AF581899

Fungal Nanotechnology

Fungal Nanotechnology

Editor

Kamel A. Abd-Elsalam

MDPI • Basel • Beijing • Wuhan • Barcelona • Belgrade • Manchester • Tokyo • Cluj • Tianjin

Editor
Kamel A. Abd-Elsalam
Plant Pathology Research
Institute
Agricultural Research Center
Giza
Egypt

Editorial Office
MDPI
St. Alban-Anlage 66
4052 Basel, Switzerland

This is a reprint of articles from the Special Issue published online in the open access journal *Journal of Fungi* (ISSN 2309-608X) (available at: www.mdpi.com/journal/jof/special_issues/fungal_nanotechnology).

For citation purposes, cite each article independently as indicated on the article page online and as indicated below:

LastName, A.A.; LastName, B.B.; LastName, C.C. Article Title. *Journal Name* **Year**, *Volume Number*, Page Range.

ISBN 978-3-0365-1744-5 (Hbk)
ISBN 978-3-0365-1743-8 (PDF)

Contents

About the Editor

Kamel A. Abd-Elsalam

Prof. Kamel Ahmed Abd-Elsalam is a research professor specializing in molecular plant pathology in the Agricultural Research Center at the Plant Pathology Research Institute, Egypt. His areas of research are plant pathogenic fungi, using a polyphasic approach based on multi-locus phylogeny (gene-barcoding) and its correlation with polyphasic characteristics. His current research interests include developing, improving, and deploying plant biosecurity diagnostic platforms and response tools, understanding and exploiting pathogen genomes, CRISPR, and developing new nanotechnology-based platforms and materials and their applications in agricultural aanotechnology.

Preface to "Fungal Nanotechnology"

A number of topics such as the application of myconanotechnology in antimicrobials, plant and food science, management of plant disease, reducing mycotoxins in antifungal nanotherapy, management of plant diseases and veterinary applications are covered in the current Special Edition. The opportunities myconanotechnology can offer in Fungal Nanotechnology 2 are becoming increasingly interesting, both in academia and industry. We sincerely believe that we have provided a balanced, fascinating, and original perspective on the subject not only for sophisticated readers, but also for industry decision makers and individuals entering the sector with limited understanding. The present book's target readership includes academics, undergraduates, postgraduate students, and others from various sectors of science and technology. We are grateful to the publishers for providing us with this collection of high-quality manuscripts. We would like to express our heartfelt gratitude to all of the writers who contributed to the book's chapters and provided comments and useful insights in this edited issue. MDPI publishers are highly commended for their high level of competence, durability, and tolerance during the project. We want to thank MDPI personnel, especially Ms. Man Lou, Ms. Sabrina Sang, and the JoF team, for their strong support and efforts in achieving this. We also thank all reviewers who extended their valuable time to review and provide guidance. We would also like to thank my family members for their continuous support and support.

Kamel A. Abd-Elsalam
Editor

Journal of *Fungi*

Editorial

Special Issue: Fungal Nanotechnology

Kamel A. Abd-Elsalam 

Plant Pathology Research Institute, Agricultural Research Centre, Giza 12619, Egypt; kamelabdelsalam@gmail.com; Tel.: +20-010-9104-9161

1. Introduction

Fungal nanotechnology (FN) or myconanotechnology is a novel word which was originally introduced in 2009 by Rai M. from India (Myco = Fungi, Nanotechnology = material production and utilization in the 1–100 nm size range). It is described as the manufacture and subsequent use of nanoparticles via fungus, especially in biomedical, environmental and agricultural commodities [1,2]. FN investigates numerous syntheses of metal nanoparticles: processing techniques, preservation of the environment and prospects of the future. Certain nanomaterials, such as silver, magnesium, gold, palladium, copper and zinc, have also been found, such as selenium, titanium dioxide, metal sulfides, cellulose, and other key fungal species, including mushrooms, Fusarium, Trichoderma, endophytic fungus and yeast. Studying the actual process of nanoparticle production and the impact of various variables on metal ion reduction might assist in developing cost-effective synthesis and nanoparticle extraction techniques. It will also address mycogenic nanoparticles, risk assessment, protection and control. Fungi have the ability to generate many extracellular enzymes that hydrolyze complicated macromolecules and to produce a hydrolyte in the wake of these enzymes. The metabolic capability of its usage in bioprocesses has been a strong source of concern for the application of fungus as a main producer for various types of metallic NPs [3,4].

For example, *Rhizopus oryaze* metabolites were utilized as a biocatalyst for the green synthetization of magnesium oxide (MgO-NPs) nanoparticles [5]. Additionally, biogenic selenium nanoparticles (Se-NPs) was produced by *Bacillus megaterium* and was used as an antifungal agent against *R. solani*, the causal organisms of damping and root rot disease in *Vicia faba*, as well as for induction of plant growth [6]. Bacillus-mediated AgNPs with an onion-isolated endophytic bacteria Bacillus endophyticus strain H3, bactriosynthesized AgNPs with a concentration of 40 μg/mL, had a high rice-blast antifungal activity with an inhibition rate of 88% mycelial. In addition, spore germination and *M. oryzae* appressorium have been considerably suppressed by AgNPs [7]. The Fusarium genus, one of the most common fungal species, plays an important role and can be considered a nanofactory for the production of various nanoparticles. Fusarium spp. is a type of fungus. This issue discusses the production of silver nanoparticles (AgNPs) from Fusarium, as well as its mechanism and uses [8]. Although the function of lichens as natural factories for synthesizing NPs has been documented, the production of NPs using lichens has remained completely unexplored until now. Lichens have the ability to produce several forms of NMs, such as metal and metal oxide NPs, bimetallic alloys and nanocomposites, through a reducible activity [9]. Mycofabrication can be described as the synthesis of various metal nanoparticles via fungal species. Several fungal species are eco-friendly, clean, non-toxic agents for the synthesis of metal nanoparticles, using both intracellular and extracellular methods.

Due to the vast spectrum and diversity of fungi, the mycogenic synthesis of nanoparticles, an essential element of myconanotechnology, leads to an intriguing new and practicable multidisciplinary science with significant results [10]. Myconanoparticles have been used extensively to detect and control pathogenic agents, to clean wastewater, to

Citation: Abd-Elsalam, K.A. Special Issue: Fungal Nanotechnology. *J. Fungi* **2021**, *7*, 583. https://doi.org/10.3390/jof7080583

Received: 19 July 2021
Accepted: 20 July 2021
Published: 21 July 2021

conserve food, as nematicides and for many other products. The mycogenic nanoparticles produced by various fungal species may be used in some potential agricultural applications to improve crop production by increasing growth and protection from infections. In addition, this will improve the toxicity to plant ecosystems of chemical pesticides, insecticides and herbicides [11]. In the near future, myconano-functional agrochemical products, crop protection pre-harvest and post-harvest, sensing systems and genetic equipment will be enhanced for direct use in farms and fields [12]. Fungal-mediated nanoparticles have shown effective inhibition against pathogens causing infectious diseases in humans, especially against those deemed multi-resistant to traditional antibacterial agents [13]. Fungi-based nanosorbents are also a novel study path in the field of heavy metal biosorption from wastewater pollutants [14].

Fungal-mediated nanoparticles have been used effectively in a wide range of scientific domains, including medicines, pharmaceuticals, agriculture, and electronics. As a result, some evaluations concentrated on the application of mycogenic nanoparticles against plant diseases, post-harvest antibiotics, mycotoxin management, and plant pests, as well as certain animal pathogens. Furthermore, fungal nanomaterials have a high potential and promise for enhanced diagnostics, biosensors, precision agriculture, and targeted smart delivery systems. For example, soil mycobiota can influence zinc mobilization from ZnO NPs in soils and thus zinc mobility and bioavailability. As a result, *Aspergillus niger,* a common soil fungus, was chosen as a test organism to evaluate fungal interactions with ZnO NPs. As expected, the *A. niger* strain had a significant effect on the stability of particulate forms of ZnO due to the acidification of its environment [15]. The macrofungi-derived NPs produced by major mushroom species such as *Agaricus bisporus, Pleurotus* spp., *Lentinus* spp. and *Ganoderma* spp. are widely recognized to have strong nutritional, immune-modulatory, antibacterial, antifungal, antiviral, antioxidant and anticancer activities [16]. In addition, the existing and potential applications of zinc-based nanostructures in plant disease diagnosis and control, as well as their safety in the agroecosystem, are discussed [17]. The development of antifungal nanohybrid agents containing conjugates of organic or inorganic compounds, biological components and biopolymers was researched in order to generate cheaper, more dependable and effective product(s) against most fungal infections of plants and animals [18,19]. Metal-bionancomposites such as Cu-Chit/NCS hydrogel are novel nano-fungicides created by metal vapor synthesis (MVS) that are used in food and feed to promote plant defense against toxigenic fungus, such as *Aspergillus flavus* linked with peanut meal and cotton seeds [20].

In conclusion, Fungal Nanotechnology 1 and 2 provides an updated and comprehensive knowledge dealing with the green and sustainable production of metal and organic-based nanostructures by various fungal species. In addition, intracellular and extracellular mechanisms will be investigated, focusing on fungal nanotechnology applications in biomedical, environmental and agri-food sectors. FN is still in its infant stage; therefore, significant studies should be focused on this area; plants, animals and humans will benefit greatly, and efficient and ecologically friendly approaches should be created.

Funding: This research received no external funding.

Conflicts of Interest: The author declares no conflict of interest.

References

1. Rai, M.; Alka, Y.; Bridge, P.; Aniket, G. Myconanotechnology: A new and emerging science. *Appl. Mycol.* **2009**, 258–267. [CrossRef]
2. Jagtap, P.; Nath, H.; Kumari, P.B.; Dave, S.; Mohanty, P.; Das, J.; Dave, S. Mycogenic fabrication of nanoparticles and their applications in modern agricultural practices & food industries. In *Fungi Bio-Prospects in Sustainable Agriculture, Environment and Nano-Technology*; Elsevier: Amsterdam, The Netherlands, 2021; pp. 475–488.
3. Alghuthaymi, M.A.; Almoammar, H.; Rai, M.; Said-Galiev, E.; Abd-Elsalam, K.A. Myconanoparticles: Synthesis and their role in phytopathogens management. *Biotechnol. Biotechnol. Equip.* **2015**, *29*, 221–236. [CrossRef] [PubMed]
4. Prasad, R. *Fungal Nanotechnology: Applications in Agriculture, Industry, and Medicine*; Springer International Publishing: Berlin/Heidelberg, Germany, 2017; ISBN 978-3-319-68423-9.

5. Hassan, S.E.-D.; Fouda, A.; Saied, E.; Farag, M.M.S.; Eid, A.M.; Barghoth, M.G.; Awad, M.A.; Hamza, M.F.; Awad, M.F. *Rhizopus oryzae*-Mediated Green Synthesis of Magnesium Oxide Nanoparticles (MgO-NPs): A Promising Tool for Antimicrobial, Mosquitocidal Action, and Tanning Effluent Treatment. *J. Fungi* **2021**, *7*, 372. [CrossRef] [PubMed]
6. Hashem, A.H.; Abdelaziz, A.M.; Askar, A.A.; Fouda, H.M.; Khalil, A.M.A.; Abd-Elsalam, K.A.; Khaleil, M.M. *Bacillus megaterium*-Mediated Synthesis of Selenium Nanoparticles and Their Antifungal Activity against *Rhizoctonia solani* in Faba Bean Plants. *J. Fungi* **2021**, *7*, 195. [CrossRef] [PubMed]
7. Ibrahim, E.; Luo, J.; Ahmed, T.; Wu, W.; Yan, C.; Li, B. Biosynthesis of Silver Nanoparticles Using Onion Endophytic Bacterium and Its Antifungal Activity against Rice Pathogen *Magnaporthe oryzae*. *J. Fungi* **2020**, *6*, 294. [CrossRef] [PubMed]
8. Rai, M.; Bonde, S.; Golinska, P.; Trzcińska-Wencel, J.; Gade, A.; Abd-Elsalam, K.A.; Shende, S.; Gaikwad, S.; Ingle, A.P. *Fusarium* as a Novel Fungus for the Synthesis of Nanoparticles: Mechanism and Applications. *J. Fungi* **2021**, *7*, 139. [CrossRef] [PubMed]
9. Hamida, R.S.; Ali, M.A.; Abdelmeguid, N.E.; Al-Zaban, M.I.; Baz, L.; Bin-Meferij, M.M. Lichens—A Potential Source for Nanoparticles Fabrication: A Review on Nanoparticles Biosynthesis and Their Prospective Applications. *J. Fungi* **2021**, *7*, 291. [CrossRef] [PubMed]
10. Rai, M.; Gade, A.; Yadav, A. Biogenic nanoparticles: An introduction to what they are, how they are synthesized and their applications. In *Metal Nanoparticles in Microbiology*; Springer: Berlin/Heidelberg, Germany, 2011; pp. 1–14.
11. Deka, D.; Rabha, J.; Jha, D.K. Application of Myconanotechnology in the Sustainable Management of Crop Production System. In *Mycoremediation and Environmental Sustainability*; Springer: Cham, Switzerland, 2018; pp. 273–305.
12. Rao, M.; Jha, B.; Jha, A.K.; Prasad, K. Fungal nanotechnology: A pandora to agricultural science and engineering. In *Fungal Nanotechnology*; Springer: Cham, Switzerland, 2017; pp. 1–33.
13. Castro-Longoria, E.; Garibo-Ruiz, D.; Martínez-Castro, S. Myconanotechnology to Treat Infectious Diseases: A Perspective. In *Fungal Nanotechnology*; Springer: Cham, Switzerland, 2017; pp. 235–261.
14. Shakya, M.; Rene, E.R.; Nancharaiah, Y.V.; Lens, P.N. Fungal-based nanotechnology for heavy metal removal. In *Nanotechnology and Food Security and Water Treatmeant*; Springer: Berlin, Germany, 2018; pp. 229–253.
15. Šebesta, M.; Urík, M.; Bujdoš, M.; Kolenčík, M.; Vávra, I.; Dobročka, E.; Kim, H.; Matúš, P. Fungus *Aspergillus niger* Processes Exogenous Zinc Nanoparticles into a Biogenic Oxalate Mineral. *J. Fungi* **2020**, *6*, 210. [CrossRef] [PubMed]
16. Bhardwaj, K.; Sharma, A.; Tejwan, N.; Bhardwaj, S.; Bhardwaj, P.; Nepovimova, E.; Shami, A.; Kalia, A.; Kumar, A.; Abd-Elsalam, K.A.; et al. *Pleurotus* Macrofungi-Assisted Nanoparticle Synthesis and Its Potential Applications: A Review. *J. Fungi* **2020**, *6*, 351. [CrossRef] [PubMed]
17. Kalia, A.; Abd-Elsalam, K.A.; Kuca, K. Zinc-Based Nanomaterials for Diagnosis and Management of Plant Diseases: Ecological Safety and Future Prospects. *J. Fungi* **2020**, *6*, 222. [CrossRef] [PubMed]
18. Alghuthaymi, M.A.; Kalia, A.; Bhardwaj, K.; Bhardwaj, P.; Abd-Elsalam, K.A.; Valis, M.; Kuca, K. Nanohybrid Antifungals for Control of Plant Diseases: Current Status and Future Perspectives. *J. Fungi* **2021**, *7*, 48. [CrossRef] [PubMed]
19. Alghuthaymi, M.A.; Hassan, A.A.; Kalia, A.; Sayed El Ahl, R.M.H.; El Hamaky, A.A.M.; Oleksak, P.; Kuca, K.; Abd-Elsalam, K.A. Antifungal Nano-Therapy in Veterinary Medicine: Current Status and Future Prospects. *J. Fungi* **2021**, *7*, 494. [CrossRef] [PubMed]
20. Abd-Elsalam, K.A.; Alghuthaymi, M.A.; Shami, A.; Rubina, M.S.; Abramchuk, S.S.; Shtykova, E.V.; Vasil'kov, A.Y. Copper-Chitosan Nanocomposite Hydrogels Against Aflatoxigenic *Aspergillus flavus* from Dairy Cattle Feed. *J. Fungi* **2020**, *6*, 112. [CrossRef]

Journal of *Fungi*

MDPI

Article

Visible-Light-Driven Ag-Modified TiO_2 Thin Films Anchored on Bamboo Material with Antifungal Memory Activity against *Aspergillus niger*

Jingpeng Li [1,*], Rumin Ma [1], Zaixing Wu [1], Sheng He [1], Yuhe Chen [1], Ruihua Bai [1] and Jin Wang [2,*]

1 Key Laboratory of High Efficient Processing of Bamboo of Zhejiang Province, China National Bamboo Research Center, Hangzhou 310012, China; bjfu140524239@163.com (R.M.); jansonwu@126.com (Z.W.); hesheng_cbrc@163.com (S.H.); yuhec@sina.com (Y.C.); oscar_bai168@aliyun.com (R.B.)

2 Zhejiang Provincial Key Lab of Biological and Chemical Utilizing of Forest Resources, Zhejiang Academy of Forestry, Hangzhou 310023, China

* Correspondence: lijp@caf.ac.cn (J.L.); whuwj@sina.com (J.W.)

Abstract: A round-the-clock photocatalyst with energy-storage ability has piqued the interest of researchers for removing microbial contaminants from indoor environments. This work presents a moderate round-the-clock method for inhibiting the growth of fungus spores on bamboo materials using Ag-modified TiO_2 thin films. Photoactivated antifungal coating with catalytic memory activity was assembled on a hydrophilic bamboo by first anchoring anatase TiO_2 thin films (TB) via hydrogen bonding and then decorating them with Ag nanoparticles (ATB) via electrostatic interactions. Antifungal test results show that the Ag/TiO_2 composite films grown on the bamboo surface produced a synergistic antifungal mechanism under both light and dark conditions. Interestingly, post-illumination catalytic memory was observed for ATB, as demonstrated by the inhibition of *Aspergillus niger* (*A. niger*) spores, in the dark after visible light was removed, which could be attributed to the transfer of photoexcited electrons from TiO_2 to Ag, their trapping on Ag under visible-light illumination, and their release in the dark after visible light was removed. The mechanism study revealed that the immobilized Ag nanoparticles served the role of "killing two birds with one stone": increasing visible-light absorption through surface plasmon resonance, preventing photogenerated electron–hole recombination by trapping electrons, and contributing to the generation of $\bullet O_2^-$ and $\bullet OH$. This discovery creates a pathway for the continuous removal of indoor air pollutants such as volatile organic compounds, bacteria, and fungus in the day and night time.

Keywords: bamboo; Ag/TiO_2; visible light photoactivity; antifungal activity; energy storage

Citation: Li, J.; Ma, R.; Wu, Z.; He, S.; Chen, Y.; Bai, R.; Wang, J. Visible-Light-Driven Ag-Modified TiO_2 Thin Films Anchored on Bamboo Material with Antifungal Memory Activity against *Aspergillus niger*. *J. Fungi* **2021**, *7*, 592. https://doi.org/10.3390/jof7080592

Academic Editor: Kamel A. Abd-Elsalam

Received: 29 June 2021
Accepted: 15 July 2021
Published: 23 July 2021

Publisher's Note: MDPI stays neutral with regard to jurisdictional claims in published maps and institutional affiliations.

1. Introduction

Indoor air pollution has been described as the most significant environmental cause of death globally, accounting for an estimated 3.8–4.3 million premature deaths each year over the past decade [1]. In major cities around the world, people spend more than 90% of their time in confined indoor environments. There is evidence that short-term exposure of human subjects to air pollution may exacerbate asthma and lead to hospitalizations, whereas long-term exposure to air pollution is repeatedly associated with a higher incidence of cardiovascular and respiratory diseases, birth defects, and neurodegenerative disorders. Fungi are ubiquitous and are a serious threat to public health in indoor environments [2].

Fungi can grow on almost all natural and synthetic materials, especially if they are hygroscopic or wet. As common indoor building materials, inorganic [3], wood-based [4], and bamboo-based materials [5] could serve as good growth substrates for fungi. In recent years, bamboo has received considerable attention because of its high strength, fast growth, renewability, and carbon sequestration potential [6]. All types of bamboo products have been developed and used in interior construction, decoration, and furniture materials worldwide. Nevertheless, bamboo is highly vulnerable to fungal attacks, especially during the rainy season. Therefore, efficient and environmentally friendly methods for fungi inhibition are highly desirable.

Semiconductor photocatalysis has been considered one of the most promising technologies for environmental purification, as additional chemical compounds such as strong oxidants are not introduced into the environment, and energy consumption is much lower than that of other advanced oxidation technologies [7]. Among semiconductors, TiO_2 has proven to be the most suitable photocatalyst because of its abundance, chemical stability, nontoxicity, and low cost [8]. Nevertheless, TiO_2 can harvest only ultraviolet (UV) light and has a high recombination rate of electron–hole pairs, leading to low photocatalytic efficiency [9]. The key issue for TiO_2-based photocatalysts is tuning their photoactive range toward the visible light region ($\lambda > 400$ nm); thus, more solar energy can be used. To overcome this problem, we recently conducted studies to enhance photocatalytic efficiency and antifungal activities, such as decorating ZnO nanoparticles (NPs) on TiO_2 film or doping Fe^{3+} into TiO_2 films [10,11].

Photocatalysis requires a continuous light source to facilitate redox reactions. From the practical application perspective, we do not want antifungal photocatalysts to be constantly exposed to light. Presently, increasing efforts are being made to develop photocatalysts for photocatalytic reactions in both light and dark conditions, termed "round-the-clock photocatalysis" or "memory catalysis" [12]. Energy-storage substances such as carbon nanotubes [13], C_3N_4 [14], Se [15], Bi [16], WO_3 [17] and MoO_3 [18] have been developed for catalytic memory reactions. In addition to these nanomaterials, Ag NPs can also store electrons because of their capacitive activity [12]. The capacitive nature of Ag NPs impedes the charge transfer of trapped electrons out of their surface. Kamat et al. [19] suggested that electron storage depends on the amount of Ag deposited on TiO_2 NPs. Choi et al. [20] investigated a sequential photocatalysis-dark reaction, where organic pollutants were degraded on Ag/TiO_2 under UV irradiation and the storage of electrons in Ag/TiO_2, which were then used to reduce Cr(VI) in the post-irradiation period. Liu et al. [21] and Jiao et al. [22] presented a new strategy to improve the catalytic memory activity of Ag/TiO_2 for organic contaminant removal under UV light. In addition, Ag nanomaterials are widely used as antimicrobials [23]. In addition to their toxicity, they could produce a synergistic antibacterial effect with other nanomaterials such as TiO_2 [24]. Chen et al. [25] also showed that the size of the Ag nanostructure is a critical factor in antibacterial capacity. Despite research in this area, few studies have focused on the use of energy-storing photocatalysts for mildew control, let alone under visible light conditions.

In this study, a photoactivated antifungal coating with catalytic memory activity was assembled on the surface of a hydrophilic bamboo by first anchoring anatase TiO_2 thin films and then decorating Ag NPs. Different characterization methods were used to analyze the structural and optical properties of Ag-modified TiO_2 thin films grown on the bamboo surface. The Ag/TiO_2 composite films grown on the bamboo surface produced a synergistic antifungal mechanism under both light and dark conditions. Remarkably, post-illumination catalytic memory was observed for ATB in the dark after visible light was removed, as demonstrated by the inhibition of *A. niger* spores. The mechanisms involved in the antifungal processes of Ag/TiO_2 under both dark and visible-light conditions are discussed and proposed.

2. Materials and Methods

2.1. Materials

Air-dried moso bamboo (*Phyllostachys edulis* (Carr.) J.Houz.) specimens with dimensions of 50 mm (longitudinal) × 20 mm (tangential) × 5 mm (radial) were purchased from Zhejiang YoYu Corporation. All chemicals used in the experiments were of analytical reagent grade. Potato dextrose agar (PDA; 1 L of water, 6 g potato, 20 g dextrose, and 20 g agar, pH = 5.6) was obtained from Qingdao Hope Bio-Technology Co., Ltd. Deionized water was prepared using a Milli-Q Advantage A10 water purification system (Millipore, Billerica, MA, USA) and used throughout all experiments.

2.2. Preparation of Ag-Modified TiO_2 Thin Films on the Bamboo Surface

The TiO_2 thin films were synthesized on the bamboo surface via a modified procedure according to our previous work [26]. In particular, the mixed solution of $(NH_4)_2TiF_6$ and H_3BO_3 was sufficiently transferred into a 50 mL Teflon-lined autoclave without pH adjustment and heated at 90 °C for 4 h in an oven. A TiO_2 thin-film-coated bamboo sample was denoted as TB. Loading of Ag NPs on the surface of TB was achieved in the dark with a simple and rapid silver mirror method. Ammonia solution (25–28%) was added dropwise into 100 mL of $AgNO_3$ solution until the brown precipitate was dissolved. Then, a TB sample was submerged in the silver ammonia solution under feeble stirring for 1 h. Subsequently, the sample was transferred into a 0.2 M glucose solution until $[Ag(NH_3)_2]^+$ ions absorbed on the TB sample were completely reduced. Finally, the samples were repeatedly washed with deionized water and dried at 50 °C for 24 h in an oven. The Ag-NP-decorated TB samples were denoted as ATB-*x*, with *x* representing the solution concentration (5, 10, 30, 50, and 200 mM) of $AgNO_3$ as one of the raw materials. For example, ATB-10 indicates that the solution concentration of $AgNO_3$ was 10 mM, which was used to load the Ag NPs on the TB surface. Ag/bamboo (AB) samples were also prepared following the above-mentioned procedure.

2.3. Characterization

The crystal structures of the samples were determined by X-ray diffraction (XRD, Bruker D8 Advance, Germany) using Cu Kα radiation (λ = 1.5418 Å) and scanning over a 2θ range of 10° to 80°. The surface morphologies of the samples were observed using a scanning electron microscope (SEM, Hitachi S3400, Tokyo, Japan) equipped with an energy-dispersive X-ray spectroscopy (EDS) system. The UV–visible (UV–Vis) absorption spectra of the samples were obtained using a Scan UV–Vis spectrophotometer (Hitachi U-3900, Tokyo, Japan). The spectra were recorded in the range of 200–800 nm at room temperature in air. The compositions of the samples were inferred following X-ray photoelectron spectroscopy (XPS, Thermo ESCALAB 250Xi, Waltham, MA, USA) results, which were obtained using an ESCALab MKII X-ray photoelectron spectrometer with Al Kα X-ray radiation as the excitation source. The photoluminescence spectra (PL) of the samples were obtained using the Edinburgh FLS 980 (Edinburgh, UK) fluorescence spectrometer. The electron spin resonance (ESR) signals of radicals trapped by 5,5-dimethyl-1-pyrroline *N*-oxide (DMPO) were detected at ambient temperature using a Bruker (E580, Rheinstetten, Germany) spectrometer under visible-light irradiation (λ > 400 nm).

2.4. Antifungal Test

Antifungal tests of the as-prepared samples were conducted according to the Chinese Standard GB/T 18261-2013, with some modifications. For all experiments, *A. niger* was used, which is a common fungus that infects bamboo. *A. niger* spores were obtained from BeNa Culture Collection (BNCC, Beijing, China) and were activated before use. After being activated, the *A. niger* spores (approximately 1×10^6 CFU/mL (CFU = colony-forming unit)) were inoculated on each PDA plate at 28 °C and 90% relative humidity for 7 days until sporulation. Prior to inoculation, the as-prepared samples and U-shaped glass rod were sterilized using a steam sterilizer at 121 °C and 0.1 MPa for 30 min using an autoclave (SANYO, MLS-3750, Osaka, Japan). A sterilized U-shape glass rod (4 mm in diameter) was placed on the PDA substrate, which was covered with mycelium, and two specimens were placed separately on the glass rod. Subsequently, the dishes were placed in a climate chamber (Boxun, BIC-400, Shanghai, China), where temperature and relative humidity were fixed at 28 °C and 90%, respectively. The tests were conducted for 28 days.

The as-prepared samples, including the original bamboo, TB, AB-10, AB-30, ATB-5, ATB-10, ATB-30, and ATB-200, were used for the antifungal tests with and without visible-light irradiation (Figure S1). One group was analyzed under visible-light radiation (Philips TLD30W/54) for 6 h every day and then the light source was turned off, whereas the other was analyzed in the dark. All experiments were sextuplicated.

3. Results and Discussion

3.1. Structural Investigations

Figure 1a presents the SEM image of the original bamboo, which comprises numerous sizable parenchyma cells. No other substances were observed on the bamboo surfaces, except the microstructure of the bamboo. After the first step, the TiO_2 thin films with an average thickness of 1.07 μm were self-aggregated by homogeneous TiO_2 NPs on the bamboo surface (Figure 1b,(b1)). Figure 1c–g display the SEM images of the as-prepared samples with different concentrations of $AgNO_3$. As shown in Figure 1c, few Ag NPs appear for an $AgNO_3$ concentration of 0.005 M. As the $AgNO_3$ concentration increased from 0.005 to 0.01 M (Figure 1d), the nanosized Ag particles with an average size of ~35 nm were uniformly deposited on the TiO_2 thin films. As shown in Figure 1e, as the $AgNO_3$ concentration increased to 0.03 M, the particles were self-aggregated together, increasing the average diameters of the Ag NPs accordingly. Most of the uniform Ag NPs gradually vanished, and the particles became denser and even dissolved one another, forming Ag thin films on the TiO_2 surface (Figure 1g). The average thickness of composite thin films is approximately 1.15 μm (Figure 1(g1)).

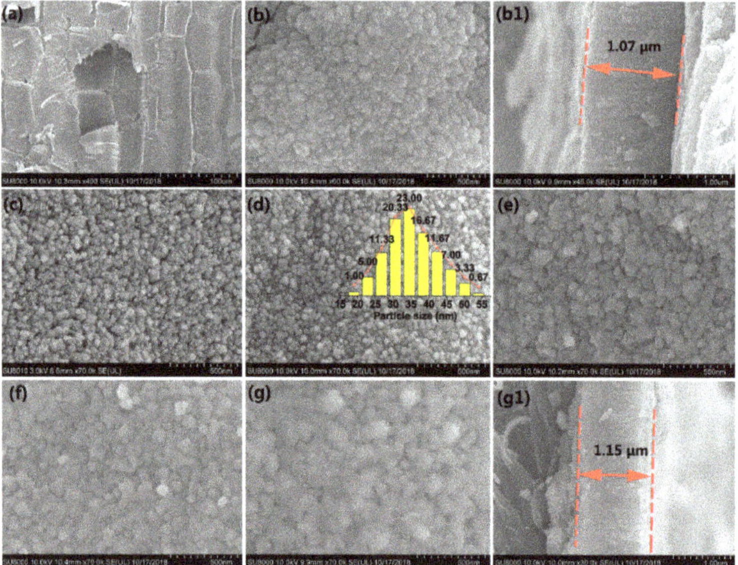

Figure 1. SEM images of (**a**) unvarnished bamboo, (**b**) TB and its corresponding cross-sectional profile (**b1**), (**c**) ATB-5, (**d**) ATB-10 and the size distribution of Ag nanocrystals (inset), (**e**) ATB-30, (**f**) ATB-50, and (**g**) ATB-200 and its corresponding cross-sectional profile (**g1**). TB: TiO_2/bamboo, ATB-x: the Ag-NP-decorated TiO_2/bamboo samples were denoted as ATB-x, with x representing the solution concentration (5, 10, 30, 50, and 200 mM) of $AgNO_3$ as one of the raw materials.

The structure of the ATB-10 sample was further studied by EDS. EDS results confirmed the presence of Ti, Ag, O, F, and C, whereas elemental mapping revealed that the Ti and Ag components were broadly and densely dispersed over the entire sample surface (Figure S2). Figure 2 shows the relative intensity of each element in the EDS spectrum measured along the thickness direction (yellow line). The signals of C, Ti, and Ag at different positions indicated that the Ag-modified TiO_2 composite thin films were successfully anchored to the bamboo surface.

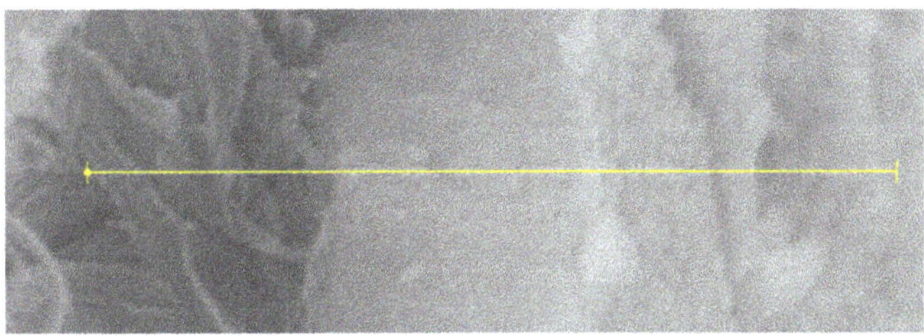

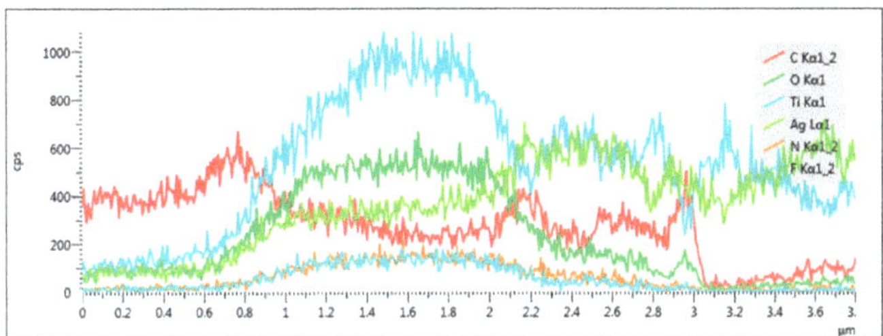

Figure 2. SEM in the line-scanning mode and the element distribution for a cross-sectional profile of ATB-10. ATB-10: Ag/TiO_2/bamboo; the solution concentration of $AgNO_3$ used is 10 mM.

The detailed crystal structures and chemical composition of the as-prepared samples were analyzed by XRD and XPS. As shown in Figure 3a, all samples exhibited similar diffraction peaks at approximately 16°, 22°, and 35°, which can be ascribed to the crystalline cellulose in bamboo. The samples all exhibited a typical anatase TiO_2 phase (JCPDS NO.71-1167), except the original bamboo. Additional diffraction peaks appeared at 38.1°, 44.4°, and 64.6°, which were assigned to the (111), (200), and (220) lattice planes of Ag, respectively [27]. No other characteristic diffraction peaks for impurities were observed in the pattern. However, the Ag diffraction peaks of the ATB-5 (green) sample could not be observed because of the relatively small amount and high dispersion of Ag metal. Notably, increasing the $AgNO_3$ concentrations from 0.01 to 0.2 M had no discernible effect on the diffraction peak intensity of the Ag metal phase. However, the diffraction peak intensity of crystalline cellulose decreased, suggesting that more Ag NPs were self-aggregated together, forming Ag thin films on the TB surface. This result is consistent with the SEM analysis, which also supported the conjecture of growth mechanism of Ag NPs on the TB surface, as shown in Figure 4b.

Research has previously suggested that only metallic Ag NPs have electron-storage ability. The chemical compositions and valence of Ag were further confirmed by XPS analysis. As shown in Figure S3, the survey spectra of ATB-10 revealed the existence of Ag, O, Ti, F, and C, which was consistent with the EDS results. As shown in Figure 3b, the XPS result of TB shows the core levels of Ti $2p_{1/2}$ and Ti $2p_{3/2}$ to be approximately at 464.6 and 458.9 eV, respectively, which was assigned to the Ti^{4+} in anatase TiO_2. However, the Ti 2p binding energy of ATB-10 is slightly shifted from 458.9 to 459.2 eV compared with that of TB. This is because the Fermi level of Ag is lower than that of TiO_2, so the conduction-band electrons of TiO_2 may be transferred to the Ag deposited on the surface of TiO_2, which decreases the outer electron cloud density of Ti ions [28]. Figure 3c shows the high-resolution XPS scans over the Ag 3d peak. The main peaks at 368.5 and 374.5 eV were ascribed to Ag metal, while the binding energies at 367.8 and 373.8 eV were attributed to Ag_2O. The two peaks detected at 368.8 and 374.7 eV could be attributed to $Ag(NH_3)_2^+$ ions [29]. This observation and XRD analysis results suggested that a small portion of Ag on the NP surface was oxidized to Ag_2O during sample drying and handling under normal ambient conditions, and the amount of Ag_2O was too small to be detected by XRD. Many researchers have reported that a small amount of Ag_2O on the Ag NP surface could enhance its stability [30].

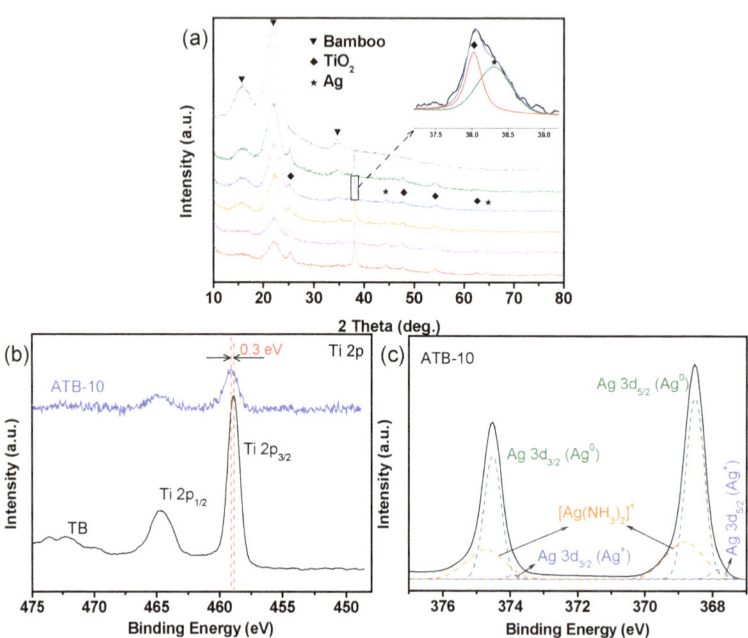

Figure 3. (**a**) XRD patterns from top to bottom are those of original bamboo, ATB-5, ATB-10, ATB-30, ATB-50, and ATB-200. The inset shows a part of the amplification of the XRD pattern (ATB-10). The high-resolution XPS spectra of (**b**) Ti 2p and (**c**) Ag 3d. ATB-*x*: the Ag-NP-decorated TiO_2/bamboo samples were denoted as ATB-*x*, with *x* representing the solution concentration (5, 10, 30, 50, and 200 mM) of $AgNO_3$ as one of the raw materials.

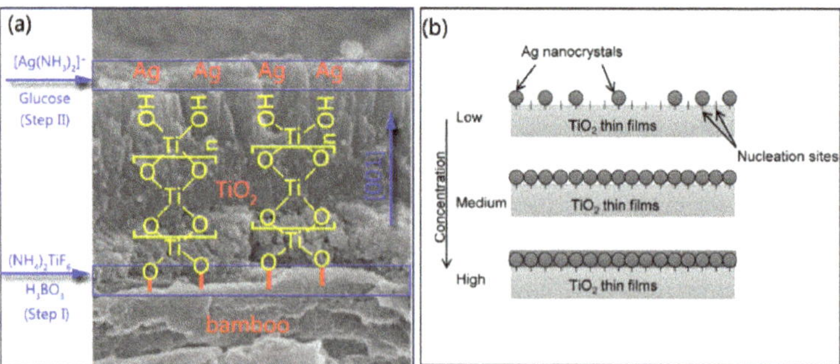

Figure 4. (**a**) Mechanism and process of nanosized Ag-modified TiO_2 thin films anchored to the bamboo surface; (**b**) schematic representation of the mineralization model proposed for the deposition of Ag nanocrystals on TiO_2 thin films with low, medium, and high densities of $[Ag(NH_3)_2]^+$ solution concentration.

3.2. Formation Mechanism of ATB Samples

Bamboo is hydrophilic, with plentiful active hydroxyl groups, and the hydroxyl groups in a bamboo substrate can react with certain metal oxides such as ZnO [31], TiO_2 [26], γ-Fe_2O_3 [32], and Cu_2O [33]. This method uses the hydrolysis of a solution containing TiF_6^{2-} in the presence of H_3BO_3 as a fluoride scavenger. The fabrication of TiO_2 thin films on the bamboo surface was accomplished by heterogeneous nucleation and homogeneous growth. For the initial heterogeneous nucleation on the bamboo surface, the existence of plentiful R–OH groups as active sites promoted the formation of R–O–Ti linkages between the bamboo surface and TiO_2 particles (Figure 4a).

$$\text{(bamboo)R-OH} + \text{HO-Ti} \rightarrow \text{(bamboo)R-O-Ti} + H_2O \quad (1)$$

The nucleated TiO_2 layer on the bamboo substrate could serve as the seed layer to further boost the homogeneous condensation of the TiO_2 NPs. For the further growth of TiO_2 NPs, the Ti–OH groups present on the surface of previous TiO_2 NPs connected with bamboo could continue to act as the active sites for the subsequent particle growth via olation and oxolation, forming Ti–O–Ti linkage (Figure 4a) [34].

$$\text{Ti-OH} + \text{HO-Ti} \rightarrow \text{Ti-O-Ti} + H_2O \quad (2)$$

From the cross-sectional profile of TiO_2 thin films, columnar crystal growth in the (001) direction can be seen on the bamboo substrate (Figure S4). This columnar morphology is consistent with the XRD measurement, which showed a significantly enhanced peak of (004) reflection (Figure S5). Previous research has demonstrated that the selective adsorption of anions on specific surfaces parallel to the (001) direction can inhibit crystal growth perpendicular to the (001) direction [34]. In our project, different types of anions, such as F^-, BO_3^{3-}, BF_4^-, and TiF_6^{2-}, were included, which could influence the growth orientation of TiO_2 crystals. Furthermore, the ζ potential of TiO_2 particles obtained using this reaction system was also confirmed to be negative owing to the strong adsorption of anions contained in the solution [34]. The XPS results also supported this standpoint because the presence of F^- anions on the surface of TB and the F^- ions on the TiO_2 surface could act as the active sites for the subsequent Ag nanocrystal growth (Figure S6).

In step II, when $[Ag(NH_3)_2]^+$ was introduced, positively charged anions were drawn to a negatively charged TiO_2 surface covered by F^- or OH groups owing to an attractive electrostatic force [35]. The silver mirror reaction generally involves the chemical reduction of the Ag compound into elemental Ag in the solution. The formed Ag subsequently nucleated on the surface of TiO_2 thin films. Figure 4b illustrates the nucleation mechanism of Ag nanocrystals that can be proposed based on SEM observations (Figure 1c–g). TB surfaces provide a certain number of nucleation sites to synthesize Ag nanocrystals. At low-level concentrations of $[Ag(NH_3)_2]^+$, the nucleation sites are sufficient to deposit Ag nanocrystals. Ag nanocrystals are uniformly deposited on nucleation sites as the concentration of the precursor solution increases. If a high concentration of the precursor solution is provided, the nucleation sites are insufficient for grafting the Ag nanocrystals, resulting in the formation of Ag thin films coated on the surface of TB, as shown in Figure 1c–g.

3.3. Optical Properties

Figure 5a presents the UV−Vis absorbance spectra of the original bamboo, TB, AB-10, and ATB prepared in the presence of $AgNO_3$: 5, 10, and 30 mM. The original bamboo exhibits strong absorption in the UV region and poor light absorption in the visible-light region from 400 to 800 nm as well as the TB sample. The samples exhibited strong visible-light absorption after the addition of Ag NPs owing to localized surface plasmon resonance. In other words, they react to visible light. Moreover, the smaller the size of Ag NPs, the greater the intensity of light absorption [36]. When comparing ATB-30 with ATB-10, the intensity of visible-light absorption decreased, implying that the Ag NPs began to grow and agglomerate. The results in Figure 5a are consistent with SEM experimental data. The efficiency of plasmon-mediated electron transfer is dominated by the size of the Ag

NPs, which plays a critical role in determining the reduction potentials of the electrons transferred to the TiO_2 conduction band [37].

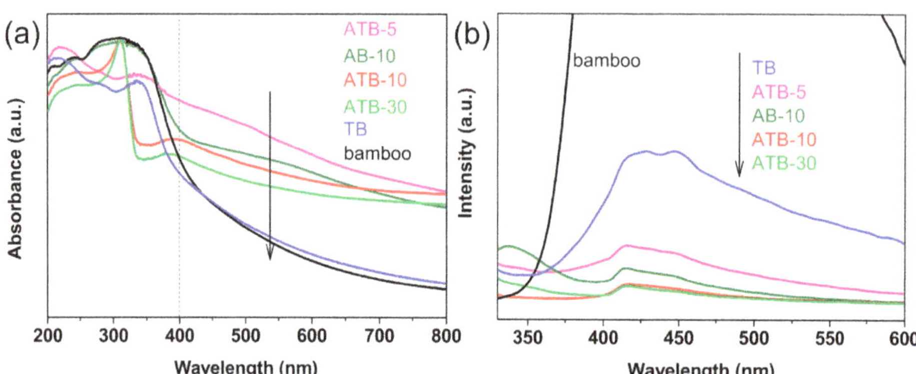

Figure 5. (**a**) UV–Visible DRS and (**b**) PL spectra (excited wavelength: 300 nm) of the original bamboo, TB, AB-10, and ATB prepared in the presence of $AgNO_3$: 5, 10, and 30 mM. TB: TiO_2/bamboo, ATB-*x*: the Ag-NP-decorated TiO_2/bamboo samples were denoted as ATB-*x*, with *x* representing the solution concentration (5, 10, and 30 mM) of $AgNO_3$ as one of the raw materials, AB-10: Ag/ bamboo; the solution concentration of $AgNO_3$ used is 10 mM.

For semiconductor nanomaterials, the PL spectra are related to the transfer behavior of the photoinduced electrons and holes, so the separation and recombination of photoinduced charge carriers can be reflected. Figure 5b shows the PL spectra of the original bamboo, TB, AB-10, and ATB prepared in the presence of $AgNO_3$: 5, 10, and 30 mM. We discovered that the original bamboo had a much higher PL intensity than other samples. Compared with TB, the intensity of the PL signal for the Ag-decorated samples was much lower, indicating that the deposition of Ag reduced the recombination rate of electrons and holes under light irradiation. The PL intensities of these samples varied in the following order: original bamboo > TB > ATB-5 > AB-10 > ATB-10 $\approx$ ATB-30. This result could be attributed to the existence of Ag NPs decorated on the TiO_2 thin films, which act as electron trappers to inhibit the recombination of photogenerated electrons and holes and decrease the PL intensity. Generally, the low PL intensity showed a high separation rate of photogenerated electron–hole pairs, resulting in a high photocatalytic activity. Therefore, a lower PL intensity indicates that the ATB samples have higher photocatalytic activities [38].

3.4. *Antifungal Performance of Ag-Modified TiO_2 Thin Films*

3.4.1. Inhibition of *A. niger* Spores in Darkness

The antifungal activity of the original bamboo, TB, AB-10, AB-30, ATB-5, ATB-10, ATB-30, and ATB-200 and their inhibition ability against *A. niger* spores in the dark are shown in Figure 6. A U-shaped glass rod was used to support the test specimens on the mycelia-covered PDA substrates, preventing their direct contact with the spores. Only five days were required for the mycelia to grow over the entire surface of the original bamboo (Figure 6a, left), indicating that the original bamboo had no resistance to *A. niger*. Peculiarly, mycelia could grow well on the bamboo surface of the AB-10 sample after incubation for five days, even though many Ag NPs were coated on the bamboo surface (Figure 6b, left). Although we increased the concentration of $[Ag(NH_3)_2]^+$ ions to prepare more Ag NPs on the bamboo surface, the AB-30 could not inhibit the growth of mycelia completely after incubation for five days (Figure 6c). These results indicate that the AB samples had poor resistance to *A. niger*.

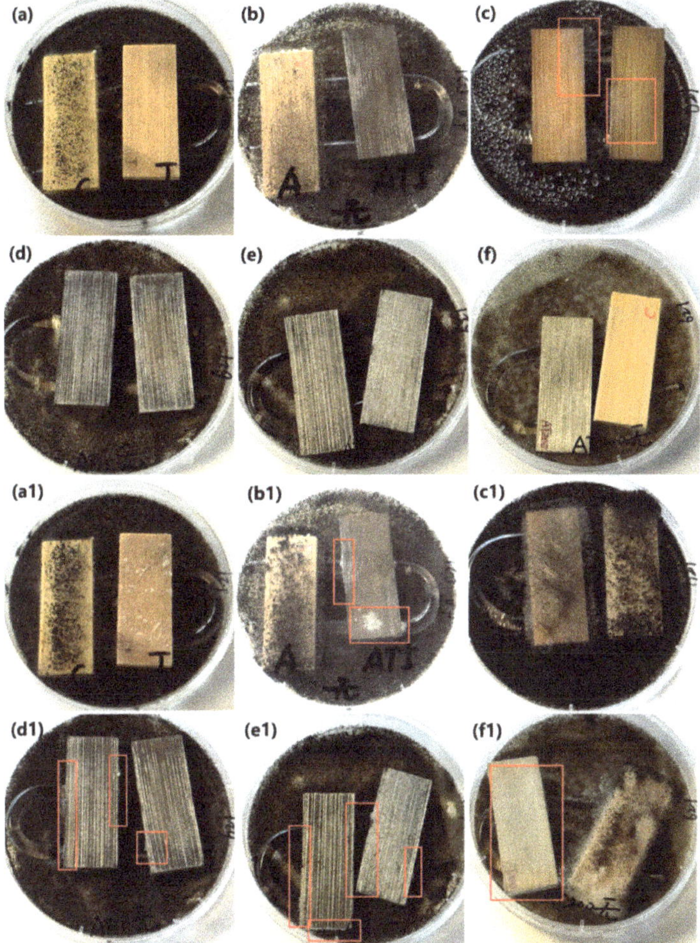

Figure 6. Antifungal properties of (**a**,**a1**) original bamboo (left) and TB (right), (**b**,**b1**) AB-10 (left) and ATB-5 (right), (**c**,**c1**) AB-30, (**d**,**d1**) ATB-10, (**e**,**e1**) ATB-30, and (**f**,**f1**) ATB-200 (left) and original bamboo (right) to inhibit *A. niger* growth in darkness. We can clearly see the mycelia in the red rectangle. Incubation period: (**a**–**f**) 5 days, (**a1**–**f1**) 28 days. TB: TiO_2/bamboo, ATB-*x*: the Ag-NP-decorated TiO_2/bamboo samples were denoted as ATB-*x*, with *x* representing the solution concentration (5, 10, 30, and 200 mM) of $AgNO_3$ as one of the raw materials, AB-*x*: the Ag-NP-decorated bamboo samples were denoted as AB-*x*, with *x* representing the solution concentration (10 and 30 mM) of $AgNO_3$ as one of the raw materials.

Our previous work similarly showed that nanosized Ag-treated bamboo samples have poor resistance to *P. citrinum and T. viride* [35]. In addition, we did not observe any mycelial growth on the surfaces of the TB, ATB-5, ATB-10, ATB-30, and ATB-200 samples after incubation for five days. After incubation for 28 days in the dark (Figure 6(a1–f1)), all samples displayed varying degrees of fungus infection. The surfaces of the original bamboo, AB-10, AB-30, TB, and ATB-200 were almost entirely covered with mycelia in the optical images (Figure 6), indicating that they had no resistance to *A. niger* in the dark. A few mycelia were directly observed on the surfaces of the ATB-5, ATB-10, and ATB-30 samples. Mycelia were mainly found on the side of the samples, especially in ATB-10 and ATB-30 samples. These results indicate that the ATB-5, ATB-10, and ATB-30 samples had a certain degree of resistance to *A. niger* in the dark. These observations suggest that the Ag/TiO_2 composite produces a synergistic antifungal effect that is unrelated to photoactivity.

3.4.2. Inhibition of *A. niger* Spores under Alternating Visible-Light Irradiation and Dark Conditions

The as-prepared samples, including the original bamboo, TB, AB-10, AB-30, ATB-5, ATB-10, ATB-30, and ATB-200 samples, were used for the antifungal test to inhibit *A. niger* spores under visible-light irradiation. The samples were tested under light radiation for 6 h every day, and then the light source was turned off. Figure 7 shows that the original bamboo and AB-10 samples were almost entirely covered with mycelia after incubation for five days, indicating poor resistance to *A. niger* under visible light. However, in addition to the original bamboo and AB-10, the TB and AB-30 samples also failed to inhibit the growth of *A. niger* after incubation for 28 days, even under visible-light irradiation. Optical images showed that their surfaces were almost entirely covered with mycelia (Figure 7(a1,c1)), indicating that they had no resistance to *A. niger*. Multiple fungal clusters were observed on the surface of ATB-5 after incubation for 28 days (Figure 7(b1), right). However, the ATB-10 and ATB-30 samples showed better antifungal activity than other samples, as *A. niger* mycelia failed to cover the entire surface of the samples after incubation for 28 days (Figure 7(d1,e1)). Note that ATB-200 exhibited better antifungal activity for *A. niger* under visible-light irradiation (Figure 7(f1), right) than under dark conditions (Figure 6(f1), left). This may be due to the plasmonic resonance effect of Ag metal under visible-light irradiation, inhibiting the growth of *A. niger* spores [39].

3.4.3. Discussion of the Antifungal Mechanisms

In this work, the hybrid Ag/TiO_2 films grown on the bamboo surface produced a synergistic antifungal mechanism under both light and dark conditions. According to data from the experiments conducted in the dark, the Ag/TiO_2 NPs showed more effectiveness at inhibiting *A. niger* growth than pure Ag NPs or TiO_2 NPs, even though they could not completely inhibit the growth of *A. niger*. The mechanism for the enhanced antimicrobial effect of Ag/TiO_2 hybrids in the absence of light is still not completely understood. Their enhanced antimicrobial qualities originated from the generation of reactive oxygen species, the release of toxic Ag ions, and cell membrane damage through their contact with the Ag NPs.

Hoek et al. [24] reported that hybrid Ag/TiO_2 NPs exhibited stronger bactericidal activity than pure Ag and TiO_2 in the absence of light. The observed synergistic effects under dark conditions were most likely caused by the variation in the dissolution and reprecipitation kinetics and equilibrium between pure Ag NPs and Ag/TiO_2 NPs. Kim et al. [40] hypothesized that the toxicity of Ag NPs is mainly caused by oxidative stress and is not related to the activity of Ag ions. Perkas et al. [41] proposed that the antibacterial activity of Ag/TiO_2 composites originates from the presence of reactive oxygen species (ROS) as well as Ag ions on the surface of TiO_2 in the dark. Chen et al. [25] reported that the antibacterial activity of Ag/TiO_2 nanocomposites under dark conditions appears to be superior to that of some pure Ag NPs. They suggested that the smaller Ag particle size should account for the higher antibacterial activity of their Ag/TiO_2. Perkas et al. [41] and Esfandiari et al. [42] both reported a similar observation, noting that the bactericidal capacity depended on the size characteristics of the Ag/TiO_2 coating. Under similar testing conditions, our previous work showed that TiO_2 thin films modified by Ag NP (diameter of 2–10 nm) have better antifungal activity for bamboo than those modified by large Ag NPs (diameter of 50–100 nm) [35]. In addition, the antifungal performance of Ag/TiO_2 nanocomposites was greater than that of AB and TiO_2/bamboo in the absence of light, indicating that the Ag/TiO_2 nanocomposite produced a synergistic antifungal effect that was unrelated to photoactivity.

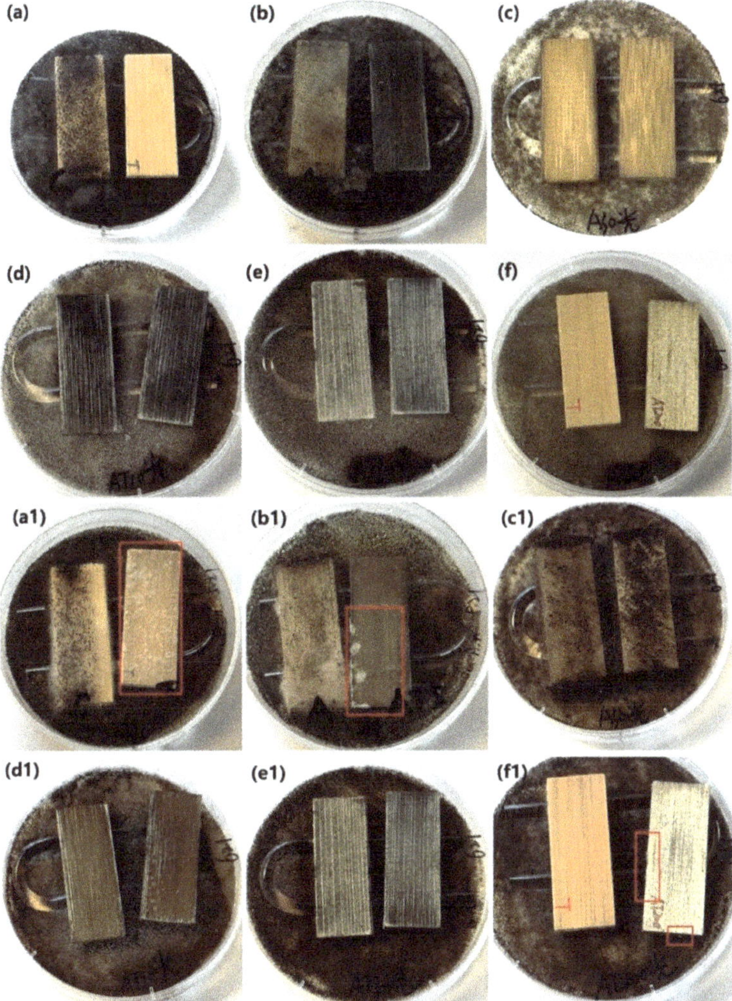

Figure 7. Antifungal properties of (**a**,**a1**) original bamboo (left) and TB (right), (**b**,**b1**) AB-10 (left) and ATB-5 (right), (**c**,**c1**) AB-30, (**d**,**d1**) ATB-10, (**e**,**e1**) ATB-30, and (**f**,**f1**) TB (left) and ATB-200 (right) to inhibit *A. niger* growth under LED light. We can clearly see the mycelia in the red rectangle. Incubation period: (**a–f**) 5 days, (**a1–f1**) 28 days. TB: TiO_2/bamboo, ATB-*x*: the Ag-NP-decorated TiO_2/bamboo samples were denoted as ATB-*x*, with *x* representing the solution concentration (5, 10, 30, and 200 mM) of $AgNO_3$ as one of the raw materials, AB-*x*: the Ag-NP-decorated bamboo samples were denoted as AB-*x*, with *x* representing the solution concentration (10 and 30 mM) of $AgNO_3$ as one of the raw materials.

As mentioned above, under dark conditions, the ATB samples could not completely inhibit *A. niger* growth on their surfaces. It is widely considered that photocatalytic microorganism disinfection depends on the interaction between microorganisms and ROS generated from photocatalysts under light illumination, such as •OH and $•O_2^-$, which can kill microorganisms [43]. Therefore, we further evaluated the antifungal activity of as-prepared samples to inhibit the growth of *A. niger* under light radiation. From the practical application perspective, photocatalysts should not be constantly exposed to light.

Therefore, we attempted to perform our experiment under visible-light irradiation for 6 h every day and then turn off the light source. Interestingly, some of the as-prepared samples could achieve complete antimicrobial activity. The ATB samples exhibited strong visible-light absorption after the addition of Ag NPs owing to the localized surface plasmon resonance. They could generate electron–hole pairs under visible-light irradiation and then migrated to the surface of the catalyst to initiate redox reactions. Most interestingly, the as-prepared ATB samples could store electrons after visible light was removed.

Figure 8 presents the ESR spectra of the ATB-10 sample. After 10-min visible-light irradiation, the strong characteristic peak DMPO-•O_2^- signals were observed, which demonstrates the formation of •O_2^- radicals by ATB-10 under light illumination (Figure 8a). When illumination was turned off, the four peaks associated with DMPO-•O_2^- adducts for ATB-10 could still be distinguished. The intensity of the DMPO-•O_2^- signals was slightly reduced after the sample was kept in the dark for 20 min. This result demonstrates that •O_2^- could be produced by ATB-10 during a dark discharge process. Similarly, we also verified the formation of •OH radicals in the dark. The ATB-10 sample exhibited slower decay kinetics of DMPO-•OH adducts after being kept in the dark for 20 min, as shown in Figure 8b. This result indicates that a considerable number of electrons in ATB-10 may remain when illumination is stopped, providing additional •OH to mitigate the decay of DMPO-•OH, which is consistent with previous work [44]. Based on the experimental data and analysis, a possible mechanism for the memory antifungal activity can be proposed as follows (Figure 8c):

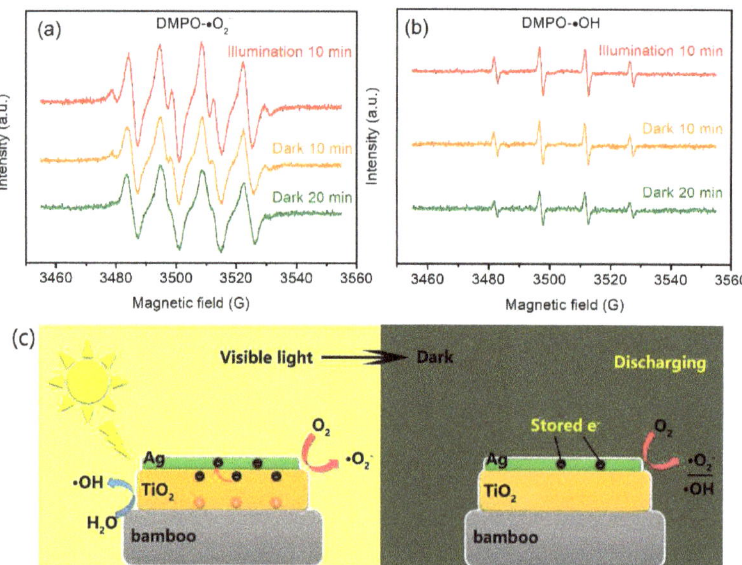

Figure 8. Time evolution of (**a**) DMPO-•O_2^- and (**b**) DMPO-•OH ESR spectra for ATB-10. (**c**) Possible mechanism of the catalytic memory reaction. ATB-10: Ag/TiO_2/bamboo; the solution concentration of $AgNO_3$ used is 10 mM.

During the photocatalytic disinfection period, excess electrons can be trapped on Ag NPs because of the capacitive nature of Ag nanomaterials. Stored electrons will be released in the dark and subsequently discharged to appropriate electron acceptors, such as O_2 and H_2O, to produce the corresponding active free radicals to inhibit the growth of fungi [44]. The combination of photocatalytic disinfection and catalytic memory reaction provides a new pathway for producing novel catalysts to achieve round-the-clock pollutant removal.

4. Conclusions

In summary, Ag-modified TiO_2 thin films were successfully anchored on bamboo material through a facile hydrothermal process, followed by an Ag mirror reaction for photoactivated antifungal coating. Upon decorating Ag NPs on anatase TiO_2 thin films, the composite films showed an enlarged optical response region and improved quantum efficiency. The antifungal test results show that the Ag/TiO_2 composite films grown on the bamboo surface produced a synergistic antifungal mechanism compared with pure Ag NPs or anatase TiO_2 film under both light and dark conditions. However, the antifungal activity of Ag/TiO_2 composite films under visible light is superior to that in the dark, owing to the transfer of photoexcited electrons from Ag to TiO_2, their trapping on TiO_2 under visible light illumination, and their release in the dark, which could give this photocatalyst a catalytic memory for producing $\bullet O_2^-$ and $\bullet OH$ radicals in the absence of light illumination. We believe that this discovery could open a door for the continuous removal of indoor air pollutants such as VOCs, bacteria, and fungus in the day and night time.

Supplementary Materials: The following are available online at https://www.mdpi.com/article/10.3390/jof7080592/s1, Figure S1: Antifungal test. Figure S2: EDS spectrum of ATB-10. Inset images are corresponding elemental mapping results. Figure S3: XPS survey spectra of (a) original bamboo, (b) TB, and (c) ATB-10. Figure S4: SEM micrograph for a cross-sectional profile of the TiO_2 thin film on bamboo substrate. Figure S5: XRD pattern of the as-prepared TB. Figure S6: High-resolution XPS spectrum of F 1s.

Author Contributions: Conceptualization, J.L. and J.W.; methodology, J.L. and Y.C.; software, J.L. and R.M.; formal analysis, Z.W.; investigation, S.H. and R.B.; resources, J.L.; writing—original draft preparation, J.L., R.M. and J.W.; writing—review and editing, J.L. All authors have read and agreed to the published version of the manuscript.

Funding: The work was financially supported by the Project of Forestry Science and Technology of the Zhejiang Province (2021SY13), Zhejiang Provincial Natural Science Foundation of China (LQ20C160002), and Fundamental Research Funds for the Central Non-profit Research Institution of CAF (CAFYBB2017MA023).

Institutional Review Board Statement: Not applicable.

Informed Consent Statement: Not applicable.

Data Availability Statement: Not applicable.

Acknowledgments: The authors would like to thank Teacher Yang from Shiyanjia Lab (http://www.shiyanjia.com, accessed on 28 May 2021) for the ESR analysis.

Conflicts of Interest: There are no conflicts to declare.

References

1. Lueker, J.; Bardhan, R.; Sarkar, A.; Norford, L. Indoor air quality among Mumbai's resettled populations: Comparing Dharavi slum to nearby rehabilitation sites. *Build. Environ.* **2020**, *167*, 106419. [CrossRef]
2. Prenafeta-Boldú, F.X.; Roca, N.; Villatoro, C.; Vera, L.; de Hoog, G.S. Prospective application of melanized fungi for the biofiltration of indoor air in closed bioregenerative systems. *J. Hazard. Mater.* **2019**, *361*, 1–9. [CrossRef]
3. Sierra-Fernandez, A.; De la Rosa-García, S.; Gomez-Villalba, L.S.; Gómez-Cornelio, S.; Rabanal, M.E.; Fort, R.; Quintana, P. Synthesis, photocatalytic, and antifungal properties of MgO, ZnO and Zn/Mg oxide nanoparticles for the protection of calcareous stone heritage. *ACS Appl. Mater. Inter.* **2017**, *9*, 24873–24886. [CrossRef]
4. Guo, H.; Bachtiar, E.V.; Ribera, J.; Heeb, M.; Schwarze, F.W.; Burgert, I. Non-biocidal preservation of wood against brown-rot fungi with a TiO2/Ce xerogel. *Green Chem.* **2018**, *20*, 1375–1382. [CrossRef]
5. Wu, Z.; Huang, D.; Wei, W.; Wang, W.; Wang, X.; Wei, Q.; Niu, M.; Lin, M.; Rao, J.; Xie, Y. Mesoporous aluminosilicate improves mildew resistance of bamboo scrimber with CuBP anti-mildew agents. *J. Clean. Prod.* **2019**, *209*, 273–282. [CrossRef]
6. Fang, C.-H.; Jiang, Z.-H.; Sun, Z.-J.; Liu, H.-R.; Zhang, X.-B.; Zhang, R.; Fei, B.-H. An overview on bamboo culm flattening. *Constr. Build. Mater.* **2018**, *171*, 65–74. [CrossRef]
7. Belver, C.; Bedia, J.; Gómez-Avilés, A.; Peñas-Garzón, M.; Rodriguez, J.J. Chapter 22—Semiconductor Photocatalysis for Water Purification. In *Nanoscale Materials in Water Purification*; Thomas, S., Pasquini, D., Leu, S.-Y., Gopakumar, D.A., Eds.; Elsevier: Amsterdam, The Netherlands, 2019; pp. 581–651.

8. Meng, A.; Zhang, L.; Cheng, B.; Yu, J. Dual cocatalysts in TiO2 photocatalysis. *Adv. Mater.* **2019**, *31*, 1807660, e1807660. [CrossRef]
9. Zhou, P.; Shen, Y.; Zhao, S.; Bai, J.; Han, C.; Liu, W.; Wei. Facile synthesis of clinoptilolite-supported Ag/TiO2 nanocomposites for visible-light degradation of xanthates. *J. Taiwan Inst. Chem. E.* **2021**, *122*, 231–240. [CrossRef]
10. Li, J.; Ren, D.; Wu, Z.; Huang, C.; Yang, H.; Chen, Y.; Yu, H. Visible-light-mediated antifungal bamboo based on Fe-doped TiO2 thin films. *RSC Adv.* **2017**, *7*, 55131–55140. [CrossRef]
11. Ren, D.; Li, J.; Xu, J.; Wu, Z.; Bao, Y.; Li, N.; Chen, Y. Efficient antifungal and flame-retardant properties of ZnO-TiO2-layered double-nanostructures coated on bamboo substrate. *Coatings* **2018**, *8*, 341. [CrossRef]
12. Cai, T.; Liu, Y.; Wang, L.; Dong, W.; Zeng, G. Recent advances in round-the-clock photocatalytic system: Mechanisms, characterization techniques and applications. *J. Photoch. Photobio. C.* **2019**, *39*, 58–75. [CrossRef]
13. Yang, Z.; Li, L.; Luo, Y.; He, R.; Qiu, L.; Lin, H.; Peng, H. An integrated device for both photoelectric conversion and energy storage based on free-standing and aligned carbon nanotube film. *J. Mater. Chem. A* **2013**, *1*, 954–958. [CrossRef]
14. Zeng, Z.; Quan, X.; Yu, H.; Chen, S.; Zhang, Y.; Zhao, H.; Zhang, S. Carbon nitride with electron storage property: Enhanced exciton dissociation for high-efficient photocatalysis. *Appl. Catal. B Environ.* **2018**, *236*, 99–106. [CrossRef]
15. Chiou, Y.-D.; Hsu, Y.-J. Room-temperature synthesis of single-crystalline Se nanorods with remarkable photocatalytic properties. *Appl. Catal. B Environ.* **2011**, *105*, 211–219. [CrossRef]
16. Dong, F.; Xiong, T.; Sun, Y.; Zhao, Z.; Zhou, Y.; Feng, X.; Wu, Z. A semimetal bismuth element as a direct plasmonic photocatalyst. *Chem. Commun.* **2014**, *50*, 10386–10389. [CrossRef]
17. Feng, F.; Yang, W.; Gao, S.; Sun, C.; Li, Q. Postillumination Activity in a Single-Phase Photocatalyst of Mo-Doped TiO2 Nanotube Array from Its Photocatalytic "Memory". *ACS Sustain. Chem. Eng.* **2018**, *6*, 6166–6174. [CrossRef]
18. Takahashi, Y.; Ngaotrakanwiwat, P.; Tatsuma, T. Energy storage TiO2–MoO3 photocatalysts. *Electrochim. Acta* **2004**, *49*, 2025–2029. [CrossRef]
19. Takai, A.; Kamat, P.V. Capture, store, and discharge. Shuttling photogenerated electrons across TiO2–silver interface. *ACS Nano* **2011**, *5*, 7369–7376. [CrossRef] [PubMed]
20. Choi, Y.; Koo, M.S.; Bokare, A.D.; Kim, D.-H.; Bahnemann, D.W.; Choi, W. Sequential process combination of photocatalytic oxidation and dark reduction for the removal of organic pollutants and Cr (VI) using Ag/TiO2. *Environ. Sci. Technol.* **2017**, *51*, 3973–3981. [CrossRef] [PubMed]
21. Cai, T.; Liu, Y.; Wang, L.; Zhang, S.; Ma, J.; Dong, W.; Zeng, Y.; Yuan, J.; Liu, C.; Luo, S. "Dark deposition" of Ag nanoparticles on TiO2: Improvement of electron storage capacity to boost "memory catalysis" activity. *ACS Appl. Mater. Inter.* **2018**, *10*, 25350–25359. [CrossRef]
22. Li, C.; Zhao, G.; Zhang, T.; Yan, T.; Zhang, C.; Wang, L.; Liu, L.; Zhou, S.; Jiao, F. A novel Ag@TiON/CoAl-layered double hydroxide photocatalyst with enhanced catalytic memory activity for removal of organic pollutants and Cr(VI). *Appl. Surf. Sci.* **2020**, *504*, 144352. [CrossRef]
23. Zheng, K.; Setyawati, M.I.; Leong, D.T.; Xie, J. Antimicrobial silver nanomaterials. *Coordin. Chem. Rev.* **2018**, *357*, 1–17. [CrossRef]
24. Li, M.; Noriega-Trevino, M.E.; Nino-Martinez, N.; Marambio-Jones, C.; Wang, J.; Damoiseaux, R.; Ruiz, F.; Hoek, E.M. Synergistic bactericidal activity of Ag-TiO2 nanoparticles in both light and dark conditions. *Environ. Sci. Technol.* **2011**, *45*, 8989–8995. [CrossRef] [PubMed]
25. Zhang, H.; Chen, G. Potent Antibacterial Activities of Ag/TiO2 Nanocomposite Powders Synthesized by a One-Pot Sol−Gel Method. *Environ. Sci. Technol.* **2009**, *43*, 2905–2910. [CrossRef] [PubMed]
26. Li, J.; Lu, Y.; Wu, Z.; Bao, Y.; Xiao, R.; Yu, H.; Chen, Y. Durable, self-cleaning and superhydrophobic bamboo timber surfaces based on TiO2 films combined with fluoroalkylsilane. *Ceram. Int.* **2016**, *42*, 9621–9629. [CrossRef]
27. Li, J.; Ma, R.; Lu, Y.; Wu, Z.; Su, M.; Jin, K.; Qin, D.; Zhang, R.; Bai, R.; He, S.; et al. A gravity-driven high-flux catalytic filter prepared using a naturally three-dimensional porous rattan biotemplate decorated with Ag nanoparticles. *Green Chem.* **2020**, *22*, 6846–6854. [CrossRef]
28. Hou, X.G.; Huang, M.; Wu, X.L.; Liu, A.D. Preparation and studies of photocatalytic silver-loaded TiO2 films by hybrid sol–gel method. *Chem. Eng. J.* **2009**, *146*, 42–48. [CrossRef]
29. Zhang, Y.; Yuan, X.; Wang, Y.; Chen, Y. One-pot photochemical synthesis of graphene composites uniformly deposited with silver nanoparticles and their high catalytic activity towards the reduction of 2-nitroaniline. *J. Mater. Chem.* **2012**, *22*, 7245–7251. [CrossRef]
30. Liu, L.; Yang, W.; Li, Q.; Gao, S.; Shang, J.K. Synthesis of Cu_2O Nanospheres Decorated with TiO2 Nanoislands, Their Enhanced Photoactivity and Stability under Visible Light Illumination, and Their Post-illumination Catalytic Memory. *ACS Appl. Mater. Inter.* **2014**, *6*, 5629–5639. [CrossRef]
31. Li, J.; Wu, Z.; Bao, Y.; Chen, Y.; Huang, C.; Li, N.; He, S.; Chen, Z. Wet chemical synthesis of ZnO nanocoating on the surface of bamboo timber with improved mould-resistance. *J. Saudi Chem. Soc.* **2017**, *21*, 920–928. [CrossRef]
32. Jin, C.; Yao, Q.; Li, J.; Fan, B.; Sun, Q. Fabrication, superhydrophobicity, and microwave absorbing properties of the magnetic γ-Fe2O3/bamboo composites. *Mater. Des.* **2015**, *85*, 205–210. [CrossRef]
33. Ma, R.; Chen, Y.; Yang, Y.; Wu, Z.; Bao, Y.; Li, N.; Li, J. Facile hydrothermal deposition of octahedral-shaped Cu2O crystallites on bamboo veneer for efficient degradation of organic aqueous solution. *Vacuum* **2021**, *185*, 110038. [CrossRef]
34. Masuda, Y.; Sugiyama, T.; Seo, W.S.; Koumoto, K. Deposition Mechanism of Anatase TiO2 on Self-Assembled Monolayers from an Aqueous Solution. *Chem. Mater.* **2003**, *15*, 2469–2476. [CrossRef]

35. Li, J.; Su, M.; Wang, A.; Wu, Z.; Chen, Y.; Qin, D.; Jiang, Z. In Situ Formation of Ag Nanoparticles in Mesoporous TiO2 Films Decorated on Bamboo via Self-Sacrificing Reduction to Synthesize Nanocomposites with Efficient Antifungal Activity. *Int. J. Mol. Sci.* **2019**, *20*, 5497. [CrossRef]
36. Mai, L.; Wang, D.; Zhang, S.; Xie, Y.; Huang, C.; Zhang, Z. Synthesis and bactericidal ability of Ag/TiO2 composite films deposited on titanium plate. *Appl. Sur. Sci.* **2010**, *257*, 974–978. [CrossRef]
37. Yu, B.; Yong, Z.; Peng, L.; Tu, W.; Ping, L.; Tang, L.; Ye, J.; Zou, Z. Photocatalytic reduction of CO2 over Ag/TiO2 nanocomposites prepared with a simple and rapid silver mirror method. *Nanoscale* **2016**, *8*, 11870–11874. [CrossRef]
38. Chen, Y.; Shen, C.; Wang, J.; Xiao, G.; Luo, G. Green Synthesis of Ag–TiO2 Supported on Porous Glass with Enhanced Photocatalytic Performance for Oxidative Desulfurization and Removal of Dyes under Visible Light. *ACS Sustain. Chem. Eng.* **2018**, *6*, 13276–13286. [CrossRef]
39. An, X.; Erramilli, S.; Reinhard, B.M. Plasmonic nano-antimicrobials: Properties, mechanisms and applications in microbe inactivation and sensing. *Nanoscale* **2021**, *13*, 3374–3411. [CrossRef] [PubMed]
40. Kim, S.; Choi, J.E.; Choi, J.; Chung, K.-H.; Park, K.; Yi, J.; Ryu, D.-Y. Oxidative stress-dependent toxicity of silver nanoparticles in human hepatoma cells. *Toxicol. In Vitro* **2009**, *23*, 1076–1084. [CrossRef] [PubMed]
41. Perkas, N.; Lipovsky, A.; Amirian, G.; Nitzan, Y.; Gedanken, A. Biocidal properties of TiO2 powder modified with Ag nanoparticles. *J. Mater. Chem. B* **2013**, *1*, 5309–5316. [CrossRef]
42. Esfandiari, N.; Simchi, A.; Bagheri, R. Size tuning of Ag-decorated TiO2 nanotube arrays for improved bactericidal capacity of orthopedic implants. *J. Biomed. Mater. Res. Part A* **2014**, *102*, 2625–2635. [CrossRef]
43. Leem, J.W.; Kim, S.-R.; Choi, K.-H.; Kim, Y.L. Plasmonic photocatalyst-like fluorescent proteins for generating reactive oxygen species. *Nano Converg.* **2018**, *5*, 8. [CrossRef]
44. Zhang, Q.; Wang, H.; Li, Z.; Geng, C.; Leng, J. Metal-free photocatalyst with visible-light-driven post-illumination catalytic memory. *ACS Appl. Mater. Inter.* **2017**, *9*, 21738–21746. [CrossRef]

Journal of *Fungi*

MDPI

Review

Antifungal Nano-Therapy in Veterinary Medicine: Current Status and Future Prospects

Mousa A. Alghuthaymi [1], Atef A. Hassan [2], Anu Kalia [3,*], Rasha M. H. Sayed El Ahl [2], Ahmed A. M. El Hamaky [2], Patrik Oleksak [4], Kamil Kuca [4,*] and Kamel A. Abd-Elsalam [5,*]

1 Biology Department, Science and Humanities College, Shaqra University, Alquwayiyah 19245, Saudi Arabia; malghuthaymi@su.edu.sa
2 Department of Mycology, Animal Health Research Institute (AHRI), Agriculture Research Center (ARC), 12611 Giza, Egypt; atefhassan2000@yahoo.com (A.A.H.); rasha_hamza2005@hotmail.com (R.M.H.S.E.A.); ahmed_elhamaky@yahoo.com (A.A.M.E.H.)
3 Electron Microscopy and Nanoscience Laboratory, Department of Soil Science, College of Agriculture, Punjab Agricultural University, Ludhiana 141004, India
4 Department of Chemistry, Faculty of Science, University of Hradec Kralove, 50003 Hradec Kralove, Czech Republic; patrik.oleksak@uhk.cz
5 Plant Pathology Research Institute, Agricultural Research Center (ARC), 9-Gamaa St., 12619 Giza, Egypt
* Correspondence: kaliaanu@pau.edu (A.K.); kamil.kuca@uhk.cz (K.K.); kamel.abdelsalam@arc.sci.eg (K.A.A.-E.); Tel.: +91-161-2401960 (A.K.)

Abstract: The global recognition for the potential of nanoproducts and processes in human biomedicine has given impetus for the development of novel strategies for rapid, reliable, and proficient diagnosis, prevention, and control of animal diseases. Nanomaterials exhibit significant antifungal and antimycotoxin activities against mycosis and mycotoxicosis disorders in animals, as evidenced through reports published over the recent decade and more. These nanoantifungals can be potentially utilized for the development of a variety of products of pharmaceutical and biomedical significance including the nano-scale vaccines, adjuvants, anticancer and gene therapy systems, farm disinfectants, animal husbandry, and nutritional products. This review will provide details on the therapeutic and preventative aspects of nanoantifungals against diverse fungal and mycotoxin-related diseases in animals. The predominant mechanisms of action of these nanoantifungals and their potential as antifungal and cytotoxicity-causing agents will also be illustrated. Also, the other theragnostic applications of nanoantifungals in veterinary medicine will be identified.

Keywords: nanoantifungal; mycotoxin degradation; theragnostic; veterinary

Citation: Alghuthaymi, M.A.; Hassan, A.A.; Kalia, A.; Sayed El Ahl, R.M.H.; El Hamaky, A.A.M.; Oleksak, P.; Kuca, K.; Abd-Elsalam, K.A. Antifungal Nano-Therapy in Veterinary Medicine: Current Status and Future Prospects. *J. Fungi* **2021**, *7*, 494. https://doi.org/10.3390/jof7070494

Academic Editor: David S. Perlin

Received: 18 May 2021
Accepted: 16 June 2021
Published: 22 June 2021

1. Introduction

Fungal diseases are manifested as active infections and/or secretion of mycotoxins on growth of fungi in different tissues of animals. The specific fungal disorders include bovine mastitis, fungal diarrhea in calve, respiratory disorders, superficial, subcutaneous, and systemic infections and mycotoxicosis [1,2]. The variability of the extent of serious public health risk effects of fungal infections in livestock and other domesticated animals spans carcinogenic, nephrotoxic, and hepatotoxic effects following their consumption in the contaminated grains/animal food products [1,3]. Animal production holds considerable economic importance for humans, particularly in low-income countries [4]. Hence, published literature searches included different studies which evaluated the extent of incidences of fungal diseases claiming morbidity and mortality in animals of economic importance besides diverse techniques that can be followed to control the growth of fungal pathogens and secretion of mycotoxins [1,5]. The global prevalence of mycosis and mycotoxicosis related diseases in livestock is about 25%. The traditional treatment procedures including the use of amphotericin (AmB) have been evaluated to be relatively ineffective in most cases due to reactivation of latent fungal infections post medication treatment [2,6].

Likewise, the treatment of fungal infections with azoles (such as fluconazole, voriconazole and itraconazole) may lead to the emergence of resistant fungal pathogens due to excessive and frequent use [7,8]. To solve these issues of fungal disorders in animals, the search for novel effective nanotechnology-enabled antifungals has gained impetus. Further, dual benefits can be reaped through use of nanomaterials for both therapy and diagnosis of disease pathogens separately and for developing conjugate systems for simultaneous diagnosis and targeted release of the therapeutic agent, theragnostic systems [9]. Moreover, novel nano-based disease diagnosis and therapeutic systems have been developed for effective treatment of different animal diseases caused by fungal, parasitic and viral pathogens [3,5]. The antifungal nanomaterials can be applied for the diagnosis of the problems related to reproductive system of the animals [10] and for the protection of the physiological activities of animal genital organs and secretions [11,12]. Also, nanomaterials can be utilized to generate effective vaccines [13]. Nanomaterials can exhibit improved killing or inhibitory activity on fungal pathogens at lower doses and can also be utilized as drug delivery vehicles to help in targeted delivery of drugs [10]. Besides, novel formulations of antifungals or new devices that increase the likelihood of the medication being administered to the site of infection tend to be important in order to boost drug efficacy [14,15]. Therefore, the aim of the present review was to investigate the types of nanoantifungals and their applications in animal health. Also, their uses for mycotoxin degradation in animal feeds, and their therapeutic and preventive aspects were illustrated. Moreover, the mechanisms of nanoantifungal actions, toxicity, and ways to overcome the suspected toxicity will also be discussed.

2. Nanoantifungals: Diversity and Relevance for Applications in Veterinary Medicine

Metal/metal oxides and their nanocomposites such as zinc, silver, selenium, copper-chitosan nanocomposite and other nanomaterials exhibit prominent fungicidal activity compared to their bulk counterparts [9,16]. These antifungal nanomaterials can be categorized into various forms according to their chemical sources and morphology [17,18]. A variety of nano-antifungals have been developed to cure different fungal diseases in animals and human beings (Figure 1).

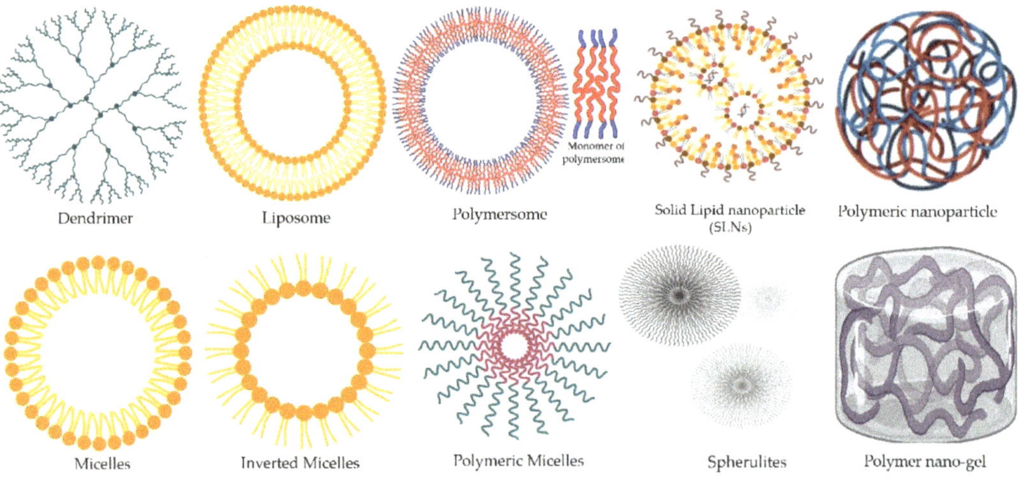

Figure 1. Nanovehicles for effective and smart delivery of therapeutic drugs and other anti-fungal agents.

2.1. Categories of Nanoantifungals in Veterinary Medicine on Basis of Their Chemical Origin, and Structure

2.1.1. Organic Synthetic and Natural Polymeric NPs

A diversity of nanoparticles and nanoscale products can be developed from synthetic and natural polymers. These NPs are formed from natural and synthetic materials including saccharides and their derivatives such as chitosan, lipids and other biomolecules. A huge variability in the size of these nanomaterials has been reported, with size dimensions spanning from 0.5 to 100 nm. These nanomaterials have high loading/conjugating capacities and have also been used for the development of hydrogel nanoformulations, particularly for the sugars and their derivatives [17]. Synthetic polymeric NPs can be composed of amphiphilic polymers such as caprolactone or PLGA which form a hydrophobic core that facilitates the transportation of hydrophobic drugs encrusted with a water-soluble coat [19]. These NPs have been used for transportation and delivery of drugs with low water-solubility such as amphotericin [20].

Solid lipid nanoparticles developed from a variety of lipids are upcoming drug delivery vehicles which exhibit great potential for lipophilic anti-cancer drugs. These can be easily combined with other materials to induce improved humoral antibody dependent immunity in the animals [21], besides their role in gene therapy by development of nucleic acid-based conjugates [22,23]. The oral, skin and parental routes of solid lipid NPs application are more effective in drug delivery and are highly absorbed [23].

Another category of polymeric nanoparticles includes the most popular forms called liposomes. These are non-toxic PEGylated NPs which are comprised of a two lipid (bilayer) cover shell having high solubility for fatty (hydrophobic) drugs. The first layer of the liposome is coated with a PEG layer to prevent any immune response towards the particles [17,24]. However, due to their vulnerability to get digested in the alimentary canal leading to loss of function, these nanoformulations are preferentially administered through parental and topical routes. Conjugating liposomes with biologically active antibodies can be useful for cancer cell treatment [23]. Further, liposomal formulations of dead pathogens can be utilized to develop vaccines [25]. The liposomes can also be conjugated with DNA to develop DNA vaccines [17]. Furthermore, the liposomes enable drug delivery and diffusion to targeted cell sites within the organism (Figure 1). Despite these benefits and potentially useful activities, these formulations are prone to changes during storage and also the encapsulated compounds may exhibit rapid destruction of their content on account of oxidation processes [26].

Similar to liposomes are the polymeric micelles with one basic difference from the former type that the latter are formed from exfoliated lipid bilayers and thus exhibit great potential to encapsulate lipophilic drugs. Therefore, micelles are hydrophobic core surrounded by a hydrophilic coat which increases their solubility in water [23].

Nano-cochleates are a specialized category of sub-micron to nanoscale solid particulate lipid-based drug carriers [27] which can be derived by the fusion of liposomes with metal cations and involve spiral rolling of continuous lipid bilayer [28,29]. These carriers can be efficiently loaded with both hydrophobic as well as hydrophilic drugs ensuring higher protection from gastrointestinal degradation of anti-fungal drugs particularly Amphotericin and thus enabling oral administration [30,31].

Synthetic polymeric nanoparticles primarily including dendrimers are derivatives of long-chain branched polymers such as polyamidoamines. Similar to micelle nanoparticles, dendrimers are water-soluble, exhibit high biological activities and possess comparatively a much smaller size than the other polymeric NPs discussed so far [26,32]. These attributes of the dendrimers do not allow stimulation of the immune response after parental administration. Dendrimers can be combined with drugs to improve their efficiency for treatment of a variety of animal disorders [26]. The dendrimer formulations have been successfully used for effective cancer treatments and may showcase multiplexed functions including detection of the tumor cells, entry through the cell membrane, targeted release of the conjugated anticancer drugs in the cytoplasm and finally the destruction and death of cancerous cells [23]. Dendrimers can also conjugate with the lipids of the cell membranes and this can create wide pores in the membrane that potentiate improved entrance of the drug containing dendrimer nanoparticles for targeted delivery leading to higher cell death rates [26].

2.1.2. Nanoemulsions

These are aqueous mixtures of oil or other hydrophobic components prepared by addition of oil to water overlaid by non-chemical surfactants [1,16,33,34]. The micelle size in the prepared nanoemulsions may vary from 0.5 to >500 nm. Nanoemulsions exhibit significantly high antifungal, bactericidal and virucidal activities. It may be attributed to greater adherence of the oil droplets on the surface of the microbial cells which facilitates the entrance of drugs to the cell [5,17,35].

2.1.3. Inorganic Metal/Non-Metal Nanomaterials

These NPs are the first-choice nanomaterials to be used as nanoantifungals due to their low cost, easy application, eco-friendly characteristics and wide viability [1,5]. They exhibit potential as antifungals [29,36,37], besides the other biomedical benefits [24]. These nanomaterials may have individual particle size dimensions ranging from 1 to 100 nm with aggregate sizes have a higher size range.

Magnetic Iron Oxide Nanoparticles

Magnetic FeO NPs mainly consist of iron core (Fe_3O_4 or Fe_2O_3) particles which have been used in several studies as significant antifungals against mycotoxigenic molds [34]. Drug delivery, heat therapy and imaging are other beneficial uses of iron core particles [23,38,39]. Moreover, the iron core can be conjugated with fluorescent shells and drugs or antibodies against targeted cancer cells [24]. Further, surface functionalization of these NPs by polyethylene glycol (PEG) can potentially help to prevent elicitation of the immune response.

Semiconductor Quantum Dots

Zinc selenide/telluride/sulphide quantum dots exhibit substantial antifungal potential [39]. QDs are core-shell aqueous materials which exhibit conjugation with drugs or other biological materials including nucleic acids (DNA/RNA), proteins and other biomolecules [17]. The biomolecule conjugated QDs have specific use for detection and diagnosis of diseases or their causative pathogens [23]. Further, QDs find peculiar applications for improved imaging and genetic analysis by observing cell activities under disease conditions, and targeted drug delivery [40].

Silicate Nanomaterials

These nanomaterials are comparatively biosafe, and do not exhibit high reactivities. Further, the silicate nanomaterials possess diverse morphologies spanning over different particle shapes and sizes which can be easily modified [23]. These silicate nanomaterials are also amenable to functionalization, and other coating treatments. Nanoshells are a specific class of silica nanomaterials which involve a thin metallic coating of the glass core [41]. These nanoparticles have been utilized for the diagnosis of the tumorous tissues [23,42] and simultaneous therapeutics applications [42–45].

2.1.4. Carbon Nanomaterials

Carbon nanomaterials have significant antifungal and antimycotoxin potential [3,5]. The carbon atom contents enable the destruction of pathogen cell walls [40]. These nanomaterials are insoluble in water and do not get digested in the alimentary tract or get excreted on oral administration [26]. The SE nanomaterials pass through the cell membranes of targeted cells to reach to the cytoplasm of the pathogens or cancer cells causing multiplexed damage resulting in the cell death [46]. Besides, buckyballs can ameliorate pH levels which help in drug delivery to targeted tissues [47]; gene therapy and DNA delivery [48].

2.1.5. Nanobubbles

These are gas core particles suspended in aqueous medium having general size dimensions ranging from 70–120 nm and function as carriers of gas molecules [49]. The nanobubbles are different from the other types of nanoparticles or nanoemulsions as these contain a shell comprised of polymer, phospholipids, proteins or anti-cancer therapeutic agent encasing a gas (generally oxygen) [50–52]. These nanomaterials are finding useful applications in diagnosis and targeted delivery of anticancer drugs [49,53].

2.1.6. Nanovaccines and Nanoadjuvants

Today, there are progressive advances in the application of nanotechnology for the production of vaccines. Nanovaccine formulations effectively activate the humoral immunity by a slow elaboration of antigens and thereby elevating the usefulness of vaccination [17,54]. These can be targeted to lymph tissues which significantly enhances the vaccine activities [55]. Nanomaterials conjugated with antibodies and other biological molecules can be used for the quick detection of pathogens and for effective treatment of the diseases caused by them [39]. However, the nanomaterials possess excellent adjuvant properties as these can bind to a variety of antigens/proteins of pathogenic origin to obtain nano-vaccines thereby replacing the use of the adjuvant material [55]. Different forms of nanomaterials used in animal antifungal nanotherapy was shown in Figure 2.

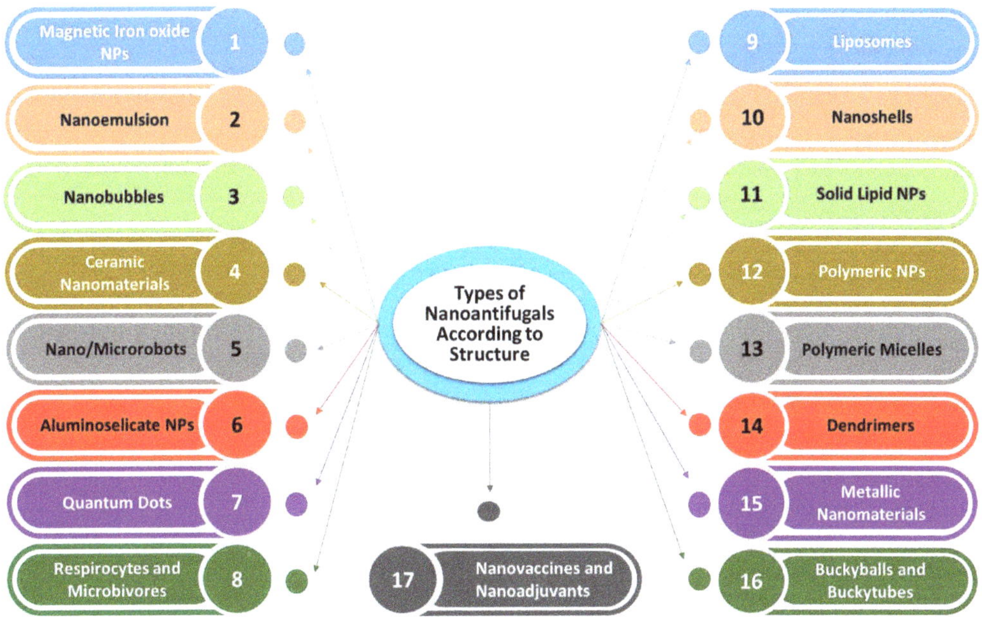

Figure 2. Various types of nano-based materials employed in antifungal nanotherapy in veterinary medicine.

3. Applications of Nanoantifungals in Veterinary Medicine

3.1. Therapeutic and Preventive Aspects of Nanomaterials

3.1.1. Metal/Metal Oxide/Non-Metal Oxide NPs and their Hybrids as Nanoantifungal Agents

The use of nanomaterials as antifungal agents is an established attribute. The nanomaterials that exhibit antifungal potentials have been evaluated in several studies with primary inhibitory impact on the vegetative growth of the fungal mycelia. The noble metal nanoparticles including the silver and gold nanoparticles possess potent antifungal properties. Nasar et al. [56] have evaluated the broad antimicrobial activity of AgNPs against human pathogenic bacteria (*Escherichia coli*, *Klebsiella pneumonia*, and *Bacillus subtilis*), and common fungal pathogen *Aspergillus niger.* The AgNPs have been found to be effective antifungals against dermal infections [57] Moreover, AgNPs can remove the human oral microbial infections caused by *S. aureus* and *C. albicans* [58], and *C. albicans*, and *Trichophyton mentagrophytes* infections in buffaloes [59]. The nanosized silver can inhibit the growth of *Fusaium* sp. at very low concentrations (<100 ppm) [16,60] and led to decreased mycotoxin production [1,61]. Also, Kischkel et al. [37] observed the antifungal activity of the AgNPs against *C. albicans*, *F. oxysporum* and *M. canis.*

Abd-Elsalam et al. [62] have discussed the fungal growth inhibitory potential of a variety of metal oxide NPs. Among the metal oxide NPs, the most promising candidates are zinc oxide NPs which inhibited the *Candida albicans* growth at very low concentrations of 1.013–296.0 µg/mL [63]. The shape and size of ZnO NPs has been an important characteristic that decides for the extent of the antifungal activity. Flower-shaped ZnO nanostructures inhibited the development of *Aspergillus flavus* and aflatoxin production at concentrations below 5 mM [64]. The next metal oxide NPs showing considerable antimicrobial potential are the iron oxide NPs. A study on magnetic NPs (Fe_2O_3 NPs) described the antifungal activity against *A. flavus* and prevention of the aflatoxin production [38]. While, Mouhamed et al. [65] documented the inhibitory effect of iron oxide NPs on ochratoxigenic *Aspergillus* sp. Moreover, Abd El-Tawab et al. [66] have detected the growth inhibitory properties of Fe_2O_3 NPs against causative pathogens of bovine skin diseases (*Trichophyton verrucosum*, *T. mentagrophytes*, and *Dermatophilus* sp.).

The coating or surface functionalization of the metal/metal oxide nanoparticles can further improve their antimicrobial properties. The chitosan NPs derived from deacetylated derivative of chitin can prevent growth of *Fusarium* sp., *Rhizopus* sp. and *Aspergillus niger* and thus can be used as an alternative to chemical pesticides [67]. Further, chitosan NPs have also been observed to inhibit fish pathogens under in vitro conditions [68]. Chitosan polymers can also be utilized to develop surface coatings on metal oxide NPs to improve their interactions and passage through the biological membranes. Recently, Abd-Elsalam et al. [69] have detected significant antifungal activity of CuNPs singly and in combination with chitosan against mycotoxigenic fungi, which also led to the prevention of aflatoxin production. The use of an acrylic resin reinforced with ZnONPs and Ag NPs can inhibit the growth of *Candida albicans* [70].

Nowadays, combinations of nanomaterials with beneficial biological active compounds are used to produce nanocomposites of significant use for animal health [1]. The conjugation and overlay of nanomaterials by other biological molecules are related to their chemical properties and used in detection of pathogens inside the body [71]. In this respect, Hassan et al. [1,5] have reported that the conjugation of metals nanomaterials with natural oils significantly improved the antifungal activity. They have detected that the composites of AgNPs, ZnONPs, and essential oils can effectively prevent the growth of fungal and bacterial pathogens. Hybrids of Ag NPs/essential oil were employed in therapy of bovine skin and udder infections [5,72] and carbon NPs [73]. Hassan et al. [5,16,74] have reported the efficient conjugation of ZnNPs and AgNPs with cinnamon and olive oils for use at low safe doses for inhibition of growth of toxigenic *A. flavus* and *E. coli* and production of respective toxins, whereas, Wang et al. [75] successfully detected that the hybrid of Au NPs with antibodies help in immune-chromatographic exploration and diagnosis of toxic AFM_1 in milk. Similar activities were obtained for QDs to observe events and activities of body cells that were found to be better than the use of traditional dyes and this helped for release of drug to the required site of infection [9,76].

Nanoparticles can also be conjugated with known standard antifungal agents or other molecules where these NPs function as nanovehicles for better delivery of the antifungal therapeutic agents at the targeted site. Therefore, common antifungal agents can also be conjugated on the metal or metal oxide NPs to enhance their antifungal activities. Kischkel et al. [29] have illustrated the potentials of different types of nanoantifungals for the treatment of mycosis caused by *Candida* sp. and *Aspergillus* sp. Hamad et al. [77] have developed a gold nanorod-fluconazole nanoconjugate which exhibited significantly high antifungal activity (9 to 12-fold) against *C. albicans* compared to either component alone, whereas, Huang et al. [36] reported the possibility of using AgNPs as antifungals singly or in conjugation with epoxiconazole (8:2 and 9:1), respectively. The concentration of AgNPs required to suppress the growth of 50% of the fungal colony was 170.20 μg/mL. The combination of AgNPs with fluconazole and florfenicol produced more antimicrobial potential against the causes of animal diseases than their single forms [5].

Inorganic mesoporous silica nanoparticles (MSNs) can also act as nanocarriers for drug delivery to target affected cells inside the body [78]. Functionalized silica NPs can be tethered to drug molecules or they can also adsorb or sequester the drug compound on the surface or inside the nanopores thereby elevating their delivery to the target organs [79,80]. Kanugala et al. [81] have developed phenazine-1-carboxamide-functionalized MSN-based antimicrobial biomaterial surfaces to prevent the formation of bioflms on medical implants. The developed MSNs exhibited superior anti-Candidal activity besides polymicrobial antibiofilm potential. Silica NPs can also be used for the development of topical cream formulations to treat skin fungal infections. Montazeri et al. [82] have synthesized and evaluated an aminopropyl functionalized MSN-econazole topical cream formulation against *Candida albicans* skin infections and observed improved antifungal activity at lower concentrations of the loaded drug.

3.1.2. Polymer Nanoparticles for Antifungal Drug Delivery

Recent drug and vaccine delivery strategies in biomedical research advocate the use of nanomaterials for successful delivery of drugs to targeted cells and tissues [10]. These strategies are beneficial as they can ensure the delivery of drugs to target tissues resulting in a decrease in the amount and required doses for the treatment of diseases. The most promising nanodelivery agents for drugs can be the polymer nanoparticles encapsulating antifungal drugs. In these respects, chitosan (CS) NPs which themselves possess considerable antifungal potential can be used for the delivery of the antibiotic drugs [68]. The encapsulation of antifungal drugs or development of their formulation as nanoemulsions can improve their action potential. Deaguero et al. [83] observed that nanoencapsulation of miconazole in cholesterol/sodium oleate vesicles have significant antifungal activity against several fungal pathogens. Siopi et al. [84] have reported that the liposome-encapsulated amphotericin B possess significant therapeutic potential against mycotic respiratory infections in animals caused by *A. fumigatus*.

Drug molecules can also be nanoformulated as nanomicelles comprised of a hydrophobic core and hydrophilic shell which improves the water solubility and therefore bioavailability of the hydrophobic drugs [85]. Further, these nanosystems can be used for the targeted delivery of the drug [86], treatment of cancer in animals [87] and to ensure drug delivery without stimulation of immunity [88].

3.1.3. Carbon Nanomaterials as Nano-Antifungals

Different forms of carbon-based nanomaterials also exhibit antimicrobial activity against bacterial and fungal pathogens causing diarrhoea [3]. These nanomaterials can inhibit the growth of *E. coli* and mycotoxigenic fungi [89]. Furthermore, conjugation of sugars with CNTs improve the ability to affect the viability of *C. albicans, A. flavus* [5]. Several benefits of nanoantifungal applications were detected, as illustrated in Figure 3.

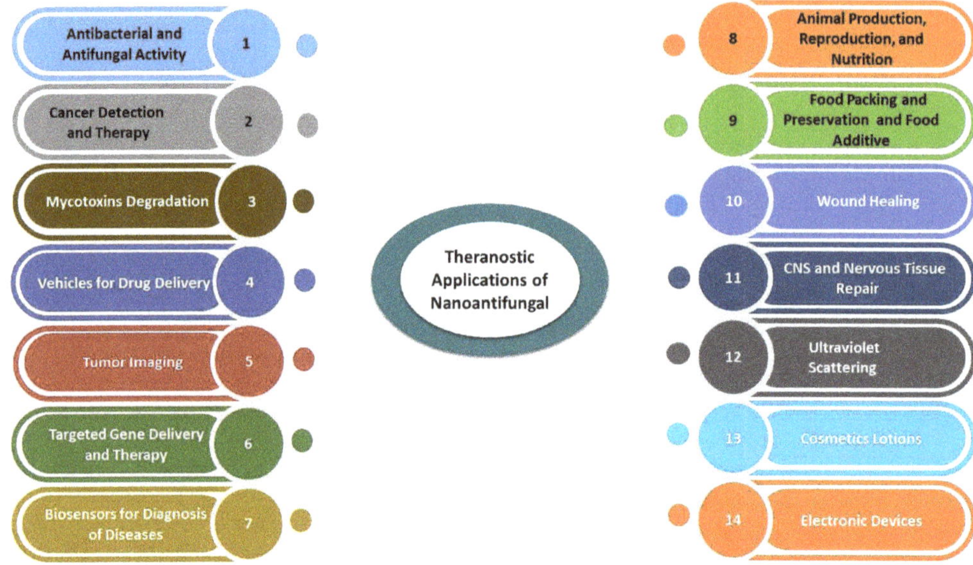

Figure 3. Theragnostic applications of nanoantifungals in animal science.

3.1.4. Nanocomposites for Antifungal Drug Delivery Agents

The nanocomposites of natural materials such as carbohydrates and proteins with polymers have the ability of effectively releasing of these materials at targeted sites [40]. However, the modified CS NPs are capable of drug delivery to diseased tissues at lower doses than traditional chemical cancer therapy. They can be used as an adjuvant for effective animal vaccination against infections. The parental administration of nanoshells comprised of silica core attached with metals NPs and drugs in animals can be useful to search and can be directed to target cancer cells [90].

The nanocomposites qualify quite uniquely considering the non-stimulation of the elaborate immune response aspect [91]. The embedding or encapsulation of the drug in polymer blend-based nanocomposite also improves its antifungal potential. Terbinafine hydrochloride was introduced into the polycaprolactone (PCL)/gelatin nanofibers generated by the hydrothermal method [92]. The resulting wound dressings were tested for their antifungal potential.The researchers were successful in inhibiting the *T. mentagrophytes* and *Aspergillus fumigatus* due to slow release of the embedded drug molecules from the nanocomposite fibers over time [92].

3.2. Antifungal Nanomaterials for Management of Mycotoxins in Animal Feeds

The morbidity and mortality caused due to global incidences of mycotoxicosis in animal and poultry industry have serious economic repercussions affecting the productivity [1,3,5,93]. Recently, it was reported that ZnO NPs and Fe_2O_3 NPs have antifungal activity against ochratoxigenic *Aspergillus* and hence prevent mycotoxin synthesis [57,65]. The supplementation of Zn NPs in aflatoxicated feed of rats and rabbits resulted in the removal of the carcinogenicity of aflatoxins on the kidney and liver [72,94]. The ZnNPs and AgNPs can inhibit the growth of *Fusarium poae* and prevent formation of trichothecenes mycotoxin [16]. The Ag NPs can eliminate aflatoxins in chickens feed [95,96]. Biosynthesized spherical SeNPs produced by *Saccharomyces cerevisiae* and originated from selenous acid and sodium sulfite were able to inhibit pathogenic saprophytes, yeasts, and dermatophytes [97]. Fadl et al. [98] have reported that CuNPs inhibit ochratoxigenic molds and prevent ochratoxin production in a fish feed.

Apart from the metal/metal oxide nanoparticles, carbon nanomaterials, particularly carbon nanodiamonds, can ameliorate the adverse effects of mycotoxins by the process of immobilization of the mycotoxins [99]. While, another report by Hassan et al. [3] detected the activity of CNTs in suppression the toxicity of *A. flavus* at a concentration of 125 μg/mL. Therefore, the primary modus operandi for the anti-mycotoxigenic effects of both nanomaterials and nanocomposites such as iron NPs [100] and MgO-SiO_2 nanocomposite [101] is through adsorption of the mycotoxins.

Nanohybrids such as polyene-functionalized magnetic NPs possess enhanced antifungal activity against opportunistic oral fungal pathogens such as *Candida* sp. [102]. A miconazole nanocarrier (MCZ) based on iron oxide nanoparticles (IONPs) functionalized with CS was prepared, characterized and screened for antifungal activity against *Candida albicans* and *Candida glabrata* biofilms. A nanocarrier with less than 50 nm dimeter presenting MIC values lower than those observed for high diameter and showed synergism against *C. albicans* [103]. Similarly, nanoformulations of known antifungal agents can improve the action spectrum of these agents and enhance the antibiofilm potential. The antimicrobial and antibiofilm effects of a colloidal nanocarrier for chlorhexidine (CHX) on yeast and bacteria such as *Candida glabrata* and *Enterococcus faecalis* were evaluated. The CHX nanocarrier has an excellent ability for the management of oral diseases linked to *C. glabrata* and *E. faecalis* [104].

The nanocomposites derived from metal oxide and carbon nanomaterials have also been evaluated for anti-mycotoxin properties. A magnetic carbon nanocomposite derived from bagasse was observed to degrade AFB1 [105], while graphene oxide nanocomposites caused a reduction in the occurrence of three prominent *Fusarium* toxins i.e., ZEA, FB, and deoxynivalenol [106] and modified halloysite nanotubes [96,107]. Also, detoxification of AFB1 by a magnetic graphene oxide nanocomposite has been reported by Ji and Xie [108]. González-Jartín et al. [109] observed that nanocomposites of carbon, bentonite, and aluminum oxide eliminated up to 87% of the mycotoxins with an adsorption efficiency of 450 μg/g. Chitosan-stabilized selenium nanoparticles have a significant ability to improve the toxic effects of aflatoxicosis in rats [35,110]. Further, the SeNPs exhibited important inhibitory effects on *A. parasiticus, A. ochraceus,* and *Aspergillus nidulans* growth at concentrations varying from 0.1–0.5 mg/L and ameliorate the dysfunction and hepatic apoptosis induced by AFB1 [30]. Chitosan-coated Fe_3O_4 particles have been reported to be substantially useful for patulin decontamination with no toxic response or histopathology in treated mice [106,111]. Recently, Hassan et al. have also assessed the efficiency of the copper-CS nanocomposites for the removal of the aflatoxins and ochratoxins in poultry (personal communication).

Conjugating metal oxide nanoparticles with other antimicrobial components such as essential oils, curcumin or ozone can improve the antimycotoxigenic activities. Also, the antimicrobial, anti-aflatoxins, and anti-shigella toxins potentials of nanoemulsion of cinnamon oil and ZnO NPs towards to fungal causes of dysentery in buffaloes were detected [16]. Hassan et al. [72] observed that conjugating ZnO-NPs along with probiotic and curcumin improved the inhibitory activities on mycotoxin producing *Fusarium* sp. besides significantly decreasing their ability for mycotoxin production. The combined application of ZnO NPs, probiotic and curcumin (ZnO NPs (100 μg/mL) + probiotic (0.5%) or curcumin (0.5%)) resulted in complete detoxification of *Fusarium* mycotoxins [72]. Hassan et al. [16] reported alteration in the gene expression profile of the ZnO and essential oil treated *E. coli* and *A. flavus* through RT-PCR studies that helped to elucidate the efficacy of the treatments. When the treatment doses of ZnO NPs, cinnamon oil, and olive oil increased, the AflR and Stx toxin genes expression efficacy, the molecular weight of DNA, and cycle threshold were decreased. The synergistic activity used lower doses of combined form than each alone. Hamza et al. [112] used hybrid β-glucan mannan lipid particles (GMLPs)-humic acid iron nanoparticles (HA-FeNPs) as an AFB1 binder provides a high binding capacity and a safe enhanced mycotoxin binding material.

3.3. Cancer Theragnostics

Nanoantifungals and their hybrids have the potential to penetrate cancer cells and accumulate in the cytoplasm of cancer cells leading to damage, followed by an inhibitory action leading to death [111]. Similar properties of nanoparticles can help in early and reliable detection of various types of cancer tissues [112]. Hybrid nanocarrier and natural body sugars-derived NPs can help in carrying and release of drugs as in cases of lung cancer [88].

3.3.1. Cancer Therapeutic Applications

Nanoantifungal Agents and Their Hybrids

Zn NPs enable killing of the tumor cells that may also help to preserve the immune cells intact and this activity can be used for both tumor detection and therapy (cancer theranostics) simultaneously [113]. Anticancer drug-functionalized iron or zinc nanoparticles can improve the adherence of the drug-NP conjugate to the target tumor cell and can ensure targeted release of the anticancer drugs to malignant or tumor cells [114–116]. Another study detected that magnetic NPs encapsulated with silica can be effectively used as antitumor drugs [117]. Polymeric nanoparticles such as solid lipid nanoparticles and dendrimers are also potent smart nanovehicles ensuring the targeted delivery of drug molecules in cancer tissues. Nanoencapsulation of 5-fluorouracil in solid lipid nanoparticles improved the specific problems associated with rapid metabolism and shorter life time [118]. This nanovehicle therefore improved the use of the 5-fluorouracil for treatment of colorectal cancer conditions. Meena et al. [40] and Hassan et al. [3] detected the major benefits of CNTs fungal and tumor infections treatment. Xie et al. [119] successfully demonstrated the relevancy of a carbon nanoparticle suspension injection for the diagnosis of thyroid carcinoma.

Nanocomposites

The combination of nanomaterials and drugs, sugars, proteins and DNA potentiated detection and control of animal tumors [120]. The parental inoculation of nanocomposite of Au NPs and gum Arabic act as fluorescent agents in canine cancer therapy [121]. Osama et al. [76] found a significant ability of liposomes to reach the targeted tumors tissues and effective drugs release. Osama et al. [76] estimated the viability of liposome hybrid NPs in detection and therapy of canine tumors in the spleen. Furthermore, dendrimers, the hybrid nanomaterials have the potentials to be conjugated with biological materials and anticancer drugs and released them in targeted tissues in the body and hence excellent tumor detection and treatment were achieved (Figure 4). The combination of MSNPs/folic acid resulted in possibilities of direction of a drug to tumor cells and hence treatment of tumors in mice [122].

3.3.2. Cancer Diagnosis Applications

The essential role for the correct disease diagnosis involves clear observation of the affected tissues activities via imaging.

Nanoantifungal-Based Diagnostic Approaches

The majority of antifungal nanomaterials such as magnetic nanoparticles (MNPs) can be employed in MRI imaging of body tissues [123]. A particular benefits of these NPs is their greater ability to penetrate through the cell membranes and reach blood supply to contrasts of targeted cells as canine stem cells [124]. Superparamagnetic iron oxide nanoparticles functionalized with PEG and ^{64}Cu exhibited encouraging PET and MRI imaging properties besides possessing good stability [125]. Packed graphene oxide also possesses significant potential for quick and sensitive detection and treatment of infections [126]. QDs, the semiconductor materials exhibit huge potential for disease diagnostic applications [127–129].

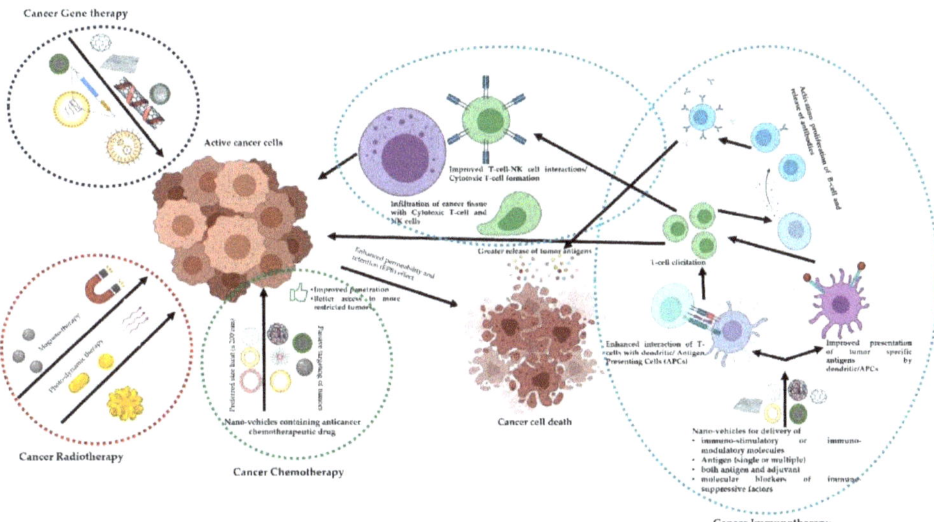

Figure 4. Immuno-modulatory and other functions of nano-antifungals for cancer therapeutics.

Nanocomposites

QDs as Co@Cd-Se core-shell nanocomposites and FePt-Zn nanosponges have fluorescence properties that help in imaging biological events [130]. However, the conjugation of QDs with biological materials (AS enzymes, antibodies and DNA) caused markers imaged by a fluorescence signal [131,132]. In addition, QDs are more photostable than traditional chemical dyes which makes them to be appropriate for use in bioimaging [133,134].

3.4. Nanoantifungal-Enabled Improved Animal Nutrition, and Breeding

Recent research focused on the use of nanomaterials for improving the efficiency of animal production has gained the attention of veterinary experts. Some relevant aspects include the reports on the supplementation of CuNPs, ZnNPs and SeNPs in chicken feed that elevated their productivity of egg and meat [135,136]. Addition of ZnO NPs to broiler chick feeds resulted in elevation of their health status and growth performance [137,138]. Moreover, multiplexed positive effects of these nanomaterials can be identified such as the fact these materials increased the growth rates, reproductive viability, and meat and egg quality of animals and poultry [4]. Also, the supplementation of coated nanomaterials kept their viability against the worst environmental conditions such as digestive enzymes, light and oxidation [40,76]. Another study on injection of Ag NPs alone or in combination with cysteine/threonine amino acids in chicken embryos increased the formation of breast tissue and also improved the chicken immunity through the immunomodulatory properties of the NPs [139]. Also, semiconductor QDs have been successfully utilized for the detection and imaging of physiological events related to functioning of the spermatozoa and female gametes [140] and imaging of fertilization events in male pig gonadal tissue [128]. QDs have the potential to determine the spermatozoon and oocyte movements, hence significant improvement in animal production occurred [12]. NPs can be used not just imaging for the elucidation of the gamete functions, but also as antibody or lectin conjugated metals for the segregation or fractionation of abnormal sperm from active healthy sperm if the functionalized antibodies can detect the defective sperms [141].

Antifungal nanomaterials showed significant activities for elevation of the efficacy of animal reproduction aspects [9,12]. Nanomaterials can be utilized to improve the life and efficacy of preserved semen specimens. The supplementation of polyethyleneimine or propyltriethoxysilane-functionalized mesoporous silica NPs did not exhibit any negative impact on any sperm activity-related properties [142]. Thus, these NPs may help in preservation of the semen quality during in vitro artificial insemination [142]. Another report showcased an improvement in the fertilization potential of the buffalo sperm on addition of titanium oxide NPs (TiO_2 NPs, 10 μg/mL) [143].

Administration of NPs of antioxidant compounds or vitamins can improve the ability of the organism to withstand and avoid oxidative stresses. Oral administration of α-tocopherol NPs in equine animals showed significantly improved rates of absorption and α-tocopherol plasma levels because of maintenance of the high oxidative status in race horses undergoing strenuous training [144]. The pig health status can be elevated by supplementation of micellar NPs conjugated with vitamin E to pigs [145]. Nanosized nutrients and vitamins used as a feed additive in feeds and pass through the alimentary tract of an animal to the blood vessels and distributed to different biological tissues cause their significant improvement [146].

4. Nanoantifungals: Can These Be the Future Innovations in Veterinary Biomedicine?

4.1. Mechanism of Action of Nanomaterials as Antifungal Agents

Nanomaterial possess a range of activities to inhibit the growth and multiplication of fungal-pathogens resulting in cell damage and loss of functions [3,5,35,38]. Nanomaterials exhibit a large surface area compared to the corresponding bulk materials [147]. These materials interact with the various biomolecules in the biological milieu eliciting formation of reactive oxygen species. The action of several nanoantifungals leads to an augmentation of intracellular ROS, an important mediator for exerting antifungal effects. The antifungal activity of nanosilver has been associated with the induction of mitochondrial dysfunctional apoptosis through an increase in oxidative stress via ROS generation especially hydroxyl radicals [148]. The ROS generation is initiated as a response to attachment of antifungal nanomaterial with targeted cells leading to elaboration of O_2 atom and metal ions [149], whereas, the elaborated O_2 increases the oxidative stress causing damage of the mitochondria proteins, leading to denaturation and loss of their functions. These potentials of ROS production have been observed on supplementation of C_{60} fullerenes, SWNTs, and QDs [1,5,150].

4.2. Cytoxicity Risks of the Use of Nanoantifungal Agents

The continuous awareness about the toxicity risk of nanomaterials to animal and the environment have led to refusals of applications of nanomaterials in animal science by several international authorities [151]. The toxicity of nanomaterials can be affected by a variety of factors such as particle size, dose level, type of animal species and the period of exposure [94] and the physico-chemical characters of the nanomaterials used [152]. Chronic exposure of buffalo sperm to ZnNPs and TiONPs (100 μg/mL) caused several abnormalities resulting in suppression of viability and diminished fertility [143], while, sperm exposure to 100–500 μg/mL of Zn NPs caused their damage and rapid death [153]. Hence, the estimation of safe doses of the used nanomaterial should be investigated in laboratory animals before application to field animals [35,72,94].

Furthermore, upon ingestion of nanomaterials by humans and animals they enter the alimentary tract, reach the circulatory system and are carried over via the liver and spleen [35,154,155], whereas, inhalation and skin exposure to nanomaterials allows for their penetration through the skin tissues and nerve cells [152,155]. Inhalation of TiO_2 NPs was identified to have an effect in the development of lung cancer [154]. When NPs reach blood vessels, pathological effects occur such as blood clots and disorders in the cardiovascular system functions [156]. Inhalation of low doses of TiO_2 NPs can cause vascular disorders in rats [157], besides inhalation of single wall and multiwall CNTs [158,159]. We have little knowledge about toxicity and the journey of the nanoparticles in the animal body from the site of administration, passage through absorption, blood vessels, distribution in body tissues and their further journey. Hence, broad toxicological studies are needed before launching commercially viable nanotechnology applications in biomedicine and animal health.

4.3. Safety Concerns of Nanoantifungals

There are many challenges related to the potentially toxic effects of nanomaterials. Incorporation of nanomaterials into polymeric hydrogel matrices may reduce the toxicity and improve its efficacy because of sustained and controlled release of the incorporated NPs. The effective delivery of the nanomaterials can be ensured by their functionalization with polymers at low doses to avoid elicitation of the cellular toxicity [1,3,5,160]. Moreover, several benefits of nanomaterials use for improvement in biomedical applications have also been realized. Although, information related to their harmful impacts is not sufficient and special attention is required for identification of their toxicity risk before practical biomedical applications can be approved for use.

5. Conclusions and Future Perspectives

Over the past decades, nanotechnology has offered progressive novel advances to improve animal health and production. Today, several nanomaterials are used as nanoantifungals, besides having other benefits such as disease detection, diagnosis and therapy, use of additives to animal feed and their products, and finally food safety. The essential therapeutic and preventive activities of nanoantifungals, particularly the zinc and copper nanomaterials, have been evaluated against a variety of fungal diseases and mycotoxicosis in animals. Also, super paramagnetic iron, semiconductor quantum dots and gold nanoparticles are finding applications for early and sensitive detection followed by detailed prognosis and therapy of cancer. Both inorganic and organic polymeric nanomaterials have also been utilized for targeted delivery of various vaccines, quick and on-site detection of pathogens or their signature protein and other biological molecules. The mechanisms of nanoantifungal activity are related to their ability to penetrate the cell membrane, damage the cytoplasmic contents, leading to loss of function and death of the cells. Therefore, further studies illustrating the cellular toxicity mechanisms that result in oxidative stress and leading to genotoxicity and cancers need a detailed evaluation to manipulate the roles of nanomaterial in animal health. Moreover, the toxicity risks of nanomaterials must be determined before application of nanomaterials in veterinary medicine for safeguarding the health of the animals and their role in animal production.

Author Contributions: Conceptualization, A.A.H., M.A.A.; resources, A.A.H. writing—original draft preparation, A.A.M.E.H., A.K.; writing—review and editing, R.M.H.S.E.A., K.K., K.A.A.-E.; supervision, K.A.A.-E.; funding acquisition, K.K., P.O. All authors have read and agreed to the published version of the manuscript.

Funding: This work was supported by the project UHK VT2019-2021.

Institutional Review Board Statement: Not applicable.

Informed Consent Statement: Not applicable.

Data Availability Statement: Not applicable.

Acknowledgments: The authors wish to express their grateful appreciation and heartfelt thanks to M.K. Refai, Professor of Microbiology, Cairo University, Egypt for his continuous assistance and advice for initiating this work.

Conflicts of Interest: The authors declare no conflict of interest.

References

1. Hassan, A.A.; Sayed-ElAhl, R.M.H.; Oraby, N.H.; El-Hamaky, A.M.A. Metal nanoparticles for management of mycotoxigenic fungi and mycotoxicosis diseases of animals and poultry. In *Nanomycotoxicology*; Elsevier: Amsterdam, The Netherlands, 2020; pp. 251–269; ISBN 9780128179987.
2. Tiew, P.Y.; Mac Aogain, M.; Ali, N.A.B.M.; Thng, K.X.; Goh, K.; Lau, K.J.X.; Chotirmall, S.H. The Mycobiome in Health and Disease: Emerging Concepts, Methodologies and Challenges. *Mycopathologia* **2020**, *185*, 207–231. [CrossRef]

3. Hassan, A.A.; Mansour, M.K.; Sayed-ElAhl, R.M.H.; El Hamaky, A.M.A.; Oraby, N.H. Toxic and beneficial effects of carbon nanomaterials on human and animal health. In *Carbon Nanomaterials for Agri-Food and Environmental Applications*; Elsevier: Amsterdam, The Netherlands, 2020; pp. 535–555; ISBN 9780128197868.
4. Fesseha, H.; Degu, T.; Getachew, Y. Nanotechnology and its Application in Animal Production: A Review. *Vet. Med. Open J.* **2020**, *5*, 43–50. [CrossRef]
5. Hassan, A.A.; Mansour, M.K.; El Hamaky, A.M.; Sayed-ElAhl, R.M.H.; Oraby, N.H. *Nanomaterials and Nanocomposite Applications in Veterinary Medicine*; Elsevier: Amsterdam, The Netherlands, 2020; ISBN 9780128213544.
6. Brunet, K.; Alanio, A.; Lortholary, O.; Rammaert, B. Reactivation of dormant/latent fungal infection. *J. Infect.* **2018**, *77*, 463–468. [CrossRef]
7. Di Mambro, T.; Guerriero, I.; Aurisicchio, L.; Magnani, M.; Marra, E. The yin and yang of current antifungal therapeutic strategies: How can we harness our natural defenses? *Front. Pharmacol.* **2019**, *10*, 1–11. [CrossRef] [PubMed]
8. Gintjee, T.J.; Donnelley, M.A.; Thompson, G.R. Aspiring Antifungals: Review of Current Antifungal Pipeline Developments. *J. Fungi* **2020**, *6*, 28. [CrossRef]
9. El-Sayed, A.; Kamel, M. Advanced applications of nanotechnology in veterinary medicine. *Environ. Sci. Pollut. Res.* **2020**, *27*, 19073–19086. [CrossRef]
10. Youssef, F.S.; El-Banna, H.A.; Elzorba, H.Y.; Galal, A.M. Application of some nanoparticles in the field of veterinary medicine. *Int. J. Vet. Sci. Med.* **2019**, *7*, 78–93. [CrossRef]
11. Saragusty, J.; Arav, A. Current progress in oocyte and embryo cryopreservation by slow freezing and vitrification. *Reproduction* **2011**, *141*, 1–19. [CrossRef]
12. Hill, E.K.; Li, J. Current and future prospects for nanotechnology in animal production. *J. Anim. Sci. Biotechnol.* **2017**, *8*, 1–13. [CrossRef]
13. Souza, A.C.O.; Amaral, A.C. Antifungal therapy for systemic mycosis and the nanobiotechnology era: Improving efficacy, biodistribution and toxicity. *Front. Microbiol.* **2017**, *8*, 1–13. [CrossRef]
14. Sousa, F.; Ferreira, D.; Reis, S.; Costa, P. Current insights on antifungal therapy: Novel nanotechnology approaches for drug delivery systems and new drugs from natural sources. *Pharmaceuticals* **2020**, *13*, 248. [CrossRef]
15. Martínez-Montelongo, J.H.; Medina-Ramírez, I.E.; Romo-Lozano, Y.; González-Gutiérrez, A.; Macías-Díaz, J.E. Development of nano-antifungal therapy for systemic and endemic mycoses. *J. Fungi* **2021**, *7*, 158. [CrossRef]
16. Hassan, A.A.; Abo-Zaid, K.F.; Oraby, N.H. Molecular and conventional detection of antimicrobial activity of zinc oxide nanoparticles and cinnamon oil against escherichia coli and aspergillus flavus. *Adv. Anim. Vet. Sci.* **2020**, *8*, 839–847. [CrossRef]
17. Torres-Sangiao, E.; Holban, A.M.; Gestal, M.C. Advanced nanobiomaterials: Vaccines, diagnosis and treatment of infectious diseases. *Molecules* **2016**, *21*, 867. [CrossRef]
18. Riley, M.K.; Vermerris, W. Recent advances in nanomaterials for gene delivery—A review. *Nanomaterials* **2017**, *7*, 94. [CrossRef]
19. Ferraz, M.P.; Mateus, A.Y.; Sousa, J.C.; Monteiro, F.J. Nanohydroxyapatite microspheres as delivery system for antibiotics: Release kinetics, antimicrobial activity, and interaction with osteoblasts. *J. Biomed. Mater. Res. Part A* **2007**, *81A*, 994–1004. [CrossRef]
20. Prabhu, R.H.; Patravale, V.B.; Joshi, M.D. Polymeric nanoparticles for targeted treatment in oncology: Current insights. *Int. J. Nanomedicine* **2015**, *10*, 1001–1018. [CrossRef]
21. Krishnan, S.R.; George, S.K. Nanotherapeutics in Cancer Prevention, Diagnosis and Treatment. *Pharmacol. Ther.* **2014**. [CrossRef]
22. Mishra, B.; Patel, B.B.; Tiwari, S. Colloidal nanocarriers: A review on formulation technology, types and applications toward targeted drug delivery. *Nanomed. Nanotechnol. Biol. Med.* **2010**, *6*, 9–24. [CrossRef]
23. Mohanty, N.N.; Palai, T.K.; Prusty, B.R.; Mohapatra, J.K. An Overview of Nanomedicine in Veterinary Science. *Vet. Res. Int.* **2014**, *2*, 90–95.
24. Elgqvist, J. Nanoparticles as theranostic vehicles in experimental and clinical applications-focus on prostate and breast cancer. *Int. J. Mol. Sci.* **2017**, *18*, 1102. [CrossRef]
25. De Serrano, L.O.; Burkhart, D.J. Liposomal vaccine formulations as prophylactic agents: Design considerations for modern vaccines. *J. Nanobiotechnology* **2017**, *15*, 1–23. [CrossRef]
26. Jurj, A.; Braicu, C.; Pop, L.A.; Tomuleasa, C.; Gherman, C.D.; Berindan-Neagoe, I. The new era of nanotechnology, an alternative to change cancer treatment. *Drug Des. Devel. Ther.* **2017**, *11*, 2871–2890. [CrossRef]
27. Nagarsekar, K.; Ashtikar, M.; Thamm, J.; Steiniger, F.; Schacher, F.; Fahr, A.; May, S. Electron microscopy and theoretical modeling of cochleates. *Langmuir* **2014**, *30*, 13143–13151. [CrossRef]
28. Pawar, A.; Bothiraja, C.; Shaikh, K.; Mali, A. An insight into cochleates, a potential drug delivery system. *RSC Adv.* **2015**, *5*, 81188–81202. [CrossRef]
29. Kischkel, B.; Rossi, S.A.; Santos, S.R.; Nosanchuk, J.D.; Travassos, L.R.; Taborda, C.P. Therapies and Vaccines Based on Nanoparticles for the Treatment of Systemic Fungal Infections. *Front. Cell. Infect. Microbiol.* **2020**, *10*. [CrossRef]
30. Aigner, M.; Lass-Flörl, C. Encochleated amphotericin B: Is the oral availability of amphotericin B finally reached? *J. Fungi* **2020**, *6*, 66. [CrossRef]
31. Faustino, C.; Pinheiro, L. Lipid systems for the delivery of amphotericin B in antifungal therapy. *Pharmaceutics* **2020**, *12*, 29. [CrossRef]
32. Vikrama Chakravarthi, P.; Balaji, S.N. Applications of nanotechnology in veterinary medicine. *Vet. World* **2010**, *3*, 477–480. [CrossRef]

33. Aboalnaja, K.O.; Yaghmoor, S.; Kumosani, T.A.; McClements, D.J. Utilization of nanoemulsions to enhance bioactivity of pharmaceuticals, supplements, and nutraceuticals: Nanoemulsion delivery systems and nanoemulsion excipient systems. *Expert Opin. Drug Deliv.* **2016**, *13*, 1327–1336. [CrossRef]
34. Rodríguez-Burneo, N.; Busquets, M.A.; Estelrich, J. Magnetic nanoemulsions: Comparison between nanoemulsions formed by ultrasonication and by spontaneous emulsification. *Nanomaterials* **2017**, *7*, 190. [CrossRef]
35. Hassan, A.A.; Mansour, M.K.; Sayed-ElAhl, R.M.H.; Tag El-Din, H.A.; Awad, M.E.A.; Younis, E.M. Influence of Selenium Nanoparticles on The Effects of Poisoning with Aflatoxins. *Adv. Anim. Vet. Sci.* **2020**, *8*. [CrossRef]
36. Huang, W.; Yan, M.; Duan, H.; Bi, Y.; Cheng, X.; Yu, H. Synergistic Antifungal Activity of Green Synthesized Silver Nanoparticles and Epoxiconazole against Setosphaeria turcica. *J. Nanomater.* **2020**, *2020*. [CrossRef]
37. Kischkel, B.; De Castilho, P.F.D.; De Oliveira, K.M.P.; Rezende, P.S.T.; Bruschi, M.L.; Svidzinski, T.I.E.; Negri, M.; Negri, M. Silver nanoparticles stabilized with propolis show reduced toxicity and potential activity against fungal infections. *Future Microbiol.* **2020**, *15*, 521–539. [CrossRef] [PubMed]
38. Nabawy, G.A.; Hassan, A.A.; Sayed-ElAhl, R.M.H.; Refai, M.K. Effect of Metal Nanoparticles in Comparison With Commercial Antifungal Feed Additives on the Growth of Aspergillus Flavus and Aflatoxin B1 Production. *J. Glob. Biosci.* **2014**, *3*, 954–971.
39. Manuja, A.; Kumar, B.; Singh, R.K. Nanotechnology developments: Opportunities for animal health and production. *Nanotechnol. Dev.* **2012**, *2*, 4. [CrossRef]
40. Meena, N.S.; Sahni, Y.P.; Singh, R.P. Applications of nanotechnology in veterinary therapeutics. *J. Entomol. Zool. Stud.* **2018**, *6*, 167–175.
41. Dahman, Y. Nanoshells. In *Nanotechnology and Functional Materials for Engineers*; Elsevier: Amsterdam, The Netherlands, 2017; pp. 175–190; ISBN 9780323512565.
42. Loo, C.; Lin, A.; Hirsch, L.; Lee, M.H.; Barton, J.; Halas, N.; West, J.; Drezek, R. Nanoshell-Enabled Photonics-Based Imaging and Therapy of Cancer. *Technol. Cancer Res. Treat.* **2004**, *3*, 33–40. [CrossRef]
43. Loo, C.; Lin, A.; Hirsch, L.; Lee, M.H.; Barton, J.; Halas, N.; West, J.; Drezek, R. Diagnostic and Therapeutic Applications of Metal Nanoshells. *Nanofabrication Towar. Biomed. Appl. Tech. Tools Appl. Impact* **2005**, 327–342. [CrossRef]
44. Nghiem, T.H.L.; Le, T.N.; Do, T.H.; Vu, T.T.D.; Do, Q.H.; Tran, H.N. Preparation and characterization of silica-gold core-shell nanoparticles. *J. Nanoparticle Res.* **2013**, *15*. [CrossRef]
45. Mochizuki, C.; Nakamura, J.; Nakamura, M. Development of non-porous silica nanoparticles towards cancer photo-theranostics. *Biomedicines* **2021**, *9*, 73. [CrossRef] [PubMed]
46. Dobrovolskaia, M.A.; Shurin, M.; Shvedova, A.A. Current understanding of interactions between nanoparticles and the immune system. *Toxicol. Appl. Pharmacol.* **2016**, *299*, 78–89. [CrossRef] [PubMed]
47. Chowdhury, A.; Kunjiappan, S.; Panneerselvam, T.; Somasundaram, B.; Bhattacharjee, C. Nanotechnology and nanocarrier-based approaches on treatment of degenerative diseases. *Int. Nano Lett.* **2017**, *7*, 91–122. [CrossRef]
48. Reilly, R.M. Carbon nanotubes: Potential benefits and risks of nanotechnology in nuclear medicine. *J. Nucl. Med.* **2007**, *48*, 1039–1042. [CrossRef] [PubMed]
49. Rapoport, N.; Gao, Z.; Kennedy, A. Multifunctional nanoparticles for combining ultrasonic tumor imaging and targeted chemotherapy. *J. Natl. Cancer Inst.* **2007**, *99*, 1095–1106. [CrossRef] [PubMed]
50. Bhandari, P.; Novikova, G.; Goergen, C.J.; Irudayaraj, J. Ultrasound beam steering of oxygen nanobubbles for enhanced bladder cancer therapy. *Sci. Rep.* **2018**, *8*, 1–10. [CrossRef] [PubMed]
51. Song, L.; Wang, G.; Hou, X.; Kala, S.; Qiu, Z.; Wong, K.F.; Cao, F.; Sun, L. Biogenic nanobubbles for effective oxygen delivery and enhanced photodynamic therapy of cancer. *Acta Biomater.* **2020**, *108*, 313–325. [CrossRef]
52. Shen, S.; Li, Y.; Xiao, Y.; Zhao, Z.; Zhang, C.; Wang, J.; Li, H.; Liu, F.; He, N.; Yuan, Y.; et al. Folate-conjugated nanobubbles selectively target and kill cancer cells via ultrasound-triggered intracellular explosion. *Biomaterials* **2018**, *181*, 293–306. [CrossRef]
53. Khan, M.S.; Hwang, J.; Lee, K.; Choi, Y.; Seo, Y.; Jeon, H.; Hong, J.W.; Choi, J. Anti-tumor drug-loaded oxygen nanobubbles for the degradation of HIF-1α and the upregulation of reactive oxygen species in tumor cells. *Cancers* **2019**, *11*, 1464. [CrossRef]
54. Underwood, C.; van Eps, A.W. Nanomedicine and veterinary science: The reality and the practicality. *Vet. J.* **2012**, *193*, 12–23. [CrossRef]
55. Moyer, T.J.; Zmolek, A.C.; Irvine, D.J. Beyond antigens and adjuvants: Formulating future vaccines. *J. Clin. Invest.* **2016**, *126*, 799–808. [CrossRef] [PubMed]
56. Qasim Nasar, M.; Zohra, T.; Khalil, A.T.; Saqib, S.; Ayaz, M.; Ahmad, A.; Shinwari, Z.K. Seripheidium quettense mediated green synthesis of biogenic silver nanoparticles and their theranostic applications. *Green Chem. Lett. Rev.* **2019**, *12*, 310–322. [CrossRef]
57. Hassan, A.A.; Howayda, M.E.; Mahmoud, H.H. Effect of Zinc Oxide Nanoparticles on the Growth of Mycotoxigenic Mould. *Stud. Chem. Process Technol.* **2013**, *1*, 66–74.
58. Refai, H.; Badawy, M.; Hassan, A.; Sakr, H.; Baraka, Y. Antimicrobial Effect of Biologically Prepared Silver Nanoparticles (AgNPs) on Two Different Obturator's Soft Linings in Maxillectomy Patients. *Eur. J. Acad. Essays* **2017**, *4*, 15–25.
59. Hassan, A.A.; Mansour, M.K.; Mahmoud, H. Biosynthesis of silver nanoparticles (Ag-Nps) (a model of metals) by Candida albicans and its antifungal activity on Some fungal pathogens (*Trichophyton mentagrophytes* and *Candida albicans*). *N. Y. Sci. J.* **2013**, *6*, 27–34.
60. Al Abboud, M.A. Fungal biosynthesis of silver nanoparticles and their role in control of Fusarium wilt of sweet pepper and soil-borne fungi in vitro. *Int. J. Pharmacol.* **2018**, *14*, 773–780. [CrossRef]

61. Pietrzak, K.; Twaruzek, M.; Czyzowska, A.; Kosicki, R.; Gutarowska, B. Influence of silver nanoparticles on metabolism and toxicity of moulds. *Acta Biochim. Pol.* **2015**, *62*, 851–857. [CrossRef]
62. Abd-Elsalam, K.A.; Hashim, A.F.; Alghuthaymi, M.A.; Said-Galiev, E. Nanobiotechnological strategies for toxigenic fungi and mycotoxin control. In *Food Preservation*; Elsevier: Amsterdam, The Netherlands, 2017; pp. 337–364; ISBN 9780128043035.
63. Hosseini, S.S.; Mohammadi, R.; Joshaghani, H.R.; Eskandari, M. Antifungal effect of Sodium Dodecil Sulfate and Nano particle ZnO on growth inhibition of standard strain of Candida albicans. *J. Gorgan Univ. Med. Sci.* **2011**, *12*, 64–69.
64. Hernández-Meléndez, D.; Salas-Téllez, E.; Zavala-Franco, A.; Téllez, G.; Méndez-Albores, A.; Vázquez-Durán, A. Inhibitory effect of flower-shaped zinc oxide nanostructures on the growth and aflatoxin production of a highly toxigenic strain of Aspergillus flavus Link. *Materials* **2018**, *11*, 1265. [CrossRef]
65. Mouhamed, A.E.; Hassan, A.A.; Hassan, A.; Hariri, M.E.; Refai, M. Effect of Metal Nanoparticles on the Growth of Ochratoxigenic Moulds and Ochratoxin A Production Isolated From Food and Feed. *Int. J. Res. Stud. Biosci.* **2015**, *3*, 1–14.
66. El-Tawab, A.A.A.; El-Hofy, F.I.; Metwally, A. A Comparative Study on Antifungal Activity of Fe_2O_3, and Fe_3O_4 Nanoparticles. *Int. J. Adv. Res.* **2018**, *6*, 189–194. [CrossRef]
67. Kheiri, A.; Moosawi Jorf, S.A.; Mallihipour, A.; Saremi, H.; Nikkhah, M. Application of chitosan and chitosan nanoparticles for the control of *Fusarium* head blight of wheat (*Fusarium graminearum*) *in vitro* and greenhouse. *Int. J. Biol. Macromol.* **2016**, *93*, 1261–1272. [CrossRef]
68. Ahmed, F.; Soliman, F.M.; Adly, M.A.; Soliman, H.A.M.; El-Matbouli, M.; Saleh, M. *In vitro* assessment of the antimicrobial efficacy of chitosan nanoparticles against major fish pathogens and their cytotoxicity to fish cell lines. *J. Fish Dis.* **2020**, *43*, 1049–1063. [CrossRef]
69. Abd-Elsalam, K.A.; Alghuthaymi, M.A.; Shami, A.; Rubina, M.S.; Abramchuk, S.S.; Shtykova, E.V.; Vasil'kov, A.Y. Copper-chitosan nanocomposite hydrogels against aflatoxigenic *Aspergillus flavus* from dairy cattle feed. *J. Fungi* **2020**, *6*, 112. [CrossRef]
70. Anaraki, M.R.; Jangjoo, A.; Alimoradi, F.; Dizaj, S.M. Comparison of Antifungal Properties of Acrylic Resin Reinforced with ZnO and Ag Nanoparticles. *Tabriz Univ. Med. Sci.* **2017**, *23*, 207–214. [CrossRef]
71. Shokrollahi, H. Structure, synthetic methods, magnetic properties and biomedical applications of ferrofluids. *Mater. Sci. Eng. C* **2013**, *33*, 2476–2487. [CrossRef] [PubMed]
72. Atef, H.A.; Mansour, M.K.; Ibrahim, E.M.; Sayed-ElAhl, R.M.H.; Al-Kalamawey, N.M.; El Kattan, Y.A.; Ali, M.A. Efficacy of Zinc Oxide Nanoparticles and Curcumin in Amelioration the Toxic Effects in Aflatoxicated Rabbits. *Int. J. Curr. Microbiol. Appl. Sci.* **2016**, *5*, 795–818. [CrossRef]
73. Sanchez, V.C.; Jachak, A.; Hurt, R.H.; Kane, A.B. Biological Interactions of Graphene-Family Nanomaterials: An Interdisciplinary Review. *Chem. Res. Toxicol.* **2012**, *25*, 15–34. [CrossRef] [PubMed]
74. Hassan, A.A.; Oraby, N.H.; Manal, M.E.-M. Detection of Mycotoxigenic *Fusarium* Species in Poultry Rations and Their Detection of Mycotoxigenic *Fusarium* Species in Poultry Rations and Their Growth Control by Zinc Nanoparticles. *Anim. Health Res. J.* **2019**, *7*, 1075–1091.
75. Wang, J.J.; Liu, B.H.; Hsu, Y.T.; Yu, F.Y. Sensitive competitive direct enzyme-linked immunosorbent assay and gold nanoparticle immunochromatographic strip for detecting aflatoxin M1 in milk. *Food Control* **2011**, *22*, 964–969. [CrossRef]
76. Osama, E.; El-Sheikh, S.M.A.; Khairy, M.H.; Galal, A.A.A. Nanoparticles and their potential applications in veterinary medicine. *J. Adv. Vet. Res.* **2020**, *10*, 268–273.
77. Hamad, K.M.; Mahmoud, N.N.; Al-Dabash, S.; Al-Samad, L.A.; Abdallah, M.; Al-Bakri, A.G. Fluconazole conjugated-gold nanorods as an antifungal nanomedicine with low cytotoxicity against human dermal fibroblasts. *RSC Adv.* **2020**, *10*, 25889–25897. [CrossRef]
78. Carvalho, G.C.; Sábio, R.M.; de Cássia Ribeiro, T.; Monteiro, A.S.; Pereira, D.V.; Ribeiro, S.J.L.; Chorilli, M. Highlights in Mesoporous Silica Nanoparticles as a Multifunctional Controlled Drug Delivery Nanoplatform for Infectious Diseases Treatment. *Pharm. Res.* **2020**, *37*. [CrossRef]
79. Vallet-Regí, M. Mesoporous Silica Nanoparticles: Their Projection in Nanomedicine. *ISRN Mater. Sci.* **2012**, *2012*, 1–20. [CrossRef]
80. Castillo, R.R.; Lozano, D.; Vallet-Regí, M. Mesoporous silica nanoparticles as carriers for therapeutic biomolecules. *Pharmaceutics* **2020**, *12*, 432. [CrossRef]
81. Kanugala, S.; Jinka, S.; Puvvada, N.; Banerjee, R.; Kumar, C.G. Phenazine-1-carboxamide functionalized mesoporous silica nanoparticles as antimicrobial coatings on silicone urethral catheters. *Sci. Rep.* **2019**, *9*, 1–16. [CrossRef] [PubMed]
82. Montazeri, M.; Razzaghi-Abyaneh, M.; Nasrollahi, S.A.; Maibach, H.; Nafisi, S. Enhanced topical econazole antifungal efficacy by amine-functionalized silica nanoparticles. *Bull. Mater. Sci.* **2020**, *43*. [CrossRef]
83. Deaguero, I.G.; Huda, M.N.; Rodriguez, V.; Zicari, J.; Al-hilal, T.A.; Badruddoza, A.Z.M.; Nurunnabi, M. Nano-vesicle based anti-fungal formulation shows higher stability, skin diffusion, biosafety and anti-fungal efficacy *in vitro*. *Pharmaceutics* **2020**, *12*, 516. [CrossRef]
84. Siopi, M.; Mouton, J.W.; Pournaras, S.; Meletiadis, J. In Vitro and In Vivo Exposure-Effect Relationship of Liposomal Amphotericin B against *Aspergillus fumigatus*. *Antimicrob. Agents Chemother.* **2019**, *63*, 1–7. [CrossRef]
85. Kunjumon, S.; Krishnakumar, K.; Nair, S.K. Nanomicelles Formulation: In Vitro Anti-Fungal Study. *Int. J. Pharm. Sci. Rev. Res.* **2020**, *62*, 78–81.
86. Lee, A.L.Z.; Wang, Y.; Pervaiz, S.; Fan, W.; Yang, Y.Y. Synergistic Anticancer Effects Achieved by Co-Delivery of TRAIL and Paclitaxel Using Cationic Polymeric Micelles. *Macromol. Biosci.* **2011**, *11*, 296–307. [CrossRef]

87. Vail, D.M.; von Euler, H.; Rusk, A.W.; Barber, L.; Clifford, C.; Elmslie, R.; Fulton, L.; Hirschberger, J.; Klein, M.; London, C.; et al. A Randomized Trial Investigating the Efficacy and Safety of Water Soluble Micellar Paclitaxel (Paccal Vet) for Treatment of Nonresectable Grade 2 or 3 Mast Cell Tumors in Dogs. *J. Vet. Intern. Med.* **2012**, *26*, 598–607. [CrossRef]
88. Malachowski, T.; Hassel, A. Engineering nanoparticles to overcome immunological barriers for enhanced drug delivery. *Eng. Regen.* **2020**, *1*, 35–50. [CrossRef]
89. Kurantowicz, N.; Strojny, B.; Sawosz, E.; Jaworski, S.; Kutwin, M.; Grodzik, M.; Wierzbicki, M.; Lipińska, L.; Mitura, K.; Chwalibog, A. Biodistribution of a High Dose of Diamond, Graphite, and Graphene Oxide Nanoparticles After Multiple Intraperitoneal Injections in Rats. *Nanoscale Res. Lett.* **2015**, *10*. [CrossRef]
90. Ajmal, M.; Yunus, U.; Matin, A.; Haq, N.U. Synthesis, characterization and in vitro evaluation of methotrexate conjugated fluorescent carbon nanoparticles as drug delivery system for human lung cancer targeting. *J. Photochem. Photobiol. B Biol.* **2015**, *153*, 111–120. [CrossRef] [PubMed]
91. Kim, J.M.; Kim, D.H.; Park, H.J.; Ma, H.W.; Park, I.S.; Son, M.; Ro, S.Y.; Hong, S.; Han, H.K.; Lim, S.J.; et al. Nanocomposites-based targeted oral drug delivery systems with infliximab in a murine colitis model. *J. Nanobiotechnology* **2020**, *18*, 1–13. [CrossRef] [PubMed]
92. Paskiabi, F.A.; Bonakdar, S.; Shokrgozar, M.A.; Imani, M.; Jahanshiri, Z.; Shams-Ghahfarokhi, M.; Razzaghi-Abyaneh, M. Terbinafine-loaded wound dressing for chronic superficial fungal infections. *Mater. Sci. Eng. C* **2017**, *73*, 130–136. [CrossRef] [PubMed]
93. El-Nahass, E.-S.; Moselhy, W.A.; Hassan, N.E.-H.Y.; Hassan, A.A. Evaluation of the Protective Effects of Adsorbent Materials and Ethanolic Herbal Extracts against Aflatoxins Hepatotoxicity in Albino Rats: Histological, Morphometric and Immunohistochemical Study. *Adv. Anim. Vet. Sci.* **2019**, *7*, 1140–1147. [CrossRef]
94. Abd El-Fatah, S.; Bakry, H.; Abo Salem, M.; Hassan, A. Comparative Study between the Use of Bulk and Nanoparticles of Zinc Oxide in Amelioration the Toxic Effects of Aflatoxins in rats. *Benha Vet. Med. J.* **2017**, *33*, 329–342. [CrossRef]
95. Gholami-Ahangaran, M.; Zia-Jahromi, N. Nanosilver effects on growth parameters in experimental aflatoxicosis in broiler chickens. *Toxicol. Ind. Health* **2013**, *29*, 121–125. [CrossRef]
96. Gholami-Ahangaran, M.; Zia-Jahromi, N. Effect of nanosilver on blood parameters in chickens having aflatoxicosis. *Toxicol. Ind. Health* **2014**, *30*, 192–196. [CrossRef]
97. Asghari-Paskiabi, F.; Imani, M.; Rafii-Tabar, H.; Razzaghi-Abyaneh, M. Physicochemical properties, antifungal activity and cytotoxicity of selenium sulfide nanoparticles green synthesized by *Saccharomyces cerevisiae*. *Biochem. Biophys. Res. Commun.* **2019**, *516*, 1078–1084. [CrossRef]
98. Fadl, S.E.; El-Shenawy, A.M.; Gad, D.M.; El Daysty, E.M.; El-Sheshtawy, H.S.; Abdo, W.S. Trial for reduction of Ochratoxin A residues in fish feed by using nano particles of hydrated sodium aluminum silicates (NPsHSCAS) and copper oxide. *Toxicon* **2020**, *184*, 1–9. [CrossRef]
99. Gibson, N.M.; Luo, T.J.M.; Brenner, D.W.; Shenderova, O. Immobilization of mycotoxins on modified nanodiamond substrates. *Biointerphases* **2011**, *6*, 210–217. [CrossRef]
100. Asghar, M.A.; Zahir, E.; Shahid, S.M.; Khan, M.N.; Asghar, M.A.; Iqbal, J.; Walker, G. Iron, copper and silver nanoparticles: Green synthesis using green and black tea leaves extracts and evaluation of antibacterial, antifungal and aflatoxin B1 adsorption activity. *LWT Food Sci. Technol.* **2018**, *90*, 98–107. [CrossRef]
101. Moghaddam, S.H.M.; Jebali, A.; Daliri, K. The Use of Mgo-Sio2 Nanocomposite for Adsorption of Aflatoxin in Wheat Flour Samples. In Proceedings of the NanoCon 2010, Olomouc, Czech Republic, 12–14 October 2010; pp. 10–15.
102. Scorzoni, L.; de Paula e Silva, A.C.A.; Marcos, C.M.; Assato, P.A.; de Melo, W.C.M.A.; de Oliveira, H.C.; Costa-Orlandi, C.B.; Mendes-Giannini, M.J.S.; Fusco-Almeida, A.M. Antifungal therapy: New advances in the understanding and treatment of mycosis. *Front. Microbiol.* **2017**, *8*, 1–23. [CrossRef] [PubMed]
103. Arias, L.S.; Pessan, J.P.; de Souza Neto, F.N.; Lima, B.H.R.; de Camargo, E.R.; Ramage, G.; Delbem, A.C.B.; Monteiro, D.R. Novel nanocarrier of miconazole based on chitosan-coated iron oxide nanoparticles as a nanotherapy to fight *Candida* biofilms. *Colloids Surf. B Biointerfaces* **2020**, *192*, 111080. [CrossRef]
104. Araujo, H.C.; da Silva, A.C.G.; Paião, L.I.; Magario, M.K.W.; Frasnelli, S.C.T.; Oliveira, S.H.P.; Pessan, J.P.; Monteiro, D.R. Antimicrobial, antibiofilm and cytotoxic effects of a colloidal nanocarrier composed by chitosan-coated iron oxide nanoparticles loaded with chlorhexidine. *J. Dent.* **2020**, *101*. [CrossRef] [PubMed]
105. Zahoor, M.; Ali Khan, F. Adsorption of aflatoxin B1 on magnetic carbon nanocomposites prepared from bagasse. *Arab. J. Chem.* **2018**, *11*, 729–738. [CrossRef]
106. Pirouz, A.A.; Selamat, J.; Iqbal, S.Z.; Mirhosseini, H.; Karjiban, R.A.; Bakar, F.A. The use of innovative and efficient nanocomposite (magnetic graphene oxide) for the reduction on of Fusarium mycotoxins in palm kernel cake. *Sci. Rep.* **2017**, *7*, 1–9. [CrossRef] [PubMed]
107. Gao, R.; Meng, Q.; Li, J.; Liu, M.; Zhang, Y.; Bi, C.; Shan, A. Modified halloysite nanotubes reduce the toxic effects of zearalenone in gestating sows on growth and muscle development of their offsprings. *J. Anim. Sci. Biotechnol.* **2016**, *7*, 1–9. [CrossRef]
108. Ji, J.; Xie, W. Detoxification of Aflatoxin B1 by magnetic graphene composite adsorbents from contaminated oils. *J. Hazard. Mater.* **2020**, *381*, 120915. [CrossRef]

109. González-Jartín, J.M.; de Castro Alves, L.; Alfonso, A.; Piñeiro, Y.; Vilar, S.Y.; Gomez, M.G.; Osorio, Z.V.; Sainz, M.J.; Vieytes, M.R.; Rivas, J.; et al. Detoxification agents based on magnetic nanostructured particles as a novel strategy for mycotoxin mitigation in food. *Food Chem.* **2019**, *294*, 60–66. [CrossRef]
110. Zhai, X.; Zhang, C.; Zhao, G.; Stoll, S.; Ren, F.; Leng, X. Antioxidant capacities of the selenium nanoparticles stabilized by chitosan. *J. Nanobiotechnology* **2017**, *15*, 1–12. [CrossRef]
111. Luo, Y.; Zhou, Z.; Yue, T. Synthesis and characterization of nontoxic chitosan-coated Fe_3O_4 particles for patulin adsorption in a juice-pH simulation aqueous. *Food Chem.* **2017**, *221*, 317–323. [CrossRef]
112. Hamza, Z.; El-Hashash, M.; Aly, S.; Hathout, A.; Soto, E.; Sabry, B.; Ostroff, G. Preparation and characterization of yeast cell wall beta-glucan encapsulated humic acid nanoparticles as an enhanced aflatoxin B1 binder. *Carbohydr. Polym.* **2019**, *203*, 185–192. [CrossRef]
113. Nikolova, M.P.; Chavali, M.S. Metal oxide nanoparticles as biomedical materials. *Biomimetics* **2020**, *5*, 27. [CrossRef]
114. Norouzi, M.; Yathindranath, V.; Thliveris, J.A.; Kopec, B.M.; Siahaan, T.J.; Miller, D.W. Doxorubicin-loaded iron oxide nanoparticles for glioblastoma therapy: A combinational approach for enhanced delivery of nanoparticles. *Sci. Rep.* **2020**, *10*, 1–18. [CrossRef] [PubMed]
115. Soetaert, F.; Korangath, P.; Serantes, D.; Fiering, S.; Ivkov, R. Cancer therapy with iron oxide nanoparticles: Agents of thermal and immune therapies. *Adv. Drug Deliv. Rev.* **2020**, *163–164*, 65–83. [CrossRef] [PubMed]
116. Nabil, A.; Elshemy, M.M.; Asem, M.; Abdel-Motaal, M.; Gomaa, H.F.; Zahran, F.; Uto, K.; Ebara, M. Zinc Oxide Nanoparticle Synergizes Sorafenib Anticancer Efficacy with Minimizing Its Cytotoxicity. *Oxid. Med. Cell. Longev.* **2020**, *2020*. [CrossRef] [PubMed]
117. Xu, C.; Sun, S. New forms of superparamagnetic nanoparticles for biomedical applications. *Adv. Drug Deliv. Rev.* **2013**, *65*, 732–743. [CrossRef]
118. Smith, T.; Affram, K.; Nottingham, E.L.; Han, B.; Amissah, F.; Krishnan, S.; Trevino, J.; Agyare, E. Application of smart solid lipid nanoparticles to enhance the efficacy of 5-fluorouracil in the treatment of colorectal cancer. *Sci. Rep.* **2020**, *10*, 1–14. [CrossRef]
119. Xie, P.; Yang, S.T.; He, T.; Yang, S.; Tang, X.H. Bioaccumulation and toxicity of carbon nanoparticles suspension injection in intravenously exposed mice. *Int. J. Mol. Sci.* **2017**, *18*, 2562. [CrossRef]
120. Frank, A.; Eric, M.P.; Robert, L.; Omid, C.F. Nanoparticles Technologies For Cancer Therapy. In *Drug Delievery*; Springer: Berlin/Heidelberg, Germany, 2010; Volume 197, ISBN 978-3-642-00476-6.
121. Axiak-Bechtel, S.M.; Upendran, A.; Lattimer, J.C.; Kelsey, J.; Cutler, C.S.; Selting, K.A.; Bryan, J.N.; Henry, C.J.; Boote, E.; Tate, D.J.; et al. Gum arabic-coated radioactive gold nanoparticles cause no short-term local or systemic toxicity in the clinically relevant canine model of prostate cancer. *Int. J. Nanomedicine* **2014**, *9*, 5001–5011. [CrossRef] [PubMed]
122. Lu, J.; Liong, M.; Li, Z.; Zink, J.I.; Tamanoi, F. Biocompatibility, Biodistribution, and Drug-Delivery Efficiency of Mesoporous Silica Nanoparticles for Cancer Therapy in Animals. *Small* **2010**, *6*, 1794–1805. [CrossRef] [PubMed]
123. Xiao, Y.D.; Paudel, R.; Liu, J.; Ma, C.; Zhang, Z.S.; Zhou, S.K. MRI contrast agents: Classification and application (Review). *Int. J. Mol. Med.* **2016**, *38*, 1319–1326. [CrossRef] [PubMed]
124. Soenen, S.J.H.; Himmelreich, U.; Nuytten, N.; Pisanic, T.R.; Ferrari, A.; De Cuyper, M. Intracellular Nanoparticle Coating Stability Determines Nanoparticle Diagnostics Efficacy and Cell Functionality. *Small* **2010**, *6*. [CrossRef]
125. Thomas, G.; Boudon, J.; Maurizi, L.; Moreau, M.; Walker, P.; Severin, I.; Oudot, A.; Goze, C.; Poty, S.; Vrigneaud, J.M.; et al. Innovative Magnetic Nanoparticles for PET/MRI Bimodal Imaging. *ACS Omega* **2019**, *4*, 2637–2648. [CrossRef]
126. Lee, C.; Kim, J.; Zhang, Y.; Jeon, M.; Liu, C.; Song, L.; Lovell, J.F.; Kim, C. Dual-color photoacoustic lymph node imaging using nanoformulated naphthalocyanines. *Biomaterials* **2015**, *73*, 142–148. [CrossRef]
127. Martynenko, I.V.; Litvin, A.P.; Purcell-Milton, F.; Baranov, A.V.; Fedorov, A.V.; Gun'Ko, Y.K. Application of semiconductor quantum dots in bioimaging and biosensing. *J. Mater. Chem. B* **2017**, *5*, 6701–6727. [CrossRef]
128. Feugang, J.M.; Youngblood, R.C.; Greene, J.M.; Willard, S.T.; Ryan, P.L. Self-illuminating quantum dots for non-invasive bioluminescence imaging of mammalian gametes. *J. Nanobiotechnology* **2015**, *13*, 1–16. [CrossRef]
129. Chen, F.; Gerion, D. Fluorescent CdSe/ZnS nanocrystal-peptide conjugates for long-term, nontoxic imaging and nuclear targeting in living cells. *Nano Lett.* **2004**, *4*, 1827–1832. [CrossRef]
130. Wagner, A.M.; Knipe, J.M.; Orive, G.; Peppas, N.A. Quantum dots in biomedical applications. *Acta Biomater.* **2019**, *94*, 44–63. [CrossRef]
131. Dey, R.; Mazumder, S.; Mitra, M.K.; Mukherjee, S.; Das, G.C. Review: Biofunctionalized quantum dots in biology and medicine. *J. Nanomater.* **2009**, *2009*. [CrossRef]
132. Matea, C.T.; Mocan, T.; Tabaran, F.; Pop, T.; Mosteanu, O.; Puia, C.; Iancu, C.; Mocan, L. Quantum dots in imaging, drug delivery and sensor applications. *Int. J. Nanomedicine* **2017**, *12*, 5421–5431. [CrossRef]
133. Abdel-Salam, M.; Omran, B.; Whitehead, K.; Baek, K.H. Superior properties and biomedical applications of microorganism-derived fluorescent quantum dots. *Molecules* **2020**, *25*, 4486. [CrossRef] [PubMed]
134. Sahoo, S.L.; Liu, C.H.; Kumari, M.; Wu, W.C.; Wang, C.C. Biocompatible quantum dot-antibody conjugate for cell imaging, targeting and fluorometric immunoassay: Crosslinking, characterization and applications. *RSC Adv.* **2019**, *9*, 32791–32803. [CrossRef]
135. Patric Joshua, P.; Valli, C.; Balakrishnan, V. Effect of in ovo supplementation of nano forms of zinc, copper, and selenium on post-hatch performance of broiler chicken. *Vet. World* **2016**, *9*, 287–294. [CrossRef]

136. Scott, A.; Vadalasetty, K.P.; Chwalibog, A.; Sawosz, E. Copper nanoparticles as an alternative feed additive in poultry diet: A review. *Nanotechnol. Rev.* **2018**, *7*, 69–93. [CrossRef]
137. Mishra, A.; Swain, R.; Mishra, S.; Panda, N.; Sethy, K. Growth performance and serum biochemical parameters as affected by nano zinc supplementation in layer chicks. *Indian J. Anim. Nutr.* **2014**, *31*, 384–388.
138. Swain, P.S.; Rajendran, D.; Rao, S.B.N.; Dominic, G. Preparation and effects of nano mineral particle feeding in livestock: A review. *Vet. World* **2015**, *8*, 888–891. [CrossRef]
139. Bhanja, S.K.; Hotowy, A.; Mehra, M.; Sawosz, E.; Pineda, L.; Vadalasetty, K.P.; Kurantowicz, N.; Chwalibog, A. In ovo administration of silver nanoparticles and/or amino acids influence metabolism and immune gene expression in chicken embryos. *Int. J. Mol. Sci.* **2015**, *16*, 9484–9503. [CrossRef] [PubMed]
140. Feugang, J.M.; Youngblood, R.C.; Greene, J.M.; Fahad, A.S.; Monroe, W.A.; Willard, S.T.; Ryan, P.L. Application of quantum dot nanoparticles for potential non-invasive bio-imaging of mammalian spermatozoa. *J. Nanobiotechnology* **2012**, *10*, 1–8. [CrossRef] [PubMed]
141. Petruska, P.; Capcarova, M.; Sutovsky, P. Antioxidant supplementation and purification of semen for improved artificial insemination in livestock species. *Turkish J. Vet. Anim. Sci.* **2014**, *38*, 643–652. [CrossRef]
142. Barkalina, N.; Jones, C.; Kashir, J.; Coote, S.; Huang, X.; Morrison, R.; Townley, H.; Coward, K. Effects of mesoporous silica nanoparticles upon the function of mammalian sperm in vitro. *Nanomedicine Nanotechnology Biol. Med.* **2014**, *10*, 859–870. [CrossRef] [PubMed]
143. Pawar, K.; Kaul, G. Toxicity of titanium oxide nanoparticles causes functionality and DNA damage in buffalo (*Bubalus bubalis*) sperm in vitro. *Toxicol. Ind. Health* **2014**, *30*, 520–533. [CrossRef] [PubMed]
144. Rey, A.I.; Segura, J.; Arandilla, E.; López-Bote, C.J. Short- and long-term effect of oral administration of micellized natural vitamin E (D-α-tocopherol) on oxidative status in race horses under intense training. *J. Anim. Sci.* **2013**, *91*, 1277–1284. [CrossRef]
145. Rey, A.; Amazan, D.; Cordero, G.; Olivares, A.; López-Bote, C.J. Lower Oral Doses of Micellized α-Tocopherol Compared to α-Tocopheryl Acetate in Feed Modify Fatty Acid Profiles and Improve Oxidative Status in Pigs. *Int. J. Vitam. Nutr. Res.* **2014**, *84*, 229–243. [CrossRef] [PubMed]
146. King, T.; Osmond-McLeod, M.J.; Duffy, L.L. Nanotechnology in the food sector and potential applications for the poultry industry. *Trends Food Sci. Technol.* **2018**, *72*, 62–73. [CrossRef]
147. Stankic, S.; Suman, S.; Haque, F.; Vidic, J. Pure and multi metal oxide nanoparticles: Synthesis, antibacterial and cytotoxic properties. *J. Nanobiotechnology* **2016**, *14*, 1–20. [CrossRef]
148. Hwang, I.S.; Lee, J.; Hwang, J.H.; Kim, K.J.; Lee, D.G. Silver nanoparticles induce apoptotic cell death in Candida albicans through the increase of hydroxyl radicals. *FEBS J.* **2012**, *279*, 1327–1338. [CrossRef] [PubMed]
149. Kalia, A.; Abd-Elsalam, K.A.; Kuca, K. Zinc-based nanomaterials for diagnosis and management of plant diseases: Ecological safety and future prospects. *J. Fungi* **2020**, *6*, 222. [CrossRef] [PubMed]
150. Dilbaghi, N.; Kaur, H.; Kumar, R.; Arora, P.; Kumar, S. Nanoscale device for veterinay technology: Trends and future prospective. *Adv. Mater. Lett.* **2013**, *4*, 175–184. [CrossRef]
151. Oberdörster, G.; Kuhlbusch, T.A.J. In vivo effects: Methodologies and biokinetics of inhaled nanomaterials. *NanoImpact* **2018**, *10*, 38–60. [CrossRef]
152. Aschberger, K.; Micheletti, C.; Sokull-Klüttgen, B.; Christensen, F.M. Analysis of currently available data for characterising the risk of engineered nanomaterials to the environment and human health—Lessons learned from four case studies. *Environ. Int.* **2011**, *37*, 1143–1156. [CrossRef]
153. Barkhordari, A.; Hekmatimoghaddam, S.; Jebali, A.; Khalili, M.A.; Talebi, A.; Noorani, M. Effect of zinc oxide nanoparticles on viability of human spermatozoa. *Int. J. Reprod. Biomed.* **2013**, *11*, 767–771.
154. Baltic, M.; Boskovic, M.; Ivanovic, J.; Dokmanovic, M.; Janjic, J.; Loncina, J.; Baltic, T. Nanotechnology and its potential applications in meat industry. *Tehnologija Mesa* **2013**, *54*, 168–175. [CrossRef]
155. Elder, A.; Lynch, I.; Grieger, K.; Chan-Remillard, S.; Gatti, A.; Gnewuch, H.; Kenawy, E.; Korenstein, R.; Kuhlbusch, T.; Linker, F.; et al. Human Health Risks of Engineered Nanomaterials. In *Nanomaterials: Risks and Benefits*; Linkov, I., Steevens, J., Eds.; Springer Science+Business Media: New York, NY, USA, 2009; pp. 3–29.
156. Pekkanen, J.; Peters, A.; Hoek, G.; Tiittanen, P.; Brunekreef, B.; De Hartog, J.; Heinrich, J.; Ibald-Mulli, A.; Kreyling, W.G.; Lanki, T.; et al. Particulate air pollution and risk of ST-segment depression during repeated submaximal exercise tests among subjects with coronary heart disease: The exposure and risk assessment for fine and ultrafine particles in ambient air (ULTRA) study. *Circulation* **2002**, *106*, 933–938. [CrossRef]
157. Nurkiewicz, T.R.; Porter, D.W.; Hubbs, A.F.; Cumpston, J.L.; Chen, B.T.; Frazer, D.G.; Castranova, V. Nanoparticle inhalation augments particle-dependent systemic microvascular dysfunction. *Part. Fibre Toxicol.* **2008**, *5*, 1–12. [CrossRef]
158. Oberdörster, G.; Castranova, V.; Asgharian, B.; Sayre, P. Inhalation exposure to carbon nanotubes (CNT) and carbon nanofibers (CNF): Methodology and Dosimetry. *J. Toxicol. Environ. Health Part B Crit. Rev.* **2015**, *18*, 121–212. [CrossRef]
159. Orsi, M.; Al Hatem, C.; Leinardi, R.; Huaux, F. Carbon nanotubes under scrutiny: Their toxicity and utility in mesothelioma research. *Appl. Sci.* **2020**, *10*, 4513. [CrossRef]
160. Chopra, M.; Bernela, M.; Kaur, P.; Manuja, A.; Kumar, B.; Thakur, R. Alginate/gum acacia bipolymeric nanohydrogels-Promising carrier for Zinc oxide nanoparticles. *Int. J. Biol. Macromol.* **2015**, *72*, 827–833. [CrossRef] [PubMed]

Journal of
Fungi

MDPI

Article

Rhizopus oryzae-Mediated Green Synthesis of Magnesium Oxide Nanoparticles (MgO-NPs): A Promising Tool for Antimicrobial, Mosquitocidal Action, and Tanning Effluent Treatment

Saad El-Din Hassan [1,*], Amr Fouda [1,*], Ebrahim Saied [1], Mohamed M. S. Farag [1], Ahmed M. Eid [1], Mohammed G. Barghoth [1], Mohamed A. Awad [2], Mohammed F. Hamza [3,4] and Mohamed F. Awad [5,6]

1. Department of Botany and Microbiology, Faculty of Science, Al-Azhar University, Nasr City, Cairo 11884, Egypt; hema_almassry2000@azhar.edu.eg (E.S.); mohamed.farag@azhar.edu.eg (M.M.S.F.); aeidmicrobiology@azhar.edu.eg (A.M.E.); mohamed_gamal.sci@azhar.edu.eg (M.G.B.)
2. Department of Zoology and Entomology, Faculty of Science, Al-Azhar University, Nasr City, Cairo 11884, Egypt; Mohamed_awad@azhar.edu.eg
3. Guangxi Key Laboratory of Processing for Non-Ferrous Metals and Featured Materials, School of Resources, Environment and Materials, Guangxi University, Nanning 530004, China; m_fouda21@hotmail.com
4. Nuclear Materials Authority, El-Maadi, Cairo POB 530, Egypt
5. Department of Biology, College of Science, Taif University, P.O. Box 11099, Taif 21944, Saudi Arabia; m.fadl@tu.edu.sa
6. Botany and Microbiology Department, Faculty of Science, Al-Azhar University, Assiut Branch, Assiut 71524, Egypt

* Correspondence: Saad.el-din.hassan@umontreal.ca (S.E.-D.H.); amr_fh83@azhar.edu.eg (A.F.); Tel.: +20-102-388-4804 (S.E.-D.H.); +20-111-335-1244 (A.F.)

check for updates

Citation: Hassan, S.E.-D.; Fouda, A.; Saied, E.; Farag, M.M.S.; Eid, A.M.; Barghoth, M.G.; Awad, M.A.; Hamza, M.F.; Awad, M.F. *Rhizopus oryzae*-Mediated Green Synthesis of Magnesium Oxide Nanoparticles (MgO-NPs): A Promising Tool for Antimicrobial, Mosquitocidal Action, and Tanning Effluent Treatment. *J. Fungi* **2021**, *7*, 372. https://doi.org/10.3390/jof7050372

Academic Editor: Kamel A. Abd-Elsalam

Received: 14 April 2021
Accepted: 7 May 2021
Published: 10 May 2021

Publisher's Note: MDPI stays neutral with regard to jurisdictional claims in published maps and institutional affiliations.

Abstract: The metabolites of the fungal strain *Rhizopus oryaze* were used as a biocatalyst for the green-synthesis of magnesium oxide nanoparticles (MgO-NPs). The production methodology was optimized to attain the maximum productivity as follows: 4 mM of precursor, at pH 8, incubation temperature of 35 °C, and reaction time of 36 h between metabolites and precursor. The as-formed MgO-NPs were characterized by UV-Vis spectroscopy, TEM, SEM-EDX, XRD, DLS, FT-IR, and XPS analyses. These analytical techniques proved to gain crystalline, homogenous, and well-dispersed spherical MgO-NPs with an average size of 20.38 ± 9.9 nm. The potentiality of MgO-NPs was dose- and time-dependent. The biogenic MgO-NPs was found to be a promising antimicrobial agent against the pathogens including *Staphylococcus aureus*, *Bacillus subtilis*, *Pseudomonas aeruginosa*, *Escherichia coli*, and *Candida albicans* with inhibition zones of 10.6 ± 0.4, 11.5 ± 0.5, 13.7 ± 0.5, 14.3 ± 0.7, and 14.7 ± 0.6 mm, respectively, at 200 µg mL^{-1}. Moreover, MgO-NPs manifested larvicidal and adult repellence activity against *Culex pipiens* at very low concentrations. The highest decolorization percentages of tanning effluents were 95.6 ± 1.6% at 100 µg/ 100 mL after 180 min. At this condition, the physicochemical parameters of tannery effluents, including TSS, TDS, BOD, COD, and conductivity were reduced with percentages of 97.9%, 98.2%, 87.8%, 95.9%, and 97.3%, respectively. Moreover, the chromium ion was adsorbed with percentages of 98.2% at optimum experimental conditions.

Keywords: MgO-NPs; optimization; antimicrobial; mosquitocidal and repellence activity; tannery effluents; chromium ion

1. Introduction

The revolution of nanoscience has been intensely grown daily. Nanoscience refers to the production of new materials at the nanoscale (1–100 nanometers). The prepared materials have unique properties that are not found in bulk materials [1]. Among these unique properties are shape, size, compatibility, surface charge, chemical stability, catalytic activity, and small size to a large surface area [2]. These properties enable them to integrate into various biotechnological and biomedical applications. The nanomaterial compounds have been fabricated using different routes including chemical, physical, and biological methods [3]. The physical methods were achieved under harsh working conditions and consumed high energy with complicated experimental devices [4], while the synthesis by chemical routes took place using deleterious organic solvents with dangerous reducing agents and produced undesirable by-products with negative impacts to the surrounding environment [5]. Therefore, the researchers' attention is directed to biological approaches to overcome or avoid the disadvantages associated with physical and chemical methods.

Biological synthesis or green synthesis has been described as cost-effectiveness, biocompatible, eco-friendly nature, and scalable, avoiding harsh conditions and not utilizing hazardous chemicals [6]. Therefore, it can be incorporated into various applications such as textiles, wastewater treatment, paper preservation, the food industry, cosmetics and pharmaceuticals, optics, and smart devices [7–11]. Various biological entities such as bacteria, fungi, yeast, actinomycetes, and plant extracts are utilized in the green synthesis of different metal and metal oxide nanoparticles, such as Ag, Cu, CuO, ZnO, TiO_2, Se, and Fe_2O_3 [12–16].

Among important metal oxide nanoparticles, magnesium oxide nanoparticles (MgO-NPs) are characterized by biocompatibility, high stability, cost-effectiveness, high ionic properties, and having a crystal structure and safe and high contaminant adsorbents [17]. Miscellaneous applications and unique properties of MgO-NPs has made it highly attractive to researchers everywhere during the past two decades over other metal oxide nanoparticles [18]. Chemically, MgO-NPs can be synthesized through numerous processes including solvent modification, co-precipitation, moist chemical, sol-gel, and hydrothermal, but these methods have drastic effects on the environment [19]. Recently, the biological synthesis of nanoparticles using fungi has become the preferable method because of the flexibility of handling, multiplication, high growth rates, species diversity (more than 1.52 million species), cost-effectiveness, novelty, and environmentally friendly nature [20,21]. The synthesis of MgO-NPs using fungi can be achieved either by intracellular method through the transportation of metal ions inside the fungal cell and then reduced by enzymes, or extracellular method through the reacting of metal ions with fungal biomass filtrate [22]. It has been reported that MgO-NPs can be used in novel applications including the removal of toxic waste, catalysis, antimicrobial property, refractory materials and wastewater treatment, ceramics, heavy fuel oils, enhancement of the potential of antioxidant and substrate in ferroelectric thin films, biomedical fields, sensing, adsorbents, lithium batteries, and agriculture sectors [18,23–25]. MgO-NPs are characterized by long-lasting antimicrobial activity, and this phenomenon can be attributed to its ability to tolerate high temperatures and low volatility [26].

The common house mosquito *Culex pipiens* L. is present in Egypt and plays a critical role in the transmission of different human pathogens such as malaria parasites, filarial worms, and viruses of Rift valley fever, West Nile, and Japanese encephalitis [27]. This insect acquired resistance against different insecticides in Egypt [28], and therefore it is necessary to urgently discover new alternative insecticides.

The current study aims to myco-synthesize MgO-NPs by harnessing metabolites of the fungal strain *Rhizopus oryaze* isolated from a soil sample. The characterization of myco-synthesized MgO-NPs was attained by UV-Vis spectroscopy and TEM, SEM-EDX, XRD, DLS, FT-IR, and XPS analyses. Moreover, factors including pH, contact times, incubation temperature, and precursor concentrations that affect the myco-synthesis of MgO-NPs were optimized. The biomedical activities of MgO-NPs including antibacterial, antifungal, larvicidal, and repellency activities were investigated. Finally, the efficacy of MgO-NPs in decolorization and chromium adsorption from tanning effluents were assessed.

2. Materials and Methods

2.1. Chemicals Used

In the current study, magnesium nitrate hexahydrate (Mg $(NO_3)_2{\cdot}6H_2O$) and sodium hydroxide (NaOH) were used as analytical grade and obtained from Sigma Aldrich, Cairo, Egypt. The Malt Extract agar (MEA) media used for fungal isolations and Mueller-Hinton agar media used for antimicrobial activities were ready-made (Oxoid, Cairo, Egypt). All biological reactions were carried out using distilled water (dis. H_2O). The tannery wastewater was collected from Robbiki Leather City, 10th of Ramadan, Cairo, Egypt (GPS: 30°17′898″ N, 31°76′840″ E).

2.2. Isolation and Identification of the Fungal Strain

The fungal strain used for biosynthesis of MgO-NPs was isolated from Qalyubia Governorate, Egypt (31°18′522.07″ E, 30°15′524.13″ N), and has a code E3. The isolation procedure was achieved according to Fouda et al. [29] as follows: approximately 1.0 g of soil sample undergo diluted in sterilized dis. H_2O. After that, 100 µL of the fourth dilution was inoculated onto MEA plates and incubated for 3–4 days at 30 ± 2 °C. The appeared colonies were picked up and re-inoculated onto the same media for purifications. The purified colony was preserved on an MEA slant for further use.

The identification was accomplished by cultural and microscopic characteristics and confirmed by molecular identification using internal transcribed spacer (ITS) sequence analysis. The ITS rDNA region was amplified using primers for ITS1 f (5-CTTG GTCATTTAGAGGAAGTAA-3) and ITS4 (5-TCCTCCGCTTATTGATATGC-3) [30]. The PCR mixture contained: 1X PCR buffer, 0.5 mM $MgCl_2$, 2.5 U Taq DNA polymerase (QIAGEN, Germantown, MD 20874, USA), 0.25 mM dNTP, 0.5 µL of each primer, and 1 µg of extracted genomic DNA. The PCR was performed in a DNA Engine Thermal Cycler (PTC-200, BIO-RAD, Hercules, CA, USA) with a program of 94 °C for 3 min, followed by 30 cycles of 94 °C for 30 s, 55 °C for 30 s, and 72 °C for 1 min, followed by a final extension performed at 72 °C for 10 min. The PCR product was checked for the expected sizes on 1% agarose gel and was sequenced by Sigma Company for Scientific Research, Egypt, with the two primers. The sequence was compared against the GenBank database using the NCBI BLAST tool. Multiple sequence alignment was done using the Clustal Omega software package (http://www.clustal.org/clustal2, accessed on 17 November 2010), and a phylogenetic tree was constructed using the neighbor-joining method with MEGA (v.6.1) software, with confidence tested by bootstrap analysis (1000 repeats).

2.3. Green Synthesis of MgO-NPs

2.3.1. Preparation of Fungal Biomass Filtrate

Three disks (0.8 cm in diameter) of the old culture of fungal strain E3 were inoculated into 100 mL of malt extract broth (MAB) media and incubated for 5 days at 30 ± 2 °C under 150 rpm shaking condition. At the end of the incubation period, the inoculated MAB was centrifuged and fungal biomass was collected. About 10.0 g of fungal biomass was resuspended in 100 mL dis. H_2O for 48 h. at 30 ± 2 °C and 150 rpm shaking condition. The previous mixture was centrifuged at 10,000 rpm for 5 minutes and the supernatant (fungal biomass filtrate) was collected and used for green synthesis of MgO-NPs.

2.3.2. Green Synthesis of MgO-NPs

Approximately 102.5 mg of $Mg(NO_3)_2{\cdot}6H_2O$ was dissolved in 10 mL dis. H_2O, then mixed with 90 mL of fungal biomass filtrate overnight to produce a final concentration of 4 mM. At the end of the incubation period, the $Mg(OH)_2$ was formed as turbid white precipitate, which was collected and rinsed with dis. H_2O to remove any impurities before being oven-dried at 100 °C for 1 h (Equation (1)).

$$Mg(NO_3)_2 \cdot 6H_2O + H_2O \underset{\text{Metabolites}}{\overset{\text{Fungal}}{\rightarrow}} Mg(OH)_2 \quad (1)$$

The formed $Mg(OH)_2$ was calcinated at 400 °C for 3 h to form MgO-NPs, as represented in Equation (2) [31]. The controls including fungal biomass filtrate and $Mg(NO_3)_2{\cdot}6H_2O$ solution ran alongside the experiment under the same conditions.

$$Mg(OH)_2 \overset{400°C}{\rightarrow} MgO \quad (2)$$

2.4. Optimizing Myco-Synthesis of MgO-NPs

The MgO-NPs production and distribution were affected by the environmental factors such as pH, contact times, incubation temperature, and precursor concentrations. These factors were investigated at maximum surface plasmon resonance (SPR) detected by using a UV-Vis spectrophotometer (Jenway 6305, Staffordshire, UK). The effect of pH values at 6, 7, 8, 9, 10, and 11 on MgO-NPs sorption properties was investigated. The contact time either between fungal biomass and distilled water to produce biomass filtrate (24, 48, 72, and 96 h) or times between biomass filtrate and precursors (6, 12, 24, 36, 48, and 72 h) were also investigated. The incubation temperature (25 °C to 40 °C with intervals of 5 degrees) and precursor concentrations (1–5 mM) were assessed. At the end of each parameter, 1.0 mL of the sample was withdrawn and measured at maximum SPR at λ_{max} = 282 nm.

2.5. Characterization of Biosynthesized MgO-NPs

The particle sizes and shapes of biosynthesized MgO-NPs were detected using Transmission Electron Microscopy (TEM) (JEOL 1010, Japan, acceleration voltage of 120 KV). A drop of NP solution was loaded on the carbon-copper grid and underwent vacuum desiccation for 24 h and placed after that onto a specimen holder [32]. The elemental compositions of biosynthesized MgO-NPs were assessed using scanning electron microscopy connected with energy dispersive X-ray (SEM-EDX) (JEOL, JSM-6360LA, Akishima, Japan). The crystallographic structure of biosynthesized MgO-NPs was determined using X-ray diffraction (XRD) analysis by X'Pert pro diffractometer (Philips, Eindhoven, Netherlands). The operating conditions were, 2θ values measured in ranges of 4° to 80°, X-ray radiation source was Ni-filtered Cu Ka and the operating voltage and current were 40 KV and 30 mA, respectively. The average NP sizes were measured using the Debye–Scherrer equation [33] as follows:

$$D = K\lambda / \beta \cos\theta \quad (3)$$

where D is average particle size, K is the Scherrer's constant (0.9), λ is the wavelength of X-ray radiation (0.154 nm), and β and θ are the half of maximum intensity and Bragg's angle, respectively. Moreover, the size distribution of MgO-NPs in colloidal solution was investigated by dynamic light scattering (DLS) analysis. The sample was subjected to measurement by Zeta sizer nano series (Nano ZS, Malvern, UK).

On the other hand, the functional groups present in fungal biomass filtrate involved in reduction, capping, and stabilization of MgO-NPs were investigated using Fourier transform infrared (FT-IR) spectroscopy (Agilent system Cary 660 FT-IR model). The MgO-NPs sample was ground with KBR pellets (1% *w*/*w*), pressure was applied to form a disk, and scanning was done in the range of 400–4000 cm^{-1}.

Finally, the X-ray photoelectron spectroscopy (XPS) analysis was analyzed by ESCALAB 250XI^{+} instrument (Thermo Fischer Scientific, Inc., Waltham, MA, USA) connected with monochromatic X-ray Al Kα radiation (1486.6 eV). The analysis was conducted under the following conditions: the size of the spot was 500 μm, the samples were prepared under a pressure adjusted at 10^{-8} mbar., the energy was calibrated with $Ag3d_{5/2}$ signal (ΔBE: 0.45 eV) and C 1s signal (ΔBE: 0.82 eV), and the full and narrow-spectrum passed energies were 50 eV and 20 eV, respectively [34,35].

2.6. Antimicrobial Activity

The antimicrobial activity of MgO-NPs synthesized by fungal metabolites was investigated against various pathogenic microbes, including *Staphylococcus aureus* ATCC 6538, *Bacillus subtilis* ATCC 6633 (Gram-positive strains), *Pseudomonas aeruginosa* ATCC 9022,

Escherichia coli ATCC 8739 (Gram-negative strains), and unicellular fungi *Candida albicans* ATCC 10231. The bacterial strains were subcultured on nutrient agar media (containing g L^{-1}: peptone, 5; beef extract, 3; NaCl, 5; agar, 20; distilled water, 1000 mL) while *C. albicans* was subcultured on yeast extract peptone dextrose (YEPD) agar media (containing g L^{-1}: glucose, 20; peptone, 20; yeast extract, 10; agar, 20; distilled water, 1000 mL) for 24 h. To check antimicrobial activity, each strain was homogenously streaked over Mueller-Hinton agar (for bacterial strains) and YEPD agar plates (for *C. albicans)* using a sterilized cotton swab. Three wells (0.7 cm diameter) were cut in the streaked Mueller-Hinton plates and filled with 100 µL of biosynthesized MgO-NPs (200 µg mL^{-1}). The minimum inhibitory concentration (MIC) was assessed by using different concentrations of MgO-NPs (150, 100, 50, and 25 µg mL^{-1}). The loaded Mueller-Hinton plates were kept in the refrigerator for 1 h before being incubated at 35 °C for 24 h. The results were recorded as a zone of inhibitions (ZOIs) around each well by mm [36]. The experiment was achieved in triplicate.

2.7. Mosquitocidal Bioassay

2.7.1. Mosquito Rearing

Culex pipiens L. mosquitos' vectors were reared in the Laboratory of Medical Entomology, Animal House, Department of Zoology and Entomology, Faculty of Science, Al-Azhar University, Cairo, Egypt. The mosquitoes were preserved at 30–35 °C and 60–80% relative humidity with a photoperiod of 12:12 h light: dark. Larvae were reared in white plastic cups (30 cm × 35 cm × 5 cm and containing 500 mL tap water). Fish was added to each tray for 2 weeks for optimum feeding of the larvae as follows: 0.1 g for 1st and 2nd instar larvae, 0.3 g for 3rd instar larvae, and 0.5 g for 4th instar larvae per day, until the pupation stage appeared. New pupae were transferred from the trays to plastic cups containing water and placed in screened cages (size 30 cm × 30 cm × 30 cm) until they emerged as adults. The adults were constantly provided with 5% sucrose solution on saturated cotton pads. The adult female mosquitoes of 5 days were destitute of sugar for 12 h, then fed on human blood by artificial membrane feeding technique [37] for 30 min. After two days of blood-feeding, plastic cups full of tap water were placed inside the cage for oviposition, which occurred on day 3 or 4 after blood feeding.

2.7.2. Larvicidal Activity

The larvicidal activity (at room temperature) was assessed by the standard method of the World Health Organization (WHO) with slight modifications using the method described by Velayutham and Ramanibai. [38]. Twenty-five 3rd instar larvae of *C. pipiens* were moved separately from their colony in the laboratory to a plastic beaker (250 mL) containing 100 mL of the biogenic MgO-NPs (10 µg mL^{-1}). The control was set up using chlorinated tap water. The same experiment was repeated with different concentrations of MgO-NPs (8, 6, 4, and 2 µg mL^{-1}). The larval mortality percentages were counted every 24 h for 72 h using the following equation [39]:

$$\text{Mortality percentages}(\%) = \frac{\text{mortality percentages of treatment} - \text{mortality percentages of control}}{100 - \text{mortality percentages of control}} \times 100 \quad (4)$$

The lethal concentrations (LC_{50}) and higher lethal concentrations (LC_{90}) values were calculated at their 95% confidence intervals, as previously reported [40].

2.7.3. Repellent Activity

The repellent study was conducted according to the method described by WHO [41]. This test was monitored by the Faculty of Science, Zoology and Entomology Department, Al-Azhar University, Cairo, Egypt, according to WHO ethics. Briefly, 3-day-old, blood-starved female *C. pipiens* were held in a net cage (45 cm × 30 cm × 45 cm). The volunteer had no contact with perfumes, lotions, perfumed soaps, or oils for one day before the experiment. From each volunteer arm, only 25 cm^2 of skin, upper side, were uncovered and exposed to female *C. pipiens*, while the residual arm area was covered with elastic gloves. The biosynthesized MgO-NPs were applied at 1.0, 2.5, 5.0, and 10.0 mg/cm^2 separately to the uncovered arm area. The commercial DEET (N, N-Diethyl-meta-toluamide) was used as a positive control. The repellent experiment was achieved during the night from 7:00 p.m. to 3:00 a.m. The volunteer introduces their arm (control and treated) simultaneously into the cage, gently tapping the sides of the experimental cages to activate the mosquitoes. The experiment was repeated three times for each concentration used. The volunteer was asked to insert their arm (control and treated) into the experiment cage simultaneously for 1 minute every 5 minutes. The mosquitoes that settled on the arm were recorded and then shaken off before sucking any blood. The percentages of repellency were calculated according to the following equation [42]:

$$\text{Repellency percentages}(\%) = \frac{T_a - T_b}{T_a} \times 100 \tag{5}$$

where T_a and T_b denote the number of mosquitoes in the control and treated groups, respectively.

2.8. Bio-Adsorption and Treatment of Tannery Effluent

The decolorization of tanning effluents using biosynthesized MgO-NPs was investigated at different concentrations (50, 75, and 100 mg/100 mL) for different contact times (60, 120, 180, and 240 min). The experiment was conducted on a 250 mL conical flask containing 100 mL of tanning effluent mixed with specific NP concentrations. The mixture was stirred for 30 min before the experiment to reach absorption/desorption equilibrium. At the end of each incubation time, 1.0 mL of the mixture (tanning effluent with NPs) was withdrawn and centrifuged at 10,000 rpm for 3 min, and its optical density was measured at the maximum absorption band (λ_{max}) of tanning effluent (550 nm) by a spectrophotometer (721 spectrophotometers, M-ETCAL). The decolorization percentages (%) of tanning effluents were calculated according to the following equation [10]:

$$\text{Decolorization percentages}(\%) = \frac{C_0 - C_t}{C_0} \times 100 \tag{6}$$

where C_0 is the absorbance at zero time and C_t is the absorbance after specific time t (min).

Based on the most suitable MgO-NPs concentration at the optimum contact time, the biological oxygen demand (BOD), chemical oxygen demand (COD), total dissolved solids (TDS), total suspended solids (TSS), and conductivity were assessed according to the standard recommended methods [43].

The major common tanning heavy metal represented by Cr was measured before and after MgO-NPs treatment using atomic adsorption spectroscopy (A PerkinElmer Analyst 800 atomic spectrometer). The Cr heavy metal was detected by atomic absorption spectroscopy according to the absorption of light by free metallic ions.

2.9. Statistical Analysis

All results presented are the means of three independent replicates. Data were subjected to statistical analysis by a statistical package SPSS v.17. The mean difference comparison between the treatments was analyzed by *t*-test or the analysis of variance (ANOVA) and subsequently by Tukey HSD test at $p < 0.05$.

3. Results and Discussion

3.1. Isolation and Identification of Fungal Isolate

The fungal isolate E3 was isolated and underwent primary identification according to standard keys. Based on morphological and culturable characteristics, the fungal isolate E3 was belonging to *Rhizopus* sp., which was subjected to molecular identification based on amplification and sequencing of the internal transcribed spacer (ITS) gene. The sequence analysis revealed that the fungal strain E3 was highly related to *Rhizopus oryaze* (accession number: NR103596), with similarity percentages of 99%. Therefore, the fungal strain obtained in this study was specifically identified as *Rhizopus oryaze* strain E3 (Figure 1). The obtained ITS sequence was deposited in the gene bank under accession number MW774584.

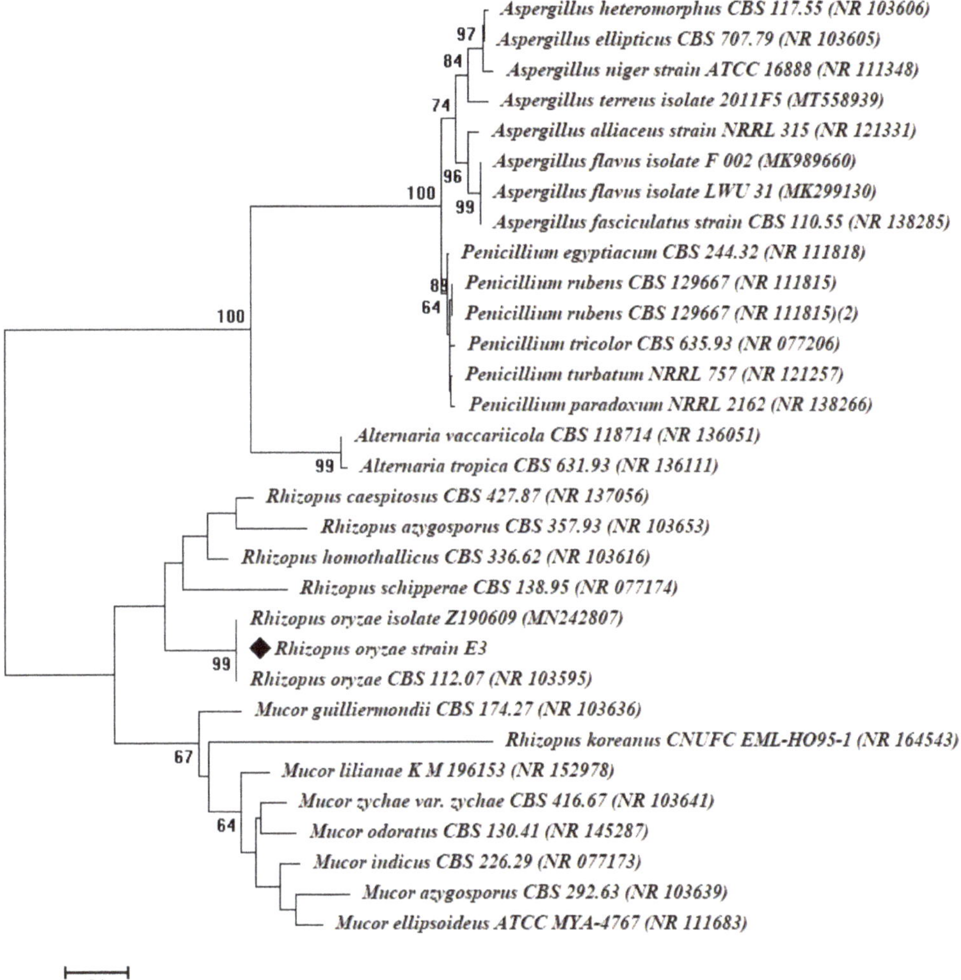

Figure 1. Phylogenetic tree of the fungal strain E3 with the sequences from NCBI. The symbol ♦ refers to ITS fragments retrieved from this study. The tree was conducted with MEGA 6.1 using the neighbor-joining method.

Rhizopus oryaze are characterized by their platform's highly secondary metabolites such as chemicals (fumaric acid, lactic acid, and ethanol), enzymes, fermentative compounds, and a wide range of by-products [44,45]. This wide range of metabolites increases the possibilities of *R. oryaze* to incorporate into various biomedical and biotechnological applications. To date, this is the first report that utilized *R. oryaze* as a biocatalyst for the green synthesis of MgO-NPs.

3.2. Myco-Synthesis of MgO-NPs

The myco-synthesis of metal and metal oxide NPs has recently drawn more attention due to high scalability, easy handling, high metabolite secretions. Moreover, fungi are characterized by high accumulation and high tolerance to metals, as well as high biomass production that is used to produce high active metabolites [46]. In this study, biomass filtrate of *R. oryaze* strain E3 was utilized as a biocatalyst for the green synthesis of MgO-NPs. The first monitor for successful fabrication was the color change of fungal biomass filtrate from colorless to turbid white after mixed with precursor (Mg $(NO_3)_2.6H_2O$). Changing of the color indicates the activity of metabolites involved in biomass filtrate to reduce nitrate (NO_3) to nitrite (NO_2) and the liberated electron used to reduce Mg^{2+} to form $Mg(OH)_2$, which was calcinated at 400 °C to form MgO-NPs (Equations (1) and (2)) [47].

To confirm MgO-NPs formation, the color change was monitored by UV-Vis spectroscopy to detect the maximum surface plasmon resonance (SPR). The size, shape, and well distribution of green synthesized NPs were usually influenced by SPR, as reported previously by Fedlheim and Foss [48]. In this regard, Jeevanandam et al. [49] reported that the size of biosynthesized MgO-NPs tends to be small when the SPR is less than 300 nm, whereas it became more anisotropic at SPR greater than 300 nm. In the present study, myco-synthesized MgO-NPs showed SPR at a wavelength of 282 nm (Figure 2), which indicates the presence of particles at the nanoscale, as reported previously [50]. The presented data are compatible with those reported by Abdallah et al. [51] and Nguyen et al. [50]; these studies showed that the maximum peaks of MgO-NPs synthesized by *Rosmarinus officinalis* L. and *Tecoma stans* (L.) were at 250 nm and 281 nm, respectively. Moreover, the chemically synthesized MgO-NPs showed a broad absorption peak at 290 nm [52]. Therefore, we can assume the efficacy of metabolites involved in biomass filtrate to reduce, cap, and stabilize MgO-NPs.

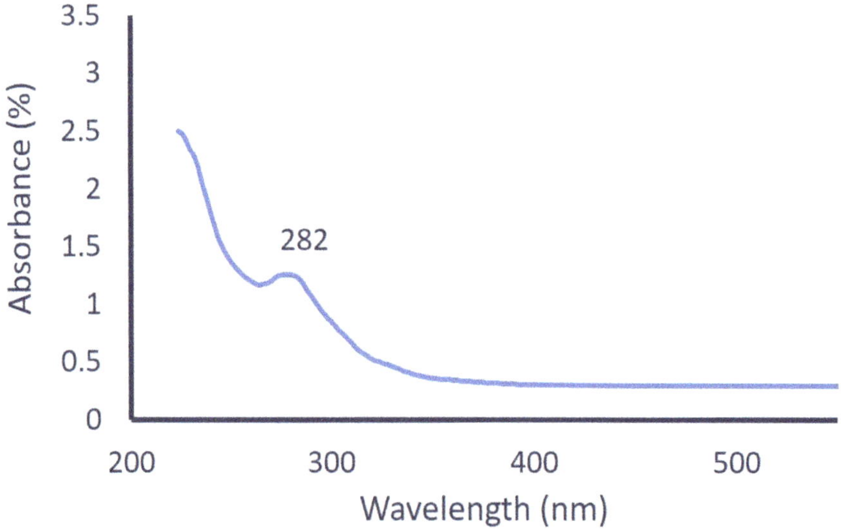

Figure 2. UV-Vis spectroscopy of myco-synthesized MgO-NPs showed maximum SPR at 282 nm.

3.3. Optimizing Myco-Synthesis of MgO-NPs

Several environmental factors affect the biosynthesis processes, the stability of NPs, and their applications. These factors can affect the reducing agents such as enzymes, proteins, carbohydrates, and others present in biomass filtrate, which directly influence the biosynthesis processes [53]. The optimization of these factors has positive impacts on reducing and capping agents, decreasing the biosynthesis time, increasing NP stability, and decrease aggregation [32]. Among these factors, contact time, pH, temperature, and concentrations of precursor are investigated in the current study.

The incubation period of fungal biomass in distilled water to secrete bioactive compounds that acts as reducing and capping agents is considered a critical factor. In the present study, the optimum contacting time between fungal biomass and distilled water was 72 h. At this time, the color intensity drastically increased, referring to the efficacy of active compounds involved in biomass filtrate to reduce the highest amount of $Mg(NO_3)_2{\cdot}6H_2O$ (Figure 3A). The concentration of Mg ions is considered an important factor that influences the myco-synthesis of MgO-NPs; therefore, different concentrations of precursor were used. In this study, the absorbance at λ_{282} was increased gradually by increasing $Mg(NO_3)_2{\cdot}6H_2O$ concentration until 4 mM, after which the absorbance was decreased (Figure 3B). This indicates the efficacy of active metabolites including enzymes and protein to reduce $Mg(NO_3)_2{\cdot}6H_2O$, whereas the MgO-NPs tend to aggregate or agglomerate by increasing precursor concentrations and hence decreasing the absorbance [54]. Compatible with our study, Ibrahim [55] reported that the color intensity of Ag-NPs synthesized by banana peel extract was changed from yellowish-brown to dark reddish-brown by increasing the precursor ($AgNO_3$) concentrations.

Another important factor affecting the myco-synthesis of MgO-NPs is the contact time or the reaction time between biomass filtrate and optimum precursor concentration (4 mM). The color changes to white precipitate offer evidence for the successful fabrication of MgO-NPs, which is monitored by detecting the absorbance at λ_{282}. Our data showed that the highest absorbance was achieved at 36 h of reaction time between fungal biomass filtrate and $Mg(NO_3)_2{\cdot}6H_2O$ by constant the previous optimum conditions (Figure 3C). At the early stages of reaction time, the SPR peak was broadened due to the slow reduction of metal ions to NPs. At the optimum reaction time, the large number of metal ions were reduced to NPs and identical SPR bands were formed. By increasing the reaction time, the absorbance gradually decreased due to the aggregation of some particles, consequently reducing the color intensity and particle size [55].

In addition, pH is considered an important factor that influences reducing agents, which affects the green synthesis. In the current study, different pH values were adjusted ranging from 6 to 11, and their impact on the reduction process was investigated through measuring the absorbance at λ_{282} after the constant of the other parameters. Data analysis showed that the alkaline solution was preferred for the myco-synthesis of MgO-NPs by *R. oryaze* strain E3. The highest absorbance at λ_{282} was attained at pH 8, which was evidence for maximum MgO-NPs production (Figure 3D). The lowest absorbance was attained at an acidic solution (pH = 6) with an absorbance value of 0.845 ± 0.01, which indicates that the activity of reducing groups present in biomass filtrate was reduced at acidic medium. Our recently published study showed that the activities of metabolites secreted by *Aspergillus carbonarious* strain D-1 were highly active at alkaline medium for reducing, cap, and stabilizing of α-Fe_2O_3-NPs and MgO-NPs [56]. The charges of biomolecules present in biomass filtrate can be altered due to differences in pH values and thus, the reducing capacity is affected [57].

Finally, the temperature is considered the main factor effect on the enzymes, proteins, carbohydrates, and other reducing agents present in biomass filtrate and consequently influence the NP synthesis process. In this study, different incubation temperatures (25–40 °C) were investigated to detect the preferable incubation temperature by measuring the absorbance intensity at λ_{282}. Data analysis showed that the capacity of metabolites as reducing, capping, and stabilizing agents were accomplished at 35 °C (Figure 3E). The intensity of color as indicated by absorbance was decreased at a temperature greater or less than 35 °C. This phenomenon could be attributed to the inactivation of reducing agents at low and high incubation temperatures. Rai et al. [58] reported that the incubation temperature of the NP synthesis process can influence the nature, size, and shape of NPs formed. Interestingly, Patra and Baek [53] reported that the synthesis of NPs using physical methods needs an incubation temperature >350 °C, whereas chemical approaches require an incubation temperature <350 °C; in contrast, green approaches required ambient temperature.

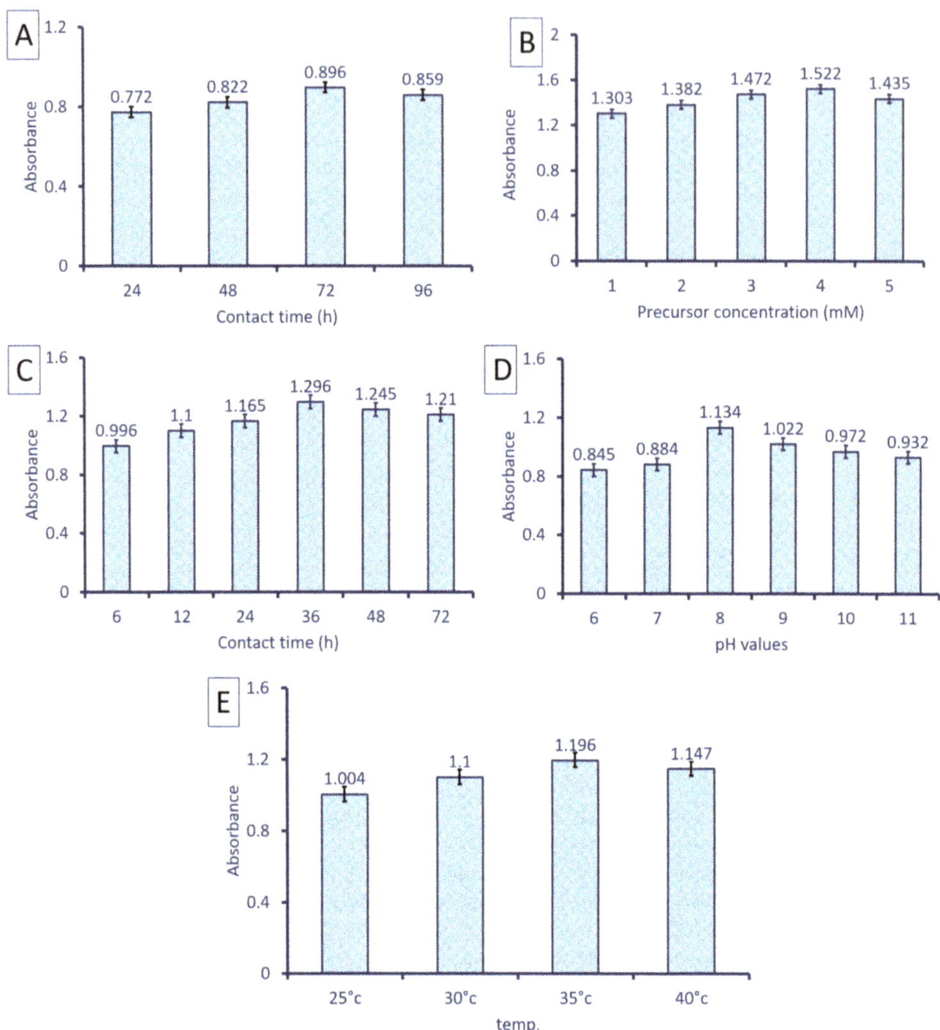

Figure 3. Optimizing factors for myco-synthesis MgO-NPs using *R. oryaze* strain E3. (**A**) is contact time between fungal biomass and distilled water, (**B**) is precursor concentrations, (**C**) is contact time between biomass filtrate and optimum precursor concentration, (**D**) is the effect of pH values, and (**E**) is the effect of incubation temperature on the biosynthesis process.

3.4. Myco-Synthesized MgO-NPs Characterization

3.4.1. Transmission Electron Microscopy

The morphological characteristics of myco-synthesized MgO-NPs including size and shape were investigated using TEM analysis. As shown in Figure 4A, the capability of metabolites secreted by *R. oryaze* strain E3 for reducing, capping, and stabilizing of well-dispersed spherical shape with sizes ranging from 8.0 to 47.5 nm with an average diameter of 20.38 ± 9.9 nm (Figure 4B). Recently, spherical MgO-NPs were successfully fabricated through harnessing metabolites of *Aspergillus carbonarious* D-1 with a size range of 20–80 nm [56]. Moreover, leaf aqueous extract of *Carica papaya* L. was used to synthesis of spherical MgO-NPs with an average size of 100 nm [59]. In this study, the obtained data evidence the potential for green synthesis of MgO-NPs with small size by harnessing metabolites of identified fungal strain. The activities of NPs are correlated to their size and shape, where the small sizes have more activities [36,60]. For instance, synthesized MgO-NPs with a varied size of 35.9, 47.3, and micron-size of 2145.9 nm exhibited bactericidal efficacy against *Bacillus subtilis* with percentages of 96.12%, 94.46%, and 75.71%, respectively [61]. The difference in the sizes and shapes of NPs synthesized by green approaches could be attributed to metabolites between different biological entities (bacteria, fungi, actinomycetes, algae, and plants), and even between the different species in the same genus [32].

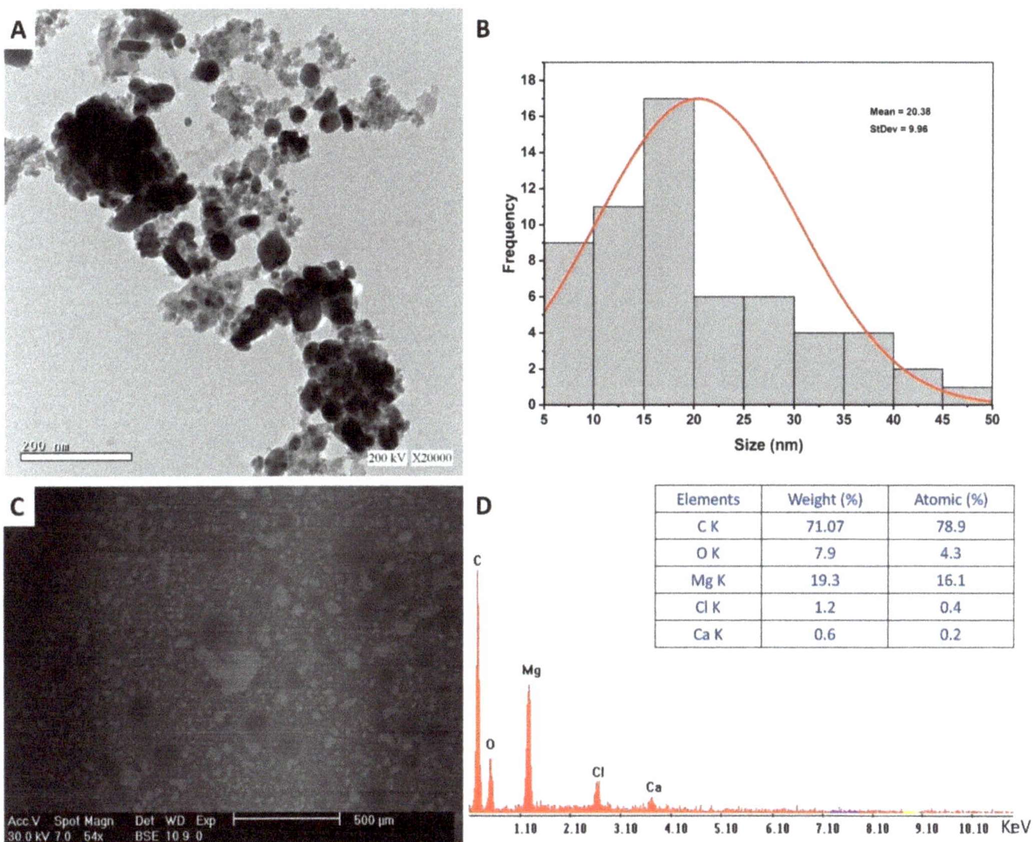

Elements	Weight (%)	Atomic (%)
C K	71.07	78.9
O K	7.9	4.3
Mg K	19.3	16.1
Cl K	1.2	0.4
Ca K	0.6	0.2

Figure 4. Characterization of myco-synthesized MgO-NPs. (**A**) TEM image, (**B**) size distribution, (**C**) SEM image, and (**D**) EDX spectrum.

3.4.2. Scanning Electron Microscopy—Energy Dispersive X-ray (SEM-EDX)

The surface morphology of myco-synthesized MgO-NPs, agglomeration, and qualitative and quantitative chemical compositions were investigated using SEM-EDX analysis. The SEM image showed well-dispersed spherical MgO-NPs without any aggregation (Figure 4C). Furthermore, the EDX profile contains Mg and O with weight percentages of 19.3% and 7.9%, respectively, and with atomic percentages of 16.1% and 4.3%, respectively (Figure 4D). Moreover, the EDX profiles showed that the weight percentages of other elements present in the sample were 71.07%, 1.2%, and 0.6% for C, Cl, and Ca, respectively. Dobrucka [62] reported that the presence of peaks at an energy between 0.5 and 1.5 KeV indicates the successful synthesis of MgO-NPs, which is confirmed by our study. Consistent with the obtained data, the EDX profile of MgO-NPs synthesized by water extract of *Artemisia abrotanum* contains Mg, O, Al, Si, K, and Ca with weight percentages of 13.9%, 39.4%, 1.4%, 0.3%, 0.8%, and 0.5%, respectively [62]. The presence of peaks other than Mg and O could be attributed to the hydrolysis of capping and stabilizing agents such as proteins, enzymes, polysaccharides, and amino acids by X-ray [63].

3.4.3. X-ray Diffraction (XRD) Analysis

The crystalline nature of myco-synthesized MgO-NPs was investigated using XRD analysis. Data showed five intense peaks at 2θ° of 36.9°, 42.6°, 62.2°, 75.4°, and 78.6° that correspond to planes of (111), (200), (220), (311), and (222), respectively (Figure 5A). These data confirmed that the as-formed MgO-NPs by *R. oryaze* strain E3 were crystallographic structures according to the JCPDS standard (JCPDS file No. 89-7746) [64]. The observed XRD peaks indicate the presence of oxide in the sample as $Mg(OH)_2$ and MgO, confirmed by XPS analysis. Lekota et al. [65] reported that the peaks observed at 2θ° of 36.9° (111) and 75.4 (311) correspond to $Mg(OH)_2$, whereas the diffraction peaks at 2θ° of 42.6° (200), 62.2° (220), and 78.6° (222) signified cubic MgO-NPs. The crystal size of myco-synthesized MgO-NPs was measured as 50 nm from the XRD pattern by the Debye–Scherrer equation.

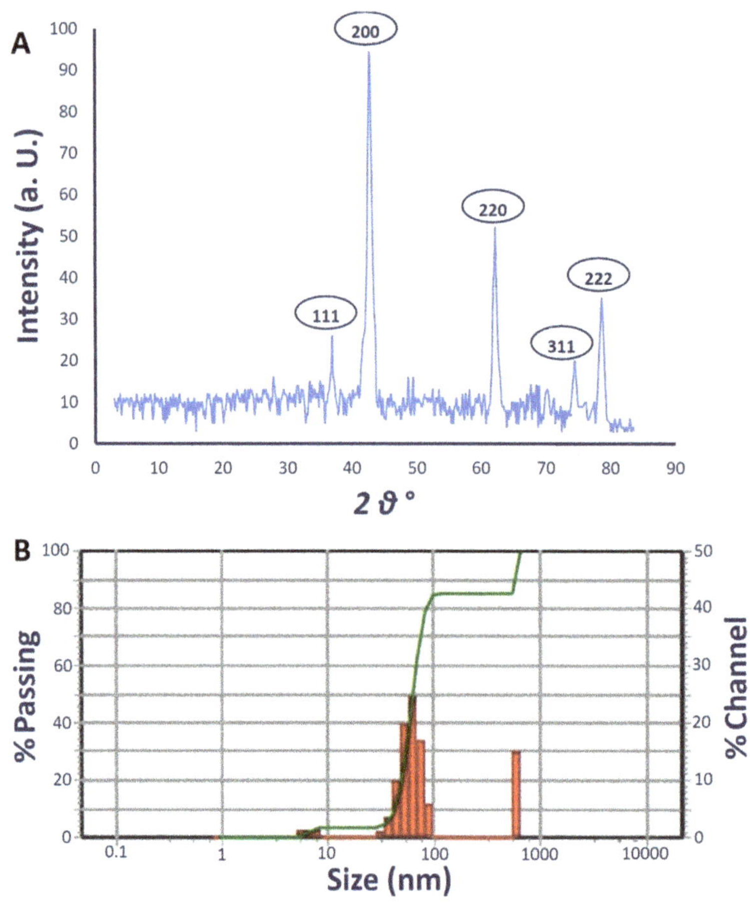

Figure 5. (**A**) XRD analysis showed the crystalline nature; (**B**) DLS analysis of myco-synthesized MgO-NPs.

3.4.4. Dynamic Light Scattering (DLS)

The size distribution of MgO-NPs in colloidal solution was assessed using the DLS technique by reacting light beams with myco-synthesized MgO-NPs [66]. In this study, the average diameter size of MgO-NPs calculated using DLS was 56.1 nm, 60.5 nm, and 54.9 nm for volume intensities of 39.6%, 55.5%, and 4.9%, respectively, of colloidal solution (Figure 5B). In most cases, the size obtained from DLS is bigger than those obtained from TEM and XRD. This phenomenon can be attributed to the metabolites coating the NP surface which act as capping and stabilizing agents [63,66]. Moreover, the non-homogenous distribution of NPs in colloidal solution and hydrodynamic residue measured by DLS can give a bigger size [67,68].

Furthermore, the polydispersity index (PDI) that indicates the homogeneity of NPs in the colloidal solution can be measured based on DLS analysis. The homogeneity increases or decreases if the PDI values are below or higher than 0.4, while the colloidal sample is considered completely heterogenous if PDI values are higher than 1. The obtained data demonstrate that the PDI value of MgO-NPs synthesized by *R. oryaze* strain E3 was 0.2, which indicates the high homogenous colloidal solution.

3.4.5. Fourier Transform Infrared (FT-IR) Spectroscopy

The bioactive compounds present in biomass filtrate of *R. oryaze* strain E3, which is responsible for the reduction of metal precursors to form MgO-NPs, are identified by FT-IR analysis which recorded a wavenumber between 400 to 4000 cm^{-1}, as shown in Figure 6. The peak observed at 3700 cm^{-1} signifies to –OH stretching band [69]. The broadness peak at 3431 cm^{-1} is corresponding to the O–H stretching vibration mode of the hydroxyl groups overlapped with the NH stretching mode of amines [62]. The medium observed peaks at 1650 cm^{-1} are signified to the bending mode of primary amine (N–H) overlapped with either amide and carboxylate salt (see XPS analysis). The medium peak at 1435 cm^{-1} corresponds to the C=O stretching of carboxylate salt as well as the adsorption of CO_3^{2-} and CO_2 at the surface of MgO-NPs [70,71]. The adsorption of such functional groups on the surface of MgO-NPs has a critical role in catalytic reactions [72]. whereas the peak at 1030 cm^{-1} matched the Mg–OH stretching [73] with C-H out-of-plane bend. The peak located at 930 cm^{-1} corresponds to C-O stretching, *trans*-C-H out-of-plane bend, and P–O which refers to phosphate-containing molecules [74]. The successful fabrication of Mg-O was confirmed by peaks observed at a wavenumber between 400 to 700 cm^{-1}, as reported in various published studies [69,75,76]. The peaks observed in FT-IR spectra reflect the role of metabolites involved in biomass filtration of *R. oryaze* strain E3 for reducing and stabilizing of MgO-NPs.

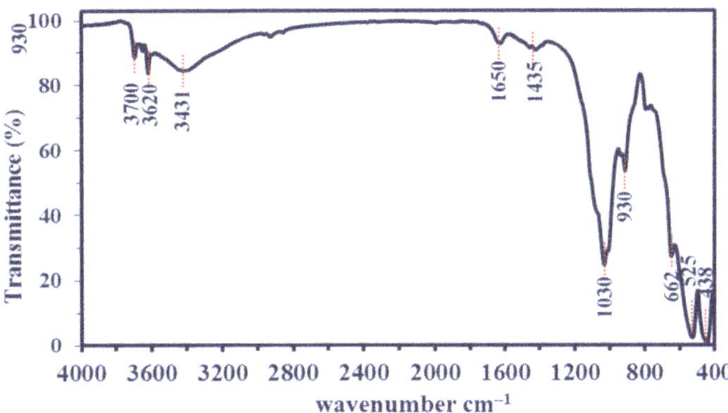

Figure 6. The FT-IR spectrum of myco-synthesized MgO-NPs fabricated by metabolites of *R. oryaze* strain E3.

3.4.6. X-ray Photoelectron Spectroscopy (XPS) Analysis

The XPS survey spectra of MgO-NPs synthesized by *R. oryaze* strain E3 is shown in Figure 7A. The magnesium was mainly characterized at different bending energies with different species Mg (1s, 2s, 2p, KL1, KL2, KL3 KL4, and KL5); other associated ions were detected as O (1s, 2s, KL1, and KL2), Cl (2p and 1s), and N 1s. The high-resolution spectra were performed for the most familiar peak, as shown below.

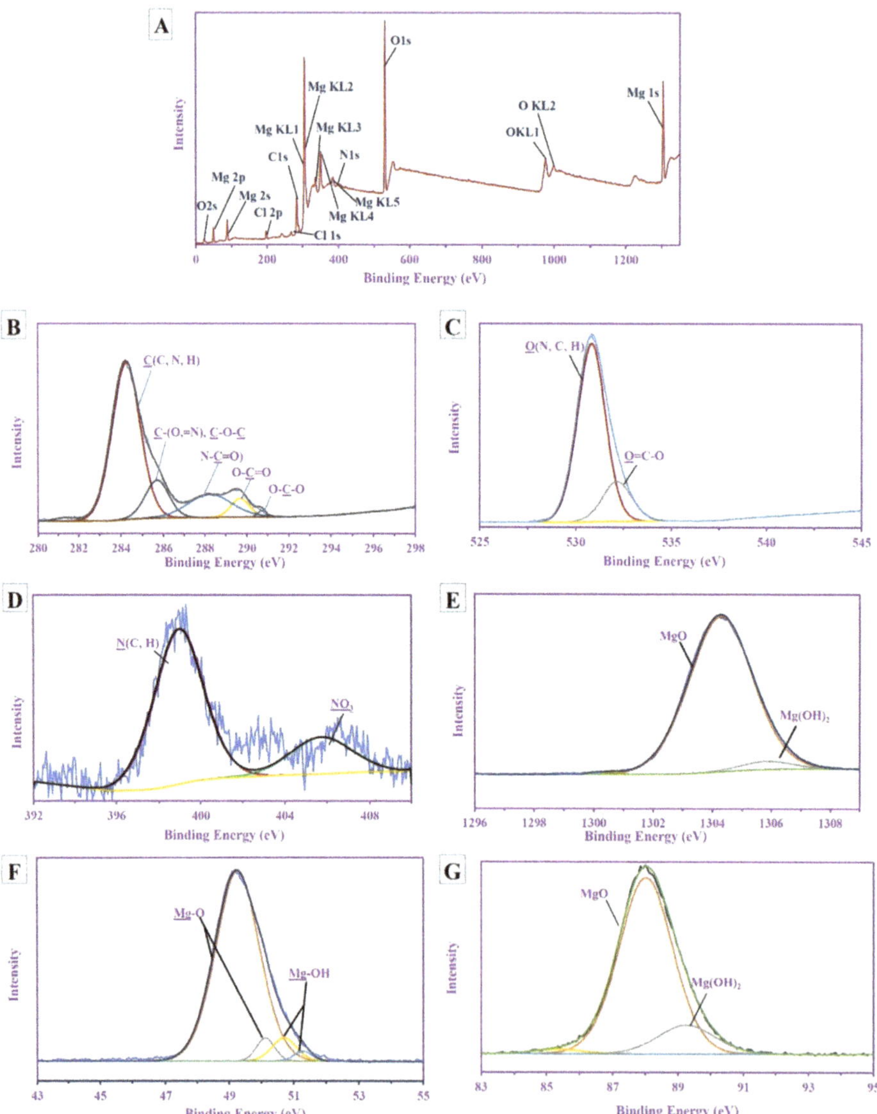

Figure 7. The X-ray photoelectron spectroscopy (XPS) analysis of biosynthesized MgO-NPs. (**A**) Overall view; (**B**) C 1s; (**C**) O 1s; (**D**) N 1s; (**E**) Mg 1s; (**F**) Mg 2p; (**G**) Mg 2s.

The nano-MgO included carbon in the main polysaccharide ring that was characterized by the presence of different peaks for O-C-O at 290.68 eV [77], O-C=O at 289.64 eV [78],

N-C=O at 288.19 eV [79], while the other two peaks at 284.2 eV and 285.65 eV for C(C, N, H) and C(=H, O), C-O-C, respectively [79,80] (Figure 7B). This indicated the presence of amide and carbonyl groups in the polysaccharide's skeletons (these results are parallel to the FT-IR analysis). Khan et al. reported that the presence of carbon labelled as C 1s in MgO-NPs could be attributed to the exposure of the sample to air before characterization [18]. The carboxyl group appeared in the O 1s (which split into two different peaks) at 532.12 eV, which is related to O-C=O, while the other peak is at 530.77 eV for O (N, C, H) [81,82] (Figure 7C). The N 1s deconvoluted into two peaks at 398.94 eV for N (C, H) [83] of the saccharide and 405.72 eV for NO_3 originated from the precursor [84] (Figure 7D).

The HRES XPS spectra for Mg show a different types of species; the Mg 1s splitting into two peaks at 1304.24 eV and 1305.8 eV with At% 95.35% and 4.65% for MgO and $Mg(OH)_2$, respectively [85], the Mg 2s has two different peaks at 87.97 eV and 89.21 eV with At% equivalent to 83.53% and 16.47% for both MgO and $Mg(OH)_2$, respectively [86,87], while the Mg 2p shows 4 peaks at 49.22 eV and 50.08 eV with At% (i.e., 87.47% and 4.42% (major), respectively) for MgO, the other two peaks at 50.66 eV and 51.23 eV with At% ranged 6.18% and 1.92% (minor), respectively, for Mg-OH [87] (Figure 7E–G). This indicates that the majority of appearances of Mg is in MgO form with a trace amount of $Mg(OH)_2$

3.5. Antimicrobial Activity

The antimicrobial activity of biosynthesized MgO-NPs was evaluated against different pathogenic Gram-positive and Gram-negative bacteria represented by *Staphylococcus aureus, Bacillus subtilis, Pseudomonas aeruginosa,* and *Escherichia coli* as well as unicellular fungi represented by *candida albicans*. The presented data showed that the activities of MgO-NPs against different pathogenic microbes were dose-dependent. Compatible with recently published studies, the activities of nanomaterials such as Ag, Fe_2O_3, Se, CuO, and ZnO were dose-dependent [88–91]. Analysis of variance showed that the zones of inhibitions (ZOIs) caused by 200 μg mL^{-1} were 14.7 ± 0.6, 14.3 ± 0.7, 13.7 ± 0.5, 10.6 ± 0.4, and 11.5 ± 0.5 mm for *C. albicans, E. coli, P. aeruginosa, S. aureus,* and *B. subtilis,* respectively (Figure 8). Similarly, the MgO-NPs fabricated by *Rhizophora lamarckii's* extract have antibacterial activity against *Staphylococcus aureus, E. coli,* and *Streptococcus pneumoniae* and with a zone of inhibitions of 26.5, 26.1, and 26.3 mm, respectively [5].

The lowest concentration of MgO-NPs that inhibits microbial growth is defined as the minimum inhibitory concentration (MIC), which differs based on the microbial used. To detect the MIC value for each tested organism, different concentrations of MgO-NPs were investigated. Data analysis showed that the MIC value for *S. aureus* was 150 μg mL^{-1} with ZOI of 8.3 ± 0.6 mm, whereas *C. albicans, E. coli, P. aeruginosa,* and *B. subtilis* have MIC values of 100 μg mL^{-1} with ZOI of 9.7 ± 0.7, 9.0 ± 0.0, 9.8 ± 0.3, and 8.3 ± 0.6 mm, respectively (Figure 8).

The antimicrobial activity of biosynthesized MgO-NPs has been reported by several researchers [5,18,69]. These activities could be attributed to different mechanisms such as the enhancement of the production of reactive oxygen species (ROS), interactions between MgO-NPs and microbial cell walls, discharge of Mg^{2+} upon the entrance of microbial cells, and the alkaline effects of MgO-NPs on the microbial cells. Rai et al. [21] reported that the toxicity of nanoparticles synthesized using fungal species are dependent on size, shape, concentration, and surface charge of nanoparticles used.

The production of ROS, primarily superoxide radicals ($^{-}O_2$), hydrogen peroxide (H_2O_2), and reactive hydroxyl radicals ($^{\bullet}OH$), interferes with nucleic acids and proteins, ultimately leading to cell death [73]. In the current study, Gram-negative bacteria are more sensitive to biosynthesized MgO-NPs than Gram-positive bacteria and this phenomenon can be related to variations between two bacterial kinds in cell wall structures. The Gram-positive bacterial cell wall contains a thick layer of peptidoglycans; on the contrary, Gram-negative bacteria contain a thin layer of peptidoglycans and possess rich lipopolysaccharides (LPS). The interaction between NPs and bacterial cells is due to the negative charge of LPS and the positive charge of NPs [92,93]. Moreover, due to the thin peptidoglycan layer in Gram-negative bacteria, the MgO-NPs can penetrate the cell wall and deposit on the cell membrane, which changes the selective permeability function and is followed by cell death [94]. Furthermore, the MgO-NPs can disrupt the quorum sensing, which is responsible for communications between microbial strains, leading to the inhibition of microbial activities and functions [94,95].

The Mg^{2+} formed due to the penetration of MgO-NPs into the microbial cells can interact with thiol groups of amino acids, leading to the disruption of protein structure and ultimately cell death [75]. Sawai et al. [96] reported that a thin layer of water was adsorbed on the surface of MgO-NPs, resulting in a higher pH value in an aqueous solution than the equilibrium value, leading to microbial cell membrane damage upon contact.

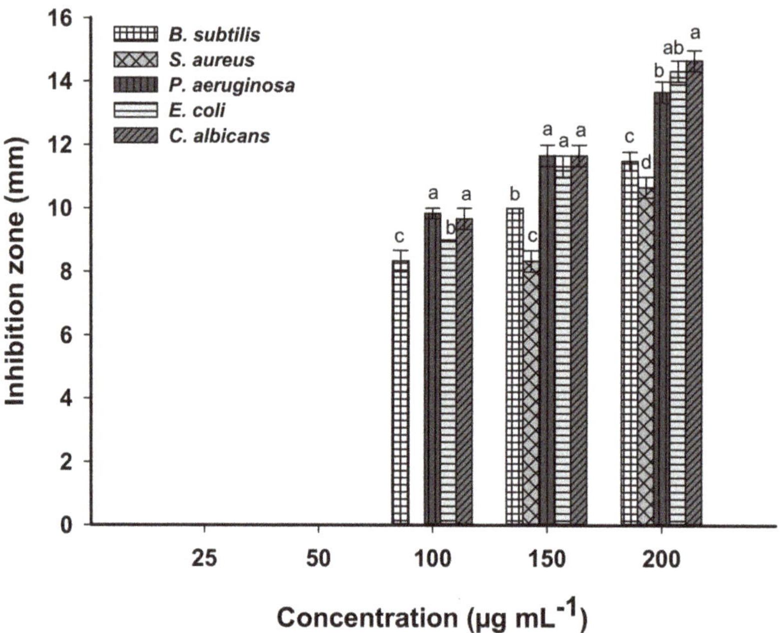

Figure 8. The antimicrobial activity of MgO-NPs at different concentrations against Gram-positive and Gram-negative bacteria, and unicellular fungi. Different letters (a, b, c, and d) on bars at the same concertation denote that mean values are significantly different ($p \leq 0.05$) ($n = 3$).

3.6. Larvicidal Activity

The diseases caused by mosquitos are considered key challenges to public health worldwide and many efforts have been established to control these diseases. The conventional methods (either biological or chemical methods) have been utilized to control the spread of mosquitos, but these methods have drawbacks with effects such as toxicity to consumers and an increase in mosquito resistance to these compounds [97]. Recently, due to the unique characteristics of NPs, they have been used as an alternative source to control mosquito-borne diseases instead of conventional methods. To date, this is the first report to investigate the efficacy of myco-synthesized MgO-NPs against *C. Pipiens*.

The mosquitocidal mechanism of MgO-NPs may be related to two mechanisms: the production of ROS and lipid peroxidation and the leakage of internal cellular contents due to cell membrane damage [73,98]. Once MgO-NPs is sprayed over one stage of mosquito life cycles such as egg, larvae, pupa, or adult, it is dissociated into Mg^{2+} and O^{2-} ions in the surrounding environment. The high concentration of O^{2-} ions forms ROS, which ultimately leads to oxidative stress and lipid peroxidation. Moreover, the small size of MgO-

NPs can react with nucleic acid and deform it, then inhibit the growth of mosquitos [99]. Moreover, the high concentration of Mg^{2+} leads to the destabilization or damage of cellular equilibrium, causing more stress, leakage of cellular components, and finally mosquito cell death [75] (Figure 9A).

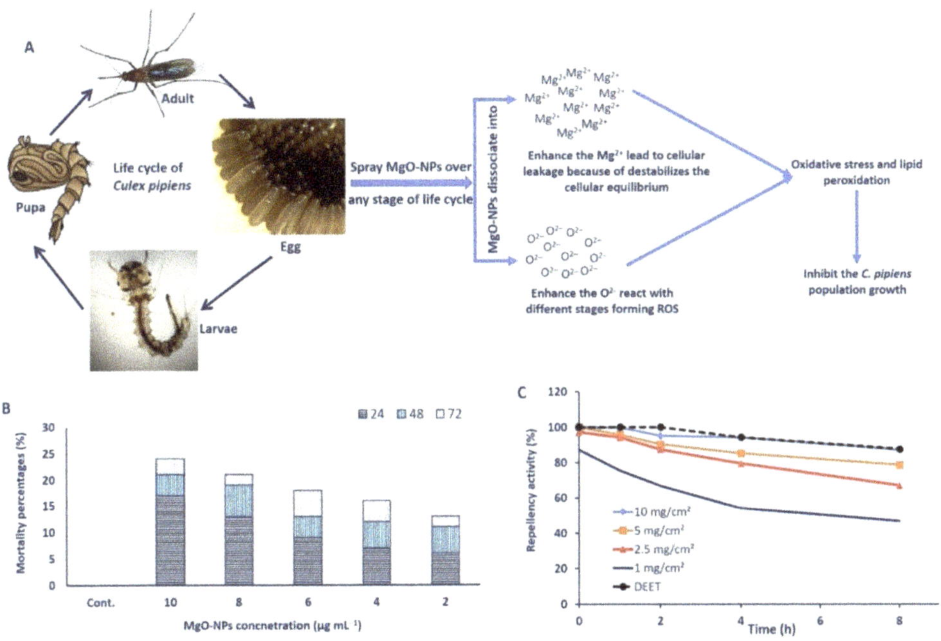

Figure 9. Mosquitocidal activity of MgO-NPs against *C. Pipiens*. (**A**) Proposal mechanism of MgO-NPs as mosquitocidal; (**B**) larvicidal activity of MgO-NPs against *C. Pipiens* at different concentrations (2, 4, 6, 8, and 10 µg mL^{-1}) after different contact times (24, 48, and 72 h); (**C**) repellence activity of different concentrations of MgO-NPs as compared to a positive control (DEET).

In the current study, data analysis showed that the mortality percentages differed based on MgO-NPs concentrations. More than 50% mortality was observed in all concentrations after 72 h. The lowest mortality percentages (52%) were recorded after 3 days of treatment with 2 µg mL^{-1} of MgO-NPs, whereas the maximum mortality percentages (96%) were obtained at 10 µg mL^{-1}. At MgO-NPs concentrations of 4, 6, and 8 µg mL^{-1}, the mortality percentages reached 64%, 72%, and 84%, respectively, after 72 h (Figure 9B). The LC_{50} (concentration of MgO-NPs that inhibit 50% of the population) and LC_{90} (concentration of MgO-NPs that inhibit 90% of the population) were 2.21 µg mL^{-1} and 10.71 µg mL^{-1}, respectively. Recently, several studies have reported the efficacy of different metal and metal oxides NPs to control or inhibit the mosquito populations [15,37,39]. The rod-shaped MgO/ hydroxyapatite nanoparticles showed potentiality to inhibit *Aedes aegypti*, *Anopheles stephensi*, and *Culex quinquefasciatus* through the accumulation of Mg^{2+}, which destroys the mosquito cells [100].

3.7. Mosquito Repellent Activity

Data analysis showed that all tested concentrations exhibit more than 60% repellency against female *C. Pipiens* mosquitoes. However, the repellent activity of MgO-NPs was decreased with time increase (Figure 9C). The MgO-NPs concentrations of 10 mg/cm^2, 5 mg/cm^2, 2.5 mg/cm^2, and DEET (as a positive control) showed repellency percentages of 95.16%, 90.34%, 87.45%, and 100%, respectively, for more than 60 min (Figure 9C).

Interestingly, the repellency percentages between 10 mg/cm^2 and the positive control do not show any difference after 240 min (94.1% and 94.2%) and 480 min (87.3% and 87.5%). After maximum time (8 h), the repellence percentages were 78.6%, 67.1%, and 46. 78% for MgO-NPs concentration of 5 mg/cm^2, 2.5 mg/cm^2, and 1.0 mg/cm^2, respectively, as compared with DEET (87.6%). Based on the obtained data, the MgO-NPs have repellence activity at high concentrations as compared to commercial substances (EDDT). The TiO_2-NPs exhibit repellence activity against *Aedes aegypti* with a value of 80.43% at 100 ppm [101]. The present study provides an eco-friendly, cost-effective, and simple approach to control the spread of C. *Pipiens* populations, as well as new repellent agents that can be used instead of commercial compounds.

3.8. Decolorization and Degradation of Tanning Effluents

The materials at the nanoscale are characterized by their eco-friendly and large surface area, and these features enable them to absorb many contaminants. The tanning wastewater is characterized by its greenish-blue color due to the extensive use of chrome ions and dyes during various processing steps [102]. Disposing of this effluent without treatment prevents the penetration of sunlight and hence decreases or hinders the oxidation of the pollutants [103]. Recently, global trends have been directed to reduce the chemical methods used for remediation and degradation of environmentally hazardous materials by newly advanced, eco-friendly, and safe methods. Hence, the efficacy of biosynthesized MgO-NPs as adsorbents to treat the tannery effluent was investigated at different concentrations (50, 75, and 100 µg/100 mL) and different contact times (60, 120, 180, and 240 min). Data analysis showed that the potentiality of MgO-NPs was dose- and time-dependent. The decolorization percentages were increased by increasing NP concentrations and contact time. This phenomenon could be attributed to the adsorption sites being increased by increasing the adsorbent concentrations [104]. The highest decolorizations were accomplished at 100 µg MgO-NPs after 180 min, a record of 95.6 ± 1.6% (Table 1). Analysis of variance showed that the difference between decolorization after 180 and 240 min is not significant and the time is considered a critical factor at a large scale. Therefore, the treatment of 100 mL tanning effluent with 100 µg of MgO-NPs for 180 min was selected as the optimum condition to study the physicochemical parameters of tanning effluents.

Table 1. Decolorization percentages (%) of tanning effluents using different concentrations (50, 75, and 100 µg/100 mL) of myco-synthesized MgO-NPs at different contact times (60, 120, 180, and 240 min).

MgO-NPs Concentration /100 mL Tanning Effluent	Decolorization Percentages (%) after Time (min)			
	60 min	120 min	180 min	240 min
Control	2.3 ± 0.2 ^{a}	3.5 ± 0.4 ^{b}	4.6 ± 0.3 ^{c}	5.4 ± 0.4 ^{c}
50 µg	34.3 ± 2.2 ^{a}	48.7 ± 3.7 ^{b}	54.2 ± 2.02 ^{c}	59.6 ± 2.5 ^{c}
75 µg	44.5 ± 3.3 ^{a}	57.7 ± 2.4 ^{b}	71.3 ± 2.3 ^{c}	75.2 ± 1.7 ^{c}
100 µg	65.4 ± 1.9 ^{a}	81.1 ± 1.6 ^{b}	95.6 ± 1.6 ^{c}	96.7 ± 0.7 ^{c}

Different letters in the same row are significantly different ($p \leq 0.05$) by the Tukey LSD test. Data are represented by mean ± SD (n = 3).

Tanning effluents are characterized by high contents of hazardous chemicals, chlorides, calcium phosphates, bicarbonates, sulfates, sodium, nitrates, potassium, and varied dissolved salts. Therefore, the physicochemical parameters including pH, TDS, TSS, BOD, COD, and conductivity are usually high [105]. Moreover, the physicochemical parameters of tanning effluents are varied based on tannery size, chemicals used according to the type of products, and the amount of water used [106]. Data presented in Table 2 show a high level of physicochemical parameters of crude tanning effluents before MgO-NPs treatment. The pH of crude tannery effluent is usually alkaline because of the usage of bicarbonates and carbonates [107]. Furthermore, the values of TSS, TDS, BOD, COD, and conductivity of crude tanning effluents are 8745.3 ± 5.5 mg L^{-1}, 15,704 ± 4.1 mg L^{-1}, 2355.7 ± 7.0 mg L^{-1}, 651.7 ± 4.7 mg L^{-1}, and 26,738.7 ± 6.0 S m^{-1}, respectively (Table 2). The high values of TSS and TDS may be attributed to the high concentrations of salts and inorganic or organic insoluble substances, which make tanning effluents unsuitable for plant irrigations [108]. The high level of electrical conductivity can be attributed to the high usage of acid and salts such as chromium salts and sodium during various tanning processes [109]. The high level of BOD and COD of untreated effluents have negative impacts on aquatic and environmental ecosystems [110]. Data analysis showed that the MgO-NPs have high efficiency to highly reduce the physicochemical parameters of tanning effluents. As shown in Table 2, the removal percentages of TSS, TDS, BOD, COD, and conductivity due to MgO-NPs treatment were 97.9%, 98.2%, 87.8%, 95.9%, and 97.3%, respectively. This activity can be due to high MgO-NPs concentration, which provides the large adsorption sites and large surface area that facilitate the adsorption of most organic ions and other pollutants, especially at high contact times [111].

Table 2. Physicochemical characterizations and chromium ion adsorption from tanning effluents by MgO-NPs.

Physicochemical Parameters	Control	After MgO-NPs Treatment	Removal Percentages (%)
pH	10.5	8	-
TSS (mg L^{-1})	8745.3 ± 5.5 [a]	177.7 ± 5.1 [b]	97.9
TDS (mg L^{-1})	15,704 ± 4.1 [a]	286.7 ± 4.2 [b]	98.2
BOD (mg L^{-1})	2355.7 ± 7.0 [a]	287.3 ± 4.9 [b]	87.8
COD (mg L^{-1})	651.7 ± 4.7 [a]	26.7 ± 2.1 [b]	95.9
Conductivity (S m^{-1})	26,738.7 ± 6.0 [a]	708.7 ± 4.0 [b]	97.3
Cr (mg L^{-1})	822.3 ± 2.5 [a]	14.5 ± 0.9 [b]	98.2

Different letters in the same row are significantly different ($p \leq 0.05$) by the Tukey LSD test. Data are represented by mean ± SD (n = 3).

Heavy metals are considered one of the most damaging environmental pollutants and can cause severe human and animal problems due to their accumulation in the food chains [112]. The tanning industry is one of the main sources of the discharge of heavy metals into the environment. Among these heavy metals is chromium ions, which, despite their presence in some daily diets, can cause skin allergies and lung cancer [113]. MgO-NPs have various advantages to be used as a bio-adsorbent for different heavy metals, such as cost-effectiveness, high adsorption capacity, nontoxicity, eco-friendly, abundance, and biocompatibility [17]. In the current study, data analysis showed the efficacy of biosynthesized MgO-NPs to reduce chromium ion in tanning effluents from 822.3 ± 2.5 mg L^{-1} to 14.5 ± 0.9 mg L^{-1} with removal percentages of 98.2%. This high removal efficacy can be attributed to the precipitation and adsorption of metals on the MgO; on the contrary, for other nanomaterials such as Al_2O_3 and TiO_2, the removal mechanism was adsorption only as mentioned previously [114]. Yang et al. [115] reported that the high adsorption efficacy of MgO-NPs is due to the dissociation of OH^- from $Mg(OH)_2$, along with the synergistic effects of the adsorption and precipitation process. Interestingly, MgO-NPs showed the adsorption efficacy of Pb (II) and Cd (II) with values of 2614 and 2294 mg g^{-1} [116], respectively. Recently, the MgO-NPs synthesized by *Aspergillus niger* F1 showed the removal of Cr with percentages of 94.2 ± 1.2% [76]. Moreover, Seif et al. reported the high potentiality of MgO-NPs to adsorb Cr^{3+} with values of 1033.8 mg g^{-1} as compared to montmorillonite nanoparticles, for which the maximum adsorption was 3.6 mg g^{-1} [117].

4. Conclusions

In this study, synthesis of MgO-NPs was pursued by using the metabolites on the biomass filtrate of *Rhizopus oryaze* with Mg $(NO_3)_2.6H_2O$ as a precursor. The production technique was optimized by studying the precursor concentration, the contact time between the fungal biomass filtrate and precursor, incubation temperatures, and pH values. The color change to turbid white and the maximum SPR at 282 nm confirmed the successful synthesis of MgO-NPs, which is characterized by TEM, SEM-EDX, XRD, DLS, FT-IR spectroscopy, and XPS analyses. These analyses revealed that the biosynthesized MgO-NPs are crystalline, spherical, and well-dispersed, with sizes ranging between 8.0 to 47.5 nm.

The fungal-induced MgO-NPs offer potential anti-microbial activity against *S. aureus*, *B. subtilis*, *P. aeruginosa*, *E. coli*, and *C. albicans*, with varied inhibition zones based on NP concentrations. Moreover, these NPs provide a modest, eco-friendly, and cost-effective way of controlling *C. Pipiens* populations with enhanced repellent activity as compared with commercial compounds. Additionally, they could be used efficiently to decolorize tanning effluents through the reduction of physicochemical parameters such as TSS, TDS, BOD, COD, and conductivity. Moreover, the biosynthesized MgO-NPs exhibit high efficacy to bio-adsorb chromium ions from tanning effluents. This study provides a simple, eco-friendly, cost-effective, and rapid approach to inhibit the growth of the microbial pathogen, prevent the spread of adverse insects, treat some of the worst environmental contaminants, and adsorb the most hazardous heavy metal.

Author Contributions: Conceptualization, S.E.-D.H. and A.F.; methodology, S.E.-D.H., A.F., E.S., M.M.S.F., A.M.E., M.G.B., and M.A.A.; software, S.E.-D.H., A.F., E.S., M.M.S.F., A.M.E., M.G.B., and M.A.A.; validation, S.E.-D.H., A.F., E.S., M.M.S.F., A.M.E., M.G.B., and M.A.A.; formal analysis, S.E.-D.H., A.F., E.S., A.M.E., M.G.B., M.A.A., M.F.H., and M.F.A.; investigation, S.E.-D.H., A.F., E.S., M.M.S.F., A.M.E., M.G.B., M.F.A., and M.A.A.; resources, S.E.-D.H., A.F., E.S., M.M.S.F., A.M.E., M.G.B., M.A.A., M.F.H., and M.F.A.; data curation, S.E.-D.H., A.F., E.S., A.M.E., M.G.B., and M.A.A.; writing—original draft preparation, S.E.-D.H., A.F., E.S., M.M.S.F., A.M.E., M.G.B., M.A.A., M.F.H., and M.F.A.; writing—review and editing, A.F., E.S., M.M.S.F., A.M.E., M.G.B., M.A.A., M.F.H., and M.F.A.; visualization, S.E.-D.H., A.F., E.S., M.M.S.F., A.M.E., M.G.B., M.A.A., M.F.H., and M.F.A.; supervision, S.E.-D.H. and A.F.; project administration, S.E.-D.H. and A.F.; funding acquisition, M.F.A. All authors have read and agreed to the published version of the manuscript.

Funding: This research received no external funding.

Institutional Review Board Statement: Not applicable.

Informed Consent Statement: Not applicable.

Data Availability Statement: The data presented in this study are available on request from the corresponding author.

Acknowledgments: Authors extend their appreciation to Mamdouh Salem El-Gamal (Head of Microbial Physiology Lab), Botany and Microbiology Department, Faculty of Science, Al-Azhar University, Cairo, Egypt, for the great help in the current study.

Conflicts of Interest: The authors declare no conflict of interest.

References

1. Soliman, A.M.; Abdel-Latif, W.; Shehata, I.H.; Fouda, A.; Abdo, A.M.; Ahmed, Y.M. Green Approach to Overcome the Resistance Pattern of *Candida* spp. Using Biosynthesized Silver Nanoparticles Fabricated by *Penicillium chrysogenum* F9. *Biol. Trace Elem. Res.* **2021**, *199*, 800–811. [CrossRef]
2. Rajeswaran, S.; Somasundaram Thirugnanasambandan, S.; Dewangan, N.K.; Moorthy, R.K.; Kandasamy, S.; Vilwanathan, R. Multifarious Pharmacological Applications of Green Routed Eco-Friendly Iron Nanoparticles Synthesized by *Streptomyces* Sp. (SRT12). *Biol. Trace Elem. Res.* **2020**, *194*, 273–283. [CrossRef]
3. Mahmoudi, F.; Mahmoudi, F.; Gollo, K.H.; Amini, M.M. Biosynthesis of Novel Silver Nanoparticles Using *Eryngium thyrsoideum* Boiss Extract and Comparison of their Antidiabetic Activity with Chemical Synthesized Silver Nanoparticles in Diabetic Rats. *Biol. Trace Elem. Res.* **2021**, *199*, 1967–1978. [CrossRef] [PubMed]
4. Alrashed, A.A.A.A.; Akbari, O.A.; Heydari, A.; Toghraie, D.; Zarringhalam, M.; Shabani, G.A.S.; Seifi, A.R.; Goodarzi, M. The numerical modeling of water/FMWCNT nanofluid flow and heat transfer in a backward-facing contracting channel. *Phys. B Condens. Matter* **2018**, *537*, 176–183. [CrossRef]
5. Prasanth, R.; Kumar, S.D.; Jayalakshmi, A.; Singaravelu, G.; Govindaraju, K.; Kumar, V.G. Green synthesis of magnesium oxide nanoparticles and their antibacterial activity. *Indian J. Geo Mar. Sci.* **2019**, *48*, 1210–1215.
6. Eid, A.M.; Fouda, A.; Niedbała, G.; Hassan, S.E.-D.; Salem, S.S.; Abdo, A.M.; Hetta, H.F.; Shaheen, T.I. Endophytic *Streptomyces laurentii* Mediated Green Synthesis of Ag-NPs with Antibacterial and Anticancer Properties for Developing Functional Textile Fabric Properties. *Antibiotics* **2020**, *9*, 641. [CrossRef]
7. Eid, A.M.; Fouda, A.; Abdel-Rahman, M.A.; Salem, S.S.; Elsaied, A.; Oelmüller, R.; Hijri, M.; Bhowmik, A.; Elkelish, A.; Hassan, S.E.-D. Harnessing Bacterial Endophytes for Promotion of Plant Growth and Biotechnological Applications: An Overview. *Plants* **2021**, *10*, 935. [CrossRef]

8. Fouda, A.; Abdel-Maksoud, G.; Abdel-Rahman, M.A.; Eid, A.M.; Barghoth, M.G.; El-Sadany, M.A.-H. Monitoring the effect of biosynthesized nanoparticles against biodeterioration of cellulose-based materials by *Aspergillus niger*. *Cellulose* **2019**, *26*, 6583–6597. [CrossRef]
9. Elfeky, A.S.; Salem, S.S.; Elzaref, A.S.; Owda, M.E.; Eladawy, H.A.; Saeed, A.M.; Awad, M.A.; Abou-Zeid, R.E.; Fouda, A. Multifunctional cellulose nanocrystal/metal oxide hybrid, photo-degradation, antibacterial and larvicidal activities. *Carbohydr. Polym.* **2020**, *230*, 115711. [CrossRef]
10. Fouda, A.; Salem, S.S.; Wassel, A.R.; Hamza, M.F.; Shaheen, T.I. Optimization of green biosynthesized visible light active CuO/ZnO nano-photocatalysts for the degradation of organic methylene blue dye. *Heliyon* **2020**, *6*, e04896. [CrossRef] [PubMed]
11. Hamza, M.F.; Ahmed, F.Y.; El-Aassy, I.; Fouda, A.; Guibal, E. Groundwater purification in a polymetallic mining area (SW Sinai, Egypt) using functionalized magnetic chitosan particles. *Water Air Soil Pollut.* **2018**, *229*, 1–14. [CrossRef]
12. Vasantharaj, S.; Sathiyavimal, S.; Senthilkumar, P.; LewisOscar, F.; Pugazhendhi, A. Biosynthesis of iron oxide nanoparticles using leaf extract of *Ruellia tuberosa*: Antimicrobial properties and their applications in photocatalytic degradation. *J. Photochem. Photobiol. B Biol.* **2019**, *192*, 74–82. [CrossRef] [PubMed]
13. Fouda, A.; Abdel-Maksoud, G.; Abdel-Rahman, M.A.; Salem, S.S.; Hassan, S.E.-D.; El-Sadany, M.A.-H. Eco-friendly approach utilizing green synthesized nanoparticles for paper conservation against microbes involved in biodeterioration of archaeological manuscript. *Int. Biodeterior. Biodegrad.* **2019**, *142*, 160–169. [CrossRef]
14. Mohamed, A.A.; Fouda, A.; Elgamal, M.S.; EL-Din Hassan, S.; Shaheen, T.I.; Salem, S.S. Enhancing of Cotton Fabric Antibacterial Properties by Silver Nanoparticles Synthesized by New Egyptian Strain *Fusarium Keratoplasticum* A1-3. *Egypt. J. Chem.* **2017**, *60*, 63–71. [CrossRef]
15. Hassan, S.E.-D.; Fouda, A.; Radwan, A.A.; Salem, S.S.; Barghoth, M.G.; Awad, M.A.; Abdo, A.M.; El-Gamal, M.S. Endophytic actinomycetes *Streptomyces* spp mediated biosynthesis of copper oxide nanoparticles as a promising tool for biotechnological applications. *JBIC J. Biol. Inorg. Chem.* **2019**, *24*, 377–393. [CrossRef]
16. Hassan, S.E.L.D.; Salem, S.S.; Fouda, A.; Awad, M.A.; El-Gamal, M.S.; Abdo, A.M. New approach for antimicrobial activity and bio-control of various pathogens by biosynthesized copper nanoparticles using endophytic actinomycetes. *J. Radiat. Res. Appl. Sci.* **2018**, *11*, 262–270. [CrossRef]
17. Cai, Y.; Li, C.; Wu, D.; Wang, W.; Tan, F.; Wang, X.; Wong, P.K.; Qiao, X. Highly active MgO nanoparticles for simultaneous bacterial inactivation and heavy metal removal from aqueous solution. *Chem. Eng. J.* **2017**, *312*, 158–166. [CrossRef]
18. Khan, A.; Shabbier, D.; Ahmad, P.; Khandaker, M.U.; Faruque, M.R.I.; Din, I. Biosynthesis and antibacterial activity of MgO-NPs produced from *Camellia-sinensis* leaves extract. *Mater. Res. Express* **2020**, *8*, 015402. [CrossRef]
19. Duong, T.H.Y.; Nguyen, T.N.; Oanh, H.T.; Dang Thi, T.A.; Giang, L.N.T.; Phuong, H.T.; Anh, N.T.; Nguyen, B.M.; Tran Quang, V.; Le, G.T.; et al. Synthesis of Magnesium Oxide Nanoplates and Their Application in Nitrogen Dioxide and Sulfur Dioxide Adsorption. *J. Chem.* **2019**, *2019*, 4376429. [CrossRef]
20. Clarance, P.; Luvankar, B.; Sales, J.; Khusro, A.; Agastian, P.; Tack, J.C.; Al Khulaifi, M.M.; Al-Shwaiman, H.A.; Elgorban, A.M.; Syed, A.; et al. Green synthesis and characterization of gold nanoparticles using endophytic fungi *Fusarium solani* and its in-vitro anticancer and biomedical applications. *Saudi J. Biol. Sci.* **2020**, *27*, 706–712. [CrossRef]
21. Rai, M.; Bonde, S.; Golinska, P.; Trzcińska-Wencel, J.; Gade, A.; Abd-Elsalam, K.A.; Shende, S.; Gaikwad, S.; Ingle, A.P. *Fusarium* as a Novel Fungus for the Synthesis of Nanoparticles: Mechanism and Applications. *J. Fungi* **2021**, 139. [CrossRef]
22. Bhardwaj, K.; Sharma, A.; Tejwan, N.; Bhardwaj, S.; Bhardwaj, P.; Nepovimova, E.; Shami, A.; Kalia, A.; Kumar, A.; Abd-Elsalam, K.A.; et al. *Pleurotus* Macrofungi-Assisted Nanoparticle Synthesis and Its Potential Applications: A Review. *J. Fungi* **2020**, *6*, 351. [CrossRef] [PubMed]
23. Din, M.I.; Najeeb, J.; Ahmad, G. Recent advancements in the architecting schemes of zinc oxide-based photocatalytic assemblies. *Sep. Purif. Rev.* **2018**, *47*, 267–287. [CrossRef]
24. Jeevanandam, J.; Chan, Y.S.; Danquah, M.K. Evaluating the Antibacterial Activity of MgO Nanoparticles Synthesized from Aqueous Leaf Extract. *Med. One* **2019**, *4*, e190011. [CrossRef]
25. Rani, P.; Kaur, G.; Rao, K.V.; Singh, J.; Rawat, M. Impact of Green Synthesized Metal Oxide Nanoparticles on Seed Germination and Seedling Growth of *Vigna radiata* (Mung Bean) and *Cajanus cajan* (Red Gram). *J. Inorg. Organomet. Polym. Mater.* **2020**, *30*, 4053–4062. [CrossRef]
26. Ammulu, M.A.; Vinay Viswanath, K.; Giduturi, A.K.; Vemuri, P.K.; Mangamuri, U.; Poda, S. Phytoassisted synthesis of magnesium oxide nanoparticles from *Pterocarpus marsupium* rox.b heartwood extract and its biomedical applications. *J. Genet. Eng. Biotechnol.* **2021**, *19*, 21. [CrossRef] [PubMed]
27. Pigeault, R.; Bataillard, D.; Glaizot, O.; Christe, P. Effect of Neonicotinoid Exposure on the Life History Traits and Susceptibility to Plasmodium Infection on the Major Avian Malaria Vector *Culex pipiens* (Diptera: Culicidae). *Parasitologia* **2021**, *1*, 3. [CrossRef]
28. Younis, M.; Hussein, A.; Aboelela, H.; Rashed, G. The Potential Role of Photosensitizers in Fight against Mosquitoes: Phototoxicity of Rose Bengal against Culex Pipiens Larvae. *Benha Med. J.* **2021**, *38*, 1–10. [CrossRef]
29. Fouda, A.; Khalil, A.; El-Sheikh, H.; Abdel-Rhaman, E.; Hashem, A. Biodegradation and detoxification of bisphenol-A by filamentous fungi screened from nature. *J. Adv. Biol. Biotechnol.* **2015**, 123–132. [CrossRef]
30. White, T.J.; Bruns, T.; Lee, S.; Taylor, J. Amplification and direct sequencing of fungal ribosomal RNA genes for phylogenetics. *PCR Protoc. Guide Methods Appl.* **1990**, *18*, 315–322.

31. Essien, E.R.; Atasie, V.N.; Okeafor, A.O.; Nwude, D.O. Biogenic synthesis of magnesium oxide nanoparticles using *Manihot esculenta* (Crantz) leaf extract. *Int. Nano Lett.* **2020**, *10*, 43–48. [CrossRef]
32. Fouda, A.; El-Din Hassan, S.; Salem, S.S.; Shaheen, T.I. In-Vitro cytotoxicity, antibacterial, and UV protection properties of the biosynthesized Zinc oxide nanoparticles for medical textile applications. *Microb. Pathog.* **2018**, *125*, 252–261. [CrossRef] [PubMed]
33. Badawy, A.A.; Abdelfattah, N.A.H.; Salem, S.S.; Awad, M.F.; Fouda, A. Efficacy Assessment of Biosynthesized Copper Oxide Nanoparticles (CuO-NPs) on Stored Grain Insects and Their Impacts on Morphological and Physiological Traits of Wheat (*Triticum aestivum* L.) Plant. *Biology* **2021**, *10*, 233. [CrossRef]
34. Wei, Y.; Salih, K.A.M.; Lu, S.; Hamza, M.F.; Fujita, T.; Vincent, T.; Guibal, E. Amidoxime Functionalization of Algal/Polyethyleneimine Beads for the Sorption of Sr(II) from Aqueous Solutions. *Molecules* **2019**, *24*, 3893. [CrossRef] [PubMed]
35. Shaheen, T.I.; Fouda, A.; Salem, S.S. Integration of Cotton Fabrics with Biosynthesized CuO Nanoparticles for Bactericidal Activity in the Terms of Their Cytotoxicity Assessment. *Ind. Eng. Chem. Res.* **2021**, *60*, 1553–1563. [CrossRef]
36. El-Belely, E.F.; Farag, M.; Said, H.A.; Amin, A.S.; Azab, E.; Gobouri, A.A.; Fouda, A. Green Synthesis of Zinc Oxide Nanoparticles (ZnO-NPs) Using *Arthrospira platensis* (Class: *Cyanophyceae*) and Evaluation of their Biomedical Activities. *Nanomaterials* **2021**, *11*, 95. [CrossRef]
37. Marimuthu, S.; Rahuman, A.A.; Rajakumar, G.; Santhoshkumar, T.; Kirthi, A.V.; Jayaseelan, C.; Bagavan, A.; Zahir, A.A.; Elango, G.; Kamaraj, C. Evaluation of green synthesized silver nanoparticles against parasites. *Parasitol. Res.* **2011**, *108*, 1541–1549. [CrossRef]
38. Velayutham, K.; Ramanibai, R. Larvicidal activity of synthesized silver nanoparticles using isoamyl acetate identified in *Annona squamosa* leaves against *Aedes aegypti* and *Culex quinquefasciatus*. *J. Basic Appl. Zool.* **2016**, *74*, 16–22. [CrossRef]
39. Fouda, A.; Hassan, S.E.; Abdo, A.M.; El-Gamal, M.S. Antimicrobial, Antioxidant and Larvicidal Activities of Spherical Silver Nanoparticles Synthesized by Endophytic *Streptomyces* spp. *Biol. Trace Elem. Res.* **2020**, *195*, 707–724. [CrossRef]
40. Postelnicu, T. Probit analysis. In *International Encyclopedia of Statistical Science*; Lovric, M., Ed.; Springer: Berlin/Heidelberg, Germany, 2011; pp. 1128–1131.
41. World Health Organization. *Guidelines for Efficacy Testing of Mosquito Repellents for Human Skin*; World Health Organization: Geneva, Switzerland, 2009.
42. Govindarajan, M.; Sivakumar, R. Mosquito adulticidal and repellent activities of botanical extracts against malarial vector, *Anopheles stephensi* Liston (Diptera: Culicidae). *Asian Pac. J. Trop. Med.* **2011**, *4*, 941–947. [CrossRef]
43. Eugene, W.R.; Baird, R.B.; Andrew, D.E.; Lenore, S.C. *Standard Methods for the Examination of Water and Wastewater*; American Public Health Association, American Water Works Association, Water Environment Federation: Washington, DC, USA, 2012.
44. Ibarruri, J.; Hernández, I. Rhizopus oryzae as Fermentation Agent in Food Derived Sub-products. *Waste Biomass Valorization* **2018**, 9. [CrossRef]
45. López-Fernández, J.; Benaiges, M.D.; Valero, F. *Rhizopus oryzae* Lipase, a Promising Industrial Enzyme: Biochemical Characteristics, Production and Biocatalytic Applications. *Catalysts* **2020**, *10*, 1277. [CrossRef]
46. Salem, S.S.; Fouda, A. Green synthesis of metallic nanoparticles and their prospective biotechnological applications: An overview. *Biol. Trace Elem. Res.* **2020**, *199*, 344–370. [CrossRef]
47. Khanra, K.; Panja, S.; Choudhuri, I.; Bhattacharyya, N. Evaluation of Antibacterial Activity and Cytotoxicity of Green Synthesized Silver Nanoparticles Using *Scoparia dulcis*. *Nano Biomed Eng.* **2015**, *7*, 128–133. [CrossRef]
48. Fedlheim, D.L.; Foss, C.A. *Metal Nanoparticles: Synthesis, Characterization, and Applications*; CRC Press: Boca Raton, FL, USA, 2001.
49. Jeevanandam, J.; Chan, Y.S.; Danquah, M.K. Biosynthesis and characterization of MgO nanoparticles from plant extracts via induced molecular nucleation. *New J. Chem.* **2017**, *41*, 2800–2814. [CrossRef]
50. Nguyen, D.T.C.; Dang, H.H.; Vo, D.-V.N.; Bach, L.G.; Nguyen, T.D.; Tran, T.V. Biogenic synthesis of MgO nanoparticles from different extracts (flower, bark, leaf) of *Tecoma stans* (L.) and their utilization in selected organic dyes treatment. *J. Hazard. Mater.* **2021**, *404*, 124146. [CrossRef] [PubMed]
51. Abdallah, Y.; Ogunyemi, S.O.; Abdelazez, A.; Zhang, M.; Hong, X.; Ibrahim, E.; Hossain, A.; Fouad, H.; Li, B.; Chen, J. The Green Synthesis of MgO Nano-Flowers Using *Rosmarinus officinalis* L. (Rosemary) and the Antibacterial Activities against *Xanthomonas oryzae* pv. *oryzae*. *BioMed Res. Int.* **2019**, *2019*, 5620989. [CrossRef]
52. Shawky, A.M.; El-Tohamy, M.F. Highly Functionalized Modified Metal Oxides Polymeric Sensors for Potentiometric Determination of Letrozole in Commercial Oral Tablets and Biosamples. *Polymers* **2021**, *13*, 1384. [CrossRef]
53. Patra, J.K.; Baek, K.-H. Green Nanobiotechnology: Factors Affecting Synthesis and Characterization Techniques. *J. Nanomater.* **2014**, *2014*, 417305. [CrossRef]
54. Rao, C.R.K.; Trivedi, D.C. Synthesis and characterization of fatty acids passivated silver nanoparticles—Their interaction with PPy. *Synth. Met.* **2005**, *155*, 324–327. [CrossRef]
55. Ibrahim, H.M.M. Green synthesis and characterization of silver nanoparticles using banana peel extract and their antimicrobial activity against representative microorganisms. *J. Radiat. Res. Appl. Sci.* **2015**, *8*, 265–275. [CrossRef]
56. Fouda, A.; Hassan, S.E.-D.; Abdel-Rahman, M.A.; Farag, M.M.S.; Shehal-deen, A.; Mohamed, A.A.; Alsharif, S.M.; Saied, E.; Moghanim, S.A.; Azab, M.S. Catalytic degradation of wastewater from the textile and tannery industries by green synthesized hematite (α-Fe_2O_3) and magnesium oxide (MgO) nanoparticles. *Curr. Res. Biotechnol.* **2021**, *3*, 29–41. [CrossRef]

57. Verma, A.; Mehata, M.S. Controllable synthesis of silver nanoparticles using Neem leaves and their antimicrobial activity. *J. Radiat. Res. Appl. Sci.* **2016**, *9*, 109–115. [CrossRef]
58. Rai, A.; Singh, A.; Ahmad, A.; Sastry, M. Role of Halide Ions and Temperature on the Morphology of Biologically Synthesized Gold Nanotriangles. *Langmuir* **2006**, *22*, 736–741. [CrossRef]
59. Chen, J.; Wu, L.; Lu, M.; Lu, S.; Li, Z.; Ding, W. Comparative Study on the Fungicidal Activity of Metallic MgO Nanoparticles and Macroscale MgO against Soilborne Fungal Phytopathogens. *Front. Microbiol.* **2020**, *11*, 365. [CrossRef] [PubMed]
60. Lashin, I.; Fouda, A.; Gobouri, A.A.; Azab, E.; Mohammedsaleh, Z.M.; Makharita, R.R. Antimicrobial and In Vitro Cytotoxic Efficacy of Biogenic Silver Nanoparticles (Ag-NPs) Fabricated by Callus Extract of *Solanum incanum* L. *Biomolecules* **2021**, *11*, 341. [CrossRef]
61. Huang, L.; Li, D.; Lin, Y.; Evans, D.G.; Duan, X. Influence of nano-MgO particle size on bactericidal action against *Bacillus subtilis* var. *niger*. *Chin. Sci. Bull.* **2005**, *50*, 514–519. [CrossRef]
62. Dobrucka, R. Synthesis of MgO Nanoparticles Using *Artemisia abrotanum* Herba Extract and Their Antioxidant and Photocatalytic Properties. *Iran. J. Sci. Technol. Trans. A Sci.* **2018**, *42*, 547–555. [CrossRef]
63. Alsharif, S.M.; Salem, S.S.; Abdel-Rahman, M.A.; Fouda, A.; Eid, A.M.; Hassan, S.E.-D.; Awad, M.A.; Mohamed, A.A. Multifunctional properties of spherical silver nanoparticles fabricated by different microbial taxa. *Heliyon* **2020**, *6*, e03943. [CrossRef] [PubMed]
64. Safaei, G.J.; Zahedi, S.; Javid, M.; Ghasemzadeh, M. MgO nanoparticles: An efficient, green and reusable catalyst for the onepot syntheses of 2, 6-dicyanoanilines and 1, 3-diarylpropyl malononitriles under different conditions. *J. Nanostruct.* **2015**, *5*, 153–160.
65. Lekota, M.W.; Dimpe, K.M.; Nomngongo, P.N. MgO-ZnO/carbon nanofiber nanocomposite as an adsorbent for ultrasound-assisted dispersive solid-phase microextraction of carbamazepine from wastewater prior to high-performance liquid chromatographic detection. *J. Anal. Sci. Technol.* **2019**, *10*, 25. [CrossRef]
66. Tomaszewska, E.; Soliwoda, K.; Kadziola, K.; Tkacz-Szczesna, B.; Celichowski, G.; Cichomski, M.; Szmaja, W.; Grobelny, J. Detection Limits of DLS and UV-Vis Spectroscopy in Characterization of Polydisperse Nanoparticles Colloids. *J. Nanomater.* **2013**, *2013*, 313081. [CrossRef]
67. Singh, T.; Jyoti, K.; Patnaik, A.; Singh, A.; Chauhan, R.; Chandel, S.S. Biosynthesis, characterization and antibacterial activity of silver nanoparticles using an endophytic fungal supernatant of *Raphanus sativus*. *J. Genet. Eng. Biotechnol.* **2017**, *15*, 31–39. [CrossRef] [PubMed]
68. Salem, S.S.; El-Belely, E.F.; Niedbała, G.; Alnoman, M.M.; Hassan, S.E.-D.; Eid, A.M.; Shaheen, T.I.; Elkelish, A.; Fouda, A. Bactericidal and In-Vitro Cytotoxic Efficacy of Silver Nanoparticles (Ag-NPs) Fabricated by Endophytic Actinomycetes and Their Use as Coating for the Textile Fabrics. *Nanomaterials* **2020**, *10*, 2082. [CrossRef]
69. Ramanujam, K.; Sundrarajan, M. Antibacterial effects of biosynthesized MgO nanoparticles using ethanolic fruit extract of *Emblica officinalis*. *J. Photochem. Photobiol. B Biol.* **2014**, *141*, 296–300. [CrossRef]
70. Coates, J. Interpretation of infrared spectra, a practical approach. *Encycl. Anal. Chem. Appl. Theory Instrum.* **2006**. [CrossRef]
71. Hamza, M.F.; Salih, K.A.M.; Abdel-Rahman, A.A.H.; Zayed, Y.E.; Wei, Y.; Liang, J.; Guibal, E. Sulfonic-functionalized algal/PEI beads for scandium, cerium and holmium sorption from aqueous solutions (synthetic and industrial samples). *Chem. Eng. J.* **2021**, *403*, 126399. [CrossRef]
72. Taourati, R.; Khaddor, M.; Laghzal, A.; El Kasmi, A. Facile one-step synthesis of highly efficient single oxide nanoparticles for photocatalytic application. *Sci. Afr.* **2020**, *8*, e00305. [CrossRef]
73. Karthik, K.; Dhanuskodi, S.; Gobinath, C.; Prabukumar, S.; Sivaramakrishnan, S. Fabrication of MgO nanostructures and its efficient photocatalytic, antibacterial and anticancer performance. *J. Photochem. Photobiol. B Biol.* **2019**, *190*, 8–20. [CrossRef] [PubMed]
74. Wei, Y.; Salih, K.A.M.; Rabie, K.; Elwakeel, K.Z.; Zayed, Y.E.; Hamza, M.F.; Guibal, E. Development of phosphoryl-functionalized algal-PEI beads for the sorption of Nd (III) and Mo (VI) from aqueous solutions–Application for rare earth recovery from acid leachates. *Chem. Eng. J.* **2020**, *412*, 127399. [CrossRef]
75. Pugazhendhi, A.; Prabhu, R.; Muruganantham, K.; Shanmuganathan, R.; Natarajan, S. Anticancer, antimicrobial and photocatalytic activities of green synthesized magnesium oxide nanoparticles (MgONPs) using aqueous extract of *Sargassum wightii*. *J. Photochem. Photobiol. B Biol.* **2019**, *190*, 86–97. [CrossRef] [PubMed]
76. Fouda, A.; Hassan, S.E.-D.; Saied, E.; Hamza, M.F. Photocatalytic degradation of real textile and tannery effluent using biosynthesized magnesium oxide nanoparticles (MgO-NPs), heavy metal adsorption, phytotoxicity, and antimicrobial activity. *J. Environ. Chem. Eng.* **2021**, *9*, 105346. [CrossRef]
77. Hamza, M.F.; Wei, Y.; Mira, H.I.; Abdel-Rahman, A.A.H.; Guibal, E. Synthesis and adsorption characteristics of grafted hydrazinyl amine magnetite-chitosan for Ni(II) and Pb(II) recovery. *Chem. Eng. J.* **2019**, *362*, 310–324. [CrossRef]
78. Sun, Y.; Wang, X.; Ding, C.; Cheng, W.; Chen, C.; Hayat, T.; Alsaedi, A.; Hu, J.; Wang, X. Direct Synthesis of Bacteria-Derived Carbonaceous Nanofibers as a Highly Efficient Material for Radionuclides Elimination. *ACS Sustain. Chem. Eng.* **2016**, *4*, 4608–4616. [CrossRef]
79. Jurado-López, B.; Vieira, R.S.; Rabelo, R.B.; Beppu, M.M.; Casado, J.; Rodríguez-Castellón, E. Formation of complexes between functionalized chitosan membranes and copper: A study by angle resolved XPS. *Mater. Chem. Phys.* **2017**, *185*, 152–161. [CrossRef]
80. Wei, Y.; Rakhatkyzy, M.; Salih, K.A.M.; Wang, K.; Hamza, M.F.; Guibal, E. Controlled bi-functionalization of silica microbeads through grafting of amidoxime/methacrylic acid for Sr(II) enhanced sorption. *Chem. Eng. J.* **2020**, *402*, 125220. [CrossRef]

81. Hamza, M.F.; Wei, Y.; Benettayeb, A.; Wang, X.; Guibal, E. Efficient removal of uranium, cadmium and mercury from aqueous solutions using grafted hydrazide-micro-magnetite chitosan derivative. *J. Mater. Sci.* **2020**, *55*, 4193–4212. [CrossRef]
82. Yang, K.; Zhong, L.; Guan, R.; Xiao, M.; Han, D.; Wang, S.; Meng, Y. Carbon felt interlayer derived from rice paper and its synergistic encapsulation of polysulfides for lithium-sulfur batteries. *Appl. Surf. Sci.* **2018**, *441*, 914–922. [CrossRef]
83. He, C.; Salih, K.A.M.; Wei, Y.; Mira, H.; Abdel-Rahman, A.A.H.; Elwakeel, K.Z.; Hamza, M.F.; Guibal, E. Efficient Recovery of Rare Earth Elements (Pr(III) and Tm(III)) From Mining Residues Using a New Phosphorylated Hydrogel (Algal Biomass/PEI). *Metals* **2021**, *11*, 294. [CrossRef]
84. Bartis, E.A.; Luan, P.; Knoll, A.J.; Hart, C.; Seog, J.; Oehrlein, G.S. Polystyrene as a model system to probe the impact of ambient gas chemistry on polymer surface modifications using remote atmospheric pressure plasma under well-controlled conditions. *Biointerphases* **2015**, *10*, 029512. [CrossRef]
85. Yao, Y.; Gao, B.; Chen, J.; Yang, L. Engineered Biochar Reclaiming Phosphate from Aqueous Solutions: Mechanisms and Potential Application as a Slow-Release Fertilizer. *Environ. Sci. Technol.* **2013**, *47*, 8700–8708. [CrossRef]
86. Le Febvrier, A.; Jensen, J.; Eklund, P. Wet-cleaning of MgO(001): Modification of surface chemistry and effects on thin film growth investigated by x-ray photoelectron spectroscopy and time-of-flight secondary ion mass spectroscopy. *J. Vac. Sci. Technol. A Vac. Surf. Film.* **2017**, *35*, 021407. [CrossRef]
87. Dang, L.; Nai, X.; Dong, Y.; Li, W. Functional group effect on flame retardancy, thermal, and mechanical properties of organophosphorus-based magnesium oxysulfate whiskers as a flame retardant in polypropylene. *RSC Adv.* **2017**, *7*, 21655–21665. [CrossRef]
88. Fouda, A.; Abdel-Maksoud, G.; Saad, H.A.; Gobouri, A.A.; Mohammedsaleh, Z.M.; El-Sadany, M.A. The efficacy of silver nitrate ($AgNO_3$) as a coating agentto protect paper against high deteriorating microbes. *Catalysts* **2021**, *11*, 310. [CrossRef]
89. Fouda, A.; Hassan, S.E.-D.; Saied, E.; Azab, M.S. An eco-friendly approach to textile and tannery wastewater treatment using maghemite nanoparticles (γ-Fe_2O_3-NPs) fabricated by *Penicillium expansum* strain (Kw). *J. Environ. Chem. Eng.* **2021**, *9*, 104693. [CrossRef]
90. Salem, S.S.; Fouda, M.M.G.; Fouda, A.; Awad, M.A.; Al-Olayan, E.M.; Allam, A.A.; Shaheen, T.I. Antibacterial, Cytotoxicity and Larvicidal Activity of Green Synthesized Selenium Nanoparticles Using *Penicillium corylophilum*. *J. Clust. Sci.* **2021**, *32*, 351–361. [CrossRef]
91. Mohamed, A.A.; Fouda, A.; Abdel-Rahman, M.A.; Hassan, S.E.-D.; El-Gamal, M.S.; Salem, S.S.; Shaheen, T.I. Fungal strain impacts the shape, bioactivity and multifunctional properties of green synthesized zinc oxide nanoparticles. *Biocatal. Agric. Biotechnol.* **2019**, *19*, 101103. [CrossRef]
92. Hamza, M.F.; Fouda, A.; Elwakeel, K.Z.; Wei, Y.; Guibal, E.; Hamad, N.A. Phosphorylation of Guar Gum/Magnetite/Chitosan Nanocomposites for Uranium (VI) Sorption and Antibacterial Applications. *Molecules* **2021**, *26*, 1920. [CrossRef]
93. Roy, A.; Bulut, O.; Some, S.; Mandal, A.K.; Yilmaz, M.D. Green synthesis of silver nanoparticles: Biomolecule-nanoparticle organizations targeting antimicrobial activity. *RSC Adv.* **2019**, *9*, 2673–2702. [CrossRef]
94. Nguyen, N.-Y.T.; Grelling, N.; Wetteland, C.L.; Rosario, R.; Liu, H. Antimicrobial Activities and Mechanisms of Magnesium Oxide Nanoparticles (nMgO) against Pathogenic Bacteria, Yeasts, and Biofilms. *Sci. Rep.* **2018**, *8*, 16260. [CrossRef] [PubMed]
95. Holm, A.; Vikström, E. Quorum sensing communication between bacteria and human cells: Signals, targets, and functions. *Front. Plant Sci.* **2014**, *5*, 309. [CrossRef] [PubMed]
96. Sawai, J.; Kojima, H.; Igarashi, H.; Hashimoto, A.; Shoji, S.; Takehara, A.; Sawaki, T.; Kokugan, T.; Shimizu, M. *Escherichia coli* damage by ceramic powder slurries. *J. Chem. Eng. Jpn.* **1997**, *30*, 1034–1039. [CrossRef]
97. Suresh, M.; Jeevanandam, J.; Chan, Y.S.S.; Danquah, M.; Kalaiarasi, J. Opportunities for Metal Oxide Nanoparticles as a Potential Mosquitocide. *BioNanoScience* **2020**, 10. [CrossRef]
98. Leung, Y.H.; Ng, A.M.; Xu, X.; Shen, Z.; Gethings, L.A.; Wong, M.T.; Chan, C.M.; Guo, M.Y.; Ng, Y.H.; Djurišić, A.B.; et al. Mechanisms of antibacterial activity of MgO: Non-ROS mediated toxicity of MgO nanoparticles towards *Escherichia coli*. *Small* **2014**, *10*, 1171–1183. [CrossRef] [PubMed]
99. Magdolenova, Z.; Collins, A.; Kumar, A.; Dhawan, A.; Stone, V.; Dusinska, M. Mechanisms of genotoxicity. A review of in vitro and in vivo studies with engineered nanoparticles. *Nanotoxicology* **2014**, *8*, 233–278. [CrossRef]
100. Gayathri, B.; Muthukumarasamy, N.; Velauthapillai, D.; Santhosh, S.B.; Asokan, V. Magnesium incorporated hydroxyapatite nanoparticles: Preparation, characterization, antibacterial and larvicidal activity. *Arab. J. Chem.* **2018**, *11*, 645–654. [CrossRef]
101. Murugan, K.; Jaganathan, A.; Rajaganesh, R.; Suresh, U.; Madhavan, J.; Senthil-Nathan, S.; Rajasekar, A.; Higuchi, A.; Kumar, S.S.; Alarfaj, A.A.; et al. Poly(Styrene Sulfonate)/Poly(Allylamine Hydrochloride) Encapsulation of TiO_2 Nanoparticles Boosts Their Toxic and Repellent Activity Against Zika Virus Mosquito Vectors. *J. Clust. Sci.* **2018**, *29*, 27–39. [CrossRef]
102. Sepehr, M.; Nasseri, S.; Mazaheri Assadi, M. Chromium bioremoval from tannery industries effluent by *Aspergillus oryzae*. *Iran. J. Environ. Health Sci. Eng.* **2005**, *2*, 273–279.
103. Verma, T.; Ramteke, P.W.; Garg, S.K. Quality assessment of treated tannery wastewater with special emphasis on pathogenic *E. coli* detection through serotyping. *Environ. Monit. Assess.* **2008**, *145*, 243–249. [CrossRef]
104. Li, S. Combustion synthesis of porous MgO and its adsorption properties. *Int. J. Ind. Chem.* **2019**, *10*, 89–96. [CrossRef]
105. Ben, A.; Amutha, E.; Pushpalaksmi, E.; Samraj, J.; Gurusamy, A. Physicochemical Characteristics, Identification of Fungi and Optimization of Different Parameters for Degradation of Dye from Tannery Effluent. *J. Appl. Sci. Environ. Manag.* **2020**, *24*, 1203–1208. [CrossRef]

106. Jahan, M.A.A.; Akhtar, N.; Khan, N.M.S.; Roy, C.K.; Islam, R.; Nurunnabi, M. Characterization of tannery wastewater and its treatment by aquatic macrophytes and algae. *Bangladesh J. Sci. Ind. Res.* **2015**, 49. [CrossRef]
107. Sharma, S.; Malaviya, P. Bioremediation of tannery wastewater by *Aspergillus niger* SPFSL 2-a isolated from tannery sludge. *J. Basic Appl. Sci* **2013**, *2*, 88–93.
108. Selim, M.T.; Salem, S.S.; Mohamed, A.A.; El-Gamal, M.S.; Awad, M.F.; Fouda, A. Biological treatment of real textile effulent using *Aspergillus flavus* and *Fusarium oxysorium* and their consortium along with the evaluation of their phytotoxicity. *J. Fungi.* **2021**, *7*, 193. [CrossRef]
109. Chowdhury, M.; Mostafa, M.G.; Biswas, T.K.; Mandal, A.; Saha, A.K. Characterization of the Effluents from Leather Processing Industries. *Environ. Process.* **2015**, *2*, 173–187. [CrossRef]
110. Chowdhury, M.; Mostafa, M.; Biswas, T.K.; Saha, A.K. Treatment of leather industrial effluents by filtration and coagulation processes. *Water Resour. Ind.* **2013**, *3*, 11–22. [CrossRef]
111. Gao, X.; Zhang, Y.; Dai, Y.; Fu, F. High-performance magnetic carbon materials in dye removal from aqueous solutions. *J. Solid State Chem.* **2016**, *239*, 265–273. [CrossRef]
112. Hamza, M.F.; Hamad, N.A.; Hamad, D.M.; Khalafalla, M.S.; Abdel-Rahman, A.A.-H.; Zeid, I.F.; Wei, Y.; Hessien, M.M.; Fouda, A.; Salem, W.M. Synthesis of Eco-Friendly Biopolymer, Alginate-Chitosan Composite to Adsorb the Heavy Metals, Cd(II) and Pb(II) from Contaminated Effluents. *Materials* **2021**, *14*, 2189. [CrossRef]
113. Alvarez, C.C.; Bravo Gómez, M.E.; Hernández Zavala, A. Hexavalent chromium: Regulation and health effects. *J. Trace Elem. Med. Biol.* **2021**, *65*, 126729. [CrossRef]
114. Mahdavi, S.; Jalali, M.; Afkhami, A. Heavy Metals Removal from Aqueous Solutions Using TiO_2, MgO, and Al_2O_3 Nanoparticles. *Chem. Eng. Commun.* **2013**, 200. [CrossRef]
115. Yang, J.; Hou, B.; Wang, J.; Tian, B.; Bi, J.; Wang, N.; Li, X.; Huang, X. Nanomaterials for the Removal of Heavy Metals from Wastewater. *Nanomaterials* **2019**, *9*, 424. [CrossRef] [PubMed]
116. Xiong, C.; Wang, W.; Tan, F.; Luo, F.; Chen, J.; Qiao, X. Investigation on the efficiency and mechanism of Cd(II) and Pb(II) removal from aqueous solutions using MgO nanoparticles. *J. Hazard. Mater.* **2015**, *299*, 664–674. [CrossRef] [PubMed]
117. Seif, S.; Marofi, S.; Mahdavi, S. Removal of Cr^{3+} ion from aqueous solutions using MgO and montmorillonite nanoparticles. *Environ. Earth Sci.* **2019**, *78*, 377. [CrossRef]

Journal of
Fungi

Review

Lichens—A Potential Source for Nanoparticles Fabrication: A Review on Nanoparticles Biosynthesis and Their Prospective Applications

Reham Samir Hamida [1], Mohamed Abdelaal Ali [2,3], Nabila Elsayed Abdelmeguid [1], Mayasar Ibrahim Al-Zaban [4,*], Lina Baz [5,*] and Mashael Mohammed Bin-Meferij [4]

1 Molecular Biology Unit, Department of Zoology, Faculty of Science, Alexandria University, Alexandria 21500, Egypt; reham.hussein@alexu.edu.eg (R.S.H.); dr_nabila_elsayed2000@yahoo.com (N.E.A.)
2 Biotechnology Unit, Department of Plant Production, College of Food and Agriculture Science, King Saud University, Riyadh 11543, Saudi Arabia; mali3@ksu.edu.sa
3 Plant Production Department, Arid Lands Cultivation Research Institute, City of Scientific Research and Technological Applications (SRTA-City), New Borg El-Arab, Alexandria 21934, Egypt
4 Department of Biology, College of Science, Princess Nourah bint Abdulrahman University, Riyadh 11543, Saudi Arabia; mmbinmufayrij@pnu.edu.sa
5 Department of Biochemistry, Faculty of Science, King Abdulaziz University, Jeddah 21589, Saudi Arabia
* Correspondence: Mialzaban@pnu.edu.sa (M.I.A.-Z.); Lbaz@kau.edu.sa (L.B.)

Citation: Hamida, R.S.; Ali, M.A.; Abdelmeguid, N.E.; Al-Zaban, M.I.; Baz, L.; Bin-Meferij, M.M. Lichens—A Potential Source for Nanoparticles Fabrication: A Review on Nanoparticles Biosynthesis and Their Prospective Applications. *J. Fungi* **2021**, *7*, 291. https://doi.org/10.3390/jof7040291

Academic Editor: Kamel A. Abd-Elsalam

Received: 16 March 2021
Accepted: 8 April 2021
Published: 12 April 2021

Publisher's Note: MDPI stays neutral with regard to jurisdictional claims in published maps and institutional affiliations.

Abstract: Green synthesis of nanoparticles (NPs) is a safe, eco-friendly, and relatively inexpensive alternative to conventional routes of NPs production. These methods require natural resources such as cyanobacteria, algae, plants, fungi, lichens, and naturally extracted biomolecules such as pigments, vitamins, polysaccharides, proteins, and enzymes to reduce bulk materials (the target metal salts) into a nanoscale product. Synthesis of nanomaterials (NMs) using lichen extracts is a promising eco-friendly, simple, low-cost biological synthesis process. Lichens are groups of organisms including multiple types of fungi and algae that live in symbiosis. Until now, the fabrication of NPs using lichens has remained largely unexplored, although the role of lichens as natural factories for synthesizing NPs has been reported. Lichens have a potential reducible activity to fabricate different types of NMs, including metal and metal oxide NPs and bimetallic alloys and nanocomposites. These NPs exhibit promising catalytic and antidiabetic, antioxidant, and antimicrobial activities. To the best of our knowledge, this review provides, for the first time, an overview of the main published studies concerning the use of lichen for nanofabrication and the applications of these NMs in different sectors. Moreover, the possible mechanisms of biosynthesis are discussed, together with the various optimization factors influencing the biological synthesis and toxicity of NPs.

Keywords: lichen; nanoparticles; green synthesis; eco-friendly; antimicrobial; antioxidant

1. Introduction

Nanotechnology has recently created a revolution in the scientific world, particularly in the industrial, medical, agricultural, and electronic sectors [1,2]. This revolution is due to the ability of this technology to generate new products in the nanoscale (at least one of their dimensions in the range 1–100 nm) with unique and desirable physicochemical and biological characteristics that are missing in their precursor forms [3,4]. These nanomaterials (NMs) have a high surface-area-to-volume ratio and can therefore be used in drug delivery [5], catalysis [6], therapeutics [7,8], theranostics [9], and detection and diagnostic fields [10]. For instance, the specific surface properties, porosity, and ability for functionalization render silica nanoparticles (NPs) appealing options for drug delivery [11]. Fullerenes can be loaded with different therapeutic agents such as antibiotics and anticancer drugs [12,13]. Silver NPs show unique reactivity, selectivity, and stability, as well as recyclability in catalytic reactions [14]. Magnetic NPs have a high magnetic moment and consequently are attractive tools for magnetic resonance imaging for cancer diagnostic [15].

Furthermore, the smaller size of NPs has facilitated the development of new therapeutic agents against serious global illnesses such as cancer and infectious and parasitic diseases [11]. Platinum [16], selenium [17], and palladium NPs acted as potent anticancer agents [18], while zinc oxide [19], copper oxide [20], and titanium dioxide NPs exhibited significant inhibitory activity against different microbes including bacteria, fungi, and viruses [21,22]. NPs also play important roles in communication and electronic fields due to high electro-optical activity, enabling them to be used in electronic and optical industries [23,24]. The thermal conductivity of NPs provides scope for researchers to develop numerous energy cells such as solar cells and batteries [25,26], while the unique photothermal properties of NPs mean they are a promising therapeutic for various types of cancers [27]. The localized surface plasmon resonance (SPR) of gold NPs enable these particles to absorb specific wavelengths, leading to photoacoustic and photothermal characteristics; consequently, these NPs are promising tools for hyperthermic cancer therapies and bioimaging [27].

These unique features of NPs facilitate the creation and development of new tools, processes, and products with applications in many sectors, including medicine, industry, and communication; thus, enthusiasm for producing novel nanoproducts has rapidly increased. However, this swift growth has resulted in harmful effects on living organisms and their environments [4,28]. This damage arises from using and yielding hazardous materials during the production of NPs via chemical synthesis approaches. Moreover, the approach of using physical methods in the fabrication of NPs consumes more energy and money than other synthesis methods. To minimize these drawbacks, eco-friendly alternatives to traditional synthesis methods (chemical and physical routes) have been sought. One of these alternatives is the green fabrication of NPs.

Biofabrication, biological synthesis, green synthesis, and biosynthesis are synonymous terms corresponding to the use of eco-friendly, rapid, simple, and low-cost technology for NP production. This technology has numerous advantages, including high scalability, variation in size/shape and chemical compositions, and high mono-dispersity of NPs [29]. Moreover, this approach uses living organisms or their products to reduce bulk materials into NPs and stabilize the NPs without needing chemical materials or producing any hazardous materials [4]. Multicellular and unicellular organisms (plants, algae, worms, lichens, fungi, bacteria, cyanobacteria, yeast, actinomycetes, etc.) and their biomolecules, such as proteins, pigments, enzymes, vitamins, polysaccharides, and lignin were used as reductants and surfactants for fabricating precursors into their nanoforms [4,30–32]. These biogenic NPs can be consumed in numerous industrial and medical processes due to their unique physicochemical and biological features such as efficiency, biocompatibility, bioactivity, and stability [33].

Although there are limited reports about the lichen-based green synthesis of NPs, this method is considered a promising technology for NP production. Lichens are composite organisms, which live in both obligate and beneficial symbiosis with fungi, algae, perennial trees, or cyanobacteria [34]. Lichen cells contain many types of secondary metabolites and other bioactive molecules, rendering them valuable for industrial, pharmaceutical, biotechnological, medical, and cosmetics applications [35]. Some researchers have demonstrated the potentiality of different species of lichens to fabricate unique NPs with different shapes, sizes, and physicochemical and biological activities [36]. Rattan et al. demonstrated the role of different lichen species to synthe-size different types of NMs and their potentiality to act as promising antimicrobial agents [36]. Alqahtani et al. reported that methanolic extracts of two lichen species, *Xanthoria parietina* and *Flavopunctelia flaventior*, were recently shown to have the po-tential to reduce silver nitrate into Ag-NPs extracellularly [37]. The resultant Ag-NPs were spherical, had a nanosize range of 1–40 nm, and reduced the proliferation of human colorectal cancer (HCT 116), breast cancer (MDA-MB-231), and pharynx can-cer (FaDu) cell lines, and the growth of methicillin-resistant Staphylococcus aureus (MRSA), vancomycin-resistant Enterococcus (VRE), Pseudomonas aeruginosa, and Esche-richia coli [37]. This review provides, for the first time, an overview of the main published studies concerning the use of lichen for nanofabrication and the applications of these nanomaterials in different sectors. Moreover, the possible mechanisms of biosynthesis are discussed together with the various optimization factors influencing the biological synthesis and toxicity of NPs.

2. Classification of Nanoparticles

Classification of NPs varies according to their origins, structures, shapes, dimensions, chemical and phase compositions, physical and chemical properties, and crystallinity [29,38,39]. For example, NPs can be obtained naturally from dust, volcanic eruptions, and living organisms such as bacteria, fungi, algae, plants, etc., and artificially by utilizing chemical, physical, and biological synthesis routes such as colloidal, chemical precipitation, laser ablation, sputtering, and micro- and macro-organism-mediated synthesis approaches [30,31,40]. Furthermore, NPs can be classified into organic, inorganic, and semi-organic types according to their chemical nature [29,41]. Similarly, NPs can be categorized according to their shape, including rod, spherical, cubic, triangular, octahedral, pentahedral, flower, star, etc. [4,39], and on the basis of the number of their dimension in nanoscale, including zero dimension (0D) such as quantum dots, one dimension (1D) such as nanowires and nanorods, two dimensions (2D) such as nanolayers and nanoplates, and three dimensions (3D) such as nanocoils and nanoflowers [42]. In terms of magnetic properties, NPs can belong to either the paramagnetic category, which includes iron oxide and zinc sulfide NPs, or the diamagnetic category comprising titanium oxide and magnesium ferrite NPs [43] (Figure 1).

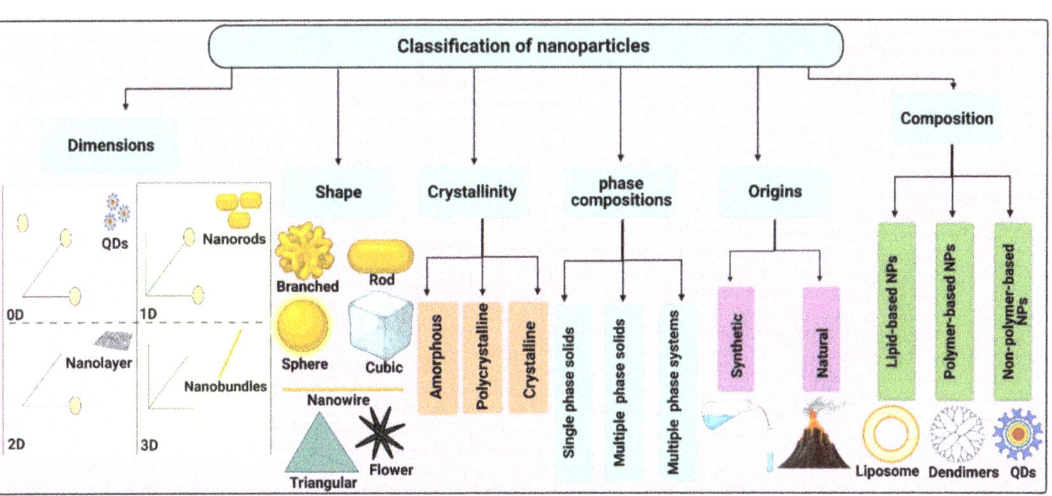

Figure 1. Classifications of nanoparticles (NPs).

3. Synthesis Routes of Nanoparticles

Nanofabrication routes can generally be classified into two main groups: top-down methods, such as physical synthesis approaches, and bottom-up synthesis methods, such as chemical and biological synthesis processes (Figure 2) [44].

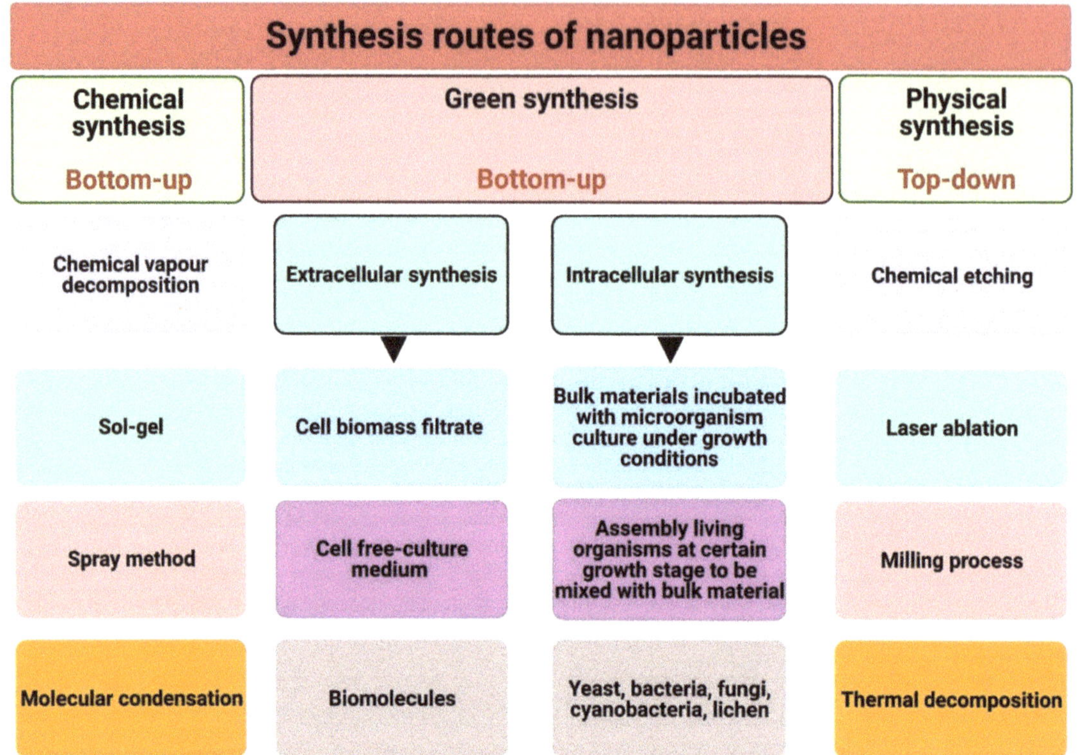

Figure 2. Synthesis routes of nanoparticles.

3.1. Physical Synthesis

This route is a top-down synthesis process in which precursors are reduced into NPs using physical approaches such as ultra-sonication, laser ablation, mechanical milling, sputtering, microwave irradiation, and electrochemical methods [31]. Quantum dot NPs have been physically synthesized using molecular beam epitaxy, ion implantation, e-beam lithography, and X-ray lithography [45,46]. Recently, carbon nanostructures were prepared from elemental graphite powders using a mechanical milling method in air [47]. Niasari et al. synthesized silica NPs using rice husk ash at ambient temperature by utilizing a high-energy planetary ball mill [48]. The scholar reported that silica NPs were synthesized after 6 h of ball milling. Fe-SEM and transmission electron microscopy (TEM) micrographs exhibited that silica NPs have a spherical shape and nanosize of 70 nm. They reported that silica NPs acted as a promising drug delivery system for controlling penicillin-G drug releasing. Recently, simple green- microwave-assisted synthesis route of fluorescent carbon quantum dots (CQDs) was conducted using roasted chickpea as the carbon source in one step without using any chemicals [49]. The study provided an eco-friendly method to fabricate CQDs with advantageous properties such as high fluorescence intensity, excellent photostability, and good water solubility. The physicochemical features of CQDs were determined using UV–Vis spectroscopy, fluorescence spectroscopy, Fourier transform infrared spectroscopy (FTIR), X-ray diffraction (XRD), transmission electron microscopy (TEM), and selected-area electron diffraction based on TEM micrographs. The data revealed that CQDs emitted blue fluorescent at a UV wavelength of 365 nm and have a spherical shape with an amorphous structure and nanodiameter less than 10 nm.

3.2. Chemical Synthesis

Chemical synthesis is a bottom-up process in which atoms are assembled into nuclei and grown to NPs [50]. The predominant components of this route are the reducing (such as sodium citrate and ascorbate) and capping agents (such as sodium carboxyl methylcellulose) [51,52]. Chemical vapor deposition, spinning, pyrolysis, and sol-gel process are examples of chemical synthesis approaches [41]. Titanium (Ti) dioxide NPs were fabricated from the precursor Ti-isopropoxide and calcined at 300, 350, 400, and 450 °C using a simple sol-gel method [53]. The chemical reduction was used to reduce silver nitrate into Ag-NPs using sodium citrate (TSC) and sodium borohydride ($NaBH_4$) as reducing agents [54]. The authors used the face-centered central composite model with four abiotic parameters including $AgNO_3$, TSC, and $NaBH_4$ concentrations and the pH of the reaction. They revealed that optimal conditions to synthesize spherical Ag-NPs with a nanosize of less than 10.3 nm were pH 8 and 0.01 M, 0.06 M, 0.01 M for the concentration of TSC, $AgNO_3$, and $NaBH_4$, respectively. Yu et al. synthesized hollow silica spheres (HSSs) through a self-templating route in acidic aqueous media under hydrothermal conditions [55]. The resultant HSSs have a spherical shape with a nanodiameter of 190 nm. They found that the hollowed-out interior space of HSSs was dependent on reaction time and silica concentration, while their porous structure in the shell can be mitigated by tuning the acidity of the silica dispersion. Moraes et al. synthesized tadpole-like gold nanowires (AuNWs) by mixing 0.1 mmol of $HAuCl_4 \cdot 3H_2O$ with 12 mL of the oleylamine as a reducing agent at 65 °C under stirring for 72 h [56]. The scholar exhibited that AuNWs were polydispersed and branched with length ranging from a few nanometers to larger than 500 nm with a diameter of 23 nm.

3.3. Biological (Green) Synthesis

Biological synthesis is a modern alternative to both physical and chemical synthesis processes and is considered a type of bottom-up route [57]. This approach utilizes natural sources such as microorganisms, macroorganisms, and biomolecules (proteins, lipids, polysaccharides, pigments, etc.) to fabricate NPs from their bulk materials without the need for toxic chemicals during the fabrication process [4,58]. Various significant properties of biosynthesis routes such as the absence of poisonous chemical compounds used as reducing or stabilizing agents, no toxic yields generated from the process, the low energy consumption, inexpensive cost, and high scalability have resulted in green synthesis methods becoming more attractive than other traditional methods [4].

Biological synthesis routes are categorized into two main approaches—extracellular and intracellular synthesis routes.

3.3.1. Extracellular Synthesis

In extracellular synthesis routes, the fabrication process occurs outside living cells [57,59]. This process can be achieved via three different patterns:

(i) Cell-biomass-filtrate synthesis of NPs: In this pattern, cells of living organisms are dried with a lyophilizer, oven, or air-based methods and then crushed into fine powders that are mixed with distilled water for boiling. The mixture is cooled, passed through a filtration system such as Whatman filter paper, and then the resulting filtrate is mixed with a defined concentration of bulk material to fabricate it into NPs [28,58,60]. An alternative procedure involves washing the natural sources such as *Streptomyces* sp., algae, etc., then soaking the cells in water for a number of days, followed by centrifugation and use of the resulting supernatant as a reducing and stabilizing agent to synthesize NPs [61,62]. Other methods achieved the extracellular synthesis of NPs by sonicating or boiling the natural sources under certain conditions, then filtering the mixture and using the filtrate in the synthesis process [33,63];

(ii) Cell-free, culture-medium-based synthesis of NPs: This method is suitable for cultured microorganisms. First, the culture is centrifuged and the supernatant used for bioreduction of bulk compounds into their NPs under suitable conditions. Keskin et al.

demonstrated that cell-free culture media of *Synechococcus* sp. had a reducible activity that resulted, under light conditions, in the formation of Ag-NPs with an average nanosize of 140 nm [64]. However, this process is sometimes unsuitable for NP synthesis because many types of media used for the culture of microorganisms contain components that act as reductants and stabilizing agents. These compounds interfere with the reducible activity of active biomolecules of the cultured microorganisms [65];

(iii) Biomolecule-mediated synthesis of NPs: This approach uses biomolecules such as pigments, carbohydrates, proteins, enzymes, etc. as reducing and capping materials to produce NPs. Briefly, target biomolecules are extracted from their micro- or macro-organisms, purified, and mixed with a defined concentration of bulk material solutions to start the NP fabrication process under specific conditions of temperature, illumination, and pH [66,67] (Figure 3).

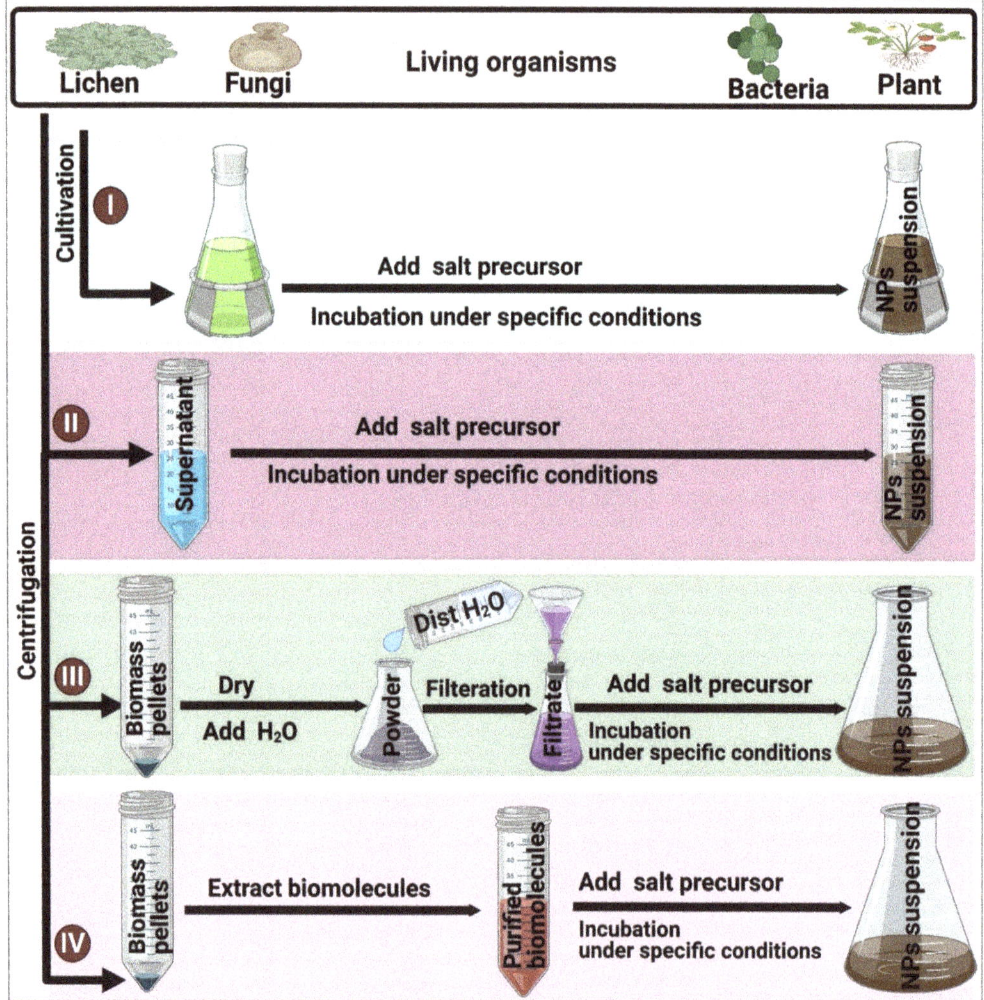

Figure 3. Green synthesis methods include intracellular synthesis route (**I**) and extracellular synthesis routes including cell-free, culture-medium-based synthesis of NPs (**II**), cell–biomass-filtrate synthesis of NPs (**III**), and biomolecule-mediated synthesis of NPs (**IV**).

3.3.2. Intracellular Synthesis

The intracellular synthesis method refers to the production of NPs inside living cells, with biological processes such as metabolic activity, respiration, and growth stage, playing crucial roles in the biosynthesis process [68]. Intracellular synthesis can be performed according to two protocols, each composed of three steps: (i) culturing the target living organism, (ii) the reaction between precursor materials and living cells, and (iii) separation and purification of NPs and subsequent characterization using different physicochemical methods [4,69]. The first protocol includes the incubation of bulk materials solution with microbe cultures during their growing period under standard culture conditions until the microbes reach a certain growth [70]. In the second method, living cells in the logarithmic phase are collected by centrifugation, washed multiple times to discard any undesired materials, and then the cleaned microbial biomass is dissolved in water and mixed with a suitable amount of bulk material solution [71].

Intracellular synthesis is more complicated than extracellular fabrication due to the additional steps required to extract and purify NPs from inside the cells [72]. Both types of biological synthesis methods are eco-friendly routes, do not usually need toxic chemical materials, and are easily performed under normal laboratory conditions [4,65] (Figure 3).

4. Green Synthesis-Based Systems

4.1. *Biomolecule-Mediated Fabrication of NPs*

Recently, natural products have become a target area for many researchers due to their promising applications in numerous sectors including biotechnology (e.g., biofuel and biofertilizer production), bioremediation, and the cosmetics industry (e.g., synthesis of natural sunblock creams). Moreover, natural biomolecules exert broad biomedicinal and therapeutic potentials for serious diseases including cancers and infectious, parasitic, and immune diseases, etc. [7,73]. For instance, scytonemin, a natural pigment extracted from cyanobacteria *Scytonema* sp., is an extremely potent modulator of mitotic spindle formation [74]. In addition, calothrixins, quinone-based natural products extracted from cyanobacteria *Calothrix* sp., exhibit potent antiproliferative activity against cancer cell lines [75]. Normavacurine-21-one, isolated from *Alstonia scholaris* leaves, displays antibacterial activities against *Enterococcus faecalis* ATCC 10541. Conversely, biomolecules exhibit significant reducible properties and thus have the ability to fabricate numerous metal precursors into their nanoforms [76].

Biomolecules such as proteins, amino acids, or secondary metabolites from microorganisms or plant extracts, can act as reduction, stabilization, functionalization, and capping agents for NPs [77]. For the green synthesis of metal NPs, aqueous extracts of dried plants or algae are commonly used [33,78]. The water extract classically contains phenolics, terpenoids, polysaccharides, flavonoids, alkaloids, lipids, proteins, and carbohydrates, which collectively represent the reducing power needed for the process. Generally, plant extracts contain enzymes and amino acids that can act as reductants for silver ions and are therefore utilized as scaffolding to facilitate the formation of silver NPs [79]. This unique synthesis strategy provides several kinds of functional groups for NP functionalization [80]. Numerous studies have attributed the synthesis mechanisms for NPs to the potentiality of biomolecules to reduce and stabilize NPs, thereby providing more provision to improve and control the shape, size, and crystallinity of nanomaterials [81].

4.1.1. Pigments

A vital constituent of most photosynthetic organisms is the pigments, including chlorophylls, carotenes, and anthocyanins [82]. Natural pigments produced by plants, algae, and microorganisms are distinctive biomolecules that have been used in the biological synthesis of NPs. Although studies on the use of biopigments for bioreduction of NPs are limited, these biopigments are known to act as potent reducing and stabilizing agents during biofabrication of NPs [80]. Photosynthetic accessory pigments, such as carotenoid, cochineal, flexirubin, fucoxanthin, melanin, phycocyanin, and C-phycoerythrin and Rphycoerythrin, are the predominant pigments in many organisms, including cyanobacteria, microalgae, actinomycete, algae, etc., and have been extensively exploited in the synthesis of NPs [4,66,83].

Actinorhodin isolated from *Streptomyces coelicolor* successfully reduced silver nitrate ($AgNO_3$) into stable Ag-NPs [84]. El-Naggar et al. synthesized Ag-NPs using phycocyanin extracted from *Nostoc linckia* and studied the anticancer, antibacterial, and antihemolytic activities of these NPs [66]. The blue pigment was observed to be an efficient reductant and surfactant material for the production of Ag-NPs. Moreover, pigment-coated Ag-NPs exhibited significant antitumor properties against MCF-7 cell lines, with an IC_{50} of 27.79 ± 2.3 μg/mL, and act as a tumor progression suppressor against Ehrlich ascites carcinoma-bearing mice. Green pigment extracted from Alfalfa plant leaves extracellularly reduced $AgNO_3$ into Ag-NPs [82]. The particle size of the resultant quasi-spherical biogenic Ag-NPs was 25 nm, and the reducible activity of the green pigments was attributed to chlorophylls and carotenes.

The pigment produced by *Talaromyces purpurogenus* was also used as a reducing agent to manufacture Ag-NPs [85]. A reaction mixture (5 mL) was prepared by mixing 0.5 g/L of extracted pigment with 2 mM $AgNO_3$ and adjusting the pH to 12 using 5 N sodium hydroxide solution. The mixture was vortexed then incubated at 28 °C with 2000 lux of light for 48 h. The formation of Ag-NPs was monitored by color change from light orange to brown and by UV–visible (UV–Vis) spectroscopy detection. The UV–Vis spectrum displayed a peak at 410 nm, the known SPR of Ag-NPs. The size of the resulting NPs was in the range of 4–41 nm. To investigate the functional groups present in the pigment, Fourier transform infrared spectroscopy (FTIR) analysis was conducted at a fixed pH of 12, the conditions in which Ag-NPs were generated. At alkaline pH, phenolic groups were reported to donate electrons that reduce the silver ions to Ag-NPs.

4.1.2. Carbohydrates

Polysaccharide-based green synthesis of NPs has been a more attractive method in nanobiotechnology due to the stability, hydrophilicity, nontoxicity, bioactivity, and biodegradable properties of these NPs [86]. Ebrahiminezhad et al. synthesized Ag-NPs using the carbohydrate secreted by *Chlorella vulgaris* [87]. The resulting green Ag-NPs were uniformly dispersed and spherical shaped, with an average size of 7 nm and positive zeta potential of +26 mV. The authors suggested the carbohydrate coat surrounding the Ag-NPs was 2 nm based on a comparison between the size of the NPs in transmission electron microscopy (TEM) micrographs (7 nm) and their hydrodynamic diameter (9 nm).

Palladium NPs (Pd-NPs) have been fabricated from palladium chloride using carboxymethyl cellulose as a reducing and capping agent at 80 °C for 30 min [88]. The Pd-NPs were spherical with a crystallinity structure and an average size of 2.5 nm. Pd-NPs have a negative zeta potential value of −52.6 mV, which is indicative of their high stability. Furthermore, the biogenic Pd-NPs showed high catalytic activity against azo-dyes.

4.1.3. Enzymes

Enzymes are complex globular proteins present in living cells where they act as catalysts to facilitate chemical changes in substances. With the development of biochemistry came a fuller understanding of the wide range of enzymes present in living cells and their modes of action [89]. Although enzymes are only formed in living cells, many can be extracted or separated from the cells and can continue to function in vitro. This unique ability of enzymes to perform their specific chemical transformations in isolation has led to the use of enzymes in industrial and food processes, bioremediation, and medicine [90]. Furthermore, enzymes are nontoxic and biodegradable, making them environmentally friendly and attractive for medical applications [91]. All these characteristics of enzymes, plus their unique and precise structure, have rendered them desirable for green synthesis of NPs [31]. A prime example is the synthesis of Au-NPs by the action of extracellular amylase from *Bacillus licheniformis* on $AuCl_4$ at pH 8 [92]. Another example is the sulfite reductase enzyme extracted from *E. coli* by ion-exchange chromatography and used for the production of Au-NPs that exhibit antifungal activity [93]. NADH and NADH-dependent enzymes were investigated for their role in the biosynthesis of metal NPs. These extracellular enzymes are highly effective reducing agents due to their ability to shuttle electrons in the reduction process of metals to produce NPs [94,95].

4.1.4. Proteins

NP biosynthesis in the presence of proteins from several biological sources can produce NPs with uniform size and shape and minimal particle aggregation. In these processes, the functional groups of proteins act as the reducing and capping agents to metal ions [96,97]. Proteins were utilized in the bioreduction, capping, and assembly of selenium oxyanion, contributing to controlling the size and morphology of selenium NPs (Se-NPs) [98]. Proteins are crucial in the reduction of selenites and selenates and the stabilization of Se-NPs, which exhibit a unique nanostructure contrary to those obtained chemically [98]. Sanghi et al. found that the production of Au-NPs was facilitated by proteins of the fungus *Coriolus versicolor* [99]. Characterization of these Au-NPs by UV–Vis spectroscopy, scanning electron microscopy (SEM), and atomic force microscopy (AFM), revealed that the NPs had high stability (they can be stored up to six months without any aggregation) and a size of 5–30 nm. FTIR data demonstrated the crucial role of different fungal proteins in the fabrication of Au-NPs. A study in 2018 reported the synthesis of Au-NPs with high stability by using the supernatant of fermented fungi containing the extracellular proteins [65]. This process resulted in the formation of Au-NPs with sizes ranging from 6 to 40 nm.

4.1.5. Lipids

Mannosylerythritol lipids were used as a reducing and stabilizing agent in the green synthesis of Ag-NPs [100]. The process commenced with the addition of 0.01 g mannosylerythritol lipids to 1 mL acetone diluted with 10 mL dechlorinated water; pH of the whole solution was adjusted to 7 utilizing 0.1 M sodium hydroxide. The solution was added dropwise to 100 mL of 2 mM silver nitrate solution and kept at room temperature with continuous stirring. The mixture changed from pale-yellow to brownish-red, and the UV–Vis absorption spectrum of the synthesized Ag-NPs was recorded at 430 nm. This confirmed that mannosylerythritol lipids were effective as reducer and stabilizer agents in the formulation of Ag-NPs. An energy dispersive spectroscopy (EDS) instrument equipped with the SEM was used to determine the chemical composition, size, and morphology of Ag-NPs. The structure of the Ag-NPs was perceived by TEM after dispersing powdered NPs in methanol and sonicating the solution. The TEM structure provided more information about the crystallinity and average size of the Ag-NPs.

X-ray diffraction (XRD) of the produced Ag-NPs showed four characteristic peaks of 28.4°, 33.2°, 47.4°, and 56.3° at 2θ, which correspond to the lattice planes (111), (200), (220), and (311), respectively, confirming the crystalline and face-centered cubic (fcc) structure of the NPs. Meanwhile, the FTIR spectrum of the mannosylerythritol lipids Ag-NPs demonstrated significant peaks at 3337, 2923, 1742, 1562, 1344, 1093, 718, and 534 cm^{-1}, which indicate the presence of various functional groups in the mannosylerythritol glycolipid capping the Ag-NPs. The peak at 3337 cm^{-1} may be due to –OH from polysaccharides, while the peak at 2923 cm^{-1} might indicate (C–H) stretching of alkanes. The strong band at 1562 cm^{-1} could be due to the carbonyl stretching vibration. The peaks at 1466 and 1344 cm^{-1} can be assigned to (C–N) and (C–C) stretching vibration of aromatic and aliphatic amines, while the band at 1093 cm^{-1} could be assigned to (C–O) of alkoxy groups, and peaks at 718 and 534 cm^{-1} to CH_2 groups.

A different study used *Lactobacillus casei* to synthesize of Au-NPs and the *L. casei* components were compared before and after the addition of auric acid (0.5 mM $K[AuCl_4]$) [101]. The levels of unsaturated lipids decreased significantly after the addition of auric acid. Moreover, the formation of Au-NPs caused a reduction in the levels of diglycosyldiacylglycerol (DGDG) and triglycosyldiacylglycerol (TGDG). DGDG extracted from *L. casei* induced the formation of Au-NPs, suggesting that these glycolipids can act as potent reducing agents for the conversion of Au(III) to Au(0) and that results in the formation of small NPs.

4.1.6. Vitamins

The utilization of vitamin B2 as a reducing and capping agent in the green synthesis of Ag and Pd nanowires and nanorods is a distinctive technique in the field of green nanotechnology [102]. Ascorbic acid (vitamin C) is used as a reducing factor in combination with chitosan as a stabilizing agent to fabricate sodium alginate-silver NPs [103]. Malassis et al. demonstrated a prompt and effective method to fabricate Au-NPs and Ag-NPs by exploiting ascorbic acid as a reducing and stabilizing agent [104]. The size of the NPs produced was 8–80 nm for Au-NPs and 20–175 nm for Ag-NPs. The method yielded versatile NP surface modification with a large variety of water-soluble surfactants that can be neutral, positively, or negatively charged. Ahmed et al. reported that ascorbic acid in *Desmodium triflorum* was the predominant biomolecule in the reduction process for Ag-NPs [105].

Production of Se-NPs coated with ascorbic acid was achieved through the bioreduction of selenite (Na_2SeO_3) [106]. Selenite was mixed with ascorbic acid and the mixture turned orange red after 30 min, confirming the fabrication of Se-NPs. The produced Se-NPs were analyzed by TEM and dynamic light scattering (DLS) and were observed to have an average size of 23 ± 5.0 nm. These NPs were shown to be an excellent candidate for radiopharmaceutical imaging techniques used in the diagnosis of liver and kidney cancers.

Another important vitamin exploited for the synthesis of NPs is vitamin B12. To synthesize Ag-NPs, Au-NPs, and Pd-NPs, vitamin B12 solution was mixed with silver nitrate, gold (III) chloride, and palladium acetate solutions, respectively [107]. All mixtures were tested in the presence and absence of microwave (MV) irradiation. The results exhibited that in the absence of MV irradiation, vitamin B12 did not reduce bulk material to their nanoform. However, MV irradiation enhanced the reduction ability of vitamin B12 to fabricate metals into NPs. XRD analysis of the resultant metallic NPs confirmed the efficiency of this vitamin as a reducing agent. The morphological features of the synthesized Ag-NPs, Au-NPs, and Pd-NPs were examined by using SEM and TEM techniques, and large aggregates with irregular shapes and diameters in the range 70–600 nm were observed. Ag samples treated with MW irradiation for 6 min produced NPs with diameters less than 30 nm. While Au samples treated with MV irradiation for 3 min showed irregular shapes and small-size particles with an average diameter of 40 ± 11.7; larger Au NPs with a diameter > 500 nm were observed after a longer period of irradiation (i.e., 6 min). Pd samples irradiated with MV for 3 min resulted in NPs with an average size of 40.2 ± 7.3 nm, whereas that irradiated with MV for 6 min produced two different diameters of 43.9 ± 7.1 and 6.6 ± 2.1 nm. The NPs were spherical, triangular, and decahedron shaped. It was concluded that MV irradiation duration is the key to mitigate noble NPs size.

4.1.7. Secondary Metabolites

Secondary metabolites of different microorganisms, plants, and animal collagen waste were noted to have several properties that enhance the synthesis of NPs and could potentially be deployed in major pharmaceutical studies. Some of the notable secondary metabolites that serve as NP stabilizers include alkaloids, cardiac glycosides, flavonoids, phenols, tannins, and terpenoids [108–110]. Of these compounds, flavonoids are the most utilized secondary metabolites for green synthesis due to their practical structure and the favorable qualities they provide for human health. Pertaining to the flavonoid family are anthocyanins, which have been thoroughly investigated for their antioxidant activity [110]. One study tested the effects of anthocyanins as secondary metabolites on the green synthesis of Ag-NPs by using an aqueous extract of saffron wastage and reported a marked reduction of silver ions and antibacterial activity against several bacterial strains [109].

A study on Ag-NP synthesized using an aqueous extract of *Pteris tripartita* proved the anti-inflammatory activity of flavonoids-coated Ag-NPs by conducting an in vivo investigation on mice with edema, and reported a success rate of nearly 60% [111]. These findings provide an optimistic outlook for the future of NPs in biotechnology and drug discovery applications since they present an efficient way of producing metal NPs without chemical stabilizers or reducers through the use of abundant and natural compounds such as flavonoids, phenols, tannins, terpenoids, reducing sugars, and proteins.

4.2. Living Organisms-Mediated Fabrication of NPs

Many micro- and macro-organisms are used as biofactories to produce NPs with unique physicochemical and biological activities.

4.2.1. Plants

Plant-mediated fabrication of NPs, or phytonanotechnology, is a recognized branch of green synthesis of NPs due to being an eco-friendly, low-cost, rapid, and simple method. Other beneficial features of phytonanotechnology processes are their scalability, bioactivity, biocompatibility, and broad medical applicability [112]. Plant extracts act as reducing and capping agents for the synthesis of many types of NPs [33]. Different parts of plants, including leaves, fruits, stems, seeds, and roots, showed their reducing ability during the synthesis of metallic NPs [113,114]. Singh et al. successfully synthesized Au-NPs and Ag-NPs using *Panax ginseng* leaf and root extracts within 3 and 45 min at 80 °C [115]. Saratale et al. fabricated silver nitrate into Ag-NPs using *Acacia nilotica* leave extract as reducing and stabilizing agents to investigate their antineoplastic, free radical scavenging activity and sensing potency for H_2O_2 [116]. The scholars reported that Ag-NPs formed within 20 min of mixing 10 mL of plant leave extract to 100 mL of 1 mM $AgNO_3$ solution. The resultant NPs have a spherical shape and nanosize range of 5 to 30 nm.

Krishnan et al. biosynthesized Ag-NPs from *Piper nigrum* extract and investigated their antitumor activity [117]. TEM images revealed that the Ag-NPs were spherical with a size of 20 nm. The cytotoxicity of Ag-NPs and *Piper nigrum* extract at various concentrations in the range of 10–100 μg/mL was investigated against breast and liver cancer cell lines (MCF-7 and HepG2 cells, respectively) and confirmed their potent cytotoxic effect. In a different study, biosurfactant extracted from corn steep liquor was used to biosynthesize Ag- and Au-NPs. The bioreduction process was completed in one step under a controlled temperature at 60 °C and resulted in a mixture of nanospheres and nanoplates. Biosurfactants were essential for the bioreduction process and also for stabilization of the produced NPs, which improved the antimicrobial activity of the NPs [118].

Green synthesis of Au-NPs by *Salicornia brachiata* (Sb) plant extract and characterization of the formed NPs revealed that mixing plant extract (50 mL) with 10 mM $NaBH_4$ was sufficient to yield the purple color that indicated the formation of Sb-Au-NPs [119]. TEM micrographs showed that the size of Sb-Au-NPs was approximately 30 nm, while XRD and EDS data proved that Sb-Au-NPs had a pure crystalline form.

4.2.2. Algae, Microalgae, Cyanobacteria, and Diatoms

Algae, microalgae, and cyanobacteria have emerged as attractive biofabrication machines for many NPs [4,68]. The synthesis and antimicrobial and antioxidant applications of Au- and Ag-NPs produced through the exploitation of cell-free extracts of the microalga *Neodesmus pupukensis* were explored [120]. Zone of inhibition tests showed that Ag-NPs were active against *Pseudomonas* sp. (43 mm), *Escherichia coli* (24.5 mm), *Klebsiella pneumoniae* (27 mm), and *Serratia marcescens* (39 mm). In contrast, Au-NPs only showed activity against *Pseudomonas* sp. (27.5 mm) and *Serratia marcescens* (28.5 mm). Antifungal tests indicated that Ag-NPs had mycelial inhibition of 80.6, 57.1, 79.4, 65.4, and 69.8% against *Aspergillus niger*, *A. fumigatus*, *A. flavus*, *Fusarium solani*, and *Candida albicans*, respectively, while Au-NPs had 79.4, 44.3, 75.4, 54.9, and 66.4% against *A. niger*, *A. fumigatus*, *A. flavus*, *F. solani*, and *C. albicans*, respectively. The free radical scavenging power of Au-NPs and Ag-NPs was 68.9 and 41.21%, respectively. The authors concluded that Au- and Ag-NPs fabricated by *Neodesmus pupukensis* have significant potential as antimicrobial and antioxidant agents and could be used for various biotechnological applications.

Colin et al. reported an eco-friendly green synthesis method to produce Au-NPs with enhanced biocompatibility [76]. The method used an extract from the alga *Egregia* sp., which naturally contains biomolecules that are important for shell formation around the Au-NPs to improve their biocompatibility. The algae extract functions as the reducing agent and as the stabilizing capping shell for the Au-NPs colloid. The yielded Au-NPs had a diameter of approximately 8 nm with a narrow size distribution.

El-Kassas et al. revealed that the formation and stabilization of Au-NPs using *Corallina officinalis* extract could be attributed to the existence of the hydroxyl functional group of polyphenols and the carbonyl group of proteins [121]. Hamida et al. extracellularly synthesized, for the first time, Ag-NPs using the novel cyanobacterial strain *Desertifilum* IPPAS B-1220 [28]. The green Ag-NPs ranged from 4.5 to 26 nm in size, were spherical, and exhibited potent anticancer and antibacterial activities. Similarly, *Nostoc Bahar M* sp. exhibited a potent reducible activity to fabricate silver nitrate into Ag-NPs at ambient temperature after 24 h under dark conditions [58]. The biogenic Ag-NPs were spherical with an average diameter of 14.9 nm and showed antiproliferative activity against colon cancer cells.

Diatoms are unicellular photosynthetic microalgae that are distinguished by hydrated amorphous silica exoskeletons of different sizes and shapes [122]. The use of live diatoms in biotechnology and their applications in ecological monitoring and biofuel production were reported in several studies [123,124]. The biosynthesis of metal NPs using live diatoms as a reducing agent has been demonstrated [125,126]. Jena et al. reported the formation of Ag-NPs by a light-dependent reaction in an aqueous cell extract of diatom *Amphora* sp. [127]. The aqueous extract of *Amphora* sp. was light yellow, indicating that only yellow pigment was extracted but not the chlorophyll. The aqueous extract was added to the silver nitrate solution for the biosynthesis of Ag-NPs. The reaction mixture started to change color from light yellow to brown within seconds and became red brown within 30 min. Ag-NPs were formed only in light conditions because no color change was observed when the reaction was conducted in dark conditions. UV–Vis spectroscopy of the Ag-NP suspension showed a peak at 413 nm. The authors reported that the increase in peak intensity at 413 nm, which was linked to the time of reaction, confirmed the rise in the number of Ag-NPs in the reaction mixture. TEM analysis revealed that Ag-NPs were polydispersed, spherical, and ranged in size from 5 to 70 nm, with an average particle size of 20–25 nm. XRD spectra revealed four intense diffraction peaks at 2θ values of 38.48°, 44°, 64.74°, and 77.4° corresponding to (111), (200), (220), and (311) planes, indicating the crystallinity of Ag-NP. These findings indicated that aqueous extract of *Amphora* sp. diatom was highly effective in reducing Ag ions to formulate scattered Ag-NPs.

4.2.3. Actinomycetes

Extracellular synthesis of Au-NPs was explored using the supernatant of *Streptomyces griseoruber*, an actinomycete culture isolated from soil [128]. The development of NPs was confirmed by UV–Vis spectroscopy, which showed a peak between 520 and 550 nm. High-resolution TEM (HRTEM) analysis revealed that the formed Au-NPs were in the range of 5–50 nm and exhibited catalytic activity to degrade methylene blue.

The marine actinomycete, *Nocardiopsis alba*, isolated from mangrove soil, was utilized to produce Ag-NPs, and several bioassays were performed to evaluate the antibacterial and antiviral activities of these NPs [129]. UV–Vis spectroscopy showed the absorption peak at 420 nm, while SEM and XRD analysis revealed that the Ag-NPs were spherical and crystalline, respectively. The Ag-NPs showed antiviral activity and significant antibacterial activity against *Pseudomonas aeruginosa*, *Klebsiella pneumonia*, *Streptococcus aureus*, and *E. coli*.

4.2.4. Bacteria

Bacteria, especially thermophilic bacteria, have huge potential in the extracellular green synthesis of Ag- and Au-NPs [130]. These extracellular mechanisms facilitate the production of metal NPs in an eco-friendly manner, which reduces the downstream processing of these metals.

Patil et al. investigated the effectiveness of the marine bacterium *Paracoccus haeundaensis* in the extracellular synthesis of Au-NPs and assayed the antioxidant and antiproliferative effects of the Au-NPs on both normal and cancer cells [131]. The formation of Au-NPs was confirmed by following the development of a ruby red color with a UV–Vis absorbance peak at about 535 nm. The resultant Au-NPs were spherical and had an average size of 20.93 ± 3.46 nm. The results showed no growth inhibition effect of the Au-NPs on normal cells, while the growth of cancer cells was inhibited in a concentration-dependent manner. These findings indicated that the biogenic Au-NPs were nontoxic to human cells and could therefore be used in biomedical applications.

Bacillus brevis (NCIM 2533) was exploited in the green synthesis of Ag-NPs [132]. The synthesized Ag-NPs, which were characterized by several spectroscopic and microscopic techniques, were spherical, had a size range of 41–68 nm, and presented with an SPR peak at 420 nm. In addition, the antibacterial effect of the Ag-NPs against multidrug-resistant pathogens, including *Salmonella typhi* and *Staphylococcus aureus*, was verified in vitro.

4.2.5. Fungi

Fungi contain a plethora of biocompounds; approximately 6400 have been extracted from filamentous fungi making these organisms attractive in many applications [133]. Furthermore, these microorganisms have a potentially reduced ability to produce NPs from many bulk materials owing to their tolerance against heavy metals and potentiality to accumulate metals [134].

Molnár et al. synthesized Au-NPs using 29 different thermophilic fungi and compared the results of the extracellular fraction to those of the intracellular fraction of the fungi [65]. The fabricated Au-NPs had a size ranging between 6 and 40 nm, and the sizes vary according to the fungal strain and experimental conditions.

Another study focused on exploring the anticancer activity of Au-NPs synthesized using *Fusarium solani* [135]. Properties of the Au-NPs were observed by UV–Vis spectroscopy, FTIR, SEM, and XRD. SEM images revealed that the average diameter of the NPs was between 40 and 45 nm. These Au-NPs demonstrated dose-dependent cytotoxicity against cervical cancer cells and human breast cancer cells by inducing apoptosis pathways. The findings of this research present a safer chemotherapeutic agent with lower systemic toxicity.

4.2.6. Lichens

Lichens are composite organisms that live in both obligate and beneficial symbiosis with fungi, algae, perennial trees, or cyanobacteria [34]. These organisms have been used globally in enceinte traditional medicine. Some lichens are recognized as an effective treatment for gastritis, diabetes, hemorrhoids, dysentery, dyspepsia, amenorrhea, vomiting, and respiratory tract illnesses such as pulmonary tuberculosis, throat irritation, bronchitis, and dry cough [136]. Many countries are using commercial lichen-derived pharmacological products. For example, usnic acid was used in antiseptic products in Germany (Camillen 60 Fudes spray and nail oil) and Italy (Gessato™ shaving) [137]. Icelandic lichens were used in cold remedies by the trade names of Isla-Moos® (Engelhard Arzneimittel GmbH & Co. KG, Germany) and Broncholind® (MCM Klosterfrau Vertriebsgesellschaft mbH, Germany). The riminophenazine was demonstrated as antimycobacterial drugs [138]. Generally, lichens contain high proportions of phenolic compounds and polysaccharides such as lichenan and isolichenan, and various secondary metabolites, including protolichesterinic acid and fumarprotocetraric acid [34,139]. These biomolecules make the lichen extracts have many biological activities such as antioxidant, antimicrobial and anticancer potencies. Moriano et al. investigated the antioxidant potency of 10 lichen species of *Parmeliaceae* spp. using oxygen radical absorbance capacity (ORAC) and 1,1-diphenyl-2-picrylhydrazyl (DPPH) radical scavenging activities and the ferric reducing antioxidant power [140]. The data exhibited that antioxidant capacities were variable between lichen species. For instance, methanolic extract of *Flavoparmelia euplecta* showed the highest ORAC value (3.30 μmol TE/mg dry extract), *Myelochroa irrugans* methanolic extract demonstrated the maximum DPPH scavenging activity (EC_{50} = 384 μg/mL), and the extract of *Hypotrachyna cirrhata* showed the highest reducing power (316 μmol of $Fe2^{+}$ eq/g sample).

Felczykowska et al. studied the antiproliferative potency and antibacterial activity of acetonic extracts of three lichen species, namely, *Caloplaca pusilla*, *Protoparmeliopsis muralis*, and *Xanthoria parietina* [141]. The scholars exhibited that *P. muralis* significantly suppressed the growth of *Bacillus subtilis*, *Enterococcus faecalis*, *Staphylococcus aureus*, and *Staphylococcus epidermidis*. Moreover, *X. parietina* showed antiproliferative activity against both Hela and MCF-7 cancer cells with IC_{50} values of 8 μg/mL, and *C. pusilla* revealed the highest potency to reduce Hela, MCF-7, and PC-3 cancer cells viability with IC_{50} values of 6.57, 7.29, and 7.96 μg/mL, respectively.

Usnic acid, along with isodivaricatic acid, 5-propylresorcinol, and divaricatinic acid derived from *Protousnea poeppigii* and *Usnea florida*, showed potent antifungal activity against *Microsporum gypseum*, *Trichophyton mentagrophytes*, and *T. rubrum* [142].

Furthermore, lichen extract has been established as an efficient reducing and capping agent for NPs due to the vast abundance, rapid growth, and most importantly, environmental sustainability of these organisms [143]. The functional groups of secondary metabolites from lichen extracts are instrumental in preventing aggregation of NPs and hence improve the fabrication and stabilization of NPs [144]. Lichen-based NPs show great potential as therapeutic agents, serving as antimicrobials, antidiabetics, and antioxidants [145,146].

5. Lichens as Biosynthesizers for Nanoparticles

5.1. Metallic Nanoparticles (MNPs)

MNPs have become the most fundamental NMs in many applied and research areas due to their unique physical, chemical, and biological properties that make them promising candidates in the fields of industry, medicine, electronics, etc. [4,147]. The most frequently studied MNPs are Ag-NPs and Au-NPs due to their significant therapeutic activity against many serious diseases and their smaller-size-to-large-surface-area ratio, which enables them to be used as drug delivery systems and catalysts [8,148].

Siddiqi et al. reported that aqueous-ethanolic extract of *Usnea longissima* has the potential to fabricate silver nitrate into Ag-NPs extracellularly under laboratory conditions [136]. The lichen samples were washed, dried at 60 °C, and then crushed into fine powder [136]. Next, 10 g lichen powder was refluxed into 100 mL ethanol-distilled water (50:50) for 3 h. The samples were centrifuged to remove debris and 10 mL of supernatant was mixed with 1 mL of 0.01 M solution of $AgNO_3$. The reduction process was completed after 72 h of incubation under stirring and dark conditions at room temperature. The resultant Ag-NPs were stable for weeks; however, the yield of NPs was extremely low at approximately 35%. TEM of the synthesized Ag-NPs indicated that the NPs were spherical with an average nanosize of 10.49 nm. FTIR peaks of the Ag-NPs exhibited distinct bands at 3500–3300 cm^{-1} that refer to primary amines, while bands at 1600–1500 cm^{-1} correspond to C–O stretching and amide bands (N–H). Additionally, at 1650 cm^{-1}, amide I and amide II bands were reported, and at frequencies of up to 1600 cm^{-1}, a COO band overlapped with the amide II band. The authors speculated that organic molecules of *U. longissima* were responsible for the reduction of silver nitrate into Ag-NPs.

The bioreduction ability of *Cetraria islandica* extract for the fabrication of silver nitrate into Ag-NPs was first studied by Yıldız et al. [139]. They reported that *C. islandica* can extracellularly fabricate Ag-NPs with diverse morphologies and sizes under different parameters such as time of exposure, the concentration of both silver nitrate and lichen extracts, and temperature. Increasing the time of exposure resulted in an increase in UV absorbance values, indicating higher production of NPs. Low $AgNO_3$/lichen ratio and low temperature caused increases in absorbance values, which again indicated higher production of NPs. The authors suggested that the higher production of Ag-NPs may be due to an increase in bioreducing agents (low $AgNO_3$/lichen ratio) represented by the lichen extract. It was also speculated that the average size of the Ag-NPs (5–29 nm) may be controlled by varying the silver nitrate and lichen concentrations, time of reaction, and temperature.

Khandel et al. studied the catalytic activity of *Parmotrema tinctorum* to form silver-NPs [149]. *P. tinctorum* could effectively synthesize Ag-NPs from silver nitrate in an eco-friendly manner. These NPs are distinguished by their high stability, spherical shape, and average diameter of 15.14 nm. Leela and Anchana reported for the first time the potentiality of aqueous extracts of *Parmelia perlata* and their purified fractions (secondary metabolites) to fabricate silver nitrate into Ag-NPs [150]. Briefly, the lichen extract was prepared with a cold extraction method utilizing methanol and water, in which 50 g pulverized lichen was mixed with 500 mL methanol and incubated under dark conditions on a rotary shaker for three days at ambient temperature and the same amount of lichen powder was mixed with 500 mL distilled water and boiled for 1 h at 65 °C. Each mixture was filtrated using Whatman filter paper No. 1, and the filtrates were used for the synthesis of NPs. Thin-layer chromatography (TLC), column chromatography (CC), and gas chromatography-mass spectroscopy were performed to obtain purified secondary metabolites to use in Ag-NP synthesis. Both aqueous extract of lichen and their secondary metabolites were potent reducing and stabilizing agents for the fabrication of Ag-NPs.

Aqueous extracts of *Parmotrema praesorediosum* and *Ramalina dumeticola* were also exploited for the extracellular fabrication of Ag-NPs after 72 h at room temperature [151]. Both lichen species could form Ag-NPs, but *R. dumeticola* showed the highest bioreduction activity. *R. dumeticola* induced the formation of spherical Ag-NPs with an average size of 20 nm, while those synthesized by *P. praesorediosum* were spherical with an average size of 42 nm. *P. praesorediosum* was recently reported to extracellularly synthesize Ag-NPs with a cubic structure and a nanodiameter of 19 nm [152]. Similarly, *Cetraria islandica* was an effective biosynthetic source for both Ag-NPs and Au-NPs [144]. This lichen could produce spherical silver-NPs and gold-NPs with a dominant nanosize of 6 and 19 nm, respectively, after 30 min at 80 °C. The authors suggested that oxidation of phenolic compounds was a result of the reduction process of metal ions into their nanoform.

Dasari et al. used in vitro cultures of four species of lichen, *Parmeliopsis ambigua*, *Punctelia subrudecta*, *Evernia mesomorpha*, and *Xanthoparmelia plitti* to synthesize Ag-NPs extracellularly [153]. These lichens were collected from Goolapalli, Ramakuppam Mandal, Chittoor (District), Andhra Pradesh, India. Five grams of lichen thalli was cut, washed with water, and then sterilized with 0.01% $HgCl_2$. Small pieces of the thalli were then inoculated on plates of malt yeast extract medium and incubated at 28 ± 5 °C for 7–10 days before transfer to fresh culture media. Four types of mycelial mat were collected separately by filtering the cultures through Whatman No. 1 filter paper. Each type of mat was separately added to 100 mL of 1 mM silver nitrate solution and incubated for 24 h at room temperature with shaking and light conditions. The solutions were then centrifuged for 10 min at 12,000 rpm to collect the synthesized Ag-NPs. UV–Vis spectrum analysis was used to examine the reduction of Ag^+ ions into Ag-NPs. The absorbance peak maximum was at 410–420 nm, which is typical for Ag-NPs, while the control solution (incubated without silver nitrate) did not show any peak of absorbance. The samples displayed a broad resonance (390–420 nm), suggestive of the aggregation of Ag-NPs. SEM analysis of the formed Ag-NPs disclosed their different sizes ranging between 150 and 200 nm and that the Ag-NPs were in a polydispersed mixture. FTIR analysis was conducted to identify the biomolecules present in the mycobiont mat and responsible for Ag-NPs synthesis. Briefly, samples were mixed with KBr at a ratio of 1:100, and the spectra were recorded at 1000–3500 cm^{-1}. Ag-NPs synthesized by *Parmeliopsis ambigua* had an IR spectra peak at 3332 cm^{-1} that confirmed the presence of polyphenolic –OH group and peaks at 1639 and 1252 cm^{-1} that reflected the presence of amide I and carboxylic groups, respectively. Similar findings were obtained for the *Punctelia subrudecta* sample. Ag-NPs of *Evernia mesomorpha* also showed the same functional groups, –OH and –NHCO, at peaks of 3248 and 1739 cm^{-1}, respectively. However, spectra of Ag-NPs synthesized by *Xanthoparmelia plittii* revealed the presence of C–N at a peak of 1015 cm^{-1} and the asymmetric mode of both the aliphatic and aromatic functional group –C–H peaking at 2923 cm^{-1}. The presence of C–H stretching was confirmed by a peak at 2853 cm^{-1}, and a peak at 3234 cm^{-1} corresponded to primary aliphatic amines. The presence of carbonyl group C=O from the phenols was indicated by the peak at 1656 cm^{-1}, and the C–O single bonds were indicated by a peak at 1000–1200 cm^{-1}, while the aromatic C–H functional group was found below 700 cm^{-1}. The predicted phenols in the samples included catechin gallate, epicatechin gallate, and gallocatechin gallate. The authors speculated that polyphenolic compounds were the essential molecules in the bioreduction process of Ag-NPs.

The efficiency of aqueous extract of the lichen *Ramalina dumeticola* as a reducing and stabilizing agent for extracellular fabrication of silver nitrate into Ag-NPs was recently explored [154]. The reaction between the lichen aqueous extract (10 mL) and 30 mL of 1 mM silver nitrate solution was conducted at room temperature for 24 h. The formation of Ag-NPs was confirmed by the solution turning yellowish brown. NPs were obtained from the solution by centrifugation at 5000 rpm for 20 min and were subsequently freeze dried. UV–Vis spectral analysis, which was taken with a resolution of 2 nm at a range of 400–450 nm, monitored the formation of Ag-NPs and revealed the characteristic SPR band of Ag-NPs at approximately 433 nm. XRD analysis to define the chemical composition and crystal structure of the sample showed four peaks of 38.1°, 44.3°, 64.4°, and 77.4° at 2θ, which corresponded to the (1 1 1), (2 0 0), (2 2 0), and (3 1 1) crystallographic planes of face-centered cubic of silver, respectively. The average of crystal size was calculated by utilizing the Debye-Scherrer equation, i.e., $D = (0.94\lambda)/(\beta \cos \theta)$, where D is the mean crystallite domain size, λ is the wavelength of $Cu_{k\alpha}$, β is the full width at half maximum (FWHM), and θ is the Bragg diffraction angle. The average size of the Ag-NP crystals was 17.1 nm. TEM revealed that Ag-NPs were polydispersed and mainly spherical with a size between 6 and 28 nm and an average diameter of 13 nm.

Rai and Gupta tested the possibility of biofabricating Ag-NPs by exploiting the reducing capacity of aqueous extracts of the lichen *Cladonia rangiferina* [145]. The lichen was collected from the Govind wildlife sanctuary in the Uttarkashi District of Uttarakhand, western Himalaya, at an altitude above 3500 m. Silver nitrate solution (45 mL of 1 mM) was mixed with 15 mL lichen aqueous extract and 2–3 drops of 0.1 M sodium hydroxide to reach an alkaline pH. After 72 h at room temperature, the reaction mixture turned yellow brown, verifying the presence of Ag-NPs. UV–Vis spectrophotometry indicated that the spectral band peak at 402 nm corresponded to the specific color change that resulted from the reduction of silver ions to Ag-NPs by secondary metabolites of the lichen. The presence of these secondary metabolites was consolidated by the detection of their functional groups by FTIR analysis. The FTIR scan taken at a range of 450–4000 cm^{-1} showed several functional groups corresponding to specific biomolecules in the extract such as polyphenols that could participate in the fabrication and stabilization of Ag-NPs. Peaks were observed in the range of 1000–4000 cm^{-1}, demonstrating the presence of O–H (3400 cm^{-1}), C–H (2853 cm^{-1}), C=O (1742 cm^{-1}), C=O (1691 cm^{-1}), C=O aldehyde (1651 cm^{-1}), C=C vibration (1573 cm^{-1}), CH_2, CH_3 (1443 cm^{-1}), and C–O (1273 cm^{-1}). For TEM analysis, the solution was sonicated for 15 min, loaded onto a carbon-coated copper grid, and incubated under a fume hood for 30 min for the solvent to evaporate. The Ag-NPs visualized by TEM were spherical and rod shaped, with a particle size ranging from 5 to 40 nm and an average diameter of 20 nm. Similarly, different studies were reported that lichen species including *Xanthoria elegans*, *Usnea antractica*, *Leptogium puberulum*, *Cetraria islandica*, *Pseudevernia furfuracea*, *Lobaria pulmonaria*, *Heterodermia boryi* and *Parmotrema stuppeum* have potentiality to reduce silver nitrate into Ag-NPs with different shapes (bimodal and cubic) and sizes [155–157].

Ethanolic extract of the lichen *Parmotrema clavuliferum* was used for the biological synthesis of Ag-NPs from silver nitrate [158]. The extraction was carried out by adding 10 g dried lichen to 100 mL ethanol and incubating with shaking at 80 °C for 24 h then filtering through 25-mm pore-sized papers. The extract was added to silver nitrate solution at room temperature and the production of Ag-NPs was indicated by brown color development. The brown color appeared immediately, which verifies the high potency of *Parmotrema clavuliferum* extract as a reducing and capping agent for NPs. For further confirmation, the excitation of SPR provided by the Ag-NPs was measured spectrophotometrically at 400–450 nm. The plasmon absorption bands showed an absorbance peak at 440 nm. The DLS and zeta potential data indicated that the particle size distribution of the biogenic Ag-NPs was in the range of 80–120 nm and the particles had negative charges suggesting their stability at room temperature. TEM and SEM revealed that the Ag-NPs were spherical and approximately 106 nm in diameter. Potential biomolecules in the lichen extract were explored using FTIR spectroscopy, and broad peaks at 3264 and 1634 cm^{-1} were recorded, which correspond to O–H of phenolic compounds stretching groups and C=O of the peptide bond, respectively. These findings imply the role of phenolic compounds and in the bioreduction of silver ions.

Abdolmaleki et al. used two lichen species, *Usnea articulata* and *Ramalina sinensis*, to reduce 1 mmol of silver nitrate solution into Ag-NPs [143]. The resulting Ag-NPs were spherical with a nanosize range of 10–50 and 50–80 nm, respectively.

Recently, aqueous extracts of two novel lichen species, *Acroscyphus sphaerophoroides Lev* and *Sticta nylanderiana*, were utilized to fabricate chloroauric acid (10^{-3} M $HAuCl_4$) into Au-NPs at room temperature for 12 h [159]. Physicochemical analyses confirmed the potentiality of both lichen species to generate gold-NPs. The UV–Vis-spectra of biogenic Au-NP was at 535 nm and the XRD pattern confirmed the face-centered cubic of Au-NPs. FTIR spectra of both types of Au-NPs featured bands at 3446 and 1041 cm^{-1} that relate to N–H and C–O stretching, respectively, bands at 2922 and 2849 cm^{-1} corresponding to C–H stretching, and bands at 1638 and 1456 cm^{-1} that relate to the amide and carboxylate groups, respectively, in the amino acid residues of the biomolecules. The authors speculated that the presence of these functional groups might help prevent agglomeration of the NPs. Moreover, TEM revealed that *A. sphaerophoroides*-mediated Au-NPs are multiply twinned quasi-spherical and prismatic with a size range between 5 and 35 nm, while *S. nylanderiana*-mediated Au-NPs were exclusively multiply twinned with a nanosize range of 20–50 nm.

An extract of the lichen *Parmelia sulcate* was used to synthesize Au-NPs; for its preparation, 90 mL of a 1 mM $HAuCl_4$ solution (to provide Au^{3+}) and 10 mL of the 5% *Parmelia sulcata* extract were heated to 60 °C and kept on a magnetic stirrer for 20 min [160]. The color change of the reaction mixture from yellow to purple was monitored to observe the formation of Au-NPs then the solution was dried in an oven (70 °C) for 48 h to obtain powdered particles. The UV–Vis spectrum from 300–700 nm confirmed the reduction of Au^{3+} to gold NPs (Au^0). The peak observed at 540 nm represents the SPR, verifying the formation of the Au-NPs. The XRD spectrum had peaks at 38.3°, 44.6°, 64.7°, and 77.7°, which revealed the crystalline feature of gold and the face-centered cubic particles depending on the angular positions of the Bragg peak. SEM and TEM demonstrated that the particles had an average size of 54 nm and were spherical. TEM with energy-dispersive spectroscopy (EDS) was utilized to determine the elemental composition. The EDS pattern displayed a strong signal for the gold peak, indicating successful fabrication of Au-NPs. FTIR spectra were measured over a range of 400–4000 cm^{-1} to follow the reaction between the *Parmelia sulcata* extract and chloroauric acid. Biomolecules in the lichen extract were confirmed to interact with Au-NPs. The peaks at 3443 cm^{-1} corresponded to the O–H strong stretch of the alcohol, while a peak at 1640 cm^{-1} corresponded to the C=C of the alkenes, and the peaks at 1544 and 1384 cm^{-1} were related to the N–H and N–O bending and stretching of the amide and nitro groups, respectively. The peaks at 1272 and 1206 cm^{-1} were related to the C–O stretch of esters. The presence of proteins in the solution, indicated by the amide and nitro group peaks, might contribute to the stabilization of the newly formed NPs. DLS and zeta potential analyses revealed that the average size of the hydrodynamic diameter of the Au-NPs was 54.14 nm and their charge was negative (−18.4). This indicates the existence of lichen biomolecules surrounding the Au-NPs, which provide stability to the Au-NPs.

Devasena et al. used the Soxhlet extraction method to obtain the lichen extract to use in magnesium nanoparticles (Mg-NPs) synthesis [161]. They reported that *Cladonia rangiferina* has the ability to reduce magnesium sulfate into Mg-NPs extracellularly. UV–Vis spectroscopy analysis revealed that the absorption peak of Mg-NPs was at 262 nm, while the DLS technique revealed that Mg-NPs have an average hydrodynamic diameter of 23 nm.

Protoparmeliopsis muralis was first used for the synthesis of different metallic and metal oxide NPs (MONPs) by Alavi et al., who utilized the aqueous extract of this lichen to extracellularly fabricate silver-NPs and copper-NPs under dark and stirred conditions for 24 h at ambient temperature [162]. Synthesis of the MNPs was confirmed by UV–Vis spectroscopy, TEM, SEM, EDAX, XRD, and FTIR analyses. The resulting data demonstrated that the maximum absorbance peak for Ag-NPs and Cu-NPs was 378 and 567 nm, respectively, and that the MNPs were spherical with an average nanosize of 33.49 ± 22.91 and 253.97 ± 57.2 nm, respectively. EDAX data demonstrated that Ag- and Cu-NPs were present in the sample at 87.72 and 26.42%, respectively. Furthermore, the 2θ degree values of both Ag- and Cu-NPs were 35.5°, 43.6°, 65.6°, and 72.1°, and 35.9°, 39.6°, 44.3°, 54.3°, and 57.2°, respectively, indicating the crystallinity of these NPs. Based on FTIR data, there were three dominant functional groups, C=C, S=O, and C-Br, in all samples (lichen extract, Ag- and Cu-NPs, and metal oxides NPs). However, O–H bond bending corresponding to secondary metabolites such as phenol was observed for the MNPs, suggesting that these secondary metabolites may act as a potential reducing and stabilizing agent during the synthesis process of both Ag- and Cu-NPs. To prove this hypothesis, the authors analyzed the total phenol, flavonoid, flavanol, and tannin contents (TPC, TFC, TFLC, and TTC, respectively) of the samples via Folin-Ciocaltue assay. The Ag-NP solution contained higher amounts of TPC, TFC, TFLC, and TTC, compared with the lichen extract and other MNPs and MONPs. The same study also screened the effect of time exposure (24, 48, 72, and 96 h) on the biosynthesis process of MNPs and demonstrated that the concentration of MNPs increased as the time of exposure increased. The authors noted that the synthesis process of MNPs using lichen aqueous extract was slower than that using plant watery extract suggesting the cause of the slow reaction is the lower reducing capacity of lichens (Table 1).

Table 1. Lichen-based synthesis of nanoparticles (NPs).

Strains	Type of NPs	Size (nm)	Shape	Illumination	Time of Exposure	pH	Temperature (°C)	Mode of Synthesis	Application	Reference
Usnea longissima	Ag-NPs	9.40–11.23	Spherical	Dark	72 h	7	RT	-	Antibacterial agent	[136]
Parmotrema praesorediosum	Ag-NPs	19	Cubic structure	NM	24 h	NM	RT	-	Antibacterial agent	[152]
Cetraria islandica (L) Ach	Ag-NPs	5-29	Spherical	NM	19.09, 60 120, 180 and 220.91 min	NM	16.48, 25, 37.5, 50, 58.52	-	NA	[139]
Parmotrema praesorediosum	Ag-NPs	42	Spherical	NM	72 h	Alkaline	RT	-	NA	[151]
Ramalina dumeticola	Ag-NPs	20	Spherical	NM	72 h	Alkaline	RT	-	NA	[151]

Table 1. *Cont.*

Strains	Type of NPs	Size (nm)	Shape	Illumination	Time of Exposure	pH	Temperature (°C)	Mode of Synthesis	Application	Reference
Ramalina dumeticola	Ag-NPs	13	Spherical	NM	24 h	NM	RT	-	NA	[154]
Cetraria islandica (L.) Ach	Ag-NPs	6	Spherical	NM	30 min	NM	80	-	Catalytic activity	[144]
	Au-NPs	19	Spherical	NM	30 min	NM	80	-		
	Ag-Au NPs	6 and 21	Polygonal and Spherical	NM	30 min	NM	80	-		
Parmelia perlata	Ag-NPs	NA	Spherical	NM	30 min	NM	60	-	Antimicrobial, antioxidant and antidiabetic agents	[150]
Ramalina sinensis	Fe_3O_4 NPs	20-40	Uniform Spherical	NM	1 h	NM	70	-	Removing heavy metals such as Pb and Cd	[163]
Lecanora muralis	ZnO@TiO2@SiO2 nanocomposites	55–90	Spherical	NM	5 h	NM	80	-	Antimicrobial agent	[164]
	Fe3O4@SiO2 nanocomposites	55–85	Spherical	NM	5 h	NM	80	-		
Ramalina sinensis	Iron oxide nanoparticles	31.74-53.91	Uniform spherical	NM	1 h	7	70	-	Antibacterial agent	[165]
Protoparmeliopsis muralis	Ag-NPs	33.49 ± 22.91	Spherical	NM	24 h	8	RT	-	Antibacterial, antibiofilm, antiquorum sensing, antimotility, and antioxidant activities	[162]
	Cu-NPs	253.97 ± 57.2	Triangular	NM	24 h	8	RT	-		
	Fe3O4 NPs	307 ± 154	Spherical	NM	24 h	8	RT	-		
	TiO_2 NPs	133.32 ± 35.33	Polyhedral	NM	24 h	8	RT	-		
	ZnO NPs	178.06 ± 49.97	Cubic	NM	24 h	8	RT	-		
Parmeliopsis ambigua	Ag-NPs	150–250	NM	Light	24 h	NM	RT	+, -	Antibacterial and antioxidant agents	[153]
Punctelia subrudecta	Ag-NPs	150–250	NM	Light	24 h	NM	RT	+, -		
Evernia mesomorpha	Ag-NPs	150–250	NM	Light	24 h	NM	RT	+, -		
Xanthoparmelia plitti	Ag-NPs	150–250	NM	Light	24 h	NM	RT	+, -		
Cladonia rangiferina	Mg-NPs	23	NM	NM	24 h	NM	NM	NM	NA	[161]
Parmotrema tinctorum	Ag-NPs	15 ± 5.1	Spherical	Dark	24 h	NA	RT	-	Antibacterial agent	[149]
Acroscyphus sphaerophoroides	Ag-NPs	5–35	Twinned quasi-spherical and prismatic shapes	NM	12 h	NM	RT	-	Antioxidant agent	[159]
Sticta nylanderiana	Ag-NPs	20–50	Multiply twinned	NM	12 h	NM	RT	-		
Parmotrema clavuliferum	Ag-NPs	106	Spherical	Dark	48 h	NM	80 °C	-	Antibacterial agent	[158]
Parmotrema perlatum	Ag-NPs	NM	NM	NM	NM	NM	NM	NM	Antibacterial agent	[166]
Xanthoria parietina	Ag-NPs	1–40	Spherical	Dark	72 h	NM	40 °C	-	Anticancer and antibacterial agents	[37]
Flavopunctelia flaventior	Ag-NPs	1–40	Spherical	Dark	72 h	NM	40 °C	-		
Parmelia perlata	Ag-NPs	NM	NM	NM	NM	NM	NM	NM	Antibacterial agent	[167]
Umbilicaria Americana	Ag-NPs	NM	NM	NM	NM	NM	NM	NM	NM	[168]
Cladonia rangiferina	Ag-NPs	20	Spherical and rods	NM	72 h	Alkaline	RT	-	Antibacterial agent	[145]

Table 1. *Cont.*

Strains	Type of NPs	Size (nm)	Shape	Illumination	Time of Exposure	pH	Temperature (°C)	Mode of Synthesis	Application	Reference
Usnea articulata	Ag-NPs	10–50	Spherical	NM	72 h	Alkaline	27 °C	-	Antibacterial agent	[143]
Ramalina sinensis	Ag-NPs	50–80	Spherical	NM	72 h	Alkaline	27 °C	-		
Parmelia sulcate	Au-NP	54	Spherical	NM	20 min	NM	60 °C	-	Antioxidant and mosquitocidal agents	[160]
Ramalina fraxinea	ZnO-NPs	21	Spherical	NM	Up to 2 h	NM	60 °C	-	Neuroprotection activity	[169]
Aspicilia lichens	Nanohyaluronic acid	29–89	Spherical	NM	48 h	Alkaline then neutralize by acid	50 °C	-	Antidiabetic agent	[146]
Xanthoria elegans	Ag-NPs	Bimodal	5-100	NM	2 h	NM	NM	-	Antibacterial agent	[155]
Usnea antarctica		Bimodal	5-100	NM	6 h	NM	NM	-	Antibacterial agent	
Leptogium puberulum		Bimodal	5-100	NM	6 h	NM	NM	-	Antibacterial agent	
Cetraria islandica		Bimodal	5-100	NM	2 h	NM	NM	-	Antibacterial agent	
Pseudevernia furfuracea	Ag-NPs	Bimodal	<10-100	NM	2 h	NM	NM	-	Antibacterial and antioxidant agents	[156]
Lobaria pulmonaria		Bimodal	<10-100	NM	2 h	NM	NM	-	Antibacterial and antioxidant agents	
Heterodermia boryi	Ag-NPs	Cubic	27.91–37.21	NM	NM	NM	NM	-	Antibacterial agent	[157]
Parmotrema stuppeum		Cubic	27.69–36.00	NM	NM	NM	NM	-	Antibacterial agent	

Abbreviation: (-), extracellular synthesis; (+), intracellular synthesis; NM, not mentioned; NA, no applications; RT, room temperature.

5.2. Metal Oxide Nanoparticles (MONPs)

MONPs are one of the widest used nanomaterials due to their unique properties including high stability, porosity, and easy functionalization with different molecules because of their negative charge; these properties mean MONPs are particularly suited to biomedical applications [170].

Alavi et al. utilized the aqueous extract of *Protoparmeliopsis muralis* to biosynthesize three different types of MONPs—ferric oxide, zinc oxide, and titanium oxide (Fe_3O_4, ZnO, and TiO_2, respectively) NPs [162]. Briefly, lichen samples collected from Kane Gonabad Mountains were washed with distilled water, air dried for six days, then crushed into a fine powder, and boiled with 250 mL distilled water at 90 °C for 30 min. The mixture was filtered through Whatman filter paper No. 40 and 10 mL of the filtrate, was mixed with 50 mL $TiO(OH)_2$ or $Zn(NO_3)_2{\cdot}6H_2O$ (0.01, and 0.001 M concentrations, respectively), and incubated for 24 h with stirring. For fabrication of Fe_3O_4 NPs, the same amount of lichen extract was added to flasks containing $FeCl_3{\cdot}6H_2O$ (0.2 M) and $FeCl_2{\cdot}4H_2O$ (0.001, 0.01, and 0.1 M), and the pH was adjusted to 8 by adding 0.1 M NaOH solution. Mixtures were kept under stirred conditions for 24 h at room temperature. The resultant NPs were collected by centrifugation at 4000 rpm for 30 min, washed, and dried at 70 °C for 8 h. Physicochemical analyses showed that the UV-spectra peaks of Fe_3O_4, ZnO, and TiO_2 NPs were 216, 328, and 283 nm, respectively. TEM and SEM images demonstrated that all the MONPs were spherical with an average nanodiameter of 307 ± 154 (Fe_3O_4 NPs), 133.32 ± 35.33 (TiO_2 NPs), and 178.06 ± 49.97 nm (ZnO NPs). The presence of Fe (84.07%), Ti (66.41%), and Zn (25.61%) in the Fe_3O_4, ZnO, and TiO_2 NPs samples was detected by EDAX analysis. XRD and FTIR analyses proved that these MONPs had nanocrystal structures and were coated with organic molecules such as secondary metabolites (phenols, O–H), which have a significant role in reducing and stabilizing NPs.

The hydrolytic capacity of aqueous extracts of a new strain of lichen, *Ramalina sinensis*, was recently reported to extracellularly fabricate ferric chloride salts into iron oxide NPs [165]. The UV-spectra curve of the NP samples appeared in the range of 280–320 nm, indicating the formation of magnetic iron oxide NPs. The XRD pattern of the biosynthesized iron oxide NPs showed distinct diffraction peaks of 30.5°, 36.1°, 43.3°, 53.9°, 57.5°, and 63.3° at 2θ, indicating the cubical nanocrystalline structure of iron oxide NPs. Furthermore, FTIR analysis demonstrated that π-electrons of carbonyl groups of flavonoid and phenolic compounds of *R. sinensis* were responsible for the reduction of iron ions into their nanoforms. Field emission scanning electron microscopy (FESEM) revealed that the particle size of the iron oxide NPs was between 31.74 and 53.91 nm and that pores existed in the iron oxide NPs structure. EDX analysis showed that Fe and O elements were the main constituents in the iron oxide nanostructure.

Similarly, Arjaghi et al. performed extracellular reduction of ferric chloride salts ($FeCl_2 \cdot 4H_2O$ and $FeCl_3 \cdot 6H_2O$) into Fe_3O_4 NPs by utilizing *R. sinensis* [163]. Sharp absorption peaks were observed between 300 and 350 nm owing to the interaction between the chemicals, and tensile vibration resulted from the formation of a new bond between iron and oxygen and the synthesis of Fe_3O_4 NPs. The authors hypothesize that the existence of biomolecules in *R. sinensis* might prevent the agglomeration of NPs, that polysaccharide sulfate acts as a potent reducing agent, and that sulfate groups have significant roles in the extracellular synthesis of iron oxide NPs by oxidizing the aldehyde group into carboxylic acids. XRD and SEM data revealed that Fe_3O_4 NPs were nanocrystalline and 20–40 nm in size.

ZnO-NPs were biologically synthesized by Koca et al. using *Ramalina fraxinea* extract [169]. Lichen samples were carefully washed, dried in a 70 °C oven overnight, and extracted in water by heating (80 °C) for 1 h, and the resulting extract was filtered through Whatman No 1 filter paper. For the synthesis of ZnO-NPs, 100 mL filtered extract was added to 5 g $Zn(NO_3)_2 \cdot 6H_2O$ and incubated at 60 °C with continuous stirring until the color changed, indicating the formation of NPs. The solution was then heated at 400 °C for approximately 2 h to obtain a fine powder of ZnO-NPs. The characteristic SPR band of the ZnO-NPs was determined by UV analysis at a wavelength range between 200 and 900 nm. Peaks were observed at 269 nm, which related to *Ramalina fraxinea* extracts, and 330 nm, suggesting synthesis of ZnO NPs was successful. FTIR analysis of the ZnO-NPs revealed O–H (alcohol) band vibrations at 3128 and 1398 cm^{-1} and stretching bands at 1620 and 1575 cm^{-1} that were correlated to the alkenes (C=C), while absorption peaks at 1480 cm^{-1} were for alkanes (C–H). Bands at 1379 and 1335 cm^{-1} related to O–H (alcohol and phenol), and the presence of amine groups (C–N) was observed at 1192 cm^{-1}. The peak at 1295 cm^{-1} represented the aromatic ester (C–O) and aromatic amine groups (C–N), while the bands at 1121 and 1087 cm^{-1} were identified as amine (C–N) and aliphatic ether (C–O) groups. Stretching bands at 1038 and 960 cm^{-1} were allotted to ether (C–O) and alkene (C–H), respectively. The bands at 871, 842, 773, 671, and 605 cm^{-1} confirmed the presence of halo compounds (C–Cl), and the peaks observed at 407, 444, and 535 cm^{-1} were assigned to Zn-O (metal-oxygen) vibration. In conclusion, FTIR analysis disclosed that functional groups in the extract of *Ramalina fraxinea* are crucial for the synthesis of ZnO-NPs. XDR spectra at 2θ showed a group of diffraction peaks of 31.7°, 34.4°, 36.2°, 47.5°, 56.5°, 62.8°, 66.4°, 67.9°, 69.1°, 72.5°, and 77.1° indicative of the crystal planes of (1 0 0), (0 0 2), (1 0 1), (1 0 2), (1 1 0), (1 0 3), (2 0 0), (1 1 2), (2 0 1), (0 0 4), (2 0 2), and (1 0 4), respectively. Collectively, the XDR analysis demonstrated that the *Ramalina fraxinea* extract delivered ZnO-NPs with a characteristic hexagonal and crystalline structure. The biogenic ZnO-NPs were spherical with a size of around 21 nm, as evidenced by SEM and FESEM imaging. The authors reported the existence of aggregation of NPs that resulted from the impact of Van der Waals forces between NPs (Table 1).

5.3. Other Nanomaterials

Abdullah et al. introduced the green synthesis method of $ZnO@TiO_2@SiO_2$ and $Fe_3O_4@SiO_2$ nanocomposites (NCs) utilizing the novel lichen species *Lecanora muralis* [164]. Lichen specimens were collected from Grdmandil mountain and the chemical compositions of the rock-inhibiting lichen samples were analyzed via XRD assay. The samples comprised quartz, hematite, magnetite, and maghemite Q. Similarly, biomolecules inside the lichen cell extract were determined using gas chromatography-mass spectroscopy, and there were a variety of secondary metabolites present that have many medical applications through their antioxidant, anticancer, analgesic, and antipyretic activities. To synthesize $ZnO@TiO_2@SiO_2$, 2 g of *L. muralis* (LM) was mixed with 30 mL distilled water and the mixture was boiled at 80 °C for 1 h then filtrated. Next, 20 mL LM filtrate was mixed with 0.5 g $ZnCl_2$, 1.5 g $TiO(OH)_2$ (titanyl hydroxide), and 2.5 g Na_2SiO_3 at pH 8 and 80 °C for 5 h under stirring conditions. Similarly, $Fe_3O_4@SiO_2$ NCs were formed by mixing 20 mL LM filtrate and 2 g Na_2SiO_3 with 0.7 g $FeCl_2$ and 1.2 g $FeCl_3$ at pH 9 and 80 °C for 5 h under stirring conditions. After the incubation period, NP precipitates were filtrated, washed with hot distilled water to discard any impurities, and then dried. Physical and chemical analyses of the $ZnO@TiO_2@SiO_2$ and $Fe_3O_4@SiO_2$ NCs showed that *L. muralis* has the potential to fabricate NCs from their bulk materials. XRD demonstrated that the biosynthesis of $ZnO@TiO_2@SiO_2$ and $Fe_3O_4@SiO_2$ NCs generates crystallinity nanoforms of 55 and 53 nm, respectively. Furthermore, the authors reported that Fe_3O_4 NPs coated the surface of the silica oxide nanoparticles. SEM micrographs of $ZnO@TiO_2@SiO_2$ and $Fe_3O_4@SiO_2$ NCs revealed that these NCs were spherical and had a nanosize range of 55–90 and 50–85 nm, respectively. Some agglomeration was observed by SEM in both types of NC. EDX and elemental mapping showed that $ZnO@TiO_2@SiO_2$ NC was synthesized from Zn, O, Ti, and Si with no further elements, indicating the purity of the formed nanostructure; however, the compositional elements of $Fe_3O_4@SiO_2$ NCs involving Fe, O, and Si indicated the binding of Fe_3O_4 NPs on the surface of SiO_2 NPs (Table 1).

Bimetallic NPs (Au–Ag NPs) were extracellularly synthesized using *Cetraria islandica* [144]. In brief, 1 mL lichen extract was mixed with 10 mL of 1.5 mm $HAuCl_4$ and $AgNO_3$ solutions and 0.5 m NaOH solution (pH 10) and incubated for 30 min at 80 °C under continuous mixing and stirring conditions. The reaction was repeated with different molar ratios of the Ag and Au solutions (1:1, 1:2, and 2:1), and the resulting bimetallic NPs are defined as Ag50Au50, Ag33Au67, and Ag67Au33, respectively. Only one absorption band appeared between the SPR of both monometallic Au- and Ag-NPs, and as Au content increased, the absorbance band redshifted. UV absorbance peaks of Ag50Au50, Ag33Au67, and Ag67Au33 were 412, 519, and 523 nm, respectively. This finding implied that bimetallic NPs may have only one alloy. FTIR spectra peaks for Ag33Au67 were at 1383 cm^{-1}, which relate to C=O of carboxylic acid and methyl interactions in large, branched molecules that play an important role in stabilizing and capping NPs. TEM images of Ag33Au67 bimetallic NPs showed that these NPs were spherical and polygonal with nanosizes of 6 and 21 nm, respectively. The narrow particle size indicates a large active surface area for catalytic activity.

Esmaeili and Rajaee explored an eco-friendly synthesis method using lichen usnic acid as a nanoparticle mediator to produce nanohyaluronic acid [146]. In brief, usnic acid (UAL) was extracted from *Aspicilia lichens* with acetone using a Soxhlet apparatus for 8 h. The FTIR spectrum of purified UAL was similar to the standard usnic acid spectrum. The stretching of O–H in the Ar–OH intramolecular hydrogen bond showed strong bands at 3421 and 3371 cm^{-1}. Additionally, ($-CH_3$) of the alkane groups in UAL had peaks at 2925 and 2854 cm^{-1} caused by stretching of the C–H bond. The presence of the C=C group in UAL was indicated by a peak at 1658 cm^{-1}, which is due to the presence of the aryl group. The aromatic methyl ketone at 1625 cm^{-1} is related to the hydrogen bonds. Similarly, the conjugate cyclic ketone group is confirmed by a peak at 1739 cm^{-1}. UAL also showed hydroxyl phenolic signals at 3095 cm^{-1}, which is likely due to the stretching of a symmetrical or nonsymmetrical group C–O–C that bonded to an aryl-alkyl-ester at

1265 and 1074 cm^{-1}. SEM micrographs demonstrated that UAL NPs were spherical with a mean size of 29–89 nm and no agglomeration. For the preparation of nanohyaluronic acid, 2 g hyaluronic acid (produced by mixing *Bifidobacterium* sp. and the solution of UAL extract) was added to 50 mL distilled water and 200 mL of a UAL solution with acetone and methanol (5:2) at 50 °C for 48 h with stirring. The NPs were then collected by centrifugation at 12,000 rpm for 30 min. SEM micrographs showed that hyaluronic NPs have an average nanosize of 55 nm. FTIR spectra of nanohyaluronic acid showed that the usnic acid extracted from *Aspicilia* sp. has strong redox activity that enables these compounds to reduce the hyaluronic acid into their nanoform (Figure 4).

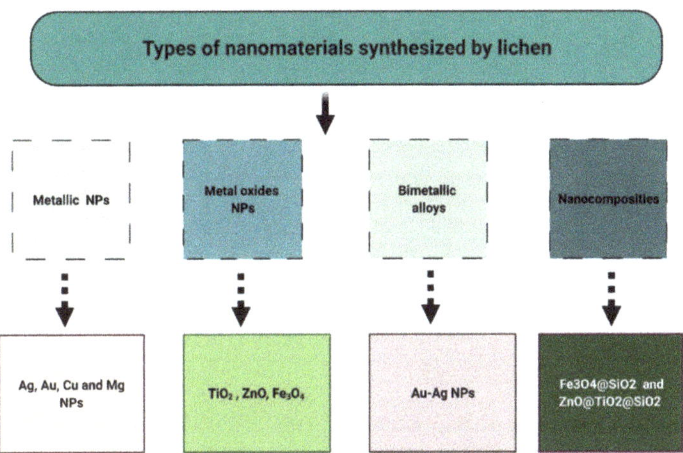

Figure 4. Types of nanoparticles (NPs) synthesized by lichen species.

6. Prospective Applications of Lichen-Based Nanoparticles

6.1. Antimicrobial Activity

Khandel et al. studied the inhibitory activity of Ag-NPs synthesized by *Parmotrema tinctorum* and the activity of silver nitrate and lichen extract against five pathogenic bacteria—*Pseudomonas aeruginosa, Staphylococcus aureus, Escherichia coli, Bacillus subtilis,* and *Klebsiella pneumoniae*—for 24 h at 35 °C using the agar well diffusion method [149]. Ag-NP (10, 30, and 50 μL) was the most potent antibacterial agent, causing greater inhibition of bacterial growth compared with silver nitrate and lichen extract. Ag-NPs suppressed the growth of both Gram-negative and Gram-positive bacteria at the three concentrations tested. At the highest concentration of Ag-NPs (50 μL), the greatest inhibition zone (IZ) was detected against *P. aeruginosa* (17 ± 0.50 mm), *K. pneumoniae* (14 ± 0.10 mm), and *E. coli* (11 ± 0.10 mm), while the lowest IZ was observed against *B. subtilis* (8 ± 0.30 mm) and *S. aureus* (7 ± 0.30 mm). The authors reported that Ag-NPs were more effective at inhibiting Gram-negative bacteria than Gram-positive bacteria due to the difference in the cell-wall structure of the bacteria; Gram-positive bacteria have a thicker cell wall than Gram-negative bacteria, and hence, penetration of their cell wall is difficult. Furthermore, the authors conclude that the mode of action of Ag-NPs against bacteria is via their ability to change the membrane structure and permeability, leading to bacterial death.

Iron oxide-NPs (0.075–0.00046875 mg/mL) bioformed by *Ramalina sinensis* significantly inhibited the bacterial growth of both *P. aeruginosa* and *S. aureus* after incubation for 24 h at 37 °C [165]. Iron oxide-NPs at 0.075 mg/mL exhibited the highest antibacterial activity, while 0.0075 and 0.000234375 mg/mL of iron oxide-NPs were the lowest inhibitory concentrations of NPs against both *P. aeruginosa* and *S. aureus*, respectively. The antibacterial activity of iron oxide-NPs was almost equivalent to that of tetracycline. The authors suggested that the expected killing mechanism of NPs against bacteria may be related to the electrostatic activity between iron oxide-NPs and the bacterial membrane. This interaction might result in the release of the iron ions by the NPs, and these ions can then interact with the thiol group on membrane proteins, causing bacterial membrane oxidation, subsequent stimulation of reactive oxygen species (ROS), loss of membrane permeability, disruption of cell membrane respiration, and ultimately, bacterial death.

Abdullah et al. studied the antibacterial and antifungal activities of both $ZnO@TiO_2@SiO_2$ and $Fe_3O_4@SiO_2$ NCs synthesized by *Lecanora muralis* against *S. aureus*, *E. coli*, *Pseudomonas* sp., and five species of fungi, i.e., *Candida albicans*, *Candida* spp., *Aspergillus flavus*, *Aspergillus niger*, and *Aspergillus terrus*, utilizing both disk diffusion and agar well diffusion assays and compared the results with those of lichen extract alone [164]. Both NCs showed higher inhibitory activity than lichen extract alone against bacterial and fungal species, with the exception of the three species of the genus *Aspergillus* (zero inhibition zone). $Fe_3O_4@SiO_2$ exhibited the highest bioactivity among the treatments, suggesting more bioactive molecules were precipitated on $Fe_3O_4@SiO_2$ NCs than on $ZnO@TiO_2@SiO_2$ NCs. The authors noted that the increased antioxidant molecules adsorbed on $Fe_3O_4@SiO_2$ NCs contributed to the long-term stabilization of NCs against decomposition and deformation conditions.

Lichen ethanolic extract (*Parmotrema clavuliferum*) and the corresponding lichen-synthesized Ag-NPs were investigated as an antibacterial treatment in a recent study by Alqahtani et al. [158]. Biogenic Ag-NPs showed a significant inhibitory effect against *P. aeruginosa* (11.5 ± 0.9 mm), *Streptococcus faecalis* (7.6 ± 1.7 mm), *B. subtilis* (8.1 ± 1.5 mm), and *S. aureus* (8.1 ± 1.5 mm). However, the ethanolic extract of the lichen caused had the highest zone of inhibition (19.8 ± 0.9 mm) against *B. subtilis* and the lowest zone of inhibition (3.6 ± 0.9 mm) against *S. aureus*. Furthermore, *S. faecalis* and *P. aeruginosa* showed inhibition zones of 15.5 ± 1.6 and 13.8 ± 0.9 mm, respectively, after spiking with lichen extract. The authors suggested that Ag-NPs could have this bactericidal effect due to one or more of the following actions: Ag-NPs may induce cell lysis, hinder transduction, change membrane permeability, or destroy the bacterial genome through DNA fragmentation.

Ag-NPs produced using four lichens, *Parmeliopsis ambigua*, *Punctelia subrudecta*, *Evernia mesomorpha*, and *Xanthoparmelia plitti*, were screened for antibacterial activity against several Gram-negative and Gram-positive bacteria, including *Pseudomonas aeruginosa*, *Escherichia coli*, *Proteus vulgaricus*, *Staphylococcus aureus*, *Streptococcus pneumoniae*, and *Bacillus subtilis* [153]. The disk diffusion method was used for screening with 0.02 mg of the produced Ag-NPs. *Pseudomonas putida* was the most susceptible to Ag-NPs synthesized by *X. plitti* (2.3 cm), followed by *Pseudomonas aeruginosa* with Ag-NPs synthesized by *E. mesomorpha* extract (1.9 cm) and *Bacillus subtilis*, which was the least susceptible with Ag-NPs synthesized by *X. plitti* extract (1.3 cm).

The antibacterial activity of Ag-NPs synthesized using an aqueous extract of *Ramalina dumeticola* were examined against four Gram-positive pathogenic bacteria (*Staphylococcus epidermidis*, methicillin-resistant *Staphylococcus aureus* (MRSA), *Bacillus subtilis*, and *Streptococcus faecalis*) and four Gram-negative strains (*Proteus vulgaris*, *Pseudomonas aeruginosa*, *Serratia marcescens*, and *Salmonella typhi*) by applying the disk diffusion method [154]. Gentamycin (30 µg disks) was used as a positive control, and the negative control was sterile distilled water. A total of 10 microliters of 100 µg/mL of the Ag-NPs solution was applied to the sterile disks on agar plates, and inhibition zones (IZs) were measured after incubation for 18–24 h at 37 °C. The results demonstrated the potential of the Ag-NPs as a bactericidal factor. The Ag-NPs were effective against both types of bacteria but showed more efficacy towards Gram-negative bacteria than Gram-positive bacteria. The largest IZ was observed in *Proteus vulgaris* (10.5 ± 0.7 mm), followed by *Pseudomonas aeruginosa* (9.5 ± 3.5 mm) and MRSA (9.5 ± 0.7 mm). The Ag-NPs were less effective in *Salmonella typhi* (IZ diameter of 8.5 ± 2.1 mm) and *Serratia marcescens* (7.5 ± 0.7 mm). *Bacillus subtilis* and *Streptococcus faecalis* had identical IZ (7.5 ± 0.0 mm), while the least inhibitory effect of the Ag-NPs was observed on *Staphylococcus epidermidis* with an IZ of 7 ± 0.0 mm. The same concentration of aqueous extract (100 µg/mL) resulted in a lower inhibition zone diameter of 6 mm against all tested microbes compared with the IZ diameters produced with Ag-NPs. This suggested that the Ag-NPs have higher antibacterial activity than the lichen extract. The authors hypothesize that the increased susceptibility of Gram-negative bacteria to Ag-NPs compared with Gram-positive bacteria is likely due to the thinner peptidoglycan layer of Gram-negative bacteria, which provides the Ag-NPs with better anchoring and penetration of the cell wall.

Ag-NPs synthesized by Usnea longissima were evaluated for antimicrobial potency against six-Gram positive bacteria (*Staphylococcus aureus*, *Streptococcus mutans*, *Streptococcus pyogenes*, *Streptococcus viridans*, *Corynebacterium diphtheriae*, and *Corynebacterium xerosis*) and three Gram-negative bacteria (*Escherichia coli*, *Klebsiella pneumoniae*, and *Pseudomonas aeruginosa*) by agar well diffusion method [136]. The bacteria were incubated with Ag-NPs for 24 h at 37 °C. A negative control (DMSO) and positive controls (ciprofloxacin (5 μg/disk) for Gram-positive bacteria and Gentamicin (10 μg/disk) for Gram-negative strains) were used to compare the inhibitory activity of NPs. The Ag-NPs displayed the highest antibacterial efficiency against *E. coli* and *K. pneumoniae* with IZ diameters of 20.8 ± 0.02 and 16 ± 0.31 mm, respectively. In contrast, *S. mutans* (6.5 ± 0.89 mm), *C. diphtheriae* (6.2 ± 0.37 mm), and *P. aeruginosa* (7 ± 0.31 mm) were not affected by the Ag-NPs. The Ag-NPs were suggested to have a low antibacterial effect on the basis that the antibacterial effect can be amplified by reducing the NP size and hence increasing the surface area. As the surface area of the Ag-NPs increases, contact with microorganisms improves, which mediates penetration of the particles into the bacterial cell membrane or attachment to the bacterial surface. When silver ions reach the bacterial cytoplasm, they can denature the ribosome, thus directing the suppression of cell enzymes and proteins. Consequently, the metabolic function of the bacterial cell will be disrupted and the cell will undergo apoptosis. The authors reported that the lethal effect of Ag-NPs against bacteria can be achieved by different mechanisms including (i) interfering with cell wall, (ii) suppression of protein synthesis, and (iii) disruption of transcription and primary metabolic processes.

Kumar et al. studied the synergistic antibacterial effect of the extracts of two lichens, *Parmotrema pseudotinctorum* and *Ramalina hossei*, combined with chemically synthesized Ag-NPs, against several strains of Gram-positive and Gram-negative bacteria known to cause food poisoning [171]. The tested strains *Staphylococcus aureus*, *Bacillus cereus*, *Escherichia coli*, and *Salmonella typhi* were treated with the lichen extracts and Ag-NPs individually and with a combination of both, utilizing the agar well diffusion method. On Muller-Hinton agar plates, bacterial broth cultures (10^8 cells/mL) were swabbed then wells of 6-mm diameter were loaded as follows: lichen extracts (10 mg/mL in DMSO), Ag-NPs (1 mg/mL in DMSO), standard (chloramphenicol, 1 mg/mL), a combination of lichen extract and Ag-NPs (1:1 ratio), and control (DMSO). The plates (two replicates of each) were incubated at 37 °C for 24 h and the IZs were measured and the mean value was calculated for each sample. According to the IZs, the lichen extracts were more effective than the Ag-NPs alone on most plates. The Ag-NPs were more effective against Gram-negative bacteria than Gram-positive bacteria. However, the combination of the lichen extracts and Ag-NPs showed more bacterial inhibition than that of the extract alone or the NPs alone. After exposure to the combined treatment, *S. typhi* had an IZ of 2.8 cm, followed by *E. coli* (2.6 cm), *B. cereus* (2.1 cm), and *S. aureus* (1.9 cm). The enhanced antibacterial activity of the combined treatment might be attributed to the presence of effective secondary metabolites in the lichen extracts, and also the smaller-size-to-large-surface-area ratio of Ag-NPs. In the lichen extracts experiments, Gram-positive bacteria were more affected than Gram-negative bacteria. However, in the combined treatment assays, the antibacterial activity was more pronounced against Gram-negative bacteria. The authors attributed this to Gram-negative bacteria being naturally more resistant due to their thick outer membrane that prevents harmful substances from entering the cell. This barrier comprises an exterior lipopolysaccharide layer and a thin layer of peptidoglycan at the interior. TEM imaging confirmed that Ag-NPs can be effective bactericidal agents by rupturing the bacterial membrane even at low concentrations.

6.2. Antioxidants

The antioxidant activity of biomatrix loaded with Au-NPs synthesized by *Acroscyphus sphaerophoroides* and *Sticta nylanderiana* was screened by Debnath et al. using a modified diphenylpicrylhydrazyl (DPPH) method. Powdered samples of 2 and 5 mg were treated in two separate test tubes with 3 mL of 100 M methanolic solution of DPPH. The surface reaction for both mixtures was amplified by sonicating them in the dark. To confirm time-dependent DPPH scavenging, centrifugation was performed and the absorbance of the supernatants over time was measured at 517 nm with DPPH as a reference and a gap of 15, 30, 45, and 60 min. Measurement of the scavenging potential (SC_{50}) of biomatrix-loaded Au-NPs synthesized by *A. sphaerophoroides* and *S. nylanderiana* is achieved via a similar process, where absorbances are documented at 30 min after administering 1, 1.5, 2, 2.5, 3 mg and 1, 3, 5, 7, 10 mg of the samples, respectively. The concentrations of gold-NPs synthesized by *A. sphaerophoroides* and *S. nylanderiana* responsible for scavenging of 50% of DPPH (SC_{50}) were 1.66 and 4.48 mg, respectively, suggesting biogenic gold-NPs were potent antioxidant agent [159].

An extract of the lichen *Parmelia sulcata* was exploited for the biological formulation of Au-NPs [160], and the resulting Au-NPs and *P. sulcata* extract were tested for their free radical scavenging potential in antioxidant bioassays involving DPPH and hydrogen peroxide. For the DPPH method, 2.96 mL of 0.1 mM solution of DPPH was added to 0.4 mL of the extract or Au-NPs at different concentrations (250, 500, 750, and 1000 μg/mL) and incubated under dark conditions at ambient temperature for 30 min. The absorbance was recorded at 517 nm and used to calculate the percentage inhibition of scavenging potential. For the hydrogen peroxide scavenging test, 40 mM H_2O_2 solution was prepared in phosphate buffer at pH 7.4 and then several concentrations (250, 500, 750, 1000 μg/mL) of extracts and Au-NPs were added and incubated for 10 min at room temperature. The absorbance was measured at 230 nm and was subsequently used to determine the percentage of inhibition. The outcomes of these bioassays consolidated the ability of the lichen extract and Au-NPs to scavenge free radicals; the IC_{50} values of DPPH were 1020 and 815 μg/mL and the IC_{50} values of H_2O_2 were 694 and 510 μg/mL, respectively. These results indicated that the Au-NPs had greater potential for free radical scavenging (FRS) compared with the lichen extract. In addition, the FRS activity of both lichen extract and Au-NPs appears to be concentration dependent.

6.3. Other Applications

A recent study used ZnO-NPs biosynthesized by *Ramalina fraxinea* extract as a cytotoxic agent for human neuroblastoma cells [169]. The study focused on evaluating the neurotoxicity and neuroprotective effect of lichen-synthesized ZnO-NPs against SHSY-5Y human neuroblastoma cells. Several concentrations of ZnO-NPs were prepared to identify the cytotoxic doses of these NPs. A concentration of 25 μg/mL ZnO-NPs significantly increased the cell viability ($p < 0.05$) when compared with the control group. However, a lower concentration (5 μg/mL) of ZnO-NPs did not affect SHSY-5Y cells. ZnO-NPs at 50, 100, 200, and 400 μg/mL caused a marked reduction ($p < 0.001$) in cell viability, compared with those of the control group. To estimate the neuroprotective effect of ZnO-NPs, the authors exploited the ability of hydrogen peroxide to induce apoptosis of SHSY-5Y cells via oxidative stress; for this purpose, 300 μM H_2O_2 was used to treat the cells. This treatment resulted in a significant reduction ($p < 0.001$) of cell viability compared with the control group. ZnO-NPs (at all tested doses) did not increase H_2O_2-induced death of the SHSY-5Y cells. Moreover, the higher doses (100, 200, and 400 μg/mL) of ZnO-NPs markedly reduced ($p < 0.001$) the cell viability compared with the H_2O_2 group. In summary, ZnO-NPs at high doses ($\geq$50 μg/mL) can induce neurotoxicity in SHSY-5Y neuroblastoma cells but provide neuroprotection against the neurotoxic effect of hydrogen peroxide at a low to moderate doses (25 μg/mL).

Iron oxide-NPs fabricated by *Ramalina sinensis* were able to remove lead and cadmium (82 and 77%, respectively) from aqueous solution at an initial concentration of 50 mg/L and with pH in the range of 5–4, indicating the potential of these NPs to be heavy metal eliminators [163].

Çıplak et al. conducted the first study on the catalytic activity of biogenic monometallic NPs (Ag- and Au-NPs) and bimetallic NPs (Au-Ag NPs) synthesized by *Cetraria islandica* [144]. Bimetallic NPs showed higher catalytic activity than monometallic NPs for the reduction of nitrophenols (4-nitrophenol; 4-NP) to aminophenols (4-aminophenol; 4-AP) with sodium borohydride ($NaBH_4$). The higher catalytic performance of Au-Ag NPs might be attributed to the higher ionization potential of Au (9.22 eV) than Ag (7.58 eV), which causes electronic charge transfer from Ag to Au and results in an increase in the electron density on the NP surface. Similarly, Au-NPs exhibited better catalytic potentiality than the Ag-NPs.

The reducing power, hydrogen peroxide scavenging ability, and antidiabetic activities of Ag-NPs synthesized by both *Parmelia perlata* aqueous extract and their purified glycoside and alkaloid fractions were screened by Leela and Anchana [150]. Biogenic Ag-NPs generated from lichen fraction biomolecules have significant antidiabetic potential, reducing power, and free radical scavenging ability, compared with the Ag-NPs fabricated by lichen aqueous extract. The antidiabetic properties of the biogenic Ag-NPs were tested using an alpha-amylase inhibition assay and the percentage inhibition of alpha-amylase was 11.11% for Ag-NPs synthesized by lichen aqueous extract, and 51.85 and 29.62% for the Ag-NPs fabricated by the glycoside fraction and alkaloid fraction, respectively. This indicated that glycoside-mediated-Ag-NPs exhibited the strongest antidiabetic activity. The authors suggest that these biogenic Ag-NPs may lead to improvements in type 2 diabetic disease. Furthermore, the reducing activity of the same Ag-NPs was explored in a reducing power assay in which Ag-NPs interact with potassium ferricyanide (Fe^{3+}), leading to the generation of potassium ferricyanide (Fe^{2+}), which then reacts with ferric chloride to form a ferric-ferrous complex that is readily detected by UV spectrophotometer. Glycoside-mediated-Ag-NP had the greatest reducing activity among the three types of Ag-NPs (absorbance of 0.771, compared with 0.639 and 0.4 for Ag-NPs fabricated utilizing lichen aqueous extract and the alkaloid fraction, respectively). The hydrogen peroxide scavenging ability of the Ag-NPs was also examined. Glycoside-mediated-Ag-NP had the highest scavenging activity (28.89%), compared with Ag-NPs biofabricated by alkaloid fraction (21.86%) and lichen aqueous extract (7.21%).

Parmelia sulcata extract (PSE) and PSE-synthesized Au-NPs were investigated for their mosquitocidal activity against *Anopheles stephensi* and *Anopheles aegypti* mosquito larvae, pupae, adults, and egg hatching [160]. Varying concentrations of the lichen extract (75, 150, 225, 300, and 375 ppm) were tested and deemed toxic against larval instars I–IV and pupae of *A. stephensi* and *Anopheles aegypti*. The registered lethal concentration$_{50}$ (LC_{50}) values of instars of *A. stephensi* were: 172.16 ppm (I), 201.39 ppm (II), 219.04 ppm (III), 243.89 ppm (IV), and 288.03 ppm (pupae), and the ones for *Anopheles aegypti* were 281.71 ppm (I), 244.46 ppm (II), 283.90 ppm (III), 330.35 ppm (IV), and 346.99 ppm (pupae). The green-synthesized Au-NPs showed exceptionally high activity against larvae and pupae. At concentrations of 15, 30, 45, 60, and 75 ppm, the Au-NPs presented LC_{50} values of 29.82 ppm (I), 33.83 ppm (II), 37.55 ppm (III), 44.26 ppm (IV), and 50.44 ppm (pupae) for *A. stephensi*, and 34.49 ppm (I), 38.72 ppm (II), 44.72 ppm (III), 51.41 ppm (IV), and 59.00 ppm (pupae) for *A. aegypti*. For the adulticidal experiments, the PSE concentrations were 25, 50, 75, 100, and 125 ppm, while those of the Au-NPs were 10, 20, 30, 40, and 50 ppm. The LC_{50} and LC_{90} values of PSE and Au-NPs for *A. stephensi* were 59.35 and 132.80 ppm, and 22.43 and 49.02 ppm, respectively. For *A. aegypti*, the LC_{50} and LC_{90} values of PSE and Au-NPs were 70.16 and 149.66 ppm, and 24.55 and 52.74 ppm, respectively. Multiple concentrations of PSE and Au-NPs (60, 120, 180, 240, 300, 360, and 420 ppm) were tested for their ovicidal effects, and it was concluded that both *A. stephensi* and *A. aegypti* hinder complete egg hatchability at 360 and 240 ppm, respectively. The Au-NPs impose a high toxicity risk on *A. stephensi* and *A. aegypti*. Importantly, the synthesized Au-NPs worked at a far higher efficacy when compared with the PSE. These experiments proved that PSE-synthesized

Au-NPs have a markedly successful mosquitocidal effect against *Anopheles*. The study was concluded to provide evidential support for the use of PSE and Au-NPs as a solution towards mosquito-manifested environments (Figure 5).

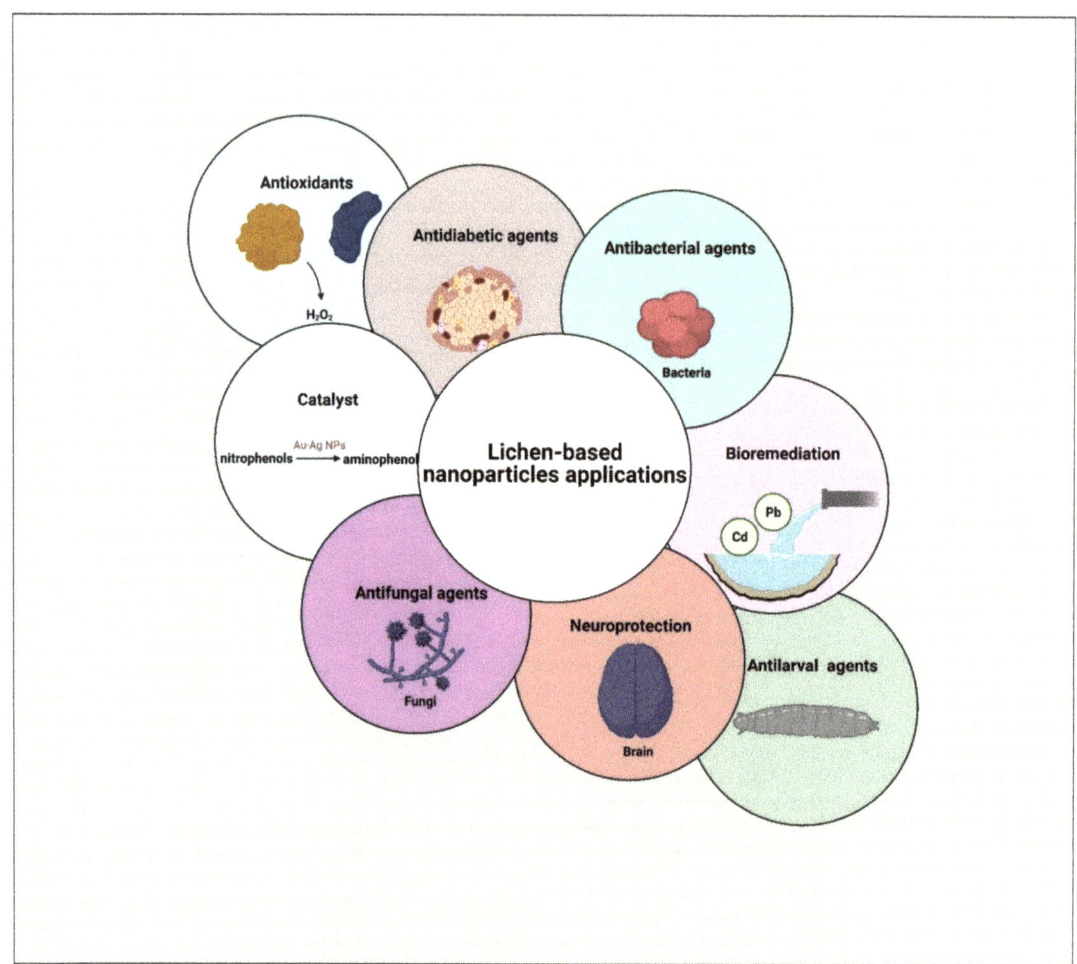

Figure 5. Application of lichen-based nanoparticles.

7. Analysis and Characterization of Nanoparticles

Physicochemical characterization analyses of NP samples are the initial and most significant step following the fabrication process of NPs. These analyses are required to confirm the synthesis of NPs and their unique properties such as increased surface area, stability, crystallinity, charge, dispersion, magnetic, thermal, and optical properties, and morphological features such as shape and size. The techniques utilized include spectroscopic analyses such as UV–visible spectroscopy, FTIR, zeta potential, dynamic light scattering, and nuclear magnetic resonance spectroscopy. These spectroscopic methods estimate the corresponding wavelength ranges of NPs, the functional groups surrounding NPs, and evaluate the charge and hydrodynamic diameter of NPs. X-ray-based analyses such as XRD, X-ray photoelectron spectroscopy (XPS), and energy-dispersive spectroscopy (EDAX or EDS) are performed to reveal the chemical composition and crystal structure and phase of the NPs. Microscopic analyses such as TEM, SEM, high-resolution TEM (HRTEM), and atomic force microscope (AFM) are used to demonstrate the morphological features of NPs [153].

8. Nanoparticle-Based Green Synthesis-Regulated Parameters: Clues to Enhance Their Activity

Different parameters should be used to optimize synthesis and obtain high efficiency, stabilized, and applicable NPs. These parameters include the type of synthesis method, temperature, pH, time of exposure, concentration of reductants and stabilizing agents, concentration of bulk materials, type of natural sources, illumination, and microorganism growth phase (Figure 6).

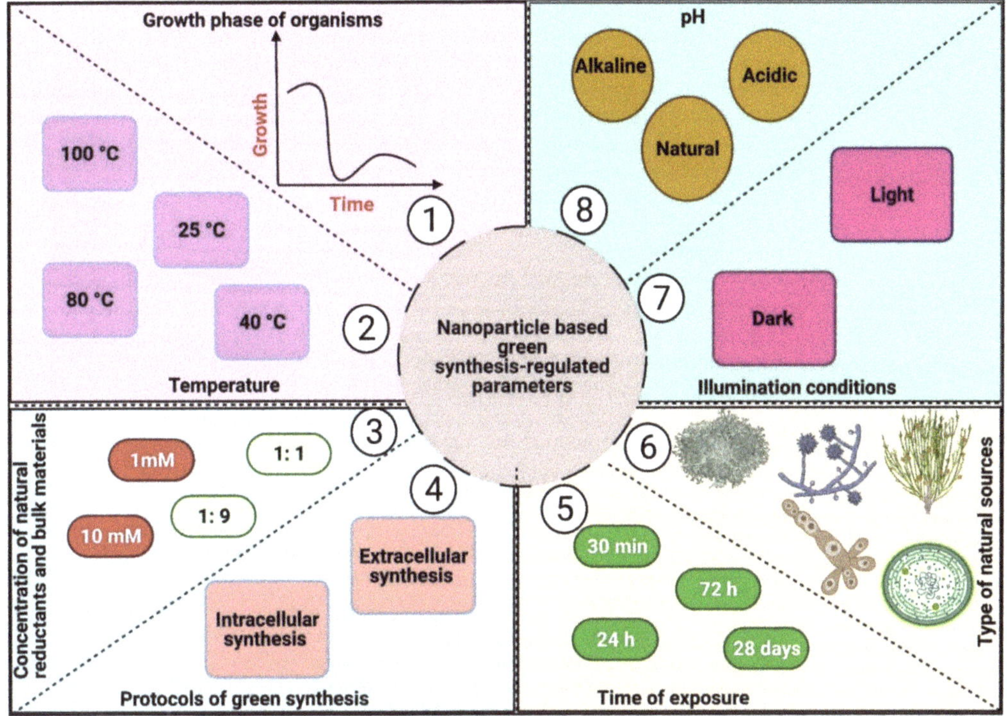

Figure 6. Nanoparticle-based, green synthesis-regulated parameters including (**1**) growth phase of organisms, (**2**) temperature, (**3**) concentrations of reductants and bulk materials, (**4**) protocols of green synthesis, (**5**) time of exposure, (**6**) type of natural sources, (**7**) illumination conditions and (**8**) pH.

8.1. Temperature

Temperature is an important factor in controlling the nature of NPs. Generally, biosynthesis processes using natural sources require temperatures ranging from room temperature to 100 °C [172]. Liu et al. synthesized Ag-NPs using *Cinnamomum Camphor* leaf extract at different temperatures including 70, 75, 80, 85, and 90 °C [173]. The temperature had impressively different effects on the nucleation kinetics constant k1 and growth kinetics constant k2 resulting in the generation of NPs with different sizes.

8.2. pH

pH is another significant parameter mitigating the green synthesis of NPs. pH influences the size and texture of NPs [174]. Mohammadi et al. synthesized zinc oxide-NPs at different pH (4, 6, 7, 8, and 10) using cherry extract and found that the optimum pH for the fabrication of hexagonal, small NPs was pH 8 (alkaline medium) [175].

8.3. Time of Exposure

Wei et al. studied the effect of reaction time (0, 1, 2, 3, and 4 h) on the yield of Ag-NPs using berry extract of sea buckthorn [63]. As the reaction time increased, the absorption intensity of NPs increased steadily, reaching a maximum at 4 h, which was indicative of a high concentration of Ag-NPs. The UV spectra also showed a slight blue shift from 415 to 413 nm with the increase in time of exposure, indicating the formation of smaller-sized NPs.

8.4. Concentration of Natural Reductants, Stabilizing Agents, and Bulk Materials

The concentration of biological entities and the salts used for the synthesis of NPs influence the size and shape of the NPs. Hamouda et al. revealed that a surge in the amount of *Oscillatoria limnetica* extract during synthesis of Ag-NPs shifted the UV-spectra peak of the Ag-NPs from 420 to 430 nm, which reflected an increase in the size of the NPs [176]. Similarly, Vellora et al. synthesized copper oxide NPs using different concentrations (1, 2, and 3 mM) of copper chloride ($CuCl_2 \cdot H_2O$) and a constant concentration of Gum karya (10 mg/mL) with incubation at 75 °C and 250 rpm for 1 h in an orbital shaker [177]. The increase in the concentration of precursors promoted the generation of NPs of increasing sizes; the nanosizes were 4.8, 5.5, and 7.8 nm, respectively.

8.5. Illumination

Light energy is critical for accelerating the reaction rate of NP fabrication. The illumination factor affects the intracellular and extracellular synthesis of NPs using photosynthetic organisms [178]. In contrast, some organisms do not need a light source to synthesize NPs. This phenomenon can be attributed to the fact that some organisms secrete different biomolecules capable of NP fabrication, only some of which need light activation [178]. Light intensity may be another important factor controlling the stability and production of NPs [179]. Recently, Ag-NPs were completely fabricated after 5 min using *Azadirachta indica* leaf extract under sunlight [180].

8.6. Protocol of Green Synthesis Method

The type of method(s) used for synthesizing NPs is a crucial parameter controlling the physicochemical and biological properties of NPs. Molnár et al. studied three different methods—extracellular fraction, autolysate of the fungal cells, and intracellular fractions—for synthesizing Au-NPs using 29 thermophilic filamentous fungi [65]. NPs were formed using all three methods; however, the extracellular methods were the most acceptable, yielding NPs with a smaller size and low polydispersity. The authors also recommended washing the fungal mycelia several times before extracellular biofabrication to avoid the influence of residual growth media components on the NP synthesis process.

8.7. Type of Natural Sources

The nature of organisms used in NP fabrication processes significantly influences the nature of the resulting NPs. Biomolecules such as pigments, proteins, polysaccharides, etc. vary between different organisms and in strains of the same species, and this leads to variations in an organism's potentiality to produce NPs [30]. Recently, Ag-NPs were produced using three different strains (*Nostoc* sp. Bahar M [58], *Nostoc* HKAR-2,98 [181], and *Nostoc muscorum* NCCU-442,56 [60]). These strains produced Ag-NPs with a range of different sizes including 8.5–26.44 mm, 51–100 mm, and 42 nm, respectively.

8.8. Growth Phase of Organisms Used for NP Fabrication

The effect of the growth phase on the fabrication process of NPs was studied by Sweeney et al. [182]. Cadmium sulfide nanocrystal production varies dramatically depending on the growth phase of *E. coli*. The formation of NPs increased approximately 20-fold in the stationary phase of *E. coli*, compared with that grown in the late logarithmic phase.

9. The Mechanism of Biological Synthesis of NPs

Different speculations about the mechanism of NP synthesis using living organisms were reported, but until now, the exact mechanism remained unclear. However, each organism was found to have its own synthesis mechanism against different metals [4,183]. One hypothesis is the ability of living organisms to synthesize NPs occurs via two general steps—(i) metal ions are trapped on the surface of an organism and/or inside their cells and (ii) these ions are reduced to NPs aided by biomolecules such as enzymes, proteins, pigments, or polysaccharides, or by the union effect of different molecules [4,67,183,184]. These biological molecules are responsible for the electron shuttle reduction and stabilization of NPs. Sneha et al. reported that metal ions, particularly Au or Ag ions, are captured on the fungal cell surface through the electrostatic force between the metal ions and cell wall, which carries a negative charge from the enzyme carboxylate groups. The enzymes then fabricate the metals into Au or Ag nuclei, which sequentially grow by reduction and accumulation [185]. Kalishwaralal et al. reported that nitrate reductase produced by *Bacillus licheniformis* facilitates the bioreduction of Ag-NPs. The nitrate ions of silver nitrate were found to activate the nitrate reductase enzyme, resulting in reducing silver ions to metallic silver via an electron shuttle enzymatic metal reduction process [186]. During the biosynthesis of MNPs, NADH and NADH-dependent nitrate reductase enzymes (especially nitrate reductase) are essential factors [187]. These enzymes are known to be secreted from *B. licheniformis* and may be linked to the biofabrication of Ag^{+} to Ag^{0} and subsequent synthesis of Ag-NPs. Divya et al. reported that the existence of NADH and NADH-based reductases in the supernatant of *Alcaligenes faecalis* was responsible for the reduction of silver nitrate into Ag-NPs [188]. Hamedi et al. studied the synthesis of Ag-NPs using *Fusarium oxysporum66* cell-free culture filtrates [189]. They reported that a surge in C:N ratio resulted in the enhancement of the nitrate reductase activity, causing an increase in the Ag-NPs fabrication process rate. Furthermore, they obtained small Ag-NPs with a narrow size distribution.

Exopolysaccharides (EPSs) are predominantly composed of carbohydrates (such as D-glucose and D-mannose) and noncarbohydrate components (such as carboxyl, phosphate, and sulfate), which characterize them with anionic properties. These organic molecules enhance the lipophilicity of EPSs and directly influence their interaction with polysaccharides and cations. It was found that EPSs chelate metal ions. Then, sugar molecules of EPSs reduce metal ions into NPs to be capping with different functional groups of EPSs [190–192].

Kang et al. synthesized Ag-NPs using EPSs in an *E. coli* biofilm. They reported that EPSs aldehyde and hemiacetal groups of rhamnose sugars were responsible for the reduction and stabilization of NPs [193].

Moreover, the synthesis of heavy MNPs can be due to the genetic and proteomic responses of a metallophilic microorganism to toxic environments [194]. Heavy metal ions such as mercury, cadmium, zinc, and copper ions are a hazard to microbial survival and consequently, microorganisms have developed genetic and proteomic responses to tackle these threats [195]. Microorganisms contain many gene clusters of metal resistance that enable cell detoxification via different mechanisms, such as complexation, efflux, or reductive precipitation [196]. Recently, a mechanism for the synthesis of magnetites using *Shewanella oneidensis* was suggested and comprised both passive and active processes [197]. Initially, Fe^{2+} is actively produced when bacteria use ferrihydrite as a terminal electron acceptor, accompanied by elevation in the pH round the cells due to the bacterial metabolism of amino acids. The passive mechanism then uses the localized concentration of Fe^{2+} and Fe^{3+} at the net negatively charged cell wall, cell structures, and/or cell debris, which enhances a local rise of supersaturation of the system with respect to magnetite, resulting in precipitation of the magnetite phase.

Another hypothesis discussed the role of c-type cytochromes redox proteins for electron transfer during the reduction of metals. Ng et al. synthesized Ag-NPs using a mutant strain of *Shewanella oneidensis* missing cytochrome genes (MtrC and OmcA) and a wild-type strain of *S. oneidensis*. They found that c-type cytochromes aid in electrons transfers to metal ions outside the cells. Similarly, Liu et al. reported that c-type cytochrome protein complexes (ombB, omaB, and omcB) in the outer membrane of metal-reducing bacterium Geobacter sulfurreducens PCA was responsible for the extracellular reduction of Fe (III)-citrate and ferrihydrite (Figure 7).

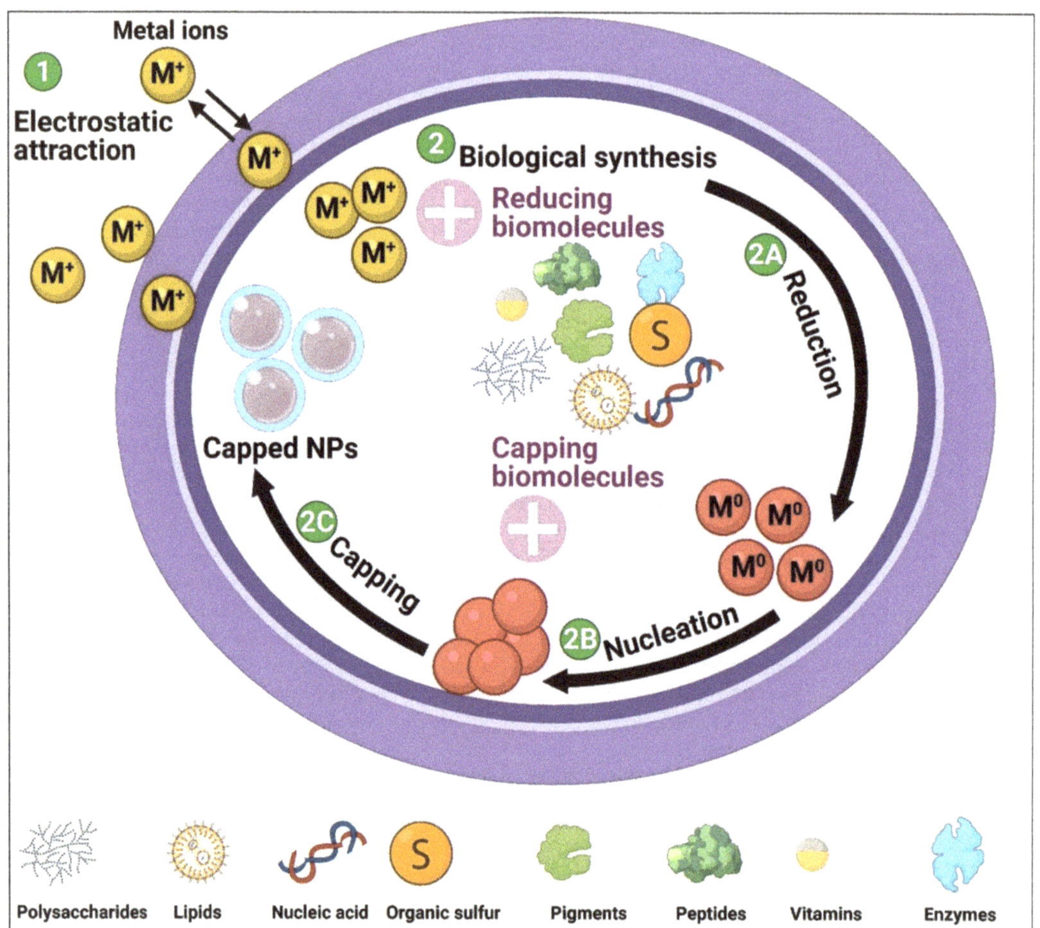

Figure 7. The potential mechanism of biological synthesis of NPs.

10. Toxicity of Nanoparticles

Although nanotechnology is rapidly growing with a wide range of applications in different areas such as industry, agriculture, medicine, biotechnology, etc., there remain many barriers with this technology such as the toxicological effects of NPs on ecology and living organisms. Many in vitro and in vivo investigations revealed that metallic and non-metallic NPs have serious side effects on human health and the environment. Senut et al. reported that mercaptosuccinic acid-capped Au-NPs (1.5 nm) at a concentration of 0.1 μg/mL enhanced cell death of human embryonic stem cells (ESCs). However, other Au-NPs (4 and 14 nm) at the same concentration showed almost no toxic effect on ESCs [198]. Chen et al. studied the relation between the toxicity of Ag-NPs and their size against fresh red blood cells. The scholars used three different characteristic sizes (15, 50, and 100 nm) of Ag-NPs [199]. They reported that smaller sizes of Ag-NPs enhancing the hemolysis and membrane damage of blood cells, compared with that of other sizes. Wan et al. investigated the genotoxicity of the chemically synthesized cobalt NPs in vivo by utilizing guanine phosphoribosyltransferase delta transgenic mice [200]. The authors reported that cobalt NPs induced oxidative stress, lung inflammation and injury, DNA damage, and mutation.

The toxic effects of TiO_2 NPs against human cells, vertebrates, and invertebrate animals might be attributed to their potency to form free radicals with water in the presence of sunlight. TiO_2 NPs caused DNA damage with or without light and induced the cell death pathway in hamster fibroblasts with stretched micronuclei [201,202]. Although biological synthesis processes have become a simple, eco-friendly, low-cost alternative to traditional methods of NPs fabrication, there are few studies in the literature discussing their toxicity on humans and the environment. Some investigations reported the biocompatibility of green NPs, compared with that synthesized by chemical and physical methods. This could be assumed to the synergetic effect between synthesized NPs and their biological coats [203]. Devasena et al. reported that Mg-NPs synthesized by *Cladonia rangiferina* extract exhibited a better antimicrobial activity and lower toxicity [161]. Khorrami et al. reported that Ag-NPs synthesized by walnut green husk (as reducing and capping agents) showed high selectivity toward breast cancer cells (MCF-7) than normal cells (L-929) [203]. However, the chemically synthesized Ag-NPs showed high toxicity against L-929 cells, compared with biologically synthesized Ag-NPs. Ag-NPs formed by biological synthesis using *Lycium chinense* fruit extract showing low cytotoxicity against normal murine macrophage RAW264.7 cells [204]. On the other hand, Krishnaraj et al. exhibited that Ag-NPs synthesized by *Malva crispa* plant caused morphological alterations in adult zebrafish gills and reduced the biological connection and the homogenous distribution of their liver parenchyma cells [205]. To fill the gap in knowledge regarding this association, many additional in vitro and in vivo investigations are required to test the toxicity of green NPs against normal cells, explore the biocompatibility of biogenic NPs, and determine the precise lethal mechanism of NPs against living cells to increase the potentiality of using these NPs as FDA-approved drugs.

11. Future Prospects and Conclusions

The current review discussed the biological synthesis methods in depth with emphasis on the lichen-mediated synthesis of NPs. The biological synthesis of NPs has recently become an increasingly active area of research. Through exhibiting several advantageous qualities and numerous potential applications, biological synthesis methods of NPs have proven to be superior to the traditional chemical and physical methods. These qualities include being cost effective, eco-friendly, and vastly applicable in biomedical fields due to biocompatibility. Lichen species are seldom considered as biomachinery for the biofabrication of NPs. In this review, we have highlighted the potential of these organisms as natural biofactories for NP formation. The symbiosis between fungi and cyanobacteria or algae or sometimes plants makes lichen a promising alternative biomachinery for NPs fabrication. Due to the variation in their biomolecule contents and structures, which are responsible for reducing metal ions into NPs. Devasena et al. reported that lichen-mediated synthesis NPs are distinguished from other alternative biological methods by being lesser toxic and needing low-processing conditions [161]. Lichen species have a reducible activity to fabricate different types of NPs, including gold (Au)-NPs, silver (Ag)-NPs, metal oxide-NPs such as iron oxide- and zinc oxide-NPs, and other nanomaterials such as bimetallic alloys (Au-Ag NPs) and nanocomposites such as $ZnO@TiO_2@SiO_2$ and $Fe_3O_4@SiO_2$. These biogenic NPs have significant antimicrobial activities against both Gram-positive and Gram-negative bacteria, and fungi, and they also display mosquitocidal activity. Additionally, these NPs act as potential catalytic materials, bioremediatory agents for heavy metals, antidiabetics, antioxidants, and neuroprotection agents against neurotoxin.

Extending the utilization of lichen-mediated green synthesis methods and exploring the optimum conditions of these processes to fabricate applicable, bioactive, scalable, and biocompatible nanoproducts may lead to the development of novel green NPs with unique physicochemical and biological features that can be applied in different sectors, including agriculture, industry, medicine, biotechnology, and pharmaceutics. Moreover, there remain many barriers against the biological synthesis process, including toxicity and agglomeration, polydispersity, stability, and the nonuniform size of NPs. These issues can be solved by increasing the optimization studies for green synthesis of NPs to obtain the desirable NPs. Additionally, exploring the synthesis mechanism of NPs using natural sources will facilitate the development and launch of nanodrugs in different fields.

Author Contributions: Conceptualization, R.S.H.; methodology, R.S.H., L.B. and M.A.A.; software, R.S.H. and M.A.A.; investigation, R.S.H. and L.B.; data curation, R.S.H., L.B. and M.A.A.; writing—original draft preparation, R.S.H. and L.B.; writing—review and editing, R.S.H. and L.B.; visualization, R.S.H.; supervision, N.E.A., M.M.B.-M. and M.I.A.-Z.; project administration, R.S.H.; funding acquisition, M.M.B.-M. and M.I.A.-Z. All authors have read and agreed to the published version of the manuscript.

Funding: This research received no external funding.

Institutional Review Board Statement: Not applicable.

Informed Consent Statement: Not applicable.

Data Availability Statement: The data supporting this article are shown in Figures 1–7 and one table. The data sets analyzed in the present study are available from the corresponding author upon reasonable request.

Acknowledgments: This research was funded by the Deanship of Scientific Research at Princess Nourah bint Abdulrahman University through the Fast-track Research Funding Program.

Conflicts of Interest: The authors declare no conflict of interest.

References

1. Saratale, R.G.; Karuppusamy, I.; Saratale, G.D.; Pugazhendhi, A.; Kumar, G.; Park, Y.; Ghodake, G.S.; Bharagava, R.N.; Banu, J.R.; Shin, H.S. A comprehensive review on green nanomaterials using biological systems: Recent perception and their future applications. *Colloids Surf. B Biointerfaces* **2018**, *170*, 20–35. [CrossRef]
2. Saratale, R.G.; Saratale, G.D.; Shin, H.S.; Jacob, J.M.; Pugazhendhi, A.; Bhaisare, M.; Kumar, G. New insights on the green synthesis of metallic nanoparticles using plant and waste biomaterials: Current knowledge, their agricultural and environmental applications. *Environ. Sci. Pollut. Res.* **2018**, *25*, 10164–10183. [CrossRef]
3. Gahlawat, G.; Choudhury, A.R. A review on the biosynthesis of metal and metal salt nanoparticles by microbes. *RSC Adv.* **2019**, *9*, 12944–12967. [CrossRef]
4. Hamida, R.S.; Ali, M.A.; Redhwan, A.; Bin-Meferij, M.M. Cyanobacteria—A Promising Platform in Green Nanotechnology: A Review on Nanoparticles Fabrication and Their Prospective Applications. *Int. J. Nanomed.* **2020**, *15*, 6033–6066. [CrossRef]
5. Mitchell, M.J.; Billingsley, M.M.; Haley, R.M.; Wechsler, M.E.; Peppas, N.A.; Langer, R. Engineering precision nanoparticles for drug delivery. *Nat. Rev. Drug Discov.* **2020**, *20*, 1–24.
6. Narayan, N.; Meiyazhagan, A.; Vajtai, R. Metal nanoparticles as green catalysts. *Materials* **2019**, *12*, 3602. [CrossRef] [PubMed]
7. Bajpai, V.K.; Shukla, S.; Kang, S.-M.; Hwang, S.K.; Song, X.; Huh, Y.S.; Han, Y.-K. Developments of cyanobacteria for nano-marine drugs: Relevance of nanoformulations in cancer therapies. *Mar. Drugs* **2018**, *16*, 179. [CrossRef]
8. Hamida, R.S.; Albasher, G.; Bin-Meferij, M.M. Oxidative Stress and Apoptotic Responses Elicited by Nostoc-Synthesized Silver Nanoparticles against Different Cancer Cell Lines. *Cancers* **2020**, *12*, 2099. [CrossRef] [PubMed]
9. Liu, R.; Hu, C.; Yang, Y.; Zhang, J.; Gao, H. Theranostic nanoparticles with tumor-specific enzyme-triggered size reduction and drug release to perform photothermal therapy for breast cancer treatment. *Acta Pharm. Sin. B* **2019**, *9*, 410–420. [CrossRef] [PubMed]
10. Agasti, S.S.; Rana, S.; Park, M.-H.; Kim, C.K.; You, C.-C.; Rotello, V.M. Nanoparticles for detection and diagnosis. *Adv. Drug Deliv. Rev.* **2010**, *62*, 316–328. [CrossRef]
11. Yetisgin, A.A.; Cetinel, S.; Zuvin, M.; Kosar, A.; Kutlu, O. Therapeutic nanoparticles and their targeted delivery applications. *Molecules* **2020**, *25*, 2193. [CrossRef] [PubMed]

12. Mroz, P.; Pawlak, A.; Satti, M.; Lee, H.; Wharton, T.; Gali, H.; Sarna, T.; Hamblin, M.R. Functionalized fullerenes mediate photodynamic killing of cancer cells: Type I versus Type II photochemical mechanism. *Free Radic. Biol. Med.* **2007**, *43*, 711–719. [CrossRef] [PubMed]
13. Tegos, G.P.; Demidova, T.N.; Arcila-Lopez, D.; Lee, H.; Wharton, T.; Gali, H.; Hamblin, M.R. Cationic fullerenes are effective and selective antimicrobial photosensitizers. *Chem. Biol.* **2005**, *12*, 1127–1135. [CrossRef] [PubMed]
14. Dong, X.-Y.; Gao, Z.-W.; Yang, K.-F.; Zhang, W.-Q.; Xu, L.-W. Nanosilver as a new generation of silver catalysts in organic transformations for efficient synthesis of fine chemicals. *Catal. Sci. Technol.* **2015**, *5*, 2554–2574. [CrossRef]
15. Wu, M.; Huang, S. Magnetic nanoparticles in cancer diagnosis, drug delivery and treatment. *Mol. Clin. Oncol.* **2017**, *7*, 738–746. [CrossRef]
16. Zeng, X.; Sun, J.; Li, S.; Shi, J.; Gao, H.; Leong, W.S.; Wu, Y.; Li, M.; Liu, C.; Li, P. Blood-triggered generation of platinum nanoparticle functions as an anti-cancer agent. *Nat. Commun.* **2020**, *11*, 1–12. [CrossRef]
17. Sonkusre, P. Specificity of biogenic selenium nanoparticles for prostate cancer therapy with reduced risk of toxicity: An in vitro and in vivo study. *Front. Oncol.* **2020**, *9*, 1541. [CrossRef]
18. Phan, T.T.V.; Huynh, T.-C.; Manivasagan, P.; Mondal, S.; Oh, J. An up-to-date review on biomedical applications of palladium nanoparticles. *Nanomaterials* **2020**, *10*, 66. [CrossRef]
19. Tiwari, V.; Mishra, N.; Gadani, K.; Solanki, P.S.; Shah, N.; Tiwari, M. Mechanism of anti-bacterial activity of zinc oxide nanoparticle against carbapenem-resistant Acinetobacter baumannii. *Front. Microbiol.* **2018**, *9*, 1218. [CrossRef]
20. Muñoz-Escobar, A.; Reyes-López, S.Y. Antifungal susceptibility of Candida species to copper oxide nanoparticles on polycaprolactone fibers (PCL-CuONPs). *PLoS ONE* **2020**, *15*, e0228864. [CrossRef]
21. Akhtar, S.; Shahzad, K.; Mushtaq, S.; Ali, I.; Rafe, M.H.; Fazal-ul-Karim, S.M. Antibacterial and antiviral potential of colloidal Titanium dioxide (TiO2) nanoparticles suitable for biological applications. *Mater. Res. Express* **2019**, *6*, 105409. [CrossRef]
22. Azizi-Lalabadi, M.; Ehsani, A.; Divband, B.; Alizadeh-Sani, M. Antimicrobial activity of Titanium dioxide and Zinc oxide nanoparticles supported in 4A zeolite and evaluation the morphological characteristic. *Sci. Rep.* **2019**, *9*, 1–10.
23. Kumbhakar, P.; Ray, S.S.; Stepanov, A.L. Optical properties of nanoparticles and nanocomposites. *J. Nanomater.* **2014**, *2014*, 1–2. [CrossRef]
24. Matsui, I. Nanoparticles for electronic device applications: A brief review. *J. Chem. Eng. Jpn.* **2005**, *38*, 535–546. [CrossRef]
25. Wei, W.; Wang, H.; Wang, C.; Luo, H. Advanced Nanomaterials and Nanotechnologies for Solar Energy. *Int. J. Photoenergy* **2019**. [CrossRef]
26. Kiani, M.; Ansari, M.; Arshadi, A.A.; Houshfar, E.; Ashjaee, M. Hybrid thermal management of lithium-ion batteries using nanofluid, metal foam, and phase change material: An integrated numerical-experimental approach. *J. Therm. Anal. Calorim.* **2020**, *141*, 1–13. [CrossRef]
27. Vines, J.B.; Yoon, J.-H.; Ryu, N.-E.; Lim, D.-J.; Park, H. Gold nanoparticles for photothermal cancer therapy. *Front. Chem.* **2019**, *7*, 167. [CrossRef]
28. Hamida, R.S.; Abdelmeguid, N.E.; Ali, M.A.; Bin-Meferij, M.M.; Khalil, M.I. Synthesis of silver nanoparticles using a novel cyanobacteria Desertifilum sp. extract: Their antibacterial and cytotoxicity effects. *Int. J. Nanomed.* **2020**, *15*, 49. [CrossRef] [PubMed]
29. Khan, I.; Saeed, K.; Khan, I. Nanoparticles: Properties, applications and toxicities. *Arab. J. Chem.* **2019**, *12*, 908–931. [CrossRef]
30. Asmathunisha, N.; Kathiresan, K. A review on biosynthesis of nanoparticles by marine organisms. *Colloids Surf. B Biointerfaces* **2013**, *103*, 283–287. [CrossRef] [PubMed]
31. Gour, A.; Jain, N.K. Advances in green synthesis of nanoparticles. *Artif. Cells Nanomed. Biotechnol.* **2019**, *47*, 844–851. [CrossRef]
32. Saratale, R.G.; Saratale, G.D.; Ghodake, G.; Cho, S.-K.; Kadam, A.; Kumar, G.; Jeon, B.-H.; Pant, D.; Bhatnagar, A.; Shin, H.S. Wheat straw extracted lignin in silver nanoparticles synthesis: Expanding its prophecy towards antineoplastic potency and hydrogen peroxide sensing ability. *Int. J. Biol. Macromol.* **2019**, *128*, 391–400. [CrossRef]
33. Noruzi, M. Biosynthesis of gold nanoparticles using plant extracts. *Bioprocess Biosyst. Eng.* **2015**, *38*, 1–14. [CrossRef]
34. Yuan, X.; Xiao, S.; Taylor, T.N. Lichen-like symbiosis 600 million years ago. *Science* **2005**, *308*, 1017–1020. [CrossRef] [PubMed]
35. Müller, K. Pharmaceutically relevant metabolites from lichens. *Appl. Microbiol. Biotechnol.* **2001**, *56*, 9–16. [CrossRef] [PubMed]
36. Rattan, R.; Shukla, S.; Sharma, B.; Bhat, M. A Mini-Review on Lichen-Based Nanoparticles and Their Applications as Antimicrobial Agents. *Front. Microbiol.* **2021**, *12*, 336.
37. Alqahtani, M.A.; Al Othman, M.R.; Mohammed, A.E. Bio fabrication of silver nanoparticles with antibacterial and cytotoxic abilities using lichens. *Sci. Rep.* **2020**, *10*, 1–17. [CrossRef] [PubMed]
38. Ahmad, S.; Munir, S.; Zeb, N.; Ullah, A.; Khan, B.; Ali, J.; Bilal, M.; Omer, M.; Alamzeb, M.; Salman, S.M. Green nanotechnology: A review on green synthesis of silver nanoparticles—An ecofriendly approach. *Int. J. Nanomed.* **2019**, *14*, 5087. [CrossRef]
39. Jeevanandam, J.; Barhoum, A.; Chan, Y.S.; Dufresne, A.; Danquah, M.K. Review on nanoparticles and nanostructured materials: History, sources, toxicity and regulations. *Beilstein J. Nanotechnol.* **2018**, *9*, 1050–1074. [CrossRef]
40. Baker, S.; Harini, B.; Rakshith, D.; Satish, S. Marine microbes: Invisible nanofactories. *J. Pharm. Res.* **2013**, *6*, 383–388. [CrossRef]
41. Ijaz, I.; Gilani, E.; Nazir, A.; Bukhari, A. Detail review on chemical, physical and green synthesis, classification, characterizations and applications of nanoparticles. *Green Chem. Lett. Rev.* **2020**, *13*, 223–245. [CrossRef]
42. Murr, L.E. Classifications and Structures of Nanomaterials. In *Handbook of Materials Structures, Properties, Processing and Performance*; Springer International Publishing: Cham, Switzerland, 2015; pp. 719–746.

43. Issa, B.; Obaidat, I.M.; Albiss, B.A.; Haik, Y. Magnetic nanoparticles: Surface effects and properties related to biomedicine applications. *Int. J. Mol. Sci.* **2013**, *14*, 21266–21305. [CrossRef] [PubMed]
44. Iqbal, P.; Preece, J.A.; Mendes, P.M. Nanotechnology: The "Top-Down" and "Bottom-Up" Approaches. In *Supramolecular Chemistry: From Molecules to Nanomaterials*; Steed, J.W., Gale, P.A., Eds.; John Wiley & Sons Ltd.: Chichester, UK, 2012; Volume 8, pp. 3589–3602.
45. Nakata, Y.; Mukai, K.; Sugawara, M.; Ohtsubo, K.; Ishikawa, H.; Yokoyama, N. Molecular beam epitaxial growth of InAs self-assembled quantum dots with light-emission at 1.3 μm. *J. Cryst. Growth* **2000**, *208*, 93–99. [CrossRef]
46. Bertino, M.F.; Gadipalli, R.R.; Martin, L.A.; Rich, L.E.; Yamilov, A.; Heckman, B.R.; Leventis, N.; Guha, S.; Katsoudas, J.; Divan, R. Quantum dots by ultraviolet and x-ray lithography. *Nanotechnology* **2007**, *18*, 315603. [CrossRef]
47. Patiño-Carachure, C.; Martínez-Vargas, S.; Flores-Chan, J.; Rosas, G. Synthesis of carbon nanostructures by graphite deformation during mechanical milling in air. *Fuller. Nanotub. Carbon Nanostruct.* **2020**, *28*, 869–876. [CrossRef]
48. Salavati-Niasari, M.; Javidi, J.; Dadkhah, M. Ball milling synthesis of silica nanoparticle from rice husk ash for drug delivery application. *Comb. Chem. High Throughput Screen.* **2013**, *16*, 458–462. [CrossRef] [PubMed]
49. Başoğlu, A.; Ocak, Ü.; Gümrükçüoğlu, A. Synthesis of Microwave-Assisted Fluorescence Carbon Quantum Dots Using Roasted–Chickpeas and its Applications for Sensitive and Selective Detection of Fe^{3+} Ions. *J. Fluoresc.* **2020**, *30*, 515–526. [CrossRef] [PubMed]
50. Salem, S.S.; Fouda, A. Green synthesis of metallic nanoparticles and their prosective biotechnological applications: An overview. *Biol. Trace Elem. Res.* **2020**, *6*, 1–27.
51. Zhang, X.-F.; Liu, Z.-G.; Shen, W.; Gurunathan, S. Silver nanoparticles: Synthesis, characterization, properties, applications, and therapeutic approaches. *Int. J. Mol. Sci.* **2016**, *17*, 1534. [CrossRef]
52. Patel, K.; Bharatiya, B.; Mukherjee, T.; Soni, T.; Shukla, A.; Suhagia, B. Role of stabilizing agents in the formation of stable silver nanoparticles in aqueous solution: Characterization and stability study. *J. Dispers. Sci. Technol.* **2017**, *38*, 626–631. [CrossRef]
53. Mushtaq, K.; Saeed, M.; Gul, W.; Munir, M.; Firdous, A.; Yousaf, T.; Khan, K.; Sarwar, H.M.R.; Riaz, M.A.; Zahid, S. Synthesis and characterization of TiO_2 via sol-gel method for efficient photocatalytic degradation of antibiotic ofloxacin. *Inorg. Nano-Met. Chem.* **2020**, *50*, 580–586. [CrossRef]
54. Quintero-Quiroz, C.; Acevedo, N.; Zapata-Giraldo, J.; Botero, L.E.; Quintero, J.; Zárate-Triviño, D.; Saldarriaga, J.; Pérez, V.Z. Optimization of silver nanoparticle synthesis by chemical reduction and evaluation of its antimicrobial and toxic activity. *Biomater. Res.* **2019**, *23*, 1–15. [CrossRef] [PubMed]
55. Yu, Q.; Wang, P.; Hu, S.; Hui, J.; Zhuang, J.; Wang, X. Hydrothermal synthesis of hollow silica spheres under acidic conditions. *Langmuir* **2011**, *27*, 7185–7191. [CrossRef]
56. Moraes, D.A.; Junior, J.B.S.; Ferreira, F.F.; Mogili, N.V.V.; Varanda, L.C. Gold nanowire growth through stacking fault mechanism by oleylamine-mediated synthesis. *Nanoscale* **2020**, *12*, 13316–13329. [CrossRef] [PubMed]
57. Khanna, P.; Kaur, A.; Goyal, D. Algae-based metallic nanoparticles: Synthesis, characterization and applications. *J. Microbiol. Methods* **2019**, *163*, 105656. [CrossRef]
58. Bin-Meferij, M.M.; Hamida, R.S. Biofabrication and antitumor activity of silver nanoparticles utilizing novel nostoc sp. Bahar M. *Int. J. Nanomed.* **2019**, *14*, 9019. [CrossRef]
59. Mata, Y.; Torres, E.; Blazquez, M.; Ballester, A.; González, F.; Munoz, J. Gold (III) biosorption and bioreduction with the brown alga Fucus vesiculosus. *J. Hazard. Mater.* **2009**, *166*, 612–618. [CrossRef]
60. Husain, S.; Sardar, M.; Fatma, T. Screening of cyanobacterial extracts for synthesis of silver nanoparticles. *World J. Microbiol. Biotechnol.* **2015**, *31*, 1279–1283. [CrossRef] [PubMed]
61. Hassan, S.E.-D.; Fouda, A.; Radwan, A.A.; Salem, S.S.; Barghoth, M.G.; Awad, M.A.; Abdo, A.M.; El-Gamal, M.S. Endophytic actinomycetes Streptomyces spp mediated biosynthesis of copper oxide nanoparticles as a promising tool for biotechnological applications. *JBIC J. Biol. Inorg. Chem.* **2019**, *24*, 377–393. [CrossRef]
62. Singaravelu, G.; Arockiamary, J.; Kumar, V.G.; Govindaraju, K. A novel extracellular synthesis of monodisperse gold nanoparticles using marine alga, Sargassum wightii Greville. *Colloids Surf. B Biointerfaces* **2007**, *57*, 97–101. [CrossRef]
63. Wei, S.; Wang, Y.; Tang, Z.; Hu, J.; Su, R.; Lin, J.; Zhou, T.; Guo, H.; Wang, N.; Xu, R. A size-controlled green synthesis of silver nanoparticles by using the berry extract of Sea Buckthorn and their biological activities. *New J. Chem.* **2020**, *44*, 9304–9312. [CrossRef]
64. Abdel-Gawad, E.I.; Hassan, A.I.; Awwad, S.A. Efficiency of calcium phosphate composite nanoparticles in targeting Ehrlich carcinoma cells transplanted in mice. *J. Adv. Res.* **2016**, *7*, 143–154. [CrossRef]
65. Molnár, Z.; Bódai, V.; Szakacs, G.; Erdélyi, B.; Fogarassy, Z.; Sáfrán, G.; Varga, T.; Kónya, Z.; Tóth-Szeles, E.; Szűcs, R. Green synthesis of gold nanoparticles by thermophilic filamentous fungi. *Sci. Rep.* **2018**, *8*, 1–12. [CrossRef] [PubMed]
66. El-Naggar, N.E.-A.; Hussein, M.H.; El-Sawah, A.A. Bio-fabrication of silver nanoparticles by phycocyanin, characterization, in vitro anticancer activity against breast cancer cell line and in vivo cytotxicity. *Sci. Rep.* **2017**, *7*, 1–20. [CrossRef] [PubMed]
67. El-Naggar, N.E.-A.; Hussein, M.H.; El-Sawah, A.A. Phycobiliprotein-mediated synthesis of biogenic silver nanoparticles, characterization, in vitro and in vivo assessment of anticancer activities. *Sci. Rep.* **2018**, *8*, 1–20. [CrossRef]
68. Sharma, A.; Sharma, S.; Sharma, K.; Chetri, S.P.; Vashishtha, A.; Singh, P.; Kumar, R.; Rathi, B.; Agrawal, V. Algae as crucial organisms in advancing nanotechnology: A systematic review. *J. Appl. Phycol.* **2016**, *28*, 1759–1774. [CrossRef]

69. Lengke, M.F.; Fleet, M.E.; Southam, G. Biosynthesis of silver nanoparticles by filamentous cyanobacteria from a silver (I) nitrate complex. *Langmuir* **2007**, *23*, 2694–2699. [CrossRef]
70. Merin, D.D.; Prakash, S.; Bhimba, B.V. Antibacterial screening of silver nanoparticles synthesized by marine micro algae. *Asian Pac. J. Trop. Med.* **2010**, *3*, 797–799. [CrossRef]
71. Jena, J.; Pradhan, N.; Dash, B.P.; Sukla, L.B.; Panda, P.K. Biosynthesis and characterization of silver nanoparticles using microalga Chlorococcum humicola and its antibacterial activity. *Int. J. Nanomater. Biostruct.* **2013**, *3*, 1–8.
72. Castro-Longoria, E.; Vilchis-Nestor, A.R.; Avalos-Borja, M. Biosynthesis of silver, gold and bimetallic nanoparticles using the filamentous fungus Neurospora crassa. *Colloids Surf. B Biointerfaces* **2011**, *83*, 42–48. [CrossRef]
73. Solárová, Z.; Liskova, A.; Samec, M.; Kubatka, P.; Büsselberg, D.; Solár, P. Anticancer potential of lichens' secondary metabolites. *Biomolecules* **2020**, *10*, 87. [CrossRef] [PubMed]
74. Stevenson, C.; Capper, E.; Roshak, A.; Marquez, B.; Grace, K.; Gerwick, W.; Jacobs, R.; Marshall, L. Scytonemin-a marine natural product inhibitor of kinases key in hyperproliferative inflammatory diseases. *Inflamm. Res.* **2002**, *51*, 112. [CrossRef]
75. Bernardo, P.H.; Chai, C.L.; Heath, G.A.; Mahon, P.J.; Smith, G.D.; Waring, P.; Wilkes, B.A. Synthesis, electrochemistry, and bioactivity of the cyanobacterial calothrixins and related quinones. *J. Med. Chem.* **2004**, *47*, 4958–4963. [CrossRef] [PubMed]
76. Colin, J.A.; Pech-Pech, I.; Oviedo, M.; Águila, S.A.; Romo-Herrera, J.M.; Contreras, O.E. Gold nanoparticles synthesis assisted by marine algae extract: Biomolecules shells from a green chemistry approach. *Chem. Phys. Lett.* **2018**, *708*, 210–215. [CrossRef]
77. Madhumitha, G.; Roopan, S.M. Devastated crops: Multifunctional efficacy for the production of nanoparticles. *J. Nanomater.* **2013**, *2013*, 1–12. [CrossRef]
78. Mohammed, A.E.; Al-Qahtani, A.; Al-Mutairi, A.; Al-Shamri, B.; Aabed, K. Antibacterial and cytotoxic potential of biosynthesized silver nanoparticles by some plant extracts. *Nanomaterials* **2018**, *8*, 382. [CrossRef] [PubMed]
79. Kesharwani, J.; Yoon, K.Y.; Hwang, J.; Rai, M. Phytofabrication of silver nanoparticles by leaf extract of Datura metel: Hypothetical mechanism involved in synthesis. *J. Bionanosci.* **2009**, *3*, 39–44. [CrossRef]
80. Adelere, I.A.; Lateef, A. A novel approach to the green synthesis of metallic nanoparticles: The use of agro-wastes, enzymes, and pigments. *Nanotechnol. Rev.* **2016**, *5*, 567–587. [CrossRef]
81. Ali, J.; Ali, N.; Wang, L.; Waseem, H.; Pan, G. Revisiting the mechanistic pathways for bacterial mediated synthesis of noble metal nanoparticles. *J. Microbiol. Methods* **2019**, *159*, 18–25. [CrossRef]
82. Baraka, A.; Dickson, S.; Gobara, M.; El-Sayyad, G.S.; Zorainy, M.; Awaad, M.I.; Hatem, H.; Kotb, M.M.; Tawfic, A. Synthesis of silver nanoparticles using natural pigments extracted from Alfalfa leaves and its use for antimicrobial activity. *Chem. Pap.* **2017**, *71*, 2271–2281. [CrossRef]
83. Manikprabhu, D.; Lingappa, K. Microwave assisted rapid and green synthesis of silver nanoparticles using a pigment produced by Streptomyces coelicolor klmp33. *Bioinorg. Chem. Appl.* **2013**, *2013*, 1–5. [CrossRef]
84. Manikprabhu, D.; Lingappa, K. Synthesis of silver nanoparticles using the Streptomyces coelicolor klmp33 pigment: An antimicrobial agent against extended-spectrum beta-lactamase (ESBL) producing Escherichia coli. *Mater. Sci. Eng. C* **2014**, *45*, 434–437. [CrossRef] [PubMed]
85. Bhatnagar, S.; Kobori, T.; Ganesh, D.; Ogawa, K.; Aoyagi, H. Biosynthesis of silver nanoparticles mediated by extracellular pigment from talaromyces purpurogenus and their biomedical applications. *Nanomaterials* **2019**, *9*, 1042. [CrossRef] [PubMed]
86. Banerjee, A.; Halder, U.; Bandopadhyay, R. Preparations and applications of polysaccharide based green synthesized metal nanoparticles: A state-of-the-art. *J. Clust. Sci.* **2017**, *28*, 1803–1813. [CrossRef]
87. Ebrahiminezhad, A.; Bagheri, M.; Taghizadeh, S.-M.; Berenjian, A.; Ghasemi, Y. Biomimetic synthesis of silver nanoparticles using microalgal secretory carbohydrates as a novel anticancer and antimicrobial. *Adv. Nat. Sci. Nanosci. Nanotechnol.* **2016**, *7*, 015018. [CrossRef]
88. Li, G.; Li, Y.; Wang, Z.; Liu, H. Green synthesis of palladium nanoparticles with carboxymethyl cellulose for degradation of azo-dyes. *Mater. Chem. Phys.* **2017**, *187*, 133–140. [CrossRef]
89. Copeland, R.A. *Enzymes: A Practical Introduction to Structure, Mechanism, and Data Analysis*, 2nd ed.; John Wiley & Sons: New York, NY, USA, 2000.
90. Adamu, S.M.; Koki, A.Y.; Adamu, S.; Musa, A.M.; Abdullahi, A.S. Biotechnology as a Cradle of Scientific Development: A Review on Historical Perspective. *J. Adv. Biol. Biotechnol.* **2016**, *10*, 1–12. [CrossRef]
91. Wong, J.K.H.; Tan, H.K.; Lau, S.Y.; Yap, P.-S.; Danquah, M.K. Potential and challenges of enzyme incorporated nanotechnology in dye wastewater treatment: A review. *J. Environ. Chem. Eng.* **2019**, *7*, 103261. [CrossRef]
92. Manivasagan, P.; Venkatesan, J.; Kang, K.-H.; Sivakumar, K.; Park, S.-J.; Kim, S.-K. Production of α-amylase for the biosynthesis of gold nanoparticles using Streptomyces sp. MBRC-82. *Int. J. Biol. Macromol.* **2015**, *72*, 71–78. [CrossRef]
93. Gholami-Shabani, M.; Shams-Ghahfarokhi, M.; Gholami-Shabani, Z.; Akbarzadeh, A.; Riazi, G.; Ajdari, S.; Amani, A.; Razzaghi-Abyaneh, M. Enzymatic synthesis of gold nanoparticles using sulfite reductase purified from Escherichia coli: A green eco-friendly approach. *Process. Biochem.* **2015**, *50*, 1076–1085. [CrossRef]
94. Subbaiya, R.; Saravanan, M.; Priya, A.R.; Shankar, K.R.; Selvam, M.; Ovais, M.; Balajee, R.; Barabadi, H. Biomimetic synthesis of silver nanoparticles from Streptomyces atrovirens and their potential anticancer activity against human breast cancer cells. *IET Nanobiotechnol.* **2017**, *11*, 965–972. [CrossRef]

95. Ovais, M.; Khalil, A.T.; Islam, N.U.; Ahmad, I.; Ayaz, M.; Saravanan, M.; Shinwari, Z.K.; Mukherjee, S. Role of plant phytochemicals and microbial enzymes in biosynthesis of metallic nanoparticles. *Appl. Microbiol. Biotechnol.* **2018**, *102*, 6799–6814. [CrossRef]
96. Tareq, F.K.; Fayzunnesa, M.; Kabir, M.S.; Parvin, R.; Zahid, M.A. Biomolecule stabilized/functionalized nanomaterials: Advanced synthesis strategies, characterization and unique properties as antimicroorganism agent. *Import Appl. Nanotechnol.* **2020**, *2*, 1–7.
97. Kim, M.; Jee, S.-C.; Shinde, S.K.; Mistry, B.M.; Saratale, R.G.; Saratale, G.D.; Ghodake, G.S.; Kim, D.-Y.; Sung, J.-S.; Kadam, A.A. Green-synthesis of anisotropic peptone-silver nanoparticles and its potential application as anti-bacterial agent. *Polymers* **2019**, *11*, 271. [CrossRef]
98. Tugarova, A.V.; Kamnev, A.A. Proteins in microbial synthesis of selenium nanoparticles. *Talanta* **2017**, *174*, 539–547. [CrossRef]
99. Sanghi, R.; Verma, P. pH dependant fungal proteins in the 'green'synthesis of gold nanoparticles. *Adv. Mater. Lett.* **2010**, *1*, 193–199. [CrossRef]
100. Ga'al, H.; Yang, G.; Fouad, H.; Guo, M.; Mo, J. Mannosylerythritol lipids mediated biosynthesis of silver nanoparticles: An eco-friendly and operative approach against chikungunya vector Aedes albopictus. *J. Clust. Sci.* **2020**, *32*, 1–9. [CrossRef]
101. Kikuchi, F.; Kato, Y.; Furihata, K.; Kogure, T.; Imura, Y.; Yoshimura, E.; Suzuki, M. Formation of gold nanoparticles by glycolipids of Lactobacillus casei. *Sci. Rep.* **2016**, *6*, 1–8. [CrossRef]
102. Nadagouda, M.N.; Varma, R.S. Green and controlled synthesis of gold and platinum nanomaterials using vitamin B2: Density-assisted self-assembly of nanospheres, wires and rods. *Green Chem.* **2006**, *8*, 516–518. [CrossRef]
103. Shao, Y.; Wu, C.; Wu, T.; Yuan, C.; Chen, S.; Ding, T.; Ye, X.; Hu, Y. Green synthesis of sodium alginate-silver nanoparticles and their antibacterial activity. *Int. J. Biol. Macromol.* **2018**, *111*, 1281–1292. [CrossRef]
104. Malassis, L.; Dreyfus, R.; Murphy, R.J.; Hough, L.A.; Donnio, B.; Murray, C.B. One-step green synthesis of gold and silver nanoparticles with ascorbic acid and their versatile surface post-functionalization. *RSC Adv.* **2016**, *6*, 33092–33100. [CrossRef]
105. Ahmad, N.; Sharma, S.; Singh, V.; Shamsi, S.; Fatma, A.; Mehta, B. Biosynthesis of silver nanoparticles from Desmodium triflorum: A novel approach towards weed utilization. *Biotechnol. Res. Int.* **2011**, *2011*, 1–8. [CrossRef]
106. Korany, M.; Mahmoud, B.; Ayoub, S.M.; Sakr, T.M.; Ahmed, S.A. Synthesis and radiolabeling of vitamin C-stabilized selenium nanoparticles as a promising approach in diagnosis of solid tumors. *J. Radioanal. Nucl. Chem.* **2020**, *325*, 237–244. [CrossRef]
107. Han, C.; Nagendra, V.; Baig, R.; Varma, R.S.; Nadagouda, M.N. Expeditious synthesis of noble metal nanoparticles using vitamin B12 under microwave irradiation. *Appl. Sci.* **2015**, *5*, 415–426. [CrossRef]
108. Suman, T.; Rajasree, S.R.; Kanchana, A.; Elizabeth, S.B. Biosynthesis, characterization and cytotoxic effect of plant mediated silver nanoparticles using Morinda citrifolia root extract. *Colloids Surf. B Biointerfaces* **2013**, *106*, 74–78. [CrossRef]
109. Bagherzade, G.; Tavakoli, M.M.; Namaei, M.H. Green synthesis of silver nanoparticles using aqueous extract of saffron (Crocus sativus L.) wastages and its antibacterial activity against six bacteria. *Asian Pac. J. Trop. Biomed.* **2017**, *7*, 227–233. [CrossRef]
110. Marslin, G.; Siram, K.; Maqbool, Q.; Selvakesavan, R.K.; Kruszka, D.; Kachlicki, P.; Franklin, G. Secondary metabolites in the green synthesis of metallic nanoparticles. *Materials* **2018**, *11*, 940. [CrossRef]
111. Baskaran, X.; Vigila, A.V.G.; Parimelazhagan, T.; Muralidhara-Rao, D.; Zhang, S. Biosynthesis, characterization, and evaluation of bioactivities of leaf extract-mediated biocompatible silver nanoparticles from an early tracheophyte, Pteris tripartita Sw. *Int. J. Nanomed.* **2016**, *11*, 5789. [CrossRef]
112. Singh, P.; Kim, Y.-J.; Zhang, D.; Yang, D.-C. Biological synthesis of nanoparticles from plants and microorganisms. *Trends Biotechnol.* **2016**, *34*, 588–599. [CrossRef] [PubMed]
113. Kuppusamy, P.; Yusoff, M.M.; Maniam, G.P.; Govindan, N. Biosynthesis of metallic nanoparticles using plant derivatives and their new avenues in pharmacological applications—An updated report. *Saudi Pharm. J.* **2016**, *24*, 473–484. [CrossRef] [PubMed]
114. Saratale, G.D.; Saratale, R.G.; Kim, D.-S.; Kim, D.-Y.; Shin, H.-S. Exploiting fruit waste grape pomace for silver nanoparticles synthesis, assessing their antioxidant, antidiabetic potential and antibacterial activity against human pathogens: A novel approach. *Nanomaterials* **2020**, *10*, 1457. [CrossRef] [PubMed]
115. Singh, P.; Kim, Y.J.; Yang, D.C. A strategic approach for rapid synthesis of gold and silver nanoparticles by Panax ginseng leaves. *Artif. Cells Nanomed. Biotechnol.* **2016**, *44*, 1949–1957. [CrossRef]
116. Saratale, R.G.; Saratale, G.D.; Cho, S.-K.; Ghodake, G.; Kadam, A.; Kumar, S.; Mulla, S.I.; Kim, D.-S.; Jeon, B.-H.; Chang, J.S. Phyto-fabrication of silver nanoparticles by Acacia nilotica leaves: Investigating their antineoplastic, free radical scavenging potential and application in H_2O_2 sensing. *J. Taiwan Inst. Chem. Eng.* **2019**, *99*, 239–249. [CrossRef]
117. Krishnan, V.; Bupesh, G.; Manikandan, E.; Thanigai, A.; Magesh, S.; Kalyanaraman, R.; Maaza, M. Green synthesis of silver nanoparticles using Piper nigrum concoction and its anticancer activity against MCF-7 and Hep-2 cell lines. *J. Antimicro* **2016**, *2*. ISSN 2472-1212.
118. Gómez-Graña, S.; Perez-Ameneiro, M.; Vecino, X.; Pastoriza-Santos, I.; Perez-Juste, J.; Cruz, J.M.; Moldes, A.B. Biogenic synthesis of metal nanoparticles using a biosurfactant extracted from corn and their antimicrobial properties. *Nanomaterials* **2017**, *7*, 139. [CrossRef]
119. Ahmed, K.B.A.; Subramanian, S.; Sivasubramanian, A.; Veerappan, G.; Veerappan, A. Preparation of gold nanoparticles using Salicornia brachiata plant extract and evaluation of catalytic and antibacterial activity. *Spectrochim. Acta Part A Mol. Biomol. Spectrosc.* **2014**, *130*, 54–58. [CrossRef]

120. Omomowo, I.; Adenigba, V.; Ogunsona, S.; Adeyinka, G.; Oluyide, O.; Adedayo, A.; Fatukasi, B. Antimicrobial and antioxidant activities of algal-mediated silver and gold nanoparticles. In *Proceedings of IOP Conference Series: Materials Science and Engineering*; IOP Publishing Ltd.: Bristol, UK, 2020; p. 012010.
121. El-Kassas, H.Y.; El-Sheekh, M.M. Cytotoxic activity of biosynthesized gold nanoparticles with an extract of the red seaweed Corallina officinalis on the MCF-7 human breast cancer cell line. *Asian Pac. J. Cancer Prev.* **2014**, *15*, 4311–4317. [CrossRef] [PubMed]
122. Parkinson, J.; Gordon, R. Beyond micromachining: The potential of diatoms. *Trends Biotechnol.* **1999**, *17*, 190–196. [CrossRef]
123. Atazadeh, I.; Sharifi, M. Algae as bioindicators. In *The Effects of Heavy Metals on Algae and Development of an Algal Index System for Assessing Water Quality*; LAP LAMBERT Academic Publishing: Saarbrücken, Germany, 2010.
124. Graham, J.M.; Graham, L.E.; Zulkifly, S.B.; Pfleger, B.F.; Hoover, S.W.; Yoshitani, J. Freshwater diatoms as a source of lipids for biofuels. *J. Ind. Microbiol. Biotechnol.* **2012**, *39*, 419–428. [CrossRef] [PubMed]
125. Gutu, T.; Gale, D.K.; Jeffryes, C.; Wang, W.; Chang, C.-H.; Rorrer, G.L.; Jiao, J. Electron microscopy and optical characterization of cadmium sulphide nanocrystals deposited on the patterned surface of diatom biosilica. *J. Nanomater.* **2009**, *2009*, 1–7. [CrossRef]
126. Jeffryes, C.; Gutu, T.; Jiao, J.; Rorrer, G.L. Metabolic insertion of nanostructured TiO_2 into the patterned biosilica of the diatom Pinnularia sp. by a two-stage bioreactor cultivation process. *Acs Nano* **2008**, *2*, 2103–2112. [CrossRef] [PubMed]
127. Jena, J.; Pradhan, N.; Dash, B.P.; Panda, P.K.; Mishra, B.K. Pigment mediated biogenic synthesis of silver nanoparticles using diatom Amphora sp. and its antimicrobial activity. *J. Saudi Chem. Soc.* **2015**, *19*, 661–666. [CrossRef]
128. Ranjitha, V.; Rai, V.R. Actinomycetes mediated synthesis of gold nanoparticles from the culture supernatant of Streptomyces griseoruber with special reference to catalytic activity. *3 Biotech* **2017**, *7*, 1–7. [CrossRef] [PubMed]
129. Avilala, J.; Golla, N. Antibacterial and antiviral properties of silver nanoparticles synthesized by marine actinomycetes. *Int. J. Pharm. Sci. Res.* **2019**, *10*, 1223–1228.
130. Menon, S.; Rajeshkumar, S.; Kumar, V. A review on biogenic synthesis of gold nanoparticles, characterization, and its applications. *Resour. Effic. Technol.* **2017**, *3*, 516–527. [CrossRef]
131. Patil, M.P.; Kang, M.-J.; Niyonizigiye, I.; Singh, A.; Kim, J.-O.; Seo, Y.B.; Kim, G.-D. Extracellular synthesis of gold nanoparticles using the marine bacterium Paracoccus haeundaensis BC74171T and evaluation of their antioxidant activity and antiproliferative effect on normal and cancer cell lines. *Colloids Surf. B Biointerfaces* **2019**, *183*, 110455. [CrossRef] [PubMed]
132. Saravanan, M.; Barik, S.K.; MubarakAli, D.; Prakash, P.; Pugazhendhi, A. Synthesis of silver nanoparticles from Bacillus brevis (NCIM 2533) and their antibacterial activity against pathogenic bacteria. *Microb. Pathog.* **2018**, *116*, 221–226. [CrossRef] [PubMed]
133. Berdy, J. Bioactive microbial metabolites. *J. Antibiot.* **2005**, *58*, 1–26. [CrossRef] [PubMed]
134. Guilger-Casagrande, M.; Lima, R.D. Synthesis of silver nanoparticles mediated by fungi: A review. *Front. Bioeng. Biotechnol.* **2019**, *7*, 287. [CrossRef]
135. Clarance, P.; Luvankar, B.; Sales, J.; Khusro, A.; Agastian, P.; Tack, J.-C.; Al Khulaifi, M.M.; Al-Shwaiman, H.A.; Elgorban, A.M.; Syed, A. Green synthesis and characterization of gold nanoparticles using endophytic fungi Fusarium solani and its in-vitro anticancer and biomedical applications. *Saudi J. Biol. Sci.* **2020**, *27*, 706–712. [CrossRef] [PubMed]
136. Siddiqi, K.S.; Rashid, M.; Rahman, A.; Husen, A.; Rehman, S. Biogenic fabrication and characterization of silver nanoparticles using aqueous-ethanolic extract of lichen (Usnea longissima) and their antimicrobial activity. *Biomater. Res.* **2018**, *22*, 23. [CrossRef]
137. Ingólfsdóttir, K.; Gudmundsdóttir, G.; Ögmundsdóttir, H.; Paulus, K.; Haraldsdóttir, S.; Kristinsson, H.; Bauer, R. Effects of tenuiorin and methyl orsellinate from the lichen Peltigera leucophlebia on 5-/15-lipoxygenases and proliferation of malignant cell lines in vitro. *Phytomedicine* **2002**, *9*, 654–658. [CrossRef]
138. Reddy, V.M.; O'Sullivan, J.F.; Gangadharam, P.R. Antimycobacterial activities of riminophenazines. *J. Antimicrob. Chemother.* **1999**, *43*, 615–623. [CrossRef] [PubMed]
139. Yıldız, N.; Ateş, Ç.; Yılmaz, M.; Demir, D.; Yıldız, A.; Çalımlı, A. Investigation of lichen based green synthesis of silver nanoparticles with response surface methodology. *Green Process. Synth.* **2014**, *3*, 259–270. [CrossRef]
140. Fernández-Moriano, C.; González-Burgos, E.; Divakar, P.; Crespo, A.; Gómez-Serranillos, M. Evaluation of the antioxidant capacities and cytotoxic effects of ten parmeliaceae lichen species. *Evid. Based Complement. Altern. Med.* **2016**, *2016*, 1–11. [CrossRef]
141. Felczykowska, A.; Pastuszak-Skrzypczak, A.; Pawlik, A.; Bogucka, K.; Herman-Antosiewicz, A.; Guzow-Krzemińska, B. Antibacterial and anticancer activities of acetone extracts from in vitro cultured lichen-forming fungi. *Bmc Complementary Altern. Med.* **2017**, *17*, 1–12. [CrossRef]
142. Schmeda-Hirschmann, G.; Tapia, A.; Lima, B.; Pertino, M.; Sortino, M.; Zacchino, S.; Arias, A.R.D.; Feresin, G.E. A new antifungal and antiprotozoal depside from the Andean lichen Protousnea poeppigii. *Phytother. Res. Int. J. Devoted Pharmacol. Toxicol. Eval. Nat. Prod. Deriv.* **2008**, *22*, 349–355.
143. Abdolmaleki, H.; Sohrabi, M. Biosynthesis of silver nanoparticles by two lichens of "Usnea articulate" and "Ramalina sinensis" and investigation of their antibacterial activity against some pathogenic bacteria. *Ebnesina* **2016**, *17*, 33–42.
144. Çıplak, Z.; Gökalp, C.; Getiren, B.; Yıldız, A.; Yıldız, N. Catalytic performance of Ag, Au and Ag-Au nanoparticles synthesized by lichen extract. *Green Process. Synth.* **2018**, *7*, 433–440. [CrossRef]
145. Rai, H.; Gupta, R.K. Biogenic fabrication, characterization, and assessment of antibacterial activity of silver nanoparticles of a high altitude Himalayan lichen-Cladonia rangiferina (L.) Weber ex FH Wigg. *Trop. Plant Res.* **2019**, *6*, 293–298. [CrossRef]

146. Esmaeili, A.; Rajaee, S. The preparation of hyaluronic acid nanoparticles from Aspicilia lichens using Bifido Bacteria for help in the treatment of diabetes in rats in vivo. *Phytother. Res.* **2017**, *31*, 1590–1599. [CrossRef]
147. Zhang, D.; Ma, X.-L.; Gu, Y.; Huang, H.; Zhang, G.-W. Green Synthesis of Metallic Nanoparticles and Their Potential Applications to Treat Cancer. *Front. Chem.* **2020**, *8*, 799. [CrossRef]
148. Farooq, M.U.; Novosad, V.; Rozhkova, E.A.; Wali, H.; Ali, A.; Fateh, A.A.; Neogi, P.B.; Neogi, A.; Wang, Z. Gold nanoparticles-enabled efficient dual delivery of anticancer therapeutics to HeLa cells. *Sci. Rep.* **2018**, *8*, 1–12. [CrossRef]
149. Khandel, P.; Kumar Shahi, S.; Kanwar, L.; Kumar Yadaw, R.; Kumar Soni, D. Biochemical profiling of microbes inhibiting Silver nanoparticles using symbiotic organisms. *Int. J. Nano Dimens.* **2018**, *9*, 273–285.
150. Leela, K.; Devi, C. A Study on The Applications of Silver Nanoparticle Synthesized Using The Aqueous Extract And The Purified Secondary Metabolites of Lichen Parmelia perlata. *Int. J. Pharm. Sci. Invent.* **2017**, *6*, 42–59.
151. Mie, R.; Samsudin, M.W.; Din, L.B.; Ahmad, A. Green synthesis of silver nanoparticles using two lichens species: Parmotrema praesorediosum and Ramalina dumeticola. *Proc. Appl. Mech. Mater.* **2012**, *229–231*, 256–259. [CrossRef]
152. Mie, R.; Samsudin, M.W.; Din, L.B.; Ahmad, A.; Ibrahim, N.; Adnan, S.N.A. Synthesis of silver nanoparticles with antibacterial activity using the lichen Parmotrema praesorediosum. *Int. J. Nanomed.* **2014**, *9*, 121. [CrossRef] [PubMed]
153. Dasari, S.; Suresh, K.; Rajesh, M.; Reddy, S.; Samba, C.; Hemalatha, C.; Wudayagiri, R.; Valluru, L. Biosynthesis, characterization, antibacterial and antioxidant activity of silver nanoparticles produced by lichens. *J. Bionanosci.* **2013**, *7*, 237–244. [CrossRef]
154. Din, L.B.; Mie, R.; Samsudin, M.W.; Ahmad, A.; Ibrahim, N. Biomimetic synthesis of silver nanoparticles using the lichen Ramalina dumeticola and the antibacterial activity. *Malays. J. Anal. Sci.* **2015**, *19*, 369–376.
155. Baláž, M.; Goga, M.; Hegedus, M.; Daneu, N.; Kováčová, M.; Tkáčiková, L.; Balážová, L.; Bakčor, M. Biomechanochemical solid-state synthesis of silver nanoparticles with antibacterial activity using lichens. *ACS Sustain. Chem. Eng.* **2020**, *8*, 13945–13955.
156. Goga, M.; Baláž, M.; Daneu, N.; Elečko, J.; Tkáčiková, L.; Marcinčinová, M.; Bakčor, M. Biological activity of selected lichens and lichen-based Ag nanoparticles prepared by a green solid-state mechanochemical approach. *Mater. Sci. Eng. C* **2021**, *119*, 111640.
157. Senthil Prabhu, S.; Ramanujam, J.R.; Sudha, S. Antibacterial Activity of Silver Nanoparticles Synthesized by using Lichens *Heterodermia boryi* and *Parmotrema stuppeum*. *Int. J. Pharm. Biol. Sci.* **2019**, *9*, 1397–1402.
158. Alqahtani, M.A.; Mohammed, A.E.; Daoud, S.I.; Alkhalifah, D.H.M.; Albrahim, J.S. Lichens (Parmotrema clavuliferum) extracts: Bio-mediator in silver nanoparticles formation and antibacterial potential. *J. Bionanosci.* **2017**, *11*, 410–415. [CrossRef]
159. Debnath, R.; Purkayastha, D.D.; Hazra, S.; Ghosh, N.N.; Bhattacharjee, C.R.; Rout, J. Biogenic synthesis of antioxidant, shape selective gold nanomaterials mediated by high altitude lichens. *Mater. Lett.* **2016**, *169*, 58–61. [CrossRef]
160. Gandhi, A.D.; Murugan, K.; Umamahesh, K.; Babujanarthanam, R.; Kavitha, P.; Selvi, A. Lichen Parmelia sulcata mediated synthesis of gold nanoparticles: An eco-friendly tool against Anopheles stephensi and Aedes aegypti. *Environ. Sci. Pollut. Res.* **2019**, *26*, 23886–23898. [CrossRef]
161. Devasena, T.; Ashok, V.; Dey, N.; Francis, A. Phytosynthesis of magnesium nanoparticles using lichens. *World J. Pharm. Res.* **2014**, *3*, 4625–4632.
162. Alavi, M.; Karimi, N.; Valadbeigi, T. Antibacterial, antibiofilm, antiquorum sensing, antimotility, and antioxidant activities of green fabricated Ag, Cu, TiO_2, ZnO, and Fe_3O_4 NPs via protoparmeliopsis muralis lichen aqueous extract against multi-drug-resistant bacteria. *ACS Biomater. Sci. Eng.* **2019**, *5*, 4228–4243. [CrossRef]
163. Arjaghi, S.K.; Alasl, M.K.; Sajjadi, N.; Fataei, E.; Rajaei, G.E. Green Synthesis of Iron Oxide Nanoparticles by RS Lichen Extract and its Application in Removing Heavy Metals of Lead and Cadmium. *Biol. Trace Elem. Res.* **2020**, *199*, 1–6. [CrossRef]
164. Abdullah, S.M.; Kolo, K.; Sajadi, S.M. Greener pathway toward the synthesis of lichen-based ZnO@ TiO_2@ SiO_2 and Fe_3O_4@ SiO_2 nanocomposites and investigation of their biological activities. *Food Sci. Nutr.* **2020**, *8*, 4044–4054. [CrossRef]
165. Safarkar, R.; Ebrahimzadeh Rajaei, G.; Khalili-Arjagi, S. The study of antibacterial properties of iron oxide nanoparticles synthesized using the extract of lichen Ramalina sinensis. *Asian J. Nanosci. Mater.* **2020**, *3*, 157–166.
166. Reveathy, M.; Mathiazhagan, A.; Annadurai, G. Biosynthesis, characterization, antibacterial activity of silver nanoparticles using the lichen Parmotrema perlatum. *Eur. J. Biomed. Pharm. Sci.* **2015**, *2*, 348–361.
167. Uthreshwaranath, K.; Sasidharan, J.; Yuvraj, P.; Aruna, V.; Johnson, A.W.; Karthik, S. Green synthesis of silver nanoparticles using Parmelia perlata. *Asian J. Microbiol. Biotechnol. Environ. Sci.* **2015**, *17*, 145–152.
168. Shakouri, M.; Amani, A.; Mollaei, S.; Jeddi, M. Biosynthesis of Silver Nanoparticles by Umbilicaria Americana. In Proceedings of the 24th Iranian Seminar of Organic Chemistry, Azarbaijan Shahid Madani University, Tabriz, Iran, 24–26 August 2016.
169. Koca, F.D.; Ünal, G.; Halici, M.G. Lichen Based Synthesis of Zinc Oxide Nanoparticles and Evaluation of its Neurotoxic Effects on Human Neuroblastoma Cells. *Proc. J. Nano Res.* **2019**, *59*, 15–24. [CrossRef]
170. Sanchez-Moreno, P.; Ortega-Vinuesa, J.L.; Peula-Garcia, J.M.; Marchal, J.A.; Boulaiz, H. Smart drug-delivery systems for cancer nanotherapy. *Curr. Drug Targets* **2018**, *19*, 339–359. [CrossRef]
171. Kumar, S.P.; Kekuda, T.P.; Vinayaka, K.; Yogesh, M. Synergistic efficacy of lichen extracts and silver nanoparticles against bacteria causing food poisoning. *Asian J. Res. Chem.* **2010**, *3*, 67–70.
172. Rai, A.; Singh, A.; Ahmad, A.; Sastry, M. Role of halide ions and temperature on the morphology of biologically synthesized gold nanotriangles. *Langmuir* **2006**, *22*, 736–741. [CrossRef]
173. Liu, H.; Zhang, H.; Wang, J.; Wei, J. Effect of temperature on the size of biosynthesized silver nanoparticle: Deep insight into microscopic kinetics analysis. *Arab. J. Chem.* **2020**, *13*, 1011–1019. [CrossRef]
174. Pandey, B. Synthesis of zinc-based nanomaterials: A biological perspective. *IET Nanobiotechnol.* **2012**, *6*, 144–148.

175. Mohammadi, F.M.; Ghasemi, N. Influence of temperature and concentration on biosynthesis and characterization of zinc oxide nanoparticles using cherry extract. *J. Nanostruct. Chem.* **2018**, *8*, 93–102. [CrossRef]
176. Hamouda, R.A.; Hussein, M.H.; Abo-elmagd, R.A.; Bawazir, S.S. Synthesis and biological characterization of silver nanoparticles derived from the cyanobacterium Oscillatoria limnetica. *Sci. Rep.* **2019**, *9*, 1–17. [CrossRef]
177. Padil, V.V.T.; Černík, M. Green synthesis of copper oxide nanoparticles using gum karaya as a biotemplate and their antibacterial application. *Int. J. Nanomed.* **2013**, *8*, 889.
178. Patel, V.; Berthold, D.; Puranik, P.; Gantar, M. Screening of cyanobacteria and microalgae for their ability to synthesize silver nanoparticles with antibacterial activity. *Biotechnol. Rep.* **2015**, *5*, 112–119. [CrossRef]
179. Chutrakulwong, F.; Thamaphat, K.; Limsuwan, P. Photo-irradiation induced green synthesis of highly stable silver nanoparticles using durian rind biomass: Effects of light intensity, exposure time and pH on silver nanoparticles formation. *J. Phys. Commun.* **2020**, *4*, 095015. [CrossRef]
180. Mankad, M.; Patil, G.; Patel, D.; Patel, P.; Patel, A. Comparative studies of sunlight mediated green synthesis of silver nanoparaticles from Azadirachta indica leaf extract and its antibacterial effect on Xanthomonas oryzae pv. oryzae. *Arab. J. Chem.* **2020**, *13*, 2865–2872. [CrossRef]
181. Sonker, A.S.; Pathak, J.; Kannaujiya, V.K.; Sinha, R.P. Characterization and in vitro antitumor, antibacterial and antifungal activities of green synthesized silver nanoparticles using cell extract of Nostoc sp. strain HKAR-2. *Can. J. Biotechnol.* **2017**, *1*, 26. [CrossRef]
182. Sweeney, R.Y.; Mao, C.; Gao, X.; Burt, J.L.; Belcher, A.M.; Georgiou, G.; Iverson, B.L. Bacterial biosynthesis of cadmium sulfide nanocrystals. *Chem. Biol.* **2004**, *11*, 1553–1559. [CrossRef]
183. Li, X.; Xu, H.; Chen, Z.-S.; Chen, G. Biosynthesis of nanoparticles by microorganisms and their applications. *J. Nanomater.* **2011**, *2011*, 1–16. [CrossRef]
184. Yin, Y.; Yang, X.; Hu, L.; Tan, Z.; Zhao, L.; Zhang, Z.; Liu, J.; Jiang, G. Superoxide-mediated extracellular biosynthesis of silver nanoparticles by the fungus Fusarium oxysporum. *Environ. Sci. Technol. Lett.* **2016**, *3*, 160–165. [CrossRef]
185. Sneha, K.; Sathishkumar, M.; Mao, J.; Kwak, I.; Yun, Y.-S. Corynebacterium glutamicum-mediated crystallization of silver ions through sorption and reduction processes. *Chem. Eng. J.* **2010**, *162*, 989–996. [CrossRef]
186. Kalishwaralal, K.; Deepak, V.; Ramkumarpandian, S.; Nellaiah, H.; Sangiliyandi, G. Extracellular biosynthesis of silver nanoparticles by the culture supernatant of Bacillus licheniformis. *Mater. Lett.* **2008**, *62*, 4411–4413. [CrossRef]
187. Husseiny, M.; Abd El-Aziz, M.; Badr, Y.; Mahmoud, M. Biosynthesis of gold nanoparticles using Pseudomonas aeruginosa. *Spectrochim. Acta Part A Mol. Biomol. Spectrosc.* **2007**, *67*, 1003–1006. [CrossRef]
188. Divya, M.; Kiran, G.S.; Hassan, S.; Selvin, J. Biogenic synthesis and effect of silver nanoparticles (AgNPs) to combat catheter-related urinary tract infections. *Biocatal. Agric. Biotechnol.* **2019**, *18*, 101037. [CrossRef]
189. Hamedi, S.; Ghaseminezhad, M.; Shokrollahzadeh, S.; Shojaosadati, S.A. Controlled biosynthesis of silver nanoparticles using nitrate reductase enzyme induction of filamentous fungus and their antibacterial evaluation. *Artif. Cells Nanomed. Biotechnol.* **2017**, *45*, 1588–1596. [CrossRef]
190. Freitas, F.; Alves, V.D.; Reis, M.A. Advances in bacterial exopolysaccharides: From production to biotechnological applications. *Trends Biotechnol.* **2011**, *29*, 388–398. [CrossRef]
191. Gutierrez, T.; Morris, G.; Green, D.H. Yield and physicochemical properties of EPS from Halomonas sp. strain TG39 identifies a role for protein and anionic residues (sulfate and phosphate) in emulsification of n-hexadecane. *Biotechnol. Bioeng.* **2009**, *103*, 207–216. [CrossRef]
192. Emam, H.E.; Ahmed, H.B. Polysaccharides templates for assembly of nanosilver. *Carbohydr. Polym.* **2016**, *135*, 300–307. [CrossRef] [PubMed]
193. Kang, F.; Alvarez, P.J.; Zhu, D. Microbial extracellular polymeric substances reduce Ag+ to silver nanoparticles and antagonize bactericidal activity. *Environ. Sci. Technol.* **2014**, *48*, 316–322. [CrossRef] [PubMed]
194. Reith, F.; Lengke, M.F.; Falconer, D.; Craw, D.; Southam, G. The geomicrobiology of gold. *ISME J.* **2007**, *1*, 567–584. [CrossRef] [PubMed]
195. Nies, D.H. Microbial heavy-metal resistance. *Appl. Microbiol. Biotechnol.* **1999**, *51*, 730–750. [CrossRef] [PubMed]
196. Mergeay, M.; Monchy, S.; Vallaeys, T.; Auquier, V.; Benotmane, A.; Bertin, P.; Taghavi, S.; Dunn, J.; Van Der Lelie, D.; Wattiez, R. Ralstonia metallidurans, a bacterium specifically adapted to toxic metals: Towards a catalogue of metal-responsive genes. *FEMS Microbiol. Rev.* **2003**, *27*, 385–410. [CrossRef]
197. Perez-Gonzalez, T.; Jimenez-Lopez, C.; Neal, A.L.; Rull-Perez, F.; Rodriguez-Navarro, A.; Fernandez-Vivas, A.; Iañez-Pareja, E. Magnetite biomineralization induced by Shewanella oneidensis. *Geochim. Cosmochim. Acta* **2010**, *74*, 967–979. [CrossRef]
198. Senut, M.C.; Zhang, Y.; Liu, F.; Sen, A.; Ruden, D.M.; Mao, G. Size-dependent toxicity of gold nanoparticles on human embryonic stem cells and their neural derivatives. *Small* **2016**, *12*, 631–646. [CrossRef]
199. Chen, L.Q.; Fang, L.; Ling, J.; Ding, C.Z.; Kang, B.; Huang, C.Z. Nanotoxicity of silver nanoparticles to red blood cells: Size dependent adsorption, uptake, and hemolytic activity. *Chem. Res. Toxicol.* **2015**, *28*, 501–509. [CrossRef]
200. Wan, R.; Mo, Y.; Zhang, Z.; Jiang, M.; Tang, S.; Zhang, Q. Cobalt nanoparticles induce lung injury, DNA damage and mutations in mice. *Part. Fibre Toxicol.* **2017**, *14*, 1–15. [CrossRef] [PubMed]

201. Lapied, E.; Nahmani, J.Y.; Moudilou, E.; Chaurand, P.; Labille, J.; Rose, J.; Exbrayat, J.-M.; Oughton, D.H.; Joner, E.J. Ecotoxicological effects of an aged TiO_2 nanocomposite measured as apoptosis in the anecic earthworm Lumbricus terrestris after exposure through water, food and soil. *Environ. Int.* **2011**, *37*, 1105–1110. [CrossRef]
202. Rahman, Q.; Lohani, M.; Dopp, E.; Pemsel, H.; Jonas, L.; Weiss, D.G.; Schiffmann, D. Evidence that ultrafine titanium dioxide induces micronuclei and apoptosis in Syrian hamster embryo fibroblasts. *Environ. Health Perspect.* **2002**, *110*, 797–800. [CrossRef]
203. Khorrami, S.; Zarrabi, A.; Khaleghi, M.; Danaei, M.; Mozafari, M. Selective cytotoxicity of green synthesized silver nanoparticles against the MCF-7 tumor cell line and their enhanced antioxidant and antimicrobial properties. *Int. J. Nanomed.* **2018**, *13*, 8013–8024. [CrossRef] [PubMed]
204. Chokkalingam, M.; Singh, P.; Huo, Y.; Soshnikova, V.; Ahn, S.; Kang, J.; Mathiyalagan, R.; Kim, Y.J.; Yang, D.C. Facile synthesis of Au and Ag nanoparticles using fruit extract of Lycium chinense and their anticancer activity. *J. Drug Deliv. Sci. Technol.* **2019**, *49*, 308–315. [CrossRef]
205. Krishnaraj, C.; Harper, S.L.; Yun, S.-I. In Vivo toxicological assessment of biologically synthesized silver nanoparticles in adult Zebrafish (Danio rerio). *J. Hazard. Mater.* **2016**, *301*, 480–491. [CrossRef] [PubMed]

Journal of
Fungi

Article

Differential Antimycotic and Antioxidant Potentials of Chemically Synthesized Zinc-Based Nanoparticles Derived from Different Reducing/Complexing Agents against Pathogenic Fungi of Maize Crop

Anu Kalia [1,*], Jashanpreet Kaur [2], Manisha Tondey [2], Pooja Manchanda [3], Pulkit Bindra [4], Mousa A. Alghuthaymi [5,*], Ashwag Shami [6,*] and Kamel A. Abd-Elsalam [7]

1 Electron Microscopy and Nanoscience Laboratory, Department of Soil Science, College of Agriculture, Punjab Agricultural University, Ludhiana 141004, Punjab, India
2 Department of Microbiology, College of Basic Sciences and Humanities, Punjab Agricultural University, Ludhiana 141004, Punjab, India; grewalj119@gmail.com (J.K.); tondeymanisha@gmail.com (M.T.)
3 School of Agricultural Biotechnology, College of Agriculture, Punjab Agricultural University, Ludhiana 141004, Punjab, India; poojamanchanda5@pau.edu
4 Institute of Nanoscience and Technology, Habitat Centre, Phase-10, Sector-64, Mohali 160062, Punjab, India; pulkitbindra207@gmail.com
5 Biology Department, Science and Humanities College, Shaqra University, Alquwayiyah 11726, Saudi Arabia
6 Biology Department, College of Sciences, Princess Nourah bint Abdulrahman University, Riyadh 11617, Saudi Arabia
7 Plant Pathology Research Institute, Agricultural Research Center (ARC), Giza 12619, Egypt; kamelabdelsalam@gmail.com
* Correspondence: kaliaanu@pau.edu (A.K.); malghuthaymi@su.edu.sa (M.A.A.); AYShami@pnu.edu.sa (A.S.); Tel.: +91-2401960 (A.K.); +966-11-823-3175 (A.S.)

Citation: Kalia, A.; Kaur, J.; Tondey, M.; Manchanda, P.; Bindra, P.; Alghuthaymi, M.A.; Shami, A.; Abd-Elsalam, K.A. Differential Antimycotic and Antioxidant Potentials of Chemically Synthesized Zinc-Based Nanoparticles Derived from Different Reducing/Complexing Agents against Pathogenic Fungi of Maize Crop. *J. Fungi* **2021**, *7*, 223. https://doi.org/10.3390/jof7030223

Academic Editor: David Perlin

Received: 8 March 2021
Accepted: 17 March 2021
Published: 18 March 2021

Abstract: The present study aimed for the synthesis, characterization, and comparative evaluation of anti-oxidant and anti-fungal potentials of zinc-based nanoparticles (ZnNPs) by using different reducing or organic complexing-capping agents. The synthesized ZnNPs exhibited quasi-spherical to hexagonal shapes with average particle sizes ranging from 8 to 210 nm. The UV-Vis spectroscopy of the prepared ZnNPs showed variation in the appearance of characteristic absorption peak(s) for the various reducing/complexing agents i.e., 210 (NaOH and $NaBH_4$), 220 (albumin, and thiourea), 260 and 330 (starch), and 351 nm (cellulose) for wavelengths spanning over 190–800 nm. The FT-IR spectroscopy of the synthesized ZnNPs depicted the functional chemical group diversity. On comparing the antioxidant potential of these ZnNPs, NaOH (as reducing agent, (NaOH (RA)) derived ZnNPs presented significantly higher DPPH radical scavenging potential compared to other ZnNPs. The anti-mycotic potential of the ZnNPs as performed through an agar well diffusion assay exhibited variability in the extent of inhibition of the fungal mycelia with maximum inhibition at the highest concentration (40 mg L^{-1}). The NaOH (RA)-derived ZnNPs showcased maximum mycelial inhibition compared to other ZnNPs. Further, incubation of the total genomic DNA with the most effective NaOH (RA)-derived ZnNPs led to intercalation or disintegration of the DNA of all the three fungal pathogens of maize with maximum DNA degrading effect on *Macrophomina phaseolina* genomic DNA. This study thus identified that differences in size and surface functionalization with the protein (albumin)/polysaccharides (starch, cellulose) diminishes the anti-oxidant and anti-mycotic potential of the generated ZnNPs. However, the NaOH emerged as the best reducing agent for the generation of uniform nano-scale ZnNPs which possessed comparably greater anti-oxidant and antimycotic activities against the three test maize pathogenic fungal cultures.

Keywords: metal oxides; nano-fungicides; pathogenic fungi; protein profiling; radical scavenging activity

1. Introduction

Fungal pathogenic diseases are responsible for yield losses in staple calorie and commercial commodity crops posing a major threat to crop productivity globally. The yield gaps have enhanced due to the emergence of new fungal crop pathogens [1] as a consequence of intensive monoculture and environment variations arising due to aberrant climatic conditions [2,3]. Therefore, the agronomic interventions, land management practices, and climate change have been the primary agents that have altered both abiotic and biotic components affecting crop growth and yield [4].

Maize, a versatile cereal-food, feed, and industrial crop, is sensitive to attacks and diseases caused by several fungal pathogens [5]. It is the second-largest considering the area under production and is ranked fourth in productivity among cereals [6] across the globe. Substantial annual yield decreases and depreciation in grain quality in maize are two quantifiable manifestations of the fungal infection and disease [5]. The primary fungal pathogens of maize include the *Macrophomina phaseolina*, *Curvularia lunata* and *Fusarium oxysporum* which cause charcoal rot, leaf spot, and stalk rot diseases respectively in maize. The infected plants are generally treated with anti-fungal compounds or fungicides to curb the spread of the pathogen. However, for continuous monoculture predominated agroecosystems, spraying of these antifungals only provides ephemeral protection to the plants due to the single target site mechanism and the emergence of resistant fungal strains [1]. This necessitates the development of effective analogous antifungals without compromising the ecological and bio-safety aspects.

The present decade has witnessed the emergence and use of nano-scale materials as potent anti-microbials particularly anti-fungal agents. The zinc nanomaterials including the nano-zinc particles possess excellent anti-mycotic properties against a variety of plant fungal pathogens [7–12]. The predominant mechanisms governing the anti-mycotic effect of ZnNPs include the reactive oxygen species enabled stress besides Zn^{2+}-based toxicity occurring due to the formation of these ions on dissolution of ZnNPs in the cell environment [7]. The size of ZnNPs and their crystal chemistry can affect the fungicidal potential as these characteristics alter the ability to trespass the fungal cell wall and membrane structures to elicit ROS response besides varying the dissolution of the ZnNPs within the fungal cell cytoplasm. The anti-fungal activity of ZnNPs have been identified against *Alternaria alternata* [13], *Aspergillus flavus* [14], *Botrytis cinerea* [11,15], *Candida albicans* [16], *Fusarium graminearum* [17], *Fusarium moniliforme* [8], *Fusarium solani* [18], *Penicillium* sp. [19–21], *Penicillium expansum* [22], *Pythium ultimum* and *P. aphanidermatum* [23], *Rhizopus stolonifera* [24] and many more fungal pathogens of plants.

This investigation aims for the wet chemistry-based synthesis of ZnNPs through the use of different (three each) reducing and complexing/capping agents. The generated ZnNPs were characterized through UV-Vis spectroscopy, transmission electron microscopy (TEM), X-ray diffraction spectroscopy (XRD), and Fourier transform Infrared Spectroscopy (FT-IRS). These ZnNPs were evaluated for anti-fungal potential against three prominent fungal pathogens of maize *viz.*, *Curvularia lunata*, *Fusarium oxysporum*, and *Macrophomina phaseolina* in an agar well diffusion assay. The variation in the anti-oxidant potential of these ZnNPs was assessed through scavenging of the DPPH radicals while the genomic DNA degradation potential was determined through a DNA-ZnNPs incubation study followed by agarose gel electrophoresis of the samples.

2. Materials and Methods

2.1. Chemicals and Microbial Cultures

The zinc precursor salts (zinc acetate dihydrate ($Zn(OAc)_2{\cdot}2H_2O$), zinc nitrate hexahydrate ($Zn(NO_3)_2{\cdot}6H_2O$), and zinc chloride ($ZnCl_2$)) utilized in the study were analytical grade and purchased from HiMedia (HiMedia Laboratories, Mumbai, India). The reducing agents (sodium hydroxide (NaOH), sodium borohydride ($NaBH_4$)) and other chemicals (ammonium hydroxide or ammonia solution (NH_4OH, 25%), and thiourea (NH_2CSNH_2)) were purchased from Sigma-Aldrich (St. Louis, MO, USA). The analytical grade complexing-capping agents (starch, cellulose, and bovine serum albumin) were procured from Himedia (Himedia Laboratories, Mumbai, India). The HPLC grade water (CAS No. 7732-18-5, Sisco Research Laboratories Pvt. Ltd., Mumbai, India) was used for the preparation of all the solutions and broth and agar-based media. The broth (CAS No. M403) and agar-based (CAS No. MH096) potato dextrose media were purchased from Himedia (Himedia Laboratories, Mumbai, India).

The maize crop-specific three major test fungal cultures, *Curvularia lunata* (ITCC 7170), *Fusarium oxysporum* (ITCC 7093), and *Macrophomina phaseolina* (ITCC 5467) were purchased from Indian Type Culture Collection, Division of Plant Pathology, Indian Agricultural Research Institute, New Delhi, India. The cultures were subcultured on Potato Dextrose Agar media and incubated at 25 °C in the dark.

2.2. Synthesis and Characterization of NMs

The ZnNMs were synthesized using two different approaches: Wet chemical and Sol-gel synthesis methods. Six different ZnNMs samples were prepared using soluble zinc salts (zinc acetate/zinc nitrate/zinc chloride) as the precursors. The reducing/complexing-capping agents used in the study were sodium hydroxide [25], thiourea [26] and natural polymers such as starch [27,28], and cellulose-nanocomposite [29], and protein (Bovine-serum albumin (BSA)) [30]. The brief protocols depicting the schematic steps have been provided in Figure 1.

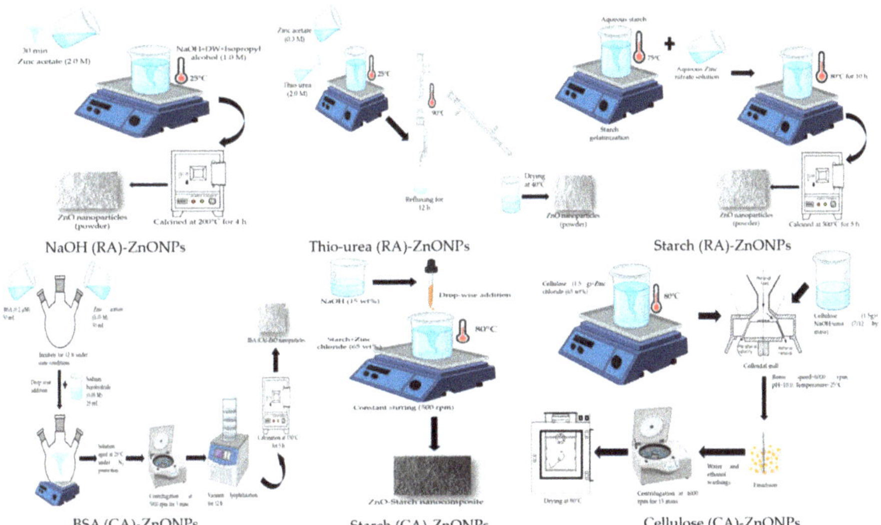

Figure 1. Schematic representation of steps for the synthesis of zinc nanomaterials using different reducing and complexing/capping agents.

2.2.1. UV-Vis Spectroscopy

The spectral absorbance behavior of the synthesized nanoparticles was analyzed on a Double Beam UV-Vis Spectrophotometer (model Elico SL 218, India) through screening over wavelengths ranging from 190 to 800 nm.

2.2.2. Transmission Electron Microscopy

The morphology of the synthesized nanoparticles was determined by the transmission electron microscopy (Hitachi H-7650, Japan) analysis. The powdered ZnNMs were ground in a polystyrene pestle mortar, homogenized, and suspended in a known volume of deionized water. The suspension was bath sonicated for 15 to 30 min and 20 µL of the suspension was then placed on a copper grid (carbon film-coated, 200-mesh size). The prepared grids were air-dried and viewed under high-resolution imaging mode in TEM operated at 80 kV acceleration voltage.

2.2.3. X-ray Diffraction Spectroscopy

The crystal structure and size of the synthesized ZnNMs were obtained through X-ray diffraction spectroscopy analysis. The X-ray diffraction patterns for the six different ZnNMs were obtained on X-ray diffractometer (Bruker D8 Advance, Germany) using Cu ($K\alpha$ λ = 0.150595 nm) radiations at specific operation conditions (voltage: 40 kV, current: 30 mA). The samples were scanned with a scanning angle (2θ) range of 5° to 80° and a step size of 0.02° respectively.

2.2.4. Fourier Transform Infra-Red Spectroscopy

The FT-IR spectroscopy (Thermo Nicolet 6700, USA) equipped with DTGS (deuterated triglycine sulfate) detector and KBr beam splitter system was used to characterize and identify the functional group diversity of the prepared nanoparticles. The summation spectra were generated in transmittance mode by placing dry, homogenized ZnNMs powdered samples on zinc selenide (flatbed configuration) crystal of the Attenuated total reflection (ATR) (Smart assembly, Thermo Fischer, USA) assembly with operational parameters of 32 scans at 4.0 cm^{-1} spectral resolution and acquired for the mid-IR region spanning over 4000 to 600 cm^{-1}.

2.3. *Anti-Oxidant Activity of the ZnNMs*

The anti-oxidant potential of the ZnNMs was determined as the free radical scavenging potential through neutralization of the 1, 1-Diphenyl-2-picrylhydrazyl (DPPH) radicals [31,32]. The aqueous suspensions of the ZnNMs were prepared by dispensing a known amount of the ZnNMs in a defined volume of HPLC-grade water. The prepared dispersions (1 mL) were bath sonicated, incubated under the same conditions with 3 mL methanolic solution of DPPH (0.1 mM) for half an hour. Variable color development in the incubated solutions indicating the radical scavenging rate was measured as absorbance at 517 nm. The percent inhibition was compared with the values obtained for the butylated hydroxytoluene (BHT) as standard.

2.4. *Anti-Mycotic Activity of the ZnNPs*

2.4.1. Agar Well Diffusion Assay

The anti-fungal activity of the ZnNPs was evaluated through agar well diffusion assay [33] involving estimation of the mycelial growth-inhibiting potential of the synthesized ZnNPs on the three test maize pathogenic fungal cultures [31]. The PDA media was poured in sterilized petri dishes (90 × 14 mm, Tarsons triple vent radiation sterile polystyrene, Code: 460091, Tarsons, Kolkata, India) and allowed to gel. The wells in the solidified media were prepared using a sterile cork borer (diameter 8.0 ± 0.2 mm, CAS No. LA737, Himedia Laboratories, Mumbai, India). The ZnNPs stock solutions were prepared in the HPLC grade water and these suspensions were bath sonicated for 30 min at room temperature. The stock solutions were then utilized for the preparation of the working concentrations (0, 5, 10, 20, and 40 mg L^{-1}). These suspensions were then given a quick bath sonication treatment for another five minutes and 20 μL of the suspensions were loaded in the agar wells. The respective fungal growth on PDA media served as the negative control. The fungal disc (8.0 mm diameter) of the freshly grown confluent growth (one-week old culture plate) was placed at the center of each plate and the inoculated petri plates were incubated in a BOD incubator at 27 ± 2 °C for seven days. The diameter of the zone of inhibition of the mycelial growth at or near the well containing the ZnNPs was taken as indicator of the decreased mycelial growth.

2.4.2. Optical Research Microscopy Studies of the Fungal Hyphae

For the optical microscopy (Leica DM5000 B, Leica Microsystems, Germany) studies, fresh slides were prepared to observe the effect of ZnNPs on the hyphal morphology and structure representing the morphological damage caused to the fungi by the ZnNPs. The slides were stained with lactophenol-cotton blue dye to visualize the variations in the hyphal morphology at 200 and 400× magnifications and imaged (Leica DFC 420C, Germany).

2.5. Fungal Genomic DNA Degrading Potential of the ZnNPs

The fungal biomass of the three fungi was generated in potato dextrose broth. The fungal mats were filtered and washed through sterilized filter paper (Whatman qualitative filter paper No. 1, Sigma-Aldrich, USA). The mycelial mat was then placed in a ceramic pestle containing liquid nitrogen and finely ground to obtain a powder. The fungal genomic DNA was extracted [34], quantified for quality and quantity and known quantity (10 µL) was incubated with 40 µg mL^{-1} of ZnNPs (NaOH as reducing agent) for 2 and 24 h at 37 °C. The incubated genomic DNA was resolved on 1.5% (*w/v*) agarose gel containing ethidium bromide (0.05 µg mL^{-1}). The resolved gel was viewed in a Gel Documentation and Analysis System (Uvitec, Cambridge, UK), and images were captured.

2.6. Statistical Analyses

The antioxidant profile data obtained for five replications were subjected to analysis of variance (ANOVA) by using the generalized linear model (Proc GLM) command for a completely randomized experimental design and results were obtained on analysis using SAS software (version 9.2, Cary, NC, USA). The mean comparisons were performed with the least significant difference (LSD, $p \leq 0.05$) approximation.

3. Results

3.1. Characterization of ZnNPs

3.1.1. UV-Visible Spectroscopy

Among the primary spectroscopy techniques utilized for the characterization of nanoparticles, this absorption spectroscopy technique is used to evaluate the light-matter interactions and has profound relevance for the determination of the optical properties of nanoparticles including key characteristics such as shape, size, and stability [35,36]. The UV-visible absorption study of the six different ZnNPs illustrated distinct and sharp absorption peaks to vary between wavelengths ranging from 210 to 350 nm (Figure 2). All the reducing and capping agents derived ZnNPs exhibited a single and sharp peak at 210 or around 210 to 220 nm except Starch (RA) (dual distinct peaks at 212 and 350 nm) and Starch (CA) (triple peaks at 212, 260, and 330 nm) derived ZnNPs.

3.1.2. Transmission Electron Microscopy

The TEM micrographs exhibit the occurrence of spherical to hexagonal-shaped ZnNPs having well-defined crystal edges and planes (Figure 3a–f). Partial (Figure 3c,e,f) to high (a, b, and d) agglomeration can also be observed. The nanoparticle size distribution was substantially variable for the reducing/capping agents used for the generation of ZnNPs. The lowest size distribution ranges of 8 to 26 nm and 6 to 22 nm were observed for thiourea (RA) and starch (CA) ZnNPs respectively. However, the starch (CA) ZnNPs appeared to be adorned on electronically less dense substrate material possibly derived from burning/charring of the starch during the calcination process. The average mean diameter of the nanoparticles (nm ± S.E.) was as follows; NaOH (RA) (31.37 ± 2.48), thiourea (RA) (15.86 ± 0.59), starch (RA) (36.82 ± 2.41), albumin (CA) (82.94 ± 3.64), starch (CA) (11.78 ± 0.74) and cellulose (CA) (209.18 ± 15.02).

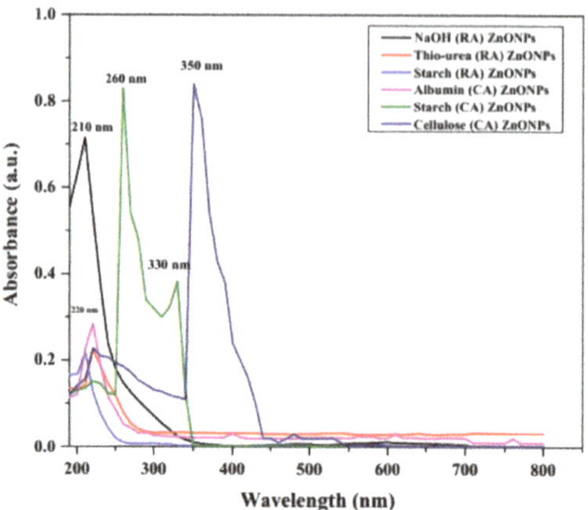

Figure 2. Variable UV-Vis absorbance spectra of the synthesized ZnO nanoparticles. RA: reducing agent, CA: capping/complexing agent.

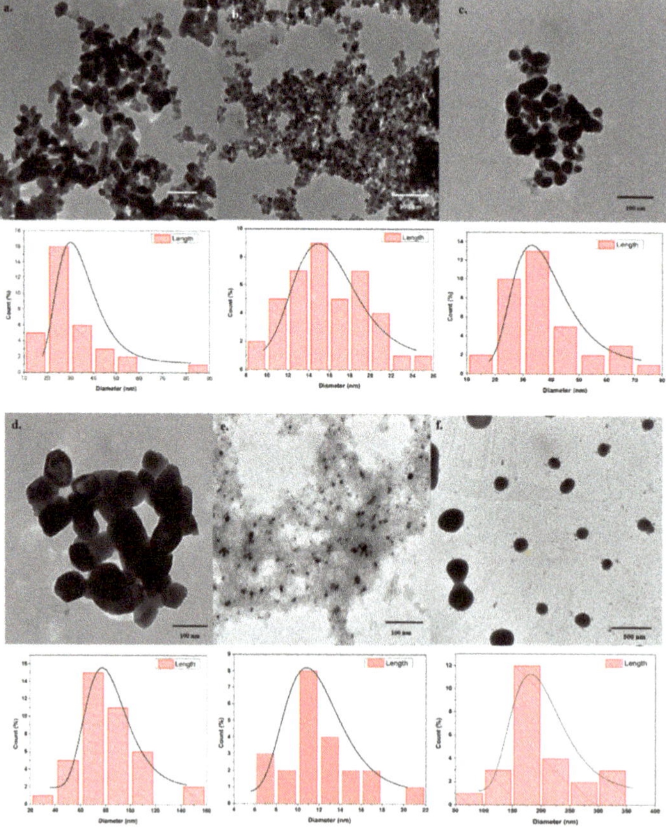

Figure 3. Transmission electron micrographs depicting the variation in the ZnO nanoparticle dimensions for the reducing (RA) and capping/complexing (CA) agents. (**a**) Sodium hydroxide (RA), (**b**) Thiourea (RA), (**c**) Starch (RA), (**d**) Bovine serum albumin (CA), (**e**) Starch (CA), and (**f**) Cellulose (CA).

3.1.3. X-ray Diffraction Spectroscopy

The XRD patterns varied in peak intensity and width according to the reducing/capping agent used. In general, the peak widening could be observed for all the six types of synthesized ZnNPs which represents the nano-scale crystalline size of the prepared ZnNPs. The XRD peaks obtained showed close agreement with the characteristic Bragg peaks of hexagonal ZnO zincite (pattern: PDF 00-036-1451), and hydro-zincite (pattern: PDF 01-072-1100). Mixed crystal phases including the simonkolleite (pattern: COD 9004683) along with zincite and hydro-zincite specific peaks can be identified in ZnNMs derived from $ZnCl_2$ salt as substrate (Figure 4). The use of starch as a reducing agent resulted in the formation of nanoscale hydrozincite crystals. While Starch (CA) ZnNPs had both zincite and hydrozincite as the predominant crystal phases. This variation may be attributed to the use of NaOH as a reducing agent during the synthesis of the Starch (CA) ZnNPs. Cellulose (CA) ZnNPs possessed a mixed crystal phase with simonkolleite as one among the predominant crystal structure. As indicated above, it can probably be due to the use of a high concentration of zinc chloride (65 wt %) as the precursor salt for the synthesis of ZnNPs.

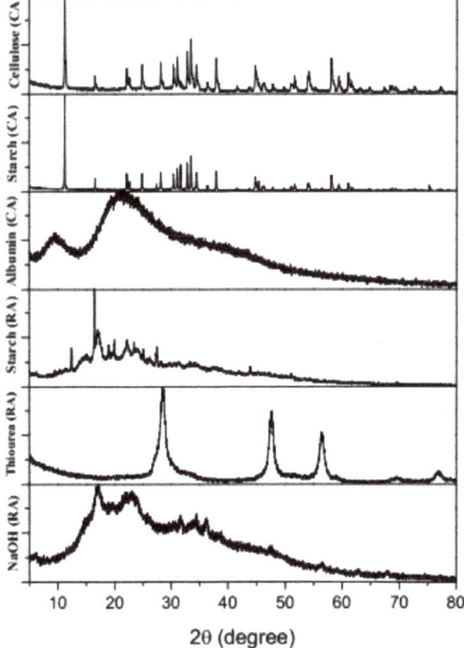

Figure 4. X-ray diffraction spectroscopy of the synthesized ZnNPs depicting the formation of wurtzite hexagonal crystal structure zincite (ZnO) on the use of various reducing and capping/complexing agents i.e., Sodium hydroxide (RA), Thiourea (RA), Starch (RA), Bovine serum albumin (CA), Starch (CA), and Cellulose (CA).

3.1.4. Fourier Transform Infra-Red Spectroscopy

The characteristic FTIR peaks for metal oxides appear in the fingerprint region of 1700 to 600 cm^{-1} due to vibrations among the metal and other atoms (O or OH) associated with it [37]. The NaOH (RA) ZnNPs exhibited specific peaks in this region at 663.0, 815.2, 952.0, 1016 (ν_1 frequency), and 1624 cm^{-1} featuring Zn-O bond deformation and stretching vibrations respectively [37]. The presence of adsorbed water molecules can be identified due to the appearance of a broad peak at 3443.0 and a sharp peak at 1103.0 cm^{-1} which can be ascribed to stretching and deformation vibrations of the O-H bond (Figure 5).

Further, the occurrence of the hydrozincite phase in the starch (RA), albumin (CA), and starch (CA) ZnNPs, the stretching vibrations of the C=O and C–O bonds in the carbonate functional group (CO_3^{2-}) can be ascribed to bands at 1360 and 1407 cm^{-1}. The conspicuous broadband (3690 to 2975 cm^{-1}) was observed in starch (RA), and albumin, starch and cellulose (CA) which can be ascribed to hydroxyl (O–H) and amine (N–H) group vibrations. Similar peaks have also been reported for thiourea derived ZnNPs [26].

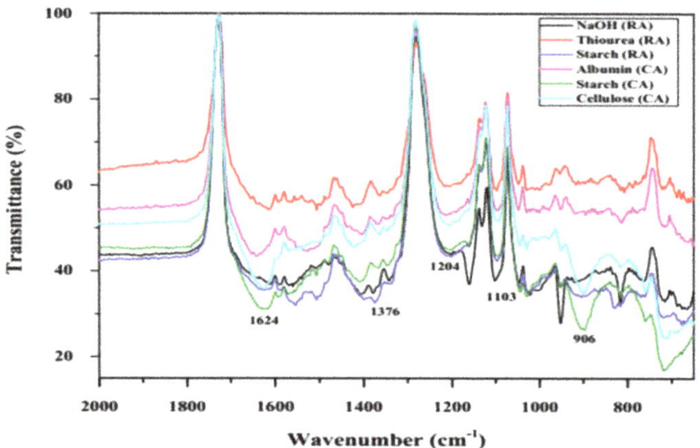

Figure 5. FT-IR cumulative spectra of the prepared ZnNPs for the mid-IR region (2000 to 650 cm^{-1} wavenumbers) indicating variability in the occurrence of chemical functional groups.

3.2. Anti-Oxidant Activity of the ZnNPs

The DPPH synthetic radicals are considered relatively stable to evaluate the radical scavenging potential of nanoparticles or other compounds. The prepared ZnNPs exhibited significant free radical scavenging activity (DPPH FRSA) which ranged from 25 to 84% inhibition (Figure 6). The NaOH (RA) ZnNPs possessed the highest antioxidant activity. The order of the FRSA was NaOH (RA) > Starch (RA)/(CA) > Albumin (CA) > Cellulose (CA) > Thiourea (RA).

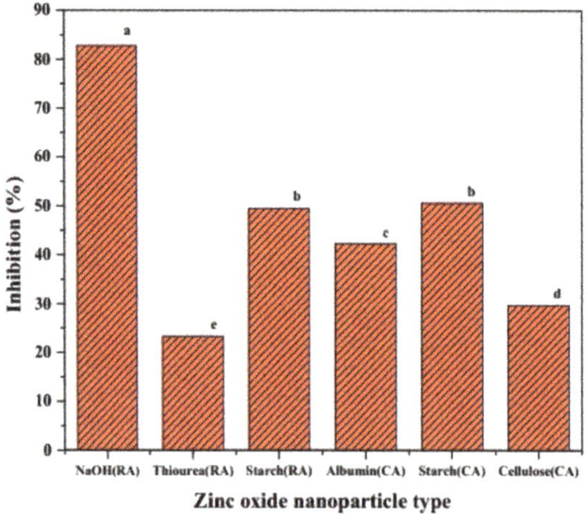

Figure 6. Comparative antioxidant potential of ZnNPs as determined through scavenging activity (%) of DPPH radicals. Different letters denote a significant difference ($p \leq 0.05$) among six different types of ZnNPs.

3.3. *Anti-Mycotic Activity of the ZnNPs*

3.3.1. Agar Well Diffusion Assay

Post three days of incubation, the prepared ZnNPs were evaluated for five different concentrations (0, 5, 10, 20, and 40 mg mL^{-1}) to exhibit an inhibitory effect on the mycelial growth for all three test fungi. Among the three fungal cultures, maximum mycelial inhibitory activity was recorded in order *Fusarium oxysporum* > *Curvularia lunata* > *Macrophomina phaseolina* (Figure 7). The efficacy of the NaOH (RA) derived ZnNPs was identified by the formation of a larger inhibition zone compared to the other ZnNPs evaluated. The control well containing only sterilized HPLC grade water did not exhibit any antifungal activity (Figure 7). The radial diameter of all the three-test fungi was the smallest for the NaOH (RA) derived ZnNPs particularly clear cottony-white hyphal inhibition could be observed for *Macrophomina phaseolina* (Figure 7d). Moreover, the sparse, aerial, and fluffy fungal growth of the *Fusarium oxysporum* indicates the response of hyphae to ZnNPs stress. This is the first report on variability in the anti-mycotic efficacy of the ZnNPs derived from different reducing/complexing agents on maize crop-specific pathogenic fungi.

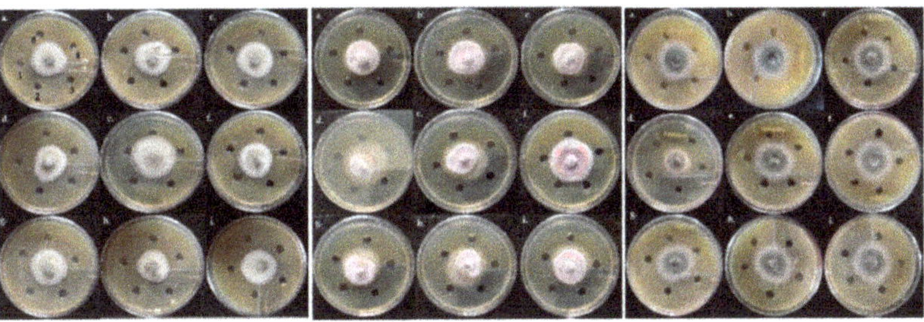

Figure 7. Effect of different ZnNPs and zinc salts on hyphal growth of three maize pathogenic cultures, Curvularia lunata, Fusarium oxysporum, and Macrophomina phaseolina. (**a**) Zinc acetate, (**b**) Zinc chloride, (**c**) Zinc sulphate, (**d**) Sodium hydroxide (RA) ZnNPs, (**e**) Thiourea (RA) ZnNPs, (**f**) Starch (RA) ZnNPs, (**g**) Bovine serum albumin (CA) ZnNPs, (**h**) Starch (CA) ZnNPs, and (**i**) Cellulose (CA) ZnNPs. The figures from 0 to 4 indicate different concentrations of the zinc salts and ZnNPs. 0 = distill water, 1 = 5 mg L^{-1}, 2 = 10 mg L^{-1}, 3 = 20 mg L^{-1}, 4 = 40 mg L^{-1}.

3.3.2. Optical Research Microscopy Studies of the Fungal Hyphae

The optical research microscopy of the mycelial growth at the fringes of the colony exhibited variation in the hyphal morphology as observed through appearance of swelling/rolling, thinning, fragmentation, and hyper-branching of the mycelia. The hyphae of all the three fungal genera sampled near the well containing NaOH (RA) (40 mg L^{-1} concentration) showed cell wall distortion, cytoplasmic shrinkage, oozing out of the cytoplasmic material, and hyphal fragmentation. The optical micrographs of the *Fusarium oxysporum* hyphae in NaOH (RA), Albumin (CA), and Starch (CA) ZnNPs (40 mg L^{-1} concentration) treatment showed extensive leakage of the cytoplasmic material from the hyphal tissue (Figure 8). The swelling of the hyphae cells can also be identified in the Thiourea (RA) ZnNPs and NaOH (RA) ZnNPs for *Curvularia lunata*, and *Macrophomina phaseolina* respectively.

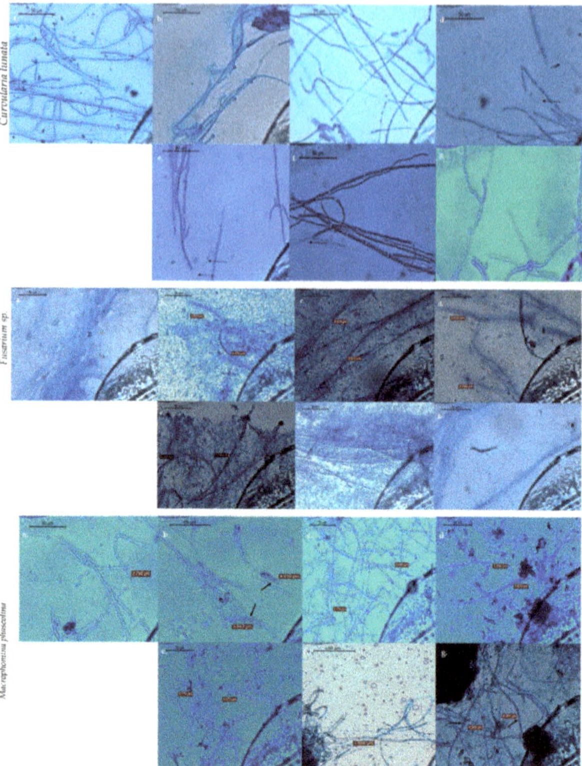

Figure 8. Optical micrographs of the three test fungal cultures depicting cytological events such as hyphal fragmentation, clearing of the cell cytoplasm, hyphal thinning, and dissolution of the fungal cell wall on incubation with ZnO nanoparticles derived from various reducing and capping/complexing agents. (**a**) Control, (**b**) Sodium hydroxide (RA), (**c**) Thiourea (RA), (**d**) Starch (RA), (**e**) Bovine serum albumin (CA), (**f**) Starch (CA), and (**g**) Cellulose (CA). Magnification-400×. The solid arrow indicates the thickening of the hyphae, dotted arrow indicates the clearing of cell cytoplasm.

3.4. Fungal Genomic DNA Degrading Potential of the ZnNPs

Incubation of the genomic fungal DNA with NaOH (RA)-derived ZnNPs resulted in fragmentation of the DNA which can be identified as smeared DNA appearance in the 1.5% agarose gel compared to the intact DNA band in the control lane of the three test fungi (Figure 9). After 2 h of incubation with the NaOH (RA) derived ZnNPs (40 $\mu g\ mL^{-1}$ concentration), a slight decrease can be noticed in the genomic DNA of all the test fungi in lanes 4, 5, and 6 (Figure 9). Therefore, the incubation was extended until 24 h to observe any further impact on the genomic DNA of these fungal genera. A clear fragmentation and decrease in the genomic DNA content can be visualized in lanes 7, 8, and 9 as compared to lanes 1, 2, and 3 respectively (Figure 9).

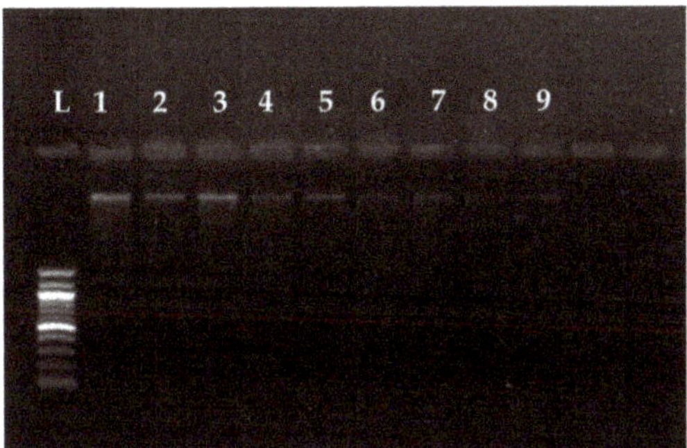

Figure 9. Fragmentation and degradation of the fungal genomic DNA on incubation with NaOH (RA)-derived ZnNPs. Lane L = Marker ladder, Lane 1 = gDNA of *Fusarium oxysporum* (*FO*), Lane 2 = gDNA of *Curvularia lunata* (*CL*), Lane 3 = gDNA of *Macrophomina phaseolina* (*MP*), Lane 4 = *FO* gDNA incubated with ZnNPs for 2 h, Lane 5 = *CL* gDNA incubated with ZnNPs for 2 h, Lane 6 = *MP* gDNA incubated with ZnNPs for 2 h, Lane 7 = *FO* gDNA incubated with ZnNPs for 24 h, Lane 8 = *CL* gDNA incubated with ZnNPs for 24 h, Lane 9 = *MP* gDNA incubated with ZnNPs for 24 h.

4. Discussion

The UV-Vis absorption peaks exhibited all the ZnNPs are way below the characteristic excitonic absorption peak value of 370 nm attributed to intrinsic band-gap absorption phenomena exhibited by the bulk ZnO under valence to conduction band electron transitions [36,38]. Similar reports of blue-shifted UV vis absorption peaks have been reported for ZnNPs synthesized through sol-gel [25,39], microemulsion route [37], and green synthesis approach [40]. Moreover, a similar UV-Vis peak at 347 nm has been observed for the starch (CA)-ZnNPs on laser ablation synthesis in an aqueous starch solution for 15 to 20 min which could be ascribed to the formation of a layered nanocomposite comprised of starch-β-$Zn(OH)_2$ sheets [41]. The occurrence of a single sharp peak indicates the formation of monodispersed ZnNPs and therefore, the narrow size distribution pattern of the synthesized ZnNPs [42]. The dual or triple peaks indicate the polydisperse nature of the aqueous nano-sol probably due to the formation of larger agglomerates by coalescence during or post-nucleation process [43]. The albumin (CA) ZnNMs showed a sharp peak at 220 nm, and multiple stout peaks at 400, 620, 700, and 760 nm wavelength which showcased the highly polydisperse nature of the albumin (CA) ZnNMs aqueous sol. However, the small area under these multiple peaks is also indicative of the presence of variable larger-sized ZnNPs in low amounts.

On the estimation of the bandgap energy of different ZnNMs according to Einstein equation; Energy = hC/λ, where h = Plank's constant (6.626×10^{-34} Joules sec), C = Velocity of light (3×10^{8} m s^{-1}) and λ = wavelength (nm), the calculated values were 5.90, 5.64, 3.76 and 3.50 eV for the sharp and distinct absorption peak wavelengths of 210, 220, 330 and 350 nm respectively. These values are quite high compared to the moderate band-gap energy (3.35 eV) of ZnO which corresponds to UV and deep blue region indicating O_{2p} to Zn_{3d} electron transitions [44]. These blue shifts in the band-gap energy may be ascribed to the inadvertent occurrence of other metal atom impurities that may have altered the electronic movement across the valence and conduction bands [45]. Similar blueshifts have been observed on doping of zinc with IIIa group metal atoms [46] and 3d transitions metals [45].

The particle size of less than 10 nm have been reported for NaOH-ethanol reaction mix [43] while ZnNPs generated from NaOH-isopropanol reaction mixture formed particles with the mean size dimensions of 189.0 ± 6.0 and 447.0 ± 22.0 nm by sonochemical and hydrothermal techniques respectively [47]. However, unlike Chen et al. [30] the ZnNPs size was larger for the albumin (CA) ZnNPs while similar size distribution and average size have been observed for thiourea (RA) [26], cellulose (CA) [29], and starch (CA) ZnNPs [27].

The use of a high concentration of $ZnCl_2$ can lead to the formation of hexagonal layered plate-like crystals of simonkolleite [18]. However, the XRD spectra for NaOH and thiourea reducing agents exhibited the presence of only hexagonal zincite phase crystals. Similar diffraction peaks have been reported for ZnNPs obtained by green synthesis from fruit parts of *Citrullus colosynthis* [14] and cotton linter pulp [11]. The XRD spectra of the albumin (CA) ZnNMs showed a very broad peak spanning over 30 to 50 2θ° diffraction region with several substantially indistinct peaks within this range indicating the occurrence of mixed crystal phase. As Bovine serum albumin (BSA) exhibits affinity to adsorb or form protein corona structure on the surface of the ZnNPs [19,20], it may have restrained the coalescence of the smaller-sized particles to form larger aggregates during the nucleation process. However, the transmission electron micrograph indicated the formation of plate-like hexagonal ZnNPs (Figure 2).

The DPPH FRSA of the ZnNMs may be ascribed to the transfer of electron density from oxygen atom in ZnNMs to N-atom odd electron in DPPH compound [48]. The antioxidant properties further depict the efficiency of the redox-catalysis reactions, electronic configuration, surface-interface effect, and biocompatibility of the ZnNMs. From the results obtained for the FRSA potential, it appears that the NPs of a particular size is critical beyond which any further reduction in the size of the NPs does not contribute significantly to the antioxidant behavior. These observations are in line with the anticipated size-dependent phenomena and larger surface area to volume ratio to be involved in improved neutralization/deactivation of the hydroxy (•OH) radicals [49]. Similar to observations of this study, the report on electron spin resonance (ESR) spectroscopy study of 4 to 15 nm AuNPs, showcased maximum antioxidant activity by 9 nm NPs and not by NPs with a size smaller than 9 nm [49]. Therefore, it can be argued that the dose [50,51] or concentration [52], specific surface, and crystallinity status [49] of the ZnNMs are relatively more critical features for altering the efficiency with which the ZnNMs interact with the DPPH radicals.

The fungal radial growth inhibition in the agar-well diffusion assay for the three test fungi indicated superior antimycotic activity of NaOH (RA) derived ZnNPs. Similar radial growth inhibition has also been reported for ZnO NPs by He et al. [11] for *Penicillium expansum* and *Botrytis cinerea*. The hyphal thinning effects can be observed for most of the ZnNMs evaluated in the study which appears to be a characteristic feature of any fungal tissue in response to nanoparticle challenge [53,54]. However, the morphological manifestations such as breakage of the cell wall and leakage of the cytoplasmic contents, and appearance of swollen hyphal cells indicate the alteration in the osmotic conditions and formation of physical pores in the cell wall/membrane of the treated fungal cells [19,23]. Coherent to the results of the mycelial inhibition in agar well assay as described in the previous Section 3.3.1, maximum impact on the hyphal morphology was recorded for the *Fusarium oxysporum*. However, the hyphal fragmentation can be observed in *Macrophomina phaseolina*, unlike the agar well assay which indicates the subtle cytological changes that occur in the fungal hyphae on treatment with ZnNMs. A substantially low concentrations of the ZnNMs have been evaluated in this study unlike the studies performed on *Pythyium* [23], *F. graminearum*, *A. flavus*, and *P. citrinum* [19], and *Aspergillus flavus* [14].

The genomic DNA fragmentation may be attributed to the physical and chemical properties of the NPs [55] including the size, concentration [56,57], chemistry [56–58], and surface functionalization [59]. Though within a fungal cell the predominant mechanism of degradation of the cellular DNA by ZnNPs is through formation of reactive oxygen species (ROS) which causes extensive DNA scissoring and fragmentation [7]. However, direct interactions of ZnNPs with DNA molecules involve the binding of DNA with ZnO nanoparticles to nucleobases [60]. The specific conjugation of the DNA on ZnO NPs surface has been further identified by Das et al. [61] evidenced through varied conductivity and mobility under electric field on an agarose gel. The results of this study effectively demonstrate that the ZnNPs can intercalate with the DNA and exhibit non-photocatalytic DNA degradation unlike the sunlight-induced fragmentation of *Leishmania* DNA on incubation with Ag-doped ZnO NPs [62].

5. Conclusions

This study provides evidence on variation in the anti-mycotic, anti-oxidant, and DNA degrading effect of the ZnO nanoparticles generated through wet chemical synthesis methods by using different reducing or complexing/capping agents. Both the anti-oxidant and anti-mycotic potentials were not observed to follow strict nanoparticle size-dependence. Thus, the NaOH (RA) derived ZnNPs had zincite crystal phase and the particle size distribution ranging from 10 to 90 nm and possessed the highest antioxidant and antimycotic activities. This study also illustrates the subtle hyphal morphological changes occurring on the use of lower concentration of the ZnNPs which may span over a variety of manifestations ranging from thinning, fragmentation, swelling, and lysis of the fungal hyphae. These alterations in the hyphal morphology could be attributed to variability in the functionalizations or chemical functional groups present on the surface of the ZnNPs. Further, the fungal DNA-NaOH (RA) ZnNPs incubation study enunciated the DNA degradation and exacerbation of the total genomic DNA of the test pathogenic fungi in a concentration and time-dependent manner with complete degradation of the *Macrophomina phaseolina* genomic DNA after 24 h of incubation. It could also be inferred from the study that the DNA damaging effect of the ZnNPs is also fungal genera/species-specific and may vary according to the fungal culture being studied. Therefore, this work establishes the impact of the appropriate nano-crystallite size dimensions and surface functionalizations as the key factors that vary the anti-mycotic, anti-oxidant, and DNA degrading potentials of ZnNPs.

Author Contributions: Conceptualization, A.K.; methodology, A.K.; validation, A.K., P.M., J.K.; formal analysis, J.K., M.T., P.B., P.M.; investigation, A.K., P.M.; resources, A.K., M.A.A.; data curation, A.K.; writing—original draft preparation, A.S., A.K.; writing—review and editing, K.A.A.-E., M.A.A.; visualization, A.K.; supervision, A.K.; project administration, A.K.; funding acquisition, K.A.A.-E., A.S., M.A.A. All authors have read and agreed to the published version of the manuscript.

Funding: This research was funded by Rashtriya Krishi Vikas Yojana (RKVY), Government of India, Grant ID (RKVY-12, PC-1320-32) to Anu Kalia. Moreover, the current research was partly supported by the Science and Technology Development Fund (STDF), Joint Egypt (STDF)-South Africa (NRF) Scientific Cooperation, Grant ID. 27837 to Kamel Abd-Elsalam.

Institutional Review Board Statement: Not applicable.

Informed Consent Statement: Not applicable.

Data Availability Statement: Data is contained within the article.

Acknowledgments: The authors thank the Director of Research for the allocation of funds in the RKVY scheme.

Conflicts of Interest: The authors declare no conflict of interest.

References

1. Fones, H.N.; Bebber, D.P.; Chaloner, T.M.; Kay, W.T.; Steinberg, G.; Gurr, S.J. Threats to global food security from emerging fungal and oomycete crop pathogens. *Nat. Food* **2020**, *1*, 332–342. [CrossRef]
2. Chaloner, T.M.; Gurr, S.J.; Bebber, D.P. The global burden of plant disease tracks crop yields under climate change. *bioRxiv* **2020**. [CrossRef]
3. Elad, Y.; Pertot, I. Climate Change Impacts on Plant Pathogens and Plant Diseases. *J. Crop Improv.* **2014**, *28*, 99–139. [CrossRef]
4. Ruffo, M.L.; Gentry, L.F.; Henninger, A.S.; Seebauer, J.R.; Below, F.E. Evaluating management factor contributions to reduce corn yield gaps. *Agron. J.* **2015**, *107*, 495–505. [CrossRef]
5. Mueller, D.S.; Pathology, P.; Wise, K.A.; Bosley, D.B.; State, C.; Collins, F.; Bradley, C.A.; Pathology, P. Corn Yield Loss Estimates Due to Diseases in the United States and Ontario, Canada from 2012 to 2015. *Plant Health Prog.* **2016**, *17*, 211–222. [CrossRef]
6. Bedford, D.; Claro, J.; Giusti, A.M.; Karumathy, G.; Lucarelli, L.; Mancini, D.; Marocco, E.; Milo, M.; Yang, D. *Food Outlook*; FAO: Rome, Itlay, 2018; ISBN 978-92-5-109782-3.
7. Kalia, A.; Abd-Elsalam, K.A.; Kuca, K. Zinc-based nanomaterials for diagnosis and management of plant diseases: Ecological safety and future prospects. *J. Fungi* **2020**, *6*, 222. [CrossRef] [PubMed]

8. Kalia, A.; Kaur, J.; Kaur, A.; Singh, N. Antimycotic activity of biogenically synthesised metal and metal oxide nanoparticles against plant pathogenic fungus Fusarium moniliforme (F. fujikuroi). *Indian J. Exp. Biol.* **2020**, *58*, 263–270.
9. Jesmin, R.; Chanda, A. Restricting mycotoxins without killing the producers: A new paradigm in nano-fungal interactions. *Appl. Microbiol. Biotechnol.* **2020**, *104*, 2803–2813. [CrossRef] [PubMed]
10. Hassan, M.; Zayton, M.A.; El-feky, S.A. Role of Green Synthesized Zno Nanoparticles as Antifungal against Post-Harvest Gray and Black Mold of Sweet Bell. *J. Biotechnol. Bioeng.* **2019**, *3*, 8–15.
11. He, L.; Liu, Y.; Mustapha, A.; Lin, M. Antifungal activity of zinc oxide nanoparticles against Botrytis cinerea and Penicillium expansum. *Microbiol. Res.* **2011**, *166*, 207–215. [CrossRef] [PubMed]
12. Abdelhakim, H.K.; El-Sayed, E.R.; Rashidi, F.B. Biosynthesis of zinc oxide nanoparticles with antimicrobial, anticancer, antioxidant and photocatalytic activities by the endophytic *Alternaria tenuissima*. *J. Appl. Microbiol.* **2020**, *128*, 1634–1646. [CrossRef]
13. Al-Dhabaan, F.A.; Shoala, T.; Ali, A.A.; Alaa, M.; Abd-Elsalam, K.; Abd-, K. Chemically-Produced Copper, Zinc Nanoparticles and Chitosan-Bimetallic Nanocomposites and Their Antifungal Activity against Three Phytopathogenic Fungi. *Int. J. Agric. Technol.* **2017**, *13*, 753–769.
14. Hernández-Meléndez, D.; Salas-Téllez, E.; Zavala-Franco, A.; Téllez, G.; Méndez-Albores, A.; Vázquez-Durán, A. Inhibitory effect of flower-shaped zinc oxide nanostructures on the growth and aflatoxin production of a highly toxigenic strain of Aspergillus flavus Link. *Materials* **2018**, *11*, 1265. [CrossRef] [PubMed]
15. Kairyte, K.; Kadys, A.; Luksiene, Z. Antibacterial and antifungal activity of photoactivated ZnO nanoparticles in suspension. *J. Photochem. Photobiol. B Biol.* **2013**, *128*, 78–84. [CrossRef] [PubMed]
16. Karimiyan, A.; Najafzadeh, H.; Ghorbanpour, M.; Hekmati-Moghaddam, S.H. Antifungal Effect of Magnesium Oxide, Zinc Oxide, Silicon Oxide and Copper Oxide Nanoparticles Against *Candida albicans*. *Zahedan J. Res. Med. Sci.* **2015**, *17*, 2–4. [CrossRef]
17. Dimkpa, C.O.; McLean, J.E.; Britt, D.W.; Anderson, A.J. Antifungal activity of ZnO nanoparticles and their interactive effect with a biocontrol bacterium on growth antagonism of the plant pathogen *Fusarium graminearum*. *BioMetals* **2013**, *26*, 913–924. [CrossRef]
18. Dos Santos, R.A.A.; D'Addazio, V.; Silva, J.V.G.; Falqueto, A.R.; Barreto da Silva, M.; Schmildt, E.R.; Fernandes, A.A. Antifungal Activity of Copper, Zinc and Potassium Compounds on Mycelial Growth and Conidial Germination of *Fusarium solani* f. sp. *piperis*. *Microbiol. Res. J. Int.* **2019**, *29*, 1–11. [CrossRef]
19. Savi, G.D.; Bortoluzzi, A.J.; Scussel, V.M. Antifungal properties of Zinc-compounds against toxigenic fungi and mycotoxin. *Int. J. Food Sci. Technol.* **2013**, *48*, 1834–1840. [CrossRef]
20. Helaly, M.N.; El-Metwally, M.A.; El-Hoseiny, H.; Omar, S.A.; El-Sheery, N.I. Effect of nanoparticles on biological contamination of in vitro cultures and organogenic regeneration of banana. *Aust. J. Crop Sci.* **2014**, *8*, 612–624.
21. Yehia, R.; Ahmed, O.F. In Vitro study of the antifungal efficacy of zinc oxide nanoparticles against *Fusarium oxysporum* and *Penicilium expansum*. *Afr. J. Microbiol. Res.* **2013**, *7*, 1917–1923. [CrossRef]
22. Sardella, D.; Gatt, R.; Valdramidis, V.P. Assessing the efficacy of zinc oxide nanoparticles against *Penicillium expansum* by automated turbidimetric analysis. *Mycology* **2018**, *9*, 43–48. [CrossRef]
23. Zabrieski, Z.; Morrell, E.; Hortin, J.; Dimkpa, C.; McLean, J.; Britt, D.; Anderson, A. Pesticidal activity of metal oxide nanoparticles on plant pathogenic isolates of *Pythium*. *Ecotoxicology* **2015**, *24*, 1305–1314. [CrossRef] [PubMed]
24. Decelis, S.; Sardella, D.; Triganza, T.; Brincat, J.P.; Gatt, R.; Valdramidis, V.P. Assessing the anti-fungal efficiency of filters coated with zinc oxide nanoparticles. *R. Soc. Open Sci.* **2017**, *4*, 1–9. [CrossRef] [PubMed]
25. Davis, D.; Singh, S. ZnO Nanoparticles Synthesis by Sol-Gel Method and Characterization. *Indian J. Nanosci.* **2016**, *4*, 1–4.
26. Wahab, R.; Ansari, S.G.; Kim, Y.S.; Dhage, M.S.; Seo, H.K.; Song, M.; Shin, H.S. Effect of annealing on the conversion of zns to zno nanoparticles synthesized by the sol-gel method using zinc acetate and thiourea. *Met. Mater. Int.* **2009**, *15*, 453–458. [CrossRef]
27. Ma, J.; Zhu, W.; Tian, Y.; Wang, Z. Preparation of Zinc Oxide-Starch Nanocomposite and Its Application on Coating. *Nanoscale Res. Lett.* **2016**, *11*. [CrossRef]
28. Raliya, R.; Avery, C.; Chakrabarti, S.; Biswas, P. Photocatalytic degradation of methyl orange dye by pristine titanium dioxide, zinc oxide, and graphene oxide nanostructures and their composites under visible light irradiation. *Appl. Nanosci.* **2017**, *7*, 253–259. [CrossRef]
29. Ma, J.; Sun, Z.; Wang, Z.; Zhou, X. Preparation of ZnO–cellulose nanocomposites by different cellulose solution systems with a colloid mill. *Cellulose* **2016**, *23*, 3703–3715. [CrossRef]
30. Chen, Z.; Shuai, L.; Zheng, B.; Wu, D. Synthesis of zinc oxide nanoparticles with good photocatalytic activities under stabilization of bovine serum albumin. *J. Wuhan Univ. Technol. Mater. Sci. Ed.* **2017**, *32*, 1061–1066. [CrossRef]
31. Kalia, A.; Manchanda, P.; Bhardwaj, S.; Singh, G. Biosynthesized silver nanoparticles from aqueous extracts of sweet lime fruit and callus tissues possess variable antioxidant and antimicrobial potentials. *Inorg. Nano-Met. Chem.* **2020**, *50*, 1053–1062. [CrossRef]
32. Azizi, S.; Mohamad, R.; Shahri, M.M.; McPhee, D.J. Green microwave-assisted combustion synthesis of zinc oxide nanoparticles with *Citrullus colocynthis* (L.) schrad: Characterization and biomedical applications. *Molecules* **2017**, *22*, 301. [CrossRef]
33. Magaldi, S.; Mata-Essayag, S.; De Capriles, C.H.; Perez, C.; Colella, M.T.; Olaizola, C.; Ontiveros, Y. Well diffusion for antifungal susceptibility testing. *Int. J. Infect. Dis.* **2004**, *8*, 39–45. [CrossRef]
34. Moslem, M.; Abd-Elsalam, K.; Yassin, M.; Bahkali, A. First morphomolecular identification of penicillium griseofulvum and penicillium aurantiogriseum toxicogenic isolates associated with blue mold on apple. *Foodborne Pathog. Dis.* **2010**, *7*, 857–861. [CrossRef]

35. Rajesh, R.W.; Jaya, L.R.; Niranjan, K.S.; Vijay, D.M.; Sahebrao, B.K. Phytosynthesis of silver nanoparticle using Gliricidia sepium (Jacq.). *Curr. Nanosci.* **2009**, *5*, 117–122.
36. Yan, J.K.; Wang, Y.Y.; Zhu, L.; Wu, J.Y. Green synthesis and characterization of zinc oxide nanoparticles using carboxylic curdlan and their interaction with bovine serum albumin. *RSC Adv.* **2016**, *6*, 77752–77759. [CrossRef]
37. Kumar, H.; Rani, R. Structural and Optical Characterization of ZnO Nanoparticles Synthesized by Microemulsion Route. *Int. Lett. Chem. Phys. Astron.* **2013**, *19*, 26–36. [CrossRef]
38. Rajiv, P.; Rajeshwari, S.; Venckatesh, R. Bio-Fabrication of zinc oxide nanoparticles using leaf extract of *Parthenium hysterophorus* L. and its size-dependent antifungal activity against plant fungal pathogens. *Spectrochim. Acta Part A Mol. Biomol. Spectrosc.* **2013**, *112*, 384–387. [CrossRef]
39. Kayani, Z.N.; Saleemi, F.; Batool, I. Synthesis and Characterization of ZnO Nanoparticles. *Mater. Today Proc.* **2015**, *2*, 5619–5621. [CrossRef]
40. Ambika, S.; Sundrarajan, M. Green biosynthesis of ZnO nanoparticles using *Vitex negundo* L. extract: Spectroscopic investigation of interaction between ZnO nanoparticles and human serum albumin. *J. Photochem. Photobiol. B Biol.* **2015**, *149*, 143–148. [CrossRef] [PubMed]
41. Zamiri, R.; Zakaria, A.; Ahangar, H.A.; Darroudi, M.; Zak, A.K.; Drummen, G.P.C. Aqueous starch as a stabilizer in zinc oxide nanoparticle synthesis via laser ablation. *J. Alloys Compd.* **2012**, *516*, 41–48. [CrossRef]
42. Panchakarla, L.S.; Govindaraj, A.; Rao, C.N.R. Formation of ZnO nanoparticles by the reaction of zinc metal with aliphatic alcohols. *J. Clust. Sci.* **2007**, *18*, 660–670. [CrossRef]
43. Talam, S.; Karumuri, S.R.; Gunnam, N. Synthesis, Characterization, and Spectroscopic Properties of ZnO Nanoparticles. *ISRN Nanotechnol.* **2012**, *2012*, 1–6. [CrossRef]
44. Yoshikawa, H.; Adachi, S. Optical constants of ZnO. *Jpn. J. Appl. Phys.* **1997**, *36*, 6237–6243. [CrossRef]
45. Joseph, D.P.; Venkateswaran, C. Bandgap Engineering in ZnO By Doping with 3d Transition Metal Ions. *J. At. Mol. Opt. Phys.* **2011**, *2011*, 1–7. [CrossRef]
46. Feng, X.; Wang, Z.; Zhang, C.; Wang, P. Electronic structure and energy band of IIIA doped group ZnO nanosheets. *J. Nanomater.* **2013**, *2013*, 181979. [CrossRef]
47. Zakirov, M.I.; Semen'ko, M.P.; Korotchenkov, O.A. A simple sonochemical synthesis of nanosized ZnO from zinc acetate and sodium hydroxide. *J. Nano-Electron. Phys.* **2018**, *10*, 4–7. [CrossRef]
48. Stan, M.; Popa, A.; Toloman, D.; Silipas, T.D.; Vodnar, D.C. Antibacterial and antioxidant activities of ZnO nanoparticles synthesized using extracts of Allium sativum, Rosmarinus officinalis and Ocimum basilicum. *Acta Metall. Sin. Engl. Lett.* **2016**. [CrossRef]
49. Yakimovich, N.O.; Ezhevskii, A.A.; Guseinov, D.V.; Smirnova, L.A.; Gracheva, T.A.; Klychkov, K.S. Antioxidant properties of gold nanoparticles studied by ESR spectroscopy. *Russ. Chem. Bull.* **2008**, *57*, 520–523. [CrossRef]
50. Watanabe, A.; Kajita, M.; Kim, J.; Kanayama, A.; Takahashi, K.; Mashino, T.; Miyamoto, Y. In Vitro free radical scavenging activity of platinum nanoparticles. *Nanotechnology* **2009**, *20*. [CrossRef]
51. Dipankar, C.; Murugan, S. The green synthesis, characterization and evaluation of the biological activities of silver nanoparticles synthesized from Iresine herbstii leaf aqueous extracts. *Colloids Surf. B Biointerfaces* **2012**, *98*, 112–119. [CrossRef]
52. Ali, A.; Ambreen, S.; Maqbool, Q.; Naz, S.; Shams, M.F.; Ahmad, M.; Phull, A.R.; Zia, M. Zinc impregnated cellulose nanocomposites: Synthesis, characterization and applications. *J. Phys. Chem. Solids* **2016**, *98*, 174–182. [CrossRef]
53. Arciniegas-Grijalba, P.A.; Patiño-Portela, M.C.; Mosquera-Sánchez, L.P.; Guerrero-Vargas, J.A.; Rodríguez-Páez, J.E. ZnO nanoparticles (ZnO-NPs) and their antifungal activity against coffee fungus Erythricium salmonicolor. *Appl. Nanosci.* **2017**, *7*, 225–241. [CrossRef]
54. Singh, S.; Kuca, K.; Kalia, A. Alterations in Growth and Morphology of Ganoderma lucidum and Volvariella volvaceae in Response to Nanoparticle Supplementation. *Mycobiology* **2020**, *48*, 383–391. [CrossRef] [PubMed]
55. Samanta, A.; Medintz, I.L. Nanoparticles and DNA-a powerful and growing functional combination in bionanotechnology. *Nanoscale* **2016**, *8*, 9037–9095. [CrossRef]
56. Hashim, A.F.; Youssef, K. Ecofriendly nanomaterials for controlling gray mold of table grapes and maintaining postharvest quality. *Eur. J. Plant Pathol.* **2019**, 377–388. [CrossRef]
57. Alghuthaymi, M.A.; Abd-Elsalam, K.A.; Shami, A.; Said-Galive, E.; Shtykova, E.V.; Naumkin, A.V. Silver/chitosan nanocomposites: Preparation and characterization and their fungicidal activity against dairy cattle toxicosis *Penicillium expansum*. *J. Fungi* **2020**, *6*, 51. [CrossRef] [PubMed]
58. Abd-Elsalam, K.A.; Alghuthaymi, M.A.; Shami, A.; Rubina, M.S.; Abramchuk, S.S.; Shtykova, E.V.; Vasil'kov, A.Y. Copper-chitosan nanocomposite hydrogels against aflatoxigenic *Aspergillus flavus* from dairy cattle feed. *J. Fungi* **2020**, *6*, 112. [CrossRef] [PubMed]
59. Das, B.; Khan, M.I.; Jayabalan, R.; Behera, S.K.; Yun, S.-I.; Tripathy, S.K.; Mishra, A. Understanding the Antifungal Mechanism of Ag@ZnO Core-shell Nanocomposites against Candida krusei. *Sci. Rep.* **2016**, *6*, 1–12. [CrossRef] [PubMed]
60. Saha, S.; Sarkar, P. Understanding the interaction of DNA-RNA nucleobases with different ZnO nanomaterials. *Phys. Chem. Chem. Phys.* **2014**, *16*, 15355–15366. [CrossRef] [PubMed]

61. Das, S.; Chatterjee, S.; Pramanik, S.; Devi, P.S.; Kumar, G.S. A new insight into the interaction of ZnO with calf thymus DNA through surface defects. *J. Photochem. Photobiol. B Biol.* **2018**, *178*, 339–347. [CrossRef]
62. Nadhman, A.; Sirajuddin, M.; Nazir, S.; Yasinzai, M. Photo-induced *Leishmania* DNA degradation by silver-doped zinc oxide nanoparticle: An in-vitro approach. *IET Nanobiotechnol.* **2016**, *10*, 129–133. [CrossRef] [PubMed]

Journal of
Fungi

MDPI

Review

Fungal–Metal Interactions: A Review of Toxicity and Homeostasis

Janelle R. Robinson, Omoanghe S. Isikhuemhen * and Felicia N. Anike

Department of Natural Resources and Environmental Design, North Carolina Agricultural and Technical State University, 1601 East Market Street, Greensboro, NC 27411, USA; jrrobin3@aggies.ncat.edu (J.R.R.); fnanike@ncat.edu (F.N.A.)
* Correspondence: omon@ncat.edu

Abstract: Metal nanoparticles used as antifungals have increased the occurrence of fungal–metal interactions. However, there is a lack of knowledge about how these interactions cause genomic and physiological changes, which can produce fungal superbugs. Despite interest in these interactions, there is limited understanding of resistance mechanisms in most fungi studied until now. We highlight the current knowledge of fungal homeostasis of zinc, copper, iron, manganese, and silver to comprehensively examine associated mechanisms of resistance. Such mechanisms have been widely studied in *Saccharomyces cerevisiae*, but limited reports exist in filamentous fungi, though they are frequently the subject of nanoparticle biosynthesis and targets of antifungal metals. In most cases, microarray analyses uncovered resistance mechanisms as a response to metal exposure. In yeast, metal resistance is mainly due to the down-regulation of metal ion importers, utilization of metallothionein and metallothionein-like structures, and ion sequestration to the vacuole. In contrast, metal resistance in filamentous fungi heavily relies upon cellular ion export. However, there are instances of resistance that utilized vacuole sequestration, ion metallothionein, and chelator binding, deleting a metal ion importer, and ion storage in hyphal cell walls. In general, resistance to zinc, copper, iron, and manganese is extensively reported in yeast and partially known in filamentous fungi; and silver resistance lacks comprehensive understanding in both.

Keywords: resistance; homeostasis; toxicity; nanoparticles; fungal–metal interaction

Citation: Robinson, J.R.; Isikhuemhen, O.S.; Anike, F.N. Fungal–Metal Interactions: A Review of Toxicity and Homeostasis. *J. Fungi* **2021**, *7*, 225. https://doi.org/10.3390/jof7030225

Received: 5 March 2021
Accepted: 17 March 2021
Published: 18 March 2021

1. Introduction

The increasing applications of fungal–metal interactions have led to the need for research on their contributions to fungal resistance [1,2]. In nature, metals serve as micronutrients required for fungal growth, however, in excess they can influence homeostatic systems. In agricultural and human medicine, there is an increasing occurrence of pathogen resistance to traditional antifungal agents which has expanded the incidence of fungal superbugs; this has led to increased research on metals as alternative fungistatic and fungicidal agents [3,4]. Fungi are also being employed in the green biosynthesis of nanoparticles due to their economic viability, high levels of natural metal resistance, and ease of mass production as antimicrobial agents [5–7]. Both instances highlight contributions to increased incidence of fungal-metal interactions, demonstrating the importance of further divulging the intricacies of their relationship.

1.1. Fungal–Metal Interactions

Metals can exist in various forms such as salts, oxides, sulfates, and nanoparticles. Fungi are able to utilize metal ions from these compounds after dissociation, which leaves unbound ions available for uptake and transport. For example, in the presence of water, copper sulfate ($CuSO_4$) hydrates to copper (II) sulfate pentahydrate ($CuSO_4$ $5H_2O$) and then dissociates into Cu^{2+} + $SO_4{}^{2-}$. Upon dissociation, Cu^{2+} can then be reduced by fungal proteins for uptake. More recently, metals in the form of nanoparticles have gained interest for use as antifungals, which has fueled the escalation of nanoparticle production [8–10].

Nanoparticles are particles that range from 1 to 100 nm in size and vary in shape, physiochemical, optical, and biological properties [11]. Ions dissociate from nanoparticles at a much lower rate, but are also available to interact with homeostatic systems [12,13].

In general, most ions have dedicated homeostatic systems to control import, export, storage, and transport within the cell (Table 1). Metal ion import and export often occurs through transmembrane channels, which are proteins that span the entirety of the membrane and protrude from both sides (e.g., transmembrane proteins Fet4, Zrt1, and Zrt2 in Figure 1) [14,15]. In some species, chelators, such as siderophores, also play a role in uptake. These organic, low molecular weight compounds have a binding capacity for certain metal ions, such as iron, and are imported into the cell through transmembrane channels [16,17]. As a mechanism of ion storage or detoxification, metallothioneins (MTs), cysteine-rich proteins that use metal ions as cofactors, bind free cytosolic ions which may be released back into the cellular environment in metal deficient conditions [18,19]. For the movement of ions to organelles for storage or as cofactors for protein functioning, intracellular transporters, such as Zrc1 (Figure 1) or Pic2 (Figure 2), are utilized [20,21]. If these systems are interfered with, homeostatic imbalance can cause toxicity.

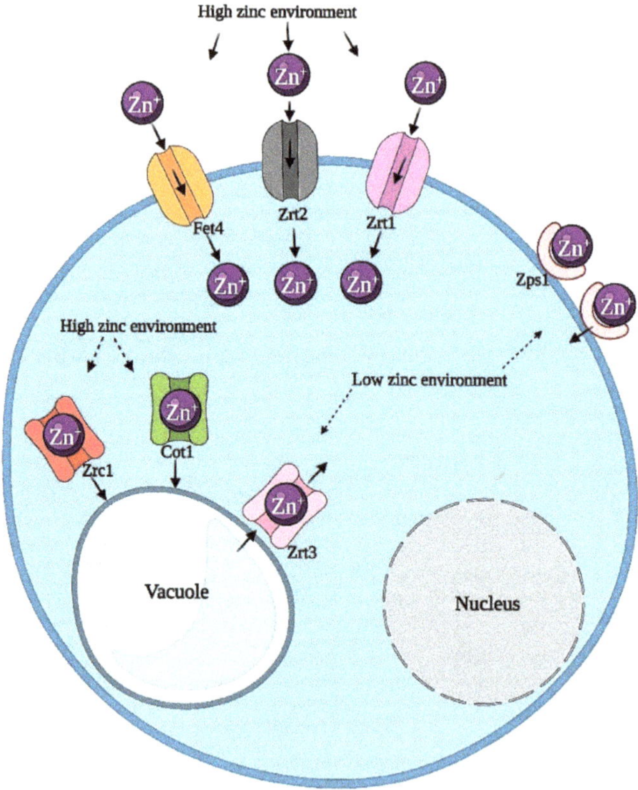

Figure 1. *S. cerevisiae* zinc homeostatic systems.

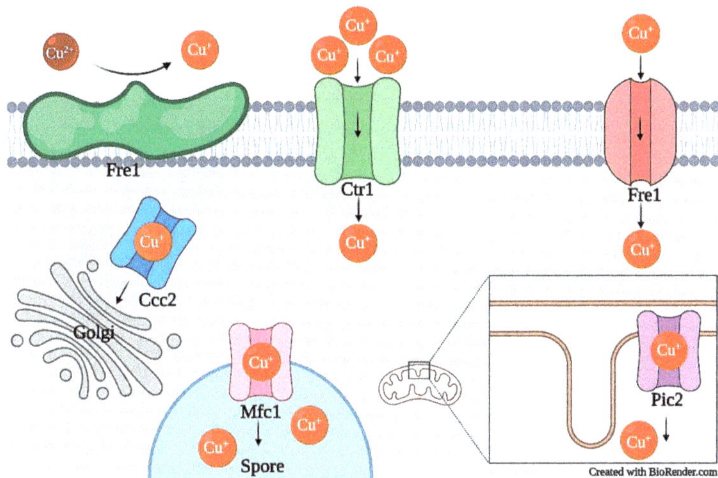

Figure 2. Yeast copper transport systems. In *S. cerevisiae*, cupric reductase, Fre1 reduces extracellular cupric oxide for transport across high and low-affinity copper membrane transports Ctr1 and Fet4. From the cytoplasm, Ccc2 shuttles Cu^+ to Golgi bodies, and Pic2 shuttles Cu^+ to the mitochondrial matrix. During meiosis in *S. pombe*, Mfc1 transports Cu^+ across the forespore membrane.

Table 1. Fungal proteins involved in metal transport.

Metal	Transport Type	Yeast Transporters	Reference	Filamentous Fungi Transporters	Reference
Zinc	Import	Zrt1, Zrt2	[15,22]	*zrfA/B/C*, UmZRT1/2, Zip1/2	[23–27]
	Vacuolar	Cot1, Zrc1	[20,28]	-	-
	Vacuole to Cytosol	Zrt3	[29]	-	-
Copper	Import	Ctr1, Ctr3, Fet4, Ctr4, Ctr5, Mfc1	[30–36]	CtrA2, CtrC, Ctr1, PaCtr2	[37–39]
	Cytosol to Golgi	Atx1, Ccc2	[34,40–42]	-	-
	Mitochondrial	Pic2, Cox17	[21,43,44]	-	-
	Cytosol to Sod1	Lys7, Pccs	[45,46]	-	-
	Mitochondrial Inner Membrane Space to Cytochrome *c* oxidase	Sco1, Sco2, Cox11	[42,47,48]	-	-
	Export	-	-	CrpA	[49]
Iron	Import	Fet4, Smf1, Fet3/Ftr1, Fip1, Str3, Shu1, Str1, Str2, Str3	[50–59]	Fer2	[60]
	Within the Nucleus	Npb35, Nar1, Cfd1, Cia1	[61,62]	-	-
	Vacuolar	Pcl1, Ccc1	[63,64]	-	-
Mangan-ese	Import	Smf1, Smf2, Pho85	[52,65–67]	PcPho84, PcSmfs	[68]
	Mitochondrial	Mtm1	[69]	PcMtm1	[68]
	Cytosol to Golgi Lumen	Pmr1, Gdt1	[70–72]	-	-
	Cytosol to Endoplasmic Reticulum Lumen	Spf1	[73]	-	-
	Vacuolar	Ccc1, Ypk9	[64,74–76]	PcCCC1	[68]
	Export	Pmr1, Hip1	[77–79]	PcMnt	[68]
Silver	Import	Ctr1	[80,81]	-	-
	Mitochondrial	Pic2	[21]	-	-

1.2. Metal Toxicity and Resistance

Metal toxicity occurs via the oligodynamic effect, which was initially described in 1893 in algae *Spirogyra nitida* and *Spirogyra dubia*, as toxicity or death in organisms due to exposure to trace amounts of metals, such as copper, lead, iron, or zinc [82]. In fungi, this exposure can have effects ranging interference in ergosterol biosynthesis to reduced MT activity (Table 2) [83,84].

Table 2. Mechanisms of toxicity in yeast and filamentous fungi.

Metal	Mechanism of Toxicity in Yeast	Reference	Mechanism of Toxicity in Filamentous Fungi	Reference
Zinc	Interference of synthesis of iron-sulfur clusters	[85,86]	increased chitin deposition within the cell wall, preventing hyphal extension	[87,88]
	Interference in ergosterol biosynthesis	[83]	increased hyphal branching and apical swelling	[88]
	Cellular leakage, polarization, and increased membrane potential	[83]	interruption of conidia and conidiophore development (interference of reproduction)	[87]
	Reduced cell wall integrity	[83]	-	-
Copper	Reduced ergosterol biosynthesis	[12,89]	Generation of reactive oxygen species	[90]
	Reduced metallothionein activity	[84]	-	-
Iron	Interference of vacuolar transport encoding gene *CCC1*	[91,92]	Inability to acquire iron	[60,93]
Manganese	Down-regulation of *HTB2, HTA1, HTA1, HTBI, HHF*	[94,95]	potentially associated to reduced functioning of manganese peroxidase	[96–98]
Silver	Interference in ergosterol biosynthesis	[80,99,100]	-	-

In an effort to counter metal toxicity, and toxicity in general, fungi develop methods of resistance which can include the alteration of the target protein to inhibit substrate binding, cellular antimicrobial efflux, antimicrobial inactivation or degradation, restricted uptake to prevent cellular interference, overproduction of targeted proteins to prevent the complete inhibition of biochemicals, and compensation for loss of function directly related to the antimicrobial [101]. Some of these resistance mechanisms are relevant to excessive metal exposure in fungi (Table 3). Presently, research utilizes yeast such as *Saccharomyces cerevisiae* to investigate cellular and molecular impacts of fungal–metal interactions, but thorough knowledge is lacking in filamentous fungi [102–104]. Due to the increase in fungal-metal interactions, we should ensure that metal resistance mechanisms in multiple types of fungi are well-understood. In this review, we summarize existing knowledge on fungal metal homeostasis of zinc, copper, iron, manganese, and silver. Conclusions and indications are presented to pave the way for further research.

Table 3. Mechanisms of metal resistance in yeast and filamentous fungi.

Metal	Mechanism of Metal Resistance in Yeast	Reference	Mechanism of Metal Resistance in Filamentous Fungi	Reference
Zinc	Up-regulation of *ZRC1* and *COT1*	[83,105–108]	storage of excess zinc in vacuoles and cell walls of spores and hyphae	[109,110]
	-	-	zinc efflux	[111]
	-	-	zinc metallothioneins	[112]
Copper	Up-regulation of *CUP1* and *CRS5*	[113]	Up-regulation of *crpA*	[81,114–116]
	Down-regulation of *FRE1* and *FRE7*, and *CTR1*	[113]	increased production of chelator copper oxalate	[117–119]
Iron	Up-regulation of *CCC1*	[64,120]	Unknown, but could associated with reduction of siderophore biosynthesis	[60,121]
	Expression of plant ferritin genes	[122–124]	-	-
Manganese	Up-regulation of *MNR1*	[65,67,125,126]	Deletion of *PcPHO84*	[68]
	Down-regulation of *PHO84, SMF1*	[67,125,126]	Expression of *PcMNT*	[68]
Silver	Expression of *CUP1-1, CUP1-2*	[81,115,116]	Expression of *crpA*	[90]
	Down-regulation of *PHO84*	[116]	-	-

2. Fungal–Metal Interactions

Metals play critical roles in fungal homeostasis. They are required for various biochemical processes, usually as enzymatic cofactors. Metals most recognized for their importance in fungi are copper, iron, zinc, and manganese. Pertaining to zinc, approximately 5% of fungal proteomes correlate to zinc-binding proteins, and 8% of yeast genomes correlate to zinc-binding proteins. In the model yeast *S. cerevisiae*, large portions of these zinc-binding proteins are related to critical functions, including DNA binding (31% of zinc-binding proteins), the regulation of transcription (25%), transcription factor activity (19%), and response to chemical stimuli (15%) [105,107,108]. Fungal–copper interactions are necessary for the activation of metalloproteins involved in biochemical processes. This includes the activation of superoxide dismutase, which is responsible for cellular detoxification of reactive oxygen species (ROS), virulence in pathogenic species, and activation of cytochrome *c* oxidate, a catalyst within the electron transport chain [39,48]. Iron is also essential for fungal virulence in pathogenic species, most importantly as an integral component of iron-sulfur clusters which are required for the activation of nuclear proteins involved in DNA repair [61]. Manganese also plays a critical role in fungi, in particular, in filamentous species where it (or copper) is required for the activation of manganese peroxidase. Dependent on nutrient availability, white-rot fungi utilize manganese peroxidase as a secondary metabolite to depolymerize lignin for nutrients; others are manipulated for increased manganese peroxidase production and extraction for use in the degradation of organo-pollutants [96,98].

Very few metals that are not considered essential have also been identified in some fungal–metal interactions; these include magnesium and molybdenum. Magnesium is a well-known micronutrient in other eukaryotic organisms, however, its homeostasis in fungi is undetermined. Only in recent years has magnesium been identified as a requirement for virulence in the agriculturally relevant fungus *Magnaporthe oryzae* [127]. Molybdenum is a metal that is discussed significantly less in eukaryotic homeostasis. It has only been identified as a cofactor for four human proteins, and in fungi it has only been suggested that it plays an unidentified role as a nitrate reductase and a xanthine dehydrogenase [128,129]. Other metals such as silver, gold, lead, nickel, and cadmium have only been implicated in fungal–metal interactions related to toxicity, nanoparticle myco-synthesis, and heavy metal myco-remediation, but information pertaining to homeostasis is limited [103,130,131].

2.1. Zinc

Zinc is a transition metal required for fungal survival and is necessary for various functions, including the structuring of nucleic acids, physical growth and, most predominately, protein folding [132,133]. In its role in DNA binding, zinc presents itself in class III zinc finger proteins, also known as zinc cluster proteins ($Zn(II)_2Cys_6$), found only in Ascomycetes (with the singular exception of *Lentinus edodes*) [107,134–136]. This protein class binds DNA, which is critical for the transcriptional activation and regulation of gene products [105,134].

In agriculture, fungal infections threaten food security by increasing global crop loss [137,138]. Traditionally, antifungal azoles have been used to combat disease, but with the emergence of azole-resistant pathogens, scientists have begun to develop possible alternatives, such as zinc-containing compounds [138,139]. Reports have demonstrated that zinc oxide nanoparticles (ZnO NPs) can control postharvest mold, plant wilts, and grey mold disease caused by *Aspergillus niger*, *Fusarium oxysporum*, and *Botrytis cinerea*, respectively [7,140–143]. It has also been demonstrated that ZnO NPs can significantly reduce the production of the mycotoxin fusaric acid from *F. oxysporum* [144]. This is significant because mycotoxins are common secondary metabolites of fungal pathogens with high rates of toxicity against cereal crops that can result in crop loss, and if consumed can result in a wide array of diseases in livestock [145,146]. Fusaric acid, in particular, can inhibit the production of dopamine-beta-hydroxylase, which acts as a messenger of signals within the nervous system and is responsible for altering the enzyme tyrosine hydrolase, which is involved in a rate-limiting step in catecholamine synthesis [147–149]. Zinc perchlorate $Zn(ClO_4)_2$ and zinc sulfate ($ZnSO_4$) also inhibit mycelial growth that produces mycotoxins and reduces the production of mycotoxins themselves [150,151].

2.1.1. Zinc Transport and Homeostasis

Many fungi have mechanisms of zinc transport similar to that of other eukaryotes, through the ZRT (zinc regulated transporter)-IRT (iron-regulated transporter)-like protein (ZIP) family and the cation diffusor facilitator (CDF) protein family [152,153]. In *S. cerevisiae*, zinc transport occurs through several protein groups; the ZIP protein family (via Zrt1, Zrt2, and Zrt3), the CDF protein family (via Zrc1, Cot1, and Msc2), the ferrous transport protein Fet4, and others (Figure 1) [105,107,108,133,154,155]. Zrt1 and Zrt2 are high and low-affinity plasma membrane transporters, respectively; both *ZRT1* and *ZRT2* are upregulated in zinc-deficient conditions and repressed when zinc conditions are favorable [15,22,156]. In an excess-zinc environment, Zrc1 and Cot1 mediate zinc transport from the cytosol into the vacuole to prevent toxicity [20,28]. In a zinc-limiting environment, zinc is released back into the cytosol from the vacuole via Zrt3 or is scavenged by zincophore Zps1 [106,157,158]. Zap1 regulates the transcription of *ZPS1* and contains two activators, Ad1 and Ad2, either independently activated or inactivated by the direct binding of zinc ions [105,108,159,160]. These mechanisms effectively control intracellular zinc uptake and help prevent excess accumulation in *S. cerevisiae*.

In filamentous Ascomycota, such as *Apergillus fumigatus*, genes in the ZIP family (*zrfA*, *zrfB*, *zrfC*, *zrfD*, *zrfE* and *zrfH*) also regulate zinc transport [23,27,161]. *zrfA* and *zrfB*, orthologues of *S. cerevisiae ZRT1* and *ZRT2*, respectively, encode zinc membrane transporters that operate in acidic, low-zinc environments and are activated by transcription factor ZafA [161,162]. Conversely, the *zrfC* gene product is an alkaline zinc transporter activated in high pH, zinc-limiting conditions [23,27]. *zrfD/E/H* are not restricted by pH and can function in either acidic or alkaline environments [23]. In *F. oxysporum*, *zrfA* and *zrfB* are also zinc importers regulated by transcription factor ZafA [163]. During infection, ZafA allows *F. oxysporum* to adapt to a zinc-limiting environment, such as if the host enacts nutritional immunity to deprive it of this essential metal [163]. Basidiomycetes have similar homology. *Ustilago maydis* UmZRT1 and UmZRT2 genes, and *Cryptococcus neoformans* Zip1 and Zip2 are homologous to *S. cerevisiae ZRT1* and *ZRT2*, respectively, with similar transport function [20,24,26]. Similarities also exist in the prevention of zinc over-accumulation. *C. neoformans* Zrc1 is homologous to *S. cerevisiae* Zrc1 and mediates zinc transport into the vacuole to prevent toxicity and decrease zinc sensitivity [20].

Mechanisms of zinc uptake and transport in fungi are mostly conserved through *S. cerevisiae* ZIP proteins and homologs. The next section will discuss how negative homeostatic interventions can result in toxicity.

2.1.2. Zinc Toxicity

Zinc-based antifungal compounds have mechanisms of toxicity that vary between species. Zinc pyrithione (ZPT), is a zinc ionophore often used to treat fungal dandruff caused by *Malassezia* spp. and induces toxicity by increasing cellular zinc uptake [164,165]. ZPT also causes partial mitochondrial malfunction by inhibiting mitochondrial synthesis of iron-sulfur clusters, which are integral in electron transport, respiration, and DNA repair and replication [165,166]. In contrast to *Malassezia* spp., ZPT toxicity in *S. cerevisiae* is not a result of increased zinc uptake, rather of increased copper uptake which overloads homeostatic systems [164,167,168]. ZnO NPs are also being explored for their antifungal properties. In *S. cerevisiae*, ZnO NPs reduce ergosterol biosynthesis which, in turn, increases cellular leakage (up to 24%) and depolarization, reduces cell wall integrity, and increases the occurrence of ROS [83]. In filamentous fungi, mechanisms of toxicity are not well-studied. In ericoid fungi, zinc ions reduced hyphal growth by increasing chitin deposition within the cell wall, preventing hyphal extension; zinc also increases hyphal branching and apical swelling, resulting in atypical hyphal morphology [88]. In the molds, excessive zinc exposure reduces hyphal growth, alters hyphal morphology and interrupts conidia and conidiophore development, limiting reproductive capabilities [87]. Zinc sensitivity can aid in the reduction of fungal pests; however, the development of tolerance and resistance can be an impedance.

2.1.3. Zinc Tolerance and Resistance

High-zinc environments can be detrimental to fungi; therefore they must possess resistance mechanisms to overcome toxicity. In yeast, resistance relies on the upregulation of Zrc1 and Cot1, which sequester Zn^{2+} to the vacuole (up to 100 mM) in *S. cerevisiae*, or the endoplasmic reticulum in *C. albicans* (Zrc1) [105–108,169]. Khouja et al. also described a resistance mechanism via OmFET in *S. cerevisiae*, though it is not yet fully understood [170]. They suggest that OmFET plays a role in Zn^{2+} uptake, and in that role increases tolerance through interactions with Mg, where Mg competes with Zn^{2+} for uptake, increasing intracellular Mg and restricting Zn [170]. In filamentous fungi, zinc resistance is not only attributed to vacuolar sequestration, but also to storage in the cytoplasm, storage in cell walls of spores and hyphae, and cellular efflux; and in ectomycorrhizal fungi, the presence of metallothionein-like peptides confers Zn^{2+} resistance [110,171–173]. To further investigate how fungi cope with toxic levels of other micronutrient metals, this review also assessed cellular interactions with copper.

2.2. *Copper*

Copper is also a transition metal and presents itself in oxidation states copper(I), Cu^{+}, and copper(II), Cu^{2+} [32,48]. It is essential to agriculture and human medicine where it can serve as a fungicidal or fungistatic agent, or be the determining factor for virulence [174,175]. Some fungal pathogens heavily rely on copper exporters to prevent host-enacted copper toxicity or import machinery to maintain virulence. In both clinical and agricultural settings, fungal exposure to excess copper can result in ionic imbalance. Therefore, homeostatic mechanisms to maintain healthy intracellular copper levels are critical.

2.2.1. Copper Transport and Homeostasis

Generally, copper cannot permeate the plasma membrane and requires membrane transporters for uptake [32,48]. Before internalization, copper must exist as Cu^{+} (cuprous oxide); however, in the environment, it often exists as Cu^{2+} (cupric oxide) and must undergo reduction. In *S. cerevisiae*, cupric reductase Fre1, transcribed by Mac1, reduces Cu^{2+} to Cu^{+}, making it readily available for uptake via high–affinity membrane transporters of the copper transporter (Ctr) protein family, Ctr1 and Ctr3 or low-affinity copper transporter Fet4 (Figure 2) [30–33,176]. Transcription of *CTR1* and *CTR3* is also regulated by transcription factor Mac1, which regulates transcription based on copper availability; copper depletion results in the upregulation of *CTR1/3*, and copper repletion results in downregulation [30,177].

After uptake, Cu^{+} serves as enzymatic cofactors. Apoproteins within the secretory pathway require copper for proper functioning, such as the multicopper oxidase Fet3, which is necessary for ferrous iron, Fe(II), uptake, and oxidation [53,178–180]. *FET3* is regulated by transcription factor Aft1 (activator of ferrous transport) in iron-deficient conditions and its gene product contains four Cu^{+} binding sites where copper serves as a cofactor for enzyme activation [53,178,181]. Unmetalated Fet3 reduces cell growth in iron-limiting conditions, demonstrating the importance of copper transport [44,182].

Another enzyme dependent on copper is the cytoplasmic Cu/Zn superoxide dismutase (Sod1). This is an antioxidant for superoxide anions ($O_2^{\bullet-}$) [183,184]. $O_2^{\bullet-}$ are ROS that cause cellular damage and toxicity and must be effectively dismutated to prevent stress; therefore, delivery of copper to Sod1 is critical [45,185,186]. In *S. cerevisiae*, the cytosolic copper chaperone Lys7 acquires Cu^{+} and delivers it to Sod1, with high specificity [45]. Once Sod1 is metalated, it is then able to catalyze the dismutation reaction that results in $O_2^{\bullet-}$ being successfully detoxified to hydrogen peroxide (H_2O_2) and molecular oxygen (O_2); H_2O_2 is now readily available for further detoxification to water via catalyst Cct1 [183,187,188]. Cu^+ transport to MTs Cup1 (also known as Cup1-1 and Cup1-2) and Crs5 is also integral to cellular detoxification [18,189]. Both MTs are regulated by transcription factor Ace1 (also known as Cup2), which activates the transcription of *CUP1* and *CRS5* at elevated copper concentrations [167,189]. Cup1 and Crs5 contain 8 and 11-12 Cu^+ binding sites, respectively, and are responsible for buffering cytosolic copper to maintain safe intracellular copper concentrations [189–191]. Though Crs5 has a greater copper binding capacity, it plays a much smaller role in detoxification due to its promoter region, which only has one recognition sequence, compared to four in *CUP1* [189–191].

S. pombe follows a pattern of copper transport similar to *S. cerevisiae*. Extracellular Cu^{2+} is reduced to Cu^+ by cell surface reductases before uptake [34,36]. Cu^+ can then be transported across the cell membrane, depending on the current cell cycle [34–36]. During mitosis, an integral membrane complex composed of proteins Ctr4 and Ctr5 are responsible for Cu^+ uptake, and during meiosis, Mfc1 (localized in the forespore membrane) is responsible [34–36]. Expression of $ctr4^+$ and $ctr5^+$ is regulated by transcription factor Cuf1, and expression of $mfc1^+$ is regulated by transcription factor Mca1, both of which are activated or deactivated by the absence or presence of sufficient copper levels, respectively [34,36]. Once inside the cell, copper chaperones such as Cox17, Pccs, and Atx1 transport Cu^+ to respective organelles [46]. Pccs is a four domain, cytosolic chaperone. The first three domains are responsible for transporting Cu^+ to unmetalated Sod1 in a copper-limited environment, activating Sod1 [46]. In high copper environments, the fourth domain acts as a copper buffering system, sequestering Cu^+ to prevent toxic cytosolic levels [46]. Atx1 in *S. pombe* plays a similar role to Atx1 in *S. cerevisiae*. In *S. pombe*, Atx1 is also located in the cytosol and responsible for carrying Cu^+ to Ccc2 [34,42]. Peter et al. and Beaudoin et al. described how Atx1 was also used for copper transport to copper amine oxidases (CAOs), a group of catalysts not present in *S. cerevisiae* [34,42]. Atx1 shuttles Cu^+ to an active site on the CAO, where copper (and another required cofactor, 2, 4, 5-trihydroxyphenylalanine quinone) activates it [34,42,192]. *S. pombe's* Cox17 is an orthologue to *S. cerevisiae* Cox17, sharing 38% identity and is located in the mitochondrial intermembrane space [42,48]. Once Cox17 acquires Cu^+ it is delivered to Sco1, Sco2, and Cox11 for copper loading to cytochrome c oxidase subunits [42,47,48].

Filamentous fungi are also important in assessing copper homeostasis, as these organisms depend on copper for growth and virulence in pathogenic species. In the pathogenic Ascomycete *Aspergillus fumigatus*, studies have shown similarities to *S. cerevisiae* and *S. pombe* in copper uptake. Cu^{2+} must also be reduced before uptake, however, there is some ambiguity regarding the reductases responsible [39]. This reductase has been referred to as unknown ferric reductase ("Fre?"), a general Fre reductase, and metallo-reductase Afu8g01310 (homolog of *S. cerevisiae FRE* or *FRE3*) [39,193,194]. After reduction, CtrA2 and CtrC (both homologs of *S. cerevisiae* Ctr1) transport Cu^+ into the cytosol and serve as enzymatic cofactors [37,39]. CtrA2 and CtrC are regulated by transcription factor MacA (also referred to as AfMac1) which senses low copper concentrations and activates CtrA2 and CtrC [39,49,195,196]. Conversely, in high copper concentrations, transcription factor AceA activates P-type ATPase CrpA as a defense mechanism for copper export and is responsible for extended life and virulence [39,49,195,196].

Limited knowledge exists on copper homeostasis in Basidiomycetes. Studies in two Basidiomycetes, the brow-rot fungus *Fibroporia radiculosa* and the edible white-rot fungus *Pleurotus ostreatus*, reported some details. In *F. radiculosa*, only the regulation of intracellular Cu^+ concentration has been unveiled, by three, unnamed copper ATPases and one gene of unknown function, CutC, [197]. In *P. ostreatus*, membrane protein Ctr1 is involved in copper uptake and shares homology with the low-affinity copper transporter PaCtr2 of the Ascomycete *Podospora anserine* (20%) and the high-affinity *S. cerevisiae* copper transporter, Ctr1 (20%) [38]. This review shows that copper homeostasis is well-studied in *S. cerevisiae* and *S. pombe*; however, more research is needed in other Ascomycetes and Basidiomycetes.

2.2.2. Copper Toxicity

Copper contains antifungals that have been investigated against various fungi. In *S. cerevisiae*, cupric sulfate ($CuSO_4$) and copper oxide nanoparticles (CuO NPs) significantly reduce growth in a dose dependent manner, with the toxicity of both potentially related to Cup2 [113,198,199]. Deletion of *CUP2* increases copper sensitivity, suggesting that a mechanism of toxicity could be reducing or inactivating its regulation, resulting in decreased Cu^+/MT binding and increased cytosolic Cu^+ [200,201]. Giannousi et al. found that CuO NPs cause DNA damage that interferes with replication and increases lipid peroxidation, reducing membrane lipid content, resulting in porous cells [202]. In *Candida* spp., CuO NPs have also shown toxic capabilities by inducing porous cell membranes [12]. Copper(II) complexes which have been shown to exhibit fungicidal and fungistatic activity in species that have a history of azole resistance appear to have a similar mechanism by reducing ergosterol content [203–207]. In filamentous fungi, copper also has dose-dependent toxicity. In the agricultural pathogen *Rhizoctonia solani*, a copper (II)–lignin hybrid had high efficacy and significantly reduced the number of plants attacked by *R. solani* [84,208]. In some instances, fungi can overcome toxicity by increasing their tolerance, which may be beneficial in the case of nanoparticle production, but can become a nuisance in pathogenic species.

2.2.3. Copper Tolerance and Resistance

Since copper is implicated as an antifungal agent, its ability to evade copper toxicity must be continuously evaluated. In *S. cerevisiae*, short-term exposure to $CuSO_4$ causes significant regulation of open reading frames (ORFs) responsible for cellular detoxification and Cu^+ uptake [113]. Exposure results in the upregulation of *CUP1* (~20-fold,) and *CRS5* (~8-fold) and the downregulation of *FRE1, FRE7,* and *CTR1* (0.07, 0.08, and 0.10-fold, respectively) [113]. This fold change, and increased $CuSO_4$ sensitivity in *cup2*Δ mutants indicates MTs, coupled with decreased Cu^{2+} reduction and decreased Cu^+ uptake, are likely to be employed as mechanisms of copper resistance [113,200].

Less is known about copper resistance in filamentous fungi. In *Aspergillus* spp., P-type ATPase CrpA has Cu^+ exporting activity that aids in cellular detoxification, increasing Cu^+ resistance [90,193,209]. High-affinity copper importers, CtrA2 and CtrC, may be involved in resistance, but are still under investigation [37,49]. In *Fusarium graminearum*, copper exposure upregulates *FgCrpA* (ATPase exporter) and the MT *FgCrdA* as a means to prevent over accumulation, with the predominate method being Cu^+ export activity [14]. In *F. oxysporum*, upregulation of oxidoreductase activity may decrease susceptibility to oxidative stress that can be induced by excessive copper exposure [210]. In Basidiomycetes, some progress has been made in identifying resistance mechanisms in *F. radiculosa*, where increased production of copper oxalate increases resistance [119]. However, this is the extent of the knowledge.

2.3. Iron

Iron (Fe) is a transition metal belonging to group eight of the periodic table and can exist as ferrous (Fe^{2+}) or ferric (Fe^{3+}) iron [211,212]. As an essential nutrient, Fe is significant for the virulence of fungi that cause disease. In *A. fumigatus* and *F. oxysporum*, survival depends on the ability to sequester iron from the host and a well-functioning homeostatic system to maintain this delicate balance [213,214]. Incapacitating the ability to do so reduces virulence and becomes a growth limiting factor, such as in the use of excessive amounts of Fe to completely overrun homeostatic systems [215–217]. Thus, homeostatic mechanisms are integral.

2.3.1. Iron Transport and Homeostasis

Generally, in *S. cerevisiae*, two iron uptake systems are described, the reductive and nonreductive systems. The reductive system recognizes Fe^{2+} salts and chelates for uptake through importers, while the nonreductive systems utilizes iron siderophores [218–221].

In the reductive system, high-affinity (aerobic) and low-affinity (anaerobic) transporters are responsible for ferric and ferrous iron transport, respectively [222,223]. For low-affinity uptake, iron must be reduced by ferric reductases Fre1 or Fre2, initially described by Lesuisse et al. in 1987 and later coined Fre1 and Fre2 by Georgatsou and Alexandrakin in 1994 [221,224]. Since then, both metallo-reductases have also been found to reduce both cupric and ferric ions, where *FRE1* expression induces the reduction of Cu^{2+} when transcription factor Mac1 is bound, and Fe^{3+} reduction occurs via binding of transcription factor Aft1 [176,181,225]. After Fe^{3+} reduction, Fe^{2+} is then ready for uptake by a six domain, transmembrane, metal transporter, Fet4 [54,55]. Fet4 can also import other metals, but is mostly responsible for Fe^{2+} uptake in iron-restricted cells [223]. In anaerobic conditions, transcription factor Aft1 is required for activation, and in aerobic conditions, expression of *FET4* is repressed by Rox1, which has two binding sites in the *FET4* promoter region [223]. This repression is necessary to prevent the unintended uptake of toxic metals, such as Cd, where it is demonstrated that *rox1Δ* mutants have increased sensitivity to Cd under aerobic conditions [223,226]. A second, less utilized iron transporter in the low-affinity uptake system is Smf1, responsible for the uptake of the Fe^{2+}/H complex [51,52]. This metal transporter is mostly known for the uptake of Cu, Mn, and Cd; however, in a study completed by Cohen et al. in 2000, it was shown that overexpression of *SMF1* also results in significant iron uptake [52,65,227]. High-affinity iron uptake is also part of the reductive system. In low-iron conditions, this system dissociates and reduces ferric iron, via Fre1 and Fre2, from a wide array of Fe^{3+} substrates such as ferric chelates, salts, and siderophores [218,219]. Fe^{2+} then transitions through the Fet3/Ftr1 complex [58]. Fet3 is activated by transcription factor Aft1 in iron-deficient conditions and contains four Cu^{+} binding domains that must be metalated for activation [53,178,181]. Activated Fet3 goes through an aerobic reaction that oxidizes Fe^{2+} to Fe^{3+} for passage to the cytosol via iron permease Ftr1 [58,178]. The final destination and the cell's utilization of Fe^{3+} is not fully elucidated.

The nonreductive system utilizes siderophores. *S. cerevisiae* is incapable of producing siderophores, but can sequester siderophores produced by other microorganisms via siderophore iron plasma membrane transporters Arn1—Arn4 [16,224]. Arn1 transports ferrichrome into the cell for iron acquisition; however, Arn1 is not always readily available in the plasma membrane because it is localized to endosomes or is routed to vacuoles for degradation when ferrichrome is unavailable [16,220,228]. When ferrichrome is present, Arn1 is routed through the plasma membrane, where ferrichrome adheres to either the low or high-affinity binding site and is transported to the cytosol [16,220,228]. It remains intact in the cytosol and serves as an intracellular Fe^{3+} storage reservoir until the cell needs iron; in this event, Fe^{3+} is reduced via metallo-reductases, or released via ferrichrome degradation [16,220,228,229]. Arn2 (also known as Taf1) is the second siderophore transporter in the *ARN* family, responsible for transporting tri-acetyl-fusarinine to the cytosol; it is unclear if Arn2 is located anywhere else aside from the plasma membrane when tri-acetyl-fusarinine is unavailable [220,230,231]. The literature is not very informative on the functions of tri-acetyl-fusarinine, but it does appear to have a similar role to ferrichrome as a store reservoir for ferric iron [230,231]. Arn3 (also known as Sit1) is a transporter for ferrioxamine B and is situated within intracellular vesicles. It appears to have a similar function to Arn1 and can progress to the plasma membrane when ferrioxamine B is available [229,232]. After ferrioxamine B is transported inside the cell, it is stored in the vacuole, likely for subsequent dissociation [232]. The first three mentioned siderophores transported by Arn1–Arn3 belong to the hydroxamate class of siderophores. However, the final transporter Arn4 (also known as Enb1) transports a siderophore of the catecholate class, ferric entero-bactin [220,233]. Unlike the other siderophore transporters, Arn4 remains at the plasma membrane regardless of the presence of its substrate [218]. Philpott and Protchenko suggested the difference in plasma membrane cycling between hydroxamate and catecholate transporters may be due to the possibility that there are toxins that can adhere to the hydroxamate transporters and not the catecholate transporters [218]. In the act of self-preservation, those transporters remove themselves as a potential source of toxicity [218]. Ferric entero-bactin is not well-studied in *S. cerevisiae*, but based on the function of other siderophores it may be reasonable to conclude that, upon cellular entry, ferric entero-bactin is also used as an Fe^{3+} storage system.

After Fe uptake, there are many intracellular destinations. Two briefly discussed here are the cytosol and the nucleus [61,62]. In the cytosol, iron–sulfur assembly (CIA) proteins Npb35 (binds two Fe–S clusters), Nar1, Cfd1 (binds one Fe–S cluster), and Cia1 form an iron–sulfur complex [61,62,234]. These complexes transfer Fe–S clusters to various apoproteins for activation [61,62,234]. In the nucleus, CIA proteins deliver Fe–S clusters to various nuclear proteins involved in DNA repair and replication [61,235].

Iron homeostasis in the fission yeast *S. pombe* is also well-studied and has three mechanisms of iron uptake [236]. One involves cell surface ferric reduction, and the other, in contrast to *S. cerevisiae*, involves the production of siderophores to capture extracellular iron and heme [236]. The first iron uptake system described here is through use of siderophore synthesis [237]. Under iron-deficient conditions, Sib2, a catalyst for ferrichrome synthesis, hydroxylates ornithine to N^5-hydroxyornithine, a newly formed hydroxy-mate group molecule, and then undergoes processing by Sib1 [236,237]. This non-ribosomal peptide synthase yields the desferri-form of ferrichrome [236,237]. Schrettl, Winkelmann, and Haas suggested that the resulting ferrichrome is excreted from the cell to capture extracellular Fe^{3+} from the surrounding environment [237]. In an iron-dependent response, transcription factor Fep1 activates ferrichrome transporters Str1, Str2, and Str3, and the iron-loaded ferrichrome re-enters the cell (predominately by way of Str1) [59,63]. *S. pombe* is also able to import exogenous iron-loaded ferrioxamine B via Str2 [63]. In addition to the previously mentioned siderophore functions, it had also been suggested that, as in *S. cerevisiae*, imported siderophores also serve as iron storage vesicles [63,237].

The second iron uptake mechanism employed by *S. pombe* is the high-affinity, reductive system that depends on cell surface ferric reductase Frp1. *frp1*$^+$ shares 27% homology with the *S. cerevisiae* Fe^{3+}/Cu^{2+} reductase encoding gene, *FRE1*, and reduces extracellular Fe^{3+} to Fe^{2+} [238]. Transcription of *frp1*$^+$ may also have some functional relation to the vacuole/cytoplasmic transporter Abc3 that transports iron from the vacuole to the cytosol in iron-deficient conditions [238,239]. Pouliot et al. found that *abc3*Δ mutants resulted in the activation of *frp1*$^+$; however, a nucleotide-based transcription factor directly linked to *frp1*$^+$ has not yet been determined and it appears to be solely activated or repressed by the absence or presence of iron, respectively [238,239]. After ferric reduction, Fe^{2+} enters an oxidase-permease complex, similar to that of the *S. cerevisiae* Fet3/Ftr1 complex, composed of proteins Fio1 and Fip1 [50]. Fio1 is a Fe^{2+} oxidase that shares 37% homology with the *S. cerevisiae* Fet3, and in an iron deprived environment, oxidizes Fe^{2+} in preparation for transfer across the plasma membrane via Fip1 [50]. fip1$^+$ is a ferrous permease having 46% homology with the *S. cerevisiae* Ftr1 [50,236].

Heme is an iron-containing compound and its acquisition and biosynthesis are the finally discussed mechanisms of iron uptake in *S. pombe*. It is notable to state that while *S. cerevisiae* does utilize heme in other processes such as respiration and ergosterol biosynthesis, it has not been determined to be used to acquire iron [240,241]. *S. pombe* imports exogenous heme for iron uptake through Str3 and Shu1 [56,57]. Shu1 is a plasma membrane protein induced during iron deprivation, when heme biosynthesis is not attainable, or if Fep1 is inactivated [56,57]. The second protein involved in heme uptake is Str3, previously mentioned as a part of a ferrichrome transporter family (Str1, Str2, and Str3). Str3 shares the lowest homology (25.1%) with Str1 when compared to Str2 (29%), and its substrate specificity is undetermined [57,59,63]. Iron release and utilization from heme is not yet fully understood in *S. pombe*; however, studies in *C. albicans* (and other fungi) show that heme degradation is catalyzed upon cellular entry via heme oxygenase [56,57,242]. *S. pombe* also biosynthesizes heme and is encoded by *hem1*$^+$, *hem2*$^+$, *hem3*$^+$, *hem12*$^+$, *hem13*$^+$, *hem14*$^+$, *hem15*$^+$, and *ups1*$^+$ [56,57]. In iron-deficient conditions, a cascade of events between the mitochondria and the cytoplasm occurs to synthesize heme for further utilization [56,57,243].

In addition to iron acquisition in *S. pombe*, regulation mechanisms must be in place to prevent over-accumulation. Mercier, Pelletier, and Labbé identified the gene *pcl1*$^+$ to play a role in vacuolar iron storage [244]. *pcl1*$^+$ shares homology to *S. cerevisiae* Ccc1, an iron vacuolar transporter, and it has been shown that *pcl1*Δ mutants have increased sensitivity to iron; this together with the study of Mercier, Pelletier, and Labbé suggests that Pcl1 might play a similar role in iron storage in *S. pombe* [63,239]. As mentioned, the final destinations of heme are somewhat unclear, but based on research in other fungi, heme may be degraded, and literature suggested that there may be a group of proteins responsible for transporting those ions to the vacuole for degradation or storage [56,57,245]. Much is known about iron homeostasis in *S. pombe;* however, there are apparent gaps in knowledge of specific processes.

In filamentous fungi, iron homeostasis is less documented. It has been investigated in *U. maydis*, a pathogenic fungus that causes corn smut disease and whose virulence is associated with iron acquisition [60,121]. There are two iron uptake mechanisms, one through hydroxamate siderophores, and the other an oxidase-permease system, similar to *S. pombe* [60,233,246,247]. In the latter, exogenous ferric iron is reduced by a seemingly unknown reductase (possibly Fer9) and then re-oxidized by ferroxidase Fer1 for uptake through the high-affinity ferric iron permease Fer2 [60,121]. In the former, siderophore iron uptake is mediated by siderophore biosynthesis encoding genes *Sdi1* and *Sid2*, and both are negatively regulated by transcription factor Urbs1 [60,121,246,247]. These siderophores play a role in iron acquisition; however, deletion mutants showed they are not necessary for virulence [121].

2.3.2. Iron Toxicity

Iron is involved in many biological processes, but can be toxic in excess. Studies have shown its toxicity in *S. cerevisiae* and fungal pathogens, but they have also demonstrated that targeting and interfering in iron acquisition mechanisms can also be detrimental. Reports indicate that iron or iron compounds are fungistatic against *F. oxysporum* and its mycotoxins in a dose dependent manner [216,217]. In discussing iron toxicity, it is also important to note that the interference of homeostatic systems can result in the inhibition of iron acquisition, which can also be toxic. Leal et al. demonstrated this with the utilization of lactoferrin, an iron-binding glycoprotein, as a topical agent to obstruct iron uptake mechanisms of *A. fumigatus* and *F. oxysporum* in mice [92,93]. Results indicated that, during corneal fungal infection, these fungi acquired iron through siderophores and that the iron-binding agent blocked the ability of the pathogen to acquire siderophore-bound iron, highlighting the inability of the fungi to proliferate without access to iron [93]. In *S. cerevisiae*, iron toxicity is related to the ability of the cell to transport cytosolic iron to the vacuole via Ccc1 [91,248]. Lin et al. showed that *ccc1*Δ mutants could not transfer cytosolic iron to the vacuole under anaerobic conditions, even with the overexpression of iron mitochondrial transporter Mrs3, effectively inducing toxicity [248]. The ability to alter and control homeostatic mechanisms are determinants of the fungal ability to resist excessive iron concentrations.

2.3.3. Iron Tolerance and Resistance

S. cerevisiae achieves iron resistance through the downregulation of iron import systems via Aft1, or activation of vacuolar transporter Ccc1 [64,249]. Ccc1 is regulated by the iron sensitive transcription factor Yap5; removal of *YAP5* increases iron sensitivity, while its overexpression dramatically reduces cytosolic iron [120]. It may be worth the effort to investigate how the overexpression of *CCC1* affects iron resistance and the vacuolar ability to store excess iron in order to prevent toxicity. Another vacuolar gene, *VMA13*, might also play a potentially novel role in iron tolerance [250]. Vma13 is commonly known as a vacuolar H^+-ATPase subunit that plays a role in vacuolar acidification; however, a study involving *vma13*Δ mutants showed that they experienced increased sensitivity to iron deprivation, suggesting Vma13 plays a role in iron import [250]. The function of

VMA13 in iron homeostasis combined with its role in vacuolar acidification should be studied to determine if mutants can also help increase iron resistance. Another method of iron resistance in *S. cerevisiae* is the expression of ferritin related genes. Ferritin is an iron storage protein found in many other eukaryotes, but is not native to fungi [122–124]. Its effects on increased iron resistance and storage capacity in yeast has been investigated and results indicate that the expression of human, soybean, and tadpole ferritin genes (*HuFH*, SFerH1/SFerH2, and *TFH*, respectively) resulted in the increased ability of yeast to store and carry higher concentration of iron [122–124]. Llanos et al. showed the ability of soybean ferritin genes, SFerH1 and SFerH2, to increase iron resistance in *ccc1*Δ mutants [122]. This is significant because, even without the natural vacuolar detoxification system, yeast cells with soybean ferritin were still able to store increased concentrations of iron and evade toxicity.

Fewer studies report on iron resistance in other fungi, but several inferences can be made based on knowledge of iron homeostasis. In *S. pombe*, ferrichome production, excretion, and subsequent uptake are used to acquire extracellular Fe^{3+} in iron-deficient conditions [59,63,236,237,251]. The engineering of cells to overexpress Sib2 and Sib1 could potentially serve as extra storage vesicles for any excess cytosolic iron acquired by the cell [59,63,236,237,251]. It is not clear how ferrichrome is excreted from the cell after production, therefore this exact mechanism would first need to be identified and well-studied to determine if inhibiting excretion would have any other adverse effects on cellular health. *U. maydis* also biosynthesizes siderophores (hydroxamate) via *sid1* for iron uptake which could also be investigated for increased production for storage of excess iron [60,121].

2.4. Manganese

Manganese (Mn) is a transition metal and also an essential micronutrient in fungi. In agriculture, Mn compounds reduce mycelial growth of fungal pathogens [252,253]. In other pathogenic fungi, Mn^{2+} is required for virulence [254]. Some lignocellulose degrading enzymes also require Mn^{2+}, such as manganese-dependent peroxidase, which white-rot fungi express during lignocellulose degradation, integral to nutrient uptake [255,256]. Many fungal species rely on Mn^{2+} and homeostatic mechanisms must exist to ensure proliferation.

2.4.1. Manganese Transport and Homeostasis

Within *S. cerevisiae*, Mn^{2+} transporters Smf1 and Smf2 (part of the Nramp metal transporter family) and phosphate transporter Pho84, have a diverging consensus on their roles in Mn^{2+} homeostasis. In the case of Smf1, it was initially determined to be a high-affinity plasma membrane transporter, which acquired extracellular Mn^{2+} in Mn^{2+} deficient environments [65,66]. Smf2 is localized in golgi-like vesicles and shares approximately 50% identity with Smf1 (at the amino acid level), but does not share functionality and is a low-affinity Mn^{2+} transporter [77,257]. Once inside the cell, the fate of Mn^{2+} is as a cofactor for proteins such as Sod2 [126,258]. Sod2 is a mitochondrial manganese superoxide dismutase that receives Mn^{2+} via Mtm1 for activation [69,126,258,259]. In *smf2*Δ mutants, the Sod2 primary protein structure accumulates in the mitochondria; however, they were mostly inactive due to inadequate Mn^{2+} transfer to the mitochondria, indicating that Smf2 is a requirement for *S. cerevisiae* Sod2 activity [126,258]. Smf1 and Smf2, unlike many other metal ion transporters discussed in this review, are not regulated at the transcriptional level, rather post-translationally by protein turnover and localization, which is directly related to Mn^{2+} availability [260]. When Mn^{2+} concentrations are stable or in excess (~100 nmol/(1 × 10^9 cells)), Smf1 and Smf2 are ubiquitinated via Rsp5 (a NEDD4 family E3 ubiquitin ligase) with the aid of Bsd2 and transferrin receptor-like proteins (Tre1 and Tre2) [260–262]. Smf1 and Smf2 are then trafficked to multivesicular bodies, which deliver the proteins to the vacuole for degradation [260,263,264]. This mechanism of action is supported by reports that *tre1*Δ, *tre2*Δ, and *bsd2*Δ mutants resulted in the accumulation of Smf1 and Smf2 [260–262]. Conversely, when Mn^{2+} starvation occurs, Bsd2 is depleted,

Smf1 is localized to the cell surface, Smf2 is localized to intracellular vesicles, and Smf1 and Smf2 resume their Mn^{2+} uptake functions [257,260,261].

The final transport system discussed here is the phosphate transporter Pho84. It was initially characterized in *S. cerevisiae* as a high-affinity, six-domain, transmembrane, inorganic phosphate transporter [265]. However, Pho84 is now also known as a low-affinity Mn^{2+} transporter, along with other metals such as cobalt, zinc, and copper [67]. Through *pho84*Δ mutants, it was shown that Mn^{2+} uptake was the most commonly affected (in relation to the other metals) when *PHO84* was removed, further proving its Mn^{2+} transporter role [67]. *PHO84* transcription is regulated by transcription factor Pho4, which inhibits Pho84 activity when it is phosphorylated in the presence of excess phosphate; Pho4 resumes transcription when phosphate levels are low [265,266].

Once Mn^{2+} is inside the cell, there are an array of destinations. Pmr1 (high-affinity Ca^{2+}/Mn^{2+} P-type ATPase) and Gdt1 (calcium/manganese transporter) both transport cytosolic Mn^{2+} to the Golgi lumen, where Mn^{2+} serves as a cofactor for mannosyl-transferases, such as Mnn1, Mnn2, Mnn5, and Mnn9, which glycosylate proteins in the secretory pathway [70–72,267–271]. This type of protein modification provides protein stability by preventing degradation, protecting against oxidative damage, and increasing thermodynamic equilibrium [272]. Concerning the ER, P-type ATPase Spf1 transports Mn^{2+} to the ER lumen; this is supported by a study showing that *spf1*Δ mutants had decreased luminal Mn^{2+}; its overexpression had the opposite effect [73]. This same study also stated that Mn^{2+} depletion observed in *spf1*Δ mutants negatively impacted luminal Mn^{2+} dependent processes. On the contrary, it positively impacted Mn^{2+} associated cytosolic processes, indicating that Spf1 is integral to *S. cerevisiae* manganese ER and cytosolic homeostasis [73].

Mn^{2+} accumulation can have severe consequences on cellular health, and systems must be in place to prevent subsequent events. We will discuss two defense mechanisms in *S. cerevisiae*, Mn^{2+} trafficking to vacuoles for storage and degradation and Mn^{2+} export. Pmr1, previously characterized as an Mn^{2+} Golgi lumen transporter, also serves as a detoxifier. Presented with toxic Mn^{2+} levels, Mn^{2+} is still transported to the Golgi lumen from the cytosol, but excess ions are delivered to secretory pathway vesicles, which ultimately exit the cell, completely removing toxic Mn^{2+} (Figure 3) [77,273]. The *HIP1* gene product also expresses export activity. Hip1 was initially characterized as a high-affinity, plasma membrane histidine permease, but has since been shown to play a role in Mn^{2+} resistance [78,274]. Farcasanu et al. investigated *S. cerevisiae* mutants having defects in Mn^{2+}transport and found that a mutation in the *HIP1* gene was responsible [78]. This mutation, originally a single base deletion, introduced a cascade of mutations that led to the protein Hip1-272 (272 amino acids long). Subsequent experiments showed that *hip1-272* mutants had significantly less cytosolic Mn^{2+} accumulation, increased Mn^{2+} efflux, and increased resistance than null mutants and wild type strains [78]. Further studies into the *hip1-272* mutant could elucidate the exact mechanisms of action of Mn^{2+} transport, determining how ions are trafficked to Hip1-272 and expelled. The second defense mechanism against Mn^{2+} toxicity in *S. cerevisiae* was Mn^{2+} trafficking to vacuoles through Ccc1 and Ypk9. Ccc1 (and possibly Cos16) is localized in the vacuolar membrane and is responsible for trafficking cytosolic Mn^{2+} to vacuoles; *CCC1* overexpression results in reduced Mn^{2+} toxicity, lower concentrations of cytosolic Mn^{2+}, and increased vacuolar concentrations (Figure 3) [64,75,77]. Ypk9 is also localized in the vacuolar membrane and shuttles Mn^{2+} to the vacuole. Gitler et al. and Schmidt et al. both demonstrated that *ypk9*Δ mutants expressed Mn^{2+} hypersensitivity when compared to wild type strains, further affirming Ypk9 involvement in Mn^{2+} homeostasis [74,76].

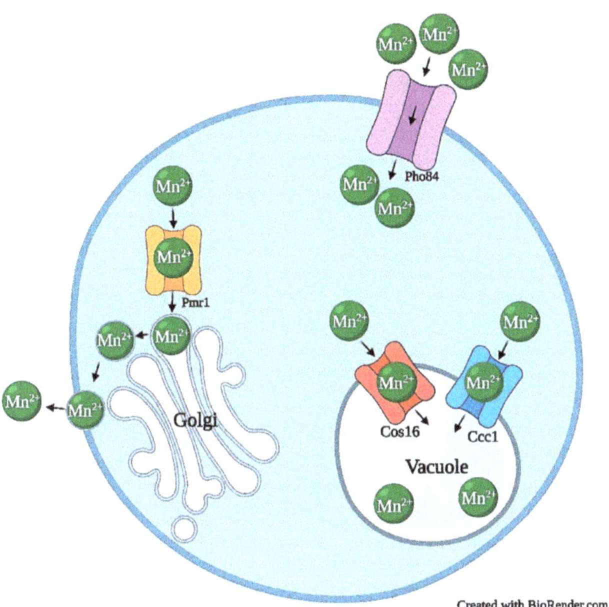

Figure 3. Mn^{2+} uptake and detoxification systems in *S. cerevisiae*.

Manganese homeostasis has not been well characterized in higher fungi, but *Phanerochaete chrysosporium* has received some attention. *P. chrysosporium* is a white-rot fungus that produces lignin-degrading enzymes, which have been useful in the biodegradation of various plant biomass and an array of organo-pollutants [275–277]. Manganese peroxidase is a common lignin depolymerizing peroxidase utilized by white-rot Basidiomycetes [278,279]. It acts in combination with other enzymes to convert various biomass to useful bio-products of commerce and agricultural operations [255,280–283]. Homologs of the *S. cerevisiae* Pho84 and Smf1/2 proteins have been found in *P. chrysosporium*, PcPho84 and PcSmfs, respectively. PcPho84 is a plasma membrane protein involved in Mn^{2+} uptake, having a similar function to its *S. cerevisiae* homolog [68]. Smf1/2 are predicted to have similar functions in *P. chrysosporium* to their *S. cerevisiae* homologs [68]. Intracellular Mn^{2+} transport has also been investigated. Yeast homolog PcAtx2, localized in the Golgi membrane, was shown to function as an antioxidant through *sod1Δ* mutants [68]. When grown on 600 μM paraquat (inducer of oxidative stress), *sod1Δ* mutants experienced almost no growth; however, in mutants expressing *PcATX2*, growth was restored, indicating that PcAtx2 exhibits similar antioxidant functionality as Sod1 [68]. In the case of mitochondrial transport, *S. cerevisiae* Mtm1 traffics Mn^{2+} to the mitochondria for Sod1 activation; however, the function of the *P. chrysosporium* homolog, PcMtm1 (localized in the mitochondrial membrane), has yet to be identified, but predicted to have a similar antioxidant activity [68]. PcMnt and PcCcc1 engage in Mn^{2+} storage and export in *P. chrysosporium*, respectively. In *Phanerochaete sordida*, PsMnt was found to be a homolog of yeast Smf2 and plays a role in Mn^{2+} uptake, suggesting that it could have dual functionality, but this is still unknown [77,284]. Limited information exists on Mn^{2+} homeostasis in other fungi; however, due to the impact of Mn^{2+} on lignin-degrading enzymes in wood-rotting fungi, more studies should be conducted. Overall, Mn^{2+} homeostasis is critical to cellular functioning to prevent toxic Mn^{2+} accumulation, detoxify cells of free radicals, and provide white-rot fungi with their capacity to degrade lignin. In the absence of such mechanisms, toxicity can impede proper functioning and cause cellular damage.

2.4.2. Manganese Toxicity

In the model yeast *S. cerevisiae*, excessive Mn^{2+} can overrun homeostatic systems and create a toxic ionic imbalance that negatively impacts survival rate [104,285,286]. Expression profiles show that high levels of Mn^{2+} down-regulate genes associated to histidine proteins (*HTB2*, *HTA1*, *HTA2*, *HTB1*, and *HHF*) that are compulsory in chromatin assembly chromosome functioning, and interface in this functioning can end in cell cycle arrest [94,95,114,287]. Filamentous fungi are often studied for their lignin degradation properties which focus on how Mn^{2+} impacts manganese peroxidase activity, but there is a lack of knowledge on how excess Mn^{2+} can be toxic towards this activity [96–98]. Due to the reliability of much of the lignin degrading properties on manganese peroxidase in many white-rot fungi, effects of Mn^{2+} over accumulation should be further investigated [96–98]. In some cases, toxicity can be avoided by resistance mechanisms.

2.4.3. Manganese Tolerance and Resistance

As with other metals, Mn^{2+} resistance is usually contingent upon homeostatic systems. In *S. cerevisiae*, several genes involved in resistance emanate from mutations. *MNR1* (also known as *HUM1* and *VCX1*), encodes a vacuolar H^+/Ca^{2+} antiporter, but has been implicated in Mn^{2+} resistance [125,288,289]. A single nucleotide alteration may affect Mnr1 function and result in increased Mn^{2+} sequestration to the vacuole [125,289,290]. A mutation in *PHO84* is also implicated in Mn^{2+} resistance, where *pho84Δ* mutants have increased resistance, likely due to the acquired inability to import and accumulate excess Mn^{2+} [67]. In filamentous fungi, Diss et al. elucidated potential resistance mechanisms through *P. chrysosporium*, where it was demonstrated that *PcPho84Δ* mutants increase Mn^{2+} resistance, as well as expression of *PcMNT*, which is likely to engage in Mn^{2+}export activity [68].

Up to this point, metals that serve as essential nutrients have been reviewed. In recent years, there has been an increase in studies on the usage of metals with no nutritional purpose, but which serve as antimicrobial agents, such as silver. This increase gives cause for further investigation into how these metals are metabolized and their intracellular functioning.

2.5. *Silver*

Silver (Ag) is a transition metal that shares similar properties to other transition metals in groups three through twelve, and closely resembles the properties of Cu and gold (Au) [291,292]. In fungi, silver is implicated in the eradication of pathogens. As part of agricultural research, silver nanoparticles (Ag NPs) and Ag ions (Ag^+) have demonstrated their ability to control plant pathogens [293–295]. As a feed additive, silver has a positive effect on the intestinal microflora, aflatoxins, and mycotoxin absorption in farm animals and in the food industry is used in food packaging for its antimicrobial properties [291,296,297]. Thus, the development of silver as an antimicrobial agent should continue to be investigated, especially on the development of fungal resistance and the impacts on non-target organisms.

2.5.1. Silver Transport and Homeostasis

Silver is a non-essential metal that has no designated cellular receptors or membrane channels for ion uptake. Much of the literature has focused on silver as an antimicrobial agent, but some studies have begun to clarify homeostatic mechanisms [81,114,116,298]. Silver has properties similar to copper, which has initiated the evaluation of copper homeostatic systems to investigate how they may contribute to silver uptake and transport [21,81,114,116,298].

In *S. cerevisiae*, Ctr1, high-affinity Cu^+ transporter, has been identified as a Ag^+ importer. This is based on observed reduced Ag^+ uptake in *ctr1Δ* mutants exposed to low silver concentrations, and transcriptional analysis that shows exposure to Ag NPs upregulates *CTR1* throughout the entire transcriptome [80,81]. The involvement of copper-related genes in Ag^+ homeostasis was also investigated by Hosiner et al. and Niazi et al.; both found that short-term exposure to silver resulted in increased expression of copper MTs Cup1-1 and Cup1-2, suggesting these MTs sequester Ag^+ in response to silver stress [114,115]. The competitiveness of Cu^+ and Ag^+ for Cup1-1 and Cup1-2 should be further investigated to determine which ion the MTs have a higher affinity for. Other metal ion transporters (Pho84, Fet3, and Smf1) have been investigated for their involvement in Ag^+ uptake, but results indicate they are not [81].

Once inside the cell, there are not many known Ag^+ destinations. $AgNO_3$ exposure results in Ag^+ accumulation in the mitochondria, which, in return, reduces Cu^+ accumulation in the mitochondrial matrix [21]. The direct result of this action is reduced copper-dependent cytochrome *c* oxidase activity, suggesting that cytosolic Ag^+ is trafficked to the mitochondria via Cu^+ mitochondrial transporter Pic2, potentially with a higher affinity, which can be toxic to cells by reducing the rate of cellular respiration [21]. No other intracellular destinations have been identified in yeast, and silver homeostasis in filamentous fungi is still unknown.

2.5.2. Silver Toxicity

Efflux systems are integral to cellular homeostasis, preventing the accumulation of toxic compounds within a cell. In *S. cerevisiae*, Ag^+ uptake can affect these systems, resulting in toxicity. Exposure to Ag^+ can increase the efflux rate of potassium ions (K^+) from *S. cerevisiae*, resulting in almost complete K^+ efflux from the cell. *S. cerevisiae* requires a minimum 30mM K^+, suggesting those events can be toxic if the ion concentration is not restored [299,300]. Another mechanism of Ag^+ toxicity is its ability to alter cellular structure [100,103]. Ionic fluids can affect cell membrane integrity of yeast *Yarrowia lipolytica*, reducing the amount of ergosterol, which fluidizes the membrane, and increases internal lateral pressures [100]. Ag^+ exposure can also deform the cell wall, which is a likely a response to the down-regulation of genes involved in ergosterol synthesis (*ERG3, ERG5, ERG6, ERG11, ERG25,* and *ERG28*) in *S. cerevisiae* [80,99]. In the aquatic fungus *Articulospora tetracladia*, transcriptome analysis via RNAseq revealed toxicity of Ag^+ and Ag NPs may result from interrupted functioning of plasma/organelle membranes and downregulation of genes associated with cellular redox [301]. Silver toxicity has also been studied in other agriculturally relevant processes and it has been determined that $AgNO_3$ and Ag NPs can be useful in pathogen control of plant diseases [174,293,295]. It may be worthwhile to investigate silver homeostasis in addressing long-term effects of exposure.

2.5.3. Silver Tolerance and Resistance

The worldwide increase of silver usage makes studies on mechanisms of silver resistance important; presently, few studies have reported on this. *CTR3* is implicated in Ag^+ resistance after an observed fold increase in its expression in a silver evolved strain of *S. cerevisiae* [116]. Insight into the expression of the Ctr3 transcription factor *MAC1* in the presence of Ag^+ may clarify its role in resistance. It is possible that MTs Cup1-1 and Cup1-2 are also involved in resistance. It was previously described that exposure to $AgNO_3$ and Ag NPs resulted in the increased expression of *CUP1-1* and *CUP1-2*, proposing that the encoded MTs may also bind Ag^+ and decrease sensitivity [81,114,115]. Similar results were observed in $AgNO_3$ exposure, where yeast had increased expression of *CUP1-1* and *CUP1-2* (4.79-fold and 4.71-fold, respectively) in an extended study that resulted in an evolved yeast strain, confirming the potential role of copper MTs in silver resistance [116]. Other Ag^+ transporters, Pho84, Fet3, and Smf1, were not implicated in Ag^+ uptake; however, significant down regulation (68.56-fold) of *PHO84* in silver evolved yeast has been observed, which may indicate that Pho84 plays a role in Ag^+ uptake, and may serve as a mechanism of Ag^+ resistance [81,116]. The effect of Ag^+ on genes involved in ergosterol biosynthesis was also investigated in a silver evolved yeast [116]. Results indicated down-regulation of those genes, suggesting that one mechanism of action of resistance against Ag^+ toxicity could be the ability to inhibit their down regulation [116]. In the filamentous fungus *A. nidulans*, silver induced expression of copper exporter crpA, indicating that it may play a role in silver export and resistance [90]. In *A. tetracladia*, resistance may be due to increased vacuolar function [301]. Overall, there has been some progress made in unveiling silver homeostasis in fungi, mostly by way of *S. cerevisiae*. Due to the increasing silver and Ag NP usage in many aspects of human life, silver–fungal interactions should be further investigated at the molecular level to decipher precise homeostatic and resistance mechanisms.

3. Omics and Metal Homeostasis

As the potential for commercial use of antifungal metals increases, so does the need to further investigate fungal homeostasis of essential and non-essential metals. Currently, research in this area is heavily reliant on assay based methods, which can be subjective and ambiguous. In this review, many of the discoveries of homeostatic mechanisms stemmed from the use of deletion libraries, microarrays, and PCR-based methods. This can restrict the scope of the research by only analyzing known genomic or transcriptiomic signatures.

The incorporation of an omics based approach is a resolution to this issue. The most popular omics utilizes bioinformatics to analyze fungal–metal interactions at a nucleotide and protein level, which can reveal novel genes and mutations. In genomics, the entirety of a genome is assessed and compared to others for similarities and differences that can contribute to an organism's characteristics [302,303]. Transcriptomics relies on RNA sequencing to survey gene expression through fold-changes in transcripts and proteomics assess fold-change in subsequent proteins. In fungi, omics is already incorporated into the identification of characteristics of multi-drug resistance, analysis of genomic divergence based on species origination, some analysis of metal tolerance due to short term exposure, and the analysis of the effects of exposure to non-metal selective pressures [210,301,304–306].

Bioinformatics analysis is used to translate omics results via computer programming methods. In nucleotide based omics, DNA or RNA is fragmented into segments or reads prior to sequencing. After sequencing, base calling assigns a nucleotide base to an intensity signal linked to a chromatogram peak and quality control measures are taken to trim reads of adapters used in the sequencing process and trim low quality bases [307]. Next, species that have a reference genome or transcriptome are mapped or aligned to that reference (resequencing). After genomic mapping, variant calling identifies distinctions between the re-sequenced organism and the reference [307]. After transcriptomic mapping, transcripts are quantified and analyzed for differential expression. Species that do not have a reference undergo de novo assembly, which constructs a genome or transcriptome from scratch. De novo assembly utilizes the fragmented reads by overlapping or matching them based on areas of similarity until the entire -ome is constructed [308]. Genome or transcriptome annotation can then be used for further interpretation of the sequencing data. In other omics, molecules produced by an organism are also analyzed and compared to chosen reference samples.

Steps within these bioinformatics pipelines require the use of computational tools written into the command line. Multiple tools with varying parameters exist to complete the same function; however, the user must decide which tools fit their scientific needs. This can result in variation between datasets and across scientific disciplines, based on accepted standards and norms. However, this limitation does not deduct from the vast amounts of data received.

With the increasing affordability of high-throughput omics, organisms can be analyzed at multiple omics levels. This is leading to a more comprehensive understanding of characteristics, especially in fungi where there is limited knowledge of their complexity. This type of research will also illuminate unique features of fungal metal homeostasis, toxicity, and resistance, especially of non-essential metals that are becoming conventional antimicrobial agents.

4. Conclusions

Fungal–metal interactions such as the synthesis of nanoparticles and metal used as antifungal agents are on the rise. Studies on metal toxicity and resistance have uncovered preserved homeostatic mechanisms. This review discussed metal homeostasis in various fungi types and has shown that essential metals have designated uptake and transport systems that regulate metal ion balance, mostly through the model organism *S. cerevisiae*. However, there was a significant lack of fundamental knowledge of such mechanisms in filamentous fungi, which play critical roles in nanoparticle biosynthesis and are targets of metal antifungals, further accentuating the need to investigate molecular systems involved in metal homeostasis. Fungal homeostasis of the non-essential metal silver was also highlighted. It showed that homeostatic mechanisms were reliant on existing copper transport systems, but were largely unclear regarding overall cellular processing. There is a need to further investigate other non-essential metals' cellular homeostasis as their commercial usage increases, due to the current lack of knowledge of future implications.

Author Contributions: J.R.R.; O.S.I.; F.N.A.; all authors contributed equally to this publication. All authors have read and agreed to the published version of the manuscript.

Funding: This research was funded by a Title III HBGI Grant sponsored by The US Department of Education.

Institutional Review Board Statement: Not applicable.

Informed Consent Statement: Not applicable.

Data Availability Statement: Data sharing not applicable.

Conflicts of Interest: The authors declare no conflict of interest.

References

1. Ahmad, A.; Mukherjee, P.; Senapati, S.; Mandal, D.; Khan, M.I.; Kumar, R.; Sastry, M. Extracellular Biosynthesis of Silver Nanoparticles Using the Fungus *Fusarium oxysporum*. *Colloids Surf. B Biointerfaces* **2003**, *28*, 313–318. [CrossRef]
2. Birla, S.S.; Gaikwad, S.C.; Gade, A.K.; Rai, M.K. Rapid Synthesis of Silver Nanoparticles from *Fusarium oxysporum* by Optimizing Physicocultural Conditions. *Sci. World J.* **2013**, *2013*, 796018. [CrossRef] [PubMed]
3. Naureen, B.; Miana, G.A.; Shahid, K.; Asghar, M.; Tanveer, S.; Sarwar, A. Iron (III) and Zinc (II) Monodentate Schiff Base Metal Complexes: Synthesis, Characterisation and Biological Activities. *J. Mol. Struct.* **2021**, *1231*, 129946. [CrossRef]
4. Mani Chandrika, K.V.S.; Sharma, S. Promising Antifungal Agents: A Minireview. *Bioorganic Med. Chem.* **2020**, *28*, 115398. [CrossRef] [PubMed]
5. Varshney, R.; Bhadauria, S.; Gaur, M.S. A Review: Biological Synthesis of Silver and Copper Nanoparticles. *Nano Biomed. Eng.* **2012**, *4*, 99–106. [CrossRef]
6. Gudikandula, K.; Charya Maringanti, S. Synthesis of Silver Nanoparticles by Chemical and Biological Methods and Their Antimicrobial Properties. *J. Exp. Nanosci.* **2016**, *11*, 714–721. [CrossRef]
7. Jamdagni, P.; Khatri, P.; Rana, J.S. Green Synthesis of Zinc Oxide Nanoparticles Using Flower Extract of Nyctanthes Arbor-tristis and Their Antifungal Activity. *J. King Saud Univ. Sci.* **2018**, *30*, 168–175. [CrossRef]
8. Bundschuh, M.; Filser, J.; Lüderwald, S.; McKee, M.S.; Metreveli, G.; Schaumann, G.E.; Schulz, R.; Wagner, S. Nanoparticles in the Environment: Where Do We Come from, Where Do We Go to? *Environ. Sci. Eur.* **2018**, *30*, 1–17. [CrossRef] [PubMed]
9. Kittler, S.; Greulich, C.; Diendorf, J.; Koller, M.; Epple, M. Toxicity of Silver Nanoparticles Increases during Storage because of Slow Dissolution under Release of Silver Ions. *Chem. Mater.* **2010**, *22*, 4548–4554. [CrossRef]
10. Mussin, J.E.; Roldán, M.V.; Rojas, F.; de los Ángeles Sosa, M.; Pellegri, N.; Giusiano, G. Antifungal Activity of Silver Nanoparticles in Combination with Ketoconazole against Malassezia Furfur. *AMB Express* **2019**, *9*, 131. [CrossRef]
11. Khan, I.; Saeed, K.; Khan, I. Nanoparticles: Properties, Applications and Toxicities. *Arab. J. Chem.* **2019**, *12*, 908–931. [CrossRef]
12. Cheong, Y.-K.; Arce, M.P.; Benito, A.; Chen, D.; Luengo Crisóstomo, N.; Kerai, L.V.; Rodríguez, G.; Valverde, J.L.; Vadalia, M.; Cerpa-Naranjo, A. Synergistic Antifungal Study of PEGylated Graphene Oxides and Copper Nanoparticles Against Candida Albicans. *Nanomaterials* **2020**, *10*, 819. [CrossRef] [PubMed]
13. Mulenos, M.R.; Liu, J.; Lujan, H.; Guo, B.; Lichtfouse, E.; Sharma, V.K.; Sayes, C.M. Copper, Silver, and Titania Nanoparticles Do not Release Ions under Anoxic Conditions and Release Only Minute Ion Levels Under Oxic Conditions in Water: Evidence for the Low Toxicity of Nanoparticles. *Environ. Chem. Lett.* **2020**, *18*, 1319–1328. [CrossRef]
14. Liu, X.; Jiang, Y.; He, D.; Fang, X.; Xu, J.; Lee, Y.-W.; Keller, N.P.; Shi, J. Copper Tolerance Mediated by FgAceA and FgCrpA in *Fusarium graminearum*. *Front. Microbiol.* **2020**, *11*, 1392. [CrossRef] [PubMed]

15. Zhao, H.; Eide, D. The Yeast ZRT1 Gene Encodes the Zinc Transporter Protein of a High-affinity uptake System Induced by Zinc Limitation. *Proc. Natl. Acad. Sci. USA* **1996**, *93*, 2454–2458. [CrossRef]
16. Moore, R.E.; Kim, Y.; Philpott, C.C. The Mechanism of Ferrichrome Transport through Arn1p and Its Metabolism in *Saccharomyces cerevisiae*. *Proc. Natl. Acad. Sci. USA* **2003**, *100*, 5664–5669. [CrossRef]
17. Voß, B.; Kirschhöfer, F.; Brenner-Weiß, G.; Fischer, R. Alternaria alternata Uses Two Siderophore Systems for Iron Acquisition. *Sci. Rep.* **2020**, *10*, 3587. [CrossRef] [PubMed]
18. Butt, T.R.; Sternberg, E.; Herd, J.; Crooke, S.T. Cloning and Expression of a Yeast Copper Metallothionein Gene. *Gene* **1984**, *27*, 23–33. [CrossRef]
19. Perinelli, M.; Tegoni, M.; Freisinger, E. Different Behavior of the Histidine Residue toward Cadmium and Zinc in a Cadmium-Specific Metallothionein from an Aquatic Fungus. *Inorg. Chem.* **2020**, *59*, 16988–16997. [CrossRef]
20. Cho, M.; Hu, G.; Caza, M.; Horianopoulos, L.C.; Kronstad, J.W.; Jung, W.H. Vacuolar Zinc Transporter Zrc1 is Required for Detoxification of Excess Intracellular Zinc in the Human Fungal Pathogen *Cryptococcus neoformans*. *J. Microbiol.* **2018**, *56*, 65–71. [CrossRef]
21. Vest, K.E.; Leary, S.C.; Winge, D.R.; Cobine, P.A. Copper Import into the Mitochondrial Matrix in *Saccharomyces cerevisiae* is Mediated by Pic2, a Mitochondrial Carrier Family Protein. *J. Biol. Chem.* **2013**, *288*, 23884–23892. [CrossRef]
22. Zhao, H.; Eide, D. The ZRT2 Gene Encodes the Low Affinity Zinc Transporter in *Saccharomyces cerevisiae*. *J. Biol. Chem.* **1996**, *271*, 23203–23210. [CrossRef]
23. Amich, J.; Vicentefranqueira, R.; Mellado, E.; Ruiz-Carmuega, A.; Leal, F.; Calera, J.A. The ZrfC Alkaline Zinc Transporter is Required for *Aspergillus fumigatus* Virulence and Its Growth in the Presence of the Zn/Mn-chelating Protein Calprotectin. *Cell. Microbiol.* **2014**, *16*, 548–564. [CrossRef]
24. Do, E.; Hu, G.; Caza, M.; Kronstad, J.W.; Jung, W.H. The ZIP Family Zinc Transporters Support the Virulence of *Cryptococcus neoformans*. *Med. Mycol.* **2016**, *54*, 605–615. [CrossRef] [PubMed]
25. Dos Santos, F.M.; Piffer, A.C.; Schneider, R.O.; Ribeiro, N.S.; Garcia, A.W.A.; Schrank, A.; Kmetzsch, L.; Vainstein, M.H.; Staats, C.C. Alterations of Zinc Homeostasis in Response to *Cryptococcus neoformans* in a Murine Macrophage Cell Line. *Future Microbiol.* **2017**, *12*, 491–504. [CrossRef]
26. Martha-Paz, A.M.; Eide, D.; Mendoza-Cózatl, D.; Castro-Guerrero, N.A.; Aréchiga-Carvajal, E.T. Zinc Uptake in the *Basidiomycota*: Characterization of Zinc Transporters in Ustilago maydis. *Mol. Membr. Biol.* **2019**, *35*, 39–50. [CrossRef] [PubMed]
27. Vicentefranqueira, R.; Amich, J.; Laskaris, P.; Ibrahim-Granet, O.; Latgé, J.P.; Toledo, H.; Leal, F.; Calera, J.A. Targeting Zinc Homeostasis to combat *Aspergillus fumigatus* Infections. *Front. Microbiol.* **2015**, *6*, 160. [CrossRef] [PubMed]
28. MacDiarmid, C.W.; Milanick, M.A.; Eide, D.J. Induction of the ZRC1 Metal Tolerance Gene in Zinc-limited Yeast Confers Resistance to Zinc Shock. *J. Biol. Chem.* **2003**, *278*, 15065–15072. [CrossRef] [PubMed]
29. MacDiarmid, C.; Gaither, L.; Eide, D. Zinc Transporters That Regulate Vacuolar Zinc Storage in *Saccharomyces cerevisiae*. *EMBO J.* **2000**, *19*, 2845–2855. [CrossRef]
30. Gross, C.; Kelleher, M.; Iyer, V.R.; Brown, P.O.; Winge, D.R. Identification of the Copper Regulon in *Saccharomyces cerevisiae* by DNA Microarrays. *J. Biol. Chem.* **2000**, *275*, 32310–32316. [CrossRef]
31. Hassett, R.; Dix, D.R.; Eide, D.J.; Kosman, D.J. The Fe(II) Permease Fet4p Functions as a Low Affinity Copper Transporter and Supports Normal Copper Trafficking in *Saccharomyces cerevisiae*. *Biochem. J.* **2000**, *351 Pt 2*, 477–484. [CrossRef]
32. Pena, M.M.; Puig, S.; Thiele, D.J. Characterization of the *Saccharomyces cerevisiae* High Affinity Copper Transporter Ctr3. *J. Biol. Chem.* **2000**, *275*, 33244–33251. [CrossRef]
33. Peña, M.M.O.; Koch, K.A.; Thiele, D.J. Dynamic Regulation of Copper Uptake and Detoxification Genes in *Saccharomyces cerevisiae*. *Mol. Cell. Biol.* **1998**, *18*, 2514–2523. [CrossRef]
34. Beaudoin, J.; Ekici, S.; Daldal, F.; Ait-Mohand, S.; Guérin, B.; Labbé, S. Copper Transport and Regulation in *Schizosaccharomyces pombe*. *Biochem. Soc. Trans.* **2013**, *41*, 1679–1686. [CrossRef] [PubMed]
35. Beaudoin, J.; Ioannoni, R.; Mailloux, S.; Plante, S.; Labbé, S. Transcriptional Regulation of the Copper Transporter Mfc1 in Meiotic Cells. *Eukaryot. Cell* **2013**, *12*, 575–590. [CrossRef] [PubMed]
36. Zhou, H.; Thiele, D.J. Identification of a Novel High Affinity Copper Transport Complex in the Fission Yeast *Schizosaccharomyces pombe*. *J. Biol. Chem.* **2001**, *276*, 20529–20535. [CrossRef]
37. Park, Y.-S.; Lian, H.; Chang, M.; Kang, C.-M.; Yun, C.-W. Identification of High-affinity Copper Transporters in *Aspergillus fumigatus*. *Fungal Genet. Biol.* **2014**, *73*, 29–38. [CrossRef]
38. Peñas, M.M.; Azparren, G.; Domínguez, Á.; Sommer, H.; Ramírez, L.; Pisabarro, A.G. Identification and Functional Characterisation of Ctr1, a *Pleurotus ostreatus* Gene Coding for a Copper Transporter. *Mol. Genet. Genom.* **2005**, *274*, 402–409. [CrossRef]
39. Raffa, N.; Osherov, N.; Keller, N. Copper Utilization, Regulation, and Acquisition by *Aspergillus fumigatus*. *Int. J. Mol. Sci.* **2019**, *20*, 1980. [CrossRef] [PubMed]
40. Lin, S.-J.; Pufahl, R.A.; Dancis, A.; O'Halloran, T.V.; Culotta, V.C. A Role for the *Saccharomyces cerevisiae* ATX1 Gene in Copper Trafficking and Iron Transport. *J. Biol. Chem.* **1997**, *272*, 9215–9220. [CrossRef]
41. Yuan, D.S.; Dancis, A.; Klausner, R.D. Restriction of Copper Export in *Saccharomyces cerevisiae* to a Late Golgi or Post-Golgi Compartment in the Secretory Pathway. *J. Biol. Chem.* **1997**, *272*, 25787–25793. [CrossRef]

42. Peter, C.; Laliberté, J.; Beaudoin, J.; Labbé, S. Copper Distributed by Atx1 is Available to Copper Amine Oxidase 1 in *Schizosaccharomyces pombe*. *Eukaryot. Cell* **2008**, *7*, 1781–1794. [CrossRef]
43. Beers, J.; Glerum, D.M.; Tzagoloff, A. Purification, Characterization, and Localization of Yeast Cox17p, a Mitochondrial Copper Shuttle. *J. Biol. Chem.* **1997**, *272*, 33191–33196. [CrossRef]
44. Glerum, D.M.; Shtanko, A.; Tzagoloff, A. Characterization of COX17, a Yeast Gene Involved in Copper Metabolism and Assembly of Cytochrome Oxidase. *J. Biol. Chem.* **1996**, *271*, 14504–14509. [CrossRef]
45. Culotta, V.C.; Klomp, L.W.J.; Strain, J.; Casareno, R.L.B.; Krems, B.; Gitlin, J.D. The Copper Chaperone for Superoxide Dismutase. *J. Biol. Chem.* **1997**, *272*, 23469–23472. [CrossRef] [PubMed]
46. Laliberté, J.; Whitson, L.J.; Beaudoin, J.; Holloway, S.P.; Hart, P.J.; Labbé, S. The *Schizosaccharomyces pombe* Pccs Protein Functions in Both Copper Trafficking and Metal Detoxification Pathways. *J. Biol. Chem.* **2004**, *279*, 28744–28755. [CrossRef] [PubMed]
47. Carr, H.S.; George, G.N.; Winge, D.R. Yeast Cox11, a Protein Essential for Cytochrome cOxidase Assembly, Is a Cu(I)-binding Protein. *J. Biol. Chem.* **2002**, *277*, 31237–31242. [CrossRef] [PubMed]
48. Smith, A.D.; Logeman, B.L.; Thiele, D.J. Copper Acquisition and Utilization in Fungi. *Annu. Rev. Microbiol.* **2017**, *71*, 597–623. [CrossRef] [PubMed]
49. Wiemann, P.; Perevitsky, A.; Lim, F.Y.; Shadkchan, Y.; Knox, B.P.; Landero Figueora, J.A.; Choera, T.; Niu, M.; Steinberger, A.J.; Wüthrich, M.; et al. *Aspergillus fumigatus* Copper Export Machinery and Reactive Oxygen Intermediate Defense Counter Host Copper-Mediated Oxidative Antimicrobial Offense. *Cell Rep.* **2017**, *19*, 1008–1021. [CrossRef] [PubMed]
50. Askwith, C.; Kaplan, J. An Oxidase-permease-based Iron Transport System in *Schizosaccharomyces pombe* and Its Expression in *Saccharomyces cerevisiae*. *J. Biol. Chem.* **1997**, *272*, 401–405. [CrossRef]
51. Chen, X.-Z.; Peng, J.-B.; Cohen, A.; Nelson, H.; Nelson, N.; Hediger, M.A. Yeast SMF1 Mediates H+-coupled Iron Uptake with Concomitant Uncoupled Cation Currents. *J. Biol. Chem.* **1999**, *274*, 35089–35094. [CrossRef] [PubMed]
52. Cohen, A.; Nelson, H.; Nelson, N. The Family of SMF Metal Ion Transporters in Yeast Cells. *J. Biol. Chem.* **2000**, *275*, 33388–33394. [CrossRef] [PubMed]
53. De Silva, D.M.; Askwith, C.C.; Eide, D.; Kaplan, J. The FET3 Gene Product Required for High Affinity Iron Transport in Yeast Is a Cell Surface Ferroxidase. *J. Biol. Chem.* **1995**, *270*, 1098–1101. [PubMed]
54. Dix, D.; Bridgham, J.; Broderius, M.; Eide, D. Characterization of the FET4 Protein of Yeast Evidence for a Direct Role in the Transport of Iron. *J. Biol. Chem.* **1997**, *272*, 11770–11777. [CrossRef] [PubMed]
55. Dix, D.R.; Bridgham, J.T.; Broderius, M.A.; Byersdorfer, C.A.; Eide, D.J. The FET4 Gene Encodes the Low Affinity Fe(II) Transport Protein of *Saccharomyces cerevisiae*. *J. Biol. Chem.* **1994**, *269*, 26092–26099. [CrossRef]
56. Mourer, T.; Jacques, J.-F.; Brault, A.; Bisaillon, M.; Labbé, S. Shu1 Is a Cell-surface Protein Involved in Iron Acquisition from Heme in *Schizosaccharomyces pombe*. *J. Biol. Chem.* **2015**, *290*, 10176–10190. [CrossRef]
57. Normant, V.; Mourer, T.; Labbé, S. The Major Facilitator Transporter Str3 is Required for Low-affinity Heme Acquisition in *Schizosaccharomyces pombe*. *J. Biol. Chem.* **2018**, *293*, 6349–6362. [CrossRef]
58. Stearman, R.; Yuan, D.S.; Yamaguchi-Iwai, Y.; Klausner, R.D.; Dancis, A. A Permease-Oxidase Complex Involved in High-Affinity Iron Uptake in Yeast. *Science* **1996**, *271*, 1552–1557. [CrossRef]
59. Pelletier, B.; Beaudoin, J.; Philpott, C.C.; Labbé, S. Fep1 Represses Expression of the Fission Yeast *Schizosaccharomyces pombe* Siderophore-iron Transport System. *Nucleic Acids Res.* **2003**, *31*, 4332–4344. [CrossRef]
60. Eichhorn, H.; Lessing, F.; Winterberg, B.; Schirawski, J.; Kämper, J.; Müller, P.; Kahmann, R. A Ferroxidation/permeation Iron Uptake System is Required for Virulence in *Ustilago maydis*. *Plant Cell* **2006**, *18*, 3332–3345. [CrossRef]
61. Lindahl, P.A. A Comprehensive Mechanistic Model of Iron Metabolism in *Saccharomyces cerevisiae*. *Met. Integr. Biometal Sci.* **2019**, *11*, 1779–1799. [CrossRef]
62. Netz, D.; Pierik, A.; Stümpfig, M.; Muhlenhoff, U.; Lill, R. The Cfd1-Nbp35 Complex Acts as a Scaffold for Iron-sulfur Protein Assembly in the Yeast Cytosol. *Nat. Chem. Biol.* **2007**, *3*, 278–286. [CrossRef] [PubMed]
63. Labbé, S.; Khan, M.G.M.; Jacques, J.-F. Iron Uptake and Regulation in *Schizosaccharomyces pombe*. *Curr. Opin. Microbiol.* **2013**, *16*, 669–676. [CrossRef] [PubMed]
64. Li, L.; Chen, O.S.; Ward, D.M.; Kaplan, J. CCC1 is a Transporter That Mediates Vacuolar Iron Storage in Yeast. *J. Biol. Chem.* **2001**, *276*, 29515–29519. [CrossRef]
65. Supek, F.; Supekova, L.; Nelson, H.; Nelson, N. A Yeast Manganese Transporter Related to the Macrophage Protein Involved in Conferring Resistance to Mycobacteria. *Proc. Natl. Acad. Sci. USA* **1996**, *93*, 5105–5110. [CrossRef] [PubMed]
66. Supek, F.; Supekova, L.; Nelson, H.; Nelson, N. Function of Metal-ion Homeostasis in the Cell Division Cycle, Mitochondrial Protein Processing, Sensitivity to Mycobacterial Infection and Brain Function. *J. Exp. Biol.* **1997**, *200*, 321–330. [PubMed]
67. Jensen, L.T.; Ajua-Alemanji, M.; Culotta, V.C. The *Saccharomyces cerevisiae* High Affinity Phosphate Transporter Encoded by PHO84 also Functions in Manganese Homeostasis. *J. Biol. Chem.* **2003**, *278*, 42036–42040. [CrossRef]
68. Diss, L.; Blaudez, D.; Gelhaye, E.; Chalot, M. Genome-wide Analysis of Fungal Manganese Transporters, with an Emphasis on *Phanerochaete chrysosporium*. *Environ. Microbiol. Rep.* **2011**, *3*, 367–382. [CrossRef]
69. Luk, E.; Carroll, M.; Baker, M.; Culotta, V.C. Manganese Activation of Superoxide Dismutase 2 in *Saccharomyces cerevisiae* Requires MTM1, a Member of the Mitochondrial Carrier Family. *Proc. Natl. Acad. Sci. USA* **2003**, *100*, 10353–10357. [CrossRef] [PubMed]

70. Dulary, E.; Yu, S.-Y.; Houdou, M.; de Bettignies, G.; Decool, V.; Potelle, S.; Duvet, S.; Krzewinski-Recchi, M.-A.; Garat, A.; Matthijs, G.; et al. Investigating the Function of Gdt1p in Yeast *Golgi glycosylation*. *Biochim. Biophys. Acta Gen. Subj.* **2018**, *1862*, 394–402. [CrossRef]
71. Lapinskas, P.J.; Cunningham, K.W.; Liu, X.F.; Fink, G.R.; Culotta, V.C. Mutations in PMR1 Suppress Oxidative Damage in Yeast Cells Lacking Superoxide Dismutase. *Mol. Cell. Biol.* **1995**, *15*, 1382–1388. [CrossRef]
72. Thines, L.; Deschamps, A.; Sengottaiyan, P.; Savel, O.; Stribny, J.; Morsomme, P. The Yeast Protein Gdt1p Transports Mn(2+) Ions and Thereby Regulates Manganese Homeostasis in the *Golgi*. *J. Biol. Chem.* **2018**, *293*, 8048–8055. [CrossRef] [PubMed]
73. Cohen, Y.; Megyeri, M.; Chen, O.C.W.; Condomitti, G.; Riezman, I.; Loizides-Mangold, U.; Abdul-Sada, A.; Rimon, N.; Riezman, H.; Platt, F.M.; et al. The Yeast P5 Type ATPase, Spf1, Regulates Manganese Transport into the Endoplasmic Reticulum. *PLoS ONE* **2014**, *8*, e85519. [CrossRef] [PubMed]
74. Gitler, A.D.; Chesi, A.; Geddie, M.L.; Strathearn, K.E.; Hamamichi, S.; Hill, K.J.; Caldwell, K.A.; Caldwell, G.A.; Cooper, A.A.; Rochet, J.-C.; et al. Alpha-synuclein is Part of a Diverse and Highly Conserved Interaction Network That Includes PARK9 and Manganese Toxicity. *Nat. Genet.* **2009**, *41*, 308–315. [CrossRef] [PubMed]
75. Lapinskas, P.J.; Lin, S.-J.; Culotta, V.C. The Role of the *Saccharomyces cerevisiae* CCC1 Gene in the Homeostasis of Manganese Ions. *Mol. Microbiol.* **1996**, *21*, 519–528. [CrossRef] [PubMed]
76. Schmidt, K.; Wolfe, D.M.; Stiller, B.; Pearce, D.A. Cd2+, Mn2+, Ni2+ and Se2+ Toxicity to *Saccharomyces cerevisiae* Lacking YPK9p the Orthologue of Human ATP13A2. *Biochem. Biophys. Res. Commun.* **2009**, *383*, 198–202. [CrossRef] [PubMed]
77. Culotta, V.C.; Yang, M.; Hall, M.D. Manganese Transport and Trafficking: Lessons Learned from *Saccharomyces cerevisiae*. *Eukaryot. Cell* **2005**, *4*, 1159–1165. [CrossRef]
78. Farcasanu, I.; Mizunuma, M.; Hirata, D.; Miyakawa, T. Involvement of Histidine Permease (Hip1p) in Manganese Transport in *Saccharomyces cerevisiae*. *Mol. Gen. Genetics MGG* **1998**, *259*, 541–548. [CrossRef]
79. Ton, V.-K.; Mandal, D.; Vahadji, C.; Rao, R. Functional Expression in Yeast of the Human Secretory Pathway Ca2+, Mn2+-ATPase Defective in Hailey-Hailey Disease. *J. Biol. Chem.* **2002**, *277*, 6422–6427. [CrossRef]
80. Horstmann, C.; Campbell, C.; Kim, D.S.; Kim, K. Transcriptome Profile with 20 nm Silver Nanoparticles in Yeast. *FEMS Yeast Res.* **2019**, *19*. [CrossRef]
81. Ruta, L.L.; Banu, M.A.; Neagoe, A.D.; Kissen, R.; Bones, A.M.; Farcasanu, I.C. Accumulation of Ag(I) by *Saccharomyces cerevisiae* Cells Expressing Plant Metallothioneins. *Cells* **2018**, *7*, 266. [CrossRef] [PubMed]
82. B, A.W. Oligodynamic Phenomena of Living Cells. *Nature* **1893**, *48*, 331. [CrossRef]
83. Galván Márquez, I.; Ghiyasvand, M.; Massarsky, A.; Babu, M.; Samanfar, B.; Omidi, K.; Moon, T.W.; Smith, M.L.; Golshani, A. Zinc Oxide and Silver Nanoparticles Toxicity in the Baker's Yeast, *Saccharomyces cerevisiae*. *PLoS ONE* **2018**, *13*, e0193111. [CrossRef] [PubMed]
84. Sinisi, V.; Pelagatti, P.; Carcelli, M.; Migliori, A.; Mantovani, L.; Righi, L.; Leonardi, G.; Pietarinen, S.; Hubsch, C.; Rogolino, D. A Green Approach to Copper-Containing Pesticides: Antimicrobial and Antifungal Activity of Brochantite Supported on Lignin for the Development of Biobased Plant Protection Products. *ACS Sustain. Chem. Eng.* **2019**, *7*, 3213–3221. [CrossRef]
85. Pasquet, J.; Chevalier, Y.; Pelletier, J.; Couval, E.; Bouvier, D.; Bolzinger, M.A. The Contribution of Zinc Ions to the Antimicrobial Activity of Zinc Oxide. *Colloids Surf. A Physicochem. Eng. Asp.* **2014**, *457*, 263–274. [CrossRef]
86. Xue, J.; Moyer, A.; Peng, B.; Wu, J.; Hannafon, B.N.; Ding, W.-Q. Chloroquine is a Zinc Ionophore. *PLoS ONE* **2014**, *9*, e109180. [CrossRef] [PubMed]
87. He, L.; Liu, Y.; Mustapha, A.; Lin, M. Antifungal Activity of Zinc Oxide Nanoparticles against *Botrytis cinerea* and *Penicillium expansum*. *Microbiol. Res.* **2011**, *166*, 207–215. [CrossRef]
88. Lanfranco, L.; Balsamo, R.; Martino, E.; Perotto, S.; Bonfante, P. Zinc Ions Alter Morphology and Chitin Deposition in an Ericoid Fungus. *Eur. J. Histochem.* **2002**, *46*, 341–350. [CrossRef] [PubMed]
89. Noor, S.; Shah, Z.; Javed, A.; Ali, A.; Hussain, S.B.; Zafar, S.; Ali, H.; Muhammad, S.A. A Fungal Based Synthesis Method for Copper Nanoparticles with the Determination of Anticancer, Antidiabetic and Antibacterial Activities. *J. Microbiol. Methods* **2020**, *174*, 105966. [CrossRef] [PubMed]
90. Antsotegi-Uskola, M.; Markina-Iñarrairaegui, A.; Ugalde, U. Copper Resistance in *Aspergillus nidulans* Relies on the PI-Type ATPase CrpA, Regulated by the Transcription Factor AceA. *Front. Microbiol.* **2017**, *8*, 912. [CrossRef]
91. Li, L.; Ward, D.M. Iron Toxicity in Yeast: Transcriptional Regulation of the Vacuolar Iron Importer Ccc1. *Curr. Genet.* **2018**, *64*, 413–416. [CrossRef]
92. Ward, P.P.; Conneely, O.M. Lactoferrin: Role in Iron Homeostasis and Host Defense against Microbial Infection. *Biometals* **2004**, *17*, 203–208. [CrossRef]
93. Leal, S.M., Jr.; Roy, S.; Vareechon, C.; Carrion, S.D.; Clark, H.; Lopez-Berges, M.S.; diPietro, A.; Schrettl, M.; Beckmann, N.; Redl, B.; et al. Targeting Iron Acquisition Blocks Infection with the Fungal Pathogens *Aspergillus fumigatus* and *Fusarium oxysporum*. *PLoS Pathog.* **2013**, *9*, e1003436. [CrossRef]
94. Dollard, C.; Ricupero-Hovasse, S.L.; Natsoulis, G.; Boeke, J.D.; Winston, F. SPT10 and SPT21 are Required for Transcription of Particular Histone Genes in *Saccharomyces cerevisiae*. *Mol. Cell. Biol.* **1994**, *14*, 5223–5228. [CrossRef]
95. Norris, D.; Osley, M.A. The Two Gene Pairs Encoding H2A and H2B Play Different Roles in the *Saccharomyces cerevisiae* Life Cycle. *Mol. Cell. Biol.* **1987**, *7*, 3473–3481. [CrossRef]

96. Janusz, G.; Kucharzyk, K.H.; Pawlik, A.; Staszczak, M.; Paszczynski, A.J. Fungal Laccase, Manganese Peroxidase and Lignin Peroxidase: Gene Expression and Regulation. *Enzym. Microb. Technol.* **2013**, *52*, 1–12. [CrossRef]
97. Manavalan, T.; Manavalan, A.; Heese, K. Characterization of Lignocellulolytic Enzymes from White-rot Fungi. *Curr. Microbiol.* **2015**, *70*, 485–498. [CrossRef]
98. Xu, H.; Guo, M.-Y.; Gao, Y.-H.; Bai, X.-H.; Zhou, X.-W. Expression and Characteristics of Manganese Peroxidase from *Ganoderma lucidum* in *Pichia pastoris* and Its Application in the Degradation of Four Dyes and Phenol. *BMC Biotechnol.* **2017**, *17*, 19. [CrossRef] [PubMed]
99. Das, D.; Ahmed, G. Silver Nanoparticles Damage Yeast Cell Wall. *J. Biotechnol.* **2012**, *3*, 36–39.
100. Walker, C.; Ryu, S.; Trinh, C. Exceptional Solvent Tolerance in *Yarrowia lipolytica* Is Enhanced by Sterols. *bioRxiv* **2018**, 324681. [CrossRef] [PubMed]
101. Ghannoum, M.A.; Rice, L.B. Antifungal Agents: Mode of Action, Mechanisms of Resistance, and Correlation of These Mechanisms with Bacterial Resistance. *Clin. Microbiol. Rev.* **1999**, *12*, 501–517. [CrossRef]
102. Avery, S.V.; Howlett, N.G.; Radice, S. Copper Toxicity towards *Saccharomyces cerevisiae*: Dependence on Plasma Membrane Fatty Acid Composition. *Appl. Environ. Microbiol.* **1996**, *62*, 3960. [CrossRef] [PubMed]
103. Babele, P.K.; Singh, A.K.; Srivastava, A. Bio-Inspired Silver Nanoparticles Impose Metabolic and Epigenetic Toxicity to *Saccharomyces cerevisiae*. *Front. Pharmacol.* **2019**, *10*, 1016. [CrossRef] [PubMed]
104. Blackwell, K.J.; Tobin, J.M.; Avery, S.V. Manganese Toxicity towards *Saccharomyces cerevisiae*: Dependence on Intracellular and Extracellular Magnesium Concentrations. *Appl. Microbiol. Biotechnol.* **1998**, *49*, 751–757. [CrossRef] [PubMed]
105. Gerwien, F.; Skrahina, V.; Kasper, L.; Hube, B.; Brunke, S. Metals in Fungal Virulence. *FEMS Microbiol. Rev.* **2017**, *42*. [CrossRef] [PubMed]
106. Simm, C.; Lahner, B.; Salt, D.; LeFurgey, A.; Ingram, P.; Yandell, B.; Eide, D.J. *Saccharomyces cerevisiae* Vacuole in Zinc Storage and Intracellular Zinc Distribution. *Eukaryot. Cell* **2007**, *6*, 1166. [CrossRef]
107. Staats, C.; Kmetzsch, L.; Schrank, A.; Vainstein, M. Fungal Zinc Metabolism and Its Connections to Virulence. *Front. Cell. Infect. Microbiol.* **2013**, *3*, 65. [CrossRef]
108. Wilson, D.; Citiulo, F.; Hube, B. Zinc Exploitation by Pathogenic Fungi. *PLoS Pathog.* **2012**, *8*, e1003034. [CrossRef]
109. González Matute, R.; Serra, A.; Figlas, D.; Curvetto, N. Copper and Zinc Bioaccumulation and Bioavailability of *Ganoderma lucidum*. *J. Med. Food* **2011**, *14*, 1273–1279. [CrossRef]
110. Ruytinx, J.; Nguyen, H.; Van Hees, M.; Op De Beeck, M.; Vangronsveld, J.; Carleer, R.; Colpaert, J.V.; Adriaensen, K. Zinc Export Results in Adaptive Zinc Tolerance in the Ectomycorrhizal Basidiomycete *Suillus bovinus*. *Metallomics* **2013**, *5*, 1225–1233. [CrossRef] [PubMed]
111. Kalsotra, T.; Khullar, S.; Agnihotri, R.; Reddy, M.S. Metal Induction of Two Metallothionein Genes in the Ectomycorrhizal Fungus *Suillus himalayensis* and Their Role in Metal Tolerance. *Microbiology* **2018**, *164*, 868–876. [CrossRef]
112. Tucker, S.L.; Thornton, C.R.; Tasker, K.; Jacob, C.; Giles, G.; Egan, M.; Talbot, N.J. A Fungal Metallothionein is Required for Pathogenicity of *Magnaporthe grisea*. *Plant Cell* **2004**, *16*, 1575–1588. [CrossRef] [PubMed]
113. Yasokawa, D.; Murata, S.; Kitagawa, E.; Iwahashi, Y.; Nakagawa, R.; Hashido, T.; Iwahashi, H. Mechanisms of Copper Toxicity in *Saccharomyces cerevisiae* Determined by Microarray Analysis. *Environ. Toxicol.* **2008**, *23*, 599–606. [CrossRef]
114. Hosiner, D.; Gerber, S.; Lichtenberg-Frate, H.; Glaser, W.; Schüller, C.; Klipp, E. Impact of Acute Metal Stress in *Saccharomyces cerevisiae*. *PLoS ONE* **2014**, *9*, e83330. [CrossRef]
115. Niazi, J.H.; Sang, B.-I.; Kim, Y.S.; Gu, M.B. Global Gene Response in *Saccharomyces cerevisiae* Exposed to Silver Nanoparticles. *Appl. Biochem. Biotechnol.* **2011**, *164*, 1278–1291. [CrossRef]
116. Terzioğlu, E.; Alkım, C.; Arslan, M.; Balaban, B.G.; Holyavkin, C.; Kısakesen, H.İ.; Topaloğlu, A.; Yılmaz Şahin, Ü.; Gündüz Işık, S.; Akman, S.; et al. Genomic, Transcriptomic and Physiological Analyses of Silver-resistant *Saccharomyces cerevisiae* Obtained by Evolutionary Engineering. *Yeast* **2020**, *37*, 413–426. [CrossRef]
117. Akgul, A.; Akgul, A. Mycoremediation of Copper: Exploring the Metal Tolerance of Brown Rot Fungi. *Bioresources* **2018**, *13*, 7155–7171.
118. Ohno, K.M.; Clausen, C.A.; Green, F.; Diehl, S.V. Insights into the Mechanism of Copper-Tolerance in *Fibroporia radiculosa*: The Biosynthesis of Oxalate. *Int. Biodeterior. Biodegrad.* **2015**, *105*, 90–96. [CrossRef]
119. Tang, J.D.; Parker, L.A.; Perkins, A.D.; Sonstegard, T.S.; Schroeder, S.G.; Nicholas, D.D.; Diehl, S.V. Gene Expression Analysis of Copper Tolerance and Wood Decay in the Brown Rot Fungus *Fibroporia radiculosa*. *Appl. Environ. Microbiol.* **2013**, *79*, 1523–1533. [CrossRef] [PubMed]
120. Li, L.; Bagley, D.; Ward, D.M.; Kaplan, J. Yap5 Is an Iron-Responsive Transcriptional Activator That Regulates Vacuolar Iron Storage in Yeast. *Mol. Cell. Biol.* **2008**, *28*, 1326–1337. [CrossRef] [PubMed]
121. Mei, B.; Budde, A.D.; Leong, S.A. sid1, a Gene Initiating Siderophore Biosynthesis in *Ustilago maydis*: Molecular Characterization, Regulation by Iron, and Role in Phytopathogenicity. *Proc. Natl. Acad. Sci. USA* **1993**, *90*, 903–907. [CrossRef]
122. de Llanos, R.; Martínez-Garay, C.A.; Fita-Torró, J.; Romero, A.M.; Martínez-Pastor, M.T.; Puig, S. Soybean Ferritin Expression in *Saccharomyces cerevisiae* Modulates Iron Accumulation and Resistance to Elevated Iron Concentrations. *Appl. Environ. Microbiol.* **2016**, *82*, 3052–3060. [CrossRef] [PubMed]
123. Kim, H.-J.; Kim, H.-M.; Kim, J.-H.; Ryu, K.-S.; Park, S.-M.; Jahng, K.-Y.; Yang, M.-S.; Kim, D.-H. Expression of Heteropolymeric Ferritin Improves Iron Storage in *Saccharomyces cerevisiae*. *Appl. Environ. Microbiol.* **2003**, *69*, 1999–2005. [CrossRef] [PubMed]

124. Shin, Y.-M.; Kwon, T.-H.; Kim, K.-S.; Chae, K.-S.; Kim, D.-H.; Kim, J.-H.; Yang, M.-S. Enhanced Iron Uptake of *Saccharomyces cerevisiae* by Heterologous Expression of a Tadpole Ferritin Gene. *Appl. Environ. Microbiol.* **2001**, *67*, 1280–1283. [CrossRef] [PubMed]
125. del Pozo, L.; Osaba, L.; Corchero, J.; Jimenez, A. A Single Nucleotide Change in the MNR1 (VCX1/HUM1) Gene Determines Resistance to Manganese in *Saccharomyces cerevisiae*. *Yeast* **1999**, *15*, 371–375. [CrossRef]
126. Luk, E.E.-C.; Culotta, V.C. Manganese Superoxide Dismutase in *Saccharomyces cerevisiae* Acquires Its Metal Co-factor through a Pathway Involving the Nramp Metal Transporter, Smf2p. *J. Biol. Chem.* **2001**, *276*, 47556–47562. [CrossRef]
127. Reza, M.H.; Shah, H.; Manjrekar, J.; Chattoo, B.B. Magnesium Uptake by CorA Transporters Is Essential for Growth, Development and Infection in the Rice Blast Fungus *Magnaporthe oryzae*. *PLoS ONE* **2016**, *11*, e0159244. [CrossRef]
128. Mendel, R.R. Molybdenum: Biological Activity and Metabolism. *Dalton Trans.* **2005**, *21*, 3404–3409. [CrossRef]
129. Novotny, J.A.; Peterson, C.A. Molybdenum. *Adv. Nutr.* **2018**, *9*, 272–273. [CrossRef]
130. Clarance, P.; Luvankar, B.; Sales, J.; Khusro, A.; Agastian, P.; Tack, J.C.; Al Khulaifi, M.M.; Al-Shwaiman, H.A.; Elgorban, A.M.; Syed, A.; et al. Green Synthesis and Characterization of Gold Nanoparticles Using Endophytic Fungi *Fusarium solani* and Its in vitro Anticancer and Biomedical Applications. *Saudi J. Biol. Sci.* **2020**, *27*, 706–712. [CrossRef]
131. Sharma, K.; Giri, R.; Sharma, R. Lead, Cadmium and Nickel Removal Efficiency of White-rot Fungus *Phlebia brevispora*. *Lett. Appl. Microbiol.* **2020**, *71*, 637–644. [CrossRef] [PubMed]
132. Brown, K.H.; Wuehler, S.E.; Peerson, J.M. The Importance of Zinc in Human Nutrition and Estimation of the Global Prevalence of Zinc Deficiency. *Food Nutr. Bull.* **2001**, *22*, 113–125. [CrossRef]
133. Feldmann, H.; Branduardi, P. *Yeast: Molecular and Cell Biology*; Wiley-Blackwell: Weinheim, Germany, 2012; p. 464.
134. Zhang, C.; Huang, H.; Deng, W.; Li, T. Genome-Wide Analysis of the $Zn(II)_2Cys_6$ Zinc Cluster-Encoding Gene Family in *Tolypocladium guangdongense* and Its Light-Induced Expression. *Genes* **2019**, *10*, 179. [CrossRef] [PubMed]
135. Todd, R.B.; Andrianopoulos, A. Evolution of a Fungal Regulatory Gene Family: The Zn(II)2Cys6 Binuclear Cluster DNA Binding Motif. *Fungal Genet. Biol.* **1997**, *21*, 388–405. [CrossRef] [PubMed]
136. Grotz, N.; Fox, T.; Connolly, E.; Park, W.; Guerinot, M.L.; Eide, D. Identification of a Family of Zinc Transporter Genes from *Arabidopsis* That Respond to Zinc Deficiency. *Proc. Natl. Acad. Sci. USA* **1998**, *95*, 7220–7224. [CrossRef] [PubMed]
137. Fisher, M.C.; Henk, D.A.; Briggs, C.J.; Brownstein, J.S.; Madoff, L.C.; McCraw, S.L.; Gurr, S.J. Emerging Fungal Threats to Animal, Plant and Ecosystem Health. *Nature* **2012**, *484*, 186–194. [CrossRef]
138. Savary, S.; Ficke, A.; Aubertot, J.-N.; Hollier, C. Crop Losses Due to Diseases and Their Implications for Global Food Production Losses and Food Security. *Food Secur.* **2012**, *4*, 519–537. [CrossRef]
139. Fisher, M.C.; Hawkins, N.J.; Sanglard, D.; Gurr, S.J. Worldwide Emergence of Resistance to Antifungal Drugs Challenges Human Health and Food Security. *Science* **2018**, *360*, 739–742. [CrossRef]
140. Fravel, D.; Olivain, C.; Alabouvette, C. *Fusarium oxysporum* and Its Biocontrol. *New Phytol.* **2003**, *157*, 493–502. [CrossRef]
141. Sadasivam, N.; Sivaraj, R. Biogenic ZnO Nanoparticles Synthesized Using L. Aculeata Leaf Extract and Their Antifungal Activity against Plant Fungal Pathogens. *Bull. Mater. Sci.* **2016**, *39*, 1–5.
142. Sharma, R. Pathogenecity of *Aspergillus niger* in Plants. *Cibtech J. Microbiol.* **2012**, *1*, 47–51.
143. WIlliamson, B.; Tudzynski, B.; Tudzynski, P.; Van Kan, J.A.L. *Botrytis cinerea*: The Cause of Grey Mould Disease. *Mol. Plant Pathol.* **2007**, *8*, 561–580. [CrossRef]
144. Yehia, R.; Ahmed, O. In vitro Study of the Antifungal Efficacy of Zinc Oxide Nanoparticles against *Fusarium oxysporum* and *Penicillium expansum*. *Afr. J. Microbiol. Res.* **2013**, *7*, 1917–1923.
145. Bryden, W.L. Mycotoxin Contamination of the Feed Supply Chain: Implications for Animal Productivity and Feed Security. *Anim. Feed Sci. Technol.* **2012**, *173*, 134–158. [CrossRef]
146. Flores-Flores, M.E.; Lizarraga, E.; López de Cerain, A.; González-Peñas, E. Presence of Mycotoxins in Animal Milk: A Review. *Food Control* **2015**, *53*, 163–176. [CrossRef]
147. Appell, M.; Jackson, M.A.; Wang, L.C.; Ho, C.-H.; Mueller, A. Determination of Fusaric Acid in Maize Using Molecularly Imprinted SPE Clean-up. *J. Sep. Sci.* **2014**, *37*, 281–286. [CrossRef]
148. Daubner, S.C.; Le, T.; Wang, S. Tyrosine Hydroxylase and Regulation of Dopamine Synthesis. *Arch. Biochem. Biophys.* **2011**, *508*, 1–12. [CrossRef]
149. Rahman, M.K.; Rahman, F.; Rahman, T.; Kato, T. Dopamine-β-Hydroxylase (DBH), Its Cofactors and Other Biochemical Parameters in the Serum of Neurological Patients in Bangladesh. *Int. J. Biomed. Sci. IJBS* **2009**, *5*, 395–401. [CrossRef] [PubMed]
150. Cuero, R.; Ouellet, T. Metal Ions Modulate Gene Expression and Accumulation of the Mycotoxins Aflatoxin and Zearalenone. *J. Appl. Microbiol.* **2005**, *98*, 598–605. [CrossRef] [PubMed]
151. Savi, G.D.; Bortoluzzi, A.J.; Scussel, V.M. Antifungal Properties of Zinc-compounds against Toxigenic Fungi and Mycotoxin. *Int. J. Food Sci. Technol.* **2013**, *48*, 1834–1840. [CrossRef]
152. Eide, D.J. Zinc Transporters and the Cellular Trafficking of Zinc. *Biochim. Biophys. Acta Mol. Cell Res.* **2006**, *1763*, 711–722. [CrossRef]
153. Gaither, L.A.; Eide, D.J. Eukaryotic Zinc Transporters and Their Regulation. *Biometals* **2001**, *14*, 251–270. [CrossRef] [PubMed]
154. Waters, B.M.; Eide, D.J. Combinatorial Control of Yeast FET4 Gene Expression by Iron, Zinc, and Oxygen. *J. Biol. Chem.* **2002**, *277*, 33749–33757. [CrossRef] [PubMed]

155. Eide, D.J. Multiple Regulatory Mechanisms Maintain Zinc Homeostasis in *Saccharomyces cerevisiae*. *J. Nutr.* **2003**, *133*, 1532S–1535S. [CrossRef] [PubMed]
156. Zhao, H.; Butler, E.; Rodgers, J.; Spizzo, T.; Duesterhoeft, S.; Eide, D. Regulation of Zinc Homeostasis in Yeast by Binding of the ZAP1 Transcriptional Activator to Zinc-responsive Promoter Elements. *J. Biol. Chem.* **1998**, *273*, 28713–28720. [CrossRef]
157. Wang, Y.; Weisenhorn, E.; MacDiarmid, C.W.; Andreini, C.; Bucci, M.; Taggart, J.; Banci, L.; Russell, J.; Coon, J.J.; Eide, D.J. The Cellular Economy of the *Saccharomyces cerevisiae* Zinc Proteome. *Met. Integr. Biometal Sci.* **2018**, *10*, 1755–1776. [CrossRef]
158. Frey, A.G.; Eide, D.J. Zinc-responsive Coactivator Recruitment by the Yeast Zap1 Transcription Factor. *Microbiologyopen* **2012**, *1*, 105–114. [CrossRef]
159. Frey, A.G.; Bird, A.J.; Evans-Galea, M.V.; Blankman, E.; Winge, D.R.; Eide, D.J. Zinc-Regulated DNA Binding of the Yeast Zap1 Zinc-Responsive Activator. *PLoS ONE* **2011**, *6*, e22535. [CrossRef]
160. Guerinot, M.L. The ZIP Family of Metal Transporters. *Biochim. Biophys. Acta Biomembr.* **2000**, *1465*, 190–198. [CrossRef]
161. Vicentefranqueira, R.; Moreno, M.Á.; Leal, F.; Calera, J.A. The zrfA and zrfB Genes of *Aspergillus fumigatus* Encode the Zinc Transporter Proteins of a Zinc Uptake System Induced in an Acid, Zinc-Depleted Environment. *Eukaryot. Cell* **2005**, *4*, 837–848. [CrossRef]
162. Moreno, M.Á.; Ibrahim-Granet, O.; Vicentefranqueira, R.; Amich, J.; Ave, P.; Leal, F.; Latgé, J.-P.; Calera, J.A. The Regulation of Zinc Homeostasis by the ZafA Transcriptional Activator is Essential for *Aspergillus fumigatus* Virulence. *Mol. Microbiol.* **2007**, *64*, 1182–1197. [CrossRef] [PubMed]
163. López-Berges, M.S. ZafA-Mediated Regulation of Zinc Homeostasis is Required for Virulence in the Plant Pathogen *Fusarium oxysporum*. *Mol. Plant Pathol.* **2020**, *21*, 244–249. [CrossRef] [PubMed]
164. Reeder, N.L.; Xu, J.; Youngquist, R.S.; Schwartz, J.R.; Rust, R.C.; Saunders, C.W. The Antifungal Mechanism of Action of Zinc Pyrithione. *Br. J. Dermatol.* **2011**, *165*, 9–12. [CrossRef] [PubMed]
165. Park, M.; Cho, Y.-J.; Lee, Y.W.; Jung, W.H. Understanding the Mechanism of Action of the Anti-Dandruff Agent Zinc Pyrithione against *Malassezia restricta*. *Sci. Rep.* **2018**, *8*, 12086. [CrossRef]
166. Stehling, O.; Lill, R. The Role of Mitochondria in Cellular Iron-sulfur Protein Biogenesis: Mechanisms, Connected Processes, and Diseases. *Cold Spring Harbor Perspect. Biol.* **2013**, *5*, a011312. [CrossRef]
167. Buchman, C.; Skroch, P.; Welch, J.; Fogel, S.; Karin, M. The CUP2 Gene Product, Regulator of Yeast Metallothionein Expression, is a Copper-activated DNA-binding Protein. *Mol. Cell. Biol.* **1989**, *9*, 4091–4095. [CrossRef]
168. Dancis, A.; Haile, D.; Yuan, D.S.; Klausner, R.D. The *Saccharomyces cerevisiae* Copper Transport Protein (Ctr1p). Biochemical Characterization, Regulation by Copper, and Physiologic Role in Copper Uptake. *J. Biol. Chem.* **1994**, *269*, 25660–25667. [CrossRef]
169. Crawford, A.C.; Lehtovirta-Morley, L.E.; Alamir, O.; Niemiec, M.J.; Alawfi, B.; Alsarraf, M.; Skrahina, V.; Costa, A.C.B.P.; Anderson, A.; Yellagunda, S.; et al. Biphasic Zinc Compartmentalisation in a Human Fungal Pathogen. *PLoS Pathog.* **2018**, *14*, e1007013. [CrossRef]
170. Khouja, H.R.; Abbà, S.; Lacercat-Didier, L.; Daghino, S.; Doillon, D.; Richaud, P.; Martino, E.; Vallino, M.; Perotto, S.; Chalot, M.; et al. OmZnT1 and OmFET, Two Metal Transporters from the Metal-tolerant Strain Zn of the Ericoid Mycorrhizal Fungus *Oidiodendron maius*, Confer Zinc Tolerance in Yeast. *Fungal Genet. Biol.* **2013**, *52*, 53–64. [CrossRef] [PubMed]
171. González-Guerrero, M.; Melville, L.H.; Ferrol, N.; Lott, J.N.A.; Azcón-Aguilar, C.; Peterson, R.L. Ultrastructural Localization of Heavy Metals in the Extraradical Mycelium and Spores of the Arbuscular Mycorrhizal Fungus *Glomus intraradices*. *Can. J. Microbiol.* **2008**, *54*, 103–110. [CrossRef] [PubMed]
172. Leonhardt, T.; Sácký, J.; Šimek, P.; Šantrůček, J.; Kotrba, P. Metallothionein-like Peptides Involved in Sequestration of Zn in the Zn-accumulating Ectomycorrhizal Fungus *Russula atropurpurea*. *Metallomics* **2014**, *6*, 1693–1701. [CrossRef]
173. Sácký, J.; Leonhardt, T.; Borovička, J.; Gryndler, M.; Briksí, A.; Kotrba, P. Intracellular Sequestration of Zinc, Cadmium and Silver in *Hebeloma mesophaeum* and Characterization of Its Metallothionein Genes. *Fungal Genet. Biol.* **2014**, *67*, 3–14. [CrossRef]
174. Ouda, S. Antifungal Activity of Silver and Copper Nanoparticles on Two Plant Pathogens, *Alternaria alternata* and *Botrytis cinerea*. *Res. J. Microbiol.* **2014**, *9*, 34–42. [CrossRef]
175. Zhang, P.; Zhang, D.; Zhao, X.; Wei, D.; Wang, Y.; Zhu, X. Effects of CTR4 Deletion on Virulence and Stress Response in *Cryptococcus neoformans*. *Antonie van Leeuwenhoek* **2016**, *109*, 1081–1090. [CrossRef]
176. Hassett, R.; Kosman, D.J. Evidence for Cu(II) Reduction as a Component of Copper Uptake by *Saccharomyces cerevisiae*. *J. Biol. Chem.* **1995**, *270*, 128–134. [CrossRef]
177. Zhu, Z.; Labbé, S.; Peña, M.M.O.; Thiele, D.J. Copper Differentially Regulates the Activity and Degradation of Yeast Mac1 Transcription Factor. *J. Biol. Chem.* **1998**, *273*, 1277–1280. [CrossRef]
178. de Silva, D.; Davis-Kaplan, S.; Fergestad, J.; Kaplan, J. Purification and Characterization of Fet3 Protein, a Yeast Homologue of Ceruloplasmin. *J. Biol. Chem.* **1997**, *272*, 14208–14213. [CrossRef]
179. Askwith, C.; Eide, D.; Van Ho, A.; Bernard, P.S.; Li, L.; Davis-Kaplan, S.; Sipe, D.M.; Kaplan, J. The FET3 Gene of S. Cerevisiae Encodes a Multicopper Oxidase Required for Ferrous Iron Uptake. *Cell* **1994**, *76*, 403–410. [CrossRef]
180. Shakoury-Elizeh, M.; Protchenko, O.; Berger, A.; Cox, J.; Gable, K.; Dunn, T.M.; Prinz, W.A.; Bard, M.; Philpott, C.C. Metabolic Response to Iron Deficiency in *Saccharomyces cerevisiae*. *J. Biol. Chem.* **2010**, *285*, 14823–14833. [CrossRef] [PubMed]
181. Yamaguchi-Iwai, Y.; Dancis, A.; Klausner, R.D. AFT1: A Mediator of Iron Regulated Transcriptional Control in *Saccharomyces cerevisiae*. *EMBO J.* **1995**, *14*, 1231–1239. [CrossRef]

182. Horng, Y.-C.; Cobine, P.A.; Maxfield, A.B.; Carr, H.S.; Winge, D.R. Specific Copper Transfer from the Cox17 Metallochaperone to Both Sco1 and Cox11 in the Assembly of Yeast Cytochrome c Oxidase. *J. Biol. Chem.* **2004**, *279*, 35334–35340. [CrossRef]
183. Garay-Arroyo, A.; Lledías, F.; Hansberg, W.; Covarrubias, A.A. Cu,Zn-superoxide Dismutase of *Saccharomyces cerevisiae* is Required for Resistance to Hyperosmosis. *FEBS Lett.* **2003**, *539*, 68–72. [CrossRef]
184. Zyrina, A.N.; Smirnova, E.A.; Markova, O.V.; Severin, F.F.; Knorre, D.A. Mitochondrial Superoxide Dismutase and Yap1p Act as a Signaling Module Contributing to Ethanol Tolerance of the Yeast *Saccharomyces cerevisiae*. *Appl. Environ. Microbiol.* **2017**, *83*, e02759–e16. [CrossRef]
185. Fridovich, I. Superoxide Anion Radical (O·2), Superoxide Dismutases, and Related Matters. *J. Biol. Chem.* **1997**, *272*, 18515–18517. [CrossRef]
186. Imlay, J.A.; Fridovich, I. Suppression of Oxidative Envelope Damage by Pseudoreversion of a Superoxide Dismutase-deficient Mutant of *Escherichia coli*. *J. Bacteriol.* **1992**, *174*, 953. [CrossRef]
187. Lushchak, V.; Semchyshyn, H.; Mandryk, S.; Lushchak, O. Possible Role of Superoxide Dismutases in the Yeast *Saccharomyces cerevisiae* under Respiratory Conditions. *Arch. Biochem. Biophys.* **2005**, *441*, 35–40. [CrossRef]
188. Seah, T.C.M.; Kaplan, J.G. Purification and Properties of the Catalase of Bakers' Yeast. *J. Biol. Chem.* **1973**, *248*, 2889–2893. [CrossRef]
189. Culotta, V.C.; Howard, W.R.; Liu, X.F. CRS5 Encodes a Metallothionein-like Protein in *Saccharomyces cerevisiae*. *J. Biol. Chem.* **1994**, *269*, 25295–25302. [CrossRef]
190. George, G.N.; Byrd, J.; Winge, D.R. X-ray Absorption Studies of Yeast Copper Metallothionein. *J. Biol. Chem.* **1988**, *263*, 8199–8203. [CrossRef]
191. Jensen, L.T.; Howard, W.R.; Strain, J.J.; Winge, D.R.; Culotta, V.C. Enhanced Effectiveness of Copper Ion Buffering by CUP1 Metallothionein Compared with CRS5 Metallothionein in *Saccharomyces cerevisiae*. *J. Biol. Chem.* **1996**, *271*, 18514–18519. [CrossRef]
192. Nagakubo, T.; Kumano, T.; Ohta, T.; Hashimoto, Y.; Kobayashi, M. Copper Amine Oxidases Catalyze the Oxidative Deamination and Hydrolysis of Cyclic Imines. *Nat. Commun.* **2019**, *10*, 413. [CrossRef] [PubMed]
193. Antsotegi-Uskola, M.; Markina-Iñarrairaegui, A.; Ugalde, U. New Insights into Copper Homeostasis in Filamentous Fungi. *Int. Microbiol.* **2020**, *23*, 65–73. [CrossRef] [PubMed]
194. Kusuya, Y.; Hagiwara, D.; Sakai, K.; Yaguchi, T.; Gonoi, T.; Takahashi, H. Transcription Factor Afmac1 Controls Copper Import Machinery in *Aspergillus fumigatus*. *Curr. Genet.* **2017**, *63*, 777–789. [CrossRef]
195. Park, Y.-S.; Kang, S.; Seo, H.; Yun, C.-W. A copper Transcription Factor, AfMac1, Regulates Both Iron and Copper Homeostasis in the Opportunistic Fungal Pathogen *Aspergillus fumigatus*. *Biochem. J.* **2018**, *475*, 2831–2845. [CrossRef] [PubMed]
196. Park, Y.-S.; Kim, T.-H.; Yun, C.-W. Functional Characterization of the Copper Transcription Factor AfMac1 from *Aspergillus fumigatus*. *Biochem. J.* **2017**, *474*, 2365–2378. [CrossRef] [PubMed]
197. Tang, J.D.; Perkins, A.D.; Sonstegard, T.S.; Schroeder, S.G.; Burgess, S.C.; Diehl, S.V. Short-Read Sequencing for Genomic Analysis of the Brown Rot Fungus *Fibroporia radiculosa*. *Appl. Environ. Microbiol.* **2012**, *78*, 2272. [CrossRef]
198. Bayat, N.; Rajapakse, K.; Marinsek-Logar, R.; Drobne, D.; Cristobal, S. The Effects of Engineered Nanoparticles on the Cellular Structure and Growth of *Saccharomyces cerevisiae*. *Nanotoxicology* **2014**, *8*, 363–373. [CrossRef]
199. Kasemets, K.; Käosaar, S.; Vija, H.; Fascio, U.; Mantecca, P. Toxicity of Differently Sized and Charged Silver Nanoparticles to Yeast *Saccharomyces cerevisiae* BY4741: A Nano-biointeraction Perspective. *Nanotoxicology* **2019**, *13*, 1041–1059. [CrossRef]
200. Jo, W.J.; Loguinov, A.; Chang, M.; Wintz, H.; Nislow, C.; Arkin, A.P.; Giaever, G.; Vulpe, C.D. Identification of Genes Involved in the Toxic Response of *Saccharomyces cerevisiae* against Iron and Copper Overload by Parallel Analysis of Deletion Mutants. *Toxicol. Sci.* **2007**, *101*, 140–151. [CrossRef]
201. Kasemets, K.; Ivask, A.; Dubourguier, H.-C.; Kahru, A. Toxicity of Nanoparticles of ZnO, CuO and TiO_2 to Yeast *Saccharomyces cerevisiae*. *Toxicol. In Vitro* **2009**, *23*, 1116–1122. [CrossRef]
202. Giannousi, K.; Sarafidis, G.; Mourdikoudis, S.; Pantazaki, A.; Dendrinou-Samara, C. Selective Synthesis of Cu_2O and Cu/Cu_2O NPs: Antifungal Activity to Yeast *Saccharomyces cerevisiae* and DNA Interaction. *Inorg. Chem.* **2014**, *53*, 9657–9666. [CrossRef]
203. Hitchcock, C.A.; Pye, G.W.; Troke, P.F.; Johnson, E.M.; Warnock, D.W. Fluconazole Resistance in *Candida glabrata*. *Antimicrob. Agents Chemother.* **1993**, *37*, 1962–1965. [CrossRef]
204. Orozco, A.S.; Higginbotham, L.M.; Hitchcock, C.A.; Parkinson, T.; Falconer, D.; Ibrahim, A.S.; Ghannoum, M.A.; Filler, S.G. Mechanism of Fluconazole Resistance in *Candida krusei*. *Antimicrob. Agents Chemother.* **1998**, *42*, 2645–2649. [CrossRef]
205. Pfaller, M.A.; Diekema, D.J.; Gibbs, D.L.; Newell, V.A.; Nagy, E.; Dobiasova, S.; Rinaldi, M.; Barton, R.; Veselov, A. Global Antifungal Surveillance, G. *Candida krusei*, a Multidrug-resistant Opportunistic Fungal Pathogen: Geographic and Temporal Trends from the ARTEMIS DISK Antifungal Surveillance Program, 2001 to 2005. *J. Clin. Microbiol.* **2008**, *46*, 515–521. [CrossRef]
206. Gomes da Silva Dantas, F.; Araújo de Almeida-Apolonio, A.; Pires de Araújo, R.; Regiane Vizolli Favarin, L.; Fukuda de Castilho, P.; de Oliveira Galvão, F.; Inez Estivalet Svidzinski, T.; Antônio Casagrande, G.; Mari Pires de Oliveira, K. A Promising Copper(II) Complex as Antifungal and Antibiofilm Drug against Yeast Infection. *Molecules* **2018**, *23*, 1856. [CrossRef]
207. Coyle, B.; Kavanagh, K.; McCann, M.; Devereux, M.; Geraghty, M. Mode of Anti-fungal Activity of 1,10-phenanthroline and Its Cu(II), Mn(II) and Ag(I) Complexes. *Biometals Int. J. Role Met. Ions Biol. Biochem. Med.* **2003**, *16*, 321–329. [CrossRef]
208. Borgatta, J.; Ma, C.; Hudson-Smith, N.; Elmer, W.; Plaza Pérez, C.D.; De La Torre-Roche, R.; Zuverza-Mena, N.; Haynes, C.L.; White, J.C.; Hamers, R.J. Copper Based Nanomaterials Suppress Root Fungal Disease in Watermelon (*Citrullus lanatus*): Role of Particle Morphology, Composition and Dissolution Behavior. *ACS Sustain. Chem. Eng.* **2018**, *6*, 14847–14856. [CrossRef]

209. Cai, Z.; Du, W.; Zhang, Z.; Guan, L.; Zeng, Q.; Chai, Y.; Dai, C.; Lu, L. The *Aspergillus fumigatus* Transcription Factor AceA is Involved not only in Cu but also in Zn Detoxification through Regulating Transporters CrpA and ZrcA. *Cell. Microbiol.* **2018**, *20*, e12864. [CrossRef]
210. Ragasa, L.R.P.; Joson, S.E.A.; Bagay, W.L.R.; Perez, T.R.; Velarde, M.C. Transcriptome Analysis Reveals Involvement of Oxidative Stress Response in a Copper-tolerant *Fusarium oxysporum* Strain. *Fungal Biol.* **2021**. [CrossRef]
211. Bolm, C. A New Iron Age. *Nat. Chem.* **2009**, *1*, 420. [CrossRef]
212. Weber, K.A.; Achenbach, L.A.; Coates, J.D. Microorganisms Pumping Iron: Anaerobic Microbial Iron Oxidation and Reduction. *Nat. Rev. Microbiol.* **2006**, *4*, 752–764. [CrossRef]
213. Hissen, A.H.T.; Wan, A.N.C.; Warwas, M.L.; Pinto, L.J.; Moore, M.M. The *Aspergillus fumigatus* Siderophore Biosynthetic Gene sidA, Encoding l-Ornithine N5-Oxygenase, Is Required for Virulence. *Infect. Immun.* **2005**, *73*, 5493–5503. [CrossRef] [PubMed]
214. Schrettl, M.; Bignell, E.; Kragl, C.; Joechl, C.; Rogers, T.; Arst, H.N., Jr.; Haynes, K.; Haas, H. Siderophore Biosynthesis but Not Reductive Iron Assimilation Is Essential for *Aspergillus fumigatus* Virulence. *J. Exp. Med.* **2004**, *200*, 1213–1219. [CrossRef] [PubMed]
215. López-Berges, M.S.; Capilla, J.; Turrà, D.; Schafferer, L.; Matthijs, S.; Jöchl, C.; Cornelis, P.; Guarro, J.; Haas, H.; Di Pietro, A. HapX-mediated Iron Homeostasis is Essential for Rhizosphere Competence and Virulence of the Soilborne Pathogen *Fusarium oxysporum*. *Plant Cell* **2012**, *24*, 3805–3822. [CrossRef] [PubMed]
216. Dong, X.; Wang, M.; Ling, N.; Shen, Q.; Guo, S. Effects of Iron and Boron Combinations on the Suppression of *Fusarium* Wilt in Banana. *Sci. Rep.* **2016**, *6*, 38944. [CrossRef]
217. Hashem, A.R. Influence of Iron on the Growth of the Tomato Wilt Pathogen, *Fusarium oxysporum*, Isolated in Saudi Arabia. *J. Plant Dis. Prot.* **1995**, *102*, 326–330.
218. Philpott, C.C. Iron Uptake in Fungi: A System for Every Source. *Biochim. Biophys. Acta Mol. Cell Res.* **2006**, *1763*, 636–645. [CrossRef]
219. Yun, C.-W.; Ferea, T.; Rashford, J.; Ardon, O.; Brown, P.O.; Botstein, D.; Kaplan, J.; Philpott, C.C. Desferrioxamine-mediated Iron Uptake in *Saccharomyces cerevisiae*: Evidence for Two Pathways of Iron Uptake. *J. Biol. Chem.* **2000**, *275*, 10709–10715. [CrossRef]
220. Yun, C.-W.; Tiedeman, J.S.; Moore, R.E.; Philpott, C.C. Siderophore-Iron Uptake in *Saccharomyces cerevisiae*: Identification of Ferrichrome and Fusarinine Transporters *J. Biol. Chem.* **2000**, *275*, 16354–16359. [CrossRef]
221. Georgatsou, E.; Alexandraki, D. Two distinctly Regulated Genes are Required for Ferric Reduction, the First Step of Iron Uptake in *Saccharomyces cerevisiae*. *Mol. Cell. Biol.* **1994**, *14*, 3065–3073. [CrossRef]
222. Martínez-Pastor, M.T.; Perea-García, A.; Puig, S. Mechanisms of Iron Sensing and Regulation in the Yeast *Saccharomyces cerevisiae*. *World J. Microbiol. Biotechnol.* **2017**, *33*, 75. [CrossRef]
223. Jensen, L.T.; Culotta, V.C. Regulation of *Saccharomyces cerevisiae* FET4 by Oxygen and Iron. *J. Mol. Biol.* **2002**, *318*, 251–260. [CrossRef]
224. Lesuisse, E.; Raguzzi, F.; Crichton, R. Iron Uptake by the Yeast *Saccharomyces cerevisiae*: Involvement of a Reduction Step. *J. Gen. Microbiol.* **1987**, *133*, 3229–3236. [CrossRef]
225. Georgatsou, E.; Mavrogiannis, L.A.; Fragiadakis, G.S.; Alexandraki, D. The Yeast Fre1p/Fre2p Cupric Reductases Facilitate Copper Uptake and Are Regulated by the Copper-modulated Mac1p Activator. *J. Biol. Chem.* **1997**, *272*, 13786–13792. [CrossRef]
226. Caetano, S.M.; Menezes, R.; Amaral, C.; Rodrigues-Pousada, C.; Pimentel, C. Repression of the Low Affinity Iron Transporter Gene FET4: A Novel Mechanism against Cadmium Toxicity Orchestrated by Yap1 via Rox1. *J. Biol. Chem.* **2015**, *290*, 18584–18595. [CrossRef]
227. Liu, X.F.; Supek, F.; Nelson, N.; Culotta, V.C. Negative Control of Heavy Metal Uptake by the *Saccharomyces cerevisiae* BSD2 Gene. *J. Biol. Chem.* **1997**, *272*, 11763–11769. [CrossRef]
228. Kim, Y.; Lampert, S.M.; Philpott, C.C. A Receptor Domain Controls the Intracellular Sorting of the Ferrichrome Transporter, ARN1. *EMBO J.* **2005**, *24*, 952–962. [CrossRef] [PubMed]
229. Kim, Y.; Yun, C.-W.; Philpott, C.C. Ferrichrome Induces Endosome to Plasma Membrane Cycling of the Ferrichrome Transporter, Arn1p, in *Saccharomyces cerevisiae*. *EMBO J.* **2002**, *21*, 3632–3642. [CrossRef]
230. Heymann, P.; Ernst, J.F.; Winkelmann, G. Identification of a Fungal Triacetylfusarinine C Siderophore Transport Gene (TAF1) in *Saccharomyces cerevisiae* as a Member of the Major Facilitator Superfamily. *Biometals* **1999**, *12*, 301–306. [CrossRef]
231. Lesuisse, E.; Blaiseau, P.-L.; Dancis, A.; Camadro, J.-M. Siderophore Uptake and Use by the Yeast *Saccharomyces cerevisiae*. *Microbiology* **2001**, *147*, 289–298. [CrossRef]
232. Froissard, M.; Belgareh-Touzé, N.; Dias, M.; Buisson, N.; Camadro, J.M.; Haguenauer-Tsapis, R.; Lesuisse, E. Trafficking of Siderophore Transporters in *Saccharomyces cerevisiae* and Intracellular Fate of Ferrioxamine B Conjugates. *Traffic* **2007**, *8*, 1601–1616. [CrossRef]
233. Kosman, D.J. Molecular Mechanisms of Iron Uptake in Fungi. *Mol. Microbiol.* **2003**, *47*, 1185–1197. [CrossRef]
234. Kohlhaw, G.B. Leucine Biosynthesis in Fungi: Entering Metabolism through the Back Door. *Microbiol. Mol. Biol. Rev.* **2003**, *67*, 1–15. [CrossRef]
235. Pokharel, S.; Campbell, J.L. Cross Talk between the Nuclease and Helicase Activities of Dna2: Role of an Essential Iron-sulfur Cluster Domain. *Nucleic Acids Res.* **2012**, *40*, 7821–7830. [CrossRef]
236. Labbé, S.; Pelletier, B.; Mercier, A. Iron Homeostasis in the Fission Yeast *Schizosaccharomyces pombe*. *Biometals* **2007**, *20*, 523–537. [CrossRef]

237. Schrettl, M.; Winkelmann, G.; Haas, H. Ferrichrome in *Schizosaccharomyces pombe*–an Iron Transport and Iron Storage Compound. *Biometals* **2004**, *17*, 647–654. [CrossRef]
238. Roman, D.G.; Dancis, A.; Anderson, G.J.; Klausner, R.D. The Fission Yeast Ferric Reductase Gene frp1+ is Required for Ferric Iron Uptake and Encodes a Protein That is Homologous to the gp91-phox Subunit of the Human NADPH Phagocyte Oxidoreductase. *Mol. Cell. Biol.* **1993**, *13*, 4342–4350. [CrossRef]
239. Pouliot, B.; Jbel, M.; Mercier, A.; Labbé, S. abc3 + Encodes an Iron-Regulated Vacuolar ABC-Type Transporter in *Schizosaccharomyces pombe*. *Eukaryot. Cell* **2010**, *9*, 59–73. [CrossRef] [PubMed]
240. Lowry, C.V.; Zitomer, R.S. ROX1 Encodes a Heme-induced Repression Factor Regulating ANB1 and CYC7 of *Saccharomyces cerevisiae*. *Mol. Cell. Biol.* **1988**, *8*, 4651–4658. [CrossRef]
241. Hickman, M.J.; Winston, F. Heme Levels Switch the Function of Hap1 of *Saccharomyces cerevisiae* between Transcriptional Activator and Transcriptional Repressor. *Mol. Cell. Biol.* **2007**, *27*, 7414–7424. [CrossRef]
242. Santos, R.; Buisson, N.; Knight, S.; Dancis, A.; Camadro, J.-M.; Lesuisse, E. Haemin Uptake and Use as an Iron Source by *Candida albicans*: Role of CaHMX1-encoded Haem Oxygenase. *Microbiology* **2003**, *149*, 579–588. [CrossRef]
243. Tsiftsoglou, A.S.; Tsamadou, A.I.; Papadopoulou, L.C. Heme as Key Regulator of Major Mammalian Cellular Functions: Molecular, Cellular, and Pharmacological Aspects. *Pharmacol. Ther.* **2006**, *111*, 327–345. [CrossRef]
244. Mercier, A.; Pelletier, B.; Labbé, S. A Transcription Factor Cascade Involving Fep1 and the CCAAT-Binding Factor Php4 Regulates Gene Expression in Response to Iron Deficiency in the Fission Yeast *Schizosaccharomyces pombe*. *Eukaryot. Cell* **2006**, *5*, 1866–1881. [CrossRef]
245. Weissman, Z.; Shemer, R.; Conibear, E.; Kornitzer, D. An Endocytic Mechanism for Haemoglobin-iron Acquisition in *Candida albicans*. *Mol. Microbiol.* **2008**, *69*, 201–217. [CrossRef]
246. Ardon, O.; Nudelman, R.; Caris, C.; Libman, J.; Shanzer, A.; Chen, Y.; Hadar, Y. Iron Uptake in *Ustilago maydis*: Tracking the Iron Path. *J. Bacteriol.* **1998**, *180*, 2021–2026. [CrossRef]
247. Wang, J.; Budde, A.D.; Leong, S.A. Analysis of Ferrichrome Biosynthesis in the Phytopathogenic Fungus *Ustilago maydis*: Cloning of an Ornithine-N5-oxygenase Gene. *J. Bacteriol.* **1989**, *171*, 2811–2818. [CrossRef]
248. Lin, H.; Li, L.; Jia, X.; Ward, D.M.; Kaplan, J. Genetic and Biochemical Analysis of High Iron Toxicity in Yeast: Iron Toxicity is Due to the Accumulation of Cytosolic Iron and Occurs under Both Aerobic and Anaerobic Conditions. *J. Biol. Chem.* **2011**, *286*, 3851–3862. [CrossRef]
249. Berthelet, S.; Usher, J.; Shulist, K.; Hamza, A.; Maltez, N.; Johnston, A.; Fong, Y.; Harris, L.J.; Baetz, K. Functional Genomics Analysis of the *Saccharomyces cerevisiae* Iron Responsive Transcription Factor Aft1 Reveals Iron-independent Functions. *Genetics* **2010**, *185*, 1111–1128. [CrossRef]
250. van Bakel, H.; Strengman, E.; Wijmenga, C.; Holstege, F.C.P. Gene Expression Profiling and Phenotype Analyses of *S. Cerevisiae* in Response to Changing Copper Reveals Six Genes with New Roles in Copper and Iron Metabolism. *Physiol. Genom.* **2005**, *22*, 356–367. [CrossRef]
251. Adamczyk, Z.; Oćwieja, M.; Mrowiec, H.; Walas, S.; Lupa, D. Oxidative Dissolution of Silver Nanoparticles: A New Theoretical Approach. *J. Colloid Interface Sci.* **2016**, *469*, 355–364. [CrossRef] [PubMed]
252. Hassan, N.; Elsharkawy, M.M.; Shimizu, M.; Hyakumachi, M. Control of Root Rot and Wilt Diseases of Roselle under Field Conditions. *Mycobiology* **2014**, *42*, 376–384. [CrossRef] [PubMed]
253. Adhilakshmi, M.; Karthikeyan, M.; Devadason, A. Effect of Combination of Bio-agents and Mineral Nutrients for the Management of Alfalfa Wilt Pathogen *Fusarium oxysporum f. sp. medicaginis*. *Arch. Phytopathol. Plant Prot.* **2008**, *47*, 514–525. [CrossRef]
254. Clark, H.L.; Jhingran, A.; Sun, Y.; Vareechon, C.; de Jesus Carrion, S.; Skaar, E.P.; Chazin, W.J.; Calera, J.A.; Hohl, T.M.; Pearlman, E. Zinc and Manganese Chelation by Neutrophil S100A8/A9 (Calprotectin) Limits Extracellular *Aspergillus fumigatus* Hyphal Growth and Corneal Infection. *J. Immunol.* **2016**, *196*, 336–344. [CrossRef] [PubMed]
255. Isikhuemhen, O.S.; Mikiashvilli, N.A. Lignocellulolytic Enzyme Activity, Substrate Utilization, and Mushroom Yield by *Pleurotus ostreatus* Cultivated on Substrate Containing Anaerobic Digester Solids. *J. Ind. Microbiol. Biotechnol.* **2009**, *36*, 1353–1362. [CrossRef] [PubMed]
256. Isikhuemhen, O.S.; Nerud, F. Preliminary Studies on the Ligninolytic Enzymes Produced by the Tropical Fungus *Pleurotus tuber-regium* (Fr.) Sing. *Antonie van Leeuwenhoek* **1999**, *75*, 257–260. [CrossRef]
257. Portnoy, M.E.; Liu, X.F.; Culotta, V.C. *Saccharomyces cerevisiae* Expresses Three Functionally Distinct Homologues of the Nramp Family of Metal Transporters. *Mol. Cell. Biol.* **2000**, *20*, 7893–7902. [CrossRef]
258. Luk, E.; Yang, M.; Jensen, L.T.; Bourbonnais, Y.; Culotta, V.C. Manganese Activation of Superoxide Dismutase 2 in the Mitochondria of *Saccharomyces cerevisiae*. *J. Biol. Chem.* **2005**, *280*, 22715–22720. [CrossRef]
259. Van Loon, A.; Pesold-Hurt, B.; Schatz, G. A Yeast Mutant Lacking Mitochondrial Manganese-superoxide Dismutase is Hypersensitive to Oxygen. *Proc. Natl. Acad. Sci. USA* **1986**, *83*, 3820–3824. [CrossRef] [PubMed]
260. Reddi, A.R.; Jensen, L.T.; Naranuntarat, A.; Rosenfeld, L.; Leung, E.; Shah, R.; Culotta, V.C. The Overlapping Roles of Manganese and Cu/Zn SOD in Oxidative Stress Protection. *Free Radic. Biol. Med.* **2009**, *46*, 154–162. [CrossRef]
261. Liu, X.F.; Culotta, V.C. Post-translation Control of Nramp Metal Transport in Yeast: Role of Metal Ions and the BSD2 Gene. *J. Biol. Chem.* **1999**, *274*, 4863–4868. [CrossRef]
262. Stimpson, H.E.; Lewis, M.J.; Pelham, H.R. Transferrin Receptor-like Proteins Control the Degradation of a Yeast Metal Transporter. *EMBO J.* **2006**, *25*, 662–672. [CrossRef]

263. Jensen, L.T.; Carroll, M.C.; Hall, M.D.; Harvey, C.J.; Beese, S.E.; Culotta, V.C. Down-regulation of a Manganese Transporter in the Face of Metal Toxicity. *Mol. Biol. Cell* **2009**, *20*, 2810–2819. [CrossRef]
264. Yang, B.; Kumar, S. Nedd4 and Nedd4-2: Closely Related Ubiquitin-protein Ligases with Distinct Physiological Functions. *Cell Death Differ.* **2010**, *17*, 68–77. [CrossRef]
265. Bun-Ya, M.; Nishimura, M.; Harashima, S.; Oshima, Y. The PHO84 Gene of *Saccharomyces cerevisiae* Encodes an Inorganic Phosphate Transporter. *Mol. Cell. Biol.* **1991**, *11*, 3229–3238. [CrossRef]
266. Wykoff, D.D.; Rizvi, A.H.; Raser, J.M.; Margolin, B.; O'Shea, E.K. Positive Feedback Regulates Switching of Phosphate Transporters in *S. cerevisiae*. *Mol. Cell* **2007**, *27*, 1005–1013. [CrossRef]
267. Reddi, A.R.; Jensen, L.T.; Culotta, V.C. Manganese Homeostasis in *Saccharomyces cerevisiae*. *Chem. Rev.* **2009**, *109*, 4722–4732. [CrossRef]
268. Rayner, J.C.; Munro, S. Identification of the MNN2 and MNN5Mannosyltransferases Required for Forming and Extending the Mannose Branches of the Outer Chain Mannans of *Saccharomyces cerevisiae*. *J. Biol. Chem.* **1998**, *273*, 26836–26843. [CrossRef]
269. Romero, P.A.; Herscovics, A. Glycoprotein Biosynthesis in *Saccharomyces cerevisiae*. Characterization of Alpha-1,6-mannosyltransferase Which Initiates outer Chain Formation. *J. Biol. Chem.* **1989**, *264*, 1946–1950. [CrossRef]
270. Wiggins, C.A.; Munro, S. Activity of the Yeast MNN1 Alpha-1,3-mannosyltransferase Requires a Motif Conserved in Many Other Families of Glycosyltransferases. *Proc. Natl. Acad. Sci. USA* **1998**, *95*, 7945–7950. [CrossRef]
271. Striebeck, A.; Robinson, D.A.; Schüttelkopf, A.W.; van Aalten, D.M.F. Yeast Mnn9 is Both a Priming Glycosyltransferase and an Allosteric Activator of Mannan Biosynthesis. *Open Biol.* **2013**, *3*, 130022. [CrossRef]
272. Solá, R.J.; Griebenow, K. Effects of Glycosylation on the Stability of Protein Pharmaceuticals. *J. Pharm. Sci.* **2009**, *98*, 1223–1245. [CrossRef]
273. Xiang, M.; Mohamalawari, D.; Rao, R. A Novel Isoform of the Secretory Pathway Ca2+,Mn(2+)-ATPase, hSPCA2, Has Unusual Properties and is Expressed in the Brain. *J. Biol. Chem.* **2005**, *280*, 11608–11614. [CrossRef]
274. Tanaka, J.; Fink, G.R. The Histidine Permease Gene (HIP1) of *Saccharomyces cerevisiae*. *Gene* **1985**, *38*, 205–214. [CrossRef]
275. Bakshi, D.K.; Saha, S.; Sindhu, I.; Sharma, P. Use of *Phanerochaete* Chrysosporium Biomass for the Removal of Textile Dyes from a Synthetic Effluent. *World J. Microbiol. Biotechnol.* **2006**, *22*, 835–839. [CrossRef]
276. Bending, G.D.; Friloux, M.; Walker, A. Degradation of Contrasting Pesticides by White Rot Fungi and Its Relationship with Ligninolytic Potential. *FEMS Microbiol. Lett.* **2002**, *212*, 59–63. [CrossRef]
277. Kirk, T.K.; Connors, W.; Zeikus, J.G. Requirement for a Growth Substrate during Lignin Decomposition by Two Wood-rotting Fungi. *Appl. Environ. Microbiol.* **1976**, *32*, 192–194. [CrossRef]
278. Kuwahara, M.; Glenn, J.K.; Morgan, M.A.; Gold, M.H. Separation and Characterization of Two Extracelluar H_2O_2-dependent Oxidases from Ligninolytic Cultures of *Phanerochaete chrysosporium*. *FEBS Lett.* **1984**, *169*, 247–250. [CrossRef]
279. Bonnarme, P.; Jeffries, T.W. Mn(II) Regulation of Lignin Peroxidases and Manganese-dependent Peroxidases from Lignin-degrading White Rot Fungi. *Appl. Environ. Microbiol.* **1990**, *56*, 210–217. [CrossRef]
280. Isikhuemhen, O.S.; Mikiashvili, N.A.; Kelkar, V. Application of Solid Waste from Anaerobic Digestion of Poultry Litter in *Agrocybe aegerita* Cultivation: Mushroom Production, Lignocellulolytic Enzymes Activity and Substrate Utilization. *Biodegradation* **2009**, *20*, 351–361. [CrossRef]
281. Isikhuemhen, O.S.; Mikiashvili, N.A.; Adenipekun, C.O.; Ohimain, E.I.; Shahbazi, G. The Tropical White Rot Fungus, *Lentinus squarrosulus Mont.*: Lignocellulolytic Enzymes Activities and Sugar Release from Cornstalks under Solid State Fermentation. *World J. Microbiol. Biotechnol.* **2012**, *28*, 1961–1966. [CrossRef]
282. Isikhuemhen, O.S.; Mikiashvili, N.A.; Senwo, Z.N.; Ohimain, E.I. Biodegradation and Sugar Release from Canola Plant Biomass by Selected White Rot Fungi. *Adv. Biol. Chem.* **2014**, *4*, 395. [CrossRef]
283. Zebulun, H.O.; Isikhuemhen, O.S.; Inyang, H. Decontamination of Anthracene-polluted Soil through White Rot Fungus-induced Biodegradation. *Environmentalist* **2011**, *31*, 11–19. [CrossRef]
284. Mori, T.; Nagai, Y.; Kawagishi, H.; Hirai, H. Functional Characterization of the Manganese Transporter smf2 Homologue Gene, PsMnt, of *Phanerochaete sordida* YK-624 via Homologous Overexpression. *FEMS Microbiol. Lett.* **2018**, *365*. [CrossRef] [PubMed]
285. Liang, Q.; Zhou, B. Copper and Manganese Induce Yeast Apoptosis via Different Pathways. *Mol. Biol. Cell* **2007**, *18*, 4741–4749. [CrossRef] [PubMed]
286. Yang, M.; Jensen, L.T.; Gardner, A.J.; Culotta, V.C. Manganese Toxicity and *Saccharomyces cerevisiae* Mam3p, a Member of the ACDP (Ancient Conserved Domain Protein) Family. *Biochem. J.* **2005**, *386*, 479–487. [CrossRef]
287. Hereford, L.; Fahrner, K.; Woolford, J., Jr.; Rosbash, M.; Kaback, D.B. Isolation of Yeast Histone Genes H2A and H2B. *Cell* **1979**, *18*, 1261–1271. [CrossRef]
288. Cunningham, K.W.; Fink, G.R. Calcineurin Inhibits VCX1-dependent H+/Ca2+ Exchange and Induces Ca2+ ATPases in *Saccharomyces cerevisiae*. *Mol. Cell. Biol.* **1996**, *16*, 2226–2237. [CrossRef]
289. Pozos, T.C.; Sekler, I.; Cyert, M.S. The Product of HUM1, a Novel Yeast Gene, is Required for Vacuolar Ca2+/H+ Exchange and is Related to Mammalian Na+/Ca2+ Exchangers. *Mol. Cell. Biol.* **1996**, *16*, 3730–3741. [CrossRef]
290. Bianchi, M.; Carbone, M.; Lucchini, G.; Magni, G. Mutants Resistant to Manganese in *Saccharomyces cerevisiae*. *Curr. Genet.* **1981**, *4*, 215–220. [CrossRef]

291. Peters, R.J.B.; Bouwmeester, H.; Gottardo, S.; Amenta, V.; Arena, M.; Brandhoff, P.; Marvin, H.J.P.; Mech, A.; Moniz, F.B.; Pesudo, L.Q.; et al. Nanomaterials for Products and Application in Agriculture, Feed and Food. *Trends Food Sci. Technol.* **2016**, *54*, 155–164. [CrossRef]
292. Wesley, A. History of the Medical Use of Silver. *Surg. Infect.* **2009**, *10*, 289–292.
293. Jo, Y.-K.; Kim, B.H.; Jung, G. Antifungal Activity of Silver Ions and Nanoparticles on Phytopathogenic Fungi. *Plant Dis.* **2009**, *93*, 1037–1043. [CrossRef]
294. Kaur, P.; Thakur, R.; Duhan, J.S.; Chaudhury, A. Management of Wilt Disease of Chickpea in vivo by Silver Nanoparticles Biosynthesized by Rhizospheric Microflora of Chickpea (*Cicer arietinum*). *J. Chem. Technol. Biotechnol.* **2018**, *93*, 3233–3243. [CrossRef]
295. Farooq, M.; Ilyas, N.; Khan, I.; Saboor, A.; Khan, K.; Khan, M.N.; Qayum, A.; Ilyas, N.; Bakhtia, M. Antifungal Activity of Plant Extracts and Silver Nano Particles Against Citrus Brown Spot Pathogen (*Alternaria citri*). *Int. J. Environ. Agric. Res.* **2018**, *4*, 118–125.
296. Gholami-Ahangaran, M.; Zia-Jahromi, N. Nanosilver Effects on Growth Parameters in Experimental Aflatoxicosis in Broiler Chickens. *Toxicol. Ind. Health* **2013**, *29*, 121–125. [CrossRef]
297. Carbone, M.; Donia, D.T.; Sabbatella, G.; Antiochia, R. Silver Nanoparticles in Polymeric Matrices for Fresh Food Packaging. *J. King Saud Univ. Sci.* **2016**, *28*, 273–279. [CrossRef]
298. Vest, K.E.; Wang, J.; Gammon, M.G.; Maynard, M.K.; White, O.L.; Cobine, J.A.; Mahone, W.K.; Cobine, P.A. Overlap of Copper and Iron Uptake Systems in Mitochondria in *Saccharomyces cerevisiae*. *Open Biol.* **2016**, *6*, 150223. [CrossRef]
299. Vagabov, V.; Yu Ivanov, A.; Kulakovskaya, T.; Kulakovskaya, E.; Petrov, V.; Kulaev, I. Efflux of Potassium Ions from Cells and Spheroplasts of *Saccharomyces cerevisiae* Yeast Treated with Silver and Copper Ions. *Biochemistry* **2008**, *73*, 1224–1227. [CrossRef]
300. Cyert, M.S.; Philpott, C.C. Regulation of Cation Balance in *Saccharomyces cerevisiae*. *Genetics* **2013**, *193*, 677–713. [CrossRef]
301. Barros, D.; Pradhan, A.; Pascoal, C.; Cássio, F. Transcriptomics Reveals the Action Mechanisms and Cellular Targets of Citrate-coated Silver Nanoparticles in a Ubiquitous Aquatic Fungus. *Environ. Pollut.* **2021**, *268*, 115913. [CrossRef]
302. Ball, B.; Langille, M.; Geddes-McAlister, J. Fun(gi)omics: Advanced and Diverse Technologies to Explore Emerging Fungal Pathogens and Define Mechanisms of Antifungal Resistance. *Mbio* **2020**, *11*, e01020–e20. [CrossRef]
303. Chitarrini, G.; Riccadonna, S.; Zulini, L.; Vecchione, A.; Stefanini, M.; Larger, S.; Pindo, M.; Cestaro, A.; Franceschi, P.; Magris, G.; et al. Two-omics Data Revealed Commonalities and Differences between Rpv12- and Rpv3-mediated Resistance in Grapevine. *Sci. Rep.* **2020**, *10*, 12193. [CrossRef] [PubMed]
304. Bazzicalupo, A.L.; Ruytinx, J.; Ke, Y.-H.; Coninx, L.; Colpaert, J.V.; Nguyen, N.H.; Vilgalys, R.; Branco, S. Fungal Heavy Metal Adaptation through Single Nucleotide Polymorphisms and Copy-number Variation. *Mol. Ecol.* **2020**, *29*, 4157–4169. [CrossRef] [PubMed]
305. Libkind, D.; Peris, D.; Cubillos, F.A.; Steenwyk, J.L.; Opulente, D.A.; Langdon, Q.K.; Rokas, A.; Hittinger, C.T. Into the Wild: New Yeast Genomes from Natural Environments and New Tools for Their Analysis. *FEMS Yeast Res.* **2020**, *20*. [CrossRef]
306. Samaras, A.; Ntasiou, P.; Myresiotis, C.; Karaoglanidis, G. Multidrug Resistance of *Penicillium expansum* to Fungicides: Whole Transcriptome Analysis of MDR Strains Reveals Overexpression of Efflux Transporter Genes. *Int. J. Food Microbiol.* **2020**, *335*, 108896. [CrossRef] [PubMed]
307. Pereira, R.; Oliveira, J.; Sousa, M. Bioinformatics and Computational Tools for Next-Generation Sequencing Analysis in Clinical Genetics. *J. Clin. Med.* **2020**, *9*, 132. [CrossRef] [PubMed]
308. Liu, P.-H.; Huang, Z.-X.; Luo, X.-H.; Chen, H.; Weng, B.-Q.; Wang, Y.-X.; Chen, L.-S. Comparative Transcriptome Analysis Reveals Candidate Genes Related to Cadmium Accumulation and Tolerance in Two Almond Mushroom (*Agaricus brasiliensis*) Strains with Contrasting Cadmium Tolerance. *PLoS ONE* **2020**, *15*, e0239617.

Journal of *Fungi*

Article

Bacillus megaterium-Mediated Synthesis of Selenium Nanoparticles and Their Antifungal Activity against *Rhizoctonia solani* in Faba Bean Plants

Amr H. Hashem [1], Amer M. Abdelaziz [1,*], Ahmed A. Askar [1], Hossam M. Fouda [1], Ahmed M. A. Khalil [1,2], Kamel A. Abd-Elsalam [3,*] and Mona M. Khaleil [2,4]

[1] Botany and Microbiology Department, Faculty of Science, Al-Azhar University, Cairo 13759, Egypt; amr.hosny86@azhar.edu.eg (A.H.H.); drahmed_askar@azhar.edu.eg (A.A.A.); hossamfouda2016@azhar.edu.eg (H.M.F.); ahmed_khalil@azhar.edu.eg (A.M.A.K.)
[2] Biology Department, College of Science, Taibah University, Yanbu 41911, Saudi Arabia; mkhaleil@taibahu.edu.sa
[3] Plant Pathology Research Institute, Agricultural Research Center (ARC), Giza 12619, Egypt
[4] Botany and Microbiology Department, Faculty of Science, Zagazig University, Zagazig 44519, Egypt
* Correspondence: amermorsy@azhar.edu.eg (A.M.A.); kamel.abdelsalam@arc.sci.eg (K.A.A.-E.); Tel.: +20-01008578963 (A.M.A.); +20-1091049161 (K.A.A.-E.)

Abstract: Rhizoctonia root-rot disease causes severe economic losses in a wide range of crops, including *Vicia faba* worldwide. Currently, biosynthesized nanoparticles have become super-growth promoters as well as antifungal agents. In this study, biosynthesized selenium nanoparticles (Se-NPs) have been examined as growth promoters as well as antifungal agents against *Rhizoctonia solani* RCMB 031001 in vitro and in vivo. Se-NPs were synthesized biologically by *Bacillus megaterium* ATCC 55000 and characterized by using UV-Vis spectroscopy, XRD, dynamic light scattering (DLS), and transmission electron microscopy (TEM) imaging. TEM and DLS images showed that Se-NPs are mono-dispersed spheres with a mean diameter of 41.2 nm. Se-NPs improved healthy *Vicia faba* cv. Giza 716 seed germination, morphological, metabolic indicators, and yield. Furthermore, Se-NPs exhibited influential antifungal activity against *R. solani* in vitro as well as in vivo. Results revealed that minimum inhibition and minimum fungicidal concentrations of Se-NPs were 0.0625 and 1 mM, respectively. Moreover, Se-NPs were able to decrease the pre-and post-emergence of *R. solani* damping-off and minimize the severity of root rot disease. The most effective treatment method is found when soaking and spraying were used with each other followed by spraying and then soaking individually. Likewise, Se-NPs improve morphological and metabolic indicators and yield significantly compared with infected control. In conclusion, biosynthesized Se-NPs by *B. megaterium* ATCC 55000 are a promising and effective agent against *R. solani* damping-off and root rot diseases in *Vicia faba* as well as plant growth inducer.

Keywords: *Vicia faba*; plant disease; root rot; *R. solani*; Se-NPs; nano-biosynthesis; plant promotion

Citation: Hashem, A.H.; Abdelaziz, A.M.; Askar, A.A.; Fouda, H.M.; Khalil, A.M.A.; Abd-Elsalam, K.A.; Khaleil, M.M. *Bacillus megaterium*-Mediated Synthesis of Selenium Nanoparticles and Their Antifungal Activity against *Rhizoctonia solani* in Faba Bean Plants. *J. Fungi* **2021**, *7*, 195. https://doi.org/10.3390/jof7030195

Academic Editor: Laurent Dufossé

Received: 26 December 2020
Accepted: 4 March 2021
Published: 9 March 2021

1. Introduction

The global population will increase to about eight billion people in 2025 and nine billion people in 2050, which requires an increase in agricultural production to feed a rapidly expanding world population [1]. Unfortunately, food security is threatened by crop losses due to attacks of pathogens, including fungi [2,3], and it is estimated that around one-third of the global crop is lost each year due to plant diseases [4]. Phytopathogenic fungi cause losing crop-yield (20–40%) annually worldwide [5]. *Vicia faba* is the main important economic legume over the world that is used as human food, livestock fodder, and silage production [6]. In Egypt, *Vicia faba* (Faba Bean) is one of the most important economic legume crops as a source of protein (18–32%), carbohydrates (55–63%), minerals (2–3.5%), fat (0.5–5.6%), phosphorus, iron, calcium, and vitamins in food [7]; also, it has an ecological role in improving soil quality by the nitrogen fixation and enhances N and P nutrition of cereals [8]. Generally, *Vicia faba* plays a vital role in crop rotation and limiting the disease cycles of various plant pathogens. Unfortunately, *Vicia faba* suffers from many abiotic and biotic stresses that have reduced crop production and led to a decrease in the cultivated area of bean plants around the world from 5 million in 1965 to 2.4 million hectares in 2016. It is susceptible to soilborne fungal pathogens, including *Rhizoctonia solani*, which causes serious root rot disease that harms the quality and quantity of crop yield [9–13], causing a significant gap between production and consumption of *Vicia faba* in Egypt [6]. Moreover, *R. solani* has a broad host range including solanaceous crops, cereals, fruits and vegetables such as potatoes, cucumbers, eggplant, peppers, sugar beet, lettuce, tomatoes, and melon, cotton, and forest trees for a long time [14,15]. *R. solani* is an aggressive fungal plant pathogen with a highly resistant structure called sclerotia, which allows the fungus to survive under environmental conditions [15]. Although fungicides are effective for controlling *R. solani*, they pollute the environment, have a high cost, and also affect other beneficial organisms in the soil [16]. Fungi are the largest group among microbes, where are used in different applications as nanotechnology, bioremediation, bio-deinking, food products, enzyme production, organic acids, and biofuels [17–23]. Dong et al. [24] reported that the management of plant diseases can be achieved by Gly-$Cu(OH)_2$ NPs by reducing the phytotoxicity to plants and improving the utilization of copper-based bactericides. Krutyakov et al. [25] proved that silver nanoparticles are an effective agent for increasing yields as well as decreasing plant diseases besides having a low harmful effect on humans and animals. The application of nanoparticles in agriculture is beneficial for improving the growth and yield of crops as well as inhibiting plant pathogens [26] by facilitating the uptake of macromolecules needed to increase resistance to plant diseases and promote growth [27]. The biological synthesis of metal nanoparticles provides an eco-friendly and cost-effective method. An alternative approach to the synthesis of metal nanoparticles is to apply biomaterials such as plants, microorganisms encompassing groups such as bacteria, yeasts, fungi, and actinomycetes as manufactories [28]. Ag-NPs can be utilized as a management and control agent against various fungal diseases of plants especially *Rhizoctonia solani* and have antifungal activity against mycelium as well as sclerotia [29]. Selenium nanoparticles (Se-NPs) synthesized from a biological source has been shown to have antimicrobial activity against pathogenic microorganisms including fungi [30]. Se-NPs is suggested to be used as a fungicide in agriculture because it has the advantage of being less toxic to humans and animals than synthetic fungicides [31]. In the same context, selenium is an essential trace element for plants growth. It is usually involved in coenzyme activation and physiological facilitation in crop plants, which contributes to food production and quality [32]. In our understanding, bacteriogenic Se-NPs antifungal action against Rhizoctonia diseases of faba bean plants is not thoroughly studied. The main aim of the current research is to (1) biosynthesize Se-NPs by *B. megaterium*, (2) characterize the physicochemical properties of the produced nanoparticles by UV-Vis spectroscopy, XRD, dynamic light scattering (DLS), and transmission electron microscopy (TEM) imaging, (3) assess and evaluate the antifungal activity of Se-NPs against *Rhizoctonia* RCMB 031001 root rot of *Vicia faba* in vitro and in vivo, (4) analyze photosynthetic pigments, metabolic indicators, protein, and phenolics compounds of *Vicia faba*, and (5) understand the antifungal mechanisms and the effects of Se-NPs on oxidative enzymes such as polyphenol oxidase (PPO) and peroxidase (POX) in *Vicia faba* under pot conditions using assays.

2. Materials and Methods

2.1. Biosynthesis of Se-NPs

Se-NPs were produced using *Bacillus megaterium* culture supernatant (as reducing and stabilizing agents). Bacteria were subcultured on nutrient broth media in conical flasks and incubated with shaking aerobically at 37 °C for 48 h. After an incubation period, the bacterial cells were removed from the suspension by filtration through a 0.44 μm PVDF filter; then, they were centrifuged at 10,000 rpm to remove occasional bacterial cells and macromolecules [33]. The next step was mixing cell-free supernatant with the selenious acid suspension (1 mM) by quotient (1:1) v/v. The mixtures were stirred at a controlled room temperature of about 25 °C. The process of selenious acid reduction was monitored by color change of the cell-free supernatant from colorless to reddish color [22–25]. The suspension of SeNPs was further centrifuged at 12,000 rpm for 30 min, and the collected precipitate pellet was dried and weighed. The concentration was calculated as follows: 1 mg of SeNPs was dissolved in 1 mL of DMSO, where the final concentration was 1000 μg/mL.

2.2. Characterization of Se-NPs

The characterization of Se-NPs was performed by using JASCO V-560, UV-Vis. spectrophotometer, Tokyo, Japan, at the wavelength range from 200–900 nm and at a resolution of 1 nm. Cell-free supernatant without SeO_2 was used as blank to adjust the baseline. Toward particle size investigation, the specimens were diluted ten times by deionized water before being estimated. To determine the morphology and size of the manufactured Se-NPs, TEM microscopy, model JEOL JEM-100 CX (Peabody, MA, USA) was used. TEM imaging was carried out by drop covering the Se-NPs upon carbon-coated TEM layers. Dynamic light scattering (DLS) was used to determine the size distribution, while the average particle size was determined by PSSNICOMP 380-ZLS particle sizing system (St. Barbara, CA, USA). For XRD analysis, the adjusted sample was centrifuged, and the precipitate was dried under vacuum and taken for XRD analysis. X-ray diffraction patterns were obtained with XRD- 6000 series, including stress analysis, residual austenite quantitation, crystallite size/lattice strain, crystallinity calculation, and materials analysis via overlaid X-ray diffraction patterns Shimadzu apparatus using nickel-filter and Cu-Ka target, Shimadzu Scientific Instruments (SSI), (Kyoto, Japan). The average crystalline size of the Se-NPs was also determined by using Debye–Scherrer equation: $D = k\lambda/\beta \cos\theta$. Here, D is the average crystalline size (nm), k is the Scherrer constant with the value from 0.9 to 1, λ is the X-ray wavelength, β is the full width of half maximum, and θ is the Bragg diffraction angle (degrees). The estimations included stress investigation, remaining austenite quantitation, crystallite capacity, crystallinity consideration, and materials examination through overlaid X-ray diffraction models. Finally, Se-NPs concentration assessment was performed using UNICAM939 Atomic Absorption Spectroscopy, Cambridge, UK, and implemented with deuterium experience improvement. All suspensions were prepared using ultra-pure water [34–41]. Furthermore, the morphology size of the manufactured NPs was read by practicing TEM microscopy, JEOL JEM-100 CX, (Peabody, MA, USA).

2.3. Control of Rhizoctonia solani by Se-NPs

2.3.1. Source of Pathogen and Culture Conditions

Rhizoctonia solani RCMB 031001 was purchased from the Regional Center for Mycology and Biotechnology (RCMB), Al-Azhar University, Cairo, Egypt. *R. solani* was cultured on potato dextrose agar medium (PDA) plates, incubated for 3–5 days at 28 ± 2 °C, and then kept at 4 °C for further use [42–45].

2.3.2. In Vitro Assessment of Antifungal Activity and Growth Inhibition

- Well diffusion method

The well diffusion method was applied to study the antifungal activity of biosynthesized Se-NPs [46] with a few modifications. *R. solani* was inoculated on PD broth medium and then incubated at 28 ± 2 °C for 3–5 days. Fungal inoculum of *R. solani* RCMB 031001 was spread thoroughly on the sterilized solidified potato dextrose agar (PDA) medium. At the same time, eight wells with 5.5 mm diameter were made using a sterile cork-borer on each agar plate (120 mm). The wells were filled with 50 μl of different concentrations of Se-NPs individually with triplicates. The culture plates were incubated at 25 °C for 7 days, and the zones of inhibition were observed and measured.

- Radial growth method

PDA medium was prepared and amended with different concentrations of Se-NPs (1, 0.5, 0.25, 0.125, and 0.0625 mM) before the pouring stage. After medium solidification, culturing of *R. solani* was carried out according to Joshi et al. [47]. The inhibition percentage of pathogen growth was calculated using the following equation:

$$\text{Inhibition of pathogen growth } (\%) = \frac{\text{Growth in the control} - \text{Growth in the treatment}}{\text{Growth in the control}} \times 100.$$

2.3.3. In Vivo Assessment Efficacy of Se-NPs on *Vicia faba*

The inoculum of the pathogenic fungus *R. solani* was prepared according to Büttner et al. [48] that comprises mixing contents of 5 pure *R. solani* culture Petri dishes with 1000 mL of distilled water using electrical blender for two minutes. This experiment was carried out in the garden of Plant and Microbiology department, Faculty of Science, Al-Azhar University, Cairo, Egypt. The source of faba bean cv. Giza 716 was obtained from the Legume Research Department, Field Crop Institute, Agricultural Research Center, Egypt. The sandy loam soil was autoclaved (1.5 atm, 121 °C for 30 min) and distributed equally in disinfected pottery pots (30 cm in diameter) with 12 sterilized seeds per pot. The *Vicia faba* seeds were washed with distilled water then sterilized using 2% sodium hypochlorite for 2 min before conducting the treatments shown in Table 1.

Table 1. Treatments used in this study.

Treatment Number	Treatment
1 (Control healthy)	The sterilized *Vicia faba* seeds submerged in distilled water for three hours and sowing in sterilized soil.
2 (Control infected)	Sowing the sterilized *Vicia faba* seeds in distilled water for three hours and sowing in inoculated soil with *R. solani*.
3 (Healthy + Nano soaking)	Soaking the sterilized *Vicia faba* seeds in Se-NPs (0.0625 mM) for three hours and sowing in sterilized soil.
4 (Infected + soaking)	Soaking the sterilized *Vicia faba* seeds in Se-NPs (0.0625 mM) for three hours and sowing in inoculated soil with *R. solani*.
5 (Healthy + soaking and spraying with Nano)	Soaking the sterilized *Vicia faba* seeds in Se-NPs (0.0625 mM) for three hours and sowing in sterilized soil, then spraying 15 mL of Se-NPs after emergence.
6 (Infected + soaking and spraying with Nano)	Soaking the sterilized *Vicia faba* seeds in Se-NPs (0.0625 mM) for three hours and sowing in inoculated soil with *R. solani*, then spraying 15 mL of Se-NPs after emergence.
7 (Healthy + spraying Nano)	Sowing the sterilized *Vicia faba* seeds in distilled water for three hours and sowing in sterilized soil, then spraying 15 mL of Se-NPs (0.0625 mM) after emergence.
8 (Infected + spraying Nano)	Sowing the sterilized *Vicia faba* seeds in distilled water for three hours and sowing in inoculated soil with *R. solani*, then spraying 15 mL of Se-NPs (0.0625 mM) after emergence.

2.3.4. Disease Symptoms and Disease Index

Pre-emergence damping-off was measured after 15 days from sowing, while post-emergence damping-off and survival were measured after 30 days from sowing according to Mousa et al. [49]. In addition, disease symptoms were assessed, and the disease index was recorded after 45 days from sowing according to Grünwald et al. [50]. The disease index scale (0–5) based on disease progress developed by the authors was used to measure the disease severity of *Rhizoctonia* root rot, in which 0 indicated no visible symptoms; 1, a few small soft lesions on a part of the root system and hypocotyls; 2, elongated, discolored lesions spread on the entire root system and hypocotyls; 3, deep brown necrosis grind the stem, partial root disintegration, and yellowing of leaves; 4, stem canker, root disintegration, yellowing of leaves, and stunting; and 5, collapse and death of the plants. Disease index = (i (rating no. × no. of plants in the rating)/(total no. of plants × highest rating) × 100. Shoot length, root length, fresh and dry weight, and pigments were also measured (one gram of fresh leaves was extracted by 100 mL of 80% aqueous acetone (*v*/*v*), filtrated, and then completed the volume to 100 mL using 80% acetone). The optical density of the plant extract was measured using the spectrophotometer of three wavelengths (470, 649, and 665 nm). Pigments were calculated using the equations mentioned Mg chlorophyll (a)/g tissue = 11.63(A665) − 2.39(A649), Mg chlorophyll (b)/g tissue = 20.11(A649) − 5.18(A665), Mg chlorophyll (a + b)/g tissue = 6.45 (A665) + 17.72(A649), and Carotenoids = $1000 \times O.D_{470} - 1.82\ C_a - 85.02\ C_b/198$ = mg/g fresh weight. "A" denotes the reading of optical density, phenol (one gram dry leaves was extracted with 80% cold methanol (*v*/*v*) three times at 0 °C. The extract was filtered; then, the volume of sample was completed to 25 mL with cold methanol. The total phenol and total soluble protein of plants were determined in the following manner: one gram of the dried leaves was added to 5 mL of 2% phenol water and 10 mL of distilled water was added; the solution was shaken and kept overnight, filtered, and completed volume to 50 mL with distilled water; then the protein content was determined according to Alhaithloul et al. [51].

2.4. Statistical Analysis

Experimental data were subjected to one-way analysis of variance (ANOVA) and the differences between means were measured using Tukey's method. The values are given as means ± SD (standard deviations). Levels of significance were considered at $p \leq 0.05$ unless otherwise stated, and the (L.S.D) at 5% level of probability using Co-state software [52].

3. Results and Discussion

3.1. Synthesis and Characterization of Se-NPs

In the current study, the supernatant of *Bacillus megaterium* ATCC 55000 was used to synthesize Se-NPs. The process of selenious acid reduction was monitored, while the cell-free extract changed from colorless to reddish color [53,54]. The UV-visible spectrum of Se-NPs synthesized indicated that it had maximum absorption at (0.860 abs) and 435 nm. DLS was performed to evaluate the particle size distribution, and the average particle size was found to be 45.9 nm, as shown in Figure 1B. On the other hand, the TEM result demonstrated that particles had a spherical shape within a nanoscale range from 29.72 to 74.36 nm with an average of the main diameter of 41.2 nm, as shown in Figure 1C. The XRD pattern for the Se-NPs was presented in Figure 1D. Several peaks were observed at nine theta (degree) as 23.2°, 30.5°, 41.7°, 44.3°, 46.4°, 52.3°, 56.7°, 62.5°, and 72.6° corresponding to the (100), (101), (110), (102), (111), (201), (113), (202), and (210) planes of the standard cubic phase of Se, respectively. The XRD pattern indicated that Se-NPs were in the face-centered cubic (FCC) structure and crystal in nature. The observation of diffraction peaks for the Se-NPs indicated that they were crystalline, while their refining was related to the particles in the nanometer size regime. The strong interaction of the Se-NPs with light was the result of the electrons conducting on the metal surface that were subjected to a collective oscillation when excited by light at specific wavelengths, which is known as surface plasma resonance (SPR) [55,56]. In another study, the culture supernatant of *A. terreus* with SeO_2 (100 µg/mL) produced Se-NPs with an average size of 47 nm [57]. *Bacillus megaterium* (a halophile strain) strongly reduced selenite (up to 0.25 mM) to Se-NPs after 40 h of incubation [58]. A microbial source *Bacillus cereus*-mediated synthesis of Se-NPs showed an absorption maxima at 590 nm, whereas nanoparticles synthesized from lemon leaf extract exhibited a maximum absorption at 395 nm [59]. The band gap energy calculated for chemically formed nano-Se was 2.1 eV, which significantly different from a biological source (band gaps for nano-Se from *Sulphurospirillum barnessi, Bacillus selenitireducens,* and *Selenihalanaerobacter shriftii* were 1.62, 1.67, and 1.52 eV, respectively [60].

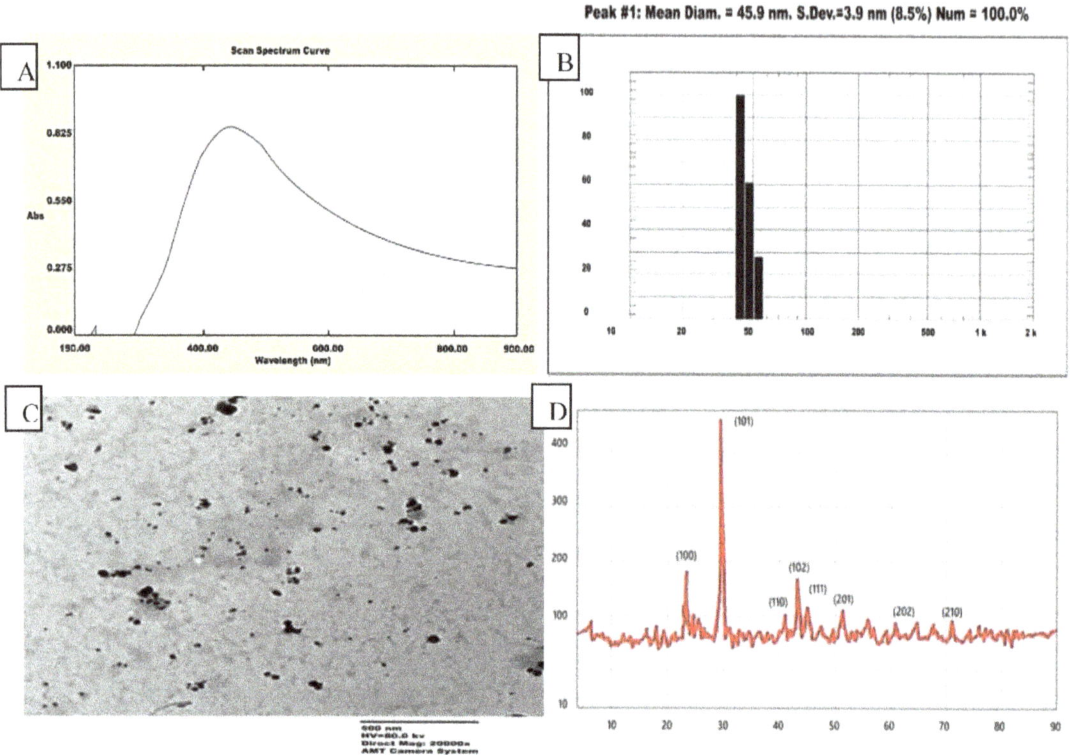

Figure 1. Characterization of bacteriogenic Se-NPs produced by *B. megaterium* (**A–D**); (**A**) UV-Visible spectrum; (**B**) dynamic light scattering (DLS); (**C**) TEM image; (**D**) XRD.

3.2. *In Vitro Control of R. solani*

3.2.1. Antifungal Activity of Se-NPs and Minimum Inhibition Concentration

Metal nanoparticles such as silver nanoparticles [61,62], copper nanoparticles [63], and zinc nanoparticles [64] are wildly used for controlling fungal plant pathogens. However, selenium nanoparticles have strong antifungal activity, while they are rarely used for controlling fungal plant pathogens. Therefore, selenium nanoparticles were biosynthesized in this study to control *R. solani*. The antifungal activity of Se-NPs was assessed against *R. solani* using the agar well diffusion method; different concentrations of Se-NPs ranging from 1 to 0.0078 mM were tested as antifungal agent, as shown in Figure 2. Results illustrated that concentrations of Se-NPs of 1, 0.5, 0.25, 0.125, and 0.0625 mM had antifungal activity against *R. solani*. Moreover, 1 mM of Se-NPs had the maximum antifungal activity and gave an inhibition zone of 45 mm, whereas 0.0625 mM had the lowest antifungal activity against *R. solani* and gave an inhibition zone of 12 mm. From these data, 0.0625 was the minimum inhibition concentration for the controlling of *R. solani*.

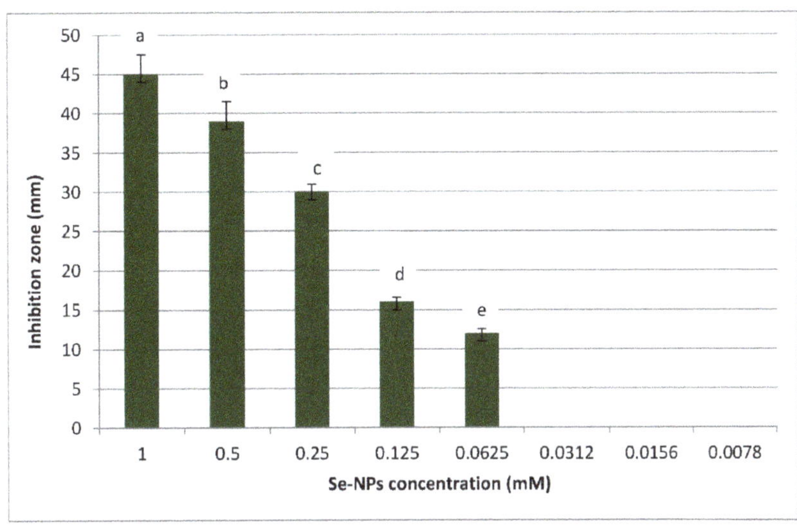

Figure 2. Antifungal activity of different concentrations of Se-NPs against *R. solani*. Data are expressed as means ± standard deviations of triplicate assays. The different alphabetic superscripts in the same column are significantly different ($p < 0.05$) based on Tukey's multiple comparison test.

3.2.2. Effect of Se-NPs on Linear Growth of *R. solani* and Minimum Fungicidal Concentration

The linear growth of *R. solani* was assessed at different concentrations of Se-NPs with different incubation periods from 1 to 7 days, as shown in Figure 3A,C. Linear growth was performed to detect the inhibition percentage for each concentration of Se-NPs against *R. solani*. Results illustrated that the inhibition percentage increased with increasing of concentration Se-NPs, while linear growth decreased, as shown in Figure 3B. At concentration 1 mM, *R. solani* could not grow on a PDA surface, as shown in Figure 3C, inhibition percentage was 100%, and this concentration had the minimum fungicidal activity. Additionally, Se-NPs at 0.5 mM gave a high inhibition percentage but less than at 1 mM, where it was 92.9%; also, inhibition percentage decreased gradually with decreasing the concentration of Se-NPs [47]. Biosynthesized Se-NPs could suppress the growth and proliferation of *Sclerospora graminicola* [65]. Moreover, selenium nanoparticles were used in controlling the leaf blight of tomato caused by *Alternaria alternate*, and Se-NPs at a concentration of 100 ppm gave an inhibition percentage 89.6% [66]. In addition, Se-NPs used against *Alternaria solani* caused Early Blight Disease on Potato, and inhibition percentage was 100% at 800 ppm [67].

3.3. *In Vivo Control of* R. solani

3.3.1. Efficacy of Se-NPs on Rhizoctonia Root Rot Disease of *Vicia faba* under Pot Conditions

The results presented in Table 2 and Figure 4 indicated that *R. solani* RCMB 031001 caused an emergence damping-off disease of 58.33% seeds and 88% Rhizoctonia root rot disease index of *Vicia faba* cultivar (treatment 2 infected control). On the other hand, the healthy control treatment pots resulted in 100% emerged and survived plants. These results confirmed that the cv. Giza 716 faba bean cultivar is susceptible to *R. solani* RCMB 031001. Application of the Se-NPs by soaking and/or spraying to *Vicia faba* infested with fungus *R. solani* showed greater potency in controlling the pathogen. The best treatment for controlling *R. solani* was treatment 6, which resulted in 83.33% survival as well as 72.27% protection followed by treatment 4 with 66.67% and 59.2%, respectively, and treatment 8 by 50% and 63.63%. These results are similar to studies by Nandini, Hariprasad, Prakash, Shetty and Geetha [65], which reported that Se-NPs had a highly effective role in controlling plant pathogenic fungus *Sclerospora graminicola* as a causative agent of downy mildew disease. However, several reports showed that the application of selenium in plants activates the defense plant mechanism against abiotic [68] and biotic stresses such as *R. solani* [69].

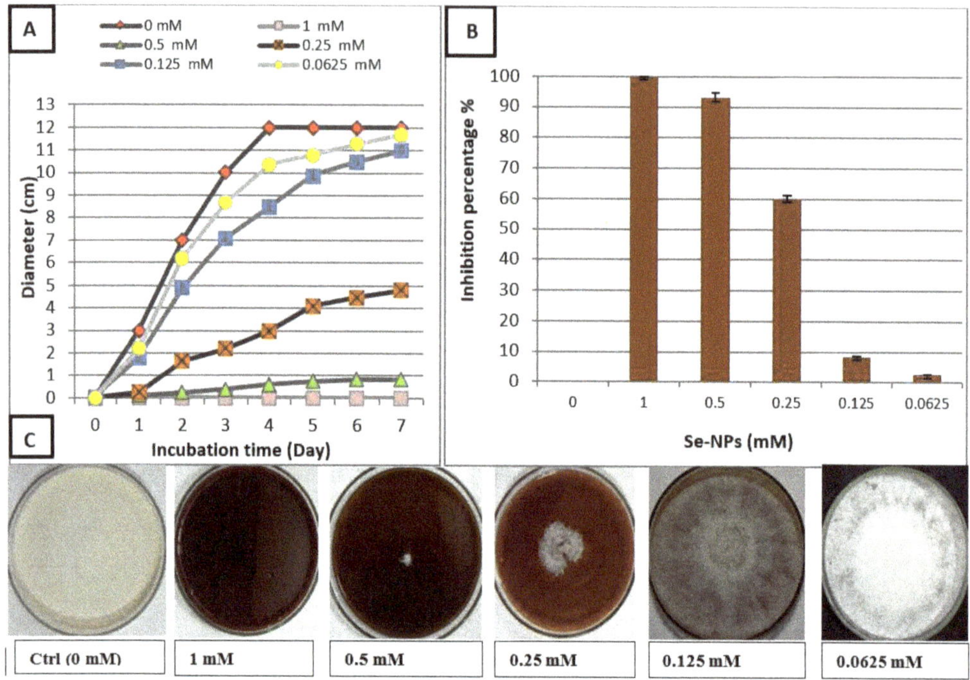

Figure 3. Effect of Se-NPs on *R. solani* (**A**–**C**): (**A**) Linear growth at different incubation periods from 0 to 7 days; (**B**) Inhibition percentages of *R. solani* at different concentration of Se-NPs; (**C**) Linear growth on potato dextrose agar medium (PDA) plates at 7 days. Data are expressed as means ± standard deviations of triplicate assays. The different alphabetic superscripts in the same column are significantly different ($p < 0.05$) based on Tukey's multiple comparison test.

Table 2. Effect of selenium nanoparticles (Se-NPs) on the disease index of *R. solani* damping-off and root rot diseases under pot conditions.

Treatment		Pre-Emergence Damping of %	Post-Emergence Damping of %	Survival Plant %	Disease Index %	Protection %
Healthy	Control	0	0	100	0	-
	Nano soaking	0	0	100	0	-
	Nano spraying	0	0	100	0	-
	Nano (soaking + spraying)	0	0	100	0	-
Infected	Control	50	8.33	41.67	88	0
	Nano soaking	16.66	16.66	66.67	36	59.2
	Nano spraying	50	0	50	32	63.63
	Nano (soaking + spraying)	16.667	0	83.33	20	77.27

Figure 4. Disease index scale (0–5); (**A**): 0, (**B**): 1, (**C**): 2, (**D**): 3, (**E**): 4, and (**F**): 5; the disease index was recorded after 45 days from sowing.

3.3.2. Growth and Yield Responses of *Vicia faba* by Se-NPs under Pot Conditions

Results presented in Tables 3 and 4 indicated that all investigated growth parameters (shoot and root length, number of leaves, fresh and dry weight plant biomass), as well as yield (number of bods per plant, number of seeds per plant, the weight of 100 seed and protein content of yield) of infected *Vicia faba* cv. Giza 716 plants with *R. solani* RCMB 031001 were significantly decreased compared with healthy control plants. The most effective treatment was treatment 6, which increased the yield and growth parameters, especially shoot dry weight 203%, root fresh weight 178.8%, root dry weight 163%, plant height, and number of seeds 116.6% compared with infected control (treatment 2). These results are similar to Abdel-Monaim [70], who reported that *R. solani* had significantly decreased fresh and dry weight compared to healthy control. Akladious et al. [71] reported that *R. solani* caused a significant decrease in shoot length, root length, number of leaves, and fresh and dry weight of shoot and root of *Vicia faba*.

The obtained results revealed that all investigated growth parameters of *Vicia faba* cv. Giza 716 plants were significantly increased in response to the application of Se-

NPs compared with the control. The simulative effects of Se-NPs on plant growth were explained by many mechanisms—firstly, the increased starch content in chloroplast [72]. Secondly, the plant cell can be protected by selenium from oxidative damage by antioxidant defenses [73]. Thirdly, selenium is a beneficial element for plants and has a bio-stimulant effect, as photocatalysis and plant growth increase plant metabolism and crop quality and stress tolerance [32,52,74]. However, the application of selenium in plants stimulates the growth and quality of fruits [75]. Our results showed that the most effective treatment was achieved by soaking and foliar spray followed by soaking and finally spraying.

Table 3. Effect of biogenic Se-NPs on morphological indicators of *Vicia faba L.* under pot conditions.

Treatments	Plant Height (cm)	Root Length (cm)	Number of Leaves	Shoot F. wt. (g)	Shoot D. wt. (g)	Root F. wt. (g)	Root F. wt. (g)
T1: Control (H.)	32 ± 1.50 ^{b}	11.33 ± 0.85 ^{cd}	16.33 ± 0.57 ^{bc}	10.48 ± 0.72 ^{bc}	3.71 ± 0.24 ^{b}	1.25 ± 0.04 ^{d}	0.37 ± 0.00 ^{bc}
T2: Control (Inf.)	19.36 ± 0.70 ^{e}	9.73 ± 0.75 ^{d}	11.66 ± 0.57 ^{d}	7.38 ± 0.33 ^{e}	1.31 ± 0.28 ^{d}	0.82 ± 0.06 ^{g}	0.19 ± 0.02 ^{d}
T3: Soaking Nano (H.)	35.83 ± 1.89 ^{b}	13.56 ± 0.51 ^{ab}	18.33 ± 1.52 ^{b}	13.56 ± 0.40 ^{a}	4.6 ± 0.35 ^{a}	1.64 ± 0.06 ^{b}	0.52 ± 0.04 ^{a}
T4: Soaking Nano (Inf.)	20.5 ± 1.80 ^{de}	10.7 ± 0.75 ^{cd}	15.33 ± 1.15 ^{c}	10.25 ± 0.67 ^{bc}	2.48 ± 0.14 ^{c}	1.07 ± 0.06 ^{ef}	0.22 ± 0.00 ^{d}
T5: Soaking + Spray Nano (H.)	42.66 ± 2.25 ^{a}	15.66 ± 1.10 ^{a}	22.33 ± 1.52 ^{a}	14.76 ± 0.92 ^{a}	4.59 ± 0.22 ^{a}	1.87± 0.03 ^{a}	0.57 ± 0.04 ^{a}
T6: Soaking + Spray Nano (Inf.)	24.5 ± 0.50 ^{cd}	10.84 ± 0.74 ^{cd}	11.66 ± 0.57 ^{d}	8.37± 0.10 ^{de}	2.67 ± 0.30 ^{c}	0.93 ± 0.06 ^{fg}	0.31 ± 0.00 ^{c}
T7: Spray Nano (H.)	34.5 ± 1.32 ^{b}	12.06 ± 0.62 ^{bc}	16.33 ± 0.57 ^{bc}	11.16 ± 0.35 ^{b}	3.92 ± 0.13 ^{ab}	1.45 ± 0.09 ^{c}	0.44 ± 0.00 ^{b}
T8: Spray Nano (Inf.)	26.66 ± 1.52 ^{c}	10.6 ± 0.65 ^{cd}	13.66 ± 0.57 ^{cd}	9.46 ± 0.47 ^{cd}	2.65 ± 0.16 ^{c}	1.13 ± 0.01 ^{de}	0.35 ± 0.00 ^{c}
L.S.D at 0.05	2.667	1.329	1.694	0.963	0.422	0.099	0.041

H. means Healthy and Inf. means infected. Data are expressed as means ± standard deviations of triplicate assays. The different alphabetic superscripts in the same column are significantly different ($p < 0.05$) based on Tukey's multiple comparison test.

Table 4. Effect of Se-NPs on the yield of (*Vicia faba* L.) plants.

Treatments	No. of Pods/Plant	No. of Seeds/Plant	wt. of 100 Seeds(g)	Protein Yield mg/g (g)
T1: Control (H.)	20.33 ± 0.57 ^{a}	51 ± 1.0 ^{bc}	101.33 ± 0.57 ^{c}	116.94 ± 0.09 ^{c}
T2: Control (Inf.)	17.66 ± 0.57 ^{b}	47 ± 2.0 ^{c}	99 ± 1.0 ^{d}	96.1 ± 0.28 ^{f}
T3: Soaking Nano (H.)	21.33 ± 0.57 ^{a}	54.66 ± 1.52 ^{ab}	105 ± 1.0 ^{b}	120.6 ± 0.46 ^{b}
T4: Soaking Nano (Inf.)	18.33 ± 0.57 ^{b}	49 ± 1.73 ^{c}	101.66 ± 0.57 ^{c}	99.02 ± 0.14 ^{e}
T5: Soaking + Spray Nano (H.)	21.66 ± 0.57 ^{a}	57.66 ± 1.15 ^{a}	107.03 ± 0.45 ^{a}	123.21 ± 0.38 ^{a}
T6: Soaking + Spray Nano (Inf.)	17.33 ± 0.57 ^{b}	49 ± 1.73 ^{c}	101.16 ± 0.28 ^{c}	101.28 ± 0.51 ^{d}
T7: Spray Nano (H.)	20.33 ± 0.57 ^{a}	54.33 ± 1.15 ^{ab}	103.5 ± 0.05 ^{b}	118.17 ± 0.32 ^{c}
T8: Spray Nano (Inf.)	18 ± 1.0 ^{b}	51 ± 1.0 ^{bc}	101.30 ± 0.02 ^{c}	100.21 ± 1.05 ^{de}
L.S.D at 0.05	1.125	2.523	1.095	0.856

Data are expressed as means ± standard deviations of triplicate assays. The different alphabetic superscripts in the same column are significantly different ($p < 0.05$) based on Tukey's multiple comparison test. LSD ($p < 0.05$) values are indicated in the data differing significantly are indicated with different letters.

3.3.3. Effect of Se-NPs on Photosynthetic Pigments of *Vicia faba* under Pot Conditions

The observed results in Figure 5 showed that chlorophyll content and carotenoids had significantly decreased by *R. solani* RCMB 031001. These results are explained with [76], which stated that phytopathogenic fungi inhibit the photosynthetic activity of plants. These reductions in chlorophyll a may be due to the more selective destruction of chlorophyll biosynthesis or degradation of chlorophyll precursors according to Saha et al. [77] or may be due to a decrease in the uptake of minerals (e.g., magnesium) that are required for chlorophyll synthesis and interfere with the photosynthesis reactions [78]. Data presented in Figure 5 indicated that the application of Se-NPs caused a significant increase in total chlorophyll content and carotenoids compared with controlled plants and the best result was achieved by soaking and foliar spraying. Several reports show that the application of selenium in plants improves photosynthesis [79].

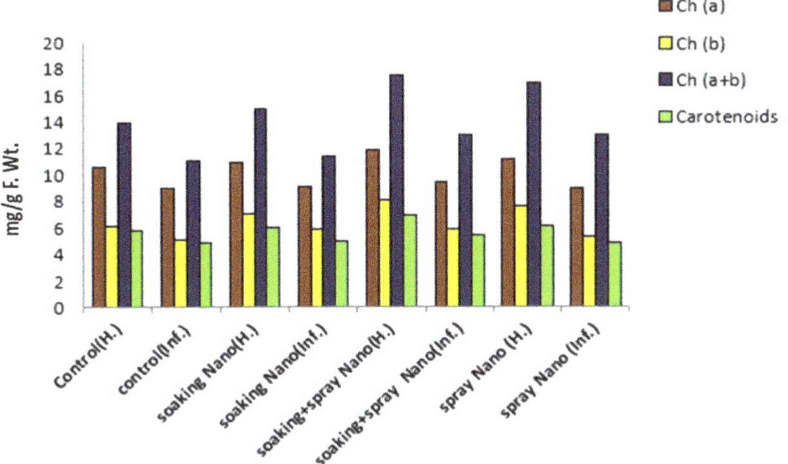

Figure 5. Effect of Se-NPs on the photosynthetic pigment's indicators of (*Vicia faba* L.).

3.3.4. Effect of Se-NPs on the Metabolic Indicators of (*Vicia faba* L.)

Effect of Se-NPs on Phenol Contents of Vicia faba under Pot Conditions

Results revealed that the contents of total phenols were significantly increased in shoots and roots of cv. Giza 716 plants in response to the infection with *R. solani* RCMB 031001, as shown in Figure 6. Moreover, results demonstrated that application of Se-NPs induced responses regarding the total contents of phenols compared with healthy control. In contrast, total phenols contents in shoots and roots-infected plants were significantly decreased in response to the treatments with Se-NPs. These results are similar to those in [80,81]; they demonstrated that the treatment of plants with NPs resulted in increasing phenolic content. This increasing in phenolic contents resulted in antifungal activity by several mechanisms including (i) cell rupture and release of intracellular proteins and carbohydrates that prevent fungal growth; (ii) inhibition of mitochondrial respiration causing reduction of ATP production, and (iii) oxidative lesions and chelation of iron ions [82,83]. Correspondingly, total phenols play a vital role in the regulation of plant metabolic process and overall plant growth as well as lignin synthesis [84]. Phenols act as free radical scavengers as well as substrates for many antioxidant enzymes [85]. Finally, Mellersh et al. [86] reported that reactive oxygen species (ROS), especially phenolic compounds, prevent penetration, restrict fungal growth, and provoke cell death and tissue necrosis, which would prevent further fungal development toward plant tissue.

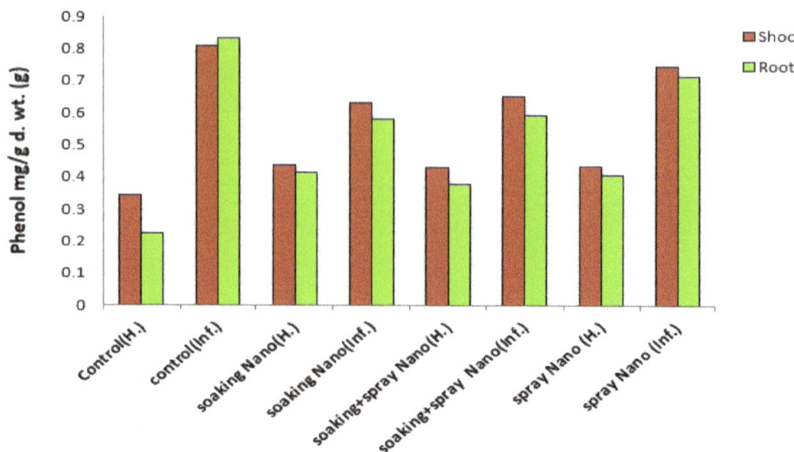

Figure 6. Effect of Se-NPs on the phenolics compounds of (*Vicia faba* L.).

Effect of Se-NPs on a Total Soluble Protein of *Vicia faba* under Pot Conditions

The presented data in Table 5 showed that the total soluble protein in shoot and root were significantly decreased in cv. Giza 716 plants in response to the infection with *R. solani* RCMB 031001. Weintraub and Jones [87] recorded that pathogen attack resulted in a reduction of several thylakoid membrane proteins and decreasing leaf soluble protein. These results are explained by several different mechanisms; firstly, the stresses may affect the process of protein synthesis, Secondly, it is also possible that the pathogens consume nitrogen, which could have been utilized for synthesizing proteins [88]. In addition, the application of Se-NPs resulted in an increase of total soluble protein compared with control. Also, the best treatment was soaking and foliar spraying, which agree with Hajiboland [89], who illustrated that the application of Se-NPs resulted in a significant increase in total soluble protein. Increasing protein content could be due to the activation of the host defense mechanisms as an indicator of resistance [88].

Table 5. Effect of Se-NPs on the total soluble protein of (*Vicia faba* L.).

Treatments	Protein Shoot mg/g d. wt. (g)	Protein Root mg/g d. wt. (g)
T1: Control (H.)	20.22 ± 0.15^{c}	18.27 ± 0.17^{c}
T2: Control (Inf.)	16.8 ± 0.24^{f}	16.63 ± 0.35^{f}
T3: Soaking Nano (H.)	21.83 ± 0.06^{b}	19.14 ± 0.17^{b}
T4: Soaking Nano (Inf.)	18.06 ± 0.48^{e}	17.39 ± 0.07^{e}
T5: Soaking + Spray Nano (H.)	22.93 ± 0.05^{a}	21.03 ± 0.07^{a}
T6: Soaking+ Spray Nano (Inf.)	19.09 ± 0.08^{d}	17.59 ± 0.07^{de}
T7: Spray Nano (H.)	21.62 ± 0.04^{b}	20.9 ± 0.09^{a}
T8: Spray Nano (Inf.)	18.08 ± 0.31^{e}	18 ± 0.17^{cd}
LSD at 0.05	0.412	0.283

Data are expressed as means ± standard deviations of triplicate assays. The different alphabetic superscripts in the same column are significantly different ($p < 0.05$) based on Tukey's multiple comparison test.

Effect of Se-NPs on Oxidative Enzymes of *Vicia faba* under Pot Conditions

Vicia faba showed variation in relative mobility and density polypeptide bands as pathogenicity indicators and or treatment with Se-NPs, where healthy control (treatment 1) gave three isozyme bands with a low density of isozymes but soaking infected (treatment 4)

and soaking + foliar spray Se-NPs (treatment 6) gave the same number of bands, three isozymes with moderate density. While infected control (treatment 2), as well as Se-NPs as a foliar spray on infected (treatment 8), gave three isozymes bands with a high density of isozymes, as shown in Figure 7, these results demonstrated that infected control recorded high activity as a high density of bands. Our results are similar to those of [90], who reported that the minimum activities of peroxidase enzymes were observed in healthy control. In this regard, Hasanuzzaman and Fujita [91] found that spraying with selenium increased the activity of many enzymes. In addition, [92] reported that nano selenium acts as a promoter and/or stressor, enhancing the antioxidant defense systems of plants, which leads to the improvement of plant tolerance under sandy soil conditions.

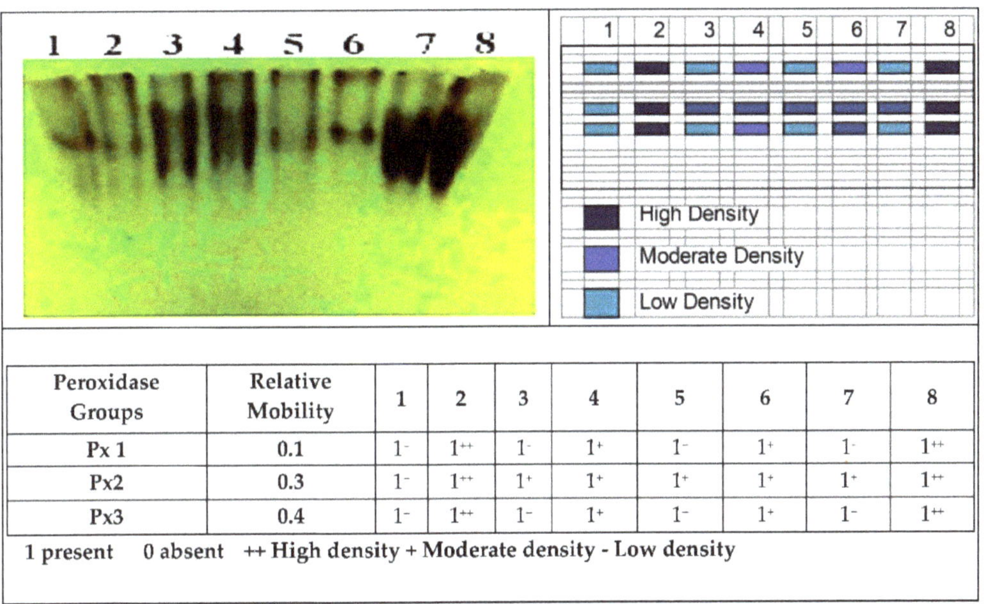

Peroxidase Groups	Relative Mobility	1	2	3	4	5	6	7	8
Px 1	0.1	1^{-}	1^{++}	1^{-}	1^{+}	1^{-}	1^{+}	1^{-}	1^{++}
Px2	0.3	1^{-}	1^{++}	1^{+}	1^{+}	1^{+}	1^{+}	1^{+}	1^{++}
Px3	0.4	1^{-}	1^{++}	1^{-}	1^{+}	1^{-}	1^{+}	1^{-}	1^{++}

1 present 0 absent ++ High density + Moderate density - Low density

Figure 7. Effect of Se-NPs on peroxidase isozymes of *Vicia faba* under pot conditions 1: Control (H.). 2: Control (Inf.). 3: Soaking Nano (H.). 4: Soaking Nano (Inf.). 5: Soaking + Spray Nano (H.). 6: Soaking + Spray Nano (Inf.). 7: Spray Nano (H.). 8: Spray Nano (Inf.).

In addition to the lowest polyphenol oxidase (PPO) activity recorded in healthy control, Se nanoparticles and *R. solani* infection application showed variation in number, relative mobility, and density polypeptide bands more than healthy ones, infected control, as well as treatment with Se-NPs as a foliar spray on infected plants (four isozyme bands) with a high density of isozymes (Figure 8). Meanwhile, plants treated with soaking or foliar spray Se-NPs (either individual or combination) gave the same number of bands (four isozymes) with moderate and high density. Variation in isozyme shows knowledge of resistant genes in the biological system to physiological changes, genetic traits, and growth in various species _ENREF_ [90]. Finally, anti-oxidative enzymes such as polyphenol oxidase (PPO) and peroxidase (POX) are most importantly involved in the scavenging system of excess reactive oxygen species (ROS) [90].

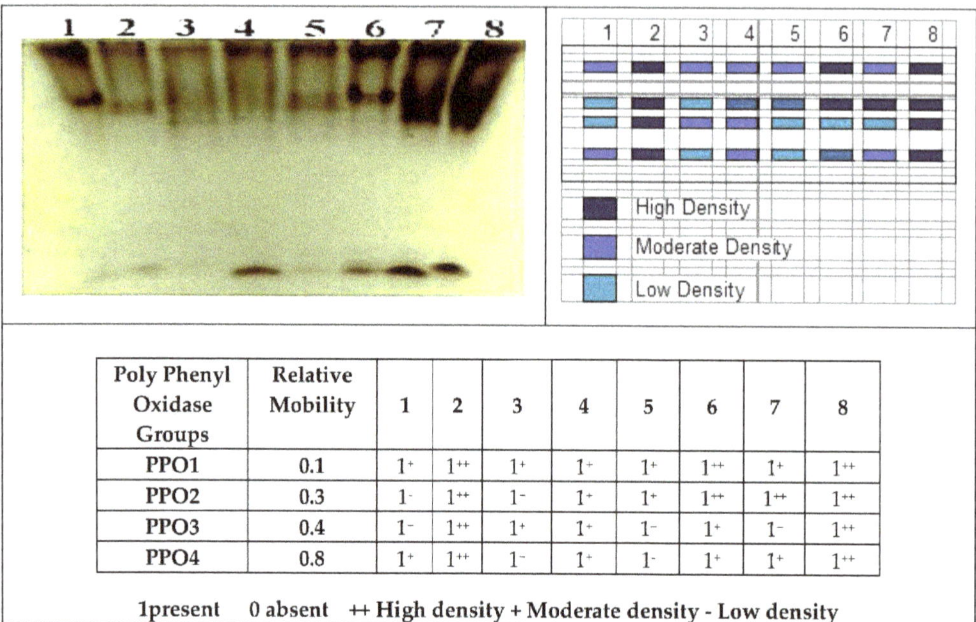

Poly Phenyl Oxidase Groups	Relative Mobility	1	2	3	4	5	6	7	8
PPO1	0.1	1^{+}	1^{++}	1^{+}	1^{+}	1^{+}	1^{++}	1^{+}	1^{++}
PPO2	0.3	1^{-}	1^{++}	1^{-}	1^{+}	1^{+}	1^{++}	1^{++}	1^{++}
PPO3	0.4	1^{-}	1^{++}	1^{+}	1^{+}	1^{-}	1^{+}	1^{-}	1^{++}
PPO4	0.8	1^{+}	1^{++}	1^{-}	1^{+}	1^{-}	1^{+}	1^{+}	1^{++}

1present 0 absent ++ High density + Moderate density - Low density

Figure 8. Effect of Se-NPs on Polyphenol oxidase isozymes of *Vicia faba* under pot conditions Control (H.). 2: Control (Inf.). 3: Soaking Nano (H.). 4: Soaking Nano (Inf.). 5: Soaking + Spray Nano (H.). 6: Soaking + Spray Nano (Inf.). 7: Spray Nano (H.). 8: Spray Nano (Inf.).

4. Conclusions

In the current study, Se-NPs were bio-synthesized by the culture supernatant of *B. megaterium* ATCC 55000, which was characterized by mono-dispersed spheres with a mean diameter of 41.2 nm. The green Se-NPs have promising antifungal activity against *R. solani* in vitro and in vivo; hence, it could use as a promising agent for the controlling of *R. solani* diseases in faba bean. Se-NPs effects on faba bean plant growth and development at the working concentration were determined. *Vicia faba* plant growth promoters in Se-NPs were the enhancement of *Vicia faba*'s morphological, metabolic and genetic parameters. Photosynthetic pigments, metabolic indicators, and phenolics compounds of *Vicia faba* were analyzed; Se-NPs caused a significant increase in total chlorophyll content and carotenoids compared with controlled plants, and the best results were achieved by soaking and foliar spraying. Moreover, results demonstrated that the application of Se-NPs induced responses regarding the total contents of phenols and total soluble protein compared with healthy control. In contrast, total phenols contents in shoots and roots-infected plants were significantly decreased in response to the treatments with Se-NPs. The effects of Se-NPs on oxidative enzymes such as polyphenol oxidase (PPO) and peroxidase (POX) in *Vicia faba* under pot conditions were assayed. Se-NPs act as a promoter and/or stressor, enhancing the antioxidant defense systems of plants, which leads to the improvement of plant tolerance. It is widely demanded that biogenic selenium NPs may be effective and economical alternatives for treating fungal plant pathogens. In the future, the adverse effects of these biogenic NPs on agriculture and ecosystems should be ascertained before their commercial use in plant disease control in the field.

Author Contributions: Conceptualization, A.H.H. and A.M.A.; methodology, A.H.H., A.M.A., A.A.A., H.M.F., A.M.A.K., K.A.A.-E. and M.M.K.; writing—review and editing, A.H.H., A.M.A., A.A.A., H.M.F., A.M.A.K., and K.A.A.-E.; All authors have read and agreed to the published version of the manuscript.

Funding: Current research was supported by the Science and Technology Development Fund (STDF), Joint Egypt (STDF)-South Africa (NRF) Scientific Cooperation, Grant ID. 27837 to Kamel Abd-Elsalam.

Institutional Review Board Statement: Not applicable.

Informed Consent Statement: Not applicable.

Data Availability Statement: Not applicable.

Acknowledgments: The authors express their sincere thanks to Faculty of Science (Boyes), Al-Azhar University, Cairo, Egypt for providing the necessary research facilities.

Conflicts of Interest: The authors declare that they have no conflict of interest.

References

1. Sekhon, B.S. Nanotechnology in agri-food production: An overview. *Nanotechnol. Sci. Appl.* **2014**, *7*, 31. [CrossRef]
2. Bebber, D.P.; Gurr, S.J. Crop-destroying fungal and oomycete pathogens challenge food security. *Fungal Genet. Biol.* **2015**, *74*, 62–64. [CrossRef]
3. Khalil, A.M.A.; Hashem, A.H.; Abdelaziz, A.M. Occurrence of toxigenic *Penicillium polonicum* in retail green table olives from the Saudi Arabia market. *Biocatal. Agric. Biotechnol.* **2019**, *21*, 101314. [CrossRef]
4. Bramhanwade, K.; Shende, S.; Bonde, S.; Gade, A.; Rai, M. Fungicidal activity of Cu nanoparticles against Fusarium causing crop diseases. *Environ. Chem. Lett.* **2016**, *14*, 229–235. [CrossRef]
5. Srivastava, S.; Bist, V.; Srivastava, S.; Singh, P.C.; Trivedi, P.K.; Asif, M.H.; Chauhan, P.S.; Nautiyal, C.S. Unraveling aspects of *Bacillus amyloliquefaciens* mediated enhanced production of rice under biotic stress of *Rhizoctonia solani*. *Front. Plant Sci.* **2016**, *7*, 587. [CrossRef] [PubMed]
6. Boghdady, M.S.; Desoky, E.; Azoz, S.N.; Abdelaziz, D.M. Effect of selenium on growth, physiological aspects and productivity of faba bean (*Vicia faba* L.). *Egypt. J. Agron.* **2017**, *39*, 83–97. [CrossRef]
7. Gupta, Y. Nutritive value of food legumes. In *Chemistry and Biochemistry of Legumes*; Arora, S.K., Ed.; Edward Arnold: London, UK, 1983.
8. Rose, T.J.; Rose, M.T.; Pariasca Tanaka, J.; Heuer, S.; Wissuwa, M. The frustration with utilization: Why have improvements in internal phosphorus utilization efficiency in crops remained so elusive? *Front. Plant Sci.* **2011**, *2*, 73. [CrossRef] [PubMed]
9. Mazen, M.; El-Batanony, N.H.; Abd El-Monium, M.; Massoud, O. Cultural filtrate of *Rhizobium* spp. and arbuscular mycorrhiza are potential biological control agents against root rot fungal diseases of faba bean. *Glob. J. Biotechnol. Biochem.* **2008**, *3*, 32–41.
10. Elhendawy, S.; Shaban, W.; Sakagami, J.-I. Does treating faba bean seeds with chemical inducers simultaneously increase chocolate spot disease resistance and yield under field conditions? *Turk. J. Agric. For.* **2010**, *34*, 475–485.
11. Köpke, U.; Nemecek, T. Ecological services of faba bean. *Field Crop. Res.* **2010**, *115*, 217–233. [CrossRef]
12. Nebiyu, A.; Diels, J.; Boeckx, P. Phosphorus use efficiency of improved faba bean (*Vicia faba*) varieties in low-input agro-ecosystems. *J. Plant Nutr. Soil Sci.* **2016**, *179*, 347–354. [CrossRef]
13. Rose, T.J.; Julia, C.C.; Shepherd, M.; Rose, M.T.; Van Zwieten, L. Faba bean is less susceptible to fertiliser N impacts on biological N_2 fixation than chickpea in monoculture and intercropping systems. *Biol. Fertil. Soils* **2016**, *52*, 271–276. [CrossRef]
14. Erper, I.; Ozkoc, I.; Karaca, G.H. Identification and pathogenicity of *Rhizoctonia* species isolated from bean and soybean plants in Samsun, Turkey. *Arch. Phytopathol. Plant Prot.* **2011**, *44*, 78–84. [CrossRef]
15. Ghosh, S.; Kanwar, P.; Jha, G. Alterations in rice chloroplast integrity, photosynthesis and metabolome associated with pathogenesis of Rhizoctonia solani. *Sci. Rep.* **2017**, *7*, 41610. [CrossRef]
16. Latz, E.; Eisenhauer, N.; Rall, B.C.; Scheu, S.; Jousset, A. Unravelling linkages between plant community composition and the pathogen-suppressive potential of soils. *Sci. Rep.* **2016**, *6*, 23584. [CrossRef]
17. Fouda, A.; Khalil, A.; El-Sheikh, H.; Abdel-Rhaman, E.; Hashem, A. Biodegradation and detoxification of bisphenol-A by filamentous fungi screened from nature. *J. Adv. Biol. Biotechnol.* **2015**, 123–132. [CrossRef]
18. Hashem, A.H.; Khalil, A.M.A.; Reyad, A.M.; Salem, S.S. Biomedical Applications of Mycosynthesized Selenium Nanoparticles Using *Penicillium expansum* ATTC 36200. *Biol. Trace Elem. Res.* **2021**, *2021*, 1–11.
19. Hashem, A.H.; Suleiman, W.B.; Abu-elreesh, G.; Shehabeldine, A.M.; Khalil, A.M.A. Sustainable lipid production from oleaginous fungus *Syncephalastrum racemosum* using synthetic and watermelon peel waste media. *Bioresour. Technol. Rep.* **2020**, *12*, 100569. [CrossRef]
20. Hashem, A.H.; Suleiman, W.B.; Abu-Elrish, G.M.; El-Sheikh, H.H. Consolidated bioprocessing of sugarcane bagasse to microbial oil by newly isolated Oleaginous Fungus: Mortierella wolfii. *Arab. J. Sci. Eng.* **2020**, *46*, 199–211. [CrossRef]
21. Hashem, A.H.; Hasanin, M.S.; Khalil, A.M.A.; Suleiman, W.B. Eco-green conversion of watermelon peels to single cell oils using a unique oleaginous fungus: Lichtheimia corymbifera AH13. *Waste Biomass Valorization* **2019**, *11*, 5721–5732. [CrossRef]
22. Hasanin, M.S.; Hashem, A.H. Eco-friendly, economic fungal universal medium from watermelon peel waste. *J. Microbiol. Methods* **2020**, *168*, 105802. [CrossRef]

23. Elbahnasawy, M.A.; Shehabeldine, A.M.; Khattab, A.M.; Amin, B.H.; Hashem, A.H. Green biosynthesis of silver nanoparticles using novel endophytic *Rothia endophytica*: Characterization and anticandidal activity. *J. Drug Deliv. Sci. Technol.* **2021**, *62*, 102401. [CrossRef]
24. Dong, H.; Xiong, R.; Liang, Y.; Tang, G.; Yang, J.; Tang, J.; Niu, J.; Gao, Y.; Zhou, Z.; Cao, Y. Development of glycine-copper (ii) hydroxide nanoparticles with improved biosafety for sustainable plant disease management. *RSC Adv.* **2020**, *10*, 21222–21227. [CrossRef]
25. Krutyakov, Y.A.; Kudrinsky, A.A.; Gusev, A.A.; Zakharova, O.V.; Klimov, A.I.; Yapryntsev, A.D.; Zherebin, P.M.; Shapoval, O.A.; Lisichkin, G.V. Synthesis of positively charged hybrid PHMB-stabilized silver nanoparticles: The search for a new type of active substances used in plant protection products. *Mater. Res. Express* **2017**, *4*, 075018. [CrossRef]
26. Kaur, P. Biosynthesis of nanoparticles using eco-friendly factories and their role in plant pathogenicity: A review. *Biotechnol. Res. Innov.* **2018**, *2*, 63–73.
27. Nair, R.; Varghese, S.H.; Nair, B.G.; Maekawa, T.; Yoshida, Y.; Kumar, D.S. Nanoparticulate material delivery to plants. *Plant Sci.* **2010**, *179*, 154–163. [CrossRef]
28. El-Batal, A.; Haroun, B.M.; Farrag, A.A.; Baraka, A.; El-Sayyad, G.S. Synthesis of silver nanoparticles and incorporation with certain antibiotic using gamma irradiation. *J. Pharm. Res. Int.* **2014**, 1341–1363. [CrossRef]
29. Kora, A.J.; Mounika, J.; Jagadeeshwar, R. Rice leaf extract synthesized silver nanoparticles: An in vitro fungicidal evaluation against *Rhizoctonia solani*, the causative agent of sheath blight disease in rice. *Fungal Biol.* **2020**, *124*, 671–681. [CrossRef] [PubMed]
30. Hariharan, H.; Al-Harbi, N.; Karuppiah, P.; Rajaram, S. Microbial synthesis of selenium nanocomposite using *Saccharomyces cerevisiae* and its antimicrobial activity against pathogens causing nosocomial infection. *Chalcogenide Lett.* **2012**, *9*, 509–515.
31. Abdelkareem, M.; Zohri, A.A. Inhibition of three toxigenic fungal strains and their toxins production using selenium nanoparticles. *Czech Mycol.* **2017**, *69*, 193–204. [CrossRef]
32. Shirsat, S.; Kadam, A.; Naushad, M.; Mane, R.S. Selenium nanostructures: Microbial synthesis and applications. *RSC Adv.* **2015**, *5*, 92799–92811. [CrossRef]
33. Liu, L.; Xiao, Z.; Niu, S.; He, Y.; Wang, G.; Pei, X.; Tao, W.; Wang, M. Preparation, characteristics and feeble induced-apoptosis performance of non-dialysis requiring selenium nanoparticles@ chitosan. *Mater. Des.* **2019**, *182*, 108024. [CrossRef]
34. Alnassar, H.S.; Helal, M.H.; Askar, A.A.; Masoud, D.M.; Abdallah, A.E. Pyridine azo disperse dye derivatives and their selenium nanoparticles (SeNPs): Synthesis, fastness properties, and antimicrobial evaluations. *Int. J. Nanomed.* **2019**, *14*, 7903. [CrossRef] [PubMed]
35. El-Sayed, M.H.; Refaat, B.M.; Askar, A.A.Z. Marine *Streptomyces cyaneus* strain Alex-SK121 mediated eco-friendly synthesis of silver nanoparticles using gamma radiation. *J. Pharm. Res. Int.* **2014**, *4*, 2525–2547.
36. El-Batal, A.; Gamal, M.; Ibrahim, M. Biosynthesis of silver nanoparticles using *Streptomyces atroverins*, strain askar-sh50 and their antimicrobial activity against foodborne pathogens and *Mycobacterium tuberculosis*. *Al Azhar Bull. Sci. March* **2017**, *9*, 87–106.
37. Tugarova, A.V.; Vetchinkina, E.P.; Loshchinina, E.A.; Burov, A.M.; Nikitina, V.E.; Kamnev, A.A. Reduction of selenite by *Azospirillum brasilense* with the formation of selenium nanoparticles. *Microb. Ecol.* **2014**, *68*, 495–503. [CrossRef] [PubMed]
38. El-Sherbiny, G.M.; El-Batal, A.; El-Sherbiny, I. Antibacterial potential with molecular docking study against multi-drug resistant bacteria and *Mycobacterium tuberculosis* of streptomycin produced by *Streptomyces atroverins*, strain Askar-SH50. *J. Chem. Pharm. Res* **2017**, *9*, 189–208.
39. Ghorab, M.M.; Alqahtani, A.S.; Soliman, A.M.; Askar, A.A. Novel N-(Substituted) Thioacetamide Quinazolinone Benzenesulfonamides as Antimicrobial Agents. *Int. J. Nanomed.* **2020**, *15*, 3161. [CrossRef]
40. Hashem, A.H.; Saied, E.; Hasanin, M.S. Green and ecofriendly bio-removal of methylene blue dye from aqueous solution using biologically activated banana peel waste. *Sustain. Chem. Pharm.* **2020**, *18*, 100333. [CrossRef]
41. Hasanin, M.S.; Hashem, A.H.; Abd El-Sayed, E.S.; El-Saied, H. Green ecofriendly bio-deinking of mixed office waste paper using various enzymes from *Rhizopus microsporus* AH3: Efficiency and characteristics. *Cellulose* **2020**, *27*, 4443–4453. [CrossRef]
42. Suleiman, W.; El-Sheikh, H.; Abu-Elreesh, G.; Hashem, A. Recruitment of *Cunninghamella echinulata* as an Egyptian isolate to produce unsaturated fatty acids. *Res. J. Pharm. Biol. Chem. Sci.* **2018**, *9*, 764–774.
43. Suleiman, W.; El-Skeikh, H.; Abu-Elreesh, G.; Hashem, A. Isolation and screening of promising oleaginous Rhizopus sp and designing of Taguchi method for increasing lipid production. *J. Innov. Pharm. Biol. Sci.* **2018**, *5*, 8–15.
44. Khalil, A.M.A.; Abdelaziz, A.M.; Khaleil, M.M.; Hashem, A.H. Fungal endophytes from leaves of Avicennia marina growing in semi-arid environment as a promising source for bioactive compounds. *Lett. Appl. Microbiol.* **2020**, *72*, 263–274. [CrossRef] [PubMed]
45. Khalil, A.M.A.; Hashem, A.H. Morphological changes of conidiogenesis in two *Aspergillus* species. *J Pure Appl Microbiol* **2018**, *12*, 2041–2048. [CrossRef]
46. Shubharani, R.; Mahesh, M.; Yogananda Murthy, V. Biosynthesis and characterization, antioxidant and antimicrobial activities of selenium nanoparticles from ethanol extract of Bee Propolis. *J. Nanomed. Nanotechnol.* **2019**, *10*, 2.
47. Joshi, S.M.; De Britto, S.; Jogaiah, S.; Ito, S.-I. Mycogenic selenium nanoparticles as potential new generation broad-spectrum antifungal molecules. *Biomolecules* **2019**, *9*, 419. [CrossRef]
48. Büttner, G.; Pfähler, B.; Märländer, B. Greenhouse and field techniques for testing sugar beet for resistance to Rhizoctonia root and crown rot. *Plant Breed.* **2004**, *123*, 158–166. [CrossRef]
49. Mousa, A.M.; El-Sayed, S.A. Effect of intercropping and phosphorus fertilizer treatments on incidence of Rhizoctonia root-rot disease of faba bean. *Int. J. Curr. Microbiol. Appl. Sci.* **2016**, *5*, 850–863. [CrossRef]

50. Grünwald, N.; Coffman, V.; Kraft, J. Sources of partial resistance to Fusarium root rot in the Pisum core collection. *Plant Dis.* **2003**, *87*, 1197–1200. [CrossRef]
51. Alhaithloul, H.A.S.; Attia, M.S.; Abdein, M.A. Dramatic biochemical and anatomical changes in eggplant due to infection with *Alternaria solani* causing early blight disease. *Int. J. Bot. Stud.* **2019**, *4*, 55–60.
52. Snedecor, G.W.; Cochran, W.G. *Statistical Methods*, 7th ed.; Iowa State University USA: Ames, IA, USA, 1980; pp. 80–86.
53. Cittrarasu, V.; Kaliannan, D.; Dharman, K.; Maluventhen, V.; Easwaran, M.; Liu, W.C.; Balasubramanian, B.; Arumugam, M. Green synthesis of selenium nanoparticles mediated from *Ceropegia bulbosa* Roxb extract and its cytotoxicity, antimicrobial, mosquitocidal and photocatalytic activities. *Sci. Rep.* **2021**, *11*, 1–15. [CrossRef]
54. Abu-Elghait, M.; Hasanin, M.; Hashem, A.H.; Salem, S.S. Ecofriendly novel synthesis of tertiary composite based on cellulose and myco-synthesized selenium nanoparticles: Characterization, antibiofilm and biocompatibility. *Int. J. Biol. Macromol.* **2021**, *175*, 294–303. [CrossRef] [PubMed]
55. Salem, S.S.; Fouda, M.M.; Fouda, A.; Awad, M.A.; Al-Olayan, E.M.; Allam, A.A.; Shaheen, T.I. Antibacterial, cytotoxicity and larvicidal activity of green synthesized selenium nanoparticles using *Penicillium corylophilum*. *J. Clust. Sci.* **2020**, *32*, 351–361. [CrossRef]
56. Zare, B.; Babaie, S.; Setayesh, N.; Shahverdi, A.R. Isolation and characterization of a fungus for extracellular synthesis of small selenium nanoparticles. *Nanomed. J.* **2013**, *1*, 13–19.
57. Ramamurthy, C.; Sampath, K.; Arunkumar, P.; Kumar, M.S.; Sujatha, V.; Premkumar, K.; Thirunavukkarasu, C. Green synthesis and characterization of selenium nanoparticles and its augmented cytotoxicity with doxorubicin on cancer cells. *Bioprocess Biosyst. Eng.* **2013**, *36*, 1131–1139. [CrossRef] [PubMed]
58. Dhanjal, S.; Cameotra, S.S. Aerobic biogenesis of selenium nanospheres by *Bacillus cereus* isolated from coalmine soil. *Microb. Cell Factories* **2010**, *9*, 1–11. [CrossRef] [PubMed]
59. Oremland, R.S.; Herbel, M.J.; Blum, J.S.; Langley, S.; Beveridge, T.J.; Ajayan, P.M.; Sutto, T.; Ellis, A.V.; Curran, S. Structural and spectral features of selenium nanospheres produced by Se-respiring bacteria. *Appl. Environ. Microbiol.* **2004**, *70*, 52–60. [CrossRef]
60. Min, J.S.; Kim, K.S.; Kim, S.W.; Jung, J.H.; Lamsal, K.; Kim, S.B.; Jung, M.; Lee, Y.S. Effects of colloidal silver nanoparticles on sclerotium-forming phytopathogenic fungi. *Plant Pathol. J.* **2009**, *25*, 376–380. [CrossRef]
61. Krishnaraj, C.; Ramachandran, R.; Mohan, K.; Kalaichelvan, P. Optimization for rapid synthesis of silver nanoparticles and its effect on phytopathogenic fungi. *Spectrochim. Acta Part A Mol. Biomol. Spectrosc.* **2012**, *93*, 95–99. [CrossRef]
62. Kanhed, P.; Birla, S.; Gaikwad, S.; Gade, A.; Seabra, A.B.; Rubilar, O.; Duran, N.; Rai, M. In vitro antifungal efficacy of copper nanoparticles against selected crop pathogenic fungi. *Mater. Lett.* **2014**, *115*, 13–17. [CrossRef]
63. Arciniegas-Grijalba, P.; Patiño-Portela, M.; Mosquera-Sánchez, L.; Guerrero-Vargas, J.; Rodríguez-Páez, J. ZnO nanoparticles (ZnO-NPs) and their antifungal activity against coffee fungus *Erythricium salmonicolor*. *Appl. Nanosci.* **2017**, *7*, 225–241. [CrossRef]
64. Nandini, B.; Hariprasad, P.; Prakash, H.S.; Shetty, H.S.; Geetha, N. Trichogenic-selenium nanoparticles enhance disease suppressive ability of Trichoderma against downy mildew disease caused by *Sclerospora graminicola* in pearl millet. *Sci. Rep.* **2017**, *7*, 1–11. [CrossRef] [PubMed]
65. El-Gazzar, N.; Ismail, A.M. The potential use of Titanium, Silver and Selenium nanoparticles in controlling leaf blight of tomato caused by *Alternaria alternata*. *Biocatal. Agric. Biotechnol.* **2020**, *27*, 101708. [CrossRef]
66. Ismail, A.-W.A.; Sidkey, N.M.; Arafa, R.A.; Fathy, R.M.; El-Batal, A.I. Evaluation of in vitro antifungal activity of silver and selenium nanoparticles against *Alternaria solani* caused early blight disease on potato. *Biotechnol. J. Int.* **2016**, 1–11. [CrossRef]
67. Feng, R.; Wei, C.; Tu, S. The roles of selenium in protecting plants against abiotic stresses. *Environ. Exp. Bot.* **2013**, *87*, 58–68. [CrossRef]
68. Freeman, J.L.; Lindblom, S.D.; Quinn, C.F.; Fakra, S.; Marcus, M.A.; Pilon-Smits, E.A. Selenium accumulation protects plants from herbivory by Orthoptera via toxicity and deterrence. *New Phytol.* **2007**, *175*, 490–500. [CrossRef]
69. Abdel-Monaim, M.F. Improvement of biocontrol of damping-off and root rot/wilt of faba bean by salicylic acid and hydrogen peroxide. *Mycobiology* **2013**, *41*, 47–55. [CrossRef] [PubMed]
70. Akladious, S.A.; Gomaa, E.Z.; El-Mahdy, O.M. Efficiency of bacterial biosurfactant for biocontrol of *Rhizoctonia solani* (AG-4) causing root rot in faba bean (*Vicia faba*) plants. *Eur. J. Plant Pathol.* **2019**, *153*, 15–35. [CrossRef]
71. Germ, M.; Kreft, I.; Osvald, J. Influence of UV-B exclusion and selenium treatment on photochemical efficiency of photosystem II, yield and respiratory potential in pumpkins (*Cucurbita pepo* L.). *Plant Physiol. Biochem.* **2005**, *43*, 445–448. [CrossRef] [PubMed]
72. Seppänen, M.; Turakainen, M.; Hartikainen, H. Selenium effects on oxidative stress in potato. *Plant Sci.* **2003**, *165*, 311–319. [CrossRef]
73. Juárez-Maldonado, A.; Ortega-Ortíz, H.; Morales-Díaz, A.B.; González-Morales, S.; Morelos-Moreno, Á.; Sandoval-Rangel, A.; Cadenas-Pliego, G.; Benavides-Mendoza, A. Nanoparticles and nanomaterials as plant biostimulants. *Int. J. Mol. Sci.* **2019**, *20*, 162. [CrossRef]
74. Pezzarossa, B.; Remorini, D.; Gentile, M.L.; Massai, R. Effects of foliar and fruit addition of sodium selenate on selenium accumulation and fruit quality. *J. Sci. Food Agric.* **2012**, *92*, 781–786. [CrossRef] [PubMed]
75. Dehghahi, R.; Subramaniam, S.; Zakaria, L.; Joniyas, A.; Firouzjahi, F.B.; Haghnama, K.; Razinataj, M. Review of research on fungal pathogen attack and plant defense mechanism against pathogen. *Int. J. Sci. Res. Agric. Sci.* **2015**, *2*, 197–208. [CrossRef]
76. Saha, P.; Raychaudhuri, S.S.; Chakraborty, A.; Sudarshan, M. PIXE analysis of trace elements in relation to chlorophyll concentration in Plantago ovata Forsk. *Appl. Radiat. Isot.* **2010**, *68*, 444–449. [CrossRef]
77. Sheng, M.; Tang, M.; Chen, H.; Yang, B.; Zhang, F.; Huang, Y. Influence of arbuscular mycorrhizae on photosynthesis and water status of maize plants under salt stress. *Mycorrhiza* **2008**, *18*, 287–296. [CrossRef]

78. Feng, T.; Chen, S.; Gao, D.; Liu, G.; Bai, H.; Li, A.; Peng, L.; Ren, Z. Selenium improves photosynthesis and protects photosystem II in pear (Pyrus bretschneideri), grape (*Vitis vinifera*), and peach (*Prunus persica*). *Photosynthetica* **2015**, *53*, 609–612. [CrossRef]
79. Ghorbanpour, M.; Hadian, J. Multi-walled carbon nanotubes stimulate callus induction, secondary metabolites biosynthesis and antioxidant capacity in medicinal plant Satureja khuzestanica grown in vitro. *Carbon* **2015**, *94*, 749–759. [CrossRef]
80. Večeřová, K.; Večeřa, Z.; Dočekal, B.; Oravec, M.; Pompeiano, A.; Tříska, J.; Urban, O. Changes of primary and secondary metabolites in barley plants exposed to CdO nanoparticles. *Environ. Pollut.* **2016**, *218*, 207–218. [CrossRef]
81. Boyer, R.F.; Clark, H.M.; LaRoche, A.P. Reduction and release of ferritin iron by plant phenolics. *J. Inorg. Biochem.* **1988**, *32*, 171–181. [CrossRef]
82. Maringgal, B.; Hashim, N.; Tawakkal, I.S.M.A.; Mohamed, M.T.M.; Hamzah, M.H.; Shukor, N.I.A. The causal agent of anthracnose in papaya fruit and control by three different Malaysian stingless bee honeys, and the chemical profile. *Sci. Hortic.* **2019**, *257*, 108590. [CrossRef]
83. Lewis, N.G.; Yamamoto, E. Lignin: Occurrence, biogenesis and biodegradation. *Annu. Rev. Plant Biol.* **1990**, *41*, 455–496. [CrossRef]
84. Martin-Tanguy, J. Metabolism and function of polyamines in plants: Recent development (new approaches). *Plant Growth Regul.* **2001**, *34*, 135–148. [CrossRef]
85. Mellersh, D.G.; Foulds, I.V.; Higgins, V.J.; Heath, M.C. H_2O_2 plays different roles in determining penetration failure in three diverse plant-fungal interactions. *Plant J.* **2002**, *29*, 257–268. [CrossRef]
86. Weintraub, P.G.; Jones, P. *Phytoplasmas: Genomes, Plant Hosts And*; Cabi: New York, NY, USA, 2009.
87. Siddique, Z.; Akhtar, K.P.; Hameed, A.; Sarwar, N.; Imran-Ul-Haq; Khan, S.A. Biochemical alterations in leaves of resistant and susceptible cotton genotypes infected systemically by cotton leaf curl Burewala virus. *J. Plant Interact.* **2014**, *9*, 702–711. [CrossRef]
88. Hajiboland, R. Selenium supplementation stimulates vegetative and reproductive growth in canola (*Brassica napus* L.) plants. *Acta Agric. Slov.* **2012**, *99*, 13. [CrossRef]
89. Mahfouze, H.A.; El-Dougdoug, N.K.; Mahfouze, S.A. Virucidal activity of silver nanoparticles against Banana bunchy top virus (BBTV) in banana plants. *Bull. Natl. Res. Cent.* **2020**, *44*, 199. [CrossRef]
90. Hasanuzzaman, M.; Fujita, M. Selenium pretreatment upregulates the antioxidant defense and methylglyoxal detoxification system and confers enhanced tolerance to drought stress in rapeseed seedlings. *Biol. Trace Elem. Res.* **2011**, *143*, 1758–1776. [CrossRef] [PubMed]
91. Hussein, H.-A.A.; Darwesh, O.M.; Mekki, B. Environmentally friendly nano-selenium to improve antioxidant system and growth of groundnut cultivars under sandy soil conditions. *Biocatal. Agric. Biotechnol.* **2019**, *18*, 101080. [CrossRef]
92. Wang, D.; Pei, K.; Fu, Y.; Sun, Z.; Li, S.; Liu, H.; Tang, K.; Han, B.; Tao, Y. Genome-wide analysis of the auxin response factors (ARF) gene family in rice (*Oryza sativa*). *Gene* **2007**, *394*, 13–24. [CrossRef] [PubMed]

Journal of *Fungi*

Article

Development of Nano-Antifungal Therapy for Systemic and Endemic Mycoses

Jorge H. Martínez -Montelongo [1], Iliana E. Medina-Ramírez [1,*], Yolanda Romo-Lozano [2], Antonio González-Gutiérrez [1] and Jorge E. Macías-Díaz [3,4]

1 Department of Chemistry, Universidad Autónoma de Aguascalientes, Aguascalientes 20131, Mexico; jorge.martmont@gmail.com (J.H.M.-M.); jangogu05@gmail.com (A.G.-G.)
2 Department of Microbiology, Universidad Autónoma de Aguascalientes, Aguascalientes 20131, Mexico; yolanda_rloza@hotmail.com
3 Department of Mathematics, School of Digital Technologies, Tallinn University, 10120 Tallinn, Estonia; jemacias@correo.uaa.mx
4 Department of Mathematics and Physics, Universidad Autónoma de Aguascalientes, Aguascalientes 20131, Mexico
* Correspondence: iemedina@correo.uaa.mx; Tel.: +52-449-9108400

Abstract: Fungal mycoses have become an important health and environmental concern due to the numerous deleterious side effects on the well-being of plants and humans. Antifungal therapy is limited, expensive, and unspecific (causes toxic effects), thus, more efficient alternatives need to be developed. In this work, Copper (I) Iodide (CuI) nanomaterials (NMs) were synthesized and fully characterized, aiming to develop efficient antifungal agents. The bioactivity of CuI NMs was evaluated using *Sporothrix schenckii* and *Candida albicans* as model organisms. CuI NMs were prepared as powders and as colloidal suspensions by a two-step reaction: first, the CuI_2 controlled precipitation, followed by hydrazine reduction. Biopolymers (Arabic gum and chitosan) were used as surfactants to control the size of the CuI materials and to enhance its antifungal activity. The materials (powders and colloids) were characterized by SEM-EDX and AFM. The materials exhibit a hierarchical 3D shell morphology composed of ordered nanostructures. Excellent antifungal activity is shown by the NMs against pathogenic fungal strains, due to the simultaneous and multiple mechanisms of the composites to combat fungi. The minimum inhibitory concentration (MIC) and minimum fungicidal concentration (MFC) of CuI-AG and CuI-Chitosan are below 50 μg/mL (with 5 h of exposition). Optical and Atomic Force Microscopy (AFM) analyses demonstrate the capability of the materials to disrupt biofilm formation. AFM also demonstrates the ability of the materials to adhere and penetrate fungal cells, followed by their lysis and death. Following the concept of safe by design, the biocompatibility of the materials was tested. The hemolytic activity of the materials was evaluated using red blood cells. Our results indicate that the materials show an excellent antifungal activity at lower doses of hemolytic disruption.

Keywords: copper (I) iodide; composites; antifungal; chitosan; atomic force microscopy

Citation: Martínez-Montelongo, J.H.; Medina-Ramírez, I.E.; Romo-Lozano, Y.; González-Gutiérrez, A.; Macías-Díaz, J.E. Development of Nano-Antifungal Therapy for Systemic and Endemic Mycoses. *J. Fungi* **2021**, *7*, 158. https://doi.org/10.3390/jof7020158

Academic Editor: Kamel A. Abd-Elsalam

Received: 10 December 2020
Accepted: 18 February 2021
Published: 23 February 2021

1. Introduction

Fungal infections and fungal contamination have become an important worldwide public health problem with a considerable impact on human morbidity and mortality. Immunosuppression, cancer treatment, and immunological dysfunctions result in disseminated or systemic fungal diseases, that might evolve into a life-threatening condition. In recent years, mycoses have increased their incidence, thus, the development of novel targeted therapeutics is an important challenge to treat them. Changes in etiology, including the emergence of new pathogens, resistance to antifungals, and opportunistic and immunosuppressive factors that many patients experience, add to the difficulties in diagnosing and treating fungal infections, a fact that increases costs in the health sector [1].

Sporothrix schenckii is a widely distributed dimorphic fungus that is mainly recovered from environmental samples (plant or soil). It can cause sporotrichosis, which is a subcutaneous mycosis, with a subacute or chronic course in humans and other animals, and like the fungus, it has a wide geographical distribution. The infection usually occurs because of the traumatic inoculation of fungal conidia or fragments of hyphae, corresponding to its mycelial morphotype, into the skin or subcutaneous tissue. The clinical outcome of the infection depends on the immune response of the host, resulting in severe disseminated disease with visceral and osteoarticular involvement in immunocompromised individuals, particularly people with AIDS [2]. In infected tissue, the fungus differentiates into the pathogenic yeast form and may spread to other tissue. In addition, it can form resistance structures such as biofilms [3]. The gold standard for the treatment of the subcutaneous disease is itraconazole, although, in Mexico and some other underdeveloped countries, potassium iodide is still used, due to its low cost and safety.

Candida albicans is a pathogenic and opportunistic fungus responsible for numerous diseases. It is one of the most common causes of hospital-acquired systemic infections due to its adhesive and invasive properties, and its capability to form biofilms. The capacity of *C. albicans* to form biofilms increases their resistance to antifungal therapy and the ability of the yeast cells within the biofilm consortium to withstand host immune response [4]. Miconazole and fluconazole are commonly used for candidiasis treatment; however, side deleterious effects, fungal resistance, and relapse cases have been observed with the use of these anti-fungal agents.

As previously mentioned, antifungal therapy for systemic mycosis is limited, most of the time expensive and causes important toxic effects. Nanotechnology has become an interesting strategy to improve the efficacy and specificity of traditional antifungal drugs, since it allows lower toxicity, better bio-distribution, drug targeting, potent activity, and broad antifungal spectrum. Nanotechnology has positively impacted the advances in the development of novel strategies for the cure of infectious diseases. For instance, it has been demonstrated that trypanocidal therapy based on nano-systems, renders higher accessibility, improved selectivity, and specific delivery of the active principle to intracellular targets [5]. Different nanomaterials have been evaluated for the development of antimicrobial agents: Metallic nanoparticles (Ag, Cu, Au, Al), metal oxides (ZnO, TiO_2, Fe_3O_4, $CoFe_2O_4$), fullerenes, carbon nanotubes, antimicrobial peptides and chitosan [6]. Lately, the use of CuI NPs was shown to inhibit the growth of Gram (+) and Gram (-) bacteria [7].

Biopolymers are valuable and flexible materials for biomedical and pharmaceutical applications. Their use as an excipient has increased since stable colloidal suspensions of different nanostructured materials can be formulated. In addition to the colloidal stability, the biopolymer might enhance the therapeutic efficacy of a compound. For example, chitosan composites are of interest due to the bioactivity of the polymer [8]. Chitosan has been investigated for its antimicrobial activity. This is shown by its multiple mechanisms: damage to peptidoglycan and/or cell membrane, damage to DNA or inhibition on the synthesis of mRNA, decreases the metalloproteinase activity of the microorganisms by metal chelation [9]. Formulations that involve the use of chitosan and nanoparticles exhibit increased surface area. Also, at physiological pH values, an increased density of positive charge on the surface of the composite favors the interaction to negatively charged biomolecules on the cell walls of microorganisms. This increases its antimicrobial activity. In this study, Copper (I) Iodide (CuI)-Arabic Gum (CuI@AG) and CuI-Chitosan (CuI@Ch) composites were prepared and evaluated for the inhibition of biofilm-forming pathogenic fungi (*S. schenckii*, *C. albicans*).

CuI is an interesting material that has found important industrial and biomedical applications [7,10,11]. Previous studies from our research group have demonstrated the antimicrobial activity of different metal oxide nanostructured materials [12–18]. We have also reported that the toxicity of metallic nanostructures is partly due to the lixiviation of ions to the medium, observing a direct correlation of toxicity and ion concentration. We explore the use of hybrid materials, to reduce toxicity and increase bio-activity (in this particular case, anti-fungal activity). In this work, we report on the synthesis, characterization, and antifungal activity of CuI composites. The composite under study contains three elements (Cu, I, chitosan) that show antifungal activity, which facilitates their administration at low doses. Moreover, the components of the composite exhibit activity under different mechanisms limiting the capacity of the pathogen to overcome their bioactivity.

2. Materials and Methods

2.1. Materials and Reagents

Copper sulfate pentahydrate (Karal, León, Mexico), potassium iodide (J.T. Baker, Estado de Mexico, Mexico), hydrazine hydrate (Sigma-Aldrich, St. Louis, MO, USA, 60%) chitosan (Sigma-Aldrich, Reykjavik, Iceland, medium molecular weight, deacetylation $\geq$ 75%, viscosity 200–800 cP, 1 wt% in 1% acetic acid, 25 °C), Arabic gum (Golden Bell, Zapopan, Mexico), Agar dextrose sabouraud (BD Bioxon, Estado de Mexico, Mexico), potato dextrose agar (BD Bioxon, Estado de Mexico, Mexico), were all analytical grade and used as received.

2.2. Synthesis of CuI Nanostructured Materials

CuI nanostructured materials were prepared by a modification in the method reported by Pramanik et al. [7]. In brief, a 6.0 mM potassium iodide solution was added to 6.0 mM of copper sulfate solution containing 5% (w/v) Arabic gum as a stabilizing agent. After that, a 30 µL concentration of hydrazine were dropwise added to the above mixture and the reaction was continued for 1 h at 60 °C. The nanostructured materials were collected by centrifugation at 4000 rpm for 10 min. These particles were washed several times with deionized water followed by centrifugation. The CuI NPs were recovered and dried at 60 °C in hot air oven for 6 h. These particles were fairly stable at room temperature for 1–2 weeks. They were stable in the sense that Cu did not oxidize and the composite did not aggregate (precipitate) during those 1–2 weeks. For the CuI@Ch composites, a 1.5% (w/v) solution of Medium Molecular Weight (MMW) chitosan was prepared in a 1% acetic acid solution. CuI nanostructured materials were prepared as powders and colloidal suspensions.

2.3. Characterization of CuI Nanostructured Materials

CuI nanostructured materials were characterized by scanning electron microscopy (SEM-EDX) (Carl Zeiss, Sigma-HDVP and EDX detector QUANTAX Brucker) and atomic force microscopy. Thin films of CuI@AG and CuI@Ch were deposited on a glass substrate using doctor blade technique. The surface roughness and morphology at micro- and nanometer scale were measured with a ScanAsyst atomic force microscope (Bruker, Dimension Edge with Scan Assist). The samples were analyzed in tapping mode in air with OTESPA-R3 tips (silicon; f_0: 300 kHz; k: 26 N/m). Data was acquired on square frames of 10 × 10, 5 × 5 and 2.5 × 2.5 µm. Images were recorded using height, amplitude and phase channels. Amplitude mode was used to evaluate the topography of the films; height mode images were used to obtain quantitative measurements, and phase images to evaluate changes in composition. A resolution of 512 × 512 pixels was used. Measurements were made by triplicate on different zones of each sample.

2.4. Evaluation of the Antifungal Activity of CuI Nanostructured Materials

Fungal strains and culture conditions. A general scheme for the antifungal evaluation of CuI@AG and/or CuI@Ch is illustrated in Figure 1. The antifungal activity of Cu@AG against *C. albicans* was used as a reference to demonstrate the enhanced activity of CuI@AG. The concentration of copper in the Cu NPs is the same as the concentration used for CuI NMs.

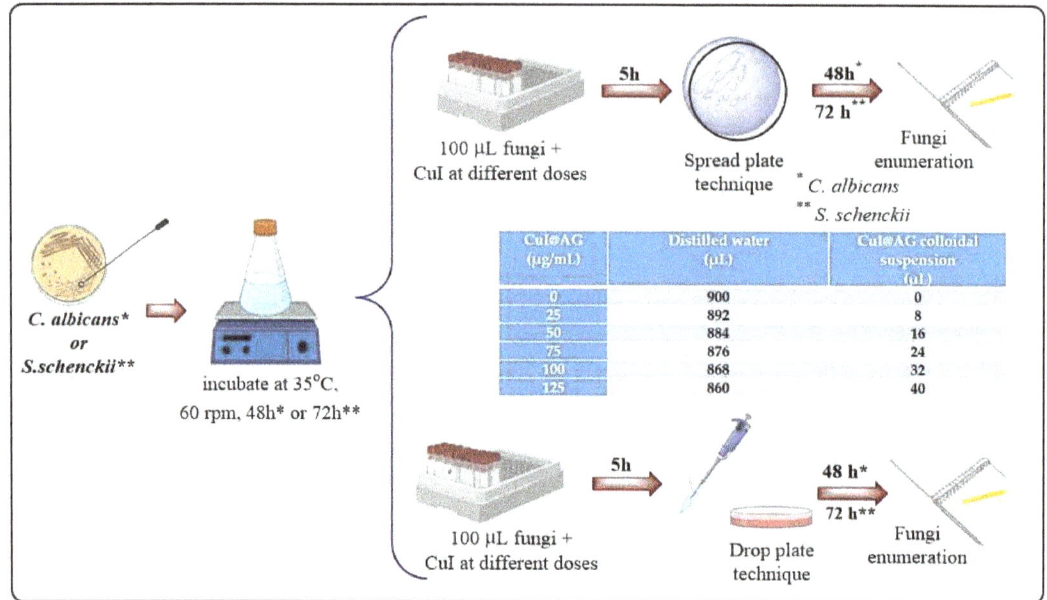

CuI@AG (µg/mL)	Distilled water (µL)	CuI@AG colloidal suspension (µL)
0	900	0
25	892	8
50	884	16
75	876	24
100	868	32
125	860	40

Figure 1. Schematic representation of the methodological approach for the evaluation of the antifungal activity of CuI@biopolymer against pathogenic fungi (*C. albicans*, *S. schenckii*).

Sporothrix schenckii. The *S. schenckii* sensu stricto wild-type strain UAA-307 was obtained from a human lymphocutaneous sporotrichosis case. The fungus identification was previously performed by biochemical, morphological, and molecular biology techniques (polymerase chain reaction sequencing of the calmodulin gene, accession number KJ921740 in GenBank) [19]. Conidial suspensions were prepared as previously reported [20]. Briefly, fungal mycelium was incubated at 28 °C for three days in sabouraud dextrose broth under orbital agitation at 150 rpm (Shaker Lumistell, IR0-60, Temperature Control). The fungal suspension was filtered on sterile filter paper (pore size 0.45 mm) and harvested by centrifugation at 750× *g*, for 20 min at 4 °C. The fungal cells were washed then with sterilized phosphate buffer solution (PBS) and counted in a Neubauer chamber before all experiments. The bioavailability of the fungi was determined by trypan blue staining. In all the experiments, a stock fungal suspension of 1×10^7 conidia/mL in distilled water was used.

Candida albicans. *C. albicans* strain used in this study was a clinical isolate, and isolates were cultured in Sabouraud Dextrose Agar (SDA), then incubated at 37 °C for 48 h. Fungal biomass was collected afterwards by a sterile loop from the surface of SDA medium and resuspended in 1 mL of PBS. An inoculum of *C. albicans* was placed in 3 mL of Potato Dextrose broth and left to grow in a shaking water bath (60 rpm) at 35 °C for 48 h (resulting in 1×10^7 CFU/mL according to the growth phase). The fungal suspension was centrifuged next at 3500 rpm for 5 min, decanted, and suspended in distilled water (stock fungal suspension).

Fusarium oxysporum. *Fusarium oxysporum* strain was isolated from samples that present fungal infection. The infected samples were placed in a PDA medium to allow fungal growth. After a small portion of the growing mycelium is reseeded in a box with fresh PDA medium and allowed to grow at 28 °C or room temperature for 5 to 7 days. Analysis of conidia production is conducted by optical microscopy. Conidia were concentrated in phosphate buffer. Fungal concentration was determined by UV-Vis spectroscopy (1×10^7 CFU/mL).

Drip dilution test. For the drip dilution test, six decimal dilutions were made in sterile distilled water (by duplicate) to evaluate the fungal growth in each dilution and to demonstrate fungal bioavailability under the experimental conditions. The initial concentration of fungal cells was 1×10^7 CFU/mL. From each dilution, an aliquot of 10 μL was laid on a Petri dish with SDA. The drops were allowed to dry to avoid movement displacement. They were left in an incubator at 28 °C for 48 h (*C. albicans*) or 72 h (*S. schenckii*).

Interaction test. 100 μL are taken from the stock fungal suspension, placed in Eppendorf tubes and diluted to 1 mL. The dilution of the fungal suspension is 1×10^6 CFU/mL. The interaction of fungi with NMs at different concentrations was performed then according to Table 1. Each experimental condition was evaluated by duplicate. The interaction was carried out for 5 h in a water bath at 37 °C. The tubes were removed next from the water bath, then 100 μL were taken from each Eppendorf tube and placed on SDA (two boxes per tube), dispersed with boiling pearls and left in an incubator for 48 h (*C. albicans*) and 72 h (*S. scenckii*) for their subsequent analysis or counting at 28 °C. The fungicidal activity of the materials was checked by determining the MIC (minimal inhibitory concentration) and MFC (minimal fungicidal concentration). The MIC was defined as the lowest concentration of the composite that significantly (visually) inhibited fungal growth with respect to CFU. The MFC was defined as the lowest concentration of the composite that completely inhibited fungal growth.

Table 1. Determination of the MIC and MFC of CuI@polymer composites. The exposure of pathogenic fungi was evaluated at different composites concentrations.

CuI (μg/mL)	Distilled Water (μL)	CuI@AG Colloidal Suspension (μL)
0	900	0
12.5	896	4
25	892	8
50	884	16
75	876	24
100	868	32
125	860	40

Evaluation of the interaction of *S. schenckii* and *C. albicans* with CuI by Atomic Force Microscopy (AFM). The analysis of the interactions of CuI@AG and CuI@Ch with the fungi was carried out by Atomic force microscopy. A 10 μL sample of the different treatments specified in the paragraph *Interaction test* was withdrawn after the exposure time was finished. The sample was deposited on a mica substrate and fixed with absolute ethanol. The samples were analyzed using a Dimension edge with a scan assyst microscope (Bruker) on tapping mode. Topography, phase, and amplitude images were collected using tapping mode in the air at room temperature. OTESPA-R3 probes (K: 26 N/m, f_0: 300 kHz) were used for the analysis.

Evaluation of the biocompatibility of CuI colloidal materials. In vitro toxicity studies were carried out using human whole blood. Heparin-stabilized human blood was freshly collected. Physiological saline solution (PSS) was added to test tubes (10 mL of PSS in each tube). Different amounts of CuI@AG were added to each tube according to Table 2. A 100 μL sample of whole blood was added to each tube. All samples were prepared in triplicate, and the suspension was briefly vortexed before putting the samples in a water bath (37 °C) for 5 h. The mixtures were centrifuged next at 3500 rpm for 5 min. The amount of hemoglobin was determined by UV–Vis spectroscopy (Thermo Scientific, Helios Omega UV-Vis) and measured with the reference wavelength of 525 nm.

Table 2. Determination of the biocompatibility of CuI@polymer composites. The exposure of RBCs was evaluated at different composites concentrations.

Tube	Concentration (µg/mL)	CuI (µg)	PSS (mL)	Distilled Water (mL)	Blood (µL)
Control −	0	–	10	0	100
1	12.5	40	9.96	0	100
2	25	80	9.92	0	100
3	50	160	9.84	0	100
4	75	240	9.76	0	100
Control +	–	–	–	10	100

3. Results and Discussions

3.1. Synthesis of CuI Nanostructured Materials

CuI powdered materials show a fine powder consistency and light creamy color. A high yield of the synthesis reaction was observed (>90%). White homogeneous colloidal suspensions of CuI were also fabricated (Figure 2). The colloids show stability for about two weeks. There are no significant differences in the appearance of the colloidal suspensions, concerning the polymer employed. The absence of greenish-blue tonalities confirms the reduction of Cu^{2+} to Cu^{1+} species.

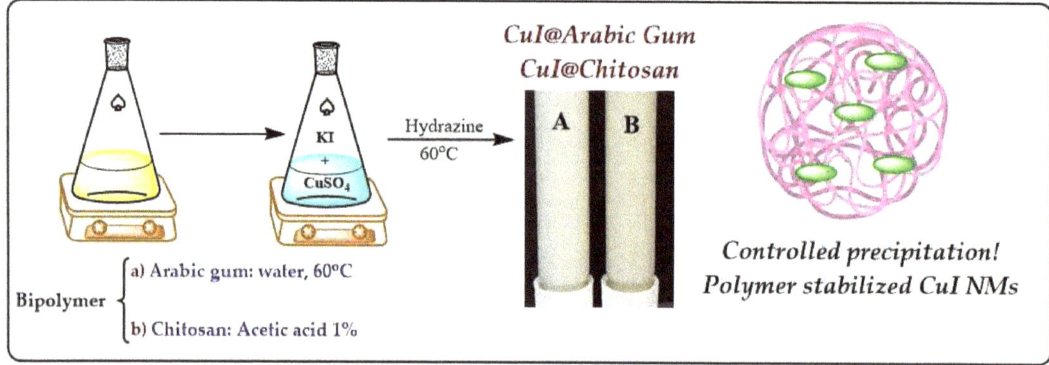

Figure 2. Synthesis of CuI nano-structured materials by a controlled precipitation method followed of reduction. Biopolymer (Arabic gum and chitosan were employed to increase colloidal stability). The materials (powders and colloids) exhibit a white color, characteristic of Cu(I) materials.

3.2. Characterization of CuI Nanostructured Materials

SEM-EDX. Figure 3 shows the SEM characterization of CuI nanostructured materials. At low magnifications, the materials exhibit a shell-like morphology (Figure 3A,C). Figure 3C clearly shows the 3D shell-like architectures with controlled internal microstructures. Figure 3B,D show the materials at higher magnifications. The materials are composed of smaller particles that are orderly aggregated in 3D shell structures. SEM-EDX analysis (Figure 3E,F) reveals the composition and purity of the materials. The weight percentages of the elements are in agreement with a CuI formulation (33.4% Cu and 66.6% I). Figure S1 (Supplementary information) shows a representative micrograph of CuI with the corresponding EDX elemental mapping for copper and iodine. Only Cu and I are present and well dispersed in the CuI lattice, demonstrating the successful preparation of the material.

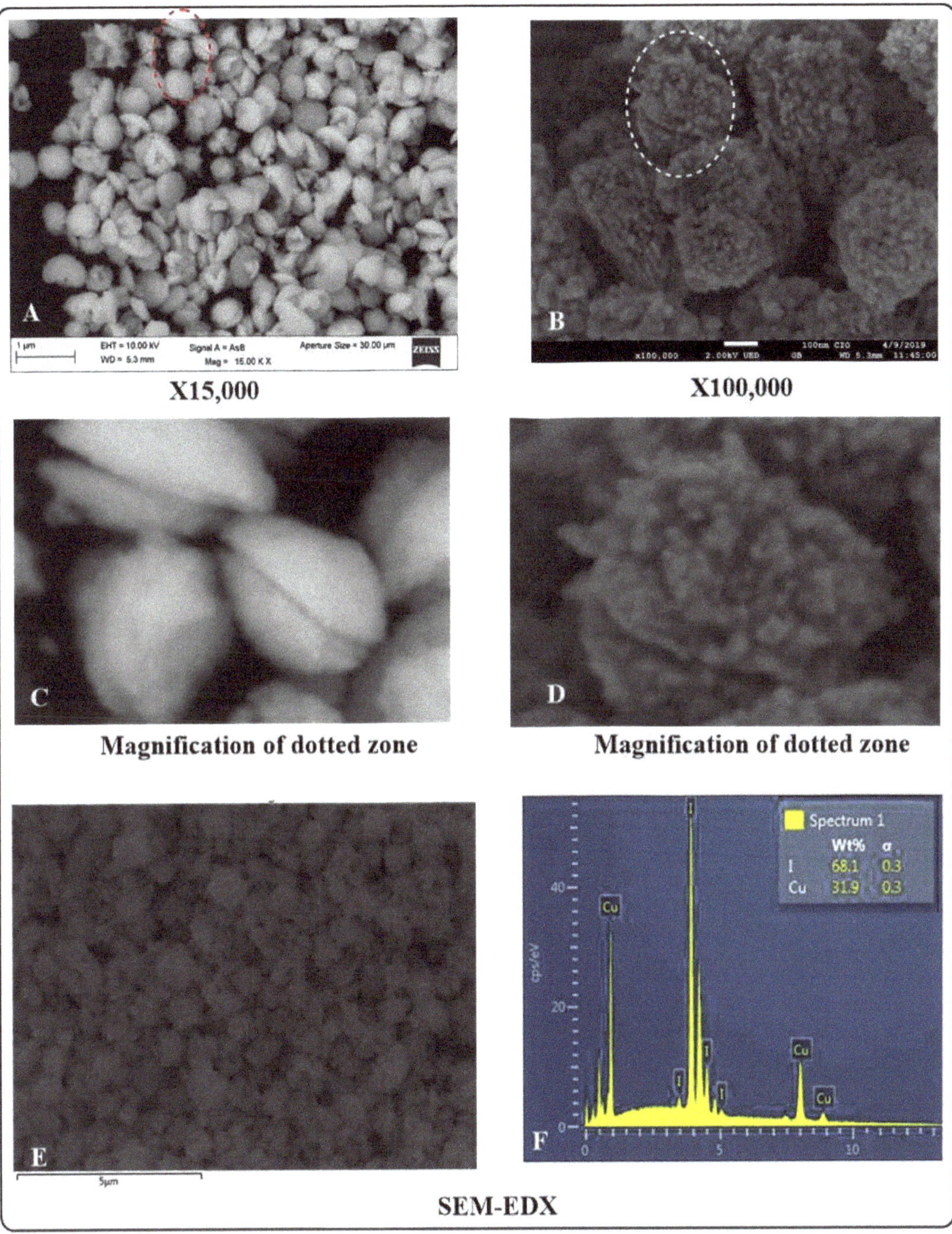

Figure 3. SEM-EDX analysis of CuI powdered nano-structured materials. (**A**) low magnification SEM micrograph of CuI powdered material, the shell like morphology of the materials is easily seen. (**B**) high magnification SEM micrograph of CuI powdered material, illustrating the ordered nanostructures that compose the 3D shell microstructures. (**C**) magnification of dotted zone in A. (**D**) magnification of dotted zone in B. (**E**) SEM micrograph for EDX analysis. (**F**) EDX analysis of CuI powdered materials.

AFM. Figure 4 portrays the AFM analyses of AG and CuI@AG thin films. Figure 4A–C show the morphology of AG films. The polymeric films (AG) are very thin (≈4 nm) and exhibit a uniform structure and composition. In comparison, CuI@AG films are thicker (≈30 nm) due to the insertion of CuI NMs in the polymeric matrix (Figure 4D–F). AFM analysis confirms the shell morphology of the CuI materials observed with SEM analysis. Tapping phase analysis (Figure 4C) of AG demonstrates the presence of pure polymer, while the CuI@AG phase analysis (Figure 4F) denotes phase changes due to inorganic (CuI) and organic (polymer) composition [21].

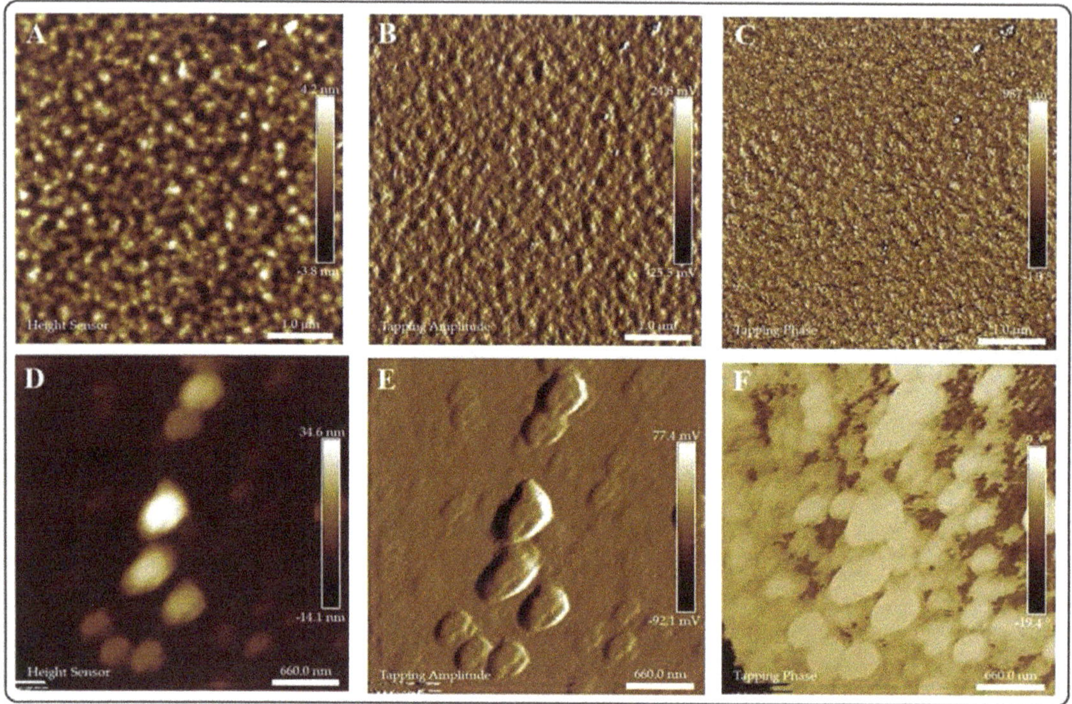

Figure 4. AFM analysis of Arabic gum (AG) and CuI@AG films. (**A–C**) Arabic gum thin films; height, amplitude and phase analysis, respectively. (**D–F**) CuI@AG thin films; height, amplitude and phase analysis, respectively.

AFM analyses of chitosan and CuI@Ch films are presented in Figure 5. Two-dimensional topography analysis of chitosan is shown in Figure 5A, illustrating its smooth surface and homogeneity. Figure 5D–L display the height, amplitude, and phase images of CuI@Ch composites at increasing magnifications. The AFM images at low magnifications reveal the presence of uniformly distributed CuI nanostructured materials with shell-like morphology. At higher magnification, a detailed inspection of the composite nanostructure denotes the presence of small (≈50 nm) ordered particles with a hierarchical 3D shell-like morphology.

In tapping mode, phase analysis is capable of imaging the nanostructure of blend or multiphase materials with high resolution of submicron length scales. We can observe evident differences in the phase images of chitosan films (Figure 5C) in comparison to CuI@Ch films. For Ch films, only one phase is observed, corresponding to the biopolymer, whereas for the composite, inorganic (CuI) and organic (polymer) phases are represented.

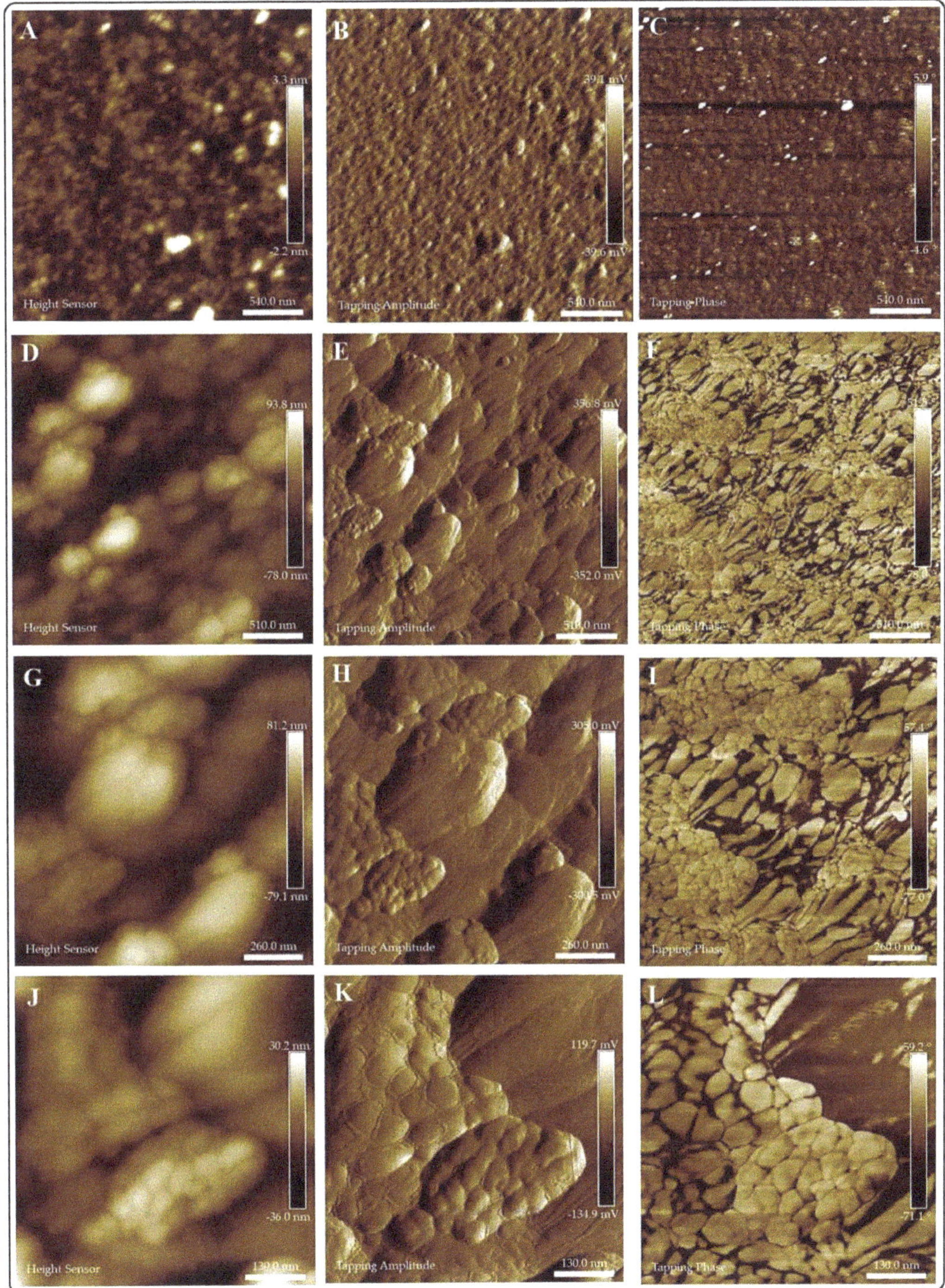

Figure 5. AFM analysis of Chitosan (Ch) and CuI@Ch films. (**A–C**) Chitosan thin films; Height, amplitude and phase analysis respectively. (**D–L**) CuI@Ch composite films; height, amplitude and phase analysis at different magnifications.

3.3. Evaluation of the Antifungal Activity of CuI

Following our methodology, different variables were evaluated to optimize the interaction of CuI materials and pathogenic fungi. Our first attempt involved the use of CuI powdered material at different doses (0.5, 2.5, and 12.5 mg/mL). Although still a common practice, the use of powdered nanomaterials for biomedical applications is cumbersome, since stable colloidal suspensions in exposition media are difficult to be obtained (Figure S2). The evaluation of the antifungal activity of CuI against *S. schenckii* was hampered by the aggregation of CuI materials in phosphate buffer, decreasing the interactions of NPs and fungi and limiting the reproducibility of observed results.

Because of the above results, stable CuI colloidal suspensions were prepared by using biopolymers as surfactants. Their antifungal activity was evaluated next at different concentrations of CuI colloids and times of exposition of pathogenic fungi. To avoid the aggregation of the NMs, their interactions with fungi were evaluated in distilled water. Initially, the following concentrations (75, 100, and 125 μg/mL) were evaluated at short exposition times (2 and 3 h). We observed good antifungal activity of the materials at all tested concentrations (Figure 6A). For instance, CuI@Ch shows antifungal activity with very high inhibition growth percentages (>90%). In addition, CuI@AG shows moderate activity after 2 h of exposition with an inhibition of 28% for the highest tested concentration. However, after 3 h of exposition, the activity of CuI@AG is very similar to CuI@Ch with inhibitions near 100%. Figure 6B shows that the fungi are viable (under the experimental conditions) in the growth controls at different serial decimal dilutions. Figure 6C illustrates the antifungal activity of the materials at different doses evaluated by the drop dilution method. Fungal growth is only registered for control groups. Figure 6D–G and Figure S4 show that longer exposition time (5 h) of fungi to these materials concentrations, result in complete growth inhibition. Figure S4 illustrates a slightly decreased antifungal activity of non-freshly prepared CuI@Ch colloidal suspension.

Freshly prepared CuI@Ch shows an enhanced antifungal activity in comparison to CuI@AG due to the synergic action of CuI and the polymer. However, its improved antifungal activity varies with time (Figure S4). A recent report remarks the importance to study the stability of biopolymers in different exposure media [22]. Accordingly, under experimental conditions, CuI@Ch tend to form aggregates, decreasing antifungal activity. Lixiviation of copper ions is one of the proposed mechanisms of CuI antifungal activity. For CuI@Ch composite, complex formation ($Cu\text{-}NH_2$) results in decreased bioavailability of Cu ions or amino groups, the chemical species responsible for antifungal activity [23,24]. In addition, decreased solubility in the exposure media lowers the stability of CuI@Ch colloidal suspension. Earlier reports point out the role of the physicochemical properties of chitosan and its antifungal activity. For example, the high molecular weight chitosan shows a better antifungal activity. Moreover, the susceptibility of the fungi specie is an important consideration [25].

Figure S5 shows the results of the antifungal activity of Cu NPs. Figure S5A illustrates the presence of fungi after their interaction with Cu NPs treatment. In contrast, CuI materials inhibit the growth of the fungi under the same experimental conditions. Figure S6 illustrates the viability of the fungal cells under experimental conditions (top row Figure S6A). It is evident the effect of Cu NPs treatment at different doses. CuI composites achieve more than 90% growth inhibition. For Cu NPs, lower efficiencies are attained under the same experimental conditions (Figure S6B). Accordingly, a recent study discusses the effect of Cu and CuO NPs to inhibit *Colletotrichum gloeoesporioides*. Cu and CuO NPs affect fungal growth at high doses (500 mg/mL) and long exposure times [26]. Cu NPs exert morphology-dependent antifungal activity. For example, the sharp edges of marigold-like petal nanostructures injure the cellular wall and membrane, and cause the death of the yeast (*C. albicans*) [27].

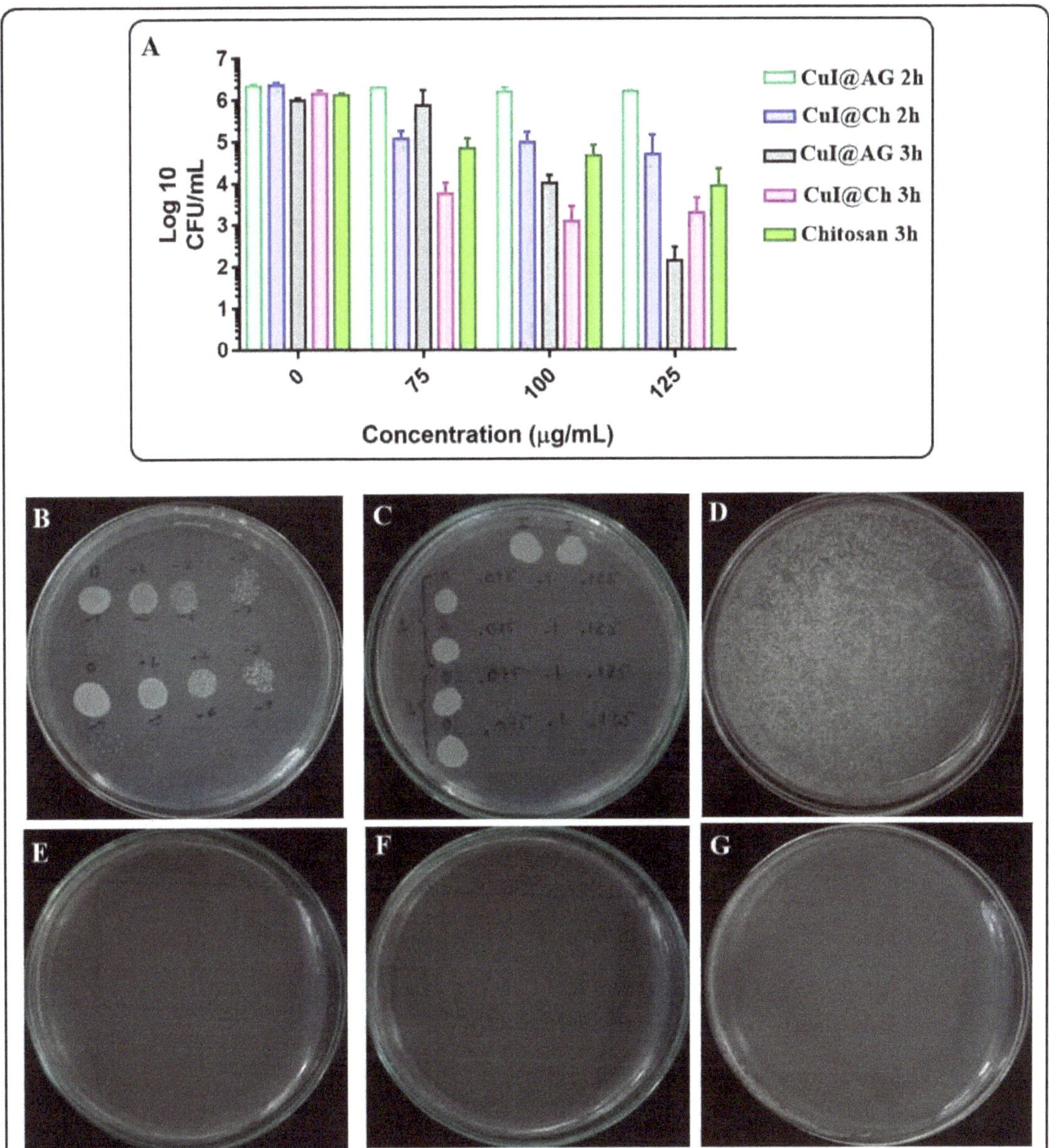

Figure 6. Evaluation of the antifungal activity of CuI@AG, CuI@Ch and chitosan to inhibit *C. albicans* growth. (**A**) Log reduction on the CFU/mL of *C. albicans* exposed to different amounts of CuI colloids or chitosan. (**B**) Viability of control group (decimal dilutions) under experimental conditions. (**C**) Evaluation of the fungicidal activity of CuI@AG against *C. albicans* (drop dilution test). (**D–G**) Enumeration of the CFU/mL of *C. albicans* exposed for 5 h to different amounts (0, 75, 100 and 125 μg/mL) of CuI NMs (spread plate procedure).

Our results indicate that the conidia of *S. schenckii* are more susceptible than the yeast of *C. albicans* to CuI treatment. For example, after 5 h of the exposition of these fungi to CuI@Ch, total growth inhibition is observed at all evaluated concentrations. Also, if the time of exposition is reduced (2 h), then the fungal growth inhibition is achieved at a concentration of 75 μg/mL of CuI@Ch. CuI@AG shows similar behavior for the treatment of *S. schenckii*. For example, after 5 h of the exposition of *S. schenckii* to CuI@AG composite, the minimum inhibitory concentration (MIC) is 12.5 μg/mL, whereas the minimum fungicidal concentration (MFC) is 25 μg/mL. At shorter exposition times (2 h), 125 μg/mL of CuI@AG composite must be used to inhibit the fungal growth (Figures S7 and S8). To our knowledge, there are not many reports regarding the development of effective and specific compounds for the treatment of *S. schenckii* infections. Potassium iodide, azoles, and amphotericin B are among the very few options for fungal infection treatment. However, these treatments result in side deleterious effects. Except for KI, the cost of the treatment is elevated, a barrier for general application in developing countries [28]. In this work, we demonstrate the effective antifungal activity of CuI materials at low doses. Their affordable cost and biocompatibility suit them as strong candidates for the development of new broad-spectrum antifungal agents.

As previously discussed, *C. albicans* is more resistant to CuI treatment. For example, after the exposition of these pathogenic fungi to CuI@AG for 5 h, fungal growth is observed at the lower evaluated doses (12.5 and 25 μg/mL). However, good inhibition activity is reached, since the fungal growth in the control group is abundant, impeding the total (CFU), direct count. On the other hand, after treatment with 12.5 μg/mL, only 50 CFU are observed. Also, treatment with 25 μg/mL decreases the growth to 5 CFU. CuI@Ch colloids exhibit the same behavior discussed for *S. schenckii*. The materials inhibit the growth of *C. albicans* at all evaluated concentrations after 5 h of exposition. However, variable effectiveness is also observed. The variability might also be a consequence of the instability of the polymer at the exposure media (pH and ionic strength).

Figure S9 shows the preliminary results on the fungicidal activity of CuI@AG against *Fusarium oxysporum*. Our results indicate fungal growth inhibition at high doses (75 μg/mL) and 20 h exposure. Future studies aim to optimize the different interaction variables to increase antifungal efficiency. Previous studies discuss the advantages of the use of NMs for the control of fungal plant diseases. Carbon, silver, silica, non-metal oxides, polymer composites and aluminosilicates show efficient activity to promote plant growth and to inhibit plant pathogens [29]. CuI@AG is a potent candidate for the development of fungicidal agents for crop protection. Also, NMs are efficient for the treatment of superficial fungal and yeast infections. Among the therapeutic agents, Ag, CuO, polymers and ZnO NPs have been investigated. NPs can also be used as antifungal carriers for the treatment of superficial fungal infections (liposomes, nanofibrous) [30] (Chapters 5, 6). From all the above, it is evident that nanotechnology offers numerous strategies to overcome the challenges that currently face the health sector.

In summary, CuI@AG shows the best antifungal activity among the different materials evaluated in this study (Cu@AG, Chitosan, CuI@Ch). Figure 7 summarizes the effective doses of the materials to inhibit the growth of pathogenic fungi. Figure 7B shows the efficiency of CuI@Ch as fungicidal at low doses. However, Figure S4 shows the variable behavior of this material. This behavior is due to its instability in the exposure medium. Figure 7C shows the enhanced antifungal activity of CuI@AG in comparison to chitosan. Although numerous reports remark the antifungal activity of this polymer, CuI@AG is more effective. Figure 7D resumes the MIC and MFC for the antifungals under study.

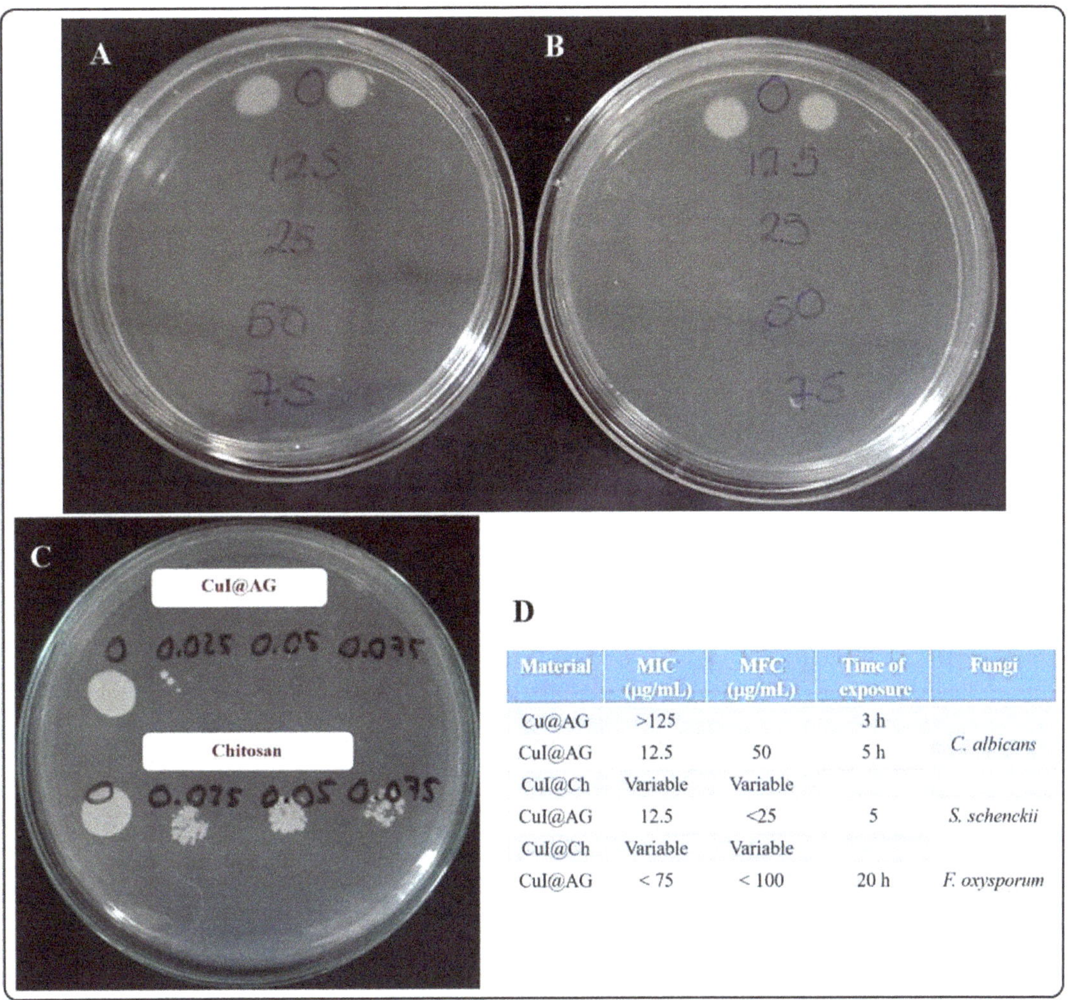

Material	MIC (µg/mL)	MFC (µg/mL)	Time of exposure	Fungi
Cu@AG	>125		3 h	*C. albicans*
CuI@AG	12.5	50	5 h	
CuI@Ch	Variable	Variable		
CuI@AG	12.5	<25	5	*S. schenckii*
CuI@Ch	Variable	Variable		
CuI@AG	< 75	< 100	20 h	*F. oxysporum*

Figure 7. Summary of the antifungal activity of CuI NMs (5 h of exposure). (**A**) CuI@AG against *S. schenckii*. (**B**) CuI@Ch against *S. schenckii*. (**C**) CuI@AG and chitosan against *C. albicans*. (**D**) MIC and MFC values.

The materials presented in this study offer numerous advantages against some previously reported antifungal compounds (Table 3). For instance: (a) the compounds can be synthesized using a green route under ambient conditions; (b) the materials show excellent antifungal activity at low doses and short exposure times; (c) the materials exhibit different mechanisms to inhibit the growth of different pathogenic fungi; (d) the materials show biocompatibility at the doses that exhibit antifungal activity; (e) Low cost.

Table 3. Comparison of the properties of non-conventional antifungal agents.

Material	Advantages	Disadvantages	Reference
CuI NPs	Broad-spectrum antibacterial and antifungal agent.	Long exposition times (24 h). The biocompatibility of the materials was not evaluated. High doses (100–150 μg/mL).	[7]
Ag NPs	Good antifungal activity at low doses (2 μg/mL)	Low fungal cell density (1×10^4/mL), Synthesis protocol indicates AgCl formation.	[31]
Chitosan NPs	Broad-spectrum antifungal activity	High doses	[25]
Cuprous iodide complexes with phenantrolines	Good antifungal and antibactericl activity at low doses (1.25–2.5 μg/mL)	Cumbersome synthetic procedure. Long exposition times (24 h). Elevated cost.	[23]
Farnesol-containing chitosan NPs	Reduces the pathogenicity of C. albicans in a murine model. Fungicidal activity at low doses	Long exposure times (48 h)	[3]
PVP-I, PEO/PVP-I complexes	Antibacterial and antifungal activity. Short time of exposure.	High doses.	[32]
PVP-I liposome hydrogel	In vitro model of oral candidosis	Irritation in epithelium. External use (ointment)	[33]
Schiff base zinc complexes with thiocyanate an iodide	Antibacterial and antifungal activity	Soluble in organic solvents, High MIC values (256 μg/mL). Expensive.	[34]
Pd@Ag Nanosheets	Broad-spectrum antifungal agent. Biocompatibility with Red Blood Cells.	Long exposure times (24 to 72 h), cumbersome synthetic procedure, expensive.	[35]
Co, CuO NPs	Antifungal activity against devastating plant fungi pathogen	High doses (500 mg/mL), long exposure times	[26]
Cu NPs	Inhibition on the growth of fluconazole resistant *C. albicans*	High doses	[36]

3.4. Evaluation of the Interaction of Fungi and CuI by AFM

Because of their physicochemical properties, NMs can display different mechanisms to destroy pathogenic microorganisms. Thus, numerous analytical techniques are applied to elucidate the antimicrobial activity of NMs. In this study, we use AFM to investigate the capacity of CuI NMs to inhibit the growth of pathogenic fungi. As previously discussed in this work, CuI NMs exhibit antifungal activity at very low doses. This capacity is not only due to their composition but also to the different and individual antifungal mechanisms shown by the elements present in the composite.

The analysis of the interaction of CuI and *Candida* by optical microscopy (Figure S10) shows the capacity of the materials to inhibit biofilm formation. The colony-forming ability of *C. albicans* decreases as the amount of CuI NMs increases. The inhibition of biofilm formation is a result of the high affinity between fungi cells and NMs. Our observations are supported by previous studies that discuss the effectiveness of iodine-containing polymers to inhibit the growth of biofilm-forming microorganisms. The materials avoid the adhesion of bacteria (*S. aureus*) or fungi (*C. albicans*) to surfaces, limiting biofilms formation [32].

AFM allows a closer examination of the interactions of CuI colloids and pathogenic fungi. There are no differences in the affinity exhibited by the colloids to adhere and penetrate the fungal cell (Figures 8–10). Short interaction times (1 h) demonstrate the adherence of the materials to fungi cell membranes. We can also observe differences in the stability of the colloids in the exposition media. For example, CuI@AG materials do not agglomerate, facilitating their interaction with fungal cells. This close interaction (CuI@AG-*C. albicans*) results in cell damage at all evaluated concentrations, as represented in Figure 8. Contrarily, CuI-Ch materials form big aggregates at the surface of the *C. albicans* membrane (Figure 9). Their increased size due to agglomeration decreases their antifungal activity. As previously discussed, the low stability of the materials renders poor reproducibility. This observation supports our results of the antifungal activity of these materials evaluated by classical microbiology methods.

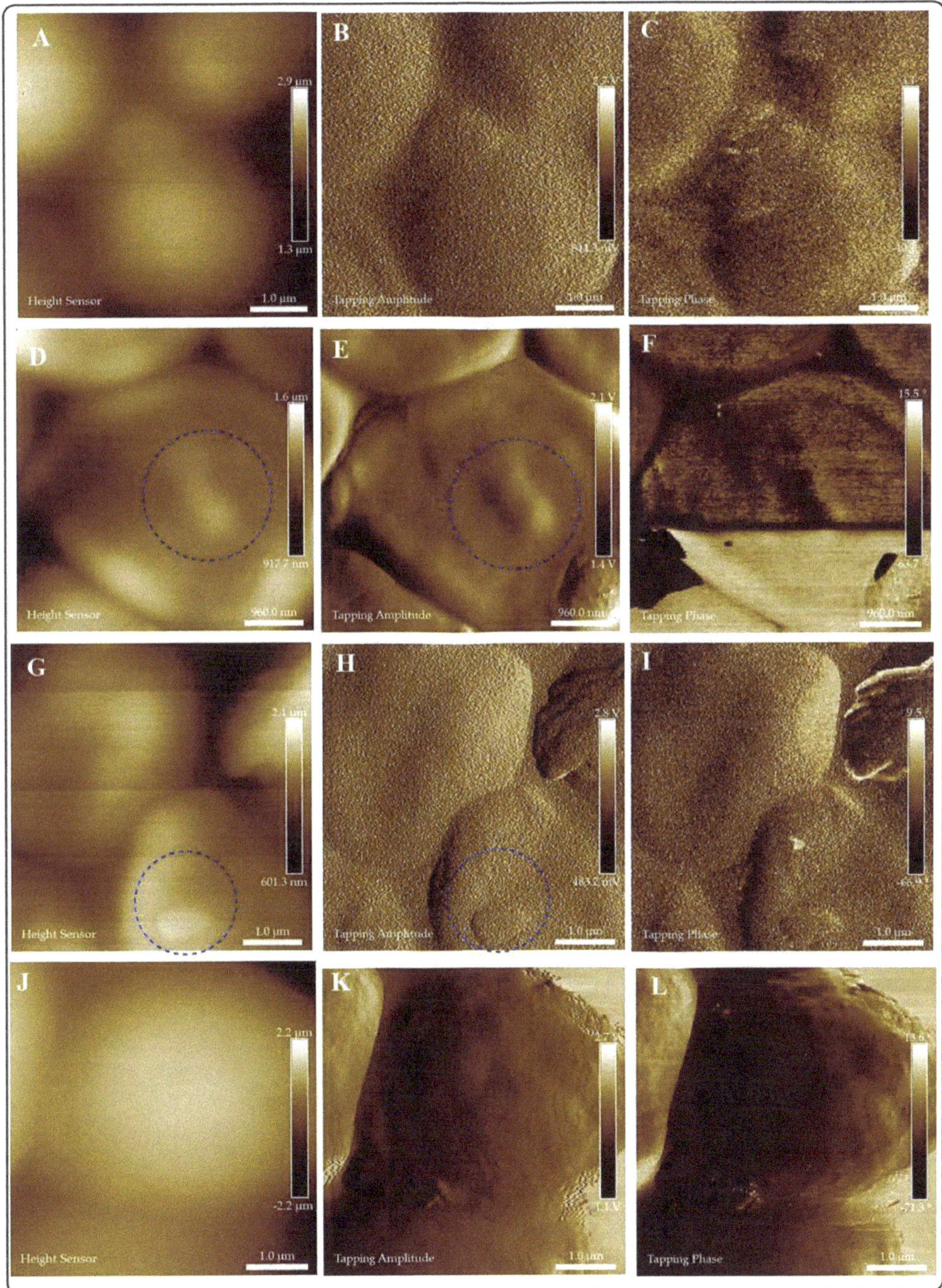

Figure 8. Evaluation of the exposition of *C. albicans* to different doses of CuI@AG for 1 h. Height, amplitude and phase images are presented for: Control group (**A–C**), and *C. albicans* exposed to different doses of CuI@AG: 75 µg/mL (**D–F**); 100 µg/mL (**G–I**); 125 µg/mL (**J–L**). Blue dotted circles indicate the presence of NMs disrupting fungal cells.

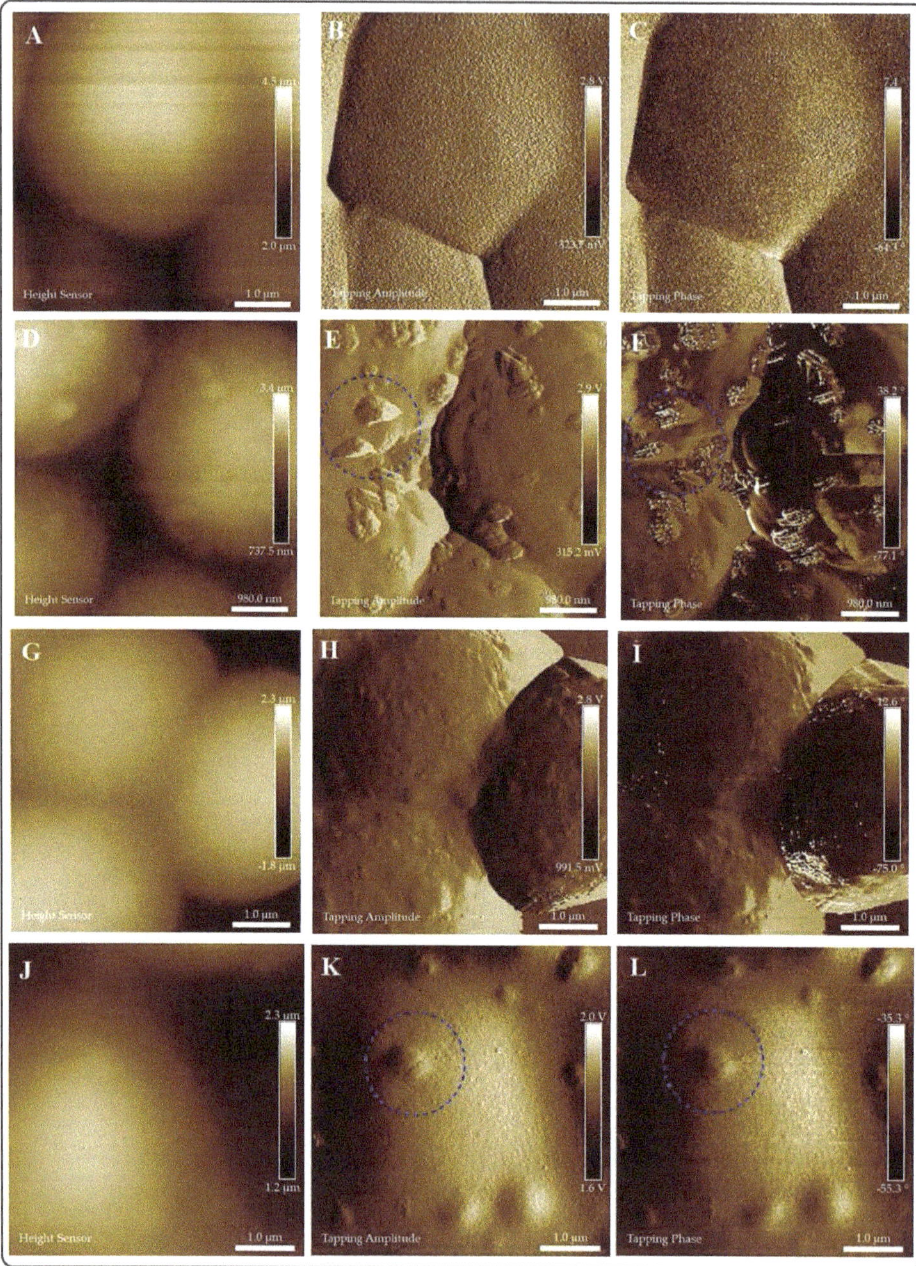

Figure 9. Evaluation of the exposition of *C. albicans* to different doses of CuI@Ch for 1 h. Height, amplitude and phase images are presented for: Control group (**A–C**), and *C. albicans* exposed to different doses of CuI@Ch: 75 µg/mL (**D–F**); 100 µg/mL (**G–I**); 125 µg/mL (**J–L**). Blue dotted circles indicate the presence of aggregated NMs on the surface (**D–F**) or penetrating the cells (**G–L**). Figures E and F denote the presence of nanomaterials on the surface of the fungi. In particular, face change in F corroborates the presence of the nanomaterials on the surface. On the other hand, K and L illustrate the penetration of nanomaterials in fungal cells.

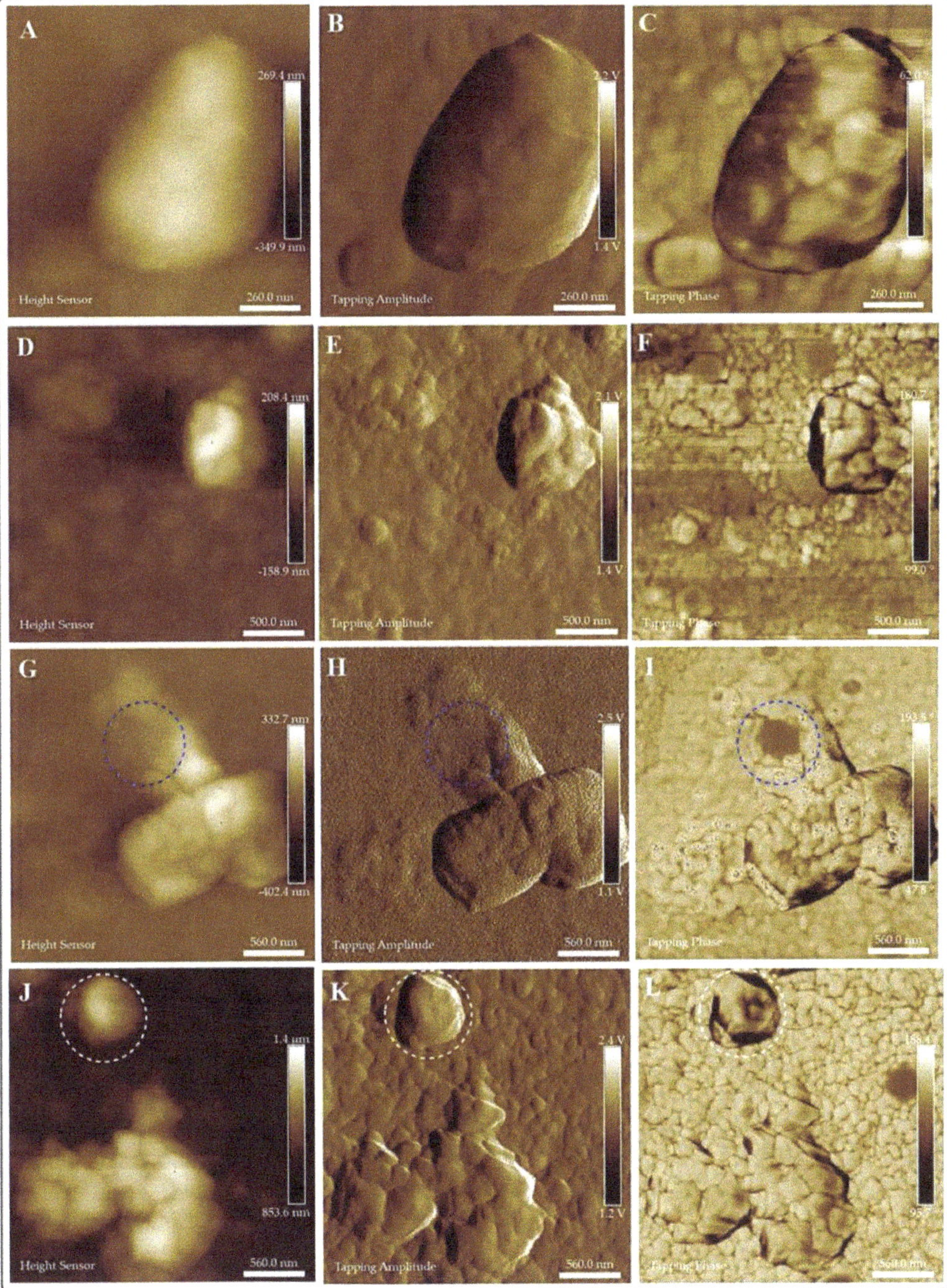

Figure 10. Evaluation of the exposition of *S. schenckii* to different doses of CuI@Ch for 1 h. Height, amplitude and phase images are presented for: Control group (**A–C**), and *S. schenckii* exposed to different doses of CuI@Ch: 75 µg/mL (**D–F**); 100 µg/mL (**G–I**); 125 µg/mL (**J–L**). Blue dotted circles indicate the penetration of the material on the fungal cell and exposition of its components. White dotted circles indicate morphological changes in *S. schenckii* due to the almost complete fungal cell coverage of the composites.

The AFM analysis of the morphological changes experimented by fungi after interacting with CuI materials indicates the penetration and disruption of the cells (Figure 8). Previous studies reported that the adherence of NMs to the membranes of microorganisms increases the lag stage of the bacterial growth period, extending the reproduction time of the microorganisms [37,38]. In accordance, in this study, the growth of pathogenic fungi was slower as a result of their exposure to CuI composites.

Kim et al. [31] reported that as a result of the interactions of *C. albicans* and nano-Ag, the fungi membranes exhibit significant changes which were manifested by the formation of "pits" on their surfaces, followed by pore development and cell death [31]. Our results indicate similar findings, the formation of a pit on the surface of *C. albicans* due to CuI penetration (Figures 8 and S11). Following the adherence of NMs to the membranes, different mechanisms of antifungal activity become active. For instance, the adhesion of CuI@Ch to fungal cell results in increased permeability of fungal membrane, then leakage of cellular contents, followed of cell death [25].

To our knowledge, there are not many reports regarding the antifungal activity of iodine based materials [32,39]. For example, the preparation of PVP-Iodine liposome hydrogels and their capacity to inhibit *C. albicans* growth was reported [33]. Human keratinocytes infected with *C. albicans* underwent antifungal treatment (PVP-iodine liposome hydrogels and commercially available Betaisodona). The iodine-based treatment results in epithelial alterations of the keratinocytes. Additionally, PVP-iodine liposome hydrogels show a multi-step antifungal mechanism. First, the adherence of the material to the cell wall of *C. albicans*, followed by membrane damage and adsorption of the active agent into the fungal cell. Our results illustrate similar findings. However, lower doses and exposure times are required.

The alteration on the morphology of *S. schenckii* due to interactions with CuI were evaluated by AFM. The results are presented in Figures 10 and S12. The high affinity of CuI colloidal materials to the *S. schenckii* cell surface results in the total coverage of fungal cells at all evaluated concentrations. AFM images show the typical morphological characteristics of these fungi by observation of the control group, demonstrating its bioavailability under test conditions (Figure 10A–C). CuI@Ch firmly adheres to fungal cells, causing damage at short exposition times (Figure 10G–L). As previously discussed by *C. albicans*, CuI@Ch forms bigger aggregates in the exposition media. Treatment of *S. schenckii* with doses of 100 and 125 μg/mL (for one hour) results in damage to fungal cells.

Figure 10G,H illustrate the formation of a cavity in *S. schenckii* (dotted blue circles). This asseveration is supported by tapping phase analysis (Figure 10I), which highlights the location of interfaces within the sample. Phase contrast is evident in the circled region, indicating the interface between NMs (light zones) and fungi (dark zone). The penetration of the materials disrupts the membrane exposing fungi cellular components. Figure 10J,K illustrate the changes in the morphology (dotted white circles) of *S. schenckii* due to interaction with the highest dose of composites. Figure 10I confirms the presence of morphologically modified fungi due to the almost complete coverage of fungi surface.

By AFM studies, we demonstrate that the highest susceptibility of *S. schenckii* to CuI materials is due to closer interactions of the materials and fungal cells. There are evident differences in the size of the fungal cells evaluated in this study (conidia of *S. schenckii* vs. *C. albicans* yeast), a variable that might influence the reported results. There are no significant differences in the activity of CuI@Ch or CuI@AG. For practical applications, CuI@AG is more suitable. Because of its abundance, low-cost, high hydrophilicity, and low toxicity. These properties suit it as an excellent candidate for the development of composites for biomedical and environmental applications.

3.5. Evaluation of the Bio-Compatibility of CuI@AG

Due to the membrane-damaging effects of CuI@AG composites in pathogenic fungi, we studied the outcomes of the interaction of composites and human red blood cells (RBCs). An *in vitro* toxicity assay to evaluate the hemolytic activity of the composites at different doses was conducted (Figure 11A). As shown in Figure 11B,C, the composites exhibited little hemolytic activity at doses where the effective antifungal activity occurs. Side effects in host cells (RBCs) result in their exposition to composite materials at 12.5 µg/mL. We can also observe a decreased hemolytic activity at higher doses of the materials. This behavior might indicate the aggregation of the composite materials in the exposure media. The homo-aggregation of the materials decreases its surface area, thus its reactivity toward surrounding RBCs.

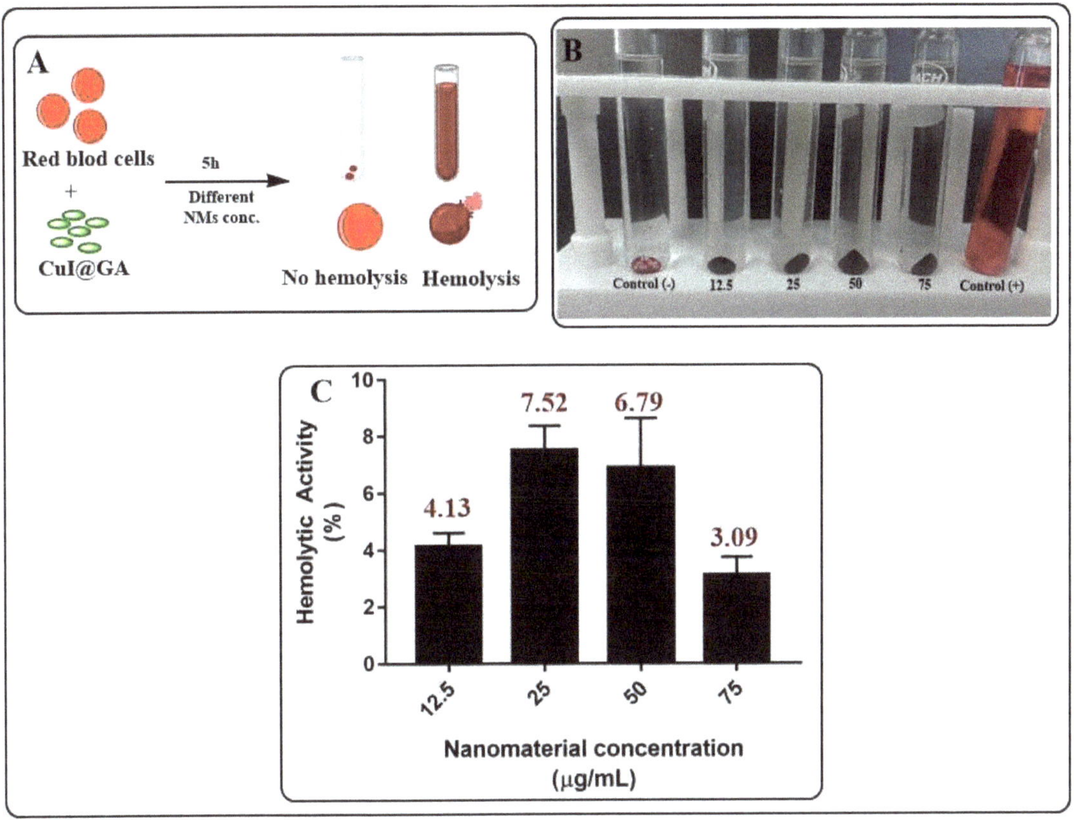

Figure 11. Evaluation of the hemolitic activity of CuI@AG. (**A**) Schematic representation of the interaction of red blood cells and CuI@AG. (**B**,**C**) Qualitative and quantitative evaluation of the hemolitic activity of CuI@AG at different doses.

It is important to remark that the interactions of NMs and living cells are complex and depend on the surface properties of the NMs and the functional groups present in the different cell membranes [40,41]. Exposure media also influence the surface properties of NMs, thus also their bio-activity. For example, Ag NPs can inhibit the growth of some, but not all, fungal strains. Another example indicates that the preparation of ketoconazole-loaded chitosan–gellan gum nanoparticles are more effective against *Aspergillus niger* than unmodified NPs and ketoconazole alone [42,43]. Current findings are not contradictory: the differences arise from the lack of uniformity in the experimental conditions among research groups.

Also, numerous reports indicate increased biocompatibility of metallic NPs after polymer functionalization. In the present study, we use Arabic gum and chitosan as CuI coatings. It is important to point up that Arabic gum is negatively charged under experimental conditions, whereas chitosan has a positive charge. RBCs are also negatively charged, avoiding electrostatic interactions with Arabic gum coated CuI materials. The AFM study of the interactions of fungal cells and CuI@polymers shows the adsorption of the composites to the fungal cells. After adsorption, the NMs enter the fungal cell, causing its death. Fungal cells contain a cell wall that can serve as a reservoir for CuI accumulation. RBS lack a cell wall. Numerous reports indicate that lixiviation of ions is favorable after a direct interaction with biological entities. CuI@AG does not interact closely due to the electrostatic repulsion of negatively charged surfaces. Penetration of NMs into the cells is not favorable under such conditions.

Although in this manuscript we present the most relevant findings related to the antifungal activity of CuI materials, previous studies show that CuI@Ch become unstable in the exposure media (aggregation), decreasing their antifungal activity. We also observe that, due to their positive charge under physiological conditions, these materials exert hemolytic activity. Our results show the same trend during the antifungal or bio-compatibility evaluation. The activity of CuI@Ch NMs is dose and time-dependent due to their instability in the exposure media. Current studies of our research group aim to improve CuI@Ch stability.

Finally, in a previous work, Pd@Ag nanosheets were studied as potential alternatives to develop efficient antifungal agents. The materials exhibit enhanced antifungal activity and biocompatibility at low doses [35]. The composites (CuI@AG) under study in the present manuscript also exhibits enhanced antifungal activity and biocompatibility in human red blood cells. CuI@AG composite materials can be a better option for large scale applications due to their lower cost and simple fabrication.

4. Conclusions

In work work, CuI@AG and CuI@Ch composites were synthesized using a facile and reproducible technique. The materials exhibit a 3D shell ordered structure that might suit them for numerous applications. We demonstrated antifungal activity at low doses and short exposure times. We also demonstrate the enhanced activity of CuI by comparing it with Cu NPs antifungal activity. The CuI NMs are bio-compatible with RBCs at the doses required to exert antifungal activity. The materials display excellent antifungal activity against biofilm-forming pathogenic fungi. The materials are also efficient to inhibit the growth of filamentous fungi (*Fusarium oxysporum*). To our knowledge, this is one of the first studies that demonstrate the efficacy of nano-antifungals to inhibit *S. schenckii* growth. Future studies will investigate the activity of the materials using *S. schenkii* yeasts. Commercially available fungal therapies show limited effectiveness due to their reduced fungal cell penetration and the development of drug-resistant strains. In this work, we demonstrate the multistep antifungal mechanism of CuI composites. Multiple studies have shown the strong antimicrobial effects of single metallic NPs on several species of microorganisms. However, some NPs are effective against certain species but have little or no effect on others. Aiming to formulate a broad-spectrum antifungal, Cu, I, and chitosan integrate the present formulation. The combination of different antifungal agents renders enhanced activity at low doses. This favors its biocompatibility with host cells and reduce the capacity of the pathogen for resistance development. It is important to remark that the synthesis of CuI@AG develops within minutes under ambient conditions. The materials are efficient, bio-compatible, broad-spectrum fungicidal agents, and of low cost. Due to all the above considerations, CuI@bio-polymer composites turn out to be potent candidates to develop efficient antifungal agents for biomedical or environmental applications.

Supplementary Materials: The following are available online at https://www.mdpi.com/2309-608X/7/2/158/s1, Figure S1. SEM-EDX analysis of CuI nanostructured materials. (A) SEM micrograph of powdered CuI NMs. (B) EDX analysis highlighting copper content. (C) EDX analysis highlighting iodide content. The mass (wt) and atomic (at) percentages of the elements confirm the formation of CuI materials; Figure S2. SEM micrographs illustrating the behavior of powdered CuI nanostructured materials at different doses in phosphate buffered medium in the presence (A*–C*) and absence (A–C) of *S. schenckii*; Figure S3. UV-Vis characterization of Cu@AG colloidal suspension.The spectra shows the characteristic plasmon of Cu NPs; Figure S4. Evaluation of the antifungal activity (*C. albicans*) of CuI@AG (A), chitosan (B) and CuI@Ch (C) at different doses; Figure S5. Evaluation of the antifungal activity (*C. albicans*) of Cu@AG (A), growth from dilution (10^{-3}) (B) at different doses; Figure S6. Evaluation of the antifungal activity of Cu@AG to inhibit C.albicans growth. (A) Viability of control group (decimal dilutions) under experimental conditions (Top row). Evaluation of the fungicidal activity of Cu@AG against *C. albicans* (drop dilution test) at differen NPs concentrations. (B) Log reduction on the CFU/mL of *C. albicans* exposed to different amounts of Cu@AG; Figure S7. Evaluation of the antifungal activity (*S. schenckii*) of CuI@AG at different concentrations and exposition of 5 h. (A) Fungal cell are viable in the growth controls at different serial decimal dilutions under experimental conditions. (B) Evaluation of the antifungal activity of CuI@AG by the drop dilution method. Fungal growth is only observed for growth control. (C–F) Evaluation of the antifungal activity of CuI@AG by the pour plate method. Fungal growth is only observed for growth control (C); Figure S8. Evaluation of the antifungal activity (*S. schenckii*) of CuI@Ch at different concentrations and exposition of 5 h. (A) Fungal cell are viable in the growth controls at different serial decimal dilutions under experimental conditions. (B) Evaluation of the antifungal activity of CuI@Ch by the drop dilution method. Fungal growth is only observed for growth control. (C–F) Evaluation of the antifungal activity of CuI@Ch by the pour plate method. Fungal growth is only observed for growth control (C); Figure S9. Evaluation of the antifungal activity (*F. oxysporumi*) of CuI@AG at different concentrations and exposition of 20 h. (A) Fungal cell are viable in the growth controls at different serial decimal dilutions under experimental conditions. (B–D) Evaluation of the antifungal activity of CuI@AG at different doses by the drop dilution method. Fungal growth is only observed for the fungi exposed to 75 μg/mL of NMs; Figure S10. Optical microscopy examination of CuI-C. albicans interaction at diferent CuI doses. (A) Big aggregates of *C. albicans* are observed for control group. (B–D) The aggregates decrease as the concentration of NMs increases (50, 75 and 100 μg/mL respectively); Figure S11. AFM analysis of the morphological changes in *C. albicans* due to exposure to CuI@AG (50 μg/mL × 1 h). Formation of a pit in fungal cell is observed due to NMs penetration (red dotted circles). A,B,C are the low-magnification images, whereas D,E,F are the high magnification images; Figure S12. AFM analysis of the morphological changes in S. schenckii due to exposure to CuI@Ch (50 μg/mL × 5 h). (A–C) Control group. The fungi maintain their typical morphology under test conditions, demonstrating the bioavailability of the MOs at the end of the experiment. (D–I) The NMs interact closely with fungal cells resulting in their destruction.

Author Contributions: Conceptualization, I.E.M.-R.; methodology, I.E.M.-R.; software, I.E.M.-R. and J.E.M.-D.; validation, I.E.M.-R.; formal analysis, J.H.M.-M., I.E.M.-R., Y.R.-L. and A.G.-G.; investigation, J.H.M.-M., I.E.M.-R. and Y.R.-L.; resources, I.E.M.-R. and J.E.M.-D.; data curation, J.H.M.-M., I.E.M.-R., Y.R.-L. and A.G.-G.; writing—original draft preparation, I.E.M.-R. and J.E.M.-D.; writing—review and editing, I.E.M.-R. and J.E.M.-D.; visualization, J.H.M.-M., I.E.M.-R. and Y.R.-L.; supervision, I.E.M.-R.; project administration, I.E.M.-R.; funding acquisition, I.E.M.-R. and J.E.M.-D. All authors have read and agreed to the published version of the manuscript.

Funding: One of the authors (I.E.M.-R.) was financially supported by the National Council for Science and Technology of Mexico (CONACyT) through the award No. 299078. Meanwhile, J.E.M.-D. was supported by CONACYT via grant A1-S-45928.

Institutional Review Board Statement: Not applicable.

Informed Consent Statement: Not applicable.

Data Availability Statement: The data presented in this study are available in insert article or supplementary material here.

Acknowledgments: This work is a submission to the special issue of Journal of Fungi on *Fungal Nanotechnology*. Beforehand, the authors wish to thank Kamel A. Abd-Elsalam for handling this submission. The authors also wish to thank the anonymous reviewers and the associate editor in charge of handlight this paper for their comments and suggestions, which helped to improve the quality of this work.

Conflicts of Interest: The authors declare no conflict of interest.

Sample Availability: Samples of all the compounds are available from the corresponding author.

References

1. Friedman, D.Z.; Schwartz, I.S. Emerging fungal infections: New patients, new patterns, and new pathogens. *J. Fungi* **2019**, *5*, 67. [CrossRef] [PubMed]
2. López-Romero, E.; Reyes-Montes, M.d.R.; Pérez-Torres, A.; Ruiz-Baca, E.; Villagómez-Castro, J.C.; Mora-Montes, H.M.; Flores-Carreón, A.; Toriello, C. Sporothrix schenckii complex and sporotrichosis, an emerging health problem. *Future Microbiol.* **2011**, *6*, 85–102. [CrossRef]
3. Fernandes Costa, A.; Evangelista Araujo, D.; Santos Cabral, M.; Teles Brito, I.; Borges de Menezes Leite, L.; Pereira, M.; Correa Amaral, A. Development, characterization, and in vitro–in vivo evaluation of polymeric nanoparticles containing miconazole and farnesol for treatment of vulvovaginal candidiasis. *Med. Mycol.* **2019**, *57*, 52–62. [CrossRef]
4. Lal, P.; Sharma, D.; Pruthi, P.; Pruthi, V. Exopolysaccharide analysis of biofilm-forming Candida albicans. *J. Appl. Microbiol.* **2010**, *109*, 128–136. [CrossRef] [PubMed]
5. Romero, E.L.; Morilla, M.J. Nanotechnological approaches against Chagas disease. *Adv. Drug Deliv. Rev.* **2010**, *62*, 576–588. [CrossRef]
6. Huh, A.J.; Kwon, Y.J. "Nanoantibiotics": A new paradigm for treating infectious diseases using nanomaterials in the antibiotics resistant era. *J. Control. Release* **2011**, *156*, 128–145. [CrossRef] [PubMed]
7. Pramanik, A.; Laha, D.; Bhattacharya, D.; Pramanik, P.; Karmakar, P. A novel study of antibacterial activity of copper iodide nanoparticle mediated by DNA and membrane damage. *Colloids Surf. B Biointerfaces* **2012**, *96*, 50–55. [CrossRef]
8. Muzzarelli, R.A. Genipin-crosslinked chitosan hydrogels as biomedical and pharmaceutical aids. *Carbohydr. Polym.* **2009**, *77*, 1–9. [CrossRef]
9. Pelgrift, R.Y.; Friedman, A.J. Nanotechnology as a therapeutic tool to combat microbial resistance. *Adv. Drug Deliv. Rev.* **2013**, *65*, 1803–1815. [CrossRef]
10. Parveen, I.; Ahmed, N. CuI mediated synthesis of heterocyclic flavone-benzofuran fused derivatives. *Tetrahedron Lett.* **2017**, *58*, 2302–2305. [CrossRef]
11. Zhang, Y.; Lin, C.; Lin, Q.; Jin, Y.; Wang, Y.; Zhang, Z.; Lin, H.; Long, J.; Wang, X. CuI-BiOI/Cu film for enhanced photo-induced charge separation and visible-light antibacterial activity. *Appl. Catal. B Environ.* **2018**, *235*, 238–245. [CrossRef]
12. Liu, J.; Medina-Ramirez, I.; Luo, Z.; Bashir, S.; Mernaugh, R. Green chemistry derived nanocomposite of silver-modified titania for disinfectant. *Nanotechnology* **2010**, *1*, 920–923.
13. Medina-Ramirez, I.; Luo, Z.; Bashir, S.; Mernaugh, R.; Liu, J.L. Facile design and nanostructural evaluation of silver-modified titania used as disinfectant. *Dalton Trans.* **2011**, *40*, 1047–1054. [CrossRef]
14. Medina-Ramírez, I.; Liu, J.L.; Hernández-Ramírez, A.; Romo-Bernal, C.; Pedroza-Herrera, G.; Jáuregui-Rincón, J.; Gracia-Pinilla, M.A. Synthesis, characterization, photocatalytic evaluation, and toxicity studies of TiO_2–Fe^{3+} nanocatalyst. *J. Mater. Sci.* **2014**, *49*, 5309–5323. [CrossRef]
15. Medina-Ramírez, I.E.; de León-Macias, C.E.D.; Pedroza-Herrera, G.; Gonzáles-Segovia, R.; Zapien, J.A.; Rodríguez-López, J.L. Evaluation of the biocompatibility and growth inhibition of bacterial biofilms by ZnO, Fe_3O_4 and ZnO@ Fe_3O_4 photocatalytic magnetic materials. *Ceram. Int.* **2020**, *46*, 8979–8994. [CrossRef]
16. Garcidueñas-Piña, C.; Medina-Ramírez, I.E.; Guzmán, P.; Rico-Martínez, R.; Morales-Domínguez, J.F.; Rubio-Franchini, I. Evaluation of the antimicrobial activity of nanostructured materials of titanium dioxide doped with silver and/or copper and their effects on Arabidopsis thaliana. *Int. J. Photoenergy* **2016**, *2016*, doi:10.1155/2016/8060847.
17. Martínez-Montelongo, J.H.; Medina-Ramírez, I.E.; Romo-Lozano, Y.; Zapien, J.A. Development of a sustainable photocatalytic process for air purification. *Chemosphere* **2020**, *257*, 127236. [CrossRef]
18. Pedroza-Herrera, G.; Medina-Ramírez, I.E.; Lozano-Álvarez, J.A.; Rodil, S.E. Evaluation of the photocatalytic activity of copper doped TiO2 nanoparticles for the purification and/or disinfection of industrial effluents. *Catal. Today* **2020**, *341*, 37–48. [CrossRef]
19. Romo-Lozano, Y.; Hernández-Hernández, F.; Salinas, E. Sporothrix schenckii yeasts induce ERK pathway activation and secretion of IL-6 and TNF-α in rat mast cells, but no degranulation. *Med. Mycol.* **2014**, *52*, 862–868. [CrossRef]
20. Curtiellas-Piñol, V.; Ventura-Juárez, J.; Ruiz-Baca, E.; Romo-Lozano, Y. Morphological changes and phagocytic activity during the interaction of human neutrophils with Sporothrix schenckii: An in vitro model. *Microb. Pathog.* **2019**, *129*, 56–63. [CrossRef]
21. Bhatt, R.; Bisen, D.; Bajpai, R.; Bajpai, A. Topological and morphological analysis of gamma rays irradiated chitosan-poly (vinyl alcohol) blends using atomic force microscopy. *Radiat. Phys. Chem.* **2017**, *133*, 81–85. [CrossRef]
22. Otulakowski, Ł.; Kasprów, M.; Strzelecka, A.; Dworak, A.; Trzebicka, B. Thermal Behaviour of Common Thermoresponsive Polymers in Phosphate Buffer and in Its Salt Solutions. *Polymers* **2021**, *13*, 90.
23. Starosta, R.; Stokowa, K.; Florek, M.; Król, J.; Chwiłkowska, A.; Kulbacka, J.; Saczko, J.; Skała, J.; Jeżowska-Bojczuk, M. Biological activity and structure dependent properties of cuprous iodide complexes with phenanthrolines and water soluble tris (aminomethyl) phosphanes. *J. Inorg. Biochem.* **2011**, *105*, 1102–1108. [CrossRef]
24. Starosta, R.; Brzuszkiewicz, A.; Bykowska, A.; Komarnicka, U.K.; Bażanów, B.; Florek, M.; Gadzała, Ł.; Jackulak, N.; Król, J.; Marycz, K. A novel copper(I) complex, [CuI(2,2′-biquinoline)P($CH_2N(CH_2CH_2)_2O)_3$]–Synthesis, characterisation and comparative studies on biological activity. *Polyhedron* **2013**, *50*, 481–489. [CrossRef]
25. Yien, L.; Zin, N.M.; Sarwar, A.; Katas, H. Antifungal activity of chitosan nanoparticles and correlation with their physical properties. *Int. J. Biomater.* **2012**, *2012*, doi:10.1155/2012/632698. [CrossRef]

26. Oussou-Azo, A.F.; Nakama, T.; Nakamura, M.; Futagami, T.; Vestergaard, M.C.M. Antifungal potential of nanostructured crystalline copper and its oxide forms. *Nanomaterials* **2020**, *10*, 1003. [CrossRef] [PubMed]
27. Martínez, A.; Apip, C.; Meléndrez, M.F.; Domínguez, M.; Sánchez-Sanhueza, G.; Marzialetti, T.; Catalán, A. Dual antifungal activity against Candida albicans of copper metallic nanostructures and hierarchical copper oxide marigold-like nanostructures grown in situ in the culture medium. *J. Appl. Microbiol.* **2020**, doi:10.1111/jam.14859. [CrossRef] [PubMed]
28. de Lima Barros, M.B.; de Almeida Paes, R.; Schubach, A.O. Sporothrix schenckii and Sporotrichosis. *Clin. Microbiol. Rev.* **2011**, *24*, 633–654. [CrossRef]
29. Alghuthaymi, M.A.; Kalia, A.; Bhardwaj, K.; Bhardwaj, P.; Abd-Elsalam, K.A.; Valis, M.; Kuca, K. Nanohybrid Antifungals for Control of Plant Diseases: Current Status and Future Perspectives. *J. Fungi* **2021**, *7*, 48. [CrossRef] [PubMed]
30. Asghari-Paskiabi, F.; Jahanshiri, Z.; Shams-Ghahfarokhi, M.; Razzaghi-Abyaneh, M. Antifungal Nanotherapy: A Novel Approach to Combat Superficial Fungal Infections. In *Nanotechnology in Skin, Soft Tissue, and Bone Infections*; Springer: Berlin/Heidelberg, Germany, 2020; pp. 93–107.
31. Kim, K.J.; Sung, W.S.; Suh, B.K.; Moon, S.K.; Choi, J.S.; Kim, J.G.; Lee, D.G. Antifungal activity and mode of action of silver nano-particles on Candida albicans. *Biometals* **2009**, *22*, 235–242. [CrossRef]
32. Ignatova, M.; Markova, N.; Manolova, N.; Rashkov, I. Antibacterial and antimycotic activity of a cross-linked electrospun poly (vinyl pyrrolidone)–iodine complex and a poly (ethylene oxide)/poly (vinyl pyrrolidone)–iodine complex. *J. Biomater. Sci. Polym. Ed.* **2008**, *19*, 373–386. [CrossRef] [PubMed]
33. Bonowitz, A.; Schaller, M.; Laude, J.; Reimer, K.; Korting, H. Comparative therapeutic and toxic effects of different povidone iodine (PVP-I) formulations in a model of oral candidosis based on in vitro reconstituted epithelium. *J. Drug Target.* **2001**, *9*, 75–83. [CrossRef]
34. Xue, L.W.; Zhao, G.Q.; Han, Y.J.; Feng, Y.X. Synthesis, structures, and antimicrobial activity of Schiff base zinc complexes with thiocyanate and iodide. *Synth. React. Inorg. Met.-Org. Nano-Met. Chem.* **2011**, *41*, 141–146. [CrossRef]
35. Zhang, C.; Chen, M.; Wang, G.; Fang, W.; Ye, C.; Hu, H.; Fa, Z.; Yi, J.; Liao, W.Q. Pd@ Ag nanosheets in combination with amphotericin B exert a potent anti-cryptococcal fungicidal effect. *PLoS ONE* **2016**, *11*, e0157000. [CrossRef] [PubMed]
36. Rasool, U.; Sah, S.K.; Hemalatha, S. Effect of Biosynthesized Copper Nanoparticles (Cunps) on the Growth And Biofilm Formation of Fluconazole-Resistant Candida Albicans. *J. Microbiol. Biotechnol. Food Sci.* **2021**, *2021*, 21–24. [CrossRef]
37. Ashrafi, M.; Bayat, M.; Mortazavi, P.; Hashemi, S.J.; Meimandipour, A. Antimicrobial effect of chitosan–silver–copper nanocomposite on Candida albicans. *J. Nanostruct. Chem.* **2020**, *10*, 87–95. [CrossRef]
38. Ashour, A.; El-Batal, A.I.; Maksoud, M.A.; El-Sayyad, G.S.; Labib, S.; Abdeltwab, E.; El-Okr, M. Antimicrobial activity of metal-substituted cobalt ferrite nanoparticles synthesized by sol–gel technique. *Particuology* **2018**, *40*, 141–151. [CrossRef]
39. Dugal, S.; Chaudhary, A. Formulation and in vitro evaluation of niosomal povidone-iodine carriers against Candida albicans. *Int. J. Pharm. Pharm. Sci.* **2013**, *5*, 509–512.
40. Akhtar, M.J.; Ahamed, M.; Alhadlaq, H.A. Challenges facing nanotoxicology and nanomedicine due to cellular diversity. *Clin. Chim. Acta* **2018**, *487*, 186–196. [CrossRef]
41. Ganguly, P.; Breen, A.; Pillai, S.C. Toxicity of nanomaterials: Exposure, pathways, assessment, and recent advances. *ACS Biomater. Sci. Eng.* **2018**, *4*, 2237–2275. [CrossRef]
42. Niemirowicz, K.; Durnaś, B.; Piktel, E.; Bucki, R. Development of antifungal therapies using nanomaterials. *Nanomedicine* **2017**, *12*, 1891–1905. [CrossRef] [PubMed]
43. Kischkel, B.; Rossi, S.A.; dos Santos Junior, S.R.; Nosanchuk, J.D.; Travassos, L.R.; Taborda, C.P. Therapies and vaccines based on nanoparticles for the treatment of systemic fungal infections. *Front. Cell. Infect. Microbiol.* **2020**, *10*, 463. [CrossRef] [PubMed]

Journal of
Fungi

MDPI

Review

Fusarium as a Novel Fungus for the Synthesis of Nanoparticles: Mechanism and Applications

Mahendra Rai [1,2,*], Shital Bonde [1], Patrycja Golinska [2], Joanna Trzcińska-Wencel [2], Aniket Gade [1], Kamel A. Abd-Elsalam [3], Sudhir Shende [1,4], Swapnil Gaikwad [5] and Avinash P. Ingle [6]

1 Department of Biotechnology, Nanobiotechnology Laboratory, Sant Gadge Baba Amravati University, Amravati 444602, India; shitalbonde@gmail.com (S.B.); aniketgade@gmail.com (A.G.); sudhirsshende13884@gmail.com (S.S.)
2 Department of Microbiology, Nicolaus Copernicus University, Lwowska, 87-100 Torun, Poland; golinska@umk.pl (P.G.); trzcinska@doktorant.umk.pl (J.T.-W.)
3 Agricultural Research Center, Plant Pathology Research Institute, Giza 12619, Egypt; kamelabdelsalam@gmail.com
4 Academy of Biology and Biotechnology, Southern Federal University, 344006 Rostov-on-Don, Russia
5 Microbial Diversity Research Centre, Dr. D. Y. Patil Biotechnology and Bioinformatics Institute, Dr. D. Y. Patil Vidyapeeth (Deemed to be University), Tathawade, Pune 411033, India; gaikwad.swapnil1@gmail.com
6 Biotechnology Centre, Department of Agricultural Botany, Dr. Panjabrao Deshmukh Krishi Vidyapeeth, Akola, Maharashtra 444104, India; Ingleavinash14@gmail.com
* Correspondence: mahendrarai7@gmail.com; Tel.: +91-779-8077-773

Abstract: Nanotechnology is a new and developing branch that has revolutionized the world by its applications in various fields including medicine and agriculture. In nanotechnology, nanoparticles play an important role in diagnostics, drug delivery, and therapy. The synthesis of nanoparticles by fungi is a novel, cost-effective and eco-friendly approach. Among fungi, *Fusarium* spp. play an important role in the synthesis of nanoparticles and can be considered as a nanofactory for the fabrication of nanoparticles. The synthesis of silver nanoparticles (AgNPs) from *Fusarium*, its mechanism and applications are discussed in this review. The synthesis of nanoparticles from *Fusarium* is the biogenic and green approach. Fusaria are found to be a versatile biological system with the ability to synthesize nanoparticles extracellularly. Different species of Fusaria have the potential to synthesise nanoparticles. Among these, *F. oxysporum* has demonstrated a high potential for the synthesis of Ag-NPs. It is hypothesised that NADH-dependent nitrate reductase enzyme secreted by *F. oxysporum* is responsible for the reduction of aqueous silver ions into AgNPs. The toxicity of nanoparticles depends upon the shape, size, surface charge, and the concentration used. The nanoparticles synthesised by different species of Fusaria can be used in medicine and agriculture.

Keywords: *Fusarium*; synthesis; nanoparticles; mechanism; medicine; agriculture; nanofactory; toxicity

Citation: Rai, M.; Bonde, S.; Golinska, P.; Trzcińska-Wencel, J.; Gade, A.; Abd-Elsalam, K.A.; Shende, S.; Gaikwad, S.; Ingle, A.P. *Fusarium* as a Novel Fungus for the Synthesis of Nanoparticles: Mechanism and Applications. *J. Fungi* **2021**, *7*, 139. https://doi.org/10.3390/jof7020139

Academic Editor: Jeffrey J. Coleman
Received: 14 December 2020
Accepted: 10 February 2021
Published: 15 February 2021

1. Introduction

Nanotechnology is an emerging branch of science having enormous applications in almost all fields related to human life. It is mainly concerned with the synthesis and applications of materials having a size in the range of 1 to 100 nanometers [1]. Nanomaterials possess exceptionally novel properties such as a high surface-area-to-volume ratio, high reactivity, enhanced catalytic and biological properties. All these unique properties make the nanomaterials appropriate for a variety of applications including in biomedicine and agriculture [2–4]. To date, a variety of nanomaterials have been developed and many more are currently under investigation to be applied in biomedicine with the emphasis on various life-threatening diseases including cancer. Therefore, some precious metals (like silver, gold and platinum) and some magnetic oxides (i.e., magnetite Fe_3O_4) nanoparticles received much attention [5]. Similarly, various nanoparticles have been reported to have many beneficial applications in agriculture which mainly include plant growth promotion, usage as nanofertilizer, and nanopesticides [6].

It is well demonstrated that various nanoparticles can be synthesized using physical, chemical, and biological methods [7]. The conventional physical and chemical methods used for the synthesis of nanoparticles usually involve the usage of toxic chemicals and also generate waste, which can cause environmental pollution [8,9]. However, the synthesis of biogenic nanoparticles using different biological agents such as plants, microbes, and their products has gained considerable attention worldwide due to the rapid synthesis and their eco-friendly nature compared to physical and chemical methods.

Among the biological agents, fungi have been preferably used for the synthesis of a variety of nanoparticles. Synthesis of nanoparticles using fungi is referred as mycosynthesis [10–13] and is being dealt with under myconanotechnology [14]. Gade et al. [15] stated the advantages of using filamentous fungi over other biological agents (e.g., bacteria) for the synthesis of nanoparticles. These mainly include high tolerance towards heavy metals, it is easy to culture fungi at mass level, synthesis of nanoparticles is extracellular which reduces the cost of down streaming, etc.

To date, several fungi have been successfully exploited for the biological synthesis of nanoparticles, but from the available literature, it is evident that different species of *Fusarium* are the prime choice for scientists. Various species of *Fusarium* such as *Fusarium oxysporum, Fusarium semitectum, Fusarium acuminatum, Fusarium solani, Fusarium culmorum*, etc. and their different strains [16–21] have been used for the synthesis of nanoparticles like silver, gold, platinum, silica, palladium, etc. Several other fungal species could also be employed in nanoparticle synthesis as described in Mahmoud et al. [22], Elamawi et al. [23], Noor et al. [24], and many more. Considering these facts, in the present review, we have discussed the importance of *Fusarium* for biosynthesis of nanoparticles. Moreover, various other key aspects such as the mechanism of nanoparticle synthesis from *Fusarium* and their applications in biomedicine and agriculture and toxicity have also been discussed.

1.1. *Diversity of* Fusarium *spp. for the Synthesis of Different Nanopartilces*

Fungi play a very important role in solving major global problems for sustainable development as compared to other biological systems. They enhance resource efficiency, converting waste to valuable food and feed ingredients, making crop plants more robust to survive in climate change conditions, and functioning as host organisms for the production of new biological drugs [25]. Fungi are the most promising hotspots for finding new drug candidates, metabolites, and antimicrobials [8]. They are also responsible for the shift from chemical processes to biological processing, achieved by fungal enzymes instead of chemical processes in industries, such as textiles, leather, paper, and pulp, which have significantly helped to make the process eco-friendly by reducing the negative impact on the environment [16,26]. The fungal system has been found to be a versatile biological system with the ability to synthesize metal nanoparticles intracellularly as well as extracellularly. Moreover, they are preferred over other biological systems because of their ubiquitous distribution in nature, and therefore, many fungi have been explored for the production of various metal nanoparticles of different shapes and size. Out of diverse fungal genera used for the synthesis of nanoparticles, the genus *Fusarium* has been the choice of many investigators [13,16]. The advantages of using *Fusarium* spp. for the synthesis of nanoparticles are enlisted in Figure 1.

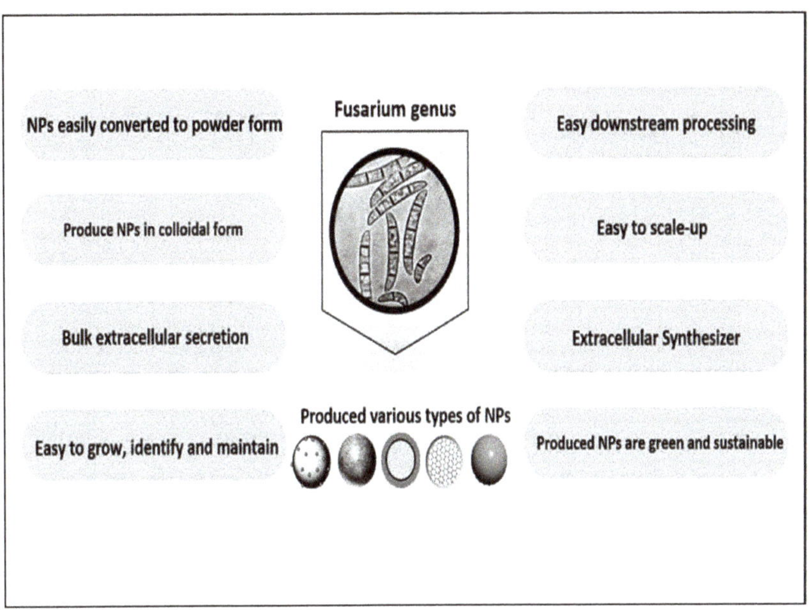

Figure 1. The advantages of using *Fusarium* spp. for the synthesis nanoparticles (NPs).

There are several reports on the synthesis of metal nanoparticles by different *Fusarium* spp. The exhaustive list of different *Fusarium* spp. involved with synthesis of different metal nanoparticles is given in Table 1.

Table 1. List of *Fusarium* spp. synthesizing metal Nanoparticles.

***Fusarium* spp.**	**Type of Nanoparticles**	**Reference**
Fusarium acuminatum	Silver,	Ingle et al. [10]
Fusarium solani	Silver	Ingle et al. [11]
Fusarium semitectum	Silver	Basavaraja et al. [19]
Fusarium acuminatum	Gold	Tidke et al. [27]
Fusarium culmorum	Silver	Bawaskar et al. [28]
Fusarium chlamydosporum NG30	Silver	Khalil et al. [29]
Fusarium equiseti, Fusarium tricinctum	Silver	Gaikwad et al. [30]
Fusarium proliferatum	Silver	Gaikwad et al. [30]
Fusarium keratoplasticum A1-3	Zinc oxide Silver	Mohamed et al. [31] Mohamed et al. [32]
Fusarium monoliforme	Silver	Gaikwad et al. [30]
Fusarium oxysporum	Silver, Gold, Lead and Cadmium Carbonate, Strontium Carbonate, Cadmium Sulfide, Silica and Titania, Silica, Barium Titanate, Zirconia Platinum Magnetite CdSe Quantum dot CdTe Quantum dot Titanium oxide Chitosan Zinc Sulfide	Birla et al. [33] Bansod et al. [34] Sanyal et al. [35] Rautaray et al. [36] Ahmad et al. [37] Bansal et al. [38] Bansal et al. [39] Bansal et al. [40] Gupta and Chundawat [41] Bharde et al. [42] Kumar et al. [43] Senapati et al. [44] Ganpathy and Siva [45] Boruah and Dutta [46] Mirzadeh et al. [47]

Table 1. *Cont.*

Fusarium spp.	Type of Nanoparticles	Reference
Fusarium oxysporum f. sp. *cubense* JT1	Gold	Thakker et al. [48]
Fusarium oxysporum 405	Silver	Rajput et al. [49]
Fusarium oxysporum f. sp. *lycopersici*	Platinum	Riddin et al. [50]
	Cadmium Sulphide	Cardenas et al. [51]
Fusarium oxysporum PTCC 5115	Silver	Karbasian et al. [52]
Fusarium graminearum	Silver	Ajah et al. [53]
Fusarium scirpi	Silver	Rodríguez-Serrano et al. [54]
Fusarium semitectum	Gold and Gold-silver alloy	Sawle et al. [55]
	Silver	Madakka et al. [56]
	Selenium	Abbas and Baker [57]
Fusarium semitectum (KSU-4)	Silver	Mahmoud et al. [58]
Fusarium solani	Zirconium Oxide	Kavitha et al. [59]
Fusarium solani ATLOY–8	Gold	Clarance et al. [60]
Fusarium verticillioides	Silver	Mekkawy Mekkawy et al. [61]

1.2. *F. oxysporum as a Novel Organism for Synthesis of Nanoparticles*

Several fungi have been used for the biosynthesis of various nanoparticles as they exhibit many advantages over other biosystems. After directing to the mycosynthesis of nanoparticles especially from *Fusarium*, nanoparticles with better size and monodispersity could be achieved. Additionally, the extracellular production of enzymes has an added benefit in the downstream handling of biomass [62] as compared to other biosystems like bacteria and plants. Consequently, using these expedient properties of *Fusarium*, it could be comprehensively used for the rapid and eco-friendly biosynthesis of nanoparticles [53]. Gaikwad and colleagues [30] have screened eleven different *Fusarium* species isolated from various infected plant materials for the synthesis of silver nanoparticles (AgNPs). All the screened species revealed the ability for synthesis of AgNPs. Based on transmission electron microscopic (TEM) analysis, six *Fusarium* species—*viz. F. graminearum, F. solani, F. oxysporum, F. culmorum, F. scirpi, F. tricinctum*—synthesized smaller-sized particles, which signifies their prominence in AgNPs synthesis (Figure 2). Moreover, AgNPs synthesis from *F. scirpi, F. graminearum, F. tricinctum* was reported for the first time.

Khalil and coworkers [29] successfully synthesized and characterized AgNPs from *F. chlamydosporum* NG30 and *P. chrysogenum* NG85 which showed promising antifungal activity. The cellular mechanism of nanoparticle synthesis is yet to be completely understood; therefore, researchers have been trying to understand the mechanism at the cellular and molecular level [63]. They have reported the synthesis of gold nanoparticles (AuNPs) using *F. oxysporum* f. sp. *cubense* JT1 in 60 min. Naimi-Shamel et al. [64] also proved that *F. oxysporum* has benefits like fast growth rate, low-cost biomass management, safety and easy processing for synthesis of AuNPs. AuNPs synthesized from *F. oxysporum* are found to have a high tendency of conjugation with β-Lactam antibiotics, and this affinity makes them a better detoxification agent as well in various areas including medicine [65]. An endophytic strain of *F. solani* isolated from *Chonemorpha fragrans* plant was used to synthesise nanoparticles with anticancer activity [60]. El-Sayed and El-Sayed [66] synthesized silver, copper and zinc biocidal nanoparticles from *F. solani* to combat multidrug-resistant pathogens. *F. oxysporum*-mediated AgNPs surface coated with different proteins and biomolecules act as a potential antimicrobial agent and proved by protein–ligand interaction in silico studies [66]. In another study by Birla et al., *F. oxysporum* produced more protein at an optimized temperature between 60° and 80 °C which showed the progressive increase in the rate of nanoparticle synthesis [33].

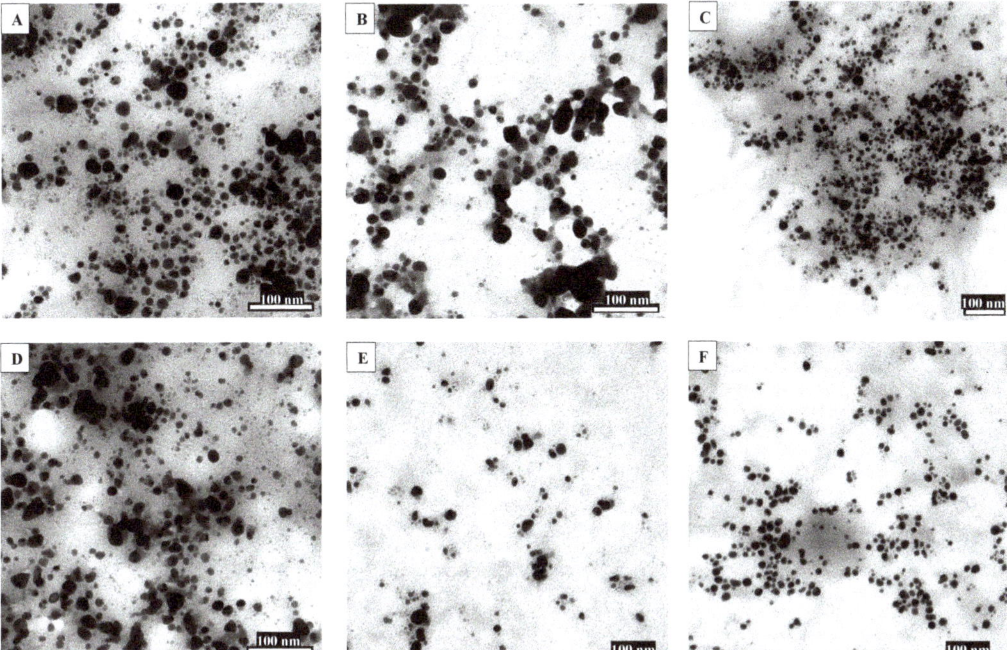

Figure 2. TEM micrographs presenting sphere-shaped AgNPs synthesized by various *Fusarium* species. (**A**) *F. graminearum;* (**B**) *F. solani;* (**C**) *F. oxysporum;* (**D**) *F. culmorum;* (**E**) *F. scirpi;* (**F**) *F. tricinctum*. (Reproduced with permission from Gaikwad et al. [30]).

2. Mechanism of Nanoparticle Synthesis from *Fusarium*

As discussed earlier, members of the genus *Fusarium* can synthesize metal nanoparticles both intracellularly and extracellularly. As far as the mechanism of extracellular mycosynthesis is concerned, it is proposed that metabolites such as enzymes, proteins, polysaccharides, flavonoids, alkaloids, phenolic and organic acids, etc. secreted by fungus-like *Fusarium* for their survival when exposed to different environmental stresses are mostly responsible for the reduction of metals ions to metallic nanoparticles through the catalytic effect [67]. Moreover, the same metabolites act as reducing and stabilizing agents which are further responsible for the growth and stabilization of biogenic metal nanoparticles [13,22]. Figure 3 represents the schematic illustration of the general mechanism involved in the synthesis, growth, and stabilization of metal nanoparticles using fungus such as *Fusarium*.

One of the hypothetical mechanisms proposed is that NADH-dependent nitrate reductase enzyme secreted by *F. oxysporum* is responsible for the reduction of aqueous silver ions into AgNPs [16]. A similar mechanism has been proposed by Ingle et al. [10] in case of synthesis of AgNPs from *F. acuminatum*, and they also pointed the involvement of cofactor NADH and nitrate reductase enzyme in the biosynthesis of AgNPs because they reported the presence of nitrate reductase in fungal cell-free extract using specific substrate utilizing discs for nitrate purchased from Hi-Media Pvt. Ltd. Mumbai, India.

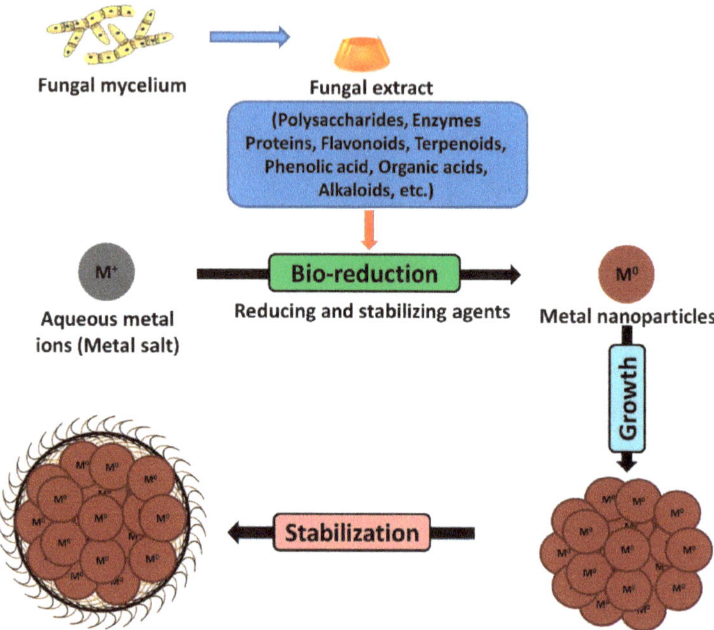

Figure 3. General mechanism involved in the synthesis, growth, and stabilization of metal nanoparticles using fungus.

In addition, Duran et al. [68] and Kumar et al. [69] proposed almost similar mechanisms for the biosynthesis of AgNPs from *F. oxysporum*. In the former study, the authors reported the role of anthraquinone and the NADPH-nitrate reductase in the biosynthesis of AgNPs, and it was hypothesized that the electron required to fulfill the deficiency of aqueous silver ions (Ag^+) and convert it into Ag neutral (Ag^0 i.e., AgNPs) was donated by both quinone and NADPH. However, in the later study, it was demonstrated that the reduction of NADPH to $NADP^+$ and the hydroxyquinoline possibly acts as an electron shuttle transferring the electron generated during the reduction of nitrate to Ag^+ ions, converting them to Ag^0 (Figure 4).

There are reports suggesting various hydroquinones to act as electron shuttles reducing the metal ions. *F. oxysporum* f. sp. *cubense* JT1 demonstrated to have the capacity to reduce the gold ions to AuNPs [48]. Moreover, as per Ahmad et al. [16], the capacity of reducing metal ions is species-specific. The reductase specific to *F. oxysporum* and *F. moniliforme* were not able to synthesize AgNPs intracellularly and extracellularly. In addition to extracellular mechanisms, there are few mechanisms proposed for intracellular mycosynthesis of metal nanoparticles. In the case of *Fusarium*-mediated intracellular mycosynthesis, metal nanoparticles usually formed below the cell surface and this may be due to the reduction of metal ions by metabolites (i.e., enzyme) present in the cell membrane. Generally, a two-step mechanism has been proposed for intracellular mycosynthesis of nanoparticles. In the first step, aqueous metal ions are attached to the fungal cell surface by the electrostatic interaction between lysine residues and metal ions (M^+). However, in the second step, the actual mycosynthesis of nanoparticles occurs by the enzymatic reduction of metal ions (M^0), which leads to the aggregation and formation of nanoparticles [49] (Figure 5).

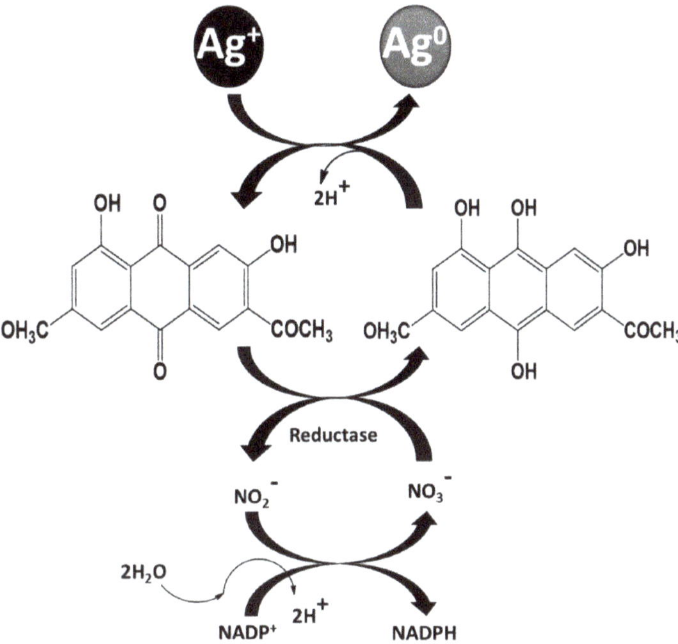

Figure 4. Hypothetical mechanism of AgNPs biosynthesis from *F. oxysporum* (Adapted from Duran et al. [67], an open-access article).

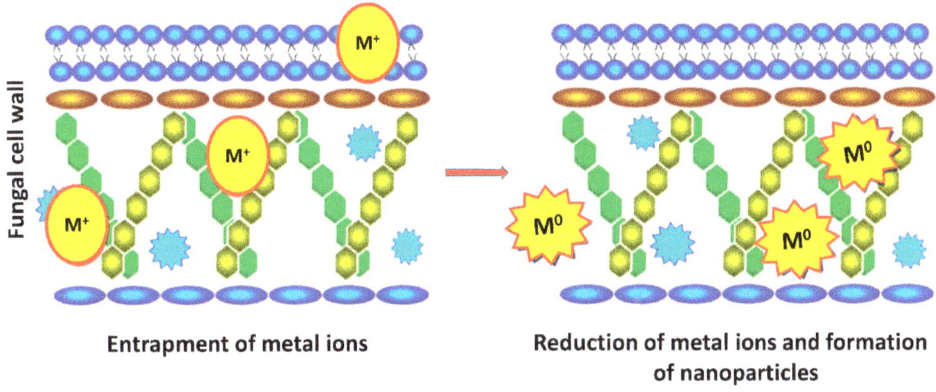

Figure 5. Hypothetical intracellular mechanisms for the synthesis of metal nanoparticles from *F. oxysporum* (Adapted and modified from Yadav et al. [13]; with copyright permission from Springer).

Moreover, various other studies performed on mycosynthesis proposed the role of different other enzymes and proteins. However, among all these mechanisms for extracellular mycosynthesis, the hypothetical mechanism involving the role of NADH-dependent nitrate reductase enzyme has been widely accepted.

2.1. Biomedical Applications of Nanoparticles Synthesized Using Fusarium Spp.

Biomedical application is an expanding field of research with tremendous prospects for the improvement of the diagnosis and treatment of human diseases [70]. The dispersed nanoparticles are usually employed in nanobiomedicine as fluorescent biological labels [71], as well as drug and gene delivery agents [72].

The biologically synthesized AgNPs could have many applications in areas such as non-linear optics, spectrally selective coating for solar energy absorption and intercalation materials for electrical batteries, as optical receptors, catalysis in chemical reactions, bio labelling [73], and as antibacterial agents [74–76]. The biomedical applications of *Fusarium*-mediated synthesized nanoparticles are shown in Figure 6.

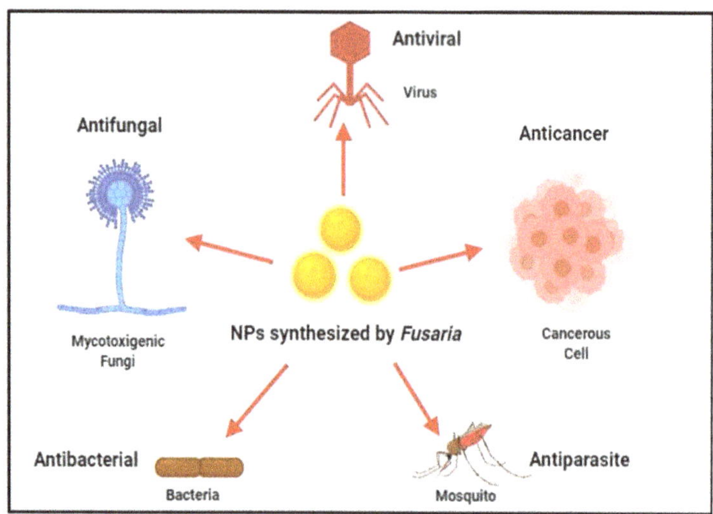

Figure 6. Biomedical applications of biogenic NPs synthesised from different *Fusaria*.

2.1.1. Antibacterial Activity of Nanoparticles

Gupta and Chundawat [41] used *Fusarium oxysporum* for the production of platinum nanoparticles. The zones of inhibition against microbes were studied by the agar well diffusion and agar dilution methods. It helps to know the minimum inhibitory concentration of platinum NPs. The minimum inhibitory concentration of platinum NPs was found to be 62.5 $\mu g\ ml^{-1}$ for *E. coli*, which is relatively better than that of commercially available drug ampicillin. Also, the antioxidant activity was studied by the α, α-diphenyl-β-picrylhydrazyl (DPPH) method and platinum nanoparticles showed 79% scavenging activity. Duran and his co-workers [68] synthesised AgNPs by *Fusarium oxysporum* and integrated into the textile fabric for the inhibition of bacterial contamination such as *Staphylococcus aureus*.

The uropathogenic *Escherichia coli* (UPEC) form biofilms. The prevalence of urinary tract infections (UTIs) is due to the inaccessibility of the antibiotics into the highly complex structure of the biofilm. However, with the appearance of antibiotic multi-resistant UPEC strains, alternatives to the treatment of UTIs are fewer. AgNPs are an effective treatment of UPEC infections due to its physicochemical properties that confer them antibacterial activity against biofilm structured cells [54].

2.1.2. Antiviral Activity of Nanoparticles

The interaction between AgNPs and viruses is attracting pronounced interest due to the potential antiviral activity of these particles. Elechiguerra et al. [77] reported that the smaller AgNPs are capable of reducing viral infectivity by inhibiting attachment to the host cells. It has been established that AgNPs undergo a size-dependent interaction with herpes simplex virus types 1 and 2, and with human parainfluenza virus type 3. Also, the authors confirmed that smaller nanoparticles were able to decrease the infectivity of the viruses [78]. Gaikwad et al. [30] synthesized AgNPs using *Fusarium oxysporum* and other fungi, which showed potential for reducing the replication of HSV-1, HSV-2, and HPIV-3 in cell cultures. The AgNPs formed by *F. oxysporum* were the most effective and presented low cytotoxicity [9].

2.1.3. Anticancer Activity of Nanoparticles

Husseiny et al. [79] reported the antibacterial and antitumor potential of AgNPs synthesized from *Fusarium oxysporum*. The nanoparticles were effective to inhibit *E. coli* and *S. aureus*, and also a tumor cell line. A low IC50 value (121.23 μg cm^{-3}) for MCF-7 cells (human breast adenocarcinoma) was gained following exposure of the cells to the nanoparticles, indicating high cytotoxicity and the potential for tumor control. The effect was recognized by the involvement of the AgNPs in the disruption of the mitochondrial respiratory chain, which led to the production of reactive oxygen species and hindered the production of adenosine triphosphate (ATP), consequently damaging the nucleic acids.

Clarance et al. [60] explored the anticancer potential of the AuNPs obtained by the green synthesis method using an endophytic strain *Fusarium solani* ATLOY—8 isolated from *Chonemorpha fragrans*. The AuNPs were tested for their cytotoxicity on cervical cancer cells (He La) and human breast cancer cells (MCF-7); the NPs exhibited dose-dependent cytotoxic effects. The results delivered an apparent and versatile biomedical application for a safer chemotherapeutic agent with little systemic toxicity.

2.1.4. Antifungal Activity of NPs

Bansod et al. [34] reported bioconjugate-nano-PCR as a rapid and specific method for the identification of *Candida* species in less time. The DNA sample of *Candida albicans* was conjugated with AuNPs and AgNPs synthesized from *F. oxysporum*. The use of this nanoparticle-altered template enhances the sensitivity and specificity of the traditional PCR assay. It is helpful in molecular diagnostics and therapeutics. It is demonstrated as an effective method for the identification of *Candida* sp. with a low concentration of DNA and less time. In another study, it was demonstrated that AgNPs synthesized by *F. oxysporum* were inhibitory to pathogenic fungi such as *Candida* and *Cryptococcus* [76].

Horky et al. [80] summarized the current findings of mycotoxins and their elimination by nanoparticles. They concluded that nanomaterials have interesting adsorption properties, which make them promising for mycotoxin elimination.

2.1.5. Antiparasitic Activity Against Vectors

Some studies showed that AgNPs synthesized from different fungal species can be employed in pest control [81]. In a study, Dhanasekaran and Thangaraj [82] evaluated the larvicidal activity of biogenically synthesized AgNPs against the larvae of *Culex* mosquito vector, which causes filariasis. The authors reported that 5 mg/L AgNPs were responsible for 100% mortality of the Culex larvae. This research has opened a new area of research that the biogenically synthesized AgNPs can be used for the control of mosquito vectors causing diseases like filaria, malaria. dengue, etc.

3. Applications in Agriculture

Agriculture plays a major role in the economy as it is the backbone of most developing countries. The worldwide population is growing day-by-day very rapidly and it is predicted that it will reach about eight billion by 2025 and 9.6 billion by 2050. It is widely recognized that global agricultural productivity must increase to feed a rapidly growing world population [83–85]. For the improvement in crop productivity, nanotechnology provides new agrochemical agents and new delivery mechanisms, and it promises to reduce pesticide use as NPs could be used in the variable applications concerned with agriculture. The applications of nanotechnology (as shown in Figure 7) can boost agricultural production, which includes the nano-formulations of agrochemicals to use as pesticides and fertilizers for crop improvement, nano-biosensors in crop protection for the identification of diseases and residues of agrochemicals, and nano-devices for the genetic manipulation of plants, etc. In agriculture, nanobiotechnology is used to improve the food production, with corresponding or even higher values of nutrition, quality and safety. Efficient application of pesticides, fertilizers, herbicides and plant growth regulators is the very critical way to get better crop production [86,87]. Nanocarriers could be used to achieve the controlled release of herbicides, pesticides, and other plant growth regulators. For example, poly (epsilon-caprolactone) nanocapsules have been recently developed as a herbicide carrier for atrazine [88]. The mustard plants (*Brassica juncea*) when treated with atrazine-loaded poly (epsilon-caprolactone) nanocapsules more significantly boosted the herbicidal activity than that of commercial atrazine, demonstrating a drastic decline in net photosynthetic rates. Moreover, the stomatal conductance and oxidative stress increased considerably, which ultimately reduced the weight and growth of the plants [88]. Likewise, other nanocarriers such as silica nanoparticles [89] as well as polymeric nanoparticles [43] have also been developed for delivering the pesticides in a prescribed manner. Nanocarriers could be employed to perfectly achieve the delivery as well as the slow release of these species, which is known as "precision farming". This helps to improve the crop yield without damaging the soil and water [90].

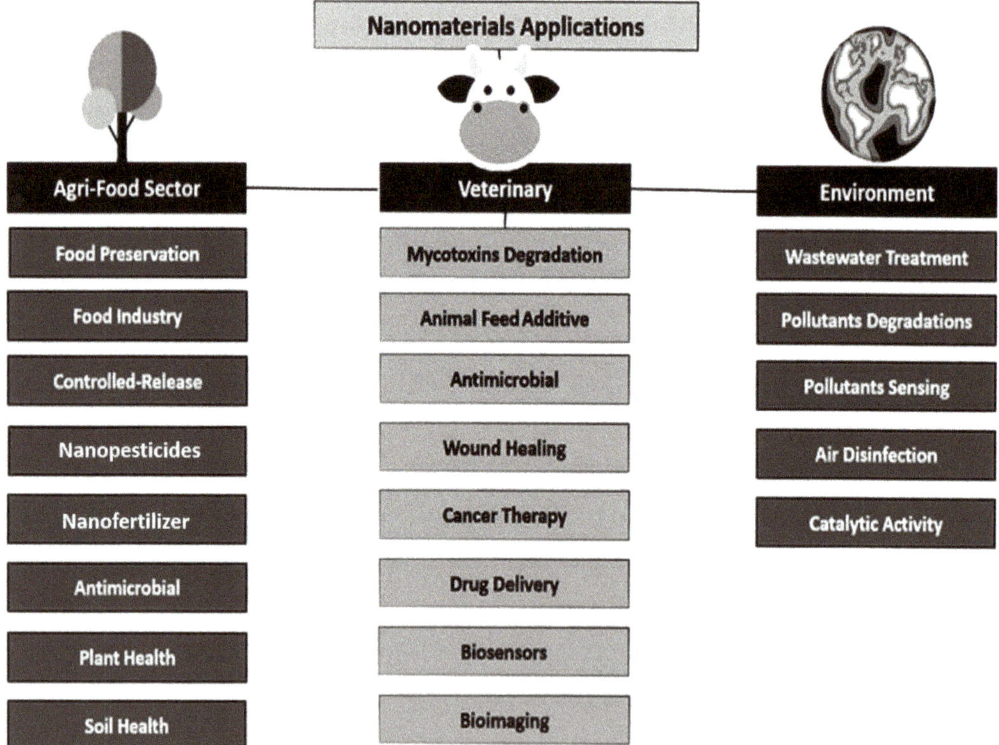

Figure 7. Applications of Nanotechnology in agri-food, veterinary medicine, and environment.

Most significantly, the application of nanoencapsulation could lower down the herbicide dosage leading to a safer environment. In addition to nanocarriers, NP-mediated gene transfer in plants was employed for the development of insect-resistant crop varieties. The detailed account for the gene or DNA transfer could be established in earlier available reviews [91,92]. Al-Askar et al. [75] demonstrated that AgNPs biosynthesized by *F. solani* isolated from wheat were shown to be effective for the treatment of wheat, barley, and maize seeds contaminated by different species of phytopathogenic fungi. Moreover, metal oxide nanomaterials such as CuO, TiO_2, and ZnO are extensively studied for plant protection from pathogen infections because of their intrinsic toxicity. For example, ZnONPs efficiently inhibited fungal growth such as *F. graminearum* [93], *Aspergillus flavus, Aspergillus fumigatus, Aspergillus niger, F. culmorum* and *F. oxysporum* [94]. The use of CuNPs as an antimicrobial agent against plant pathogens has been reported in several publications [95,96]. Mineral fertilizers used conventionally undergo substantially high losses besides lower uptake efficiencies of the nutrients. Those economic losses will be overcome by the development of nanofertilizers, which could be the novel solution. Nanofertilizers can reduce the nutrient loss as well as increasing nutrient adaptation by soil microbes and crops [97]. Nanofertilizers are mainly the micro-nutrients at nanoscale for Mn, Cu, Fe, Zn, Mo, N, and B, and are commercialized and available under different brand names in the market such as Nano-Ag Answer®, NanoPro™, NanoRise™, NanoGro™, NanoPhos™, NanoK™, NanoPack™, NanoStress™, NanoZn™, pH5®, etc. [87]. The use of other nanomaterials as an alternative for the typical conventional crop fertilizers, such as carbon nano-onions [98] and chitosan nanoparticles [99], was noted to boost the crop growth and quality. Shende et al. [100] reported the plant growth-promoting activity of biogenic CuNPs on pigeon pea (*Cajanus cajan* L.) crops that suggested the use of these nanoparticles as a nanofertilizer for the development of sustainable agriculture. It is estimated that the novel nanofertilizers will encourage and makeover current fertilizer production industries in the next decade [87].

Because of several advantageous characteristics of nanomaterials, nanosensors, particularly wireless nanosensors, have also been developed to monitor nutrient efficiency in crop plants, crop diseases, and growth, along with the environmental conditions in the field. Particularly, engineered nanosensors are capable of detecting chemicals like pesticides, herbicides, and pathogens at trace amounts in food and agricultural systems. Such an in situ and real-time monitoring system facilitates the remediation of probable crop losses as well as perking up the crop production, accompanied by the suitable application of nanopesticides, nanoherbicides, and nanofertilizers. Abbacia et al. [101] reported that the copper-doped montmorillonite will possibly be used for on-line monitoring of propineb fungicide in an aquatic environment i.e., in both fresh and salty water, with a low detection limit of about 1 mM [101]. In another study, it was demonstrated that the nanomaterials such as graphene could be developed for the detection of the pathogen in wastewater [102] and purification of drinking water can be carried out [103], signifying its potential application in aquaculture. Moreover, various other nanomaterials like carbon nanotube [104], CuNPs [105], AgNPs [106], and AuNPs [107] can be used as nanosensors designed for the real-time monitoring of crop health and growth along with the environmental conditions in the field. Table 2 showed different *Fusarium* isolates used in the green synthesis of nanoparticles.

Table 2. *Fusarium* isolates used in the green synthesis of different NPs.

Name of Fungi	Synthesised NPs	Localization	Size (in nm) and Shape	References
Fusarium oxysporum	Ag	Extracellular	10–20 and Spherical	Birla et al. [33]
			25–50 and almost spherical	Korbekandi et al. [108]
			-	Ishida et al. [76]
			5–13 and Spherical	Husseiny et al. [79]
			23 and Spherical	Hamedi et al. [109]
Fusarium verticillioides	Ag	Extracellular	-	Mekkawy et al. [61]
Fusarium semitectum	Au	-	25 and Spherical	Sawle et al. [54]
Fusarium oxysporum	Au	-	2–50 and Spherical, monodispered	Zhang et al. [110]
Fusarium oxysporum	Au-Ag bimetallic	Extracellular	8–14 and Quasi-spherical	Senapati et al. [111]
Fusarium semitectum	Au-Ag alloy	-	25 and Spherical	Sawle et al. [55]
Fusarium oxysporum	Fe_3O_4	Extracellular	20–50 and Irregular, quasi-spherical	Bharde et al. [42]

Table 2. *Cont.*

Name of Fungi	Synthesised NPs	Localization	Size (in nm) and Shape	References
Fusarium oxysporum	Pt	-	70–180 and Rectangular, triangular, spherical and aggregates	Govender et al. [112]
Fusarium oxysporum f. sp. *lycopersici*	Pt	Extra-and intracellular	10–100 and Hexagonal, pentagonal, circular, squares, rectangles	Riddin et al. [50]
Fusarium spp.	Zn	Intracellular	100–200 and Irregular, some spherical	Velmurugan et al. [113]

4. Toxicity of *Fusarium* Nanoparticles

Nanoparticles are unique materials as they have property combinations compared with conventional materials [114]. There is a wide range of applications of NPs such as in human health appliances, industrial, medical and biomedical fields, engineering, electronics, and environmental applications [115]. Among all nanomaterials, AgNPs are the most widely used in medicine, medicinal devices, pharmacology, biotechnology, electronics, engineering, energy, magnetic fields, and also in environmental remediation [116]. Their highly effective antibacterial activity has found applications in industrial sectors including textiles, food, consumer products, medicine, etc. [117].

The unique physical, chemical and biological (e.g., antimicrobial, anticancer, antiparasitic) properties of nanoparticles differ largely from corresponding bulk materials and make them a high-demand material in different sectors. However, the widespread and increased use of nanoparticles may pose a risk to both the environment and living organisms by increasing the level of toxicity [118]. To date, several studies have used different model cell lines to exhibit the cytotoxicity of nanoparticles from *Fusarium* species, mainly AgNPs, and their underlying molecular mechanisms [31,60,65,119]. Biogenic nanoparticles are capped with natural molecules like proteins [120–122]. This capping is defined as corona. This nanoparticle corona significantly affects the biological response [123]. Based on the surface affinity and exchange rate, the corona can be divided into two forms: hard corona and soft corona. The soft corona proteins are 'vehicles' for the silver ions, whereas the hard coronas are rigid for the trespass into the cellular system [124]. The functional groups of the corona play a key role in the formation of the nanoparticle–protein corona [125]. These functional groups along with the protein charges also regulate the cytotoxic properties of the nanoparticle corona [123,125]. The surface charge of nanoparticles plays an important role in their bactericidal activity against both Gram-positive and Gram-negative bacteria, as exemplified by AgNPs [126]. It was confirmed using transmission electron microscopy (TEM) that AgNPs can penetrate cellular compartments such as endosomes, lysosomes, and mitochondria [127]. Several reports indicate that the proteins contained in the nanoparticle corona interact with the cells and not the nanoparticles themselves [128,129]. Thus, the corona formation and composition have important implications for both toxicity [130] and internalization [131].

To sum up, particle size [132], particle shape [133], particle surface properties [134], biological fluid properties, and composition affect the corona structure and thus the adverse effects on human health and the environment [131,135].

4.1. *Effect of Size and Shape of Nanoparticles on Cytotoxicity*

It is claimed that apart from the size, the shape of nanoparticles affects their toxicity to cells. Mohamed et al. [31] investigated the cytotoxicity of the two different shapes of zinc oxide nanoparticles (ZnONPs) biosynthesized from *Fusarium keratoplasticum* (A1-3) and *Aspergillus niger* (G3-1). These nanoparticles displayed a similar size (10–42 and 8–38 nm) but different shapes, namely hexagonal and nanorods, respectively. It was reported that a safe dose of these nanoparticles for applications in animal cells should be lower than 20.1 and 57.6 ppm, respectively. Therefore, the rod ZnONPs were more applicable to safety at high concentrations in contrast to hexagonal ZnONPs [31]. Soleimani and co-authors [136] studied biological activity of different shapes (cube, sphere, rice and rod) of AgNPs synthesized using chitosan in acetic acid solution and 0.2 M of $AgNO_3$ (cubes) and *F. oxysporum*, starch and 0.08 M of $AgNO_3$ at pH 6.8 (spheric) or 1.0 M of $AgNO_3$ at pH 3.0 (nanorice), and finally, *F. oxysporum*, starch and 1.2 M $AgNO_3$ at pH 3.0 and 30 °C for 3 days (blunt ends rods) or with 10-days incubation (sharp-ends rods). They found that silver nanostructures with different shapes are not inherently toxic to human cells at concentrations lower than 10 $\mu g\ ml^{-1}$, whereas for higher concentrations cell viability decreased in a shape and dose-dependent manner. Nanocubes, nanorice and sharp-nanorods were found to be more toxic than spheres and blunt-nanorods. The former significantly decreased cell viability at concentrations 25 $\mu g\ mL^{-1}$ or higher, and the latter at concentrations of 50 and 75 $\mu g\ mL^{-1}$. Moreover, they showed that cubic nanoparticles inhibited the growth of all tested bacteria (*Bacillus subtilis, Staphylococcus aureus, Escherichia coli,* and *Pseudomonas aeruginosa*) at the lowest-used concentrations (10 ppm). The toxic effect of nanostructures against bacterial cells increased in order spheres, blunt-rods, nanorice, sharp-rods, and nanocubes [136]. It was concluded that due to stronger vertex in sharp ends of silver nanostructures, some geometries of AgNPs, especially cubic structures, have more interaction with the bacterial cell membrane, resulting in stronger biological activity [136]

4.2. Mechanisms of Nanoparticle Cytotoxicity

The increased use of AgNPs also increases the concentration of silver ions in soil and water that can adversely affect human and animal health or the environment [137,138].

Little is known about the diversified mechanisms of the action of nanoparticle cytotoxicity, as well as their short- or long-term exposure outcomes on human physiology [139,140]. Many authors claim that the silver ions released from AgNPs through the surface oxidation are the main mechanism that induces cytotoxicity, genotoxicity, immunological responses in biological systems and even cell death [141–145].

Several recent studies revealed that cytotoxicity of AgNPs occur due to the minimum release of silver ions [146–148]. Sambale et al. [149] showed that the mechanism of cell death is different when cells are cultivated with silver particles or ions. Cells treated with particles suffered apoptosis while those cultivated with ions die due to necrosis. They also reported that AgNPs that translocate into cytoplasm through diffusion or channel proteins are oxidized by cytoplasmic enzymes, thereby releasing silver ions. These ions may interact with thiol groups of mitochondrial membrane proteins, causing mitochondrial dysfunction and generating reactive oxygen species (ROS) production [149].

4.3. Effect of Nanoparticles on Cell Membranes

Nanosilver can interact with cellular membranes and cause toxicity. In particular, nanosilver can interact with the bacterial membrane and this is considered the main mechanism for the antimicrobial toxicity of nanosilver. The AgNPs damaged and destroyed bacterial cells by penetrating and accumulating in the bacterial membrane [150–152]. Khan et al. [153] studied the interaction of nanosilver with five types of bacteria. They found that the adsorption of nanosilver on the bacterial surface or interaction with extracellular proteins is dependent on pH, zeta potential, and NaCl concentration. Baruwati et al. [154] studied "green" synthesized nanosilver from Chinese green tea (*Camellia sinensis*) and found that exposure to these particles alters the membrane permeability of barrier cells (intestinal, brain endothelial) and stimulates oxidative stress pathways in neurons [154].

Metal nanoparticles (copper, silver and zinc oxide) from *Fusarium solani* KJ 623702 were tested against multidrug-resistant bacteria (*P. aeruginosa* and *S. aureus*), *E. coli*, *Klebsiella pneumoniae* and *Entreococcus* sp. and mycotoxigenic *Aspergillus awamori, A. fumigatus* and *F. oxysporum*. The antimicrobial activity of metal nanoparticles increased as follows: CuNPs, ZnONPs, and AgNPs. Besides the internalization of AgNPs with the cell wall resulting in the presence of pits, the fragmentation, complete disappearance of cellular contents, disorganization and leakage of internal components were observed after treatment of microorganisms with mycogenic AgNPs. Authors claimed that AgNPs attach to the microbial cell membrane and alter its structure, transport activity, penetrability, prompt neutralization of the surface electric charge and produce cracks and pits through which internal cell contents are effluxed [44].

Moreover, AgNPs synthesized using *F. oxysporum* filtrate containing reductase enzymes and quinones showed a toxic effect on the human liver cell line (Huh-7). The IC_{50} values were found to be 11.12 μM for propidium iodide assay which is an intercalating fluorescent dye and does not permeant to live cells [155]. Similarly, cell viability (hamster lung fibroblasts, V79) after the treatment with AgNPs synthesized from *F. oxysporum* was studied using neutral red uptake (NRU) assay [156]. These small (around 8 nm) nanoparticles showed a non-cytotoxic effect on fibroblasts at concentrations until 16 μM. The IC_{50} of mycogenic AgNPs was recorded at a concentration of 22 μM [156].

Vijayan et al. [157] estimated the release of haemoglobin from red blood cells due to the disruption of the cell membrane by the AgNPs from *F. oxysporum*. The percentage of haemolysis increased from 3.8 to 32.0 with an increase in AgNPs concentration from 25 to 150 μg ml^{-1} (at 25 μg ml^{-1} interval unit). The authors concluded that these biogenic AgNPs are safe to use as a drug because at safe concentrations they showed antibacterial effect against *Escherichia coli, Klebsiella pneumoniae, Pseudomonas aeruginosa* and *Salmonella typhi* [157].

4.4. *Effect of Nanoparticles on Mitochondria and/or Metabolic Activity*

Nanoparticles by mitochondrial membrane damage disturb respiratory chain activity and ATP synthesis may generate ROS production, leading to oxidative stress and eventually apoptosis [158,159]. Many authors reported the effect of metal (Ag and Au) and non-metal (selenium) nanoparticles from *Fusarium* species on the viability of human or animal cells on the basement of activity of cell/mitochondrial enzymes using mainly spectrophotometrical MTT (3-(4,5-dimethylthiazol-2-yl)-2,5-diphenyltetrazolium bromide) assay [155,156,160,161].

The toxicity of AuNPs produced by *F. oxysporum* with sizes of around 20 to 50 nm was measured by MTT assay [161]. The biologically produced AuNPs had a dose-dependent toxic effect on the human fibroblast cell line (CIRC-HLF). The IC_{50} value of the NPs was determined as 2.5 mg mL^{-1} [161]. The AuNPs from *Fusarium solani* ATLOY-8 strain were studied against cancer cell lines, namely HeLa na MCF-7 (breast cancer cells) and the human embryonic kidney (HEK) cell line using MTT assay. The anticancer potential of NPs against both cancer cell lines (IC_{50} values of 1.3 and 0.8 mg mL^{-1}, respectively) and their insignificant activity on the HEK cell line was reported [60].

AgNPs from *Fusarium oxysporum* were studied on mouse fibroblasts (3T3), hamster lung fibroblasts (V79 line), and human liver (Huh-7) cells using MTT and calcein assays [155,156,161]. AgNPs showed dose-dependent toxic effects on the mouse fibroblast. IC_{50} of mycogenic nanoparticles using MTT assay were found to be 0.312 mg mL^{-1} (= 1.187 ppm mL^{1}). The applied concentrations of the silver NPs equal to 0.625 mg mL^{-1} (2.375 ppm mL^{-1}) was determined as the toxic dose [60]. Marcato and coauthors [156] reported that AgNPs synthesized using *Fusarium oxysporum* showed a non-cytotoxic effect toward the V79 fibroblast cell line until 16 μM, evaluated by MTT assays with an IC_{50} of 22 μM; this was similar in NRU assay. AgNPs synthesized using *F. oxysporum* filtrate demonstrated a comparable dose-response relationship and similar IC_{50} values of 10.61 μM for MTT and 9.10 μM for calcein assay [155]. Small AgNPs (5–13 nm) produced by *F. oxysporum* were also found to be toxic to the human breast carcinoma cell line (MCF-7). The IC_{50} value of these NPs was found to be 121.23 μg mL^{-1} [78].

Another species of the genus *Fusarium*, namely *F. semitectum*, was used for the synthesis of selenium nanoparticles (SeNPs) which were tested for their anticancer potential on Caco-2 human colon cancer, A431 skin cancer, and SNU16 stomach gastric cancer cell lines, as well as toward THLE2 normal liver and Vero normal kidney cell lines [57]. These biogenic SeNPs showed anticancer potential toward colon, skin, and stomach gastric cancer cells (IC_{50} of 10.24, 13.27 and 20.44 μg mL^{-1}, respectively), no cytotoxic effects on normal liver cells, and weak toxicity on normal kidney cells [57].

4.5. Effect of Nanoparticles on Cell Proteins

Many authors have reported that silver ions and AgNPs can interact with various chemical groups, including sulfide and chloride [162,163]. Thiol molecules are found to be conjugated to several membrane proteins in the cell membrane, cytoplasm and mitochondria, which may serve as targets for AgNPs or Ag^+ ions [164]. AgNPs also bind to thiol groups in enzymes, such as NADH dehydrogenase, and disrupt the respiratory chain, finally generating ROS. As shown in studies by Soleimani et al. [136] the cubic AgNPs from *F. oxysporum* at a concentration of 30 ppm not only caused a toxic effect in MCF-7 cells and inhibited growth of the Gram-positive and Gram-negative bacteria, but also completely degraded proteins (albumin), whereas other shaped nanoparticles (rods, rice and spheres) had no denaturing effect on protein structure. They concluded that cubic nanosilver displays stronger biological activity when compared to rod and spherical AgNPs [136].

4.6. Effect of Nanoparticles on Cell Nucleic Acids

AgNPs, especially those smaller than 10 nm, have been shown to diffuse through the nuclear pores into the nucleus, causing DNA damage, chromosomal aberrations, and cell cycle arrest, resulting in genotoxicity in human cell lines (e.g., fibroblasts and glioblastoma cells) [162,163].

Clarance et al. [60] studied the effect of mycogenic AuNPs on cancer cells using dual AO/EtBr (acridin orange/ethidium bromide) staining. The cell line appeared as a dense red colour after treatment with nanogold, which was mainly due to apoptotic cell death. The DAPI staining revealed nuclear fragmentation and condensation in MCF-7 cells treated with AuNPs. It is well-proven that when cells undergo an apoptotic death, their DNA becomes dense, fragmented and with condensed chromatin [165,166].

It was also reported that the interaction of AgNPs synthesized from *Fusarium mangiferae* with DNA in *Staphylococcus aureus* cells resulted in an enormous reduction or degradation of both chromosomal and plasmid DNA in bacterial cells when compared with untreated cells. This gel electrophoresis study of AgNPs-treated nucleic acids also revealed that the RNA smearing was immensely decreased. These results suggest that AgNPs may prevent DNA replication via binding, and lead to cell death [167].

Spherical AgNPs synthesized by using aqueous extracts of green *Calligonum comosum* stem and *Fusarium* sp. (mean size of 105.8 and 228.4 nm, respectively) were tested for genotoxicity in bacterial cells (*S. aureus*). The agarose gel electrophoresis and (UV)–Vis spectrophotometer were used to evaluate the antimicrobial activity of AgNPs by their influence on DNA. A low concentration of DNA was observed in bacterial cells treated with plant-synthesized and mycosynthesized AgNPs when compared to untreated cells. Based on the obtained results, it was concluded that the smaller-sized nanoparticles diffuse more easily than the larger ones, which partially explained the higher toxicity of plant-synthesized than myco-synthesized nanoparticles on *S. aureus* cells. The destruction of the membrane integrity or inhibition of DNA replication were proposed by authors as AgNPs mechanisms of antibacterial action [58].

4.7. Other Nanoparticle Activity

Salaheldin et al. [119] reported that AgNPs synthesized by *F. oxysporum* caused remarkable vacuolation in the human breast carcinoma cell line (MCF-7), thus indicating potent cytotoxic activity. Al-Sharqi [168] tested these mycogenic AgNPs on mice for 21 days and showed dilation in the collecting tubule and dilation in the renal corpuscle with haemorrhage in the interstitial space between the tubules. Based on performed studies, the authors conclude that AgNPs can enter and translocate within the cell, and that the size of the AgNPs varies with its toxic effects on the cell and the cell organelles. Therefore, it was assumed that all nanoparticles are toxic and most likely only free nanoparticles that can penetrate small organelles such as mitochondria may trigger adverse health effects [168].

5. Conclusions

The potential of applications of nanotechnology particularly as nanomedicine and in agriculture has generated revolution. The bioinspired synthesis of nanoparticles using fungi is a novel and emerging field of green and sustainable nanotechnology. Among these, *Fusarium* spp. are the most studied fungi for the biosynthesis of nanoparticles. The process is green, eco-friendly, and economically viable. The applications of different nanoparticles as antimicrobials, and antiparasitic and anticancer agents have attracted the interest of the researchers globally. Moreover, these biogenic nanoparticles, synthesized by Fusaria, have huge potential in agriculture for plant growth promotion, as nanofertilisers and as fungicidal agents of the new generation. However, the mechanism of the synthesis of nanoparticles remains unclear and warrants further studies to unravel the mechanism. Looking at the potential applications of Fusarium-mediated nanoparticles, the toxicity is a major issue that depends upon shape, size, surface charge, and the dose of nanoparticles used. Thus, it can be recommended that *Fusarium* is a promising and novel fungus for the biosynthesis of nanoparticles and its potential biomedical and agricultural applications.

Author Contributions: The idea was conceived and designed by M.R. and J.T.-W. The review was written by S.B., P.G., A.G., K.A.A.-E., S.S., S.G. and A.P.I. Furthermore, M.R. supervised and edited the manuscript.

Funding: This work was supported additionally through grants from NAWA Programme, Polan.

Institutional Review Board Statement: Not applicable.

Informed Consent Statement: Not applicable.

Data Availability Statement: Not applicable.

Acknowledgments: M.R. and P.G. would like to thank NAWA Programme, Poland for financial support under the grant PPN/ULM/2019/1/00117/DEC/1 2019-10-02.

Conflicts of Interest: The authors have no relevant affiliations or financial involvement with any organization or entity with a financial interest in or financial conflict with the subject matter or materials discussed in the manuscript. This includes employment, consultancies, honoraria, stock ownership or options, expert testimony, grants or patents received or pending, or royalties.

References

1. Ranjani, S.; Shariq, A.M.; Adnan, M.; Senthil-Kumar, N.; Ruckmani, K.; Hemalatha, S. Synthesis, characterization and applications of endophytic fungal nanoparticles. *Inorg. Nano Metal Chem.* **2020**, *51*, 280–287. [CrossRef]
2. Upadhyay, P.K.; Jain, V.K.; Sharma, K.; Sharma, R. Synthesis and applications of ZnO nanoparticles in biomedicine. *Res. J. Pharm. Technol.* **2020**, *13*, 1636. [CrossRef]
3. Usman, M.; Farooq, M.; Wakeel, A.; Nawaz, A.; Cheema, S.A.; Rehman, H.U.; Ashraf, I.; Sanaullah, M. Nanotechnology in agriculture: Current status, challenges and future opportunities. *Sci. Total. Environ.* **2020**, *721*, 137778. [CrossRef] [PubMed]
4. Yaqoob, A.A.; Ahmad, H.; Parveen, T.; Ahmad, A.; Oves, M.; Ismail, I.M.I.; Qari, H.A.; Umar, K.; Ibrahim, M.N.M. Recent advances in metal decorated nanomaterials and their various biological applications: A Review. *Front. Chem.* **2020**, *8*, 341. [CrossRef]
5. Bououdina, M.; Rashdan, S.; Bobet, J.L.; Ichiyanagi, Y. Nanomaterials for biomedical applications: Synthesis, Characterization, and applications. *J. Nanomater.* **2013**, *2013*, 1–4. [CrossRef]
6. Paramo, L.A.; Feregrino-Pérez, A.A.; Guevara, R.; Mendoza, S.; Esquivel, K. Nanoparticles in agroindustry: Applications, toxicity, challenges, and trends. *Nanomaterials* **2020**, *10*, 1654. [CrossRef]
7. Ramos, M.M.; Morais, E.D.S.; Sena, I.D.S.; Lima, A.L.; De Oliveira, F.R.; De Freitas, C.M.; Fernandes, C.P.; De Carvalho, J.C.T.; Ferreira, I.M. Silver nanoparticle from whole cells of the fungi *Trichoderma* spp. isolated from Brazilian Amazon. *Biotechnol. Lett.* **2020**, *42*, 833–843. [CrossRef]
8. Ahmed, S.; Ahmad, M.; Swami, B.L.; Ikram, S. A review on plants extract mediated synthesis of silver nanoparticles for antimicrobial applications: A green expertise. *J. Adv. Res.* **2016**, *7*, 17–28. [CrossRef] [PubMed]
9. Guilger-Casagrande, M.; De Lima, R. Synthesis of silver nanoparticles mediated by fungi: A Review. *Front. Bioeng. Biotechnol.* **2019**, *7*, 287. [CrossRef]
10. Ingle, A.; Gade, A.; Pierrat, S.; Sonnichsen, C.; Rai, M. Mycosynthesis of silver nanoparticles using the fungus *Fusarium acuminatum* and its activity against some human pathogenic bacteria. *Curr. Nanosci.* **2008**, *4*, 141–144. [CrossRef]

11. Ingle, A.; Rai, M.; Gade, A.; Bawaskar, M. *Fusarium solani*: A novel biological agent for the extracellular synthesis of silver nanoparticles. *J. Nanopart. Res.* **2009**, *11*, 2079–2085. [CrossRef]
12. Rai, M.; Yadav, A.; Gade, A. Silver nanoparticles as a new generation of antimicrobials. *Biotechnol. Adv.* **2009**, *27*, 76–83. [CrossRef]
13. Yadav, A.; Kon, K.; Kratosova, G.; Duran, N.; Ingle, A.P.; Rai, M. Fungi as an efficient mycosystem for the synthesis of metal nanoparticles: Progress and key aspects of research. *Biotechnol. Lett.* **2015**, *37*, 2099–2120. [CrossRef]
14. Rai, M.; Yadav, A.; Bridge, P.; Gade, A. Myconanotechnology: A new and emerging science. In *Applied Mycology*; Rai, M.K., Bridge, P.D., Eds.; CABI: Wallingford, UK, 2009; pp. 267–285. [CrossRef]
15. Gade, A.; Ingle, A.; Whiteley, C.; Rai, M. Mycogenic metal nanoparticles: Progress and applications. *Biotechnol. Lett.* **2010**, *32*, 593–600. [CrossRef] [PubMed]
16. Ahmad, A.; Mukherjee, P.; Senapati, S.; Mandal, D.; Khan, M.I.; Kumar, R.; Sastry, M. Extracellular biosynthesis of silver nanoparticles using the fungus *Fusarium oxysporum*. *Coll. Surf. B Biointerfaces* **2003**, *28*, 313–318. [CrossRef]
17. Bansal, V.; Rautaray, D.; Ahmad, A.; Sastry, M. Biosynthesis of zirconia nanoparticles using the fungus *Fusarium oxysporum*. *J. Mater. Chem.* **2004**, *14*, 3303–3305. [CrossRef]
18. Reyes, L.R.; Gómez, I.; Garza, M.T. Biosynthesis of cadmium sulfide nanoparticles by the fungi *Fusarium* sp. *Int. J. Green Nanotechnol. Biomed.* **2009**, *1*, B90–B95. [CrossRef]
19. Basavaraja, S.; Balaji, S.D.; Lagashetty, A.; Rajasab, A.H.; Venkataraman, A. Extracellular biosynthesis of silver nanoparticles using the fungus *Fusarium semitectum*. *Mater. Res. Bull.* **2008**, *43*, 1164–1170. [CrossRef]
20. Siddiqi, K.S.; Husen, A. Fabrication of metal nanoparticles from fungi and metal salts: Scope and application. *Nanoscale Res. Lett.* **2016**, *11*, 98. [CrossRef]
21. Khan, N.T.; Jameel, M.; Jameel, J. Silver nanoparticles biosynthesis by *Fusarium oxysporum* and determination of its antimicrobial potency. *J. Nanomed. Biotherapeutic. Discov.* **2017**, *7*, 1. [CrossRef]
22. Mahmoud, M.A.; Al-Sohaibani, S.A.; Al-Othman, M.R.; El-Aziz, A.M.A.; Eifan, S.A. Synthesis of extracellular silver nanoparticles using *Fusarium semitectum* (KSU-4) isolated from Saudi Arabia. *Dig. J. Nanomat. Biostruct.* **2013**, *8*, 589–596. [CrossRef]
23. Elamawi, R.M.; Al-Harbi, R.E.; Hendi, A.A. Biosynthesis and characterization of silver nanoparticles using *Trichoderma longibrachiatum* and their effect on phytopathogenic fungi. *Egypt. J. Biol. Pest Control* **2018**, *28*, 28. [CrossRef]
24. Noor, S.; Shah, Z.; Javed, A.; Ali, A.; Hussain, S.B.; Zafar, S.; Ali, H.; Muhammad, S.A. A fungal based synthesis method for copper nanoparticles with the determination of anticancer, antidiabetic and antibacterial activities. *J. Microbiol. Methods* **2020**, *174*, 105966. [CrossRef]
25. Lange, K.; Swift, P.G.F.; Pankowska, E.; Danne, T. Diabetes education in children and adolescents. *Pediatric Diabetes* **2014**, *15*, 77–85. [CrossRef]
26. Solomon, L.; Tomii, V. Dick Importance of fungi in the petroleum, agro-allied, agriculture and pharmaceutical industries. *N. Y. Sci. J.* **2019**, *12*, 5. [CrossRef]
27. Tidke, P.R.; Gupta, I.; Gade, A.K.; Rai, M. Fungus-mediated synthesis of gold nanoparticles and standardization of parameters for its biosynthesis. *IEEE Trans. NanoBiosci.* **2014**, *13*, 397–402. [CrossRef]
28. Bawaskar, M.; Gaikwad, S.; Ingle, A.; Rathod, D.; Gade, A.; Durán, N.; Marcato, P.D.; Rai, M. A New Report on mycosynthesis of silver nanoparticles by *Fusarium culmorum*. *Curr. Nanosci.* **2010**, *6*, 376–380. [CrossRef]
29. Khalil, N.M.; El-Ghany, M.N.A.; Rodríguez-Couto, S. Antifungal and anti-mycotoxin efficacy of biogenic silver nanoparticles produced by *Fusarium chlamydosporum* and *Penicillium chrysogenum* at non-cytotoxic doses. *Chemosphere* **2019**, *218*, 477–486. [CrossRef] [PubMed]
30. Gaikwad, S.C.; Birla, S.S.; Ingle, A.P.; Gade, A.K.; Marcato, P.D.; Rai, M.K.; Durán, N. Screening of different *Fusarium* species to select potential species for the synthesis of silver nanoparticles. *J. Braz. Chem. Soc.* **2013**, *24*, 1974–1982. [CrossRef]
31. Mohamed, A.A.; Fouda, A.; Abdel-Rahman, M.A.; Hassan, S.E.-D.; El-Gamal, M.S.; Salem, S.S.; Shaheen, T.I. Fungal strain impacts the shape, bioactivity and multifunctional properties of green synthesized zinc oxide nanoparticles. *Biocatal. Agric. Biotechnol.* **2019**, *19*, 101103. [CrossRef]
32. Mohamed, A.A.; Fouda, A.; Elgamal, M.S.; Hassan, S.; Shaheen, T.I.; Salem, S.S. Enhancing of cotton fabric antibacterial properties by silver nanoparticles synthesized by new egyptian strain *Fusarium keratoplasticum* A1–3. In *Egyptian Journal of Chemistry 60, Proceedings of the 8th International Conference of the Textile Research Division (ICTRD 2017), Cairo, Egypt, 25–27 September 2017*; National Research Centre: Dokki, Egypt, 2017; pp. 63–71. [CrossRef]
33. Birla, S.S.; Gaikwad, S.C.; Gade, A.K.; Rai, M.K. Rapid Synthesis of silver nanoparticles from *Fusarium oxysporum* by optimizing physicocultural conditions. *Sci. World J.* **2013**, *2013*, 1–12. [CrossRef]
34. Bansod, S.; Bonde, S.; Tiwari, V.; Bawaskar, M.; Deshmukh, S.; Gaikwad, S.; Gade, A.; Rai, M. Bioconjugation of gold and silver nanoparticles synthesized by *Fusarium oxysporum* and their use in rapid identification of *Candida* species by using bioconjugate-nano-polymerase chain reaction. *J. Biomed. Nanotechnol.* **2013**, *9*, 1962–1971. [CrossRef]
35. Sanyal, A.; Rautaray, D.; Bansal, V.; Ahmad, A.; Sastry, M. Heavy-metal remediation by a fungus as a means of production of lead and cadmium carbonate crystals. *Langmuir* **2005**, *21*, 7220–7224. [CrossRef]
36. Rautaray, D.; Sanyal, A.; Adyanthaya, S.D.; Ahmad, A.; Sastry, M. Biological synthesis of strontium carbonate crystals using the fungus *Fusarium oxysporum*. *Langmuir* **2004**, *20*, 6827–6833. [CrossRef]
37. Ahmad, A.; Mukherjee, P.; Mandal, D.; Senapati, S.; Khan, M.I.; Kumar, R.; Sastry, M. Enzyme mediated extracellular synthesis of CdS nanoparticles by the fungus, *Fusarium oxysporum*. *J. Am. Chem. Soc.* **2002**, *124*, 12108–12109. [CrossRef] [PubMed]

38. Bansal, V.; Rautaray, D.; Bharde, A.; Ahire, K.; Sanyal, A.; Ahmad, A.; Sastry, M. Fungus-mediated biosynthesis of silica and titania particles. *J. Mater. Chem.* **2005**, *15*, 2583–2589. [CrossRef]
39. Bansal, V.; Poddar, P.; Ahmad, A.A.; Sastry, M. Room-temperature biosynthesis of ferroelectric barium titanate nanoparticles. *J. Am. Chem. Soc.* **2006**, *128*, 11958–11963. [CrossRef]
40. Bansal, V.; Sanyal, A.; Rautaray, D.; Ahmad, A.; Sastry, M. Bioleaching of sand by the fungus *Fusarium oxysporum* as a means of producing extracellular silica nanoparticles. *Adv. Mater.* **2005**, *17*, 889–892. [CrossRef]
41. Gupta, K.; Chundawat, T.S. Bio-inspired synthesis of platinum nanoparticles from fungus *Fusarium oxysporum*: Its characteristics, potential antimicrobial, antioxidant and photocatalytic activities. *Mater. Res. Express* **2019**, *6*, 1–10. [CrossRef]
42. Bharde, A.; Rautaray, D.; Bansal, V.; Ahmad, A.; Sarkar, I.; Yusuf, S.M.; Sanyal, M.; Sastry, M. Extracellular biosynthesis of magnetite using fungi. *Small* **2006**, *2*, 135–141. [CrossRef]
43. Kumar, S.; Kumar, D.; Dilbaghi, N. Preparation, characterization, and bio-efficacy evaluation of controlled release carbendazim-loaded polymeric nanoparticles. *Environ. Sci. Pollut. Res.* **2017**, *24*, 926–937. [CrossRef]
44. Senapati, S.; Syed, A.; Khan, S.; Pasricha, R.; Khan, M.I.; Kumar, R.; Ahmad, A. Extracellular biosynthesis of metal sulfide nanoparticles using the fungus *Fusarium oxysporum*. *Curr. Nanosci.* **2014**, *10*, 588–595. [CrossRef]
45. Ganapathy, S.; Siva, K. Bio-synthesis, characterisation and application of titanium oxide nanoparticles by *Fusarium oxysporum*. *Int. J. Life Sci. Res.* **2016**, *4*, 69–75. [CrossRef]
46. Boruah, S.; Dutta, P. Fungus mediated biogenic synthesis and characterization of chitosan nanoparticles and its combine effect with *Trichoderma asperellum* against *Fusarium oxysporum*, *Sclerotium rolfsii* and *Rhizoctonia solani*. *Indian Phytopathol.* **2020**, 1–13. [CrossRef]
47. Mirzadeh, S.; Darezereshki, E.; Bakhtiari, F.; Fazaelipoor, M.H.; Hosseini, M.R. Characterization of zinc sulfide (ZnS) nanoparticles biosynthesized by *Fusarium oxysporum*. *Mater. Sci. Semiconduct. Process.* **2013**, *16*, 374–378. [CrossRef]
48. Thakker, J.N.; Dalwadi, P.; Dhandhukia, P.C. Biosynthesis of gold nanoparticles using *Fusarium oxysporum* f. sp. *cubense* JT1, a plant pathogenic fungus. *ISRN Biotechnol.* **2013**, *2013*, 1–5. [CrossRef]
49. Rajput, S.; Werezuk, R.; Lange, R.M.; McDermott, M.T. Fungal isolate optimized for biogenesis of silver nanoparticles with enhanced colloidal stability. *Langmuir* **2016**, *32*, 8688–8697. [CrossRef]
50. Riddin, T.L.; Gericke, M.; Whiteley, C.G. Analysis of the inter- and extracellular formation of platinum nanoparticles by *Fusarium oxysporum* f. sp. lycopersiciusing response surface methodology. *Nanotechnology* **2006**, *17*, 3482–3489. [CrossRef] [PubMed]
51. Cárdenas, D.I.S.; Gomez-Ramirez, M.; Rojas-Avelizapa, N.G.; Vidales-Hurtado, M.A. Synthesis of cadmium sulfide nanoparticles by biomass of *Fusarium oxysporum* f. sp. *lycopersici*. *J. Nano Res.* **2017**, *46*, 179–191. [CrossRef]
52. Karbasian, M.; Atyabi, S.M.; Siadat, S.D.; Momen, S.B.; Norouzian, D. Optimizing nanosilver formation by *Fusarium oxysporum* PTCC 5115 employing response surface methodology. *AJABS* **2008**, *3*, 433–437. [CrossRef]
53. Ajah, A.H.; Hassan, A.S.; Aja, H.A. Extracellular biosynthesis of silver nanoparticles using *Fusarium graminearum* and their antimicrobial activity. *J. Global Pharma Technol.* **2018**, *10*, 683–689.
54. Rodríguez-Serrano, C.; Guzmán-Moreno, J.; Ángeles-Chávez, C.; Rodríguez-González, V.; Ortega-Sigala, J.J.; Ramírez-Santoyo, R.M.; Vidales-Rodríguez, L.E. Biosynthesis of silver nanoparticles by *Fusarium scirpi* and its potential as antimicrobial agent against uropathogenic *Escherichia coli* biofilms. *PLoS ONE* **2020**, *15*, e0230275. [CrossRef] [PubMed]
55. Sawle, B.D.; Salimath, B.; Deshpande, R.; Bedre, M.D.; Prabhakar, B.K.; Venkataraman, A. Biosynthesis and stabilization of Au and Au–Ag alloy nanoparticles by fungus, *Fusarium semitectum*. *Sci. Technol. Adv. Mater.* **2008**, *9*, 035012. [CrossRef]
56. Madakka, M.; Jayaraju, N.; Rajesh, N. Mycosynthesis of silver nanoparticles and their characterization. *MethodsX* **2018**, *5*, 20–29. [CrossRef]
57. Abbas, H.; Baker, D.A. Biological evaluation of selenium nanoparticles biosynthesized by *Fusarium semitectum* as antimicrobial and anticancer agents. *Egypt. J. Chem.* **2020**, *4*, 18–19. [CrossRef]
58. Mohammed, A.E.; Baz, F.F.B.; Albrahim, J.S. *Calligonum comosum* and *Fusarium* sp. extracts as bio-mediator in silver nanoparticles formation: Characterization, antioxidant and antibacterial capability. *3 Biotech* **2018**, *68*, 72. [CrossRef] [PubMed]
59. Kavitha, N.S.; Venkatesh, K.S.; Palani, N.S.; Ilangovan, R. Synthesis and characterization of zirconium oxide nanoparticles using *Fusarium solani* extract. In Proceedings of the 64th DAE Solid State Physics Symposium, Mumbai, India, 18–22 December 2019; AIP Publishing LLC.: Melville, NY, USA, 2020; Volume 2265, p. 030057. [CrossRef]
60. Clarance, P.; Luvankar, B.; Sales, J.; Khusro, A.; Agastian, P.; Tack, J.-C.; Al Khulaifi, M.M.; Al-Shwaiman, H.A.; Elgorban, A.M.; Syed, A.; et al. Green synthesis and characterization of gold nanoparticles using endophytic fungi *Fusarium solani* and its in-vitro anticancer and biomedical applications. *Saudi J. Biol. Sci.* **2020**, *27*, 706–712. [CrossRef]
61. Mekkawy, A.I.; El-Mokhtar, M.A.; Nafady, N.A.; Yousef, N.; Hamad, M.A.; El-Shanawany, S.M.; Ibrahim, E.H.; Elsabahy, M. In vitro and in vivo evaluation of biologically synthesized silver nanoparticles for topical applications: Effect of surface coating and loading into hydrogels. *Int. J. Nanomed.* **2017**, *12*, 759–777. [CrossRef]
62. Gade, A.; Bonde, P.; Ingle, A.P.; Marcato, P.D.; Duran, N.; Rai, M. Exploitation of *Aspergillus niger* for synthesis of silver nanoparticles. *J. Biobased Mater. Bioenergy* **2008**, *2*, 243–247. [CrossRef]
63. Slavin, Y.N.; Asnis, J.; Häfeli, U.O.; Bach, H. Metal nanoparticles: Understanding the mechanisms behind antibacterial activity. *J. Nanobiotechnol.* **2017**, *15*, 65. [CrossRef]
64. Naimi-Shamel, N.; Pourali, P.; Dolatabadi, S. Green synthesis of gold nanoparticles using *Fusarium oxysporum* and antibacterial activity of its tetracycline conjugant. *J. Mycologie Médicale* **2019**, *29*, 7–13. [CrossRef]

65. Pourali, P.; Yahyaei, B.; Afsharnezhad, S. Bio-synthesis of gold nanoparticles by *Fusarium oxysporum* and assessment of their conjugation possibility with two types of β-lactam antibiotics without any additional linkers. *Microbiology* **2018**, *87*, 229–237. [CrossRef]
66. El Sayed, M.T.; El-Sayed, A.S.A. Biocidal Activity of Metal Nanoparticles Synthesized by *Fusarium solani* against Multidrug-Resistant Bacteria and Mycotoxigenic Fungi. *J. Microbiol. Biotechnol.* **2020**, *30*, 226–236. [CrossRef]
67. Srivastava, S.; Bhargava, A.; Pathak, N.; Srivastava, P. Production, characterization and antibacterial activity of silver nanoparticles produced by *Fusarium oxysporum* and monitoring of protein-ligand interaction through in-silico approaches. *Microb. Pathog.* **2019**, *129*, 136–145. [CrossRef]
68. Durán, N.; Marcato, P.D.; Alves, O.L.; De Souza, G.I.H.; Esposito, E. Mechanistic aspects of biosynthesis of silver nanoparticles by several *Fusarium oxysporum* strains. *J. Nanobiotechnol.* **2005**, *3*, 8. [CrossRef]
69. Kumar, S.A.; Abyaneh, M.K.; Gosavi, S.W.; Kulkarni, S.K.; Pasricha, R.; Ahmad, A.; Khan, M.I. Nitrate reductase-mediated synthesis of silver nanoparticles from $AgNO_3$. *Biotechnol. Lett.* **2007**, *29*, 439–445. [CrossRef]
70. Fadeel, B.; Garcia-Bennett, A.E. Better safe than sorry: Understanding the toxicological properties of inorganic nanoparticles manufactured for biomedical applications. *Adv. Drug Deliv. Rev.* **2009**, *62*, 362–374. [CrossRef] [PubMed]
71. Tian, F.; Prina-Mello, A.; Estrada, G.; Beyerle, A.; Möller, W.; Schulz, H.; Kreyling, W.; Stoeger, T. A novel assay for the quantification of internalized nanoparticles in macrophages. *Nanotoxicology* **2008**, *2*, 232–242. [CrossRef]
72. Cui, D.; Tian, F.; Coyer, S.R.; Wang, J.; Pan, B.; Gao, F.; He, R.; Zhang, Y. Effects of antisense-myc-conjugated single-walled carbon nanotubes on HL-60Cells. *J. Nanosci. Nanotechnol.* **2007**, *7*, 1639–1646. [CrossRef] [PubMed]
73. Kowshik, M.; Ashtaputre, S.; Kharrazi, S.; Vogel, W.; Urban, J.; Kulkarni, S.K.; Paknikar, K.M. Extracellular synthesis of silver nanoparticles by a silver-tolerant yeast strain MKY3. *Nanotechnol.* **2003**, *14*, 95–100. [CrossRef]
74. Loo, Y.Y.; Rukayadi, Y.; Nor-Khaizura, M.R.; Kuan, C.H.; Chieng, B.W.; Nishibuchi, M.; Radu, M. In vitro antimicrobial activity of green synthesized silver nanoparticles against selected gram-negative foodborne pathogens. *Front. Microbiol.* **2018**, *9*, 1555. [CrossRef]
75. Al-Askar, A.; Hafez, E.E.; Kabeil, S.A.; Meghad, A. Bioproduction of silver-nano particles by *Fusarium oxysporum* and their antimicrobial activity against some plant pathogenic bacteria and fungi. *Life Sci. J.* **2013**, *10*, 2470–2475.
76. Ishida, K.; Cipriano, T.F.; Rocha, G.M.; Weissmüller, G.; Gomes, F.; Miranda, K.; Rozental, S.; Cruz, M.; de Janeiro, R. Silver nanoparticle production by the fungus *Fusarium oxysporum*: Nanoparticle characterisation and analysis of antifungal activity against pathogenic yeasts. *Mem. Inst. Oswaldo Cruz* **2014**, *109*, 220–228. [CrossRef] [PubMed]
77. Elechiguerra, J.L.; Burt, J.L.; Morones, J.R.; Camacho-Bragado, A.; Gao, X.; Lara, H.H.; Yacaman, M.J. Interaction of silver nanoparticles with HIV-1. *J. Nanobiotechnol.* **2005**, *3*, 1–8. [CrossRef]
78. Akbarzadeh, A.; Kafshdooz, L.; Razban, Z.; Tbrizi, A.D.; Rasoulpour, S.; Khalilov, R.; Kavetskyy, T.; Saghfi, S.; Nasibova, A.N.; Kaamyabi, S.; et al. An overview application of silver nanoparticles in inhibition of herpes simplex virus. *Artif. Cells Nanomed. Biotechnol.* **2018**, *46*, 263–267. [CrossRef] [PubMed]
79. Husseiny, S.M.; Salah, T.A.; Anter, H.A. Biosynthesis of size controlled silver nanoparticles by *Fusarium oxysporum*, their antibacterial and antitumor activities. *Beni Suef Univ. J. Basic Appl. Sci.* **2015**, *4*, 225–231. [CrossRef]
80. Horky, P.; Skalickova, S.; Baholet, D.; Skladanka, J. Nanoparticles as a solution for eliminating the risk of mycotoxins. *Nanomater.* **2018**, *8*, 727. [CrossRef]
81. Sayed, A.M.M.; Kim, S.; Behle, R.W. Characterisation of silver nanoparticles synthesised by *Bacillus thuringiensis* as a nanobiopesticide for insect pest control. *Biocontrol Sci. Technol.* **2017**, *27*, 1308–1326. [CrossRef]
82. Dhanasekaran, D.; Thangaraj, R. Evaluation of larvicidal activity of biogenic nanoparticles against filariasis causing Culex mosquito vector. *Asian Pac. J. Trop. Dis.* **2013**, *3*, 174–179. [CrossRef]
83. Jo, Y.K.; Kim, B.H.; Jung, G. Antifungal activity of silver ions and nanoparticles on phytopathogenic fungi. *Plant Dis.* **2009**, *93*, 1037–1043. [CrossRef]
84. U.N. Department of Economic and Social Affairs. *Population Division 2019, World Population Prospects*; United Nations: New York, NY, USA, 2019.
85. Hietzschold, S.; Walter, A.; Davis, C.; Taylor, A.A.; Sepunaru, L. Does nitrate reductase play a role in silver nanoparticle synthesis? evidence for NADPH as the sole reducing agent. *ACS Sustain. Chem. Eng.* **2019**, *7*, 8070–8076. [CrossRef]
86. Partila, A.M. Bioproduction of silver nanoparticles and its potential applications in agriculture. In *Nanotechnology for Agriculture*; Panpatte, D.G., Jhala, Y.K., Eds.; Springer: Singapore, 2019. [CrossRef]
87. He, X.; Deng, H.; Hwang, H.-M. The current application of nanotechnology in food and agriculture. *J. Food Drug Anal.* **2019**, *27*, 1–21. [CrossRef] [PubMed]
88. Oliveira, H.C.; Stolf-Moreira, R.; Martinez, C.B.R.; Grillo, R.; De Jesus, M.B.; Fraceto, L.F.U. Nanoencapsulation enhances the post-emergence herbicidal activity of atrazine against mustard plants. *PLoS ONE* **2015**, *10*, e0132971. [CrossRef]
89. Cao, L.; Zhou, Z.; Niu, S.; Cao, C.; Li, X.; Shan, Y.; Huang, Q. Positive-charge functionalized mesoporous silica nanoparticles as nanocarriers for controlled 2,4-dichlorophenoxy acetic acid sodium salt release. *J. Agric. Food Chem.* **2018**, *66*, 6594–6603. [CrossRef]
90. Duhan, J.S.; Kumar, R.; Kumar, N.; Kaur, P.; Nehra, K.; Duhan, S. Nanotechnology: The new perspective in precision agriculture. *Biotechnol. Rep.* **2017**, *15*, 11–23. [CrossRef] [PubMed]

91. Khot, L.R.; Sankaran, S.; Maja, J.M.; Ehsani, R.; Schuster, E.W. Applications of nanomaterials in agricultural production and crop protection: A review. *Crop. Prot.* **2012**, *35*, 64–70. [CrossRef]
92. Sekhon, B.S. Nanotechnology in agri-food production: An overview. *Nanotechnol. Sci. Appl.* **2014**, *7*, 31–53. [CrossRef] [PubMed]
93. Dimkpa, C.O.; McLean, J.E.; Britt, D.W.; Anderson, A.J. Antifungal activity of ZnO nanoparticles and their interactive effect with a biocontrol bacterium on growth antagonism of the plant pathogen *Fusarium graminearum*. *BioMetals* **2013**, *26*, 913–924. [CrossRef]
94. Rajiv, P.; Rajeshwari, S.; Venckatesh, R. Bio-Fabrication of zinc oxide nanoparticles using leaf extract of *Parthenium hysterophorus* L. and its size-dependent antifungal activity against plant fungal pathogens. *Spectrochim. Acta Part A* **2013**, *112*, 384–387. [CrossRef] [PubMed]
95. Bramhanwade, K.; Shende, S.; Bonde, S.; Gade, A.; Rai, M. Fungicidal activity of Cu nanoparticles against *Fusarium* causing crop diseases. *Environ. Chem. Lett.* **2016**, *14*, 229–235. [CrossRef]
96. Dimkpa, C.O.; Bindraban, P.S. Nanofertilizers: New Products for the industry? *J. Agric. Food Chem.* **2018**, *66*, 6462–6473. [CrossRef]
97. Tripathi, K.M.; Bhati, A.; Singh, A.; Sonker, A.K.; Sarkar, S.; Sonkar, S.K. Sustainable changes in the contents of metallic micronutrients in first generation gram seeds imposed by carbon nano-onions: Life cycle seed to seed study. *ACS Sustain. Chem. Eng.* **2017**, *5*, 2906–2916. [CrossRef]
98. Khalifa, N.S.; Hasaneen, M.N. The effect of chitosan-PMAA-NPK nanofertilizer on *Pisum sativum* plants. *3 Biotech* **2018**, *8*, 193. [CrossRef]
99. Abdel-Aziz, H.M.M.; Hasaneen, M.N.A.; Omer, A.M. Nano chitosan-NPK fertilizer enhances the growth and productivity of wheat plants grown in sandy soil. *Spanish J. Agri. Res.* **2016**, *14*, e0902. [CrossRef]
100. Shende, S.; Rathod, D.; Gade, A.; Rai, M. Biogenic copper nanoparticles promote the growth of pigeon pea (*Cajanus cajan* L.). *IET Nanobiotechnol.* **2017**, *11*, 773–781. [CrossRef]
101. Abbacia, A.; Azzouz, N.; Bouznit, Y. A new copper doped montmorillonite modified carbon paste electrode for propineb detection. *Appl. Clay Sci.* **2014**, *90*, 130–134. [CrossRef]
102. Wibowo, K.; Sahdan, M.; Ramli, N.; Muslihati, A.; Rosni, N.; Tsen, V.; Saïm, H.; Ahmad, S.; Sari, Y.; Mansor, Z. Detection of *Escherichia coli* bacteria in wastewater by using graphene as a sensing material. In *Journal of Physics: Conference Series, Proceedings of the 2017 International Seminar on Mathematics and Physics in Sciences and Technology (ISMAP 2017), Batu Pahat, Malaysia, 28–29 October 2017*; IOP Publishing Ltd.: Bristol, UK, 2018; Volume 995, p. 012063. [CrossRef]
103. Deng, H.; Gao, Y.; Dasari, T.P.S.; Ray, P.C.; Yu, H. A facile 3D construct of graphene oxide embedded with silver nanoparticles and its potential application as water filter. *J. Miss. Acad. Sci.* **2016**, *61*, 190–197. Available online: https://www.researchgate.net/publication/312115792 (accessed on 13 October 2020).
104. Esser, B.; Schnorr, J.M.; Swager, T.M. Selective detection of ethylene gas using carbon nanotube-based devices: Utility in determination of fruit ripeness. *Angew. Chem. Int. Ed.* **2012**, *51*, 5752–5756. [CrossRef]
105. Geszke-Moritz, M.; Clavier, G.; Lulek, J.; Schneider, R. Copper- or manganese-doped ZnS quantum dots as fluorescent probes for detecting folic acid in aqueous media. *J. Lumin.* **2012**, *132*, 987–991. [CrossRef]
106. Jokar, M.; Safaralizadeh, M.H.; Hadizadeh, F.; Rahmani, F.; Kalani, M.R. Design and evaluation of an apta-nano-sensor to detect acetamiprid in vitro and *in silico*. *J. Biomol. Struct. Dyn.* **2016**, *34*, 2505–2517. [CrossRef]
107. Lin, Y.W.; Huang, C.C.; Chang, H.T. Gold nanoparticle probes for the detection of mercury, lead and copper ions. *Analyst* **2011**, *136*, 863–871. [CrossRef] [PubMed]
108. Korbekandi, H.; Ashari, Z.; Iravani, S.; Abbasi, S. Optimization of biological synthesis of silver nanoparticles using *Fusarium oxysporum*. *Iran. J. Pharm. Res.* **2013**, *12*, 289–298. [CrossRef]
109. Hamedi, S.; Ghaseminezhad, M.; Shokrollahzadeh, S.; Shojaosadati, S.A. Controlled biosynthesis of silver nanoparticles using nitrate reductase enzyme induction of filamentous fungus and their antibacterial evaluation. *Artif. Cells Nanomed. Biotechnol.* **2017**, *45*, 1588–1596. [CrossRef]
110. Zhang, X.; He, X.; Wang, K.; Yang, X. Different active biomolecules involved in biosynthesis of gold nanoparticles by three fungus species. *J. Biomed. Nanotechnol.* **2011**, *7*, 245–254. [CrossRef]
111. Senapati, S.; Ahmad, A.; Khan, M.I.; Sastry, M.; Kumar, R. Extracellular biosynthesis of bimetallic Au-Ag alloy nanoparticles. *Small* **2005**, *1*, 517–520. [CrossRef]
112. Govender, Y.; Riddin, T.; Gericke, M.; Whiteley, C.G. Bioreduction of platinum salts into nanoparticles: A mechanistic perspective. *Biotechnol. Lett.* **2009**, *31*, 95–100. [CrossRef]
113. Velmurugan, P.; Shim, J.; You, Y.; Choi, S.; Kamala-Kannan, S.; Lee, K.J.; Kim, H.J.; Oh, B.T. Removal of zinc by live, dead, and dried biomass of *Fusarium* spp. isolated from the abandoned-metal mine in South Korea and its perspective of producing nanocrystals. *J. Hazard. Mater.* **2010**, *182*, 317–324. [CrossRef]
114. Camargo, P.H.C.; Satyanarayana, K.G.; Wypych, F. Nanocomposites: Synthesis, structure, properties and new application opportunities. *Mater. Res.* **2009**, *12*, 1–39. [CrossRef]
115. Hamzeh, M.; Sunahara, G.I. In vitro cytotoxicity and genotoxicity studies of titanium dioxide (TiO_2) nanoparticles in Chinese hamster lung fibroblast cells. *Toxicol. Vitro* **2013**, *27*, 864–873. [CrossRef] [PubMed]
116. Yu, S.J.; Yin, Y.G.; Liu, J.F. Silver nanoparticles in the environment. *Environ. Sci. Process. Impacts* **2013**, *15*, 78–92. [CrossRef] [PubMed]
117. Naidu, K.S.B.; Adam, J.K.; Govender, P. Biomedical applications and toxicity of nanosilver: A review. *Med Technol. SA* **2015**, *29*, 13–19. [CrossRef]

118. Akter, M.; Sikder, T.; Rahman, M.; Ullah, A.A.; Hossain, K.F.B.; Banik, S.; Hosokawa, T.; Saito, T.; Kurasaki, M. A systematic review on silver nanoparticles-induced cytotoxicity: Physicochemical properties and perspectives. *J. Adv. Res.* **2018**, *9*, 1–16. [CrossRef]
119. Salaheldin, T.A.; Husseiny, S.M.; Al-Enizi, A.M.; Elzatahry, A.; Cowley, A.H. Evaluation of the Cytotoxic Behavior of Fungal extracellular synthesized Ag nanoparticles using confocal laser scanning microscope. *Int. J. Mol. Sci.* **2016**, *17*, 329. [CrossRef]
120. Akhtar, M.S.; Panwar, J.; Yun, Y.-S. Biogenic synthesis of metallic nanoparticles by plant extracts. *ACS Sustain. Chem. Eng.* **2013**, *1*, 591–602. [CrossRef]
121. Karatoprak, G.Ş.; Aydin, G.; Altinsoy, B.; Altinkaynak, C.; Koşar, M.; Ocsoy, I. The Effect of *Pelargonium endlicherianum* Fenzl. root extracts on formation of nanoparticles and their antimicrobial activities. *Enzyme Microb. Technol.* **2017**, *97*, 21–26. [CrossRef]
122. Siddiqi, K.S.; Husen, A.; Rao, R.A.K. A review on biosynthesis of silver nanoparticles and their biocidal properties. *J. Nanobiotechnol.* **2018**, *16*, 14. [CrossRef] [PubMed]
123. Barbalinardo, M.; Caicci, F.; Cavallini, M.; Gentili, D. Protein corona mediated uptake and cytotoxicity of silver nanoparticles in mouse embryonic fibroblast. *Small* **2018**, *14*, e1801219. [CrossRef]
124. Ritz, S.; Schöttler, S.; Kotman, N.; Baier, G.; Musyanovych, A.; Kuharev, J.; Landfester, K.; Schild, H.; Jahn, O.; Tenzer, S.; et al. Protein corona of nanoparticles: Distinct proteins regulate the cellular uptake. *Biomacromolecules* **2015**, *16*, 1311–1321. [CrossRef]
125. Nguyen, V.H.; Lee, B.J. Protein corona: A new approach for nanomedicine design. *Int. J. Nanomed.* **2017**, *12*, 3137–3151. [CrossRef] [PubMed]
126. Abbaszadegan, A.; Ghahramani, Y.; Gholami, A.; Hemmateenejad, B.; Dorostkar, S.; Nabavizadeh, M.; Sharghi, H. The effect of charge at the surface of silver nanoparticles on antimicrobial activity against Gram-positive and Gram-negative bacteria: A preliminary study. *J. Nanomater.* **2015**, *2015*, 1–8. [CrossRef]
127. Asharani, P.V.; Hande, M.P.; Valiyaveettil, S. Anti-proliferative activity of silver nanoparticles. *BMC Cell Biol.* **2009**, *10*, 65. [CrossRef] [PubMed]
128. Walczyk, D.; Bombelli, F.B.; Monopoli, M.P.; Lynch, I.; Dawson, K.A. What the Cell "Sees" in Bionanoscience. *J. Am. Chem. Soc.* **2010**, *132*, 5761–5768. [CrossRef] [PubMed]
129. Monopoli, M.P.; Åberg, C.; Salvati, A.; Dawson, K.A. Biomolecular coronas provide the biological identity of nanosized materials. *Nat. Nanotechnol.* **2012**, *7*, 779–786. [CrossRef]
130. Lee, Y.K.; Choi, E.-J.; Webster, T.J.; Kim, S.-H.; Khang, D. Effect of the protein corona on nanoparticles for modulating cytotoxicity and immunotoxicity. *Int. J. Nanomed.* **2014**, *10*, 97–113. [CrossRef]
131. Lesniak, A.; Fenaroli, F.; Monopoli, M.P.; Åberg, C.; Dawson, K.A.; Salvati, A. Effects of the presence or absence of a protein corona on silica nanoparticle uptake and impact on cells. *ACS Nano* **2012**, *6*, 5845–5857. [CrossRef] [PubMed]
132. Yilma, A.N.; Singh, S.R.; Dixit, S.; Dennis, V.A. Anti-inflammatory effects of silver-polyvinyl pyrrolidone (Ag-PVP) nanoparticles in mouse macrophages infected with live *Chlamydia trachomatis*. *Int. J. Nanomed.* **2013**, *8*, 2421–2432. [CrossRef]
133. Yin, N.; Liu, Q.; Liu, J.; He, B.; Cui, L.; Li, Z.; Yun, Z.; Qu, G.; Liu, S.; Zhou, Q.; et al. Silver nanoparticle exposure attenuates the viability of rat cerebellum granule cells through apoptosis coupled to oxidative stress. *Small* **2013**, *9*, 1831–1841. [CrossRef] [PubMed]
134. Braydich-Stolle, L.; Hussain, S.; Schlager, J.J.; Hofmann, M.C. In vitro cytotoxicity of nanoparticles in mammalian germline stem cells. *Toxicol. Sci.* **2005**, *88*, 412–419. [CrossRef]
135. Navya, P.N.; Daima, H.K. Rational engineering of physicochemical properties of nanomaterials for biomedical applications with nanotoxicological perspectives. *Nano Converg.* **2016**, *3*, 1–14. [CrossRef]
136. Soleimani, F.F.; Saleh, T.; Shojaosadati, S.A.; Poursalehi, R. Green synthesis of different shapes of silver nanostructures and evaluation of their antibacterial and cytotoxic activity. *BioNanoScience* **2018**, *8*, 72–80. [CrossRef]
137. Aueviriyavit, S.; Phummiratch, D.; Maniratanachote, R. Mechanistic study on the biological effects of silver and gold nanoparticles in Caco-2 cells—Induction of the Nrf2/HO-1 pathway by high concentrations of silver nanoparticles. *Toxicol. Lett.* **2014**, *224*, 73–83. [CrossRef]
138. Benn, T.; Cavanagh, B.; Hristovski, K.; Posner, J.D.; Westerhoff, P. The release of nanosilver from consumer products used in the home. *J. Environ. Qual.* **2010**, *39*, 1875–1882. [CrossRef]
139. Nel, A.; Xia, T.; Mädler, L.; Li, N. Toxic potential of materials at the nanolevel. *Science* **2006**, *311*, 622–627. [CrossRef]
140. Mishra, A.R.; Zheng, J.; Tang, X.; Goering, P.L. Silver nanoparticle-induced autophagic-lysosomal disruption and NLRP3-inflammasome activation in HepG2 cells is size-dependent. *Toxicol. Sci.* **2016**, *150*, 473–487. [CrossRef] [PubMed]
141. Chen, X.; Schluesener, H.J. Nanosilver: A nanoproduct in medical application. *Toxicol. Lett.* **2008**, *176*, 1–12. [CrossRef] [PubMed]
142. Simon-Deckers, A.; Gouget, B.; Mayne-L'Hermite, M.; Herlin-Boime, N.; Reynaud, C.; Carrière, M. In vitro investigation of oxide nanoparticle and carbon nanotube toxicity and intracellular accumulation in A549 human pneumocytes. *Toxicol.* **2008**, *253*, 137–146. [CrossRef]
143. Xiu, Z.M.; Zhang, Q.B.; Puppala, H.L.; Colvin, V.L.; Alvarez, P.J. Negligible particle-specific antibacterial activity of silver nanoparticles. *Nano Lett.* **2012**, *12*, 4271–4275. [CrossRef]
144. Chernousova, S.; Epple, M. Silver as antibacterial agent: Ion, nanoparticle, and metal. *Angew. Chem. Int. Ed.* **2013**, *52*, 1636–1653. [CrossRef] [PubMed]
145. Cho, J.G.; Kim, K.T.; Ryu, T.K.; Lee, J.W.; Kim, J.E.; Kim, J.; Lee, B.C.; Jo, E.H.; Yoon, J.; Eom, I.C.; et al. Stepwise embryonic toxicity of silver nanoparticles on *Oryzias latipes*. *BioMed Res. Int.* **2013**, *2013*, 1–7. [CrossRef]

146. Gorth, D.J.; Rand, D.M.; Webster, T.J. Silver nanoparticle toxicity in Drosophila: Size does matter. *Int. J. Nanomed.* **2011**, *6*, 343–350. [CrossRef]
147. Kruszewski, M.; Brzoska, K.; Brunborg, G.; Asare, N.; Dobrzyńska, M.; Dušinská, M.; Fjellsbø, L.M.; Georgantzopoulou, A.; Gromadzka-Ostrowska, J.; Gutleb, A.C.; et al. Toxicity of silver nanomaterials in higher eukaryotes. In *Advances in Molecular Toxicology*; Fishbein, J.C., Ed.; Elsevier: Amsterdam, The Netherlands, 2011; Volume 5, pp. 179–218.
148. Van Der Zande, M.; Vandebriel, R.J.; Van Doren, E.; Kramer, E.; Rivera, Z.H.; Serrano-Rojero, C.S.; Gremmer, E.R.; Mast, J.; Peters, R.J.B.; Hollman, P.C.H.; et al. Distribution, elimination, and toxicity of silver nanoparticles and silver ions in rats after 28-Day oral exposure. *ACS Nano* **2012**, *6*, 7427–7442. [CrossRef] [PubMed]
149. Sambale, F.; Wagner, S.; Stahl, F.; Khaydarov, R.R.; Scheper, T.; Bahnemann, D.W. Investigations of the toxic effect of silver nanoparticles on mammalian cell lines. *J. Nanomater.* **2015**, *2015*, 1–9. [CrossRef]
150. El Badawy, A.M.; Silva, R.G.; Morris, B.; Scheckel, K.G.; Suidan, M.T.; Tolaymat, T.M. Surface charge-dependent toxicity of silver nanoparticles. *Environ. Sci. Technol.* **2011**, *45*, 283–287. [CrossRef]
151. Joshi, N.; Ngwenya, B.T.; French, C.E. Enhanced resistance to nanoparticle toxicity is conferred by overproduction of extracellular polymeric substances. *J. Hazard. Mater.* **2012**, *241*, 363–370. [CrossRef] [PubMed]
152. Khan, S.S.; Mukherjee, A.; Chandrasekaran, N. Studies on interaction of colloidal silver nanoparticles (SNPs) with five different bacterial species. *Coll. Surf. B Biointerfaces* **2011**, *87*, 129–138. [CrossRef]
153. Khan, S.S.; Srivatsan, P.; Vaishnavi, N.; Mukherjee, A.; Chandrasekaran, N. Interaction of silver nanoparticles (SNPs) with bacterial extracellular proteins (ECPs) and its adsorption isotherms and kinetics. *J. Hazard. Mater.* **2011**, *192*, 299–306. [CrossRef]
154. Baruwati, B.; Simmons, S.O.; Varma, R.S.; Veronesi, B. "Green" synthesized and coated nanosilver alters the membrane permeability of barrier (intestinal, brain endothelial) cells and stimulates oxidative stress pathways in neurons. *ACS Sustain. Chem. Eng.* **2013**, *1*, 753–759. [CrossRef]
155. Ferreira, L.A.; Dos Reis, S.B.; da Silva, E.D.N.; Cadore, S.; da Silva Bernardes, J.; Durán, N.; de Jesus, M.B. Thiol-antioxidants interfere with assessing silver nanoparticle cytotoxicity. *Nanomed. Nanotechnol. Biol. Med.* **2020**, *24*, 102130. [CrossRef] [PubMed]
156. Marcato, P.D.; Nakasato, G.; Brocchi, M.; Melo, P.S.; Huber, S.C.; Ferreira, I.R.; Alves, O.L.; Durán, N. Biogenic Silver Nanoparticles: Antibacterial and cytotoxicity applied to textile fabrics. *J. Nano Res.* **2012**, *20*, 69–76. [CrossRef]
157. Vijayan, S.; Divya, K.; George, T.K.; Jisha, M.S. Biogenic synthesis of silver nanoparticles using endophytic fungi *Fusarium oxysporum* isolated from *Withania somnifera* (L.), its antibacterial and cytotoxic activity. *J. Bionanosci.* **2016**, *10*, 369–376. [CrossRef]
158. Asharani, P.V.; Mun, G.L.K.; Hande, M.P.; Valiyaveettil, S. Cytotoxicity and genotoxicity of silver nanoparticles in human cells. *ACS Nano* **2009**, *3*, 279–290. [CrossRef] [PubMed]
159. Gurunathan, S.; Park, J.H.; Han, J.W.; Kim, J.H. Comparative assessment of the apoptotic potential of silver nanoparticles synthesized by Bacillus tequilensis and Calocybe indica in MDA-MB-231 human breast cancer cells: Targeting p53 for anticancer therapy. *Int. J. Nanomed.* **2015**, *10*, 4203–4223. [CrossRef]
160. Supino, R. MTT Assays: In Vitro Toxicity Testing Protocols. In *Methods in Molecular Biology*; O'Hare, S., Atterwill, C.K., Eds.; Humana Press: Totowa, NJ, USA, 1995; Volume 43.
161. Pourali, P.; Badiee, S.H.; Manafi, S.; Noorani, T.; Rezaei, A.; Yahyaei, B. Biosynthesis of gold nanoparticles by two bacterial and fungal strains, Bacillus cereus and *Fusarium oxysporum*, and assessment and comparison of their nanotoxicity in vitro by direct and indirect assays. *Electron. J. Biotechnol.* **2017**, *29*, 86–93. [CrossRef]
162. Andersson, L.O. Study of some silver-thiol complexes and polymers: Stoichiometry and optical effects. *J. Polym. Sci. Part A-1 Polym. Chem.* **1972**, *10*, 1963–1973. [CrossRef]
163. Li, X.; Lenhart, J.J. Aggregation and dissolution of silver nanoparticles in natural surface water. *Environ. Sci. Technol.* **2012**, *46*, 5378–5386. [CrossRef]
164. Almofti, M.R.; Ichikawa, T.; Yamashita, K.; Terada, H.; Shinohara, Y. Silver ion induces a cyclosporine a-insensitive permeability transition in rat liver mitochondria and release of apoptogenic cytochrome c. *J. Biochem.* **2003**, *134*, 43–49. [CrossRef]
165. Manivasagan, P.; Oh, J. Production of a novel fucoidanase for the green synthesis of gold nanoparticles by *Streptomyces* sp. and its cytotoxic effect on HeLa Cells. *Mar. Drugs* **2015**, *13*, 6818–6837. [CrossRef]
166. Park, H.Y.; Park, S.H.; Kim, G.Y.; Choi, I.W.; Kim, S.; Kim, H.S.; Cha, H.J.; Choi, Y.H.; Jeong, J.W.; Yoon, D.; et al. Induction of p53-independent apoptosis and G1 cell cycle arrest by fucoidan in HCT116 human colorectal carcinoma cells. *Mar. Drugs* **2017**, *15*, 154–159. [CrossRef] [PubMed]
167. Hamzah, H.M.; Salah, R.F.; Maroof, M.N. *Fusarium mangiferae* as new cell factories for producing silver nanoparticles. *J. Microbiol. Biotechnol.* **2018**, *28*, 1654–1663. [CrossRef]
168. Al-Sharqi, S.A.H. Histological and biometric study of the effects of *Fusarium graminarum* silver nanoparticles on the kidney in male albino mice. *Med. Leg. Update* **2020**, *20*, 1028–1035. [CrossRef]

Journal of *Fungi*

MDPI

Review

Nanohybrid Antifungals for Control of Plant Diseases: Current Status and Future Perspectives

Mousa A. Alghuthaymi [1], Rajkuberan C. [2], Rajiv P. [2], Anu Kalia [3,*], Kanchan Bhardwaj [4], Prerna Bhardwaj [4], Kamel A. Abd-Elsalam [5,*], Martin Valis [6] and Kamil Kuca [7,8,*]

1 Biology Department, Science and Humanities College, Shaqra University, Alquwayiyah 11971, Saudi Arabia; mosa-4507@hotmail.com
2 Department of Biotechnology, Karpagam Academy of Higher Education, Coimbatore, Tamil Nadu 641021, India; kuberan87@gmail.com (R.C.); rajivsmart15@gmail.com (R.P.)
3 Electron Microscopy and Nanoscience Laboratory, Department of Soil Science, College of Agriculture, Punjab Agricultural University, Ludhiana 141004, Punjab, India
4 School of Biological and Environmental Sciences, Shoolini University of Biotechnology and Management Sciences, Solan 173229, Himachal Pradesh, India; kanchankannu1992@gmail.com (K.B.); prernabhardwaj135@gmail.com (P.B.)
5 Plant Pathology Research Institute, Agricultural Research Center (ARC), Giza 12619, Egypt
6 Department of Neurology of the Medical Faculty of Charles University and University Hospital in Hradec Kralove, Sokolska 581, 50005 Hradec Kralove, Czech Republic; martin.valis@fnhk.cz
7 Department of Chemistry, Faculty of Science, University of Hradec Kralove, 50003 Hradec Kralove, Czech Republic
8 Biomedical Research Center, University Hospital in Hradec Kralove, Sokolska 581, 50005 Hradec Kralove, Czech Republic
* Correspondence: kaliaanu@pau.edu (A.K.); kamel.abdelsalam@arc.sci.eg (K.A.A.-E.); kamil.kuca@uhk.cz (K.K.); Tel.: +20-10-910-49161 (K.A.A.-E.); +420-603-289-166 (K.K.)

Citation: Alghuthaymi, M.A.; C., R.; P., R.; Kalia, A.; Bhardwaj, K.; Bhardwaj, P.; Abd-Elsalam, K.A.; Valis, M.; Kuca, K. Nanohybrid Antifungals for Control of Plant Diseases: Current Status and Future Perspectives. *J. Fungi* **2021**, *7*, 48. https://doi.org/10.3390/jof7010048

Received: 25 December 2020
Accepted: 11 January 2021
Published: 13 January 2021

Abstract: The changing climatic conditions have led to the concurrent emergence of virulent microbial pathogens that attack crop plants and exhibit yield and quality deterring impacts on the affected crop. To counteract, the widespread infections of fungal pathogens and post-harvest diseases it is highly warranted to develop sustainable techniques and tools bypassing traditional agriculture practices. Nanotechnology offers a solution to the problems in disease management in a simple lucid way. These technologies are revolutionizing the scientific/industrial sectors. Likewise, in agriculture, the nano-based tools are of great promise particularly for the development of potent formulations ensuring proper delivery of agrochemicals, nutrients, pesticides/insecticides, and even growth regulators for enhanced use efficiency. The development of novel nanocomposites for improved management of fungal diseases can mitigate the emergence of resilient and persistent fungal pathogens and the loss of crop produce due to diseases they cause. Therefore, in this review, we collectively manifest the role of nanocomposites for the management of fungal diseases.

Keywords: chitosan; nanohybrids; polymer-metal composites; antifungal; postharvest

1. Introduction

Agriculture is the major backbone of a country's economy; in recent years, the agricultural production and yields have been increased along with pesticides and insecticides usage [1–3]. But in the present scenario, the practices of agriculture methods are fretful about the extensive usage of chemical pesticides used to combat microbes/insects. Among microbes, the fungal genera belonging to class Ascomycetes (*Verticillium*, *Alternaria*, and *Fusarium*) and Basidiomycetes (*Rhizoctonia*, *Sclerotium*) can cause growth as well as yield-deterring effects on plants leading to significant economic losses to the farmers [4]. To curb these disease manifestations, a stringent dynamic approach should be implemented with novel technology encompassing smart materials with biological ingredients for sustainable delivery with prolonged efficiency.

Recently, nanoscale engineering approaches have posed a new advanced entity derived from a biological source and self-assembling systems [5]. At present, the nano concept is playing an immense role in medicine and pharmacology; in both these fields' nanotechnology has attained a decisive role in drug delivery, diagnosis, imaging, antimicrobial agents, and sensors [6]. In the agriculture sector, nanotechnology products and devices are being eventually utilized in plant hormone delivery, nano-barcoding, development of rapid and sensitive nano-sensor systems for easy diagnosis of diseases, pests, and nutritional deficiencies, the targeted/controlled/slow-release of agrochemicals, seed germination enhancers, nano-vectors for efficient gene transfer and several other applications [7].

Engineered nanoparticles (NPs) possess the desired size and shape with specific optical properties that enable them to be used for various agricultural applications particular instance is as novel pesticide formulations exhibiting improved pest and pathogen control efficiencies [8–10]. The most widely used nanoparticles for the control of plant diseases are carbon, silver, silica, and non-metal oxides or alumino-silicates. The research studies performed on carbon nanomaterials have shown diverse and promising agri-applications including the promotion of plant growth and development, besides effective control of several plant pathogens such as *Xanthomonas*, *Aspergillus* spp., *Botrytis cinerea*, and *Fusarium* spp. A study revealed that silica nanoparticles were effective in maize conferring resistance to the phytopathogens such as *Fusarium oxysporum* and *Aspergillus niger* [11]. However, the action spectrum and pest/pathogen control efficacies of the nano-enabled pesticides can be improved through the development of nano-hybrids or composites [12,13]. The components of the nano-hybrids or nanocomposites can have diverse chemical origins spanning over biological-inorganic, as well as natural/synthetic organic-inorganic materials. These composites do not involve physical mixing of the components and therefore, possess peculiar properties which may or may not essentially represent additive or augmenting effect considering the properties of the individual components [12]. Escalated research interest for the development of potent, effective and multi-functional anti-microbial nano-hybrids has been witnessed in the present decade for instance, alumino-silicate nanoplates have been used for the development of pesticide formulations that exhibit twin benefits of improved biological activity and better environmental safety compared to use of engineered NPs [14]. Therefore, nano-formulated particles/composites have the potential to tackle disease outbreaks caused by fungal pathogens effectively. In this review, we emphasize the development of antifungal nano-hybrids encompassing conjugates of organic or inorganic molecules, biological components, and biopolymers to develop the cheaper, reliable, and effective product(s) against most fungal pathogens of plants (Figure 1).

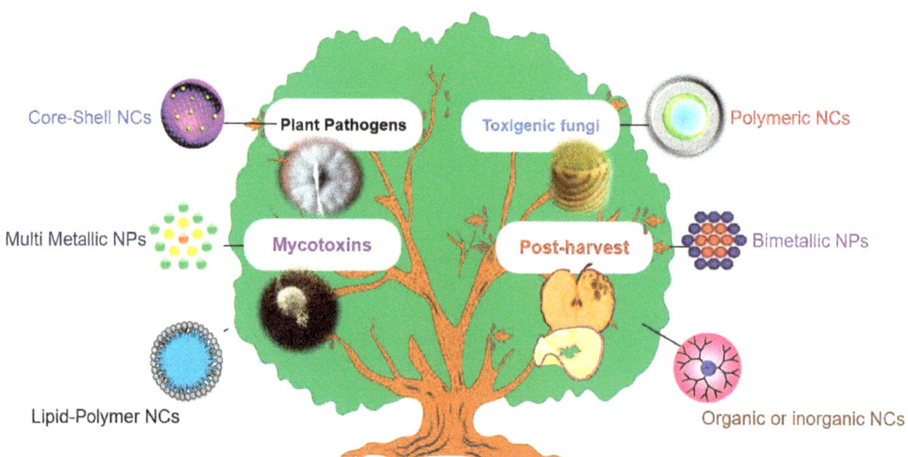

Figure 1. Nanohybrid antifungals for plant diseases control.

2. Nanotechnology Advances for Improved Pathogen Diagnosis

In agriculture, the outbreak of pathogenic infections is to be monitored at the earliest, else the crop yield might get heavily compromised. Therefore, prompt diagnostic methods are urgently required as the old conventional methods such as immunological techniques and other molecular tools require specialized skilled man-power and are not cost-effective. To pace up with the rapid spread of the plant pathogenic infections, rapid, robust, sensitive, and low-cost smart material based diagnostic protocols are required to be designed for counteracting bacterial and fungal infections in agriculture crops.

Nanotechnology-based pathogen diagnosis is gaining overwhelming attention from the research community due to the functional optical properties and ease of handling technology of these materials [15]. The added advantage of nanotechnology is that nanoparticles can be conjugated with nucleic acids, proteins, and other biomolecules, a feature that enables rapid, sensitive, and reliable diagnosis of pathogens [16]. Among the various nanomaterials, quantum dots are a special class of nanocrystals that exhibit tunable size-dependent fluorescence characteristics for which these are explored widely in agriculture and allied sectors. A specific quantum dot-based nano-sensor has been developed for diagnosing *Candidatus Phytoplasma aurantifolia* in the infected lime even at low occurrence of 5 phytoplasma cells μL^{-1} [17]. Fluorescent silica nanoparticles conjugated with antibody molecules can rapidly detect the *Xanthomonas axonopodis* pv. *vesicatoria* a causative agent for spot diseases in tomato and pepper [18].

Gold nanoparticles are widely used in pathogen diagnosis due to the unique optical or electrochemical properties enabling simple and easy protocols for the quick diagnosis of pathogens. Singh et al. [19] developed an immunosensor based on nanogold using Surface Plasmon Resonance (SPR) that could detect the Karnal bunt disease in wheat (*Tilletia indica*). Wang et al. [20] developed an electrochemical sensor comprised of copper nanoparticles to detect the fungus *Sclerotinia sclerotiorum* in oilseeds. They have utilized this electrochemical sensor to measure the level of salicyclic acid accurately. Schwenkbier et al. [21] developed a chip-based hybridization technique incorporating silver nanoparticles for the detection of *Phytophthora* species. Copper oxide nanoparticles and nanolayers have been synthesized and applied for easy detection of *Aspergillus niger* in crop plants. In another approach, portable equipment was used to detect bacterial, fungal species in stored food grains. Likewise, Ariffin et al. [22] formulated a nanowire biosensor to detect Cauliflower Mosaic Virus and Papaya Ring Spot Virus. Thus, with the above scientific evidence, it can be identified that nano-based sensors and kits play a vital role in crop health care including products for rapid testing, disease diagnostics, and environmental monitoring aspects.

3. Nanocomposites and Their Mode of Action on the Fungal Phytopathogens

Nanocomposite materials include multi-phase components. These materials may be comprised of components with variable phase domains with atleast one continuous phase while another having nano-scale dimensions [23]. These hybrid nanomaterials can be generated through co-synthesis/impregnation of diverse inorganic and organic components [24]. Nanocomposites have been extensively studied due to the properties of inorganic and organic materials that enact concurrently to perform the desired activity [25]. Generally, nanocomposites are derived by the addition of nano-particulate materials in long-chain or short-chain polymeric matrices. The derived nanocomposites exhibit improved properties not observed for any of the individual components. Most likely, the combination of polymers with nanoparticles is anticipated to increase the properties of the polymer significantly [26]. Such kind of nanocomposites are now widely being used in food processing, pest detection and management, food health screening, water treatment, disease detection, drug-delivery systems, and improvement of sustainable agriculture [27–30]. Likewise, the polymer composites act as fertilizers which increase the nutrients uptake, decrease soil toxicity [31,32]. Moreover, nanocomposites are well being used to increase the shelf life of food materials by acting as antimicrobial dispositions and as sensors [33].

Plant disease management using hybrid polymer nanocomposites is focused on making mulch films to control weeds; as nano pesticides and as a biostatic agent [34]. Min et al. [35] demonstrated that silver nanoparticles can effectively inhibit the phytopathogens such as *Rhizoctonia solani*, *Sclerotinia sclerotiorum*, and *S. minor*. Further, silver nanoparticles can cause extensive damage by breaking the hyphal wall membrane followed by internal damage of the hyphae. Sepiolite, a magnesium silicate, was blended with MgO to form a (SE-MgO) nanocomposite that exhibited excellent antifungal activity against rice pathogen *Fusarium verticillioides*, *Bipolaris oryzae*, and *Fusarium fujikuroi* with ED90 > 249 μg/mL compared with MgONPs [36].

Carbon nanomaterials have shown to possess strong antifungal activity against pathogens *Fusarium graminearum* and *Fusarium poae* [37]. Six carbon nanomaterials (CNMs), single-walled carbon nanotubes (SWCNTs), multi-walled carbon nanotubes (MWCNTs), graphene oxide (GO), reduced graphene oxide (rGO), fullerene (C_{60}), and activated carbon (AC) were evaluated against the phytopathogens. The outcome imposes that SWCNTs, MWCNTs, GO and rGO at varying concentrations decreased the biomass, mycelia growth and inhibited spore germination of the fungal pathogens. However, the C_{60} and AC didn't exert any activity against the tested pathogens.

In another study, GO-AgNPs nanocomposite was synthesized through interfacial self-assembly and was evaluated for its antifungal activity against the pathogen *Fusarium graminearum* under in vitro and in vivo conditions [38]. The fungicidal activity of GO-AgNPs against the *F. graminearum* spores was 4.68 μg/mL; the nanocomposite also constrained the germination of spores and hyphae. Moreover, the authors demonstrated response to GO-AgNPs through the SEM analysis of the events of spore germination of *F. graminearum*. The images portray that *F. graminearum* conidia were crumpled, widened, and damaged heavily as seen in Figure 2. The fungicidal activity of nanocomposite is due to the adsorption of nanoparticles to the fungal cell wall membranes and oxygen groups of GO form hydrogen bond with lipopolysaccharide subunits of the cell membrane which contains sugars, phosphates, and lipids.

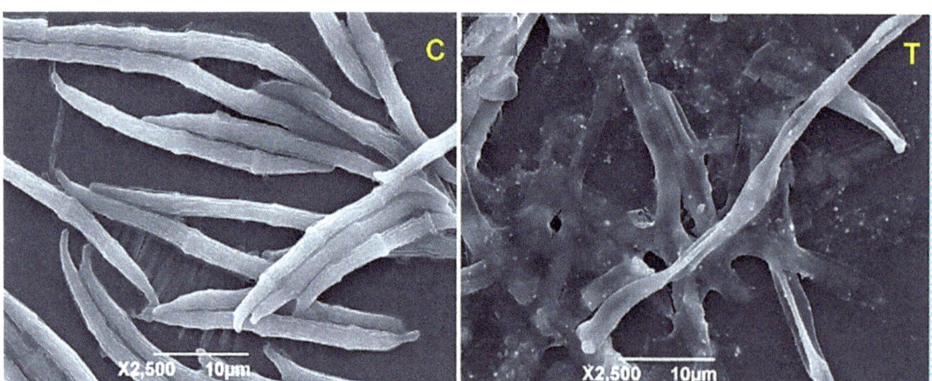

Figure 2. SEM images of *F. graminearum* spores incubated with sterile water (**C**) control and treated with GO-AgNPs nanocomposite (**T**). Images obtained with permission from Chen et al. [38]. (Cited from Chen et al., with permission from ACS).

Downy mildew is a disease caused by *Plasmopara viticola* in grapes plants that leads to extremely heavy yield loss. To combat the disease, a nanocomposite construct consisting of Graphene oxide (GO) and Iron oxide (Fe_3O_4) as GO-Fe_3O_4 nanocomposite was developed [39]. The study demonstrated that the pretreatment of the leaf discs with nanocomposite followed by inoculation with *P. viticola* sporangium suspension strongly inhibited the spore germination. This effect may be attributed to the blockage of the water channels of sporangia by surface adsorption of the nanocomposites. From the images, it is evident that bare GO and Fe_3O_4 possessed moderate spore germination inhibition activity while the nanocomposites triggered stupendous activity [39].

Silver-Titanate nanotubes (AgTNTs) nanocomposite was synthesized through a one-pot chemical method and functionalization with AgNPs. These nanocomposites were further evaluated against the phytopathogenic fungi *Botrytis cinerea* by the photoinactivation method. The nanocomposite stimulated Reactive Oxygen Species (ROS) cascades and damaged the conidia which eventually led to cell death [40]. In a microwave-assisted method, a magnetically separable Fe_3O_4/ZnO/AgBr nanocomposite was synthesized. These synthesized nanocomposites inactivated the *Fusarium graminearum* and *Fusarium oxysporum* within a short period of 120 and 60 min. Thus, the efficacy of nanocomposites can be identified as the combined aggregation of the inorganic metal complexes [41].

The leaf extract of *Adhatoda vasica* was utilized as a reducing agent to synthesize Copper oxide nanoparticles/Carbon nanocomposite through the green chemistry approach [42]. These nanocomposites exhibited effective growth inhibitory activity against *Aspergillus niger*. Thus, from these reports, it is evident that inorganic metals possessing inherent properties when combined as nanocomposites as a result of synergistic effects result in the fabrication of nanocomposites which exhibit activities on comparison with bare metals. However, the behavior of metal alloys during nanocomposite synthesis will differ according to the synthesis methods; elemental composition, and the applications.

Nanocomposites can be manufactured from any combination of materials like polymers, metals, and ceramics [43]. Among the materials, polymers and inorganic/organic materials will have a high aspect ratio and surface properties which enable them to be widely used in a different range of industries. In agriculture, polymers play an indispensable role in the release of chemical moieties such as fungicides, insecticides, growth stimulants and germicides [44]. The promising advantage of polymers is to control the release rate and rate of biodegradability of the embedded or encapsulated compound. These features make polymers widely used as a delivery agent in medicine and agrochemicals [45,46]. Polymers such as cellulose acetate phthalate, gelatin, chitosan, gum Arabic, polylactic acid, poly-butadiene, poly-lactic-glycolic acid, polyhydroxyalkanoates, polyvinyl alcohol (PVA), polyacrylamide, and polystyrene are widely used as delivery agents for drugs and agrochemicals [13,23].

In agriculture, the usage of polymer nanocomposites is to be critically chosen according to the application. Further, natural polymers are preferably chosen for the agri-applications owing to the nature of degradability and controlled release behavior. Chitosan polymer is gaining a new avenue in the plant protection field due to its outstanding properties. Nanochitosan exhibits anti-microbial potentials against bacteria and fungi at varying levels of concentrations. Combined formulation comprised of metal or metal oxide nanomaterials encapsulated or embedded in chitosan exhibit improved antimicrobial potentials. For instance, nanocomposite Ag-chitosan prominently exhibited antibacterial activity to a higher extent. Likewise, the fungicidal activity of clay chitosan nanocomposite was evaluated against *Penicillium digitatum* under in vivo and in vitro conditions [47]. The results were promising in terms of the superior activity exhibited by the nanocomposite formulation. Likewise, copper nanoparticles (Cu NPs), zinc oxide nanoparticles (ZnNPs), and chitosan, zinc oxide, and copper nanocomposites (CS–Zn-Cu NCs) were chemically fabricated and evaluated against plant pathogenic fungi *A. alternata*, *R. solani*, and *B. cinerea* [48]. The results intrinsically acclaimed that nanocomposite displayed higher activity at a concentration of 90 $\mu g\ mL^{-1}$. Antifungal activities of Bimetallic blends and Zn-Chitosan, and Cu-Chitosan at concentrations of 30, 60, 90 $\mu g\ mL^{-1}$ effectively inhibited the growth of *Rhizoctonia solani*. Further, it indicated effective control of cotton seedling damping-off under greenhouse conditions [49]. Silver/chitosan nanocomposite portrayed incremental growth-inhibitory effect against phytopathogens isolated from chickpea seeds [50]. Moreover, the individual metal and polymer components exhibited inhibitory activity lower than the Silver/Chitosan nanocomposite against the test pathogen, *Aspergillus niger*. A combination of chitosan/silica nanocomposite was evaluated against *Botrytis cinerea* under in vitro and in vivo (natural and artificial infections) conditions [51]. The in vitro study revealed complete reduction of the fungal growth by the nanocomposite compared to 72% and 76% inhibition potential of chitosan and silica nanoparticles respectively. Moreover, under natural conditions, the Chitosan/silica nanocomposite effectively hampered the gray mold disease in Italian grapes by 59% and in Benitaka grapes by 83% without affecting the grape quality.

The chitosan conjugated Ag nanoparticles functionalized with 4(*E*)-2-(3-hydroxynaphthalene-2-yl) diazenyl-1-benzoic acid were prepared which demonstrated improved effectiveness against *A. flavus* and *A. niger* forming larger inhibition zone of 20.2 mm and 27.0 mm respectively [52]. Likewise, chitosan hydrogel with cinnamic acid encapsulating *Mentha piperita* essential oil markedly inhibits the growth of mycelia of *A. flavus* at a concentration of 800 mg/mL [53]. Thus, with the above appropriate scientific evidence, it can be inferred that organic/inorganic-metal-polymer hybrid nanocomposites exhibit exceptionally superior anti-fungal activities under in vitro and in vivo conditions.

3.1. Antifungal Mechanism of Nanoparticles/Nanocomposites

A promising nano-fungicide should possess an equivalent or superior activity corresponding to the bulk metal at relatively lower concentrations. Moreover, it is desirable to understand the phyto and eco-toxicity issues due to the release of metal ions. Multifarious mechanisms were involved in the antifungal activity executed by nanomaterials. The generalized antifungal activity is provided in the Figure 3. The antifungal activity of nanomaterials can be accomplished by the following events. Generally, fungal cell wall and cell membrane architecture involves chitin, lipids, phospholipids and polysaccharides with specific predominance of mannoproteins, β-1,3-D-glucan and β-1,6-D-glucan proteins [54]. Internalization of the nanomaterials occurs through three mechanisms; (i) direct internalization of nanoparticles in the cell wall, (ii) specific receptor-mediated adsorption followed by internalization, (iii) internalization of nanomaterials through ion transport proteins [55]. Post-internalization, the nanomaterials may inhibit the enzyme β-glucan synthase thereby affecting the N-acetylglucosamine [N-acetyl-D-glucose-2-amine] synthesis in the cell wall of fungi. As a consequence of enzyme inhibition, abnormalities like enhanced thickening of the cell wall, liquefaction of cell membrane, dissolution or disorganization of the cytoplasmic organelles, hyper-vacuolization, and detachment of cell wall from cytoplasmic contents indicating incipient plasmolysis might occur [56].

At the molecular level, the nanomaterials interact with various biomolecules and form complexes with different biomolecules thereby causing structural deformation in the biomolecules, inactivation of the catalytic proteins, and nucleic acid abnormalities like DNA breakage, and chromosomal aberrations [57,58]. Reactive Oxygen Species (ROS) play a critical role in antifungal activity mechanism of nano-composite materials. The metal ions trigger ROS and damage the biomolecules leading to cell death. Further, to authenticate the role of ROS in antifungal mechanism; Lipovsky et al. [59] deciphered that the increased expression of lipid peroxidation is a clear indicator of ROS generation. Meanwhile, stress enzymes like superoxide dismutase, glutathione dismutase, ascorbate peroxidase were upregulated/downregulated upon nanomaterials treatment in fungi [60].

3.2. Nano-Hybrid Antifungals for Control of Toxigenic Fungi and Mycotoxins Degradations

Mycotoxins are the natural contaminants in food and feed worldwide. Changing climatic patterns have significantly affected the agricultural production due to limiting water, and land resources, temperature extremes and elevated humidity conditions [61]. The elevated humidity and temperature allowed for the proficient growth of a variety of mycotoxigenic fungal genera, *Aspergillus* spp., *Penicillium* spp., and *Fusarium* spp. which produce a variety of mycotoxins including aflatoxins (AFs), fumonisins (FBs), ochratoxins (OTs), trichothecenes (TCs), and zearalenone (ZEA). These mycotoxins cause detrimental health impacts in humans manifested as liver cancer, aflatoxicosis, malabsorption syndrome and reduction in bone strength [62].

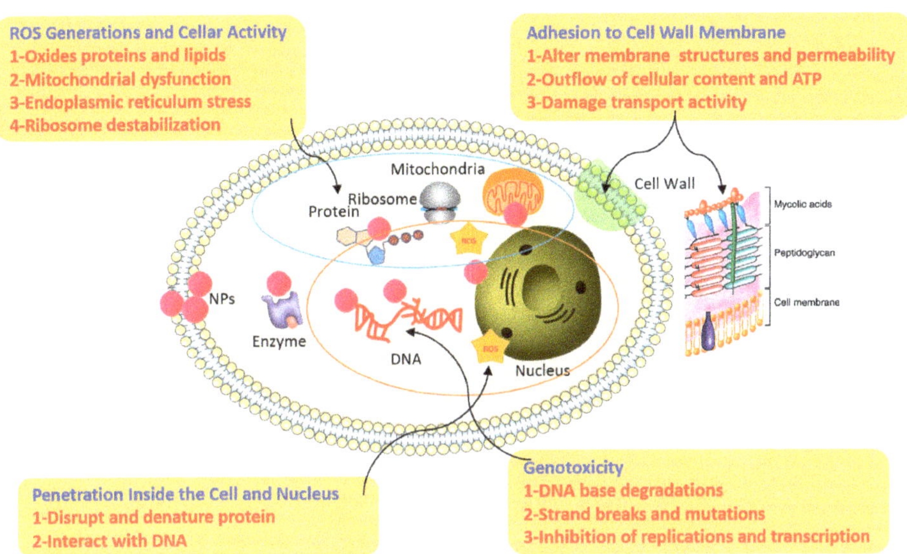

Figure 3. Antifungal activity mechanisms of hybrid nanomaterials.

Conventionally, mycotoxins can be detected by chromatographic techniques like High-Pressure Liquid Chromatography (HPLC), Gas chromatography-Mass Spectrometry (GC-MS), and Liquid Chromatography-Mass Spectrometry (LC-MS) [62,63]. These techniques are robust, sensitive and specific but have high cost of analysis per sample and are therefore, expensive and time consuming. Further, to advance with innovation a simple, cheap and sensitive technique can be achieved with the aid of nanotechnology. Nano-based mycotoxin detection, and management involves specific properties such as selectivity, sensitivity, simplicity and multiple capabilities [64]. Hybrid nanomaterials are a new paradigm to counteract mycotoxin management. Generically, hybrid nanomaterials are having superior properties and multimodality (simultaneous detection, detoxification, and management abilities) when compared with polymers/metals, organic molecules when used individually in mycotoxicology [65].

Hybrid nanomaterials consisting of polymers/metals/organic molecules can synergistically interact with each other and accelerate the reaction kinetics (Figure 4). In mycotoxin management, nanohybrid materials are used for detection, detoxification and management [61]. For instance, Bhardwaj et al. [66] developed an immunosensor comprising graphene quantum dots (GOD), gold nanoparticles (AuNPs). Further, GOD-AuNPs were fabricated onto an indium tin oxide (ITO) electrode modified with an antibody (anti-AFB1) (anti-AFB1/GQDs-AuNPs/ITO) to detect Aflatoxin (AFB1). The hybrid immunosensor detected the AFB1 with high sensitivity for the presence of aflatoxin B1 even at very low concentrations (0.1 to 3.0 ng/mL) in the food sample.

Another important concern is mycotoxin detoxification. Hybrid nanomaterials are smart detoxifying agents. The hybrid nanomaterials can be incorporated in feed to sequester mycotoxin by forming a complex in the gastrointestinal tract so that the severity of the toxin gets ceased. To detoxify mycotoxin, Hamza et al. [67] devised a hybrid nanomaterial comprised of β-glucan mannan lipid particles (GMLPs) encapsulating the humic acid nanoparticles (HA-FeNPs). The specificity of this hybrid material was that the β-glucan molecules produced 3 to 4 µm hollow porous microspheres. Moreover, the addition of humic acid increased the binding affinity of Aflatoxin B1. The bare GMLPs and HA showed a moderate binding affinity for aflatoxin (10.8 µg AFB1/mg HA for GMLP HA). However, the addition of Fe increased the adsorbent capacity for GMLP HA-FeNPs AFB1 mass to

13.5 μg AFB1/mg HA and 16.8 μg AFB1/mg. This study showcased that this nanohybrid material can be used as a safe detoxification agent.

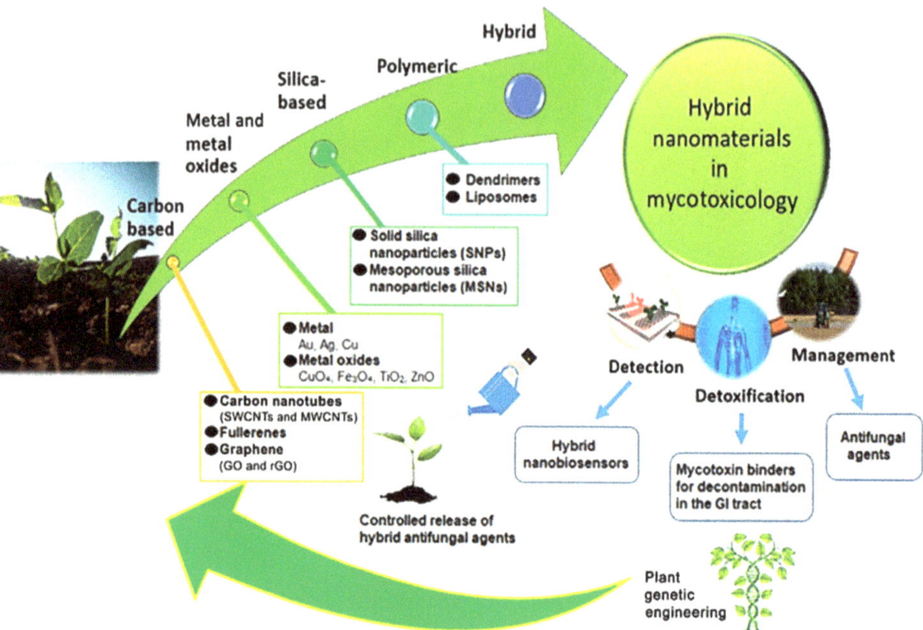

Figure 4. Hybrid-nanomaterial categories used in plant science: carbon nanotubes, including single-walled carbon nanotubes (SWCNTs) and multiwalled carbon nanotubes (MWCNTs), fullerenes); metallic and metal oxide NPs, silica-based nanostructures, and polymeric (dendrimers and liposomes) for the fabrication of biosensors for mycotoxin detection, detoxification of mycotoxin-contaminated food and feedstuffs through binders and management for sustainable control over fungal growth and mycotoxin contamination. (Cited from Thipe et al. with permission from Elsevier).

Recent findings related to nanocomposites (NCPs) based on organic polymeric and inorganic matrices or hybrid materials as effective antifungal agents against mycotoxigenic fungi and mycotoxin reduction have been summarized by Jampílek and Králová, [68]. Spadola et al. [13] have identified an interesting alternative technique to inhibit aflatoxin production in *Aspergillus flavus*. They have formulated nanoparticles of poly-ε-caprolactone polymer and loaded the generated nanoparticles with two thiosemicarbazone (benzophenone or valerophenone) compounds to curb mycotoxin production in *A. flavus*. Pirouz et al. [69] investigated the use of hybrid magnetic graphene oxides (MGOs) as an adsorbent for DON, ZEA, HT-2, and T-2 in naturally contaminated palm kernel cakes (PKC). At optimum reduction conditions at pH 6.2 for 5.2 h at 40.6 °C, the MGO was able to reduce the amount of DON, ZEA, HT-2. In the same direction, MGO adsorbents have been used to detoxify polluted AFB1 oils, their absorbents are made of MGO and magnetic reduced graphene oxides (MrGO) both of which are incorporated with Fe_3O_4 nanoparticles. The MGO and MrGO were renewable, however, after seven cycles, with no major losses in the adsorption activities [70]. Copper-chitosan nanocomposite-based chitosan hydrogels (Cu-Chit/NCs hydrogel) have been prepared using metal vapor synthesis (MVS). Also, SEM measurements revealed damage to *A. flavus* cell membranes. Current findings indicate that the antifungal activity of nanocomposites in vitro can be beneficial depending on the type of fungal strain and the concentration of nanocomposites (Figure 5A). Cu-Chit/NCS hydrogel is a revolutionary nanobiopesticide developed by MVS used in food and feed to induce plant protection against mycotoxigenic fungi [71]. The fungicidal behavior of chitosan-silver nanocomposites (Ag-Chit-NCs) against *Penicillium expansum* from the feed samples was investigated. Ag-Chit-NCs < 10 nm in size have an important antifungal inhibitory effect against *P. expansum*, the causative agent of blue mold-contaminated dairy cattle feed [72]. *P. expansum* treated with Ag-Chit *P. expansum* treated with Ag-Chit NCs was investigated by HR-SEM, alterations in conidiophores, metulae, phialides, and mature conidia characteristics had been observed to obtain information about the mode of action of Ag-Chit-NCs (Figure 5B). Therefore, nanocomposites can be utilized as viable alternatives to the already available arsenal of fungicides (Table 1).

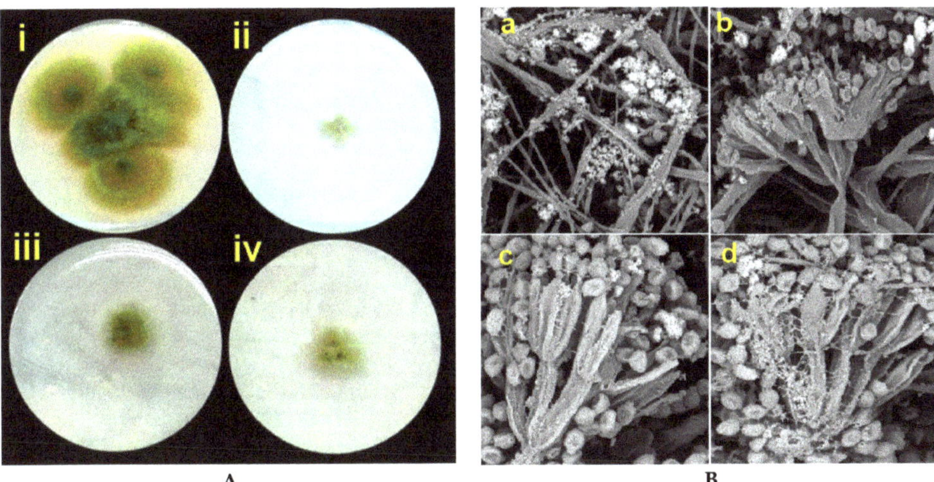

Figure 5. (**A**) Antifungal activity of *Ag-Chit NCs against A. flavus* collected from feed samples. (i), control (without nanocomposite treatment), (ii), (iii), and (iv) fungal mat treated with 30, 60, and 90 milligrams of nanocomposites. All petri dishes treatment was incubated at 28 °C for 10 days. (**B**) Fungal mycelium of *P. expansum* treated with Ag-Chit NCs referred to the morphological changes in fungal hyphae, SEM images depicted markedly shriveled, crinkled cell walls, and flattened hyphae of the fungi (a), hyphal cell wall and vesicle damaged (b), irregular branching (a and b), and collapsed cell, formation of a layer of extruded material (d) Source (Abd-Elsalam KA. unpublished data).

To prevent fungal growth and mycotoxin production in food materials and packaging; hybrid nanofiber mats composed of cellulose acetate encapsulated with AgNPs were prepared by electrospinning. Due to close tight assembly of packaging in food materials, these nanofiber mats allowed low penetration of air permeability preventing fungal growth, and silver nanoparticles inherently inhibited the growth of yeast and molds [73]. Thus, these nanocomposite materials have shown great potential for future applications in the food packaging and preservation industry.

3.3. Postharvest Management of Nanocomposite Against Pathogenic Fungi

Perishable vegetables and fruits are spoiled due to transport, storage and growth of spoilage and opportunistic microbes. Therefore, microbial decay of fruits and vegetables is a great concern for researchers to formulate a driven strategy control measures with long-standing efficiency. Nanotechnology is an alternative solution to develop sustainable horticulture in preserving and managing post-harvest diseases of fruits and vegetables [74,75]. This technology offers various products such as packaging thin films; helping for labeling fresh products using the multiple chips (nanobiosensors), improvement of packaging appearance and prevention of the impact of gases and unsafe rays.

Conventionally, fungicides like imazalil, thiabendazole, pyrimethanil, fludioxonil and chloride-based chemicals have been used for management of post-harvest diseases of horticultural produce [61]. Though effective but prolonged use of these fungicides has led to development of among the fungal genera. Further, the active ingredients of the fungicides are toxic to humans and also to the ecosystem. Post-harvest diseases can be classified into two groups (i) diseases from field infection (ii) diseases due to post-harvest infection ([76,77]).

Citrus orange fruit, *Citrus sinensis* L. Osb., is often spoiled by *P. digitatum* during the post-harvest storage/transportation periods. To counteract against the pathogen a nanocomposite clay-chitosan nanocomposite (CCNC) was synthesized and evaluated under in-vitro conditions; at 20 µg/mL the nanocomposite completely inhibited the growth of *P. digitatum*. In *in-vivo* trials, the nanocomposite reduced the lesions by 70% and inhibited the disease in orange [47]. The CNCC coated orange were observed to be free from the disease, and exhibited high pH, chroma, peel moisture, and firmness in comparison with the control [47].

A chitosan-Titanium dioxide composite film (70 µm thickness) was used as a packaging material to extend shelf life during the postharvest storage of grapes by preventing spoilage microbial infection. The composite film enhanced the shelf life and resisted the mildew infection in stored grapes for up to 22 days [78]. Similarly, a nanocomposite of silver/gelatin/chitosan was applied as a hybrid film in grapes to improve the storage shelf-life under cold conditions. The hybrid film stored grapes didn't show signs of infection, had a fresh appearance and showed no leakage of the grapes [79].

Banana (*Musa acuminata* L.) is a famed fruit consumed by all peoples and is considered to be one of the high-valued fruit in horticulture. During post-harvest, the banana fruits get deteriorated on storage and transportation periods due to its climacteric nature i.e., increased respiration rate. To overcome this and ensure delayed ripening, maintenance of fruit firmness and reduced mass loss of fruits coating of a nanocomposite containing soybean protein isolate, cinnamaldehyde, and ZnO NPs on the banana fruit was observed to be very effective [80]. In another study, a predominant disease (anthracnose) caused by *Colletotrichum musae* in banana causing a major loss to the farmers was prevented by use of metallic nanoparticles (silver, nickel, copper and magnesium) prepared from ajwain and neem leaf extract. In postharvest period, silver nanoparticles were sprayed on the banana at different concentrations ranging from 0.02 to 0.2 percent resulting in reduction in the anthracnose infection (6.67 percent disease index) on use of the least concentration of 0.2% AgNPs [81].

Apple (*Malus domestica* Bork) is a perishable fruit consumed by all ages worldwide. Since it is a climacteric fruit, it is indeed useful to pro-act for development of a sustainable methodology to prevent post-harvest losses of apple fruit. Polylactic acid incorporated with ZnO nanoparticles was applied as a thin film in a fresh-cut apple stored at 4 °C for 14 days. PLA-ZnO NPs exhibited effective inhibition of yeast and molds in fresh-cut apples. This showed the possibility that PLA nanocomposite can be used as a packaging material in apples during the storage period [82].

Mango (*Mangifera Indica* L.) is also supposed to be highly prone to infections by post-harvest pathogen(s) causing anthracnose diseases. A chitosan-silver NPs composite was prepared and evaluated for the antifungal effect against conidial germination of *C. gloeosporioides*. The prepared chitosan-silver nanocomposite (at 100 µg mL^{-1}) completely suppressed the spore germination. Under in vivo conditions, mango fruits were inoculated with the fungal spores. The infected mango fruit were then coated with the aforementioned nanocomposite and evaluated for the disease incidence. The nanocomposite prominently lowered the disease incidence with 45.7% and 71.3% at 0.5 and 1.0% of nanocomposite respectively [83]. Likewise, a nanocomposite containing aloe vera gel, ZnOPs and glycerol was coated on an edible mango and stored for 9 days at room temperature. After the storage days, the edible mango didn't show any sign of infection/diseases [84]. Similarly, nanoemulsion containing chitosan was effectively inhibited *C. musae* and *C. gloeosporioides*. However, the chitosan nanoemulsion showed better prospective results than chitosan nanocomposite in banana, papaya and dragon fruits [85]. New biopolymers composite oligochitosan (OCS) and OCS/nanosilica (OCS/nSiO_2) hybrid materials with impressive synergistic action are likely to be considered potential protections for plants infected with

Colletotrichum sp. Conjugated nanomaterials may be considered possible biotic elicitors that not only avoid anthracnose disease but also efficiently enhance plant growth [86]. The same team found that silica and hybrid material had good antifungal properties against *P. infestans*, the causal organism for late blight in tomato and potato, but the antifungal properties of hybrid materials, due to their synergistic effect, had better antifungal capacities than that of each individual component. Interestingly, the inhibition zone diameters of OCS/nSiO_2 were approximately 4–5 mm and 7–9 mm larger than those of OCS and nSiO_2, respectively [87]. Bio-synthetized MgO nanoparticles are produced using the native bacterial strain like *Bacillus* sp. The RNT3 strain was used to render CS-Mg nanocomposite. CS-Mg nanocomposite has demonstrated impressive antimicrobial activity against *Acidovorax oryzae* and *R. solani* and substantially inhibited development compared to the non-treated control [88]. In vivo assays with two plant hosts including tomatoes and peppers affected by *Fusarium* wilt and root rot diseases in which traditional chemical fungicides were used for comparative purposes displayed better antifungal activity of rGO-CuO NPs and a long-lasting impact at a very low concentration of 1 mg/mL. Interestingly, as CuO is a plant nutrient, the study of treated plants showed a positive impact on flowering, plant height and dry weight, as well as the aggregation of photosynthetic pigments [89].

Table 1. Nano-composite formulation and their application for curbing the plant fungal pathogens.

Type of Nanocomposite Applied	Method of Synthesis	Effective Working Concentration	Application Method	Pathogen Studied	Remarks	References
			Inorganic-inorganic composites			
AgNPs-titanate nanotubes	Photo-assisted functionalization of AgNPs on hydrothermal micro-wave-assisted synthesis of titanate nanotubes	30 mg/30 mL	In vitro study performed in PDB supplemented with 30 mg photo-activated AgNP-titanate nanotube composite	*Botrytis cinerea*	Fungal conidial death due to ROS damage	[40]
Fe_3O_4/ZnO/AgBr	Microwave-assisted synthesis	1:8 weight ratio nanocomposite	In vitro spore broth incubation study performed in a cylindrical Pyrex reactor	*Fusarium graminearum*, *Fusarium oxysporum*	Complete inactivation of test fungi within one hour of incubation with the nanocomposite	[41]
Bimetallic (Au/Ag) NPs with metal oxide NPs(ZnO NPs)	Physical mixture technique	50:10 µg/mL	In vitro poison food study involving the addition of NP suspension in SDB	*Aspergillus flavus*/ *A. fumigatus*	-Augmented inhibition of fungal growth by bimetallic and metal oxide NPs	[90]
ZnO:$Mg(OH)_2$ composite	Hydrothermal/co-precipitation technique	5 to 0.002 mg mL^{-1}	In vitro study involving DMSO dissolved NPs supplemented in PDB	*Colletotrichum gloeosporioides*	-Addition of MgO diminished the antifungal potential of ZnO NPs	[91]
			Inorganic-carbon composites			
CuO NPs functionalized graphene-like carbon composite	Green synthesis using *Adhatoda vasica* leaf extract and 0.01 M $CuSO_4$	5:4 ratio proportion of leaf extract: $CuSO_4$	In vitro agar well diffusion assay on PDA media	*Aspergillus niger*, *Candida albicans*	growth inhibition due to the disruption of the cell membranes	[42]
GO-AgNPs	Interfacial electrostatic self-assembly synthesis	9.37 µg/mL-MIC value	In vitro assay using growth media and detached wheat leaf bioassay	*Fusarium graminearum*	Improved anti-fungal efficiency [>3-fold for AgNPs and >7-fold over pure GO] through two mechanisms (physical injury and ROS mediated chemical injury)	[38]

Table 1. *Cont.*

Type of Nanocomposite Applied	Method of Synthesis	Effective Working Concentration	Application Method	Pathogen Studied	Remarks	References
			Inorganic-organic composites			
ZnO NPs/CS-Zn-CuNPs	Wet chemical method	0 to 90 μg mL^{-1}	In vitro study involved supplementation of nanocomposite in PDA media	*Alternaria alternata*, *B. cinerea*, *R. solani*	-Highest mycelial inhibition by chitosan mixed Zn-Cu nanocomposite	[48]
Cu-/Zn-chitosan and bimetallic nanocomposites	Wet chemical synthesis	30, 60, and 100 μg mL^{-1}	-In vitro study using agar based media -In vivo seed priming assay for damping-off disease in cotton cultivar Giza 92 seedlings	*Rhizoctonia solani*	-highest hyphal inhibition at 100 μg mL^{-1} -Augmented effect of bimetallic NC along with biocontrol fungus (*Trichoderma*)to suppress disease in vivo	[49]
Clay-chitosan nanocomposite	Anion-exchange technique	5 to 60 μg mL^{-1}	-In vitro study using PDA -In vivo assay in mature fruits of *Citrus sinensis* (L. Osbeck) cv. Valencia	*Penicillium digitatum*	-Complete inhibition of fungal hyphae in different weight ratios of clay/chitosan nanocomposite (1:0.5, 1:1, 1:2) (conc.20 μg mL^{-1})	[47]

4. Challenges

Food production and safety is a primary concern for all researchers to ensure that what we are consuming is a safe food free from contamination and an assurity of maintenance of the characteristic quality traits of the food. The primary problem of food is contamination by microbes affecting the plants at varying degree levels at seedling, rooting, flowering, and fruit stage, post-harvesting stage. Though we are practicing, routine intensive usage of chemical fertilizers, pesticides, inoculants, agrochemicals, and chemical sprayers during the postharvest period, but concerning the toxicity and accumulation of toxicants in food products causes major health allied problems in humans and also to the environment [3].

In this modern age of science, every decade is witnessing the advent of innovative scientific concepts and applications catering to the well-being of the human population and the environment. Likewise, this decade is comforted by the emergence of nanotechnology. This technology can resolve daunting problems persisting in all the sectors of physical, chemical, and biological sciences. In agriculture, the use of nanotechnology is indispensable owing to the reasons of superior properties.

Despite the usage of nanotechnology in agriculture, certain challenges have existed that need to be rectified or eliminated. In antifungal management, nanohybrid materials are being used. These consist of elements like silver, gold, copper, iron, graphene, silica; polymers like chitosan, PVC, PLGA, and other organic molecules that are incorporated to obtain composites or nano-hybrids. The nanohybrid production methods are cost-intensive since these involve the use of expensive chemicals, reagents, and physical energy. Thus, the produced nanohybrids may exhibit high effectivity against phytopathogens but on application under field conditions, these nanohybrids may exhibit off-target movements and may enter into the plant system or may get accumulated in the vegetative parts of the plants.

The effect of nanoparticles on crop plants must be understood before developing any kind of nano-formulations for antibacterial/fungicidal applications. Many researchers have suggested that nanoparticles can potentially harm plant growth and development. Dimpka et al. [92] reported that CuONPs can affect the root and shoot growth in wheat. Further, the chlorophyll content and enzyme activities peroxidase and catalase activities were reduced in CuONPs treated plants. Likewise, AgNPs at higher concentrations (10 mg/L) can have altered on the metabolism and cell defense mechanism in wheat [93]. TiO_2 when exposed to *Oryza sativa* has shown to result in reduced biomass contents, changes in metabolite concentrations, and alterations in the respiration pathways [94]. Likewise, Carbon nanotubes reduced the length of root and shoot of a rice plant besides inducing the DNA damage [95].

Nanoparticles not only exert negative effects in plants but also may have altering effects on the soil microbial communities. All metallic/metal oxide nanoparticles reduce the microbial abundance at varying levels. *Pseudomonas putida* an important bacterium in nitrogen recycling was affected by carbon nanotubes and ZnONPs [96]. The same ZnONPs were also reported to have negatively affected the soil beneficial fungi also. Soil Microbial Biomass Carbon (SMBC) was observed to be drastically very low in fields where the soil microbes were exposed to AgNPs, NiNPs, CuONPs and carbon nanotubes [96]. The induction of reactive oxygen species (ROS) by Ag-SiO_2 core-shell nanocomposites were responsible for radial hyphal growth inhibition of a few plant pathogenic fungi [97].

Another important challenge is that nanoparticles can trespass through food chains and accumulate at higher trophic levels. A report of Vittori Antisari et al. [98] documented that tomato plants exposed to engineered metal oxides resulted in increased concentrations of K and decreased in the Mg, P, and S contents in the fruits of tomato. When such fruit will be consumed by humans, their fate remains questionable. Therefore, the usage of nanoparticles remains a big and unresolved challenge for agricultural applications particularly involving their use in open field conditions.

5. Future Perspective

With the pros and cons of nanotechnology challenges; it is a time to explore the correct and beneficial usage of nanotechnology in agriculture management. To obtain the nanocomposite as a safe, clean, and eco-friendly agent for the management of fungal diseases in plants and postharvest period the following criteria can be accomplished. At present, chemical-based synthesis of nanocomposite is reported so far. However, the researchers are now endeavoring to fabricate nanocomposite through a green chemistry approach involving the utilization of agri-wastes such as banana or orange peels, wheat whiskers, straw, cotton or corn stalks, coconut or almond shells, corn silk, rice husks for the production of nanoscale carbon, silica, graphene, cellulose, and chitosan polymers. Further, tremendous opportunities of use and application of green synthesized inorganic metal/metal oxide nanoparticles can be identified in agriculture and particularly for the control and management of plant diseases caused by several fungal phytopathogens [99–101]. The biological synthesis protocols can improve the cost, time and energy requirements besides will help decrease the amounts of environment-corrosive chemicals required for the industrial production of nanomaterials and their composites through the most prevalent physical/chemical synthesis techniques [100]. The most striking benefit of the use of biodegradable polymers for development of nanocomposites by conjugation of metal/metal oxide nanoparticles is their ease of translocation within the plant tissues and the ability to exert in planta antifungal activity. Therefore, the use of biodegradable polymers must be encouraged in future research due to their biocompatible and eco-friendly characteristics.

Size and stability are two important factors for designing a novel nanocomposite. Producing a size-controlled nanocomposite will be a key success in antifungal management. Stability should be maintained till the end of the period. Before applying a nanocomposite material under field conditions; researchers should ensure the toxicity of the applied nanocomposite to the non-target organisms. Many nanocomposite materials are developed from toxic nanomaterials, for example, TiO_2 is reported to produce colon cancer. Therefore, the derived nanocomposites can be toxic to plants, microbes, and the environment, and hence, a careful preparation of nanocomposite with minimal toxicity must be preceded.

The prepared nanocomposite must not exhibit undesirable effects in plants and fruits. In certain cases, the fruit ripening process may get delayed more than the expected period. Prompt application of nanocomposites as spray/emulsion can be encouraged during the post-harvest period (storage) (Figure 6). Novel nanomaterials (sensors, kits) should be developed to detect, quantify, and analyze the fungal pathogen during the post-harvest period. The toxicity of nano-composites depends on the concentration used. Compared to chemical-based pesticides/fungicides, the working concentrations of the nanocomposites are relatively very low. Another pressing challenge for the nano-products is hurdles faced in the marketing of these products possibly due to production cost, unclear technical benefits, public opinion, and legislative uncertainties. Compared to other sectors, the usage of nanotechnology in agriculture is marginal and needs attention.

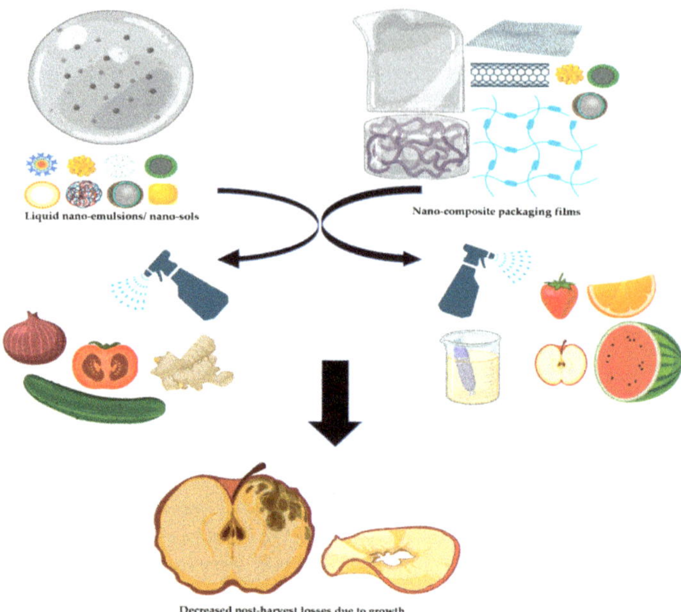

Figure 6. Application of nano-composite formulations to decrease post-harvest losses of horticultural produce.

6. Conclusions

A product that can attribute positive outcomes to our intended purpose must be welcome. Likewise, identification of the boons and banes of the nanocomposite smart materials as effective antagonistic agents to curb the fungal pathogens is critical. In this review, we have clearly emphasized the significance of nanocomposites in fungal disease management in a comprehensive approach. Post-harvest management of fruits by nanocomposites offers a successful tool to combat diseases and infections leading to produce loss through spoilage and decay.

Our findings have indicated that the control of toxigenic fungi and the detoxification of mycotoxins are not adequate for sustainable agricultural ergonomics. Therefore, novel treatment methods for improving the food safety and protection must be applied. Nanohybrid antifungals are thus, of primary importance for a synergistic approach to resolve diverse problems in the management of fungal pathogens causing agricultural/post-harvest diseases in the 21st century, with a focus on Green Nanotechnology, which is environmentally sustainable and provides a continuum for the plant, animal and human health. The nanohybrid anti-fungals are anticipated to cater to the need of the growers, consumers as well as the environment activists through rapid, effective, and comparatively improved eco-safety attributes for controlling the yield and produce quality deterring potential of the fungal phytopathogens.

Author Contributions: Conceptualization, K.A.A.-E. and R.P.; resources, M.V., K.K.; writing—original draft preparation, K.A.A.-E.; R.C.; M.A.A., writing—review and editing, A.K., K.A.A.-E., M.V., K.K.; visualization, K.B.; P.B.; funding acquisition, M.V., K.K. All authors have read and agreed to the published version of the manuscript.

Funding: The authors would like to acknowledge the funding received from UHK VT2019-2021 and the Ministry of Health of the Czech Republic (FN HK 00179906) and the Charles University in Prague, Czech Republic (PROGRES Q40).

Conflicts of Interest: The authors declare no conflict of interest.

References

1. Hassaan, M.A.; El Nemr, A. Pesticides pollution: Classifications, human health impact, extraction and treatment techniques. *Egypt. J. Aquat. Res.* **2020**, *46*, 207–220. [CrossRef]
2. Meena, R.S.; Kumar, S.; Datta, R.; Lal, R.; Vijayakumar, V.; Brtnicky, M.; Sharma, M.P.; Yadav, G.S.; Jhariya, M.K.; Jangir, C.K.; et al. Impact of agrochemicals on soil microbiota and management: A review. *Land* **2020**, *9*, 34. [CrossRef]
3. Kalia, A.; Gosal, S.K. Effect of pesticide application on soil microorganisms. *Arch. Agron. Soil Sci.* **2011**, *57*, 569–596. [CrossRef]
4. Shuping, D.S.S.; Eloff, J.N. The use of plants to protect plants and food against fungal pathogens: A review. *Afr. J. Tradit. Complement. Altern. Med.* **2017**, *14*, 120–127. [CrossRef] [PubMed]
5. Yadav, S.; Sharma, A.K.; Kumar, P. Nanoscale Self-Assembly for Therapeutic Delivery. *Front. Bioeng. Biotechnol.* **2020**, *8*, 1–24. [CrossRef] [PubMed]
6. Patra, J.K.; Das, G.; Fraceto, L.F.; Campos, E.V.R.; Rodriguez-Torres, M.D.P.; Acosta-Torres, L.S.; Diaz-Torres, L.A.; Grillo, R.; Swamy, M.K.; Sharma, S.; et al. Nano based drug delivery systems: Recent developments and future prospects. *J. Nanobiotechnol.* **2018**, *16*, 1–33. [CrossRef] [PubMed]
7. Shang, Y.; Kamrul Hasan, M.; Ahammed, G.J.; Li, M.; Yin, H.; Zhou, J. Applications of nanotechnology in plant growth and crop protection: A review. *Molecules* **2019**, *24*, 2558. [CrossRef] [PubMed]
8. Worrall, E.A.; Hamid, A.; Mody, K.T.; Mitter, N.; Pappu, H.R. Nanotechnology for plant disease management. *Agronomy* **2018**, *8*, 285. [CrossRef]
9. Pestovsky, Y.S.; Martínez-Antonio, A. The Use of Nanoparticles and Nanoformulations in Agriculture. *J. Nanosci. Nanotechnol.* **2017**, *17*, 8699–8730. [CrossRef]
10. Camara, M.C.; Campos, E.V.R.; Monteiro, R.A.; Do Espirito Santo Pereira, A.; De Freitas Proença, P.L.; Fraceto, L.F. Development of stimuli-responsive nano-based pesticides: Emerging opportunities for agriculture. *J. Nanobiotechnol.* **2019**, *17*, 1–19. [CrossRef]
11. Rangaraj, S.; Gopalu, K.; Muthusamy, P.; Rathinam, Y.; Venkatachalam, R.; Narayanasamy, K. Augmented biocontrol action of silica nanoparticles and *Pseudomonas fluorescens* bioformulant in maize (*Zea mays* L.). *RSC Adv.* **2014**, *4*, 8461–8465. [CrossRef]
12. Paek, S.M.; Oh, J.M.; Choy, J.H. A lattice-engineering route to heterostructured functional nanohybrids. *Chem. Asian J.* **2011**, *6*, 324–338. [CrossRef] [PubMed]
13. Spadola, G.; Sanna, V.; Bartoli, J.; Carcelli, M.; Pelosi, G.; Bisceglie, F.; Restivo, F.M.; Degola, F.; Rogolino, D. Thiosemicarbazone nano-formulation for the control of *Aspergillus flavus*. *Environ. Sci. Pollut. Res.* **2020**. [CrossRef] [PubMed]
14. Iavicoli, I.; Leso, V.; Beezhold, D.H.; Shvedova, A.A. Nanotechnology in agriculture: Opportunities, toxicological implications, and occupational risks. *Toxicol. Appl. Pharmacol.* **2017**, *329*, 96–111. [CrossRef] [PubMed]
15. Wang, P.; Lombi, E.; Zhao, F.; Kopittke, P.M. Nanotechnology: A New Opportunity in Plant Sciences. *Trends Plant Sci.* **2016**, *21*, 699–712. [CrossRef]
16. Kashyap, P.L.; Kumar, S.; Srivastava, A.K. Nanodiagnostics for plant pathogens. *Environ. Chem. Lett.* **2017**, *15*, 7–13. [CrossRef]
17. Rad, F.; Mohsenifar, A.; Tabatabaei, M.; Safarnejad, M.R.; Shahryari, F.; Safarpour, H.; Foroutan, A.; Mardi, M.; Davoudi, D.; Fotokian, M. Detection of *Candidatus Phytoplasma aurantifolia* with a quantum dots fret-based biosensor. *J. Plant Pathol.* **2012**, *94*, 525–534. [CrossRef]
18. Yao, K.S.; Li, S.J.; Tzeng, K.C.; Cheng, T.C.; Chang, C.Y.; Chiu, C.Y.; Liao, C.Y.; Hsu, J.J.; Lin, Z.P. Fluorescence Silica Nanoprobe as a Biomarker for Rapid Detection of Plant Pathogens. *Adv. Mater. Res.* **2009**, *79–82*, 513–516. [CrossRef]
19. Singh, S.; Singh, M.; Agrawal, V.V.; Kumar, A. An attempt to develop surface plasmon resonance based immunosensor for Karnal bunt (*Tilletia indica*) diagnosis based on the experience of nano-gold based lateral flow immuno-dipstick test. *Thin Solid Films* **2010**, *519*, 1156–1159. [CrossRef]
20. Wang, Z.; Wei, F.; Liu, S.Y.; Xu, Q.; Huang, J.Y.; Dong, X.Y.; Yu, J.H.; Yang, Q.; Di Zhao, Y.; Chen, H. Electrocatalytic oxidation of phytohormone salicylic acid at copper nanoparticles-modified gold electrode and its detection in oilseed rape infected with fungal pathogen *Sclerotinia sclerotiorum*. *Talanta* **2010**, *80*, 1277–1281. [CrossRef]
21. Schwenkbier, L.; Pollok, S.; König, S.; Urban, M.; Werres, S.; Cialla-May, D.; Weber, K.; Popp, J. Towards on-site testing of *Phytophthora* species. *Anal. Methods* **2015**, *7*, 211–217. [CrossRef]
22. Ariffin, S.A.B.; Adam, T.; Hashim, U.; Faridah, S.; Zamri, I.; Uda, M.N.A. Plant diseases detection using nanowire as biosensor transducer. *Adv. Mater. Res.* **2014**, *832*, 113–117. [CrossRef]
23. Ashfaq, M.; Talreja, N.; Chuahan, D.; Srituravanich, W. Polymeric Nanocomposite-Based Agriculture Delivery System: Emerging Technology for Agriculture. *Genet. Eng. Glimpse Tech. Appl.* **2020**, 1–16. [CrossRef]
24. Winter, J.; Nicolas, J.; Ruan, G. Hybrid nanoparticle composites. *J. Mater. Chem. B* **2020**, *8*, 4713–4714. [CrossRef]
25. Adnan, M.M.; Dalod, A.R.M.; Balci, M.H.; Glaum, J.; Einarsrud, M.A. In situ synthesis of hybrid inorganic-polymer nanocomposites. *Polymers* **2018**, *10*, 1129. [CrossRef] [PubMed]
26. Fahmy, H.M.; Salah Eldin, R.E.; Abu Serea, E.S.; Gomaa, N.M.; AboElmagd, G.M.; Salem, S.A.; Elsayed, Z.A.; Edrees, A.; Shams-Eldin, E.; Shalan, A.E. Advances in nanotechnology and antibacterial properties of biodegradable food packaging materials. *RSC Adv.* **2020**, *10*, 20467–20484. [CrossRef]
27. Idumah, C.I.; Zurina, M.; Ogbu, J.; Ndem, J.U.; Igba, E.C. A review on innovations in polymeric nanocomposite packaging materials and electrical sensors for food and agriculture. *Compos. Interfaces* **2020**, *27*, 1–72. [CrossRef]
28. Sadeghi, R.; Rodriguez, R.J.; Yao, Y.; Kokini, J.L. Advances in Nanotechnology as They Pertain to Food and Agriculture: Benefits and Risks. *Annu. Rev. Food Sci. Technol.* **2017**, *8*, 467–492. [CrossRef]

29. Padmanaban, V.C.; Giri Nandagopal, M.S.; Madhangi Priyadharshini, G.; Maheswari, N.; Janani Sree, G.; Selvaraju, N. Advanced approach for degradation of recalcitrant by nanophotocatalysis using nanocomposites and their future perspectives. *Int. J. Environ. Sci. Technol.* **2016**, *13*, 1591–1606. [CrossRef]
30. Wei, X.; Wang, X.; Gao, B.; Zou, W.; Dong, L. Facile Ball-Milling Synthesis of CuO/Biochar Nanocomposites for Efficient Removal of Reactive Red 120. *ACS Omega* **2020**, *5*, 5748–5755. [CrossRef]
31. Kalia, A.; Sharma, S.P.; Kaur, H.; Kaur, H. Novel nanocomposite-based controlled-release fertilizer and pesticide formulations: Prospects and challenges. In *Multifunctional Hybrid Nanomaterials for Sustainable Agri-Food and Ecosystem*; Abd-Elsalam, K.A., Ed.; Elsevier: Amsterdam, The Netherlands, 2020; pp. 99–134.
32. Guha, T.; Gopal, G.; Kundu, R.; Mukherjee, A. Nanocomposites for Delivering Agrochemicals: A Comprehensive Review. *J. Agric. Food Chem.* **2020**, *68*, 3691–3702. [CrossRef] [PubMed]
33. Usman, M.; Farooq, M.; Wakeel, A.; Nawaz, A.; Cheema, S.A.; Ur Rehman, H.; Ashraf, I.; Sanaullah, M. Nanotechnology in agriculture: Current status, challenges and future opportunities. *Sci. Total Environ.* **2020**, *721*, 137778. [CrossRef] [PubMed]
34. Adisa, I.O.; Pullagurala, V.L.R.; Peralta-Videa, J.R.; Dimkpa, C.O.; Elmer, W.H.; Gardea-Torresdey, J.L.; White, J.C. Recent advances in nano-enabled fertilizers and pesticides: A critical review of mechanisms of action. *Environ. Sci. Nano* **2019**, *6*, 2002–2030. [CrossRef]
35. Min, J.S.; Kim, K.S.; Kim, S.W.; Jung, J.H.; Lamsal, K.; Kim, S.B.; Jung, M.; Lee, Y.S. Effects of colloidal silver nanoparticles on sclerotium-forming phytopathogenic fungi. *Plant Pathol. J.* **2009**, *25*, 376–380. [CrossRef]
36. Sidhu, A.; Bala, A.; Singh, H.; Ahuja, R.; Kumar, A. Development of MgO-sepoilite Nanocomposites against Phytopathogenic Fungi of Rice (*Oryzae sativa*): A Green Approach. *ACS Omega* **2020**, *5*, 13557–13565. [CrossRef]
37. Wang, X.; Liu, X.; Chen, J.; Han, H.; Yuan, Z. Evaluation and mechanism of antifungal effects of carbon nanomaterials in controlling plant fungal pathogen. *Carbon* **2014**, *68*, 798–806. [CrossRef]
38. Chen, J.; Sun, L.; Cheng, Y.; Lu, Z.; Shao, K.; Li, T.; Hu, C.; Han, H. Graphene Oxide-Silver Nanocomposite: Novel Agricultural Antifungal Agent against *Fusarium graminearum* for Crop Disease Prevention. *ACS Appl. Mater. Interfaces* **2016**, *8*, 24057–24070. [CrossRef]
39. Wang, X.; Cai, A.; Wen, X.; Jing, D.; Qi, H.; Yuan, H. Graphene oxide-Fe_3O_4 nanocomposites as high-performance antifungal agents against *Plasmopara viticola*. *Sci. China Mater.* **2017**, *60*, 258–268. [CrossRef]
40. Rodríguez-González, V.; Domínguez-Espíndola, R.B.; Casas-Flores, S.; Patrón-Soberano, O.A.; Camposeco-Solis, R.; Lee, S.W. Antifungal nanocomposites inspired by titanate nanotubes for complete inactivation of *Botrytis cinerea* isolated from tomato infection. *ACS Appl. Mater. Interfaces* **2016**, *8*, 31625–31637. [CrossRef]
41. Hoseinzadeh, A.; Habibi-Yangjeh, A.; Davari, M. Antifungal activity of magnetically separable Fe_3O_4/ZnO/AgBr nanocomposites prepared by a facile microwave-assisted method. *Prog. Nat. Sci. Mater. Int.* **2016**, *26*, 334–340. [CrossRef]
42. Bhavyasree, P.G.; Xavier, T.S. Green synthesis of Copper Oxide/Carbon nanocomposites using the leaf extract of *Adhatoda vasica* Nees, their characterization and antimicrobial activity. *Heliyon* **2020**, *6*, e03323. [CrossRef] [PubMed]
43. Olad, A. Polymer–Clay nanocomposites. In *Advanced Polymeric Materials: Structure Property Relationships*; CRC Press: Boca Raton, FL, USA, 2011; pp. 349–396. ISBN 9780203492901.
44. Yao, J.; Yang, M.; Duan, Y. Chemistry, biology, and medicine of fluorescent nanomaterials and related systems: New insights into biosensing, bioimaging, genomics, diagnostics, and therapy. *Chem. Rev.* **2014**, *114*, 6130–6178. [CrossRef]
45. Sampathkumar, K.; Tan, K.X.; Loo, S.C.J. Developing Nano-Delivery Systems for Agriculture and Food Applications with Nature-Derived Polymers. *iScience* **2020**, *23*, 101055. [CrossRef] [PubMed]
46. Vega-Vásquez, P.; Mosier, N.S.; Irudayaraj, J. Nanoscale Drug Delivery Systems: From Medicine to Agriculture. *Front. Bioeng. Biotechnol.* **2020**, *8*, 1–16. [CrossRef]
47. Youssef, K.; Hashim, A.F. Inhibitory effect of clay/chitosan nanocomposite against *Penicillium digitatum* on citrus and its possible mode of action. *Jordan J. Biol. Sci.* **2020**, *13*, 349–355.
48. Al-Dhabaan, F.A.; Shoala, T.; Ali, A.A.M.; Alaa, M.; Abd-Elsalam, K. Chemically-Produced Copper, Zinc Nanoparticles and Chitosan-Bimetallic Nanocomposites and Their Antifungal Activity against Three Phytopathogenic Fungi. *Int. J. Agric. Technol.* **2017**, *13*, 753–769.
49. Abd-Elsalam, K.A.; Vasil'kov, A.Y.; Said-Galiev, E.E.; Rubina, M.S.; Khokhlov, A.R.; Naumkin, A.V.; Shtykova, E.V.; Alghuthaymi, M.A. Bimetallic blends and chitosan nanocomposites: Novel antifungal agents against cotton seedling damping-off. *Eur. J. Plant Pathol.* **2018**, *151*, 57–72. [CrossRef]
50. Kaur, P.; Thakur, R.; Choudhary, A. An In Vitro Study of The Antifungal Activity of Silver/Chitosan Nanoformulations Against Important Seed Borne Pathogens. *Int. J. Sci. Technol. Res.* **2012**, *1*, 83–86.
51. Youssef, K.; de Oliveira, A.G.; Tischer, C.A.; Hussain, I.; Roberto, S.R. Synergistic effect of a novel chitosan/silica nanocomposites-based formulation against gray mold of table grapes and its possible mode of action. *Int. J. Biol. Macromol.* **2019**, *141*, 247–258. [CrossRef]
52. Mathew, T.V.; Kuriakose, S. Photochemical and antimicrobial properties of silver nanoparticle-encapsulated chitosan functionalized with photoactive groups. *Mater. Sci. Eng. C* **2013**, *33*, 4409–4415. [CrossRef]
53. Beyki, M.; Zhaveh, S.; Khalili, S.T.; Rahmani-Cherati, T.; Abollahi, A.; Bayat, M.; Tabatabaei, M.; Mohsenifar, A. Encapsulation of *Mentha piperita* essential oils in chitosan-cinnamic acid nanogel with enhanced antimicrobial activity against *Aspergillus flavus*. *Ind. Crops Prod.* **2014**, *54*, 310–319. [CrossRef]

54. Patil, S.; Chandrasekaran, R. Biogenic nanoparticles: A comprehensive perspective in synthesis, characterization, application and its challenges. *J. Genet. Eng. Biotechnol.* **2020**, *18*. [CrossRef] [PubMed]
55. Kalia, A.; Abd-Elsalam, K.A.; Kuca, K. Zinc-based nanomaterials for diagnosis and management of plant diseases: Ecological safety and future prospects. *J. Fungi* **2020**, *6*, 222. [CrossRef] [PubMed]
56. Arciniegas-Grijalba, P.A.; Patiño-Portela, M.C.; Mosquera-Sánchez, L.P.; Guerrero-Vargas, J.A.; Rodríguez-Páez, J.E. ZnO nanoparticles (ZnO-NPs) and their antifungal activity against coffee fungus *Erythricium salmonicolor*. *Appl. Nanosci.* **2017**, *7*, 225–241. [CrossRef]
57. Patra, P.; Mitra, S.; Debnath, N.; Goswami, A. Biochemical-, biophysical-, and microarray-based antifungal evaluation of the buffer-mediated synthesized nano zinc oxide: An in vivo and in vitro toxicity study. *Langmuir* **2012**, *28*, 16966–16978. [CrossRef]
58. Shoeb, M.; Singh, B.R.; Khan, J.A.; Khan, W.; Singh, B.N.; Singh, H.B.; Naqvi, A.H. ROS-dependent anticandidal activity of zinc oxide nanoparticles synthesized by using egg albumen as a biotemplate. *Adv. Nat. Sci. Nanosci. Nanotechnol.* **2013**, *4*. [CrossRef]
59. Lipovsky, A.; Nitzan, Y.; Gedanken, A.; Lubart, R. Antifungal activity of ZnO nanoparticles-the role of ROS mediated cell injury. *Nanotechnology* **2011**, *22*. [CrossRef]
60. Yu, Z.; Li, Q.; Wang, J.; Yu, Y.; Wang, Y.; Zhou, Q.; Li, P. Reactive Oxygen Species-Related Nanoparticle Toxicity in the Biomedical Field. *Nanoscale Res. Lett.* **2020**, *15*. [CrossRef]
61. Fones, H.N.; Bebber, D.P.; Chaloner, T.M.; Kay, W.T.; Steinberg, G.; Gurr, S.J. Threats to global food security from emerging fungal and oomycete crop pathogens. *Nat. Food* **2020**, *1*, 332–342. [CrossRef]
62. Thipe, V.C.; Thatyana, M.; Ajayi, F.R.; Njobeh, P.B.; Katti, K.V. Hybrid nanomaterials for detection, detoxification, and management mycotoxins. In *Multifunctional Hybrid Nanomaterials for Sustainable Agri-Food and Ecosystems*; Elsevier: Amsterdam, The Netherlands, 2020; pp. 271–285. ISBN 9780128213544.
63. Rhouati, A.; Bulbul, G.; Latif, U.; Hayat, A.; Li, Z.H.; Marty, J.L. Nano-aptasensing in mycotoxin analysis: Recent updates and progress. *Toxins* **2017**, *9*, 349. [CrossRef]
64. Santos, A.O.; Vaz, A.; Rodrigues, P.; Veloso, A.C.A.; Venâncio, A.; Peres, A.M. Thin films sensor devices for mycotoxins detection in foods: Applications and challenges. *Chemosensors* **2019**, *7*, 3. [CrossRef]
65. Ananikov, V.P. Organic-inorganic hybrid nanomaterials. *Nanomaterials* **2019**, *9*, 1197. [CrossRef] [PubMed]
66. Bhardwaj, H.; Pandey, M.K.; Sumana, R.; Sumana, G. Electrochemical Aflatoxin B1 immunosensor based on the use of graphene quantum dots and gold nanoparticles. *Microchim. Acta* **2019**, *186*, 1–12. [CrossRef] [PubMed]
67. Hamza, Z.; El-Hashash, M.; Aly, S.; Hathout, A.; Soto, E.; Sabry, B.; Ostroff, G. Preparation and characterization of yeast cell wall beta-glucan encapsulated humic acid nanoparticles as an enhanced aflatoxin B1 binder. *Carbohydr. Polym.* **2019**, *203*, 185–192. [CrossRef] [PubMed]
68. Jampílek, J.; Kráľová, K. Nanocomposites: Synergistic nanotools for management of mycotoxigenic fungi. In *Nanomycotoxicology*; Academic Press: Cambridge, MA, USA, 2020; pp. 349–383.
69. Pirouz, A.A.; Selamat, J.; Iqbal, S.Z.; Mirhosseini, H.; Karjiban, A.R.; Bakar, F.A. The use of innovative and efficient nanocomposite (magnetic graphene oxide) for the reduction on of *Fusarium* mycotoxins in palm kernel cake. *Sci. Rep.* **2017**, *7*, 1–9. [CrossRef] [PubMed]
70. Ji, J.; Xie, W. Detoxification of Aflatoxin B 1 by magnetic graphene composite adsorbents from contaminated oils. *J. Hazard. Mater.* **2020**, *381*, 120915. [CrossRef] [PubMed]
71. Alghuthaymi, M.A.; Abd-Elsalam, K.A.; Shami, A.; Said-Galive, E.; Shtykova, E.V.; Naumkin, A.V. Silver/Chitosan Nanocomposites: Preparation and Characterization and Their Fungicidal Activity against Dairy Cattle Toxicosis *Penicillium expansum*. *J. Fungi* **2020**, *6*, 51. [CrossRef] [PubMed]
72. Abd-Elsalam, K.A.; Alghuthaymi, M.A.; Shami, A.; Rubina, M.S.; Abramchuk, S.S.; Shtykova, E.V.; Vasil'kov, A. Copper-Chitosan Nanocomposite Hydrogels Against Aflatoxigenic *Aspergillus flavus* from Dairy Cattle Feed. *J. Fungi* **2020**, *6*, 112. [CrossRef]
73. Tarus, B.K.; Mwasiagi, J.I.; Fadel, N.; Al-Oufy, A.; Elmessiry, M. Electrospun cellulose acetate and poly(vinyl chloride) nanofiber mats containing silver nanoparticles for antifungi packaging. *SN Appl. Sci.* **2019**, *1*, 1–12. [CrossRef]
74. González-Estrada, R.; Blancas-Benítez, F.; Velázquez-Estrada, R.M.; Montaño-Leyva, B.; Ramos-Guerrero, A.; Aguirre-Güitrón, L.; Moreno-Hernández, C.; Coronado-Partida, L.; Herrera-González, J.A.; Rodríguez-Guzmán, C.A.; et al. Alternative Eco-Friendly Methods in the Control of Post-Harvest Decay of Tropical and Subtropical Fruits. *Mod. Fruit Ind.* **2020**. [CrossRef]
75. Roberto, S.R.; Youssef, K.; Hashim, A.F.; Ippolito, A. Nanomaterials as alternative control means against postharvest diseases in fruit crops. *Nanomaterials* **2019**, *9*, 1752. [CrossRef] [PubMed]
76. Singh, D.; Sharma, R.R. Postharvest Diseases of Fruits and Vegetables and Their Management. In *Postharvest Disinfection of Fruits and Vegetables*; Elsevier: Amsterdam, The Netherlands, 2018; pp. 1–52. ISBN 9780128126981.
77. Pétriacq, P.; López, A.; Luna, E. Fruit decay to diseases: Can induced resistance and priming help? *Plants* **2018**, *7*, 77. [CrossRef] [PubMed]
78. Zhang, X.; Xiao, G.; Wang, Y.; Zhao, Y.; Su, H.; Tan, T. Preparation of chitosan-TiO_2 composite film with efficient antimicrobial activities under visible light for food packaging applications. *Carbohydr. Polym.* **2017**, *169*, 101–107. [CrossRef] [PubMed]
79. Kumar, S.; Shukla, A.; Baul, P.P.; Mitra, A.; Halder, D. Biodegradable hybrid nanocomposites of chitosan/gelatin and silver nanoparticles for active food packaging applications. *Food Packag. Shelf Life* **2018**, *16*, 178–184. [CrossRef]
80. Li, J.; Sun, Q.; Sun, Y.; Chen, B.; Wu, X.; Le, T. Improvement of banana postharvest quality using a novel soybean protein isolate/cinnamaldehyde/zinc oxide bionanocomposite coating strategy. *Sci. Hortic.* **2019**, *258*, 108786. [CrossRef]

81. Jagana, D.; Hegde, Y.R.; Lella, R. Green Nanoparticles—A Novel Approach for the Management of Banana Anthracnose Caused by *Colletotrichum musae*. *Int. J. Curr. Microbiol. Appl. Sci.* **2017**, *6*, 1749–1756. [CrossRef]
82. Li, W.; Li, L.; Cao, Y.; Lan, T.; Chen, H.; Qin, Y. Effects of PLA film incorporated with ZnO nanoparticle on the quality attributes of fresh-cut apple. *Nanomaterials* **2017**, *7*, 207. [CrossRef]
83. Chowdappa, P.; Shivakumar, G.; Chettana, C.S.; Madhura, S. Antifungal activity of chitosan-silver nanoparticle composite against *Colletotrichum gloeosporioides* associated with mango anthracnose. *African J. Microbiol. Res.* **2014**, *8*, 1803–1812. [CrossRef]
84. Dubey, P.K.; Shukla, R.N.; Srivastava, G.; Mishra, A.A.; Pandey, A. Study on Quality Parameters and Storage Stability of Mango Coated with Developed Nanocomposite Edible Film. *Int. J. Curr. Microbiol. Appl. Sci.* **2019**, *8*, 2899–2935. [CrossRef]
85. Zahid, N.; Ali, A.; Manickam, S.; Siddiqui, Y.; Maqbool, M. Potential of chitosan-loaded nanoemulsions to control different *Colletotrichum* spp. and maintain quality of tropical fruits during cold storage. *J. Appl. Microbiol.* **2012**, *113*, 925–939. [CrossRef]
86. Nguyen, N.T.; Nguyen, D.H.; Pham, D.D.; Nguyen, Q.H.; Hoang, D.Q. New oligochitosan-nanosilica hybrid materials: Preparation and application on chili plants for resistance to anthracnose disease and growth enhancement. *Polym. J.* **2017**, *49*, 861–869. [CrossRef]
87. Nguyen, T.N.; Huynh, T.N.; Hoang, D.; Nguyen, D.H.; Nguyen, Q.H.; Tran, T.H. Functional Nanostructured Oligochitosan–Silica/Carboxymethyl Cellulose Hybrid Materials: Synthesis and Investigation of Their Antifungal Abilities. *Polymers* **2019**, *11*, 628. [CrossRef] [PubMed]
88. Ahmed, T.; Noman, M.; Luo, J.; Muhammad, S.; Shahid, M.; Ali, M.A.; Zhang, M.; Li, B. Bioengineered chitosan-magnesium nanocomposite: A novel agricultural antimicrobial agent against *Acidovorax oryzae* and *Rhizoctonia solani* for sustainable rice production. *Int. J. Biol. Macromol.* **2020**, *24*, S0141–S8130. [CrossRef]
89. El-Abeid, S.E.; Ahmed, Y.; Daròs, J.A.; Mohamed, M.A. Reduced Graphene Oxide Nanosheet-Decorated Copper Oxide Nanoparticles: A Potent Antifungal Nanocomposite against Fusarium Root Rot and Wilt Diseases of Tomato and Pepper Plants. *Nanomaterials* **2020**, *10*, 1001. [CrossRef] [PubMed]
90. Auyeung, A.; Casillas-Santana, M.Á.; Martínez-Castañón, G.A.; Slavin, Y.N.; Zhao, W.; Asnis, J.; Häfeli, U.O.; Bach, H. Effective Control of Molds Using a Combination of Nanoparticles. *PLoS ONE* **2017**, *12*, e0169940. [CrossRef]
91. De La Rosa-García, S.C.; Martínez-Torres, P.; Gómez-Cornelio, S.; Corral-Aguado, M.A.; Quintana, P.; Gómez-Ortíz, N.M. Antifungal activity of ZnO and MgO nanomaterials and their mixtures against *Colletotrichum gloeosporioides* strains from tropical fruit. *J. Nanomater.* **2018**, *2018*. [CrossRef]
92. Dimkpa, C.O.; McLean, J.E.; Latta, D.E.; Manangón, E.; Britt, D.W.; Johnson, W.P.; Boyanov, M.I.; Anderson, A.J. CuO and ZnO nanoparticles: Phytotoxicity, metal speciation, and induction of oxidative stress in sand-grown wheat. *J. Nanopart. Res.* **2012**, *14*. [CrossRef]
93. Vannini, C.; Domingo, G.; Onelli, E.; De Mattia, F.; Bruni, I.; Marsoni, M.; Bracale, M. Phytotoxic and genotoxic effects of silver nanoparticles exposure on germinating wheat seedlings. *J. Plant Physiol.* **2014**, *171*, 1142–1148. [CrossRef]
94. Wu, B.; Zhu, L.; Le, X.C. Metabolomics analysis of TiO_2 nanoparticles induced toxicological effects on rice (*Oryza sativa* L.). *Environ. Pollut.* **2017**, *230*, 302–310. [CrossRef]
95. Shen, C.X.; Zhang, Q.F.; Li, J.; Bi, F.C.; Yao, N. Induction of programmed cell death in Arabidopsis and rice by single-wall carbon nanotubes. *Am. J. Bot.* **2010**, *97*, 1602–1609. [CrossRef]
96. Rienzie, R.; Adassooriya, N.M. Nanomaterials: Ecotoxicity, Safety, and Public Perception. In *Nanomaterials: Ecotoxicity, Safety, and Public Perception*; Springer International Publishing: Berlin, Germany, 2018; pp. 207–234. ISBN 9783030051440.
97. Zheng, L.P.; Zhang, Z.; Zhang, B.; Wang, J.W. Antifungal properties of Ag-SiO_2 core-shell nanoparticles against phytopathogenic fungi. *Adv. Mater. Res.* **2012**, *476*, 814–818. [CrossRef]
98. Vittori Antisari, L.; Carbone, S.; Gatti, A.; Vianello, G.; Nannipieri, P. Uptake and translocation of metals and nutrients in tomato grown in soil polluted with metal oxide (CeO_2, Fe_3O_4, SnO_2, TiO_2) or metallic (Ag, Co, Ni) engineered nanoparticles. *Environ. Sci. Pollut. Res.* **2015**, *22*, 1841–1853. [CrossRef] [PubMed]
99. Saratale, R.G.; Karuppusamy, I.; Saratale, G.D.; Pugazhendhi, A.; Kumar, G.; Park, Y.; Ghodake, G.S.; Bharagava, R.N.; Banu, J.R.; Shin, H.S. A comprehensive review on green nanomaterials using biological systems: Recent perception and their future applications. *Colloids Surf. B Biointerfaces* **2018**, *170*, 20–35. [CrossRef] [PubMed]
100. Saratale, R.G.; Saratale, G.D.; Shin, H.S.; Jacob, J.M.; Pugazhendhi, A.; Bhaisare, M.; Kumar, G. New insights on the green synthesis of metallic nanoparticles using plant and waste biomaterials: Current knowledge, their agricultural and environmental applications. *Environ. Sci. Pollut. Res.* **2018**, *25*, 10164–10183. [CrossRef] [PubMed]
101. Saratale, R.G.; Benelli, G.; Kumar, G.; Kim, D.S.; Saratale, G.D. Bio-fabrication of silver nanoparticles using the leaf extract of an ancient herbal medicine, dandelion (*Taraxacum officinale*), evaluation of their antioxidant, anticancer potential, and antimicrobial activity against phytopathogens. *Environ. Sci. Pollut. Res.* **2018**, *25*, 10392–10406. [CrossRef] [PubMed]

Journal of *Fungi*

Review

Pleurotus Macrofungi-Assisted Nanoparticle Synthesis and Its Potential Applications: A Review

Kanchan Bhardwaj [1], Anirudh Sharma [2], Neeraj Tejwan [2], Sonali Bhardwaj [3], Prerna Bhardwaj [1], Eugenie Nepovimova [4], Ashwag Shami [5], Anu Kalia [6], Anil Kumar [7], Kamel A. Abd-Elsalam [8,*] and Kamil Kuča [4,*]

1 School of Biological and Environmental Sciences, Shoolini University of Biotechnology and Management Sciences, Solan 173229, India; kanchankannu1992@gmail.com (K.B.); prernabhardwaj135@gmail.com (P.B.)
2 Advance School of Chemical Sciences, Shoolini University of Biotechnology and Management Sciences, Solan 173229, India; anirai3024@gmail.com (A.S.); neerajtejwan@gmail.com (N.T.)
3 School of Bioengineering and Biosciences, Lovely Professional University, Phagwara 144411, India; sonali.bhardwaj1414@gmail.com
4 Department of Chemistry, Faculty of Science, University of Hradec Kralove, 50003 Hradec Kralove, Czech Republic; eugenie.nepovimova@uhk.cz
5 Biology Department, College of Sciences, Princess Nourah bint Abdulrahman University, Riyadh 11671, Saudi Arabia; AYShami@pnu.edu.sa
6 Electron Microscopy and Nanoscience Laboratory, Punjab Agricultural University, Ludhiana 141004, India; kaliaanu@pau.edu
7 School Bioengineering and Food Technology, Shoolini University of Biotechnology and Management Sciences, Solan 173229, India; kumaranil@shooliniuniversity.com
8 Agricultural Research Center (ARC), Plant Pathology Research Institute, Giza 12619, Egypt
* Correspondence: kamelabdelsalam@gmail.com (K.A.A.-E.); kamil.kuca@uhk.cz (K.K.); Tel.: +420-603-289-166 (K.K.)

Received: 19 November 2020; Accepted: 7 December 2020; Published: 9 December 2020

Abstract: Research and innovation in nanoparticles (NPs) synthesis derived from biomaterials have gained much attention due to their unique characteristics, such as low-cost, easy synthesis methods, high water solubility, and eco-friendly nature. NPs derived from macrofungi, including various mushroom species, such as *Agaricus bisporus*, *Pleurotus* spp., *Lentinus* spp., and *Ganoderma* spp. are well known to possess high nutritional, immune-modulatory, antimicrobial (antibacterial, antifungal and antiviral), antioxidant, and anticancerous properties. Fungi have intracellular metal uptake ability and maximum wall binding capacity; because of which, they have high metal tolerance and bioaccumulation ability. Primarily, two methods have been comprehended in the literature to synthesize metal NPs from macrofungi, i.e., the intracellular method, which refers to NP synthesis inside fungal cells by transportation of ions in the presence of enzymes; and the extracellular method, which refers to the treatment of fungal biomolecules aqueous filtrate with a metal precursor. *Pleurotus* derived metal NPs are known to inhibit the growth of numerous foodborne pathogenic bacteria and fungi. To the best of our knowledge, there is no such review article reported in the literature describing the synthesis and complete application and mechanism of NPs derived from macrofungi. Herein, we intend to summarize the progressive research on macrofungi derived NPs regarding their synthesis as well as applications in the area of antimicrobial (antibacterial & antifungal), anticancer, antioxidant, catalytic and food preservation. Additionally, the challenges associated with NPs synthesis will also be discussed.

Keywords: oyster mushroom; application; antibacterial; anticancer; antioxidant

1. Introduction

The enormous impact of nanobiotechnology on almost all life forms has intrigued researchers globally [1]. In 1959, Richard Feynman introduced the theoretical concept of miniaturization, and, for the first time, provided hidden hints on nanotechnology (directing towards technology, using materials that have dimensions of approximately 1–100 nm) [2]. Nowadays, microorganisms (bacterium, fungi, including mushroom, yeast) and green plants are used for green synthesis of metallic nanoparticles [3].

Pleurotus mushrooms, commonly known as oyster mushrooms, belong to the family of genus *Pleurotus*, and they are edible and nutritious in nature [4]. Oyster mushrooms are readily available and naturally grow in nearly all latitudes, tropical and subtropical forestry, except Antarctica [5]. The worldwide geographical distribution of different *Pleurotus* species, with optimum growth temperature, is shown in Table 1. The primary role of the fruiting bodies of oyster mushrooms is to absorb amino acids, proteins, vitamin B (niacin, thiamine, and riboflavin), vitamin D, carbohydrates, and mineral salts (iron, calcium, and phosphorus) [6,7]. Additionally, they show important antifungal, anti-inflammatory, antibacterial, and immunomodulatory activities.

Table 1. Geographical Distribution of *Pleurotus* spp.

Species	Geographical Distribution	Optimum Growth Temperature °C	Reference
Pleurotus citrinopileatus	Russia, China, Japan	21–29	[8,9]
P. cornucopiae	Europe, USA, and Mexico	>25	[8,9]
P. djamor	Tropical region, Indonesia, Malaysia, Japan, Mexico	21–35	[8,9]
P. eous	Sub-tropical part of the world	23–28	[8,9]
P. eryngii	Europe, Asia, Africa	20–25	[8,9]
P. flabellatus	India, Mauritius	25–28	[8,9]
P. florida	Hungary, Kenya	20–28	[8,9]
P. giganteus	Thailand, Sri Lanka	15–35	[8,10]
P. ostreatus	Widespread around the world	18–22	[8,9]
P. sajor-caju	Kenya, India, Philippines, Australia, Mauritius	20–28	[8,9]
P. tuber-regium	Africa, Australia, Asia	25–30	[8,9]
P. platypus	India	15–25	[8,9]
P. pulmonarius	Warm tropical area	20–28	[8,9]

Moreover, they have the ability to reduce sugar and cholesterol levels in the blood [11]. *P. ostreatus* contains β-1,3-D-glucan and pectin, which are water-soluble gel-forming substances, having the ability to bind with bile acids. They inhibit cholesterol-bile micelle formation, cholesterol absorption, and endogenous synthesis, while increase the removal of plasma cholesterol by reducing the production and secretion of very low density lipoproteins (VLDL) [12]. The content of the nutrients depends on the nature, age, and size of the fungus, as well as their growing conditions [13]. Several nutrients, such as carbohydrates (40–46%), protein (20–25%), fiber (10–21%), and amino acids (20–41%) are present in considerable amounts, while the content of fat is very low, ranging from 10–20% of the dry matter. Fungi fruiting bodies are a rich source of micro- and macro-elements, such as sodium, magnesium, phosphorus, calcium, manganese, potassium, iron, copper, and zinc. Mushrooms of the genus *Pleurotus* have their own importance, more than the commercially employed basidiomycetes, because they possess better superior nutritional, gastronomic, and medical properties than mushrooms (which can be easily cultivated on a broad range of substrates) [14]. Nowadays, mushrooms show significant potential in metal nanoparticle (NP) synthesis and multifaceted applications [15].

In recent decades, several reports on mycogenesis derived NPs have been published [11]. However, the precise mechanisms of synthesis of mycogenic nanomaterials with variable size dimensions and topologies are not well understood (yet). Members of the fungi kingdom include various heterotrophic multicellular eukaryotic organisms. These microbes play an essential role in diverse ecosystems, particularly in the nutrient cycling paradigms. Fungi can be reproduced by both processes, i.e., sexually as well as asexually, and have shown symbiotic relations with bacteria and plants. Fungi groups mostly consist of mildew, mold, rust, yeast, and mushrooms [16].

The benefits and relevant use of fungal cells, such as NP factories, are attributed to the release of high amounts of extracellular enzymes that can serve as bio-reducing (as well as stabilizing) agents for NP synthesis. Moreover, fungal-derived NPs are much better than the bacteria-derived NPs. Constituents, such as enzymes and metabolites secreted by fungal cells, play an important role in synthesizing metal NPs, which reduce the toxicity of substances [15]. Ions are often liable for toxicity. When metal ions in solution are exposed to bacterial cells, they become uniformly distributed in the environment surrounding the bacterial cell, with no specific localization. In contrast, NPs that interact with the bacterial cell wall produce a focal source of ions through continuous release of ions, and cause enhanced toxicity to the cells [17]. Positively charged metal ions can easily bind with the fungal cell surface containing negative charge through electrostatic and cell receptor-specific interactions. Both types of intracellular (as well as extracellular) fungi could be used to synthesize NPs. Fungi exhibit high metal-binding capabilities in comparison to bacteria, and, hence, fungal biomass has gained the attention of researchers, for the production of NPs, at a large-scale. Various metal NPs, such as PdNPs, AgNPs, AuNPs, CuNPs, FeNPs, ZnNPs, TiNPs, and PtNPs could be synthesized using their oxides, nitrides, sulfides, and fungal biomasses [18].

NPs derived from fungal biomasses exhibit distinct optical, physical, and chemical properties, such as high quantum yield, excellent biocompatibility, high photostability, and adequate near-infrared (NIR) light-absorbing capacity, owing to which, they can be used in various chemical and medical fields, such as sensing, medicine, catalysis, and food packaging [19,20]. Oyster mushrooms mediate myco NPs by using spent mushroom substrate (SMS), and have medical importance towards many pathogenic microorganisms, as reported for the first time in 2007 [3].

To the best of our knowledge, many review articles, based on the synthesis of metal NPs derived from diverse sources and their applications, have been reported. Still, few review articles have discussed the synthesis of distinct metal nanoparticles (MNPs) and several applications of oyster mushroom derived NPs, and we attempted to fill the gap. In this article, we intend to discuss NP synthesis derived from oyster mushrooms, using various techniques and different applications. First section discuses the synthesis of nanoparticles by using intra-and extra-cellular methods. Later, different applications involving antioxidant, anticancer, antibacterial, and catalysis, with possible mechanisms of action, have also been discussed.

2. Green Synthesis of Metal-Based Nanoparticles Mediated by Genus *Pleurotus*

Fungal exploration and implications in the area of nanotechnology are very significant. In previous literature, it was reported that microorganisms, including bacteria, fungi, and yeast, could be used for the synthesis of metal NPs (metal = calcium, gold, silicon, iron, silver, lead, and gypsum) [21]. We observed that fungi have received immense attention owing to their metal bioaccumulation properties, to produce metal NPs [22]. The fungal material includes mycelia, polysaccharides, and proteins are used in the formation of metal nanoparticles; metal NPs of oyster mushroom species were synthesized using mineral salts [23]. Fungi have intracellular metal uptake capabilities and maximum wall binding abilities because they have high metal tolerance plus bioaccumulation abilities [24–27]. In comparison with other plants and microbes, the mycelia of fungi provides effective hold ability in the bioreactor, as well as in agitation and high flow pressure [28]. Moreover, fungi secrete extracellular enzymes in high amounts, leading to the massive production of enzymes [29]. Reduction of the enzyme, using both intracellular and extracellular ways, help in metal NP synthesis, nanostructure, and biomimetic mineralization [30,31].

During synthesis, fungi extracts serve the function of capping and reducing agents. At the same time, the fungal mycelium exposed to the metal precursor induces fungus to liberate metabolites and enzymes for its survival [11]. Both the fruiting bodies and mycelium of the mushrooms can be utilized for the synthesis of NPs. It has been reported that the *Pleurotus* species, such as *P. ostreatus*, are capable of synthesizing NPs, both intracellularly and extracellularly, while other species, such as *P. florida*, *P. cornucopiae var. citrinopileatus*, *P. platypus*, *P. ostreatus*, *P. sajor-caju*, *P. eous*, and *P. djamor var. roseus* synthesize NPs extracellularly [32–37]. Synthesis of *Pleurotus* derived metal NPs is shown in Table 2 and Figure 1.

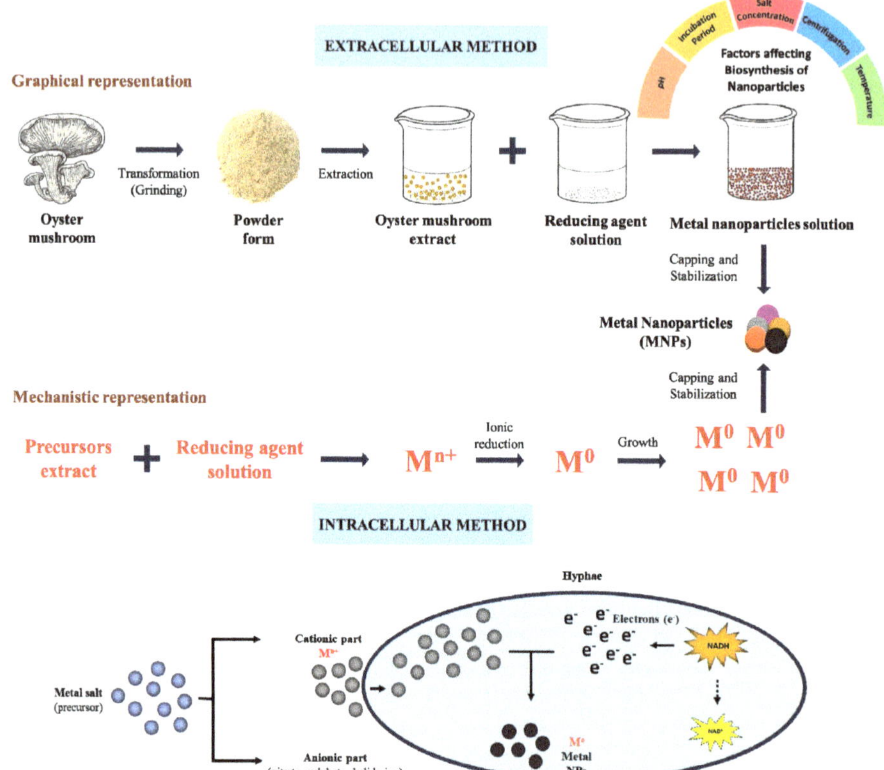

Figure 1. Graphical representation of green-synthesis of nanoparticles from *Pleurotus* (Oyster) mushroom.

Table 2. Varieties of *Pleurotus* spp. and their nanoparticles.

Species	Types of Nanoparticles Synthesize and Their Size (nm)	Chemical Used	Reaction Time (hour)	Reducing Agent	Stabilizing Agent	Specific Temperature (°C)	Morphology	References
Pleurotus citrinopileatus	Ag, 6–10	$AgNO_3$	24	mushroom extract, Nitrate	mushroom extract	60	spherical	[38]
P. cornucopiae (citrinopileatus)	Ag, 20–30	$AgNO_3$	24	aqueous extract	aqueous extract	25	spherical	[37]
P. cystidiosus	Ag, 2–100	$AgNO_3$	24	aqueous extract	aqueous extract	25	ND	[39]
P. cystidiosus	Au, ND	$HAuCl_4$	24	aqueous extract	aqueous extract	29	ND	[39]
P. djamor	Ag, 5–50	$AgNO_3$	48, 24	aqueous extract	aqueous extract	RT	spherical	[40–42]
P. djamor	ZnO, 70–80	Zn(NO3)2.5H2O	24	ND	ND	RT	spherical	[43]
P. djamor	TiO2	TiCl4	20 min	aqueous extract	aqueous extract	RT	spherical	[44]
P. eryngii	Ag, 18.45	$AgNO_3$	5 days	aqueous extract	aqueous extract	RT	spherical	[45]
P. flabellatus	Ag, 2–100	$AgNO_3$	24	aqueous extract	aqueous extract	25	ND	[39–41]
P. flabellatus	Au, ND	$HAuCl_4$	24	aqueous extract	aqueous extract	29	ND	[39]
P. florida	Ag, 20	$AgNO_3$	Overnight; 72	aqueous extract	aqueous extract	RT	spherical	[33,34,40,41, 46–48]
P. florida	Au, 2–14	$HAuCl_4$	1.5	aqueous extract, glucan	glucan	70	spherical	[49]
P. florida	Au, 20	$HAuCl_4$	24	aqueous extract	aqueous s extract	RT	spherical	[50]
P. giganteus	Ag, 5–25	$AgNO_3$	3 days	aqueous extract	aqueous extract	37	spherical	[51]

Table 2. *Cont.*

Species	Types of Nanoparticles Synthesize and Their Size (nm)	Chemical Used	Reaction Time (hour)	Reducing Agent	Stabilizing Agent	Specific Temperature (°C)	Morphology	References
P. ostreatus	Ag, 4,28,50	$AgNO_3$	24; 72; 1	aqueous extract; mushroom broth	aqueous extract	28; 75	spherical;	[36,39,52,53]
P. ostreatus	Au, 22.9	$HAuCl_4$	24 h	aqueous extract	aqueous extract	29	spherical	[39,54]
P. ostreatus	ZnS, 2–5	$ZnCl_2$	Over night	mushroom	mushroom extract	70	spherical with crystalline	[55]
P. ostreatus	Zn, 15	$ZnS\text{-}N_3$	1	aqueous extract	aqueous extract, sodium azide	4	uniform	[56]
P. platypus	Ag, 0.56µm	$AgNO_3$	72	aqueous extract	aqueous extract	37	spherical	[34]
P. pulmonarius	Ag, 2–100	$AgNO_3$	24	aqueous extract	aqueous extract	25	ND	[40,41]
P. pulmonarius	Au, ND	$HAuCl_4$	24	aqueous extract	aqueous extract	29	ND	[39]
P. sajor-caju	Ag, 5–50	$AgNO_3$	48	aqueous extract	aqueous extract	25	spherical	[32,57–60]
P. sajor-caju	Au, 16–18	$HAuCl_4{\cdot}3H_2O$	Over night	aqueous extract	aqueous extract	RT	spherical	[60]
P. tuber-regium	Ag, 50	$AgNO_3$	2	aqueous extract	aqueous extract	80	spherical and cubical	[61]

Note: RT—room temperature; ND—not detected.

2.1. Intracellular Method

This method includes synthesis of NPs inside the fungal cells by transporting ions during the exposure of enzymes [62,63]. First, the mycelia cultures are treated with a metal precursor and then they are incubated in the dark for 24 h. For intracellular identification, mycelia are resuspended in phosphate buffer saline (PBS, pH 7.4) and homogenized with a sonicator. NPs formed by the intracellular technique have a smaller size when compared with the NPs fabricated by the extracellular method [64,65]. Nucleation of particles inside the fungus could be the cause behind the variation in sizes. This technique is slower when compared with the extracellular method for synthesizing metal NPs [64]. As the NPs synthesis starts within the cell, their downstream processing becomes complicated, increasing cost of synthesizing NPs [66–68]. However, this type of synthesis technique is suitable for making composite films [69].

2.2. Extracellular Method

The extracellular synthesis method is a facile and cost-effective approach that involves the treatment of fungal biomolecule aqueous filtrate with a metal precursor, where these metal ions are adsorbed on the surface of the cells [31,70–73]. In this technique, downstream processing is not required, which makes this approach more effective in comparison to the intracellular method. Therefore, the extracellular approach is predominantly used for NP synthesis [74]. Extracellular metabolites synthesized by fungi play a crucial function in their survival when exposed to various environmental stresses, such as temperature variations, toxic materials (e.g., metallic ions), and predators [75]. Moreover, this synthesis method shows the capability of immobilization of metallic ions in a suitable carrier [69].

The accepted mechanism for the metallic NP synthesis is the enzymatic reduction via enzyme reductase, within the fungal cell or on the cell membrane [76]. This probable mechanism proposes fungus-mediated NP synthesis, i.e., the action of electron shuttle quinones, nitrate reductase, or by both. It is observed that, in bacteria and fungi, mainly two forms of enzymes: (1) nitrate reductase, and (2) α-NADPH-dependent reductases, are responsible for the metal and metal oxide NP synthesis [69]. Extracellularly synthesized NPs were stabilized by the enzymes and proteins formed by the fungi. Moreover, it has been observed that high molecular weight protein is associated with the synthesis of NPs, such as NADH-dependent reductase [69]. Furthermore, the phytochemicals found in plants play a vital role in the bioreduction of NPs [76]. In the mushroom extract of *Pleurotus* spp., phytochemicals, including alkaloids, saponins, anthraquinones, flavonoids, tannins, and steroids are present [14].

3. Different Types of Nanoparticles Derived from Oyster Mushroom

3.1. Silver Nanoparticles (AgNPs)

AgNPs play a significant character in the areas of biological and medical sciences. These NPs could be synthesized by various methods, such as physical, chemical, ionizing radiation methods, etc. [70]. However, all of these methods possess potential drawbacks; particularly, the chemicals utilized in AgNP synthesis through wet chemistry routes are less eco-friendly, expensive, and have high toxicity [46,77,78]. However, fabrication of AgNPs by green synthesis methods can be a better alternative as it is cost effective, non-toxic, and ecologically safe than the other synthesis methods [79]. Various studies on the biosynthesis of AgNPs using powdered basidiocarps and mycelia of different oyster mushroom species, such as *P. ostreatus*, *P. sajor-caju*, *P. florida*, *P. cornucopiae var. citrinopileatus*, *P. giganteus*, *P. platypus*, and *P. eous* have been reported. These basidiocarps and mycelia were soaked in distilled water, boiled, and then filtered [23–33]. The filtrate was freeze-dried to prepare aqueous extract. Various concentrations of this aqueous extract were incubated with $AgNO_3$ solution to synthesize AgNPs by the reduction of Ag^+ ions to Ag° (metal). Unboiled mycelia extract of *Pleurotus* has also been used to synthesize AgNPs [38]. In their report, they crushed the fruiting bodies and mixed them with deionized water. The content was filtered with filter paper, and then the filtrate was used to synthesize

AgNPs, with a size of 6–10 nm, with a spherical shape. Moreover, the synthesized AgNPs were further assessed for antibacterial potential against *Escherichia coli* and *Staphylococcus aureus*.

Synthesis of AgNPs was carried using *P. tuber-regium* mushroom extract and 1 mM $AgNO_3$ solution. The mixture of solutions was stirred at 90 °C for 2 h. Cubical and spherical shaped AgNPs, with an average size of 50 nm, were obtained as a black powder [61]. Debnath et al. synthesized spherical shaped AgNPs with the help of aqueous extract of mushroom (5 mL) and mixed with 95 mL silver nitrate (1 mM, $AgNO_3$) solution to reduce Ag^+ to Ag^o. This solution was kept in an incubator for 3 days at 37 °C, resulting in color change from light yellow to yellowish-brown. The obtained AgNPs were crystalline with a size ranging from 5 to 25 nm. Authors evaluated the antibacterial activity of AgNPs against *E. coli*, *B. subtilis*, *P. aeruginosa*, and *S. aureus* [51]. Similarly, the synthesis of predominantly spherical shaped AgNPs with a size ranging from 2 to 100 nm was carried by various researchers using mushroom extract and $AgNO_3$ solution [34,37,52,57,80].

3.2. Gold Nanoparticles (AuNPs)

AuNPs synthesis was performed by using edible *P. florida* mushroom by the photo-irradiation method, and evaluated for anticancer potential against A-549, HeLa, K-562, and MDA-MB cell lines. Initially, 5 g of fresh biomass of *P. florida* mushroom was washed with deionized water and then cut to small pieces. Later, the chopped pieces were added in 500 mL of double-deionized water, under stirring, for half an hour. These contents were then incubated overnight. That content was then filtered via filter paper. Later, the filtrate of mushroom was used to reduce Au^+ into Au° in the presence of bright sunlight to form spherical to triangular-shaped AuNPs in the range of 10–50 nm [48].

3.3. Zinc Sulfide Nanoparticles (ZnS) and Zinc Oxide Nanoparticles (ZnO)

ZnS NPs were fabricated using *P. ostreatus* extract, $ZnCl_2$, and Na_2S solution as the precursor material [55]. Firstly, small pieces of mushrooms were boiled and filtered. Then, different concentrations of the resultant filtrate were mixed with aqueous solutions of $ZnCl_2$ and Na_2S solution, and resulting solutions were dried at 120 °C for 2 h. Here, the resultant filtrate was used as a stabilizing (as well as a capping) agent for the fabrication of spherical shaped ZnS NPs. Obtained ZnS NPs was highly crystalline with sizes varying from 2.30 nm to 4.04 nm. The author observed that the diameter of those spherical ZnS NPs was decreasing with the increase in extract amount [55]. ZnONPs were synthesized by using *P. djamor* extract, 20 mL of mushroom extract added into 80 mL of Zn $(NO_3)_2$. The $5H_2O$ (5 mM) solution was continuously mixed for 24 h at room temperature until the color transformed into light pink, which confirmed the synthesis of ZnONPs [43].

3.4. Cadmium Sulfide Nanoparticles Quantum Dots (CdS QDs)

In contrast to traditional fluorescent organic dyes and green fluorescent proteins, CdS QDs seem to be superior as they overcame the limitations associated with different factors, such as spectral overlapping, weak signal intensity, and photobleaching [81]. The multiple characteristics of QDs are high photostability, symmetric, slow decay rates, fine emission spectra, wide absorption cross-sections, and broad absorption spectra. The emission color of QDs depends upon their size and surface chemistry; chemical composition used can be altered from the UV to visible or near NIR wavelengths. The increasing interest in the use of CdS QDs is because they act as luminescent probes and labels for biological imaging, disease diagnosis, and molecular histopathology. The studies revealed that the QDs derived from plants did not aggregate [82]. Borovaya et al. synthesized CdS NPs with the help of aqueous extract of roots of *Linaria maroccana*, $CdSO_4$, and Na_2S. First, the mixture solution was incubated for 4 days at 28 °C resulting in the formation of the clear homogeneous solution with a bright yellow color. This indicates the formation of CdS NPs, which are water-soluble and spherical, with sizes of 5–7 nm [82]. In 2015, again biosynthesis of luminescent CdS NPs using mycelium of *P. ostreatus*, $CdSO_4$ and Na_2S. In brief, $CdSO_4$ solution was mixed with mycelium followed by the incubation for 10 days at 26 °C, followed by the addition of Na_2S solution. Obtained NPs were spherically shaped, having the size in the range of 4 to 7 nm. In particular, cadmium sulfide QDs are highly useful in investigating the biomolecules interaction and cellular signaling pathway with the help of fluorescent microscopy [81].

3.5. Titanium Dioxide Nanoparticles (TiO_2)

TiO_2 NPs were synthesized by using edible *P. djamor* mushroom and evaluated for anticancer potential against A-549 (human lung carcinoma) cell lines, as well as for larvicidal and bactericidal activity. Initially, 10 g of fresh biomass of *P. djamor* mushroom was washed with deionized water for 10 min and then cut to small pieces. Later, the chopped pieces were added in 100 mL of double-deionized water, boiled at 60 °C for 15 min, and then filtered. Then, 20 mL of filtrate was added to 80 mL of $TiCl_4$ (5 mM) solution, stirred for 2 h, and kept to room temperature for 20 min until the color changed to brown. The intensity of the color of the extract was determined at the wavelength of 345 nm. The synthesized TiO_2 NPs formed, spherical in shape, with sizes of 31 nm [44].

3.6. Synthesis of Other Nanoparticles

3.6.1. Iron Nanoparticles (FeNPs)

FeNPs were intracellularly synthesized by using hypha of *Pleurotus* sp. The reduction process is involved in uptake of FeNPs via the fungal cell membrane, in which reduction of ferric ion (Fe^{+3}) to ferrous ion (Fe^{+2}) takes place. The reduction process is involved during the iron uptake by fungi [83].

3.6.2. Selenium Nanoparticle (SeNP)

SeNPs were synthesized via mushroom polysaccharide-protein complexes (PSPs) isolated from *P. tuber-regium* sclerotia. These NPs have anticancer activity, excellent bioavailability, and low toxicity. SeNPs have been recorded for inhibiting the proliferation of human breast carcinoma MCF-7 cells by apoptosis; results obtained from the study revealed that cytotoxicity was cancer-specific [84]. PSP–SeNPs have the efficiency to enhance the reactive oxygen species (ROS) generation, and inhibit dose dependently the growth of MCF-7 human breast carcinoma cells, through induction of apoptosis, with the involvement of Poly (ADP-ribose) polymerase (PARP) cleavage and caspase activation. The size of PSP–SeNPs with an average diameter < 50 nm spherical in shape [85]. The synthesis of SeNPs from *P. ostreatus* extract have been reported for in vitro anticancer activity [86].

3.6.3. Copper Nanoparticles (CuNPs)

Monodispersed copper nanoparticles (CuNPs) were synthesized from aqueous fermented fenugreek powder (FFP), polysaccharides, such as chitosan, sodium alginate, citrus, and pectin, with the help of fungal strains of *P. ostreatus*, under the exposure of gamma radiation. The CuNPs synthesized have size ranges from 25.0 to 36.0 nm. Because of the stability and the minute sizes, these synthesized CuNPs show antioxidant and antimicrobial activity, and were found beneficial in cosmetics, medical, pharmaceutical, and industrial applications [87].

4. Applications of Pleurotus Derived Nanoparticles

Different types of metal NPs synthesized from oyster mushrooms are discussed above. Recently, they have been considered as valuable in various fields of medicine and industries. Schematic representation of different applications of metal NPs derived from *Pleurotus* is summarized in Figure 2.

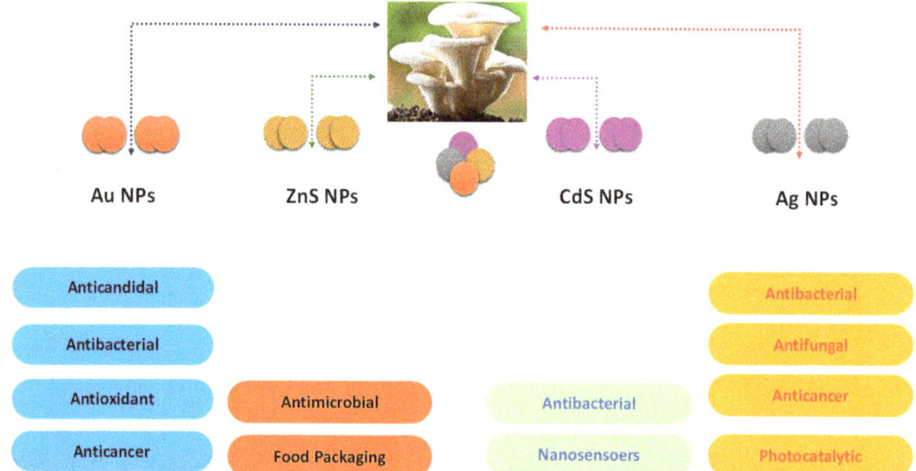

Figure 2. Schematic diagram showing the applications of oyster mushroom derived nanoparticles.

4.1. Antimicrobial Activity

In the past few decades, the microbial infection has become a major health issue across the world, due to its continuously evolving nature and ability to develop resistance against the existing regime [64]. Therefore, search for a strong alternative candidate that can kill or inhibit multidrug resistance microbes is needed [88]. Metal nanoparticles (MNPs) are known to possess potent antimicrobial activity against a wide variety of microbes, including bacteria (Gram-negative and Gram-positive) and fungi, via their photodynamic effects and strong oxidative stress [89]. Furthermore, the direct physical contact of MNPs to bacterial membrane results in the release of intercellular material, loss of cell membrane integrity, and cell death [88]. In literature, several reports are offered on MNPs' (AgNPs, AuNPs, ZnSNPs) exhibiting bactericidal activity. However, only a few reports focus on the support of their antifungal and antibacterial activity with appropriate mechanisms [88–91]. Moreover, many reports available in literature that favor the mechanism involved in bacterial and fungi cells are almost similar.

In the year 2009, Nithya and Ragunathan synthesized AgNPs (5–50 nm, spherical) derived from *Pleurotus sajor-caju* fungi as the starting material and evaluated their bactericidal activity against the both Gram-negative (*Pseudomonas aeruginosa*, *E. coli*) and Gram-positive (*S. aureus*) bacteria. Authors claimed that the inhibitory action of AgNPs was attributed to the generation of Ag^+ ions by NPs that resulted in DNA damage, protein denaturation, enzymes inhibitions [32]. Two years later, in 2011, Bhat et al. also fabricated AgNPs (20 ± 5 nm, spherical) using *P. florida* mushroom, and studied their antibacterial activity against *S. aureus, Salmonella typhi, Providencia alcalifaciens,* and *Proteus mirabilis.* The prepared NPs show higher activity against the Gram-positive microbes than Gram-negative, especially in the case of *P. mirabilis* [33]. A similar kind of study was presented by Shivashankar et al. (2013) using AgNPs derived from *P. pulmonarius, P. djamor,* and *Hypsizygus (pleurotus) ulmarius* as the precursors [40,41].

Yehia and Sheikh (2014) synthesized AgNPs (4–15 nm, spherical) derived from *P. ostreatus* extracted via the green synthesis route and evaluated their antifungal activity toward the various *Candida* species, i.e., *C. tropicalis, C. albicans, C. parapsilosis, C. krusei,* and *C. glabrata.* The (minimum inhibitory concentration) MIC (IC_{80}) results demonstrated that AgNPs showed higher toxicity against all of the *candida* species (5–28 μg/mL) than the amphotericin B (5–8 μg/mL) and fluconazole (13–33 μg/mL) [36]. The tiny size and capping ability of the bioactive white NPs derived from *P. tuber-regium* extract and silver nitrate showed higher therapeutic efficacy against the various diseases and disorders [61]. Devi and Joshi (2015) synthesized AgNPs derived from three different endophytic fungi, i.e., *Aspergillus niger*, *Aspergillus tamarii*, and *Penicillium ochrochloron* isolated from the ethnomedicinal plant *Potentilla fulgens* leaves via the green synthesis method. The electron microscopy results revealed that all of the AgNPs derived from different fungi were spherically shaped. However, NPs synthesized from *A. tamarii* showed the smallest size (~3.5 nm) than *A. niger* (~8.7 nm) and *P. ochrochloron* (~7.7 nm), respectively [74]. In the year 2018, Bawadekji et al. fabricated Au NPs (~22.9 nm, spherical) from *P. ostreatus* extract and evaluated their antimicrobial activity toward the bacteria *Enterococcus faecalis*, *E. coli*, *Klebsiella pneumonia*, *S. aureus*, *P. aeruginosa*, and *C. albicans*. The results demonstrated that synthesized NPs showed significant toxicity against *C. albicans*, *P. aeruginosa*, and *S. aureus*. In contrast, no toxicity was observed in the case of *E. faecalis*, *E. coli*, and *K. pneumonia* [54].

Acay and Baran (2019) reported the green synthesis of AgNPs derived from *Pleurotus eryngii* (PE) extract and their antimicrobial activity against the various human pathogen microorganisms, such as *E. coli*, *S. aureus*, *Streptococcus pyogenes*, *P. aeruginosa*, and *C. albicans*. The authors used drug vancomycin, colistin, and fluconazole as the control over the gram-positive, gram-negative, and fungus microorganisms. The observed MIC values for *S. aureus*, *E. coli*, *S. pyogenes*, *C. albicans*, and *P. aeruginosa* were 0.035, 0.07, 0.018, 0.07, and 0.035 mg/L, respectively. Authors claimed that AgNPs derived from (PE) *Pleurotus eryngii* extract could be used as a better alternative, as an antibiotic, compared to the other silver nitrates and antibiotics [45]. After that, Debnath et al. (2019) also fabricated AgNPs from *Pleurotus giganteus* and analyzed their antibacterial activity [51]. The TiO_2 NPs mediated from extract of *P. djamor* exhibited significant bactericidal activity against human pathogenic bacteria with maximum zone of inhibition *P. fluorescens* (33 ± 0.2 mm), *Corynebacterium diphtheria* (32 ± 0.1 mm), *S. aureus* (32 ± 0.4 mm), and showed higher levels of the inhibitory effect [44]. The *P. djamor* ZnONPs showed a maximum zone of inhibition against *C. diphtheriae* (28.6 ± 0.3 mm), *P. fluorescens* (27 ± 0.5 mm), and *S. aureus* (26.6 ± 1.5 mm) [43]. The general mechanism of microbial cell death is summarized, as below and in Figure 3.

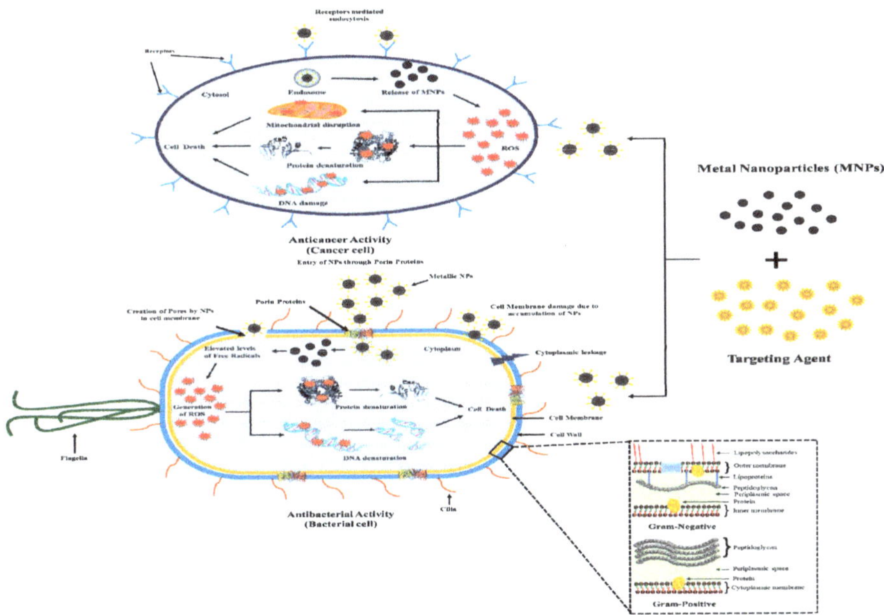

Figure 3. Graphical representation of mechanism showing anticancer and antimicrobial activity.

4.1.1. Antimicrobial Mechanisms

Physical Destruction

In this case, positively charged metal NPs can easily bind to the negatively charged components (i.e., porins, peptidoglycans, and proteins) of the cell membrane via electrostatic interaction, leading to damage of bacterial/fungi membrane, intercellular leakage and, finally, cell inhibition [92].

Oxidative Stress

Another primary mechanism for the antimicrobial activity is based upon the occurrence of oxidative stress, either in the presence of light or under dark conditions. In the microbial cell, the metal NPs can generate ROS, such as •OH and •O2, leading to protein denaturation, DNA damage, enzyme activation, ribosome disassemble and, finally, cell death [92,93]. Furthermore, metal NPs can also act as photoabsorber material upon excitation of light (most often NIR), resulting in cell death. The photothermal effect comes in origin when the emitted electrons from a higher energy state returns to a low energy state, and release their energy in the form of heat and vibrational energy [94].

4.2. *Antioxidant Activity*

In the human body, excessive reactive free radicals are formed from various sources, such as low diet, mental stress, smoking, and other ailments [95]. Metal NPs exhibited profound antioxidant activities in both intracellular and extracellular environments, as summarized in Figure 4.

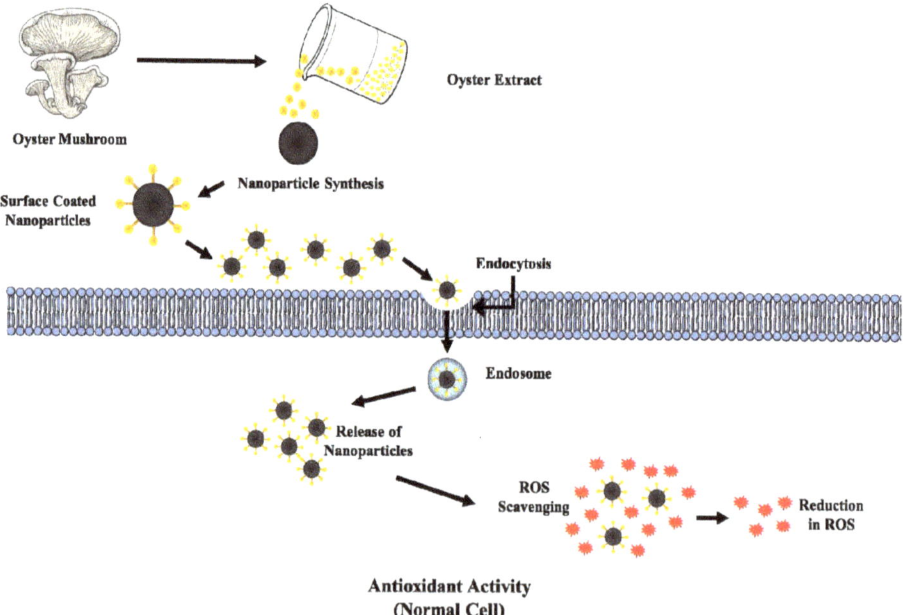

Figure 4. Schematic diagram of the antioxidant mechanism.

In the year 2012, Adebayo et al. synthesized metabolite derived from *P. pulmonarius* extract and evaluated their radical scavenging ability via α, α-diphenyl-β-picrylhydrazyl (DPPH) radical scavenging assay and the β-carotene-linoleate model method. The metabolite derived from *P. pulmonarius* extract showed dose-dependent radical scavenging activity. It was found that the existence of glutathione, ascorbic acid, cysteine, tocopherol, polyhydroxy compounds, and aromatic amines in metabolite reduces and decolorizes the violet color of DPPH via hydrogen transferability. Authors claimed that, at a concentration of 2 mg/mL, metabolite showed butylated hydroxyanisole (BHA) (75%), LAU 09 (80%), and α-tocopherol (90%) of inhibition, which attributes to of the presence of phenolic compounds in the extract [14]. A few years later, in 2017, Madhanraj et al. synthesized gold (Au) and silver (Ag) nanoparticles derived from edible mushroom (basidiomycetes) and studied their antioxidant activity via various radical scavenging assays. Both the prepared NPs (Au & Ag) showed significant antioxidant activity in a cell-free system [39]. Acay and Baran (2020), synthesized *P. eryngii* AgNPs and evaluated their radical scavenging ability via DPPH, chelation of ferrous ions reducing power, and the β-carotene-linoleate model method, and found that, at a concentration of 10 mg/mL, antioxidant activities were 85%, 82%, and 77%, respectively [96]. Zinc plays a role in protecting cells from oxidative stress and acts as an antioxidant. The ZnONPs derived with the help of *P. djamor* possess strong antioxidant properties (DPPH 59%, H_2O_2 59.65%, and ABTS 59.30%), with IC_{50} values of 428.35 μg/mL, 417.22 lg/mL, and 500 lg/mL), respectively [43].

Two possible primary mechanisms for the antioxidant activity are; (i) hydrogen atom transfer, and (ii) single electron transfer [92]. Excessive free radicals could be neutralized or terminated via donating a hydrogen atom that includes total oxyradical scavenging capacity assay, inhibition of induced low-density lipoprotein oxidation, oxygen radical absorbance capability, and radical-trapping antioxidant parameters [78]. On the other hand, the single-electron transfer involves the reduction of compounds, such as radicals, metals and carbonyls by transferring one electron, including change in the color when the compound is reduced, such as Ferric Reducing Antioxidant Potential (FRAP), 2,2-diphenyl-1-picrylhydrazyl radical (DPPH) [92]. However, in the intracellular level, metal NPs enter inside the cells via endocytosis and decrease the ROS levels generated by any probe; for example, 2, 7′-dichlorodihydrofluorescein diacetate (DCFDA) [92].

4.3. Anticancer Activity

In addition to antibacterial and antioxidant activity, metal NPs derived from fungi and other sources have been known to possess outstanding anticancer activity because of their profound ROS generation ability under the dark and light exposure [92,97–99]. Sankar et al. (2013) studied the anticancer activity of AgNPs (~136 nm) derived from *Origanum vulgare* extract against the human lung epithelial cells (A549 cells). AgNPs exhibited dose-dependent toxicity against the A549 cells by 85% inhibition at the dose of 500 μg/mL [100]. Bhat et al. (2013) fabricated Au NPs (12–15 nm, spherical) derived from *P. florida* mushroom extract via the photo-irradiated method and evaluated their anticancer activity against the A-549, MDA-MB, HeLa, and K-562 cell lines. The prepared AuNPs showed concentration-dependent activity against all cell lines in between 10 and 30 μg/mL [48].

Gliga et al. (2014) attempted to understand the coating and size-dependent toxicity of the AgNPs toward the human lung cells (BEAS-2B cells) with an appropriate mechanism. The results confirmed that prepared NPs with size <10 nm showed the highest toxicity against the BEAS-2B cells, which attributes for its aggregation in cell medium, intracellular localization, cellular uptake, and formation of Ag ions intracellularly. However, it is confirmed that all of the AgNPs showed toxicity against BEAS-2B cells via an increase in overall DNA damage within 24 h [101]. In the same year, Yehia and Sheikh (2014) used *P. ostreatus* derived AgNPs (4–15 nm, spherical) as an anticancer agent against the MCF-7 cells. The prepared AgNPs showed dose-dependent cell inhibition ranging from 5% to 78% at concentration 10 to 640 μg/mL [36].

Similarly, in the year 2015, Ismail et al., fabricated AgNPs derived from *P. ostreatus* extract and studied their anticancer effect against the HepG2 and MCF-7 adenocarcinoma cancer cell lines. The authors claimed that NPs induced cytotoxicity toward cancer cells attributes for the formation of ROS species, apoptosis, necrosis, and cell death. ROS are the highly reactive species that result in oxidative damage of proteins, DNA, and induce mitochondrial dysfunction, as summarized in Figure 3 [102]. Similarly, Raman et al. (2015) used *P. djamor* var. roseus derived AgNPs as anticancer agent toward the human prostrate carcinoma PC3 cells [42]. In the year 2013 and 2014, Priyaragini and

Kim et al. demonstrated that metal NPs are harmless at a lower concentration and may be lethal at a higher dose toward normal healthy cells. Many reports revealed the biosynthetic routes to synthesize AgNPs as an anticancer agent against various cell lines. However, AgNPs synthesized using green methods also showed a sort of cytotoxicity against tumor cells [103,104]. Even the extensive use of artificial AgNPs has been already reported, but still, there are limited studies to regulate the cytotoxic effects of AgNPs [36]. Studies on *P. eryngii* (PE) AgNPs showed cytotoxic activity of HeLa with maximum inhibitory effect 73.46% at 60 μg/mL concentration, PC-3 99.02% at 10 μg/mL and MCF-7 cells 93.89% at 20 μg/mL concentration with IC_{50} values of 46.594, 2.185, and 6.169 μg/mL, respectively, during a 24-h incubation period [90]. Chaturvedi et al. (2020) studied cytotoxic activity and revealed that the AgNPs and AuNPs mediated from *P. sajor-caju* extract (PS) showed effective results against HCT-116 cancer cell line. HCT-116 cancer cells viability showed inhibition by *P. sajor-caju* extract, Au NPs as well as Ag NPs showing IC_{50} value of 60, 80, and 50 μg/mL respectively. The study revealed that the green synthesized AgNPs showed high antiproliferative activity in contrast to other PS extract and Au NPs, and the reason behind the mechanism was due to the generation of more ROS, leading to oxidative stress, resulting in undeviated damage of protein functionality and integrity [60]. The anticancer activity of TiO_2 NPs showed potential toxic effect against human lung cancer (A549) cell lines with maximum inhibited growth of 64% at concentration of 100 μg/mL, after 24 h of exposure [44]. The anticancer activity evaluated from *P. djamor* ZnONPs showed potent inhibitory on A549 cancerous cells with (LC50(Lethal concentration required to kill 50% of population) value as 42.26 μg/mL) in a dose-dependent manner [43].

4.4. Histopathological Study and Larvicidal Activity

The histopathological profile of TiO_2NPs mediated from a *P. djamor* extract treated mosquito (*Aedes aegypti and Culex quinquefasciatus*) resulted in the complete collapse of caeca, digestive tract, and desertion of the cuticle and epithelial layer, with harsh damage to the mid-and hind-gut, muscles, as well as nerve ganglia of the brush border. The treating of TiO_2 NPs on IVth instar larvae of *Ae. aegypti* and *Cx. quinquefasciatus* resulted in larvicidal activity with LC50 (5.88 and 4.84 μg/L) and LC90 (22.80 and 19.33 μg/L) [44]. The *Ae. aegypti* larvae treated with ZnONPs showed morphological alteration in the digestive tract, wrecked membrane, midgut, and severe damaging of the brush border, cortex with hyperplasia of gut epithelial cells, and variations in the cytoplasmic masses. The larvae of *Cx. quinquefasciatus* showed the complete putrefaction of abdominal parts, specifically in the caeca, mid-gut, and epithelial layer [43].

4.5. Antidiabetic Activity

The antidiabetic activity was investigated in vitro through the inhibition of α- amylase, an enzyme that digests starch. AgNPs synthesized from *P. giganteus* possess good α-amylase inhibition activity, which helps in making diabetic drugs; inhibition percentage can be increased with increasing concentration of biosynthesized AgNPs [51].

4.6. Removal of Dyes

El-Batal et al. (2014) extracted fungal laccase derived from *P. ostreatus* via solid fermentation. The authors demonstrated that this enzyme could be used to decolorize/degrade numerous dyes, i.e., methyl orange, trypan blue, ramazol brilliant red, and ramazol brilliant yellow with more than fifty percent decolorization in their color within 3 h, confirming the laccase degrading ability. The highest reduction was observed for the methyl orange and trypan blue. Furthermore, laccase enzyme was used to synthesize gold NPs, proving that laccase obtained from *P. ostreatus* had strong potential in many significant industrial applications, for example, in biological pretreatment processes [105,106].

4.7. Catalytic Activity

The recent use of the 4-nitrophenol and derivatives in the manufacturing of insecticides, herbicides, and dyestuffs cause harm to the environment as common wastewater pollutants. Because of their high toxicity, it is challenging to eliminate these pollutants, which is a primary environmental concern. In the year 2007, Panigrahi et al. prepared citrate-capped negatively charged Au NPs (8–55 nm, spherical) for the catalytic degradation of aromatic nitro compounds. The authors claimed that the rate of the reaction rose with the rise in the loading of the catalyst, and decreased in particles size, clearly reflecting the catalytic behavior of gold nanoparticles against aromatic compounds, resulting in amino-compounds [107]. Similarly, Lim et al. (2016) prepared gold nanoparticles (AC-Au NPs, $16.88 \pm 5.47\sim29.93 \pm 9.80$ nm, spherical) from *Agrostis capillaris* extract, and studied their catalytic efficacy in the presence of $NaBH_4$ against the 4-nitrophenol. They demonstrated that particle size falls with the rise in extract concentration during the synthesis process. It was observed that the catalytic degradation of 4-nitrophenol rises as the particles size decreases [108].

In the same year, 2016, Rostami-Vartooni et al. developed AgNPs (8–35 nm, spherical) loaded on perlite (sheet-like) using *Hamamelis virginiana* leaf extract and evaluated their catalytic activity against the 4-nitrophenol and Congo red (CR) dye. The authors demonstrated that, with the rise in the concentration of $NaBH_4$ and AgNPs/perlite, the degradation time of 4-nitrophenol decreases, respectively. The AgNPs supported on the surface of perlite facilitate the electron relay from BH_4- to 4-nitrophenol as well as CR dye. Furthermore, they claimed that AgNPs/perlite showed high stability and could be used up to 4 times with significant degradation efficacy [109]. Later, Gopalakrishnan et al. (2017) reported the catalytic degradation of 4-nitrophenol to 4-aminophenol via $NaBH_4$ in the presence of PdNPs (<20 nm, spherical) derived from seed extract of *Silybum marianum*. The total reduction action was attained within 27 min and is attributed to the relay of electrons from BH4- to 4-nitrophenol, resulting in 4-aminophenol. However, the authors claimed that no reduction was detected in the case of bare $NaBH_4$ [110]. A similar kind of 4-nitrophenol reduction was also performed by Sen et al. (2013) using *P. florida* derived AuNPs and $NaBH_4$ [49].

4.8. Food Packaging and Preservation

Biocompatible fabricated zinc NPs might be efficiently applied in the biomedical and food packaging fields. The potential antimicrobial action of mushroom against many foodborne bacteria, such as *Escherichia coli*, *Streptococcus faecalis*, *Bacillus subtilis*, *Micrococcus luteus*, and *Listeria innocua* could be considered as a boon for the food industry since, by using metal NPs, the contamination of foodstuffs can be avoided, besides for long-time preservation [55,111].

5. Conclusions

Nanomaterials derived from oyster mushrooms have been found to possess great potential over a wide range of applications, especially in the biomedical field. In the present review, we discussed the progress of research, to date, on metal nanoparticles and other nanomaterials derived from oyster mushrooms, regarding their synthesis and applications, particularly in the areas of antimicrobial, larvicidal, antioxidant, anticancer, and catalysis. Generally, AgNPs derived from *Pleurotus* spp. have a higher synthesis and biomedical applications among mushrooms. The importance of derived nanoparticles is due to their unique characteristics, such as cost effective, crystalline nature, nanosize, and non-hazardous nature. Mainly there are two well-known methods of synthesis, i.e., intracellular and extracellular. It is noticed that nanomaterials derived from oyster mushrooms showed profound applications in the areas of biomedicine and catalysis, but some areas of research are needed to be addressed, which are as follows:

- To date, oyster mushroom derived NPs are not directly applied to the live samples. Hence, the progress can be made in this direction.

- Available literature provides evidence that considerable work has been carried out for ascertaining the efficacy of oyster mushroom derived NPs under in vitro conditions against the various cancer cell lines. As less information is available regarding in vivo studies, there is a need for further exploration.
- More studies are needed to define oyster mushrooms that can be genetically engineered to produce more enzymes primarily involved in NP synthesis, and to expand the knowledge and functions of nanomaterial, so that significant achievements could be attained in the fields of medicine, electronics, cosmetics, agriculture, the environment, and many more.

Author Contributions: Conceptualization, P.B., K.A.A.-E., and K.K.; manuscript writing, K.B., A.S. (Anirudh Sharma); manuscript editing, N.T., S.B., A.K. (Anu Kalia), A.K. (Anil Kumar), A.S. (Ashwag Shami). Critical revising, P.B., K.A.A.-E., E.N., and K.K. All authors have read and agreed to the published version of the manuscript.

Funding: This research was funded by the University of Hradec Kralove (Faculty of Science VT2019-2021) and Excellence project UHK.

Acknowledgments: We acknowledge the University of Hradec Kralove (Faculty of Science, VT2019-2021) and Excellence project UHK.

Conflicts of Interest: The authors declare no conflict of interest.

References

1. Bhardwaj, K.; Dhanjal, D.S.; Sharma, A.; Nepovimova, E.; Kalia, A.; Thakur, S.; Bhardwaj, S.; Chopra, C.; Singh, R.; Verma, R.; et al. Conifer-derived metallic nanoparticles: Green synthesis and biological applications. *Int. J. Mol. Sci.* **2020**, *21*, 9028. [CrossRef]
2. Kumar, H.; Bhardwaj, K.; Kuča, K.; Kalia, A.; Nepovimova, E.; Verma, R.; Kumar, D. Flower-Based Green Synthesis of Metallic Nanoparticles: Applications beyond Fragrance. *Nanomaterials* **2020**, *10*, 766. [CrossRef]
3. Owaid, M.N. Biomedical Applications of Nanoparticles Synthesized from Mushrooms. In *Green Nanoparticles*; Patra, J., Fraceto, L., Das, G., Campos, E., Eds.; Springer: Cham, Switzerland, 2020; pp. 289–303.
4. Xu, H.; Yao, L.; Sun, H.; Wu, Y. Chemical composition and antitumor activity of different polysaccharides from the roots of Actinidia eriantha. *Carbohydr. Polym.* **2009**, *78*, 316–322. [CrossRef]
5. Sykes, E.A.; Chen, J.; Zheng, G.; Chan, W.C.W. Investigating the impact of nanoparticle size on active and passive tumor targeting efficiency. *ACS Nano* **2014**, *8*, 5696–5706. [CrossRef] [PubMed]
6. Shankar, S.S.; Rai, A.; Ankamwar, B.; Singh, A.; Ahmad, A.; Sastry, M. Biological synthesis of triangular gold nanoprisms. *Nat. Mater.* **2004**, *3*, 482–488. [CrossRef] [PubMed]
7. Philip, D. Biosynthesis of Au, Ag and Au-Ag nanoparticles using edible mushroom extract. *Spectrochim. Acta Part A Mol. Biomol. Spectrosc.* **2009**, *73*, 374–381. [CrossRef] [PubMed]
8. Maftoun, P.; Johari, H.; Soltani, M.; Malik, R.; Othman, N.Z.; El Enshasy, H.A. The edible mushroom Pleurotus spp.: I. Biodiversity and nutritional values. *Int. J. Biotechnol. Wellness Ind.* **2015**, *4*, 67–83.
9. Raman, J.; Jang, K.-Y.; Oh, Y.-L.; Oh, M.; Im, J.-H.; Lakshmanan, H.; Sabaratnam, V. Cultivation and Nutritional Value of Prominent *Pleurotus* spp.: An Overview. *Mycobiology* **2020**, 1–14. [CrossRef]
10. Kumla, J.; Suwannarach, N.; Jaiyasen, A.; Bussaban, B.; Lumyong, S. Development of an edible wild strain of Thai oyster mushroom for economic mushroom production. *Chiang Mai J. Sci.* **2013**, *40*, 161–172.
11. Mirunalini, S.; Arulmozhi, V.; Deepalakshmi, K.; Krishnaveni, M. Intracellular biosynthesis and antibacterial activity of silver nanoparticles using edible mushrooms. *Not. Sci. Biol.* **2012**, *4*, 55–61. [CrossRef]
12. Mowsurni, F.R.; Chowdhury, M.B.K. Oyster mushroom: Biochemical and medicinal prospects. *Bangladesh J. Med. Biochem.* **2010**, *3*, 23–28. [CrossRef]
13. Lok, C.N.; Ho, C.M.; Chen, R.; He, Q.Y.; Yu, W.Y.; Sun, H.; Tam, P.K.H.; Chiu, J.F.; Che, C.M. Silver nanoparticles: Partial oxidation and antibacterial activities. *J. Biol. Inorg. Chem.* **2007**, *12*, 527–534. [CrossRef] [PubMed]
14. Adebayo, E.A.; Oloke, J.K.; Ayandele, A.A.; Adegunlola, C. Phytochemical, antioxidant and antimicrobial assay of mushroom metabolite from Pleurotus pulmonarius—LAU 09 (JF736658). *J. Microbiol. Biotechnol. Res.* **2012**, *2*, 366–374.
15. Owaid, M.N.; Ibraheem, I.J. Mycosynthesis of nanoparticles using edible and medicinal mushrooms. *Eur. J. Nanomed.* **2017**, *9*, 5–23. [CrossRef]

16. Duhan, J.S.; Kumar, R.; Kumar, N.; Kaur, P.; Nehra, K.; Duhan, S. Nanotechnology: The new perspective in precision agriculture. *Biotechnol. Rep.* **2017**, *15*, 11–23. [CrossRef] [PubMed]
17. McQuillan, J.S.; Groenaga Infante, H.; Stokes, E.; Shaw, A.M. Silver nanoparticle enhanced silver ion stress response in Escherichia coli K12. *Nanotoxicology* **2012**, *6*, 857–866. [CrossRef]
18. Gade, A.; Ingle, A.; Whiteley, C.; Rai, M. Mycogenic metal nanoparticles: Progress and applications. *Biotechnol. Lett.* **2010**, *32*, 593–600. [CrossRef]
19. Kreibig, U.; Vollmer, M. *Optical Properties of Metal Clusters*; Springer Series in Materials Science; Springer: Berlin/Heidelberg, Germany, 1995; Volume 25, ISBN 978-3-642-08191-0.
20. Narayanan, R.; El-Sayed, M.A. Effect of Catalytic Activity on the Metallic Nanoparticle Size Distribution: Electron-Transfer Reaction between Fe(CN)6 and Thiosulfate Ions Catalyzed by PVP—Platinum Nanoparticles. *J. Phys. Chem. B* **2003**, *107*, 12416–12424. [CrossRef]
21. Asmathunisha, N.; Kathiresan, K. A review on biosynthesis of nanoparticles by marine organisms. *Colloids Surf. B Biointerfaces* **2013**, *103*, 283–287. [CrossRef]
22. Sastry, M.; Ahmad, A.; Khan, M.I.; Kumar, R. Biosynthesis of metal nanoparticles using fungi and actinomycete. *Curr. Sci.* **2003**, *85*, 162–170.
23. Owaid, M.N. Green synthesis of silver nanoparticles by Pleurotus (oyster mushroom) and their bioactivity: Review. *Environ. Nanotechnol. Monit. Manag.* **2019**, *12*, 100256. [CrossRef]
24. Naveen, H.K.; Kumar, G.; Rao, B.K. Extracellular biosynthesis of silver nanoparticles using the filamentous fungus *Penicillium* sp. *Arch. Appl. Sci. Res.* **2010**, *2*, 161–167.
25. Ingle, A.; Gade, A.; Pierrat, S.; Sonnichsen, C.; Rai, M. Mycosynthesis of Silver Nanoparticles Using the Fungus Fusarium acuminatum and Its Activity against Some Human Pathogenic Bacteria. *Curr. Nanosci.* **2008**, *4*, 141–144. [CrossRef]
26. Jain, N.; Bhargava, A.; Majumdar, S.; Tarafdar, J.C.; Panwar, J. Extracellular biosynthesis and characterization of silver nanoparticles using Aspergillus flavus NJP08: A mechanism perspective. *Nanoscale* **2011**, *3*, 635–641. [CrossRef]
27. Sanghi, R.; Verma, P. Biomimetic synthesis and characterisation of protein capped silver nanoparticles. *Bioresour. Technol.* **2009**, *100*, 501–504. [CrossRef]
28. Soni, N.; Prakash, S. Synthesis of gold nanoparticles by the fungus Aspergillus niger and its efficacy against mosquito larvae. *Rep. Parasitol.* **2012**, *2*, 7. [CrossRef]
29. Castro-Longoria, E.; Moreno-Velásquez, S.D.; Vilchis-Nestor, A.R.; Arenas-Berumen, E.; Avalos-Borja, M. Production of platinum nanoparticles and nanoaggregates using Neurospora crassa. *J. Microbiol. Biotechnol.* **2012**, *22*, 1000–1004. [CrossRef]
30. Ahmad, A.; Senapati, S.; Khan, M.I.; Kumar, R.; Ramani, R.; Srinivas, V.; Sastry, M. Intracellular synthesis of gold nanoparticles by a novel alkalotolerant actinomycete, Rhodococcus species. *Nanotechnology* **2003**, *14*, 824–828. [CrossRef]
31. Durán, N.; Marcato, P.D.; Alves, O.L.; De Souza, G.I.H.; Esposito, E. Mechanistic aspects of biosynthesis of silver nanoparticles by several Fusarium oxysporum strains. *J. Nanobiotechnol.* **2005**, *3*, 8. [CrossRef]
32. Nithya, R.; Ragunathan, R. Synthesis of Silver Nanoparticle using Pleurotus Sajor Caju and Its Antimicrobial Study. *Dig. J. Nanomater. Biostruct.* **2009**, *4*, 623–629.
33. Bhat, R.; Deshpande, R.; Ganachari, S.V.; Huh, D.S.; Venkataraman, A. Photo-irradiated biosynthesis of silver nanoparticles using edible mushroom Pleurotus florida and their antibacterial activity studies. *Bioinorg. Chem. Appl.* **2011**, *2011*, 650979. [CrossRef] [PubMed]
34. Sujatha, S.; Tamilselvi, S.; Subha, K.; Panneerselvam, A. Studies on biosynthesis of silver nanoparticles using mushroom and its antibacterial activities. *Int. J. Curr. Microbiol. Appl. Sci.* **2013**, *2*, 605–614.
35. Owaid, M.N. Testing Efficiency of Different Agriculture Media in Growth and Production of Four Species of Oyster Mushroom Pleurotus and Evaluation the Bioactivity of Tested Species. Ph.D. Thesis, College of Science, University of Anbar, Ramadi, Iraq, 2013.
36. Yehia, R.S.; Al-Sheikh, H. Biosynthesis and characterization of silver nanoparticles produced by pleurotus ostreatus and their anticandidal and anticancer activities. *World J. Microbiol. Biotechnol.* **2014**, *30*, 2797–2803. [CrossRef] [PubMed]
37. Owaid, M.N.; Raman, J.; Lakshmanan, H.; Al-Saeedi, S.S.S.; Sabaratnam, V.; Ali Abed, I. Mycosynthesis of silver nanoparticles by Pleurotus cornucopiae var. citrinopileatus and its inhibitory effects against *Candida* sp. *Mater. Lett.* **2015**, *153*, 186–190. [CrossRef]

38. Maurya, S.; Bhardwaj, A.K.; Gupta, K.K.; Agarwal, S.; Kushwaha, A.; Vk, C.; Pathak, R.K.; Gopal, R.; Uttam, K.N.; Singh, A.K. Green synthesis of silver nanoparticles using Pleurotus and its bactericidal activity. *Cell. Mol. Biol.* **2016**, *62*, 131.
39. Madhanraj, R.; Eyini, M.; Balaji, P. Antioxidant Assay of Gold and Silver Nanoparticles from Edible Basidiomycetes Mushroom Fungi. *Free Radic. Antioxid.* **2017**, *7*, 137–142. [CrossRef]
40. Shivashankar, M.; Premkumari, B.; Chandan, N. Comparative studies on biosynthesis, partial characterization of silver nanoparticles and antimicrobial activities in some edible mushroom. *Nano Sci. Nano Technol. Indian J.* **2013**, *8*, 8–14.
41. Shivashankar, M.; Premkumari, B.; Chandan, N. Biosynthesis, partial characterization and antimicrobial activities of silver nanoparticles from pleurotus species. *Int. J. Integr. Sci. Innov. Technol.* **2013**, *2*, 13–23.
42. Raman, J.; Reddy, G.R.; Lakshmanan, H.; Selvaraj, V.; Gajendran, B.; Nanjian, R.; Chinnasamy, A.; Sabaratnam, V. Mycosynthesis and characterization of silver nanoparticles from Pleurotus djamor var. roseus and their in vitro cytotoxicity effect on PC3 cells. *Process Biochem.* **2015**, *50*, 140–147. [CrossRef]
43. Manimaran, K.; Balasubramani, G.; Ragavendran, C.; Natarajan, D.; Murugesan, S. Biological Applications of Synthesized ZnO Nanoparticles Using Pleurotus Djamor against Mosquito Larvicidal, Histopathology, Antibacterial, Antioxidant and Anticancer Effect. *J. Clust. Sci.* **2020**, 1–13. [CrossRef]
44. Manimaran, K.; Murugesan, S.; Ragavendran, C.; Balasubramani, G.; Natarajan, D.; Ganesan, A.; Seedevi, P. Biosynthesis of TiO_2 Nanoparticles Using Edible Mushroom (Pleurotus djamor) Extract: Mosquito Larvicidal, Histopathological, Antibacterial and Anticancer Effect. *J. Clust. Sci.* **2020**, 1–12. [CrossRef]
45. Acay, H.; Baran, M.F. Biosynthesis and characterization of silver nanoparticles using king oyster (Pleurotus eryngii) extract: Effect on some microorganisms. *Appl. Ecol. Environ. Res.* **2019**, *17*, 9205–9214. [CrossRef]
46. Manzoor-ul-Haq, R.V.; Patil, S.; Singh, D.; Krishnaveni, R. Isolation and Screening of Mushrooms for Potent Silver Nanoparticles Production from Bandipora District (Jammu and Kashmir) and their characterization. *Int. J. Curr. Microbiol. Appl. Sci* **2014**, *3*, 704–714.
47. Kaur, T.; Kapoor, S.; Kalia, A. Synthesis of Silver Nanoparticles from Pleurotus florida, Characterization and Analysis of their Antimicrobial Activity. *Int. J. Curr. Microbiol. Appl. Sci* **2018**, *7*, 4085–4095. [CrossRef]
48. Bhat, R.; Sharanabasava, V.G.; Deshpande, R.; Shetti, U.; Sanjeev, G.; Venkataraman, A. Photo-bio-synthesis of irregular shaped functionalized gold nanoparticles using edible mushroom Pleurotus florida and its anticancer evaluation. *J. Photochem. Photobiol. B Biol.* **2013**, *125*, 63–69. [CrossRef]
49. Sen, I.K.; Maity, K.; Islam, S.S. Green synthesis of gold nanoparticles using a glucan of an edible mushroom and study of catalytic activity. *Carbohydr. Polym.* **2013**, *91*, 518–528. [CrossRef]
50. Gurunathan, S.; Han, J.W.; Park, J.H.; Kim, J.H. A green chemistry approach for synthesizing biocompatible gold nanoparticles. *Nanoscale Res. Lett.* **2014**, *9*, 248. [CrossRef]
51. Debnath, G.; Das, P.; Saha, A.K. Green Synthesis of Silver Nanoparticles Using Mushroom Extract of Pleurotus giganteus: Characterization, Antimicrobial, and α-Amylase Inhibitory Activity. *Bionanoscience* **2019**, *9*, 611–619. [CrossRef]
52. Al-Bahrani, R.; Raman, J.; Lakshmanan, H.; Hassan, A.A.; Sabaratnam, V. Green synthesis of silver nanoparticles using tree oyster mushroom Pleurotus ostreatus and its inhibitory activity against pathogenic bacteria. *Mater. Lett.* **2017**, *186*, 21–25. [CrossRef]
53. Devika, R.; Elumalai, S.; Manikandan, E.; Eswaramoorthy, D. Biosynthesis of Silver Nanoparticles Using the Fungus Pleurotus ostreatus and their Antibacterial Activity. *Open Access Sci. Rep.* **2012**, *1*, 557. [CrossRef]
54. Bawadekji, A.; Oueslati, M.H.; Ali, A.; Basha, J. Biosynthesis of Gold Nanoparticles using Pleurotus ostreatus (Jacq. ex. Fr.) Kummer Extract and their Antibacterial and Antifungal Activities. *J. Appl. Environ. Biol. Sci.* **2018**, *8*, 142–147.
55. Senapati, U.S.; Sarkar, D. Characterization of biosynthesized zinc sulphide nanoparticles using edible mushroom Pleurotuss ostreatu. *Indian J. Phys.* **2014**, *88*, 557–562. [CrossRef]
56. Wu, H.F.; Kailasa, S.K.; Shastri, L. Electrostatically self-assembled azides on zinc sulfide nanoparticles as multifunctional nanoprobes for peptide and protein analysis in MALDI-TOF MS. *Talanta* **2010**, *82*, 540–547. [CrossRef] [PubMed]
57. Musa, S.F.; Yeat, T.S.; Kamal, L.Z.M.; Tabana, Y.M.; Ahmed, M.A.; El Ouweini, A.; Lim, V.; Keong, L.C.; Sandai, D. Pleurotus sajor-caju can be used to synthesize silver nanoparticles with antifungal activity against Candida albicans. *J. Sci. Food Agric.* **2018**, *98*, 1197–1207. [CrossRef] [PubMed]

58. Devi, M.R.; Krishnakumari, S. Qualitative screening of phytoconstituents of Pleurotus sajor caju (Fries sing) and comparison between hot and cold-aqueous and silver nanoparticles extracts. *J. Med. Plants Stud.* **2015**, *3*, 172–176.
59. Vigneshwaran, N.; Kathe, A.A.; Varadarajan, P.V.; Nachane, R.P.; Balasubramanya, R.H. Silver-protein (core-shell) nanoparticle production using spent mushroom substrate. *Langmuir* **2007**, *23*, 7113–7117. [CrossRef]
60. Chaturvedi, V.K.; Yadav, N.; Rai, N.K.; Abd Ellah, N.H.; Bohara, R.A.; Rehan, I.F.; Marraiki, N.; Batiha, G.E.S.; Hetta, H.F.; Singh, M.P. Pleurotus sajor-caju-Mediated Synthesis of Silver and Gold Nanoparticles Active against Colon Cancer Cell Lines: A New Era of Herbonanoceutics. *Molecules* **2020**, *25*, 3091. [CrossRef]
61. Dandapat, S.; Kumar, M.; Sinha, M.P. Synthesis of White Nanopartials mediate by Pleurotus tuber-regium (Rumph. ex Fr.) Extract and Silver Nitrate. In *Proceedings of the Exploring Basic and Applied Sciences (EBAS-2014), Jalandhar, India, 14–15 November 2014*; Elsevier: Amsterdam, The Netherlands, 2014; pp. 98–101.
62. Chan, S.; Mashitah, M. Instantaneous biosynthesis of silver nanoparticles by selected macro fungi. *Aust. J. Basic Appl. Sci.* **2012**, *6*, 86–88.
63. Vala, A.K.; Shah, S.; Patel, R. Biogenesis of silver nanoparticles by marine-derived fungus Aspergillus flavus from Bhavnagar Coast, Gulf of Khambhat, India. *J. Mar. Biol. Ocean.* **2014**, *3*, 1–3.
64. Narayanan, K.B.; Sakthivel, N. Biological synthesis of metal nanoparticles by microbes. *Adv. Colloid Interface Sci.* **2010**, *156*, 1–13. [CrossRef]
65. Thakkar, K.N.; Mhatre, S.S.; Parikh, R.Y. Biological synthesis of metallic nanoparticles. *Nanomed. Nanotechnol. Biol. Med.* **2010**, *6*, 257–262. [CrossRef] [PubMed]
66. Dhillon, G.S.; Brar, S.K.; Kaur, S.; Verma, M. Green approach for nanoparticle biosynthesis by fungi: Current trends and applications. *Crit. Rev. Biotechnol.* **2012**, *32*, 49–73. [CrossRef] [PubMed]
67. Gade, A.K.; Gadge Baba, S.; Ingle, A.; Marcato, P.D.; Duran, N.; Gade, A.K.; Bonde, P.; Ingle, A.P.; Marcato, P.D.; Durán, N.; et al. Exploitation of Aspergillus niger for fabrication of silver nanoparticles. *J. Biobased Mater. Bioenergy* **2008**, *2*, 1–5. [CrossRef]
68. Zhang, X.; Yan, S.; Tyagi, R.D.; Surampalli, R.Y. Synthesis of nanoparticles by microorganisms and their application in enhancing microbiological reaction rates. *Chemosphere* **2011**, *82*, 489–494. [CrossRef] [PubMed]
69. Khandel, P.; Shahi, S.K. Mycogenic nanoparticles and their bio-prospective applications: Current status and future challenges. *J. Nanostruct. Chem.* **2018**, *8*, 369–391. [CrossRef]
70. Durán, N.; Marcato, P.D.; De Souz, G.I.; Alves, O.L.; Esposito, E. Antibacterial effect of silver nanoparticles produced by fungal process on textile fabrics and their effluent treatment. *J. Biomed. Nanotech.* **2007**, *3*, 203–208. [CrossRef]
71. Mukherjee, P.; Roy, M.; Mandal, B.P.; Dey, G.K.; Mukherjee, P.K.; Ghatak, J.; Tyagi, A.K.; Kale, S.P. Green synthesis of highly stabilized nanocrystalline silver particles by a non-pathogenic and agriculturally important fungus T. asperellum. *Nanotechnology* **2008**, *19*, 075103. [CrossRef] [PubMed]
72. Nanda, A.; Majeed, S. Enhanced antibacterial efficacy of biosynthesized AgNPs from *Penicillium* glabrum (MTCC1985) pooled with different drugs. *Int. J. PharmTech Res.* **2014**, *6*, 217–223.
73. Guilger-Casagrande, M.; de Lima, R. Synthesis of Silver Nanoparticles Mediated by Fungi: A Review. *Front. Bioeng. Biotechnol.* **2019**, *7*, 287. [CrossRef]
74. Devi, L.; Joshi, S. Ultrastructures of silver nanoparticles biosynthesized using endophytic fungi. *J. Microsc. Ultrastruct.* **2015**, *3*, 29. [CrossRef]
75. Mehra, R.K.; Winge, D.R. Metal ion resistance in fungi: Molecular mechanisms and their regulated expression. *J. Cell. Biochem.* **1991**, *45*, 30–40. [CrossRef] [PubMed]
76. Singh, J.; Dutta, T.; Kim, K.H.; Rawat, M.; Samddar, P.; Kumar, P. "Green" synthesis of metals and their oxide nanoparticles: Applications for environmental remediation. *J. Nanobiotechnol.* **2018**, *16*, 84. [CrossRef] [PubMed]
77. Kumar, H.; Bhardwaj, K.; Sharma, R.; Nepovimova, E.; Kuča, K.; Dhanjal, D.S.; Verma, R.; Bhardwaj, P.; Sharma, S.; Kumar, D. Fruit and Vegetable Peels: Utilization of High Value Horticultural Waste in Novel Industrial Applications. *Molecules* **2020**, *25*, 2812. [CrossRef]
78. Kumar, H.; Bhardwaj, K.; Nepovimova, E.; Kuča, K.; Singh Dhanjal, D.; Bhardwaj, S.; Bhatia, S.K.; Verma, R.; Kumar, D. Antioxidant Functionalized Nanoparticles: A Combat against Oxidative Stress. *Nanomaterials* **2020**, *10*, 1334. [CrossRef]

79. Kumar, H.; Bhardwaj, K.; Dhanjal, D.S.; Nepovimova, E.; Şen, F.; Regassa, H.; Singh, R.; Verma, R.; Kumar, V.; Kumar, D.; et al. Fruit Extract Mediated Green Synthesis of Metallic Nanoparticles: A New Avenue in Pomology Applications. *Int. J. Mol. Sci.* **2020**, *21*, 8458. [CrossRef]
80. Latha, S. Extracellular Biosynthesis, Characterization and In Vitro Antimicrobial Potential of Silver Nanoparticles Using Myconanofactoris. Ph.D. Thesis, Bharathidasan University, Tiruchirappalli, India, 2010.
81. Borovaya, M.; Pirko, Y.; Krupodorova, T.; Naumenko, A.; Blume, Y.; Yemets, A. Biosynthesis of cadmium sulphide quantum dots by using Pleurotus ostreatus (Jacq.) P. Kumm. *Biotechnol. Biotechnol. Equip.* **2015**, *29*, 1156–1163. [CrossRef]
82. Borovaya, M.N.; Naumenko, A.P.; Matvieieva, N.A.; Blume, Y.B.; Yemets, A.I. Biosynthesis of luminescent CdS quantum dots using plant hairy root culture. *Nanoscale Res. Lett.* **2014**, *9*, 686. [CrossRef]
83. Mazumdar, H.; Haloi, N. A study on Biosynthesis of Iron nanoparticles by *Pleurotus* sp. *J. Microbiol. Biotechnol. Res. Sch. Res. Libr. J. Microbiol. Biotech. Res.* **2011**, *1*, 39–49.
84. Wong, K. Preparation of highly stable selenium nanoparticles with strong anti-tumor activity using tiger milk mushroom. *Apoptosis* **2012**, *134*, 253–261.
85. Wu, H.; Li, X.; Liu, W.; Chen, T.; Li, Y.; Zheng, W.; Man, C.W.Y.; Wong, M.K.; Wong, K.H. Surface decoration of selenium nanoparticles by mushroom polysaccharides-protein complexes to achieve enhanced cellular uptake and antiproliferative activity. *J. Mater. Chem.* **2012**, *22*, 9602–9610. [CrossRef]
86. Kaur, G.; Kalia, A.; Kapoor, S.; Sodhi, H.S.; Khanna, P.K. Scanning electron microscopy of Pleurotus ostreatus in response to inorganic selenium. In Proceedings of the Indian Mushroom Conference, Ludhiana, India, 16–17 April 2013; pp. 17–18.
87. El-Batal, A.I.; Al-Hazmi, N.E.; Mosallam, F.M.; El-Sayyad, G.S. Biogenic synthesis of copper nanoparticles by natural polysaccharides and Pleurotus ostreatus fermented fenugreek using gamma rays with antioxidant and antimicrobial potential towards some wound pathogens. *Microb. Pathog.* **2018**, *118*, 159–169. [CrossRef] [PubMed]
88. Vittal, R.R.; Aswathanarayan, J.B. Nanoparticles and their potential application as antimicrobials. In *Science against Microbial Pathogens: Communicating Current Research and Technological Advances*; Méndez-Vilas, A., Ed.; Formatex: Badajoz, Spain, 2011; pp. 197–209.
89. Warnes, S.L.; Caves, V.; Keevil, C.W. Mechanism of copper surface toxicity in Escherichia coli O157:H7 and Salmonella involves immediate membrane depolarization followed by slower rate of DNA destruction which differs from that observed for Gram-positive bacteria. *Environ. Microbiol.* **2012**, *14*, 1730–1743. [CrossRef] [PubMed]
90. Beyth, N.; Houri-Haddad, Y.; Domb, A.; Khan, W.; Hazan, R. Alternative antimicrobial approach: Nano-antimicrobial materials. *Evidence-Based Complement. Altern. Med.* **2015**, *2015*, 246012. [CrossRef] [PubMed]
91. Chiriac, V.; Stratulat, D.N.; Calin, G.; Nichitus, S.; Burlui, V.; Stadoleanu, C.; Popa, M.; Popa, I.M. Antimicrobial property of zinc based nanoparticles. In *IOP Conference Series: Materials Science and Engineering*; IOP Publishing: Bristol, UK, 2016; Volume 133, p. 012055.
92. Wang, L.; Hu, C.; Shao, L. The antimicrobial activity of nanoparticles: Present situation and prospects for the future. *Int. J. Nanomed.* **2017**, *12*, 1227–1249. [CrossRef]
93. Mehta, M.; Sharma, P.; Kaur, S.; Dhanjal, D.S.; Singh, B.; Vyas, M.; Gupta, G.; Chellappan, D.K.; Nammi, S.; Singh, T.G.; et al. Plant-based drug delivery systems in respiratory diseases. In *Targeting Chronic Inflammatory Lung Diseases Using Advanced Drug Delivery Systems*; Elsevier: Amsterdam, The Netherlands, 2020; pp. 517–539.
94. Addae, E.; Dong, X.; McCoy, E.; Yang, C.; Chen, W.; Yang, L. Investigation of antimicrobial activity of photothermal therapeutic gold/copper sulfide core/shell nanoparticles to bacterial spores and cells. *J. Biol. Eng.* **2014**, *8*, 11. [CrossRef]
95. Dhanjal, D.S.; Bhardwaj, S.; Sharma, R.; Bhardwaj, K.; Kumar, D.; Chopra, C.; Nepovimova, E.; Singh, R.; Kuca, K. Plant Fortification of the Diet for Anti-Ageing Effects: A Review. *Nutrients* **2020**, *12*, 3008. [CrossRef]
96. Acay, H.; Baran, M. Determination of Antioxidant and cytotoxic activities of king oyster mushroom mediated AgNPs synthesized with environmentally friendly methods. *Medicine* **2020**, *9*, 760–765. [CrossRef]
97. Sharma, P.; Mehta, M.; Dhanjal, D.S.; Kaur, S.; Gupta, G.; Singh, H.; Thangavelu, L.; Rajeshkumar, S.; Tambuwala, M.; Bakshi, H.A.; et al. Emerging trends in the novel drug delivery approaches for the treatment of lung cancer. *Chem. Biol. Interact.* **2019**, *309*, 108720. [CrossRef]

98. Mehta, M.; Dhanjal, D.S.; Paudel, K.R.; Singh, B.; Gupta, G.; Rajeshkumar, S.; Thangavelu, L.; Tambuwala, M.M.; Bakshi, H.A.; Chellappan, D.K.; et al. Cellular signalling pathways mediating the pathogenesis of chronic inflammatory respiratory diseases: An update. *Inflammopharmacology* **2020**, *28*, 795–817. [CrossRef]
99. Mehta, M.; Dhanjal, D.S.; Satija, S.; Wadhwa, R.; Paudel, K.R.; Chellappan, D.K.; Mohammad, S.; Haghi, M.; Hansbro, P.M.; Dua, K. Advancing of Cellular Signaling Pathways in Respiratory Diseases Using Nanocarrier Based Drug Delivery Systems. *Curr. Pharm. Des.* **2020**, *26*. [CrossRef]
100. Sankar, R.; Karthik, A.; Prabu, A.; Karthik, S.; Shivashangari, K.S.; Ravikumar, V. Origanum vulgare mediated biosynthesis of silver nanoparticles for its antibacterial and anticancer activity. *Colloids Surfaces B Biointerfaces* **2013**, *108*, 80–84. [CrossRef] [PubMed]
101. Gliga, A.R.; Skoglund, S.; Odnevall Wallinder, I.; Fadeel, B.; Karlsson, H.L. Size-dependent cytotoxicity of silver nanoparticles in human lung cells: The role of cellular uptake, agglomeration and Ag release. *Part. Fibre Toxicol.* **2014**, *11*, 11. [CrossRef] [PubMed]
102. Ismail, A.F.M.; Ahmed, M.M.; Salem, A.A.M. Biosynthesis of silver nanoparticles using mushroom extracts: Induction of apoptosis in HepG2 and MCF-7 Cells via caspases stimulation and regulation of BAX and Bcl-2 gene expressions. *J. Pharm. Biomed. Sci.* **2015**, *5*, 1–9.
103. Singh, P.; Sathishkumar, S.R.; Bhaskararao, K.V. Biosynthesis of Silver Nanoparticles using Actinobacteria and Evaluating Its Antimicrobial and Cytotoxicity Activity. *Int. J. Pharm. Pharm. Sci.* **2013**, *5*, 709–712.
104. Kim, J.H.; Lee, Y.; Kim, E.J.; Gu, S.; Sohn, E.J.; Seo, Y.S.; An, H.J.; Chang, Y.S. Exposure of iron nanoparticles to Arabidopsis thaliana enhances root elongation by triggering cell wall loosening. *Environ. Sci. Technol.* **2014**, *48*, 3477–3485. [CrossRef] [PubMed]
105. El-Batal, A.I.; Elkenawy, N.M.; Yassin, A.S.; Amin, M.A. Laccase production by Pleurotus ostreatus and its application in synthesis of gold nanoparticles. *Biotechnol. Rep.* **2015**, *5*, 31–39. [CrossRef]
106. Bagewadi, Z.K.; Mulla, S.I.; Ninnekar, H.Z. Optimization of laccase production and its application in delignification of biomass. *Int. J. Recycl. Org. Waste Agric.* **2017**, *6*, 351–365. [CrossRef]
107. Panigrahi, S.; Basu, S.; Praharaj, S.; Pande, S.; Jana, S.; Pal, A.; Ghosh, S.K.; Pal, T. Synthesis and size-selective catalysis by supported gold nanoparticles: Study on heterogeneous and homogeneous catalytic process. *J. Phys. Chem. C* **2007**, *111*, 4596–4605. [CrossRef]
108. Lim, S.H.; Ahn, E.Y.; Park, Y. Green Synthesis and Catalytic Activity of Gold Nanoparticles Synthesized by Artemisia capillaris Water Extract. *Nanoscale Res. Lett.* **2016**, *11*, 474. [CrossRef]
109. Rostami-Vartooni, A.; Nasrollahzadeh, M.; Alizadeh, M. Green synthesis of perlite supported silver nanoparticles using Hamamelis virginiana leaf extract and investigation of its catalytic activity for the reduction of 4-nitrophenol and Congo red. *J. Alloy. Compd.* **2016**, *680*, 309–314. [CrossRef]
110. Gopalakrishnan, R.; Loganathan, B.; Dinesh, S.; Raghu, K. Strategic Green Synthesis, Characterization and Catalytic Application to 4-Nitrophenol Reduction of Palladium Nanoparticles. *J. Clust. Sci.* **2017**, *28*, 2123–2131. [CrossRef]
111. Sharma, D.; Dhanjal, D.S. Bio-Nanotechnology for Active Food Packaging. *J. Appl. Pharm. Sci.* **2016**, *6*, 220–226. [CrossRef]

Journal of *Fungi*

Article

Biosynthesis of Silver Nanoparticles Using Onion Endophytic Bacterium and Its Antifungal Activity against Rice Pathogen *Magnaporthe oryzae*

Ezzeldin Ibrahim [1,2], Jinyan Luo [3], Temoor Ahmed [1], Wenge Wu [4], Chenqi Yan [5,*] and Bin Li [1,*]

1 State Key Laboratory of Rice Biology, Ministry of Agriculture Key Laboratory of Molecular Biology of Crop Pathogens and Insects, Institute of Biotechnology, Zhejiang University, Hangzhou 310058, China; ezzelbehery8818@yahoo.com (E.I.); temoorahmed@zju.edu.cn (T.A.)

2 Department of Vegetable Diseases Research, Plant Pathology Research Institute, Agriculture Research Centre, Giza 12916, Egypt

3 Department of Plant Quarantine, Shanghai Extension and Service Center of Agriculture Technology, Shanghai 201103, China; toyanzi@126.com

4 Rice Research Institute, Anhui Academy of Agricultural Sciences, Hefei 230001, China; wuwenge@aaas.org.cn

5 Institute of Biotechnology, Ningbo Academy of Agricultural Sciences, Ningbo 315040, China

* Correspondence: yanchengqi@zaas.ac.cn (C.Y.); libin0571@zju.edu.cn (B.L.)

Received: 30 September 2020; Accepted: 14 November 2020; Published: 18 November 2020

Abstract: Biosynthesis of silver nanoparticles (AgNPs) using endophytic bacteria is a safe alternative to the traditional chemical method. The purpose of this research is to biosynthesize AgNPs using endophytic bacterium *Bacillus endophyticus* strain H3 isolated from onion. The biosynthesized AgNPs with sizes from 4.17 to 26.9 nm were confirmed and characterized by various physicochemical techniques such as Fourier transform infrared spectroscopy (FT-IR), X-ray diffraction (XRD), UV-visible spectroscopy, transmission electron microscopy (TEM) and scanning electron microscopy (SEM) in addition to an energy dispersive spectrum (EDS) profile. The biosynthesized AgNPs at a concentration of 40 μg/mL had a strong antifungal activity against rice blast pathogen *Magnaporthe oryzae* with an inhibition rate of 88% in mycelial diameter. Moreover, the biosynthesized AgNPs significantly inhibited spore germination and appressorium formation of *M. oryzae*. Additionally, microscopic observation showed that mycelia morphology was swollen and abnormal when dealing with AgNPs. Overall, the current study revealed that AgNPs could protect rice plants against fungal infections.

Keywords: endophytic bacteria; silver nanoparticles; *Magnaporthe oryzae*; antifungal activity

---

1. Introduction

Rice (*Oryza sativa* L.) is the largest food crop in the world [1,2]. One of the basic hindrances to the growth of rice crops is the infection of various fungal diseases, particularly rice blast disease caused by *Magnaporthe oryzae*, which poses a serious threat to global food safety through the loss of 10–30% of rice production, enough rice for about 60 million people [1,3,4]. Current control of rice blast disease mainly depends on the use of fungicides. However, the desired goal of controlling the disease has not so far been achieved and there are serious consequences of the excessive use of fungicide on humans, ecosystems and the production of fungicide-resistant strains [5,6]. For all these risks, it is extremely important to find an alternative way to control rice blast disease. Nanotechnology can revolutionize the agricultural and food industry with new tools such as molecular plant disease management, fast disease revelation and the improvement of plants' ability to uptake nutrients. Furthermore, nanotechnology can enhance our biological knowledge of a variety of plants and therefore can improve crops or nutritional values as well as develop improved systems to monitor environmental conditions and enhance plants' ability to uptake nutrients or pesticides [7]. Therefore, the application of nanotechnology in agriculture for the control of plant diseases is a safe and eco-friendly alternative to synthetic chemical fungicides [8,9]. It has been instrumental in suppressing many of the fungal pathogens that attack the plants, causing them huge loss. For example, silver nanoparticles (AgNPs) had a significant effect in suppressing many of the air, seed and soil-borne fungal plant pathogens [9,10]. AgNP synthesis was documented using various methods including physical, chemical, and biological [11,12]. However, biological methods are safer than conventional physical and chemical methods [13,14]. It is a simple process (single vessel installation), rapid, cheap and eco-friendly. Furthermore, the polyphenols and various proteins existent in bio-sources work as a reducing agent, decreasing the use of dangerous external chemical reducing agents and, thus, toxicity. The green synthesis procedure does not require any additional capping agents, which further reduces the cost and simplifies the synthetic process. In contrast, chemical and physical methods are highly restricted in large scale applications, and also have high cost, use high energy, waste time and have difficulty in removing waste [15]. One of the most important biological methods to biosynthesize silver nanoparticles is the use of microorganisms such as bacteria, fungi and algae [11]. In this study, for the first time we biosynthesized new AgNPs using cell-free supernatants (CFSs) of endophytic bacterium *B. endophyticus* strain H3 and examined their characterization as a fungicide to inhibit *M. oryzae*.

2. Materials and Methods

2.1. Microorganisms

The virulent strain Gry of *M. oryzae*, was obtained from the Institute of Biotechnology, Zhejiang University, Hangzhou, China. Strain H3 of the endophytic bacterium isolated from onion plants was used for biosynthesis of AgNPs.

2.2. Isolation of Endophytic Bacterium

The entophytic bacterium was isolated from healthy onion plants collected from different locations in Hangzhou, China, according to [16] with slight modification. In brief, onion plants were collected at the seedling stage. The plants were washed carefully with tap water to remove the soil, then dried, and classified into leaves and roots. Onion leaves were cut into small pieces (2–3 cm) and disinfected with alcohol (70%) for 1 min followed by sodium hypoxy-chloride (1%) for 5 min and rinsed three times with sterile water under sterile conditions. One gram of sterile tissues was crushed separately in 9 mL saline water (0.85% NaCl), followed by serial dilution up to 10^{-5}. For each of these dilutions, 0.1 mL was spread on nutrient agar (NA) medium [8,11,12] bought from Sangon Biotech, Shanghai, and incubated for two days at 30 °C. The colonies were purified by transferring single colonies to a new NA plate. The isolated strains were stored in 30% (*v*/*v*) glycerol at −80 °C until use.

2.3. Identification of Endophytic Bacterium

The isolated endophytic bacterium was identified through 16S rRNA gene sequence analysis. The isolate was grown for 24 h in NA medium and a single colony was transferred to NA broth and incubated in a shaker at 30 °C overnight. Bacterial DNA was isolated using a genomic bacterial DNA isolation Kit (Sangon Biotech (Shanghai) Co, Ltd., Shanghai, China) following the instructions in the protocol. The 16S rRNA gene was amplified by the bacterial-specific primer pairs 27F (AGAGTTTGATCGCTGCTCAG) and 1492F (GGTTACCTTGTTACGACTT) [17,18]. Complete volumes of PCR amplification in 50 µL were performed using the Bioer XP Thermal Cycler (Hangzhou Bioer Tech. Co., Ltd., Hangzhou, China) in 2× TSINGKE PCR Master Mix (TsingKe Biotechnology, Beijing, China). PCR parameters including the following cycles: the initial denaturation stage was 95 °C for 5 min and 30 cycles followed, each of which consisted of denaturation at 94 °C for 30 s, annealing at

53 °C for 30 s, and extending at 72 °C for 1 min. The final extension step was carried out for 5 min at 72 °C. PCR amplicons were verified by using Agarose Gel electrophoresis (1%). PCR products were purified by StarPrep Gel Extraction Kit (GeneStar, Beijing, China) and finally submitted for DNA sequencing in TsingKe Biological Technology, Beijing, China. The 16S rRNA gene sequence of the endophytic bacterium was aligned against a reference database using the BLAST server at National Center for Biotechnology Information (NCBI) (http://www.ncbi.nlm.nih.gov). The phylogenetic tree was constructed using the MEGA 6.0 program and the neighbour-joining method [19]. The sequence was deposited in the NCBI database.

2.4. Preparation of the Cell-Free Supernatants of Endophytic Bacterium

The cell-free supernatants (CFSs) of the selected endophytic bacterium were prepared according to [20], with slight modification. In brief, *Bacillus endophyticus* strain H3 was inoculated in NA liquid medium and incubated at 30 °C and 200 rpm for 2 days. The CFSs were purified by centrifugation of bacterial culture (approximately ~1×10^8 CFU/mL) at 10,000 rpm, at 4 °C for 20 min, and twice by using filter sterilization 0.22 μm. To rule out possible contamination, 100 μL of CFSs were spread on NA agar for one day. The CFSs were kept at 4 °C till their use in the biosynthesis of AgNPs.

2.5. Biosynthesis of AgNPs

Silver nitrate (Cat. no. 10018461; Sinopharm, Shanghai, China) was used to synthesize AgNPs according to [9], with slight modification. Briefly, 10 mL of CFSs were added to 90 mL of 3 mM $AgNO_3$ aqueous solution in a 250 mL Erlenmeyer flask and then incubated at 30 °C in a rotary shaker at 200 rpm for 3 d in the dark. As a control, 10 mL of NA broth with the same volume of $AgNO_3$ was used. The successful biosynthesis of AgNPs will convert the colour from yellow light to dark brown, which can be determined by using UV–Visible spectrometry (Shimadzu UV-2550 spectrometer) in the wavelength range of 200–800 nm at a resolution of 1 nm. The resulting pellets were collected from AgNPs by centrifugation at 10,000× *g* for 20 min, purified by washing twice with sterile double-distilled water (ddH_2O) and stored at −80 °C.

2.6. Characterization of the Biosynthesized AgNPs

Characterization of the biosynthesized AgNPs was performed by using several techniques. Fourier transform infrared (FTIR) was performed to identify the functional groups of the CFSs responsible for reducing Ag ion to AgNPs; transmission electron microscopy (TEM), scanning electron microscopy (SEM) and energy dispersive spectrum (EDS) as well as X-ray diffraction (XRD) were performed to study the size and morphology and to ensure the presence of silver ion in the resulting pellets of AgNPs.

2.6.1. FTIR

The functional group of the biosynthesized AgNPs was planned by FTIR as described by [14]. Briefly, 1 mg (freeze-dried) of AgNPs powders were blended with KBr (300 mg) and the FTIR were measured with an AVATAR 370 FTIR spectrometer (Thermo Nicolet, MA, USA) at a spectral range of 500–4000 cm.

2.6.2. XRD

The crystalline phase of AgNPs was determined based on XRD analysis, which was carried out on an XPert PRO diffractometer (Holland) with a detector voltage of 45 kV and a current of 40 mA using CuKα radiation as described in the methods of [13].

2.6.3. TEM, SEM and EDS

The structural morphology and AgNP sizes were studied by TEM and SEM observation using a Transmission Electron Microscopy (JEM-1230, JEOL, Akishima, Japan) and Scanning Electron Microscopy (TM-1000, Hitachi, Japan) according to the method of [12]. In short, the AgNPs powder was equipped with a copper-coated grid and a carbon-coated grid, respectively, for one day at room temperature to form a film of the sample. The existence of silver ion was confirmed by energy dispersive spectrum (EDS).

2.7. Antifungal Activity of the Biosynthesized AgNPs

2.7.1. Effect of AgNPs on Mycelium Growth

The inhibitory effect of AgNPs at four concentrations (10, 20, 30 and 40 μg/mL) on mycelium growth of *M. oryzae* strain Gry was determined using an agar medium test as described by [9], with slight modification. In brief, a disk (10-mm in diameter) of 7-day-old mycelium was inoculated in the centre of the petri dishes (9 cm in diameter), containing the blend of Potato Dextrose Agar (PDA) medium (pH 7.0), at several concentrations of AgNPs. The PDA plates were used without AgNPs as a control. The diameter of the fungus colony was measured after 7 d of incubation at 27 °C and then the inhibition of mycelium growth was calculated.

2.7.2. Effect of AgNPs on Cell Wall Morphology

Damage to the cell wall of *M. oryzae* strain Gry was determined using AgNPs according to the method of [21], with minor modification. In short, a mycelial disc of *M. oryzae* strain Gry (10-mm in diameter) was brought from PDA medium, treated and not treated by AgNPs, and examined by using both SEM (TM-1000, Hitachi, Japan) and TEM (JEM-1230, JEOL, Akishima, Japan).

2.7.3. Effect of AgNPs on Spore Germination and Length of Germ Tubes

The influence of four concentrations (10, 20, 30 and 40 μg/mL) of AgNPs on the germination of spores and length of germ tubes of *M. oryzae* strain Gry were identified as described by [22], with slight modification. Briefly, 500 μL of spore suspension (1×10^5 spores/mL) that were prepared according to the method of [23], were added to the same volume of AgNPs in 1 mL tubes with a final concentration of 10, 20, 30 and 40 μg/mL. Mixed spore suspensions with ddH_2O were used as the control. The germination rate of spores and length of germ tubes of *M. oryzae* strain Gry was recorded after 24 h of incubation at 28 °C in the dark, using the light microscope. The experiment was repeated twice with three replicates.

2.7.4. Effect of AgNPs on Appressorium Formation

The influence of four concentrations (10, 20, 30 and 40 μg/mL) of AgNPs on the appressorium formation of *M. oryzae* strain Gry was assessed according to [6], with slight modification. In brief, 500 μL of spore suspension (1×10^5 spores/mL) was added to the same volume of AgNPs in 1 mL tubes with a final concentration of 10, 20, 30 and 40 μg/mL. Spore suspensions with ddH_2O were added as a control. The inhibitory effect of AgNPs on numbers and sizes of appressoria were recorded after 72 h of incubation at 28 °C in the dark, using the light microscope.

2.8. Statistical Analysis

All experiments were performed randomly and the results were shown as the mean ± SD (standard deviation). Statistical analysis was performed with the SPSS version software package 16.0 (SPSS Inc., Chicago, IL, USA). Differences between the groups were estimated using different test analyses. When the p value is < 0.05 or 0.01, the result is statistically significant.

3. Results and Discussion

3.1. Isolation and Identification of the Endophytic Bacterium

Endophytic bacteria live inside of plants without causing harm to the host plant [24]. In our study, we obtained a total of 17 endophytic bacterial isolates from the interior of onion leaves. One isolate, namely, H3 was selected based on its ability in the biosynthesis of AgNPs. This isolate was subjected to molecular identification. Partial 16S rRNA gene sequence analysis revealed that the selected isolate was belonging to the genus of *Bacillus* (Figure 1), and subjected to GenBank under accession number MN611116.

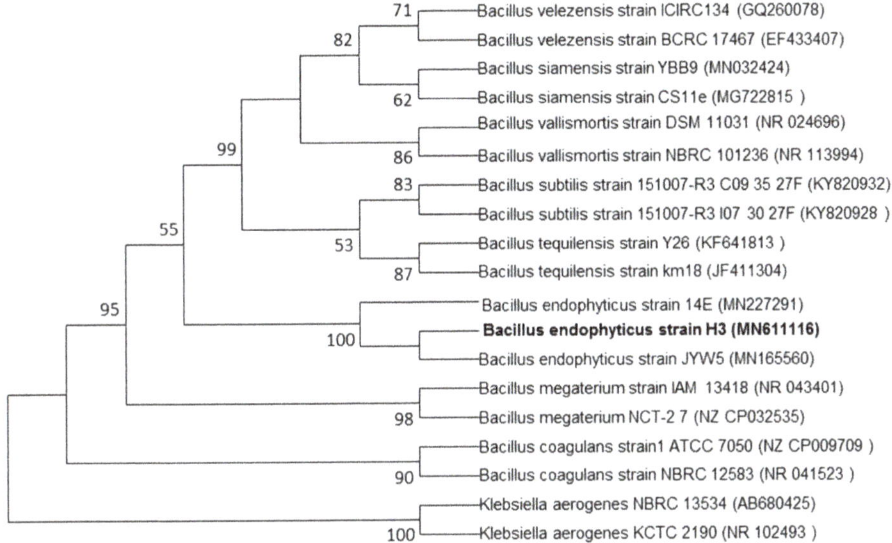

Figure 1. Phylogenetic tree of endophytic bacterium *B. endophyticus* strain H3 isolated from onion leaves in the seedling stage constructed by using 16S rRNA gene sequences. Bootstrap analysis (1000 replicates) for node values greater than 50% are given. Bar 0.02 substitutions per nucleotide position.

3.2. Biosynthesis and Characterization of AgNPs

CFSs obtained from endophytic bacterium *B. endophyticus* strain H3 were tested in the biosynthesis of AgNPs in 3 mM of $AgNO_3$ solution. Incubation of 10 mL of CFSs with 90 mL of $AgNO_3$ for three days resulted in color conversion from light yellow to dark brown, demonstrating the development of nanoparticles in the reaction mix. Differences in the color of AgNPs have been reported due to the formation of the biomolecules responsible for the synthesis of nanoparticles and the reduction of Ag^+ to Ag^0 [9,12,13,25]. The reduction of silver ion (Ag^+) of $AgNO_3$ has been found in many bacterial species, such as the *Bacillus siamensis* strain C1 [14] and *Pseudomonas rhodesiae* [13]. During the reduction of $AgNO_3$, the nitrate ions (NO_3^-) are reduced to nitrite (NO_2^-) by firstly accepting two protons and then releasing two electrons and water. The electrons emitted in the reduction reaction are transferred to the Ag^+ to form the silver element Ag^0 [26,27].

Furthermore, the formation of nanoparticles in the mixture was confirmed by a UV spectrophotometer, which showed a spectrum of surface plasmon resonance (SRP) at 412 nm (Figure 2), which is within the range reported earlier [13,28,29]. Similarly, Ahmed et al. [30] confirmed the formation of biogenic AgNPs in the reaction mixture by the presence of the peak at 418 nm.

TEM and SEM observations showed the nanoparticle to have spherical shape with sizes ranging from 4.17 to 26.9 nm in the reaction mixture (Figure 3). The present observations are consistent with the results from previous reports [13,31,32]. The toxicity of AgNPs depends on the variation of particle size.

AgNPs have an important influence on fungal cell's viability and ROS generation in a size-dependent manner. It is evident that the surface area, the volume ratio and the interaction of the surface with the particle size can be changed. Furthermore, sedimentation rate, mass diffusion, binding efficiency and sedimentation rate of NPs on biological or solid surfaces are highly influenced by particle size [33]. For example, Carlson et al. found that the 15 nm AgNPs can produce more ROS compared to 55 nm AgNPs in a macrophage cell line [34]. Furthermore, the EDS result showed that the element peak of silver, silica and sulfur are 92.77, 5.53 and 1.70%, respectively, in the reaction mixture (Figure 4). The results of this study are in agreement with the literature related to silver nanoparticles, where the silver ions peak was confirmed at 3 KeV [30,35].

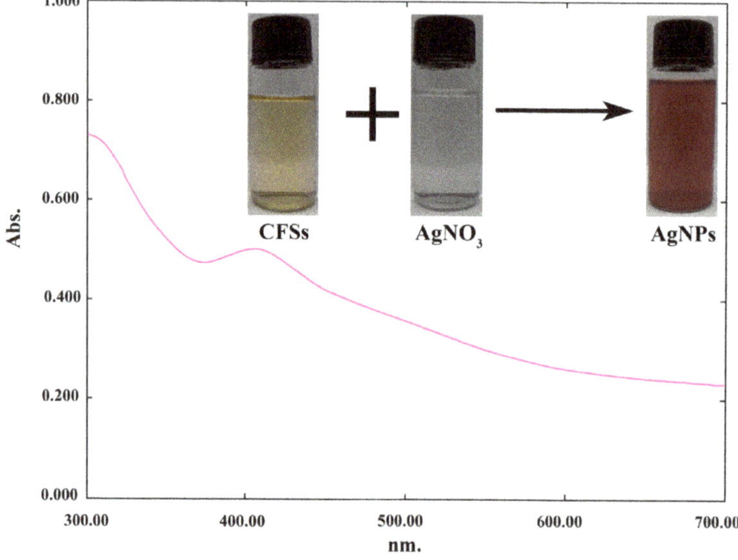

Figure 2. UV-vis absorption spectra of green silver nanoparticles (AgNPs) in the reaction mixture.

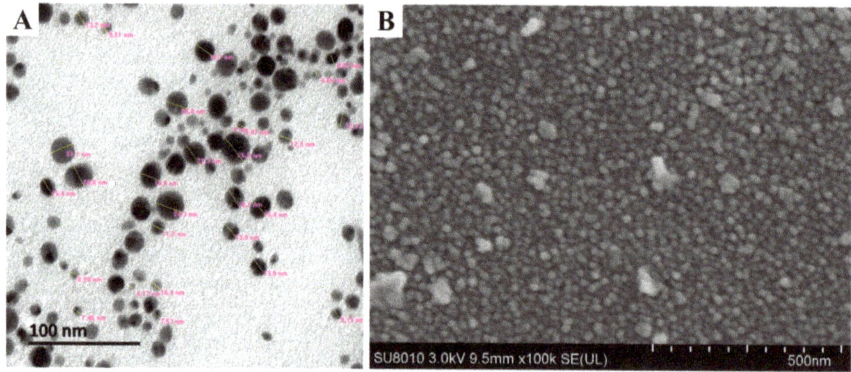

Figure 3. Characterization of AgNPs biosynthesized by using culture filtrates of *B. endophyticus* strain H3 isolated from onion. (**A**) Transmission electron microscopy and (**B**) scanning electron microscopy.

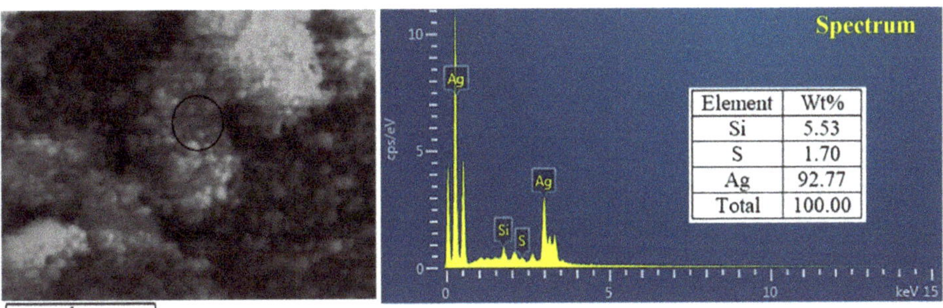

Element	Wt%
Si	5.53
S	1.70
Ag	92.77
Total	100.00

Figure 4. Characterization of biosynthesized AgNPs by using energy dispersive spectrum (EDS).

The functional groups of the biosynthesized AgNPs were confirmed by using FTIR analysis. In fact, 6 peaks at 3416, 2924, 1635, 1395, 1062 and 516 cm^{-1} were observed in the FTIR spectra of the synthesized AgNPs, which are shown in (Figure 5A). The main peak of 3416 cm^{-1} is due to the N–H stretching vibrations; a characteristic peak at 1635 cm^{-1} represents C=O carbonyl group and C=C stretching vibrations; the peaks at 2924, 1395 and 1062 cm^{-1} represent the C–H stretching vibrations, C=N bond of Amide II, O–H deformation vibrations, and C–N stretching amine vibrations, respectively; the peak at 516 cm^{-1} represents C–Br stretching. The existence of such groups in the chlorofluorocarbons (CFCs) from endophytic bacterium confirms the presence of proteins and indicates that these functional groups have a major role in reducing Ag^{+} to Ag^{0} [13,32,36]. Similarly, various studies reported the presence of functional groups representing different macromolecules, such as nucleic acids, proteins, lipids, carbohydrates and sugars, surrounding the green NPs which prevents the oxidation and deterioration of nanoparticles [30,37,38].

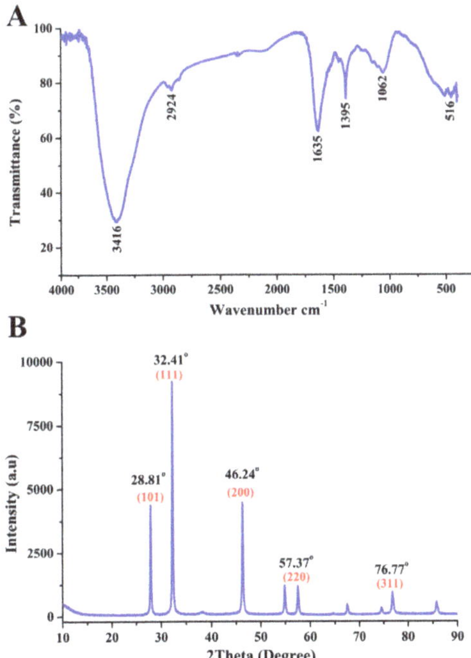

Figure 5. Characterization of the biosynthesized AgNPs. (**A**) Fourier transform infra-red (FTIR) spectra. (**B**) X-ray diffraction (XRD) spectra.

The crystalline structure of the biogenic AgNPs was determined through the XRD analysis result, that showed five emission peaks of 2θ = 28.81°, 32.41°, 46.24°, 57.37° and 76.77°, compatible with crystalline silver planes (101), (111), (200), (220), (311), respectively (Figure 5B). Similar results have been reported in other studies [11,14,38].

3.3. Antifungal Activity of AgNPs

The result of antifungal activity revealed that the mycelium growth of *M. oryaze* strain Gry was robustly suppressed by AgNPs, shown in (Figure 6). The inhibitory effect on the growth of mycelium increased with the increase of AgNP concentrations. In fact, AgNPs at 10, 20, 30 and 40 μg/mL caused an 18%, 49%, 65% and 88% reduction, respectively, in mycelial diameter. Similarly, previous studies have shown that AgNPs can be used as an antifungal agent to prevent plants from fungal infection [9,10,39–43]. Although there are different hypotheses available, the antimicrobial mechanisms of AgNPs have not yet been clearly defined. The proposed mechanisms were summarized based on the current literature, as follows: attachment of AgNPs to the surface of the cell membrane, altering the lipid bilayer or increasing permeability of the cell membrane, microbial cell intrusion of AgNPs causing damage to intracellular micro organelles (such as mitochondria, vacuoles and ribosomes) and biomolecules including DNA, protein and lipids, and modulation of the intracellular signal transduction method towards apoptosis [44].

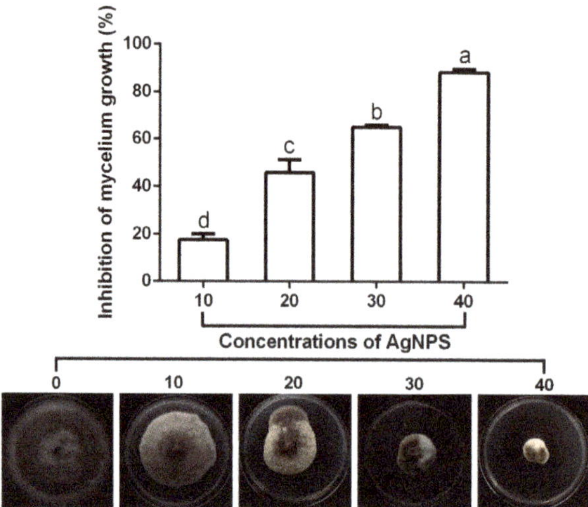

Figure 6. Effect of the biosynthesized AgNPs at four different concentrations (10, 20, 30 and 40 μg/mL) on the mycelial growth of *M. oryaze* strain Gry. The AgNPs at 10, 20, 30 and 40 μg/mL caused an 18%, 49%, 65% and 88% reduction, respectively, in mycelial diameter. Data are a mean value ± standard error of three replicates, and bars with different letters (a–d) are significantly different in LSD test.

The fungal cell wall is a flexible structure that performs several functions in determining the shape of the cell. In addition, the cell wall can protect the fungal cells from environmental stresses, such as pH, temperature and changes in osmolality [9,45]. Therefore, interference by causing harm to the fungal cell wall may lead to loss of content and death. In our results, the hyphae of *M. oryzae* strain Gry that were treated with AgNPs showed abnormal structural, swelling and damage to their cell walls causing some loss of contents. In contrast, the cell walls of *M. oryzae* strain Gry had normal structural characteristics in the absence of the AgNPs (Figure 7). Similarly, the antifungal activity of AgNPs and Cu-NPs have been found on the cell walls of many pathogenic fungi such as *Fusarium graminearum*, *Fusarium oxysporum*, *Fusarium solani* and *Colletotrichum gloeoesporioides* [9,45,46].

The germinated spores of pathogenic fungi are known to perform a major role in colonizing and infecting plants [9,47]. Therefore, inhibition rate of spore germination will significantly reduce the threat of rice fungal pathogens. The biosynthesized AgNPs were able to efficiently inhibit the spore germination and germ tube growth of *M. oryzae* strain Gry, and the antifungal activity increased with the increase in AgNP concentration. In fact, the germination rate of spores was 83%, while the length of the germ tubes was 77.63 μm in the negative control. However, the germination rate of spores was 68%, 53%, 31% and 13%, respectively (Figure 8A), while the germ tubes length was 54.62 μm, 33.73 μm, 21.74 μm and 8.11 μm, respectively, (Figure 8B) in the presence of AgNPs at four different concentrations (10, 20, 30 and 40 μg/mL). Similar results have also been reported in other studies [9,48].

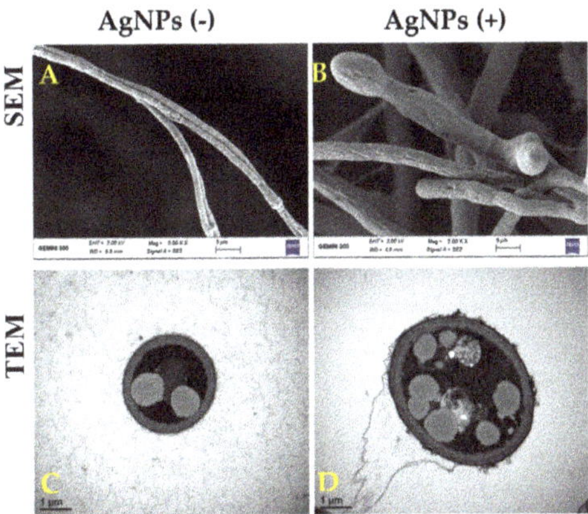

Figure 7. Scanning electron microscopy (SEM) [scale bar = 5 μm] and transmission electron microscopy (TEM) [scale bar = 1 μm] images of *M. oryaze strain Gry* in the absence of the biosynthesized AgNPs; the hyphae showed a normal structural property (**A**,**C**), and in the presence of the biosynthesized AgNPs, it had abnormal structure, swelling and damage to the cell walls' contents (**B**,**D**).

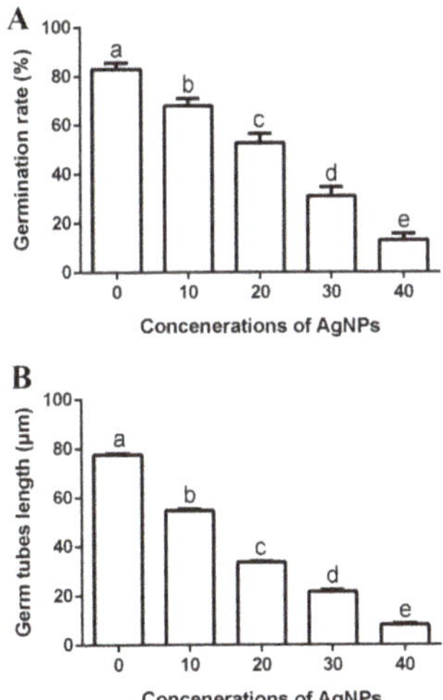

Figure 8. Effect of the biosynthesized AgNPs of four different concentrations (10, 20, 30 and 40 μg/mL) on spores' germination rate (**A**) and germ tube growth (**B**). Data are a mean value ± standard error of three replicates, and bars with different letters (a–e) are significantly different in LSD test.

Many fungi form a specialized infection structure called an appressorium that is necessary and required to penetrate the plant cell walls [1,6]. Therefore, appressorium inhibition significantly reduces the risk of rice fungal pathogens. The biosynthesized AgNPs were able to effectively inhibit the appressorium formation and appressorium diameter of *M. oryzae* strain Gry, while the inhibitory effect increased along with the increase in AgNP concentration. In fact, the appressorium formation rate was 83%, while appressorium diameter was 14.00 μm in the negative control. However, the appressorium formation rate was 66%, 34%, 11% and 0%, respectively (Figure 9A), while the appressorium diameter was 11.00 μm, 8.00 μm, 5.38 μm and 0.00 μm, respectively (Figure 9B) in the presence of AgNPs at different concentrations (10, 20, 30 and 40 μg/mL).

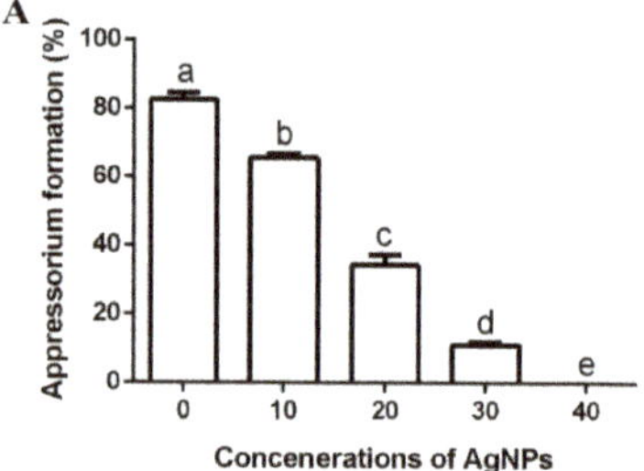

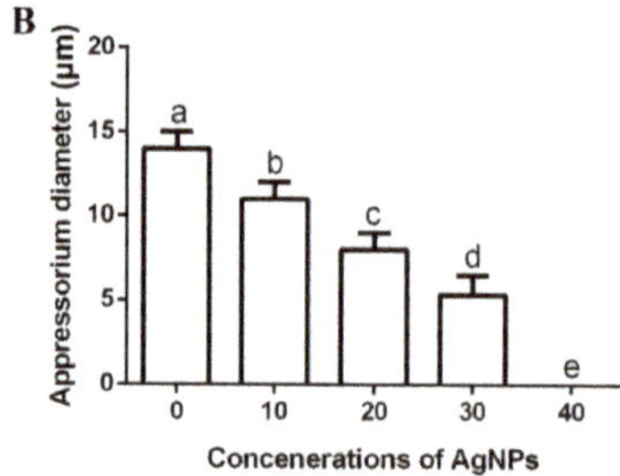

Figure 9. Effect of the biosynthesized AgNPs of four different suspensions (10, 20, 30 and 40 μg/mL) on appressorium formation (**A**) and appressorium diameter (**B**). Data are a mean value ± standard error of three replicates, and bars with different letters (a–e) are significantly different in LSD test.

4. Conclusions

The green synthesis of AgNPs is a safe alternative to physical and chemical methods. The present study reports for the first time the biosynthesis of AgNPs by using endophytic bacterium isolated from onion. The formation of biogenic AgNPs was further confirmed through UV-vis spectroscopy, FTIR, XRD, SEM, TEM and EDS. In addition, the biosynthesized AgNPs exhibited robust antifungal activity against rice blast pathogen *M. oryaze* strain Gry, which may be mainly attributed to their ability to inhibit spore germination, germ tube growth, appressorium formation and damage of cell well. Overall, these results suggest that biosynthesized AgNPs have the potential to protect rice plants from fungal diseases.

Author Contributions: Conceptualization, E.I. and J.L.; Methodology, E.I. and T.A.; Software, E.I. and J.L.; data Investigation, E.I. and J.L.; Supervision, C.Y., W.W. and B.L.; writing—original draft, E.I., T.A. and B.L.; Writing—review & editing, all the authors contribute to this part. All authors have read and agreed to the published version of the manuscript.

Funding: The work is partially supported by Shanghai Agriculture Applied Technology Development Program (2019-02-08-00-08-F01150), Zhejiang Provincial Natural Science Foundation of China (LZ19C140002), National Natural Science Foundation of China (31872017, 32072472, 31571971, 31801787, 31901925), Zhejiang Provincial Project (2017C02002, 2019C02006, 2020C02006), Key Scientific Technological Project of Ningbo (2016C11017; 2019B10004), National Key Research and Development Program of China (2018YFD0300900; 2017YFD0201104),

Dabeinong Funds for Discipline Development and Talent Training in Zhejiang University, State Key Laboratory for Managing Biotic and Chemical Treats to the Quality and Safety of Agro-products (2010DS700124-ZZ1907), the Fundamental Research Funds for the Central Universities.

Acknowledgments: We thank for the help in data investigation and analysis from Muchen Zhang, Arshad Ali, Mengju Liu and Solabomi Olaitan Ogunyemi.

Conflicts of Interest: The authors declare no conflict of interest.

References

1. Cui, S.; Wang, J.; Gao, F.; Sun, G.; Liang, J. Quantitative proteomic analyses reveal that RBBI3. 3, a trypsin inhibitor protein, plays an important role in *Magnaporthe oryzae* infection in rice. *Plant Growth Regul.* **2018**, *86*, 365–374. [CrossRef]
2. Kennedy, D. The importance of rice. *Science* **2002**, *296*, 13. [CrossRef] [PubMed]
3. Skamnioti, P.; Gurr, S.J. Against the grain: Safeguarding rice from rice blast disease. *Trends Biotechnol.* **2009**, *27*, 141–150. [CrossRef] [PubMed]
4. Dean, R.; Van Kan, J.A.; Pretorius, Z.A.; Hammond-Kosack, K.E.; Di Pietro, A.; Spanu, P.D.; Rudd, J.J.; Dickman, M.; Kahmann, R.; Ellis, J. The Top 10 fungal pathogens in molecular plant pathology. *Mol. Plant Pathol.* **2012**, *13*, 414–430. [CrossRef]
5. Gisi, U.; Sierotzki, H.; Cook, A.; McCaffery, A. Mechanisms influencing the evolution of resistance to Qo inhibitor fungicides. *Pest Manag. Sci.* **2002**, *58*, 859–867. [CrossRef]
6. Wang, N.; Cai, M.; Wang, X.; Xie, Y.; Ni, H. Inhibitory action of biofungicide physcion on initial and secondary infection of *Magnaporthe oryzae*. *J. Phytopathol.* **2016**, *164*, 641–649. [CrossRef]
7. Prasad, R.; Kumar, V.; Prasad, K.S. Nanotechnology in sustainable agriculture: Present concerns and future aspects. *Afr. J. Biotechnol.* **2014**, *13*, 705–713.
8. Abdallah, Y.; Ogunyemi, S.O.; Abdelazez, A.; Zhang, M.; Hong, X.; Ibrahim, E.; Hossain, A.; Fouad, H.; Li, B.; Chen, J. The green synthesis of MgO nano-Flowers using *Rosmarinus officinalis* L.(Rosemary) and the antibacterial activities against *Xanthomonas oryzae* pv. *oryzae*. *Biomed Res. Int.* **2019**, *2019*, 5620989. [CrossRef]
9. Ibrahim, E.; Zhang, M.; Zhang, Y.; Hossain, A.; Qiu, W.; Chen, Y.; Wang, Y.; Wu, W.; Sun, G.; Li, B. Green-Synthesization of Silver Nanoparticles Using Endophytic Bacteria Isolated from Garlic and Its Antifungal Activity against Wheat *Fusarium* Head Blight Pathogen *Fusarium graminearum*. *Nanomaterials* **2020**, *10*, 219. [CrossRef]
10. Gupta, N.; Upadhyaya, C.P.; Singh, A.; Abd-Elsalam, K.A.; Prasad, R. Applications of silver nanoparticles in plant protection. In *Nanobiotechnology Applications in Plant Protection*; Springer: Berlin/Heidelberg, Germany, 2018; pp. 247–265.
11. Masum, M.; Islam, M.; Siddiqa, M.; Ali, K.A.; Zhang, Y.; Abdallah, Y.; Ibrahim, E.; Qiu, W.; Yan, C.; Li, B. Biogenic synthesis of silver nanoparticles using Phyllanthus emblica fruit extract and its inhibitory action against the pathogen *Acidovorax oryzae* Strain RS-2 of rice bacterial brown stripe. *Front. Microbiol.* **2019**, *10*, 820. [CrossRef]
12. Ogunyemi, S.O.; Abdallah, Y.; Zhang, M.; Fouad, H.; Hong, X.; Ibrahim, E.; Masum, M.M.I.; Hossain, A.; Mo, J.; Li, B. Green synthesis of zinc oxide nanoparticles using different plant extracts and their antibacterial activity against *Xanthomonas oryzae* pv. *oryzae*. *Artif. Cells Nanomed. Biotechnol.* **2019**, *47*, 341–352. [CrossRef] [PubMed]
13. Hossain, A.; Hong, X.; Ibrahim, E.; Li, B.; Sun, G.; Meng, Y.; Wang, Y.; An, Q. Green synthesis of silver nanoparticles with culture supernatant of a bacterium *Pseudomonas rhodesiae* and their antibacterial activity against soft rot pathogen *Dickeya dadantii*. *Molecules* **2019**, *24*, 2303. [CrossRef] [PubMed]
14. Ibrahim, E.; Fouad, H.; Zhang, M.; Zhang, Y.; Qiu, W.; Yan, C.; Li, B.; Mo, J.; Chen, J. Biosynthesis of silver nanoparticles using endophytic bacteria and their role in inhibition of rice pathogenic bacteria and plant growth promotion. *RSC Adv.* **2019**, *9*, 29293–29299. [CrossRef]
15. Mukherjee, S.; Patra, C.R. Biologically synthesized metal nanoparticles: Recent advancement and future perspectives in cancer theranostics. *Future Sci.* **2017**, *3*, FSO203. [CrossRef] [PubMed]
16. Yan, X.; Wang, Z.; Mei, Y.; Wang, L.; Wang, X.; Xu, Q.; Peng, S.; Zhou, Y.; Wei, C. Isolation, diversity, and growth-promoting activities of endophytic bacteria from tea cultivars of Zijuan and Yunkang-10. *Front. Microbiol.* **2018**, *9*, 1848. [CrossRef]

17. Masum, M.; Liu, L.; Yang, M.; Hossain, M.; Siddiqa, M.; Supty, M.; Ogunyemi, S.; Hossain, A.; An, Q.; Li, B. Halotolerant bacteria belonging to operational group *Bacillus amyloliquefaciens* in biocontrol of the rice brown stripe pathogen *Acidovorax oryzae*. *J. Appl. Microbiol.* **2018**, *125*, 1852–1867. [CrossRef]
18. Li, B.; Xu, L.; Lou, M.; Li, F.; Zhang, Y.; Xie, G. Isolation and characterization of antagonistic bacteria against bacterial leaf spot of Euphorbia pulcherrima. *Lett. Appl. Microbiol.* **2008**, *46*, 450–455. [CrossRef]
19. Tamura, K.; Peterson, D.; Peterson, N.; Stecher, G.; Nei, M.; Kumar, S. MEGA5: Molecular evolutionary genetics analysis using maximum likelihood, evolutionary distance, and maximum parsimony methods. *Mol. Biol. Evol.* **2011**, *28*, 2731–2739. [CrossRef]
20. Almoneafy, A.A.; Kakar, K.U.; Nawaz, Z.; Li, B.; Yang, C.-L.; Xie, G.-L. Tomato plant growth promotion and antibacterial related-mechanisms of four rhizobacterial Bacillus strains against *Ralstonia solanacearum*. *Symbiosis* **2014**, *63*, 59–70. [CrossRef]
21. Gao, T.; Zhou, H.; Zhou, W.; Hu, L.; Chen, J.; Shi, Z. The fungicidal activity of thymol against *Fusarium graminearum* via inducing lipid peroxidation and disrupting ergosterol biosynthesis. *Molecules* **2016**, *21*, 770. [CrossRef]
22. Chen, J.; Peng, H.; Wang, X.; Shao, F.; Yuan, Z.; Han, H. Graphene oxide exhibits broad-spectrum antimicrobial activity against bacterial phytopathogens and fungal conidia by intertwining and membrane perturbation. *Nanoscale* **2014**, *6*, 1879–1889. [CrossRef] [PubMed]
23. Xing, K.; Liu, Y.; Shen, X.; Zhu, X.; Li, X.; Miao, X.; Feng, Z.; Peng, X.; Qin, S. Effect of O-chitosan nanoparticles on the development and membrane permeability of *Verticillium dahliae*. *Carbohydr. Polym.* **2017**, *165*, 334–343. [CrossRef] [PubMed]
24. Duan, J.-L.; Li, X.-J.; Gao, J.-M.; Wang, D.-S.; Yan, Y.; Xue, Q.-H. Isolation and identification of endophytic bacteria from root tissues of *Salvia miltiorrhiza* Bge. and determination of their bioactivities. *Ann. Microbiol.* **2013**, *63*, 1501–1512. [CrossRef]
25. Lateef, A.; Ojo, S.A.; Akinwale, A.S.; Azeez, L.; Gueguim-Kana, E.B.; Beukes, L.S. Biogenic synthesis of silver nanoparticles using cell-free extract of *Bacillus safensis* LAU 13: Antimicrobial, free radical scavenging and larvicidal activities. *Biologia* **2015**, *70*, 1295–1306. [CrossRef]
26. Ming, L.J. Structure and function of "metalloantibiotics". *Med. Res. Rev.* **2003**, *23*, 697–762. [CrossRef]
27. Roy, A.; Bulut, O.; Some, S.; Mandal, A.K.; Yilmaz, M.D. Green synthesis of silver nanoparticles: Biomolecule-nanoparticle organizations targeting antimicrobial activity. *RSC Adv.* **2019**, *9*, 2673–2702. [CrossRef]
28. Kasithevar, M.; Saravanan, M.; Prakash, P.; Kumar, H.; Ovais, M.; Barabadi, H.; Shinwari, Z.K. Green synthesis of silver nanoparticles using *Alysicarpus monilifer* leaf extract and its antibacterial activity against MRSA and CoNS isolates in HIV patients. *J. Interdiscip. Nanomed.* **2017**, *2*, 131–141. [CrossRef]
29. Sastry, M.; Mayya, K.; Bandyopadhyay, K. pH Dependent changes in the optical properties of carboxylic acid derivatized silver colloidal particles. *Colloids Surf. A Physicochem. Eng. Asp.* **1997**, *127*, 221–228. [CrossRef]
30. Ahmed, T.; Shahid, M.; Noman, M.; Niazi, M.B.K.; Mahmood, F.; Manzoor, I.; Zhang, Y.; Li, B.; Yang, Y.; Yan, C. Silver Nanoparticles Synthesized by Using *Bacillus cereus* SZT1 Ameliorated the Damage of Bacterial Leaf Blight Pathogen in Rice. *Pathogens* **2020**, *9*, 160. [CrossRef]
31. Kuppurangan, G.; Karuppasamy, B.; Nagarajan, K.; Sekar, R.K.; Viswaprakash, N.; Ramasamy, T. Biogenic synthesis and spectroscopic characterization of silver nanoparticles using leaf extract of *Indoneesiella echioides*: In vitro assessment on antioxidant, antimicrobial and cytotoxicity potential. *Appl. Nanosci.* **2016**, *6*, 973–982. [CrossRef]
32. Anthony, K.J.P.; Murugan, M.; Gurunathan, S. Biosynthesis of silver nanoparticles from the culture supernatant of *Bacillus marisflavi* and their potential antibacterial activity. *J. Ind. Eng. Chem.* **2014**, *20*, 1505–1510. [CrossRef]
33. Akter, M.; Sikder, M.T.; Rahman, M.M.; Ullah, A.A.; Hossain, K.F.B.; Banik, S.; Hosokawa, T.; Saito, T.; Kurasaki, M.A. Systematic review on silver nanoparticles-induced cytotoxicity: Physicochemical properties and perspectives. *J. Adv. Res.* **2018**, *9*, 1–16. [CrossRef] [PubMed]
34. Carlson, C.; Hussain, S.M.; Schrand, A.M.; Braydich-Stolle, L.K.; Hess, K.L.; Jones, R.L.; Schlager, J.J. Unique cellular interaction of silver nanoparticles: Size-dependent generation of reactive oxygen species. *J. Phys. Chem. B* **2008**, *112*, 13608–13619. [CrossRef] [PubMed]
35. Suganya, G.; Karthi, S.; Shivakumar, M.S. Larvicidal potential of silver nanoparticles synthesized from *Leucas aspera* leaf extracts against dengue vector Aedes aegypti. *Parasitol. Res.* **2014**, *113*, 875–880. [CrossRef] [PubMed]

36. Jyoti, K.; Baunthiyal, M.; Singh, A. Characterization of silver nanoparticles synthesized using Urtica dioica Linn. leaves and their synergistic effects with antibiotics. *J. Radiat. Res. Appl. Sci.* **2016**, *9*, 217–227. [CrossRef]
37. Ahmed, T.; Shahid, M.; Noman, M.; Niazi, M.B.K.; Zubair, M.; Almatroudi, A.; Khurshid, M.; Tariq, F.; Mumtaz, R.; Li, B. Bioprospecting a native silver-resistant *Bacillus safensis* strain for green synthesis and subsequent antibacterial and anticancer activities of silver nanoparticles. *J. Adv. Res.* **2020**, *24*, 475–483. [CrossRef]
38. Velmurugan, P.; Iydroose, M.; Mohideen, M.H.A.K.; Mohan, T.S.; Cho, M.; Oh, B.-T. Biosynthesis of silver nanoparticles using *Bacillus subtilis* EWP-46 cell-free extract and evaluation of its antibacterial activity. *Bioprocess Biosyst. Eng.* **2014**, *37*, 1527–1534. [CrossRef]
39. Wang, Z.; Xu, C.; Zhao, M.; Zhao, C. One-pot synthesis of narrowly distributed silver nanoparticles using phenolic-hydroxyl modified chitosan and their antimicrobial activity. *RSC Adv.* **2014**, *4*, 47021–47030. [CrossRef]
40. Khalil, N.M.; Abd El-Ghany, M.N.; Rodríguez-Couto, S. Antifungal and anti-mycotoxin efficacy of biogenic silver nanoparticles produced by *Fusarium chlamydosporum* and *Penicillium chrysogenum* at non-cytotoxic doses. *Chemosphere* **2019**, *218*, 477–486. [CrossRef]
41. Pham, D.C.; Nguyen, T.H.; Ngoc, U.T.P.; Le, N.T.T.; Tran, T.V.; Nguyen, D.H. Preparation, characterization and antifungal properties of chitosan-silver nanoparticles synergize fungicide against pyricularia oryzae. *J. Nanosci. Nanotechnol.* **2018**, *18*, 5299–5305. [CrossRef]
42. Nicomrat, D.; Janlapha, W.; Singkran, N. In vitro susceptibility of silver nanoparticles on Thai isolated fungus *Pyricularia oryzae* from *oryza sativa* L. Leaves. *Appl. Mech. Mater.* **2017**, *866*, 148–151. [CrossRef]
43. Akter, R. *Efficacy of Silver Nanoparticles Against Rice Blast Disease and Farmers Perception about Its Management in Bangladesh*; Swedish University of Agricultural Sciences: Uppsala, Sweden, 2019.
44. Lee, S.H.; Jun, B.-H. Silver nanoparticles: Synthesis and application for nanomedicine. *Int. J. Mol. Sci.* **2019**, *20*, 865. [CrossRef] [PubMed]
45. Pariona, N.; Mtz-Enriquez, A.I.; Sánchez-Rangel, D.; Carrión, G.; Paraguay-Delgado, F.; Rosas-Saito, G. Green-synthesized copper nanoparticles as a potential antifungal against plant pathogens. *RSC Adv.* **2019**, *9*, 18835–18843. [CrossRef]
46. Oussou-Azo, A.F.; Nakama, T.; Nakamura, M.; Futagami, T.; Vestergaard, M.C.M. Antifungal Potential of Nanostructured Crystalline Copper and Its Oxide Forms. *Nanomaterials* **2020**, *10*, 1003. [CrossRef]
47. Wu, D.; Lu, J.; Zhong, S.; Schwarz, P.; Chen, B.; Rao, J. Influence of nonionic and ionic surfactants on the antifungal and mycotoxin inhibitory efficacy of cinnamon oil nanoemulsions. *Food Funct.* **2019**, *10*, 2817–2827. [CrossRef]
48. Kriti, A.; Ghatak, A.; Mandal, N. Inhibitory potential assessment of silver nanoparticle on phytopathogenic spores and mycelial growth of bipolaris sorokiniana and alternaria brassicicola. *Int. J. Curr. Microbiol. Appl. Sci.* **2020**, *9*, 692–699. [CrossRef]

Publisher's Note: MDPI stays neutral with regard to jurisdictional claims in published maps and institutional affiliations.

Journal of
Fungi

Review

Zinc-Based Nanomaterials for Diagnosis and Management of Plant Diseases: Ecological Safety and Future Prospects

Anu Kalia [1,*], Kamel A. Abd-Elsalam [2] and Kamil Kuca [3,*]

1 Electron Microscopy and Nanoscience Laboratory, Department of Soil Science, College of Agriculture, Punjab Agricultural University, Ludhiana 141004, Punjab, India
2 Agricultural Research Center (ARC), Plant Pathology Research Institute, Giza 12619, Egypt; kamelabdelsalam@gmail.com
3 Department of Chemistry, Faculty of Science, University of Hradec Králové, 500 03 Hradec Králové, Czech Republic
* Correspondence: kaliaanu@pau.edu (A.K.); kamil.kuca@uhk.cz (K.K.); Tel.: +91-2401960 (A.K.); +420-603-289-166 (K.K.)

Received: 22 September 2020; Accepted: 10 October 2020; Published: 13 October 2020

Abstract: A facet of nanorenaissance in plant pathology hailed the research on the development and application of nanoformulations or nanoproducts for the effective management of phytopathogens deterring the growth and yield of plants and thus the overall crop productivity. Zinc nanomaterials represent a versatile class of nanoproducts and nanoenabled devices as these nanomaterials can be synthesized in quantum amounts through economically affordable processes/approaches. Further, these nanomaterials exhibit potential targeted antimicrobial properties and low to negligible phytotoxicity activities that well-qualify them to be applied directly or in a deviant manner to accomplish significant antibacterial, antimycotic, antiviral, and antitoxigenic activities against diverse phytopathogens causing plant diseases. The photo-catalytic, fluorescent, and electron generating aspects associated with zinc nanomaterials have been utilized for the development of sensor systems (optical and electrochemical biosensors), enabling quick, early, sensitive, and on-field assessment or quantification of the test phytopathogen. However, the proficient use of Zn-derived nanomaterials in the management of plant pathogenic diseases as nanopesticides and on-field sensor system demands that the associated eco- and biosafety concerns should be well discerned and effectively sorted beforehand. Current and possible utilization of zinc-based nanostructures in plant disease diagnosis and management and their safety in the agroecosystem is highlighted.

Keywords: ecotoxicity; nanomaterial; nanosensors; phytopathogens; zinc

---

1. Introduction

Microbial pathogenic diseases of crop plants account for substantial annual loss (in the relative manner depicted as a percentage), approximately 16–40%, of production tonnage [1]. The bacterial and fungal pathogens of various crops exhibit enormous yield and productivity losses during production and postharvest storage as well as during transportation of the crop produce [2]. To safeguard the crop from crop health and yield deterring pathogens, pesticides- organic or inorganic compounds, or their composites have been used by agriculturists or farmers. Among the diverse pesticidal agents utilized to curb weeds and plant pathogens, zinc and copper formulations have emerged as the best performers.

1.1. Use of Zinc Element as a Pesticide

Zinc alone or in combination with copper has been widely used for the development of several commercially available agricultural bio-/pesticides [3]. In the early 1970s, zinc salts for pesticide use were first registered in the United States. Later in the 1990s, the US Environment Protection Agency (US-EPA) approved three zinc salts, namely, zinc chloride, zinc oxide, and zinc sulfate for use as herbicide and the industrial preservative (to control spoilage by bacterial and fungal contaminants in carpets) [4]. Zinc phosphide, another Zn-salt, is applied as an effective rodenticide [5,6]. Further, zinc oxide has been approved to be used as a stabilizer in pesticide formulations with concentration not exceeding 15% (*w/w* or *w/v*) of the formulation [7]. Later, the zinc formulations have got popularized for the antimicrobial activity against various phytopathogens. The antimicrobial potential of the zinc formulations render its use as a considerably low cost, less environmentally toxic, and effective microbicide exhibiting broad-spectrum activities including bactericide [8], fungicide [9–11], or algaecide [12] and other activities. A growing interest exists for the development of novel zinc formulations possessing enhanced efficacy and action specificity. Generation and use of nanoenabled formulations of pesticides are one among the emerging and pertinent alternatives to manage plant diseases causing phytopathogens.

1.2. Status of Use of Nanomaterials in Plant Pathology

Changing climatic patterns and intensive agriculture has contributed enormously to the development of more fastidious and virulent pathogens, which exhibit resistance to several pesticides (bactericides, fungicides, and similar action compounds) [13–15]. These strains of microbes can survive through higher concentrations of the -cidal compounds/composites besides requiring multiple applications and therefore, have become a big menace for the farmers to avoid or control the yield losses caused by these pathogens [16]. The use of nanomaterials for control of phytopathogens has been envisioned by agriscientists after the evidence for -static to -cidal properties of various types of nanomaterials that appeared for human/livestock pathogens in journals of repute of biomedicine or pharmacology [17–21]. Amenability to fabrication/alteration of size and surface morphology and functionalization of nanomaterials is of tremendous significance considering the quick and sustainable eradication of pesticide-resistant phytopathogens [22–27].

Various categories of nanomaterials have been evaluated for their diverse agriapplications such as nanofertilizers, nanopesticides, and pesticides degradation to achieve plant growth promotion and protection [28] (Figure 1). Thus, the current manuscript entails the published research on the use of zinc nanomaterials for management and early diagnosis of phytopathogens. Further, the application of zinc nanomaterials as potent antimicrobial agents and their use for curbing the growth, virulence, and diseases caused by plant pathogens have been elaborated. The use of zinc nanomaterials as functional elements in biosensor systems for robust and sensitive identification of phytopathogens is also discussed.

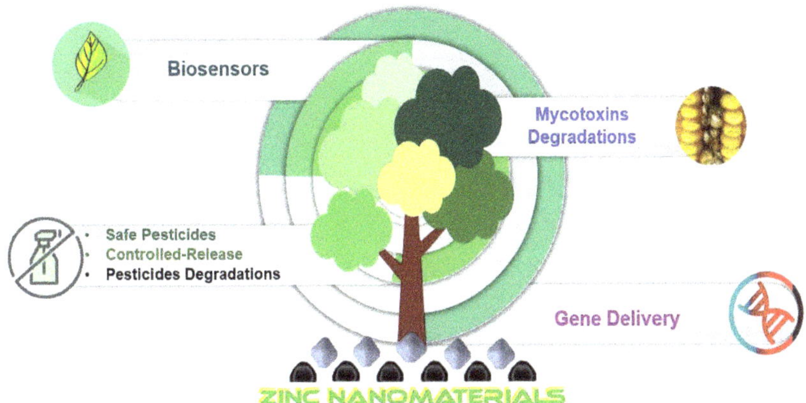

Figure 1. Zinc-based nanomaterials applications in plant pathology.

2. Nanomaterials: Can Nanosizing Matter Alter Its Properties?

Nanomaterials (NMs) exhibit enormous chemical diversity and can be categorized on basis of their chemical origin as natural, organic, synthetic, metal/nonmetal, or their oxides, sulfides, nitrides, and other forms [29]. These are considered as an intermediate state of matter with at least one of the size dimensions existing between size scales of 1–100 nm. The dimensionality classification of NMs segregates these as zero-, one-, two- and three-dimensional materials [30,31]. The nanomaterials exhibit novel physical, chemical, and biological properties [32,33]. The reason for the unusual properties of nanomaterials may be attributed to the basic phenomena of "quantum confinement" and "surface-interface effects" [34–36]. These two characteristics may alter the mechanical, optical, electrical, magnetic, and chemical catalysis properties of nanoscale materials compared to their bulk counterparts [37,38]. Thus, nanomaterials exhibit properties that are size dependent, i.e., the size of grain or particles, phase inclusions, pores, or other morphological features affect the properties exhibited by the substance [39].

2.1. Mechanism of Antimicrobial Activity

The antimicrobial potential of the nanomaterials gets improved possibly due to enhanced surface of contact with the microbial surfaces or biomolecules [17,40,41]. On interaction with the microbial cells, NMs can adsorb to oppositely charged functional groups [42] and exhibit the advantage of trespassing the intact cell boundaries/membranes. Further, NMs can generate photocatalytic or redox driven electron/hole or electron–hole pair leading to the formation of reactive oxygen moieties (superoxide anion radicals, hydroxyl radicals, singlet ion, and hydrogen peroxide), which can cause random and rapid oxidation of diverse biomolecules of critical structural, functional, and hereditary role in the cell such as proteins, enzymes, lipids, and nucleic acids [25]. Alternatively, NMs may form complexes with the biomolecules leading to damage and inactivation of biomolecules particularly the proteins [27,43]. These interactions and transformations of the biomolecules result in inhibition of cell growth and division [44]. The distortion of the cell morphology and topography is a common feature epitomized by disruption of cellular membrane including exfoliation or erosion of the membrane bilayer structure, appearance of pits due to preferential dissolution of extrinsic proteins, and leakage of cell cytoplasm or even bursting of the cell [17] (Figure 2). Therefore, the complex cascades, diversity, and multiplicity of these interactions may not allow the pathogen to develop the neutralizing or counter-acting mechanisms to address all these interactions. Thus, NM-based antimicrobials will exhibit durable efficacy as there are fewer chances of development of profound resistance in the pathogen [25].

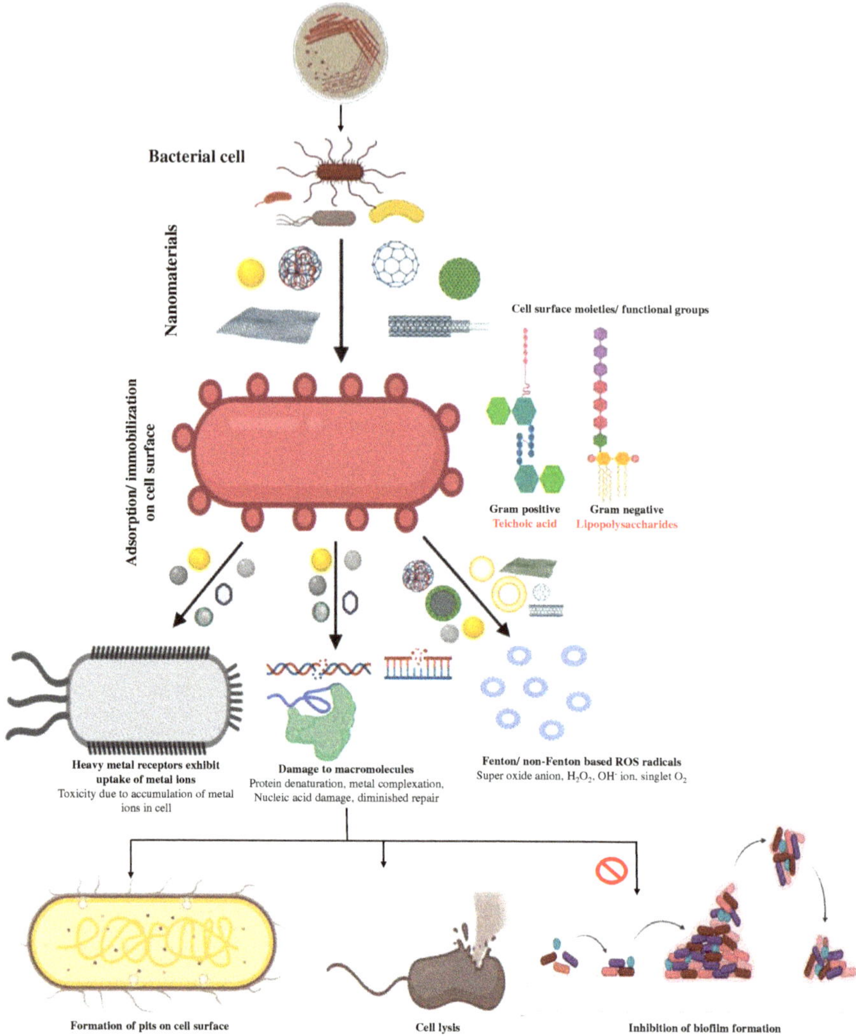

Figure 2. Mechanisms governing the antibacterial potential of different types of nanomaterials.

2.1.1. Metal/Metal Oxides, Metalloid, and Nonmetal Nanomaterials

Plants are affected by diverse biotic stress agents, particularly the phytopathogens that cause various diseases and claim the growth and yield losses in crop plants. The incidences of quick emergence of novel pesticide-resistant phytopathogens and reduced efficacy of already available arsenal of antipathogenic compounds/formulations have led towards a possibility of use of antimicrobial potentials of the nanomaterials to curb plant pathogen, which cause diseases culminating to high economic losses due to crop failure. Metal/metal oxide nanoparticles exhibit appreciable antimicrobial activities, which may span over -cidal to static potentials and help in curbing bacteria (bactericide) [17,18], fungi (fungicide) [40], virus (viricidal) [45], and algae (algicidal) [46].

The antimicrobial effect of metal/metal oxide, metalloid, and nonmetal nanomaterials on the test pathogens have been reported to be size and dose dependent [26,47,48]. Further, substantially low concentrations of nanomaterials are required to achieve significantly improved antimicrobial efficacy as compared to the standard reference antimicrobial agent (such as antibiotics and pesticides) [8,49]. Interestingly, the combinatorial use of nanomaterials along with the conventional antimicrobial agents [50] or a combination of metal/metal oxide/nonmetal oxide NPs can enhance the action-spectrum and reduces the minimum inhibitory concentrations (MIC) values [51].

Among the various inorganic nanomaterials, the antimicrobial activity including the antimycotic potential of the noble metal nanoparticles (Au/Ag NPs) against plant pathogenic microbes was identified initially [51–55]. Later, nanoparticles/nanomaterials of Group IIa metals including magnesium [56,57]; calcium [58]; other transition metals such as copper [57,59–62], iron [61], manganese [57], nickel [63,64], titanium [61,65], zinc [56,57,60,62,66–68], and zirconium [21,69]; and nonmetals such as silicon [57], selenium [70–73], and tellurium [74,75] have been evaluated for their antimicrobial potentials. However, chemically, physically, and biologically synthesized noble metal NPs (Au/Ag NPs), copper/copper oxide, zinc/zinc oxide NPs, and magnesium NPs have been mostly reported for the plant pathogenic microbes, whereas the rest of the NP-microbe studies involved evaluation of antimicrobial activity against human or food pathogenic microbial cultures.

Mechanism of Antibacterial Activity of Nanomaterials

Nanomaterials exhibit antibacterial potentials manifested as disintegration of the cell membrane leading to leakage of the cytoplasmic contents followed by the lysis of the bacterial cells [47,76,77]. Passive internalization of the NPs can occur through porin-ion channels in Gram-negative bacteria [78], whereas in Gram-positive bacteria, presence of thick cell wall hinders passive internalization and therefore, dissolved ionic species (e.g., Zn^{2+}, Cu^{2+}, and Fe^{2+} ions) released by the nanoparticles in vicinity of the cell surface get chelated by lipoteichoic acid [79]. Once inside the cell, the internalized NPs may elicit Fenton- or non-Fenton-based ROS-mediated damage of the plasma membrane, internal macromolecules, and other soluble and catalytic biomolecules [78]. Eventually or simultaneous release of ions by the dissolution of NPs leads to metal/nonmetal ion toxicity culminating to cell death [25,76]. Another interesting mechanism involves inhibited expression of the quorum-sensing regulated genes or functions in bacteria leading to inhibition of the biofilm formation [41,80]. The nanostructured materials can also help in the inhibition of the preformed biofilms of the plant pathogens, which is of great significance for the eradication of resistant bacterial pathogens [81] or pathogens related to food spoilage [80,82].

Mechanism of Antimycotic Activity of Nanomaterials

Enormous literature on antifungal potential of nanoscale silver [52–54,59,83,84], copper/copper oxide [59,62], and zinc/zinc oxide [40] materials exists (Figure 3). The fundamental benefit of the nanoparticulate fungicide is the performance of these formulations equivalent or superior to the respective bulk salt formulations at relatively lower application doses thereby effectively addressing the phyto- and ecotoxicity issues posed due to the release of the metal cations [85]. There exist multiplexed nanomaterial–fungal cell interactions. The nanomaterial internalization in the fungal cell occurs through three possible mechanisms; (i) nonspecific but direct internalization of the small-sized, mostly, spherical nanoparticles, (ii) specific receptor-mediated adsorption followed by internalization of the NPs, or (iii) internalization of dissolved metal/nonmetal ions through membrane-spanning ion transport proteins (Figure 3). Nanomaterials particularly the metal/metal oxide and nonmetal oxide nanoparticles can curb fungal growth through mechanisms that can be dichotomized as (a) antimycotic effect due to generation of ROS and dissolution of the nanoparticles in the cell environment to release specific ions leading to metal/nonmetal ion toxicity and (b) regulation of the mycotoxin-producing genes for decreased or no expression. A detailed illustration of the same for zinc nanomaterials will be incorporated in Section 3.

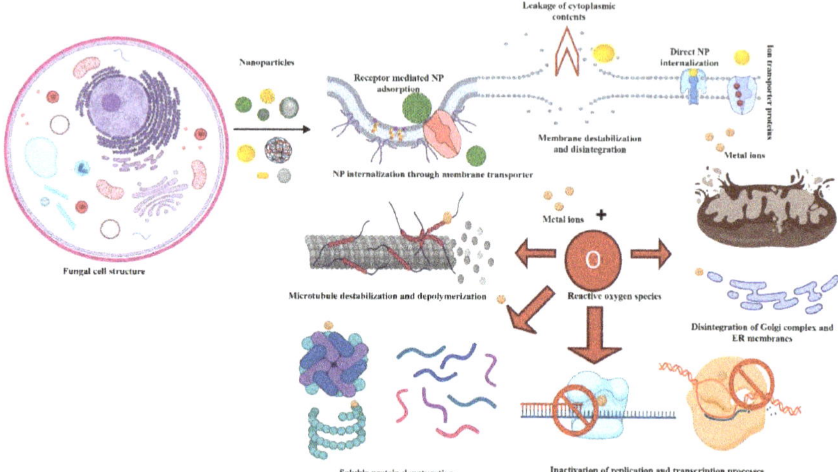

Figure 3. Effect of application of different types of nanoparticles on cellular components and organelles in a fungal cell.

Mechanism of Antiviral Activity of Nanomaterials

The M/MO/NM/NMO nanomaterials possess antiviral activity against microbial [86], animal [87–90], and human viruses [91–97] as depicted in several published reports. The green synthesized (microbial/plant cell extract-derived) nanoparticles particularly silver [98] and gold nanoparticles [99] or their composites [98] have been documented to exhibit virus-neutralizing or -inhibiting properties. Likewise, the role of zinc nanomaterials for the virostatic effect [100], virus neutralization, and for immune-modulatory significance against the emerging COVID-19 causative agent [101] has been well identified.

The application of nanomaterials for the control and treatment of viral disorders in crop plants has also been evaluated and established through molecular biology and in planta assays [45]. One-week preapplication of silver NPs at low concentration (50 ppm) on tomato plants decreased the disease severity and induced systemic resistance against two common tomato viruses, namely, Tomato mosaic virus, and Potato virus Y [102]. However, another in planta study showed significant inhibition of Tomato spotted wilt virus on foliar spray of AgNPs (200 ppm) 1 day after artificial inoculation of the TSWV, whereas the lowest and substantially low inhibition was recorded when AgNPs were applied along with and before the virus inoculation, respectively [103]. Similar results have been documented by Elbeshehy et al. [104] on foliar spray treatment of biogenically synthesized AgNPs derived from cell-free extracts of three *Bacillus* bacteria species (*B. pumilus, B. persicus*, and *B. licheniformis*). Complete inhibition of typical disease symptoms was recorded when the AgNPs were applied (concentration: 0.1 $\mu g\ \mu L^{-1}$) 24 h postinoculation with bean yellow mosaic virus in fava bean cv. Giza 3 variety, whereas weak symptoms were recorded when AgNPs formulation was sprayed on foliage simultaneously to that of swab inoculation of the fava bean plants. Low concentration of fungus generated AgNPs formulation (derived from *Curvularia lunata* cell extracts, concentration: 100 ppm) on spray treatment on the foliage of approximately 1 month (35 days) old tobacco plants (*Nicotiana tabacum* cv. *Xanthi* nc) followed by mechanical inoculation of two leaves (5th and 6th true leaf) with PVY-Ros1 virus after 2 days resulted in 2.67-fold decrease in the appearance of characteristic red lesions/infection loci in AgNP-treated plants. Development of nano-Ag composites can further improve the antiviral activity, for instance, graphene oxide-AgNP composite treatment (at 1 $\mu g\ mL^{-1}$) reduced the visible symptoms of disease caused by Tomato bushy stunt virus in test lettuce plants [105].

Apart from silver NPs, daily foliar spraying treatment for approximately 2 weeks (12 days) of micronutrient iron oxide NPs (Fe_3O_4 NPs, size: 20 nm, concentration: 100 μg mL^{-1}) enhanced the resistance of tobacco plants against Tobacco mosaic virus [106]. Another report involved daily foliar spray treatment on *Nicotiana benthamiana* plants with Fe_2O_3 (concentration: 50 mg L^{-1}) or TiO_2 NPs (concentration: 200 mg L^{-1}) (amount: 5 mL) for 21 days. When these plants were challenged with Turnip mosaic virus (green fluorescent protein-tagged TuMV), the plants exhibited significant inhibition in the proliferation of the inoculated TuMV, particularly decrease in coat protein content as identified through a decrease in the fluorescent intensity of GFP marker in new emerging leaves [107].

3. Zinc Nanomaterials and Their Use for Curbing Plant Disease-Causing Pathogens

Metal oxides exhibit substantially high antimicrobial activities. However, the eco- and cytotoxicity aspects associated with the application of these novel antimicrobial formulations have hampered their quick commercial applications. Among the various metal oxides, ZnO nanoparticles appear to be one of the most propitious candidates as these NPs can be generated through low-cost synthesis techniques in bulk amounts. Further, their better biosafety and lower cytotoxicity indices for mammalian cells have been proven through several cell line studies [108–110] including the report on the preferential killing of human cancer cells compared to normal cells by ZnO NPs [109]. The antimicrobial action spectrum of Zn nanomaterials includes antibacterial, antifungal, and antiviral characteristics [111]. Therefore, the research insights on relative multifunctional properties of the zinc nanomaterials exhibiting antimicrobial actions are based on a fundamental hypothesis of spontaneous generation of ROS species and intracellular oxidative stress leading to killing of the microbial cells [79,112].

3.1. Antibacterial and Mollicute Controlling Potential

The studies involving zinc nanomaterial-antibacterial assay against plant pathogenic bacteria are scarcely reported as the majority published research includes the antibacterial activity against pathogenic bacterial genera/species causing human or animal health diseases [113–115]. However, plant pathogenic bacteria-Zn nanomaterial interactions have been studied including the reports showcasing the inhibitory effect on the causative agent of citrus canker (*Xanthomonas citri* subsp. *citri*) [116], rice leaf blight pathogen (*Xanthomonas oryzae* pv. *oryzae*) [81], tomato bacterial spot pathogen (copper-tolerant strains of *Xanthomonas perforans*) [117], the causative agent of lentil bacterial leaf spot (*Xanthomonas axonopodis* pv. *phaseoli*) [118], the causative agent of bacterial blight of lentil (*Pseudomonas syringae* pv. *syringae*) [118], and eggplant bacterial wilt pathogen (*Ralstonia solanacearum*) [119].

On the evaluation of the relative antibacterial potential of the Zn-nanomaterials, studies established higher efficacy in comparison to the absolute or conventional bulk controls. Among the green synthesized ZnO NPs derived from three different plant extracts, *Olea europaea* extract-derived ZnO NPs exhibited the highest inhibition zone (2.2 cm at 16.0 mg mL^{-1}) for *Xanthomonas oryzae* pv. *oryzae* [81]. Likewise, Graham et al. [108] have compared the relative efficacy of nano-ZnO formulations, Zinkicide SG4 and SG6, in an in vitro assay and showed twofold and eightfold lower MIC for SG4 and SG6, respectively, against *X. alfalfae* subsp. *citrumelonis*.

The antibiofilm forming potential of nanozinc material is of remarkable significance for commercial application. The specific benefit of the antibiofilm property of the zinc nanomaterials [82] spans over the decontamination of the food articles [82], surfaces [120,121], produce processing equipment [122], and packaging systems [80,123–125].

Apart from the bacterial pathogens, the crop plants are also affected by obligate parasitic, axenically unculturable prokaryotic cell wall lacking eubacterial plant pathogens [126], the "phytoplasma" or "mollicutes" [127], which are associated with >600 plant diseases across the globe [128–131]. These initially classified as wall-less bacteria possess a trilaminated unit membrane, a small genome (~680 to 1600 kb), exhibit morphological pleomorphism (size ranging between 0.2 and 0.8 μm, and shapes varying from helical, filamentous, beaded, or simply spheroid), dwell in sieve tubes [132] and therefore, are mainly transmitted by phloem sap-feeding or sucking pest vectors, particularly planthoppers and psyllids, and by vegetatively propagated grafts or tissues [133,134]. Being obligate parasites, phytoplasma diseases can be effectively controlled by managing the vector pest population. Therefore, research efforts to develop RNAi- or dsRNA-based nanoenabled pesticides have been initiated that can effectively control the psyllids and/or leafhopper population [135,136]. However, a few reports have appeared including the development and use of nanoemulsion formulations of antibiotics [137], essential oil or aldehyde compounds (such as cinnamaldehyde), and silver nanoparticles [138] for management or eradication of *Candidatus liberibacter asiaticus* causing Huanglongbing or citrus greening disease. Foliar spray and trunk injection treatments of zinc oxide and zinc sulfide nanoparticles alone as an isopropanol-based emulsion or in combination with cinnamaldehyde-isopropanol have been reported to effectively decrease the occurrence of this bacteria in the phloem tissue [139]. Likewise, published reports indicated in planta inhibition of *Candidatus liberibacter asiaticus* by trunk injection application of aqueous formulations of 4 nm-sized zinc oxide nanoparticles and ZnONP-2S albumin protein composite [140]. A qPCR assay revealed that 1:1 proportion of ZnONPs: 2S albumin (concentration of 330 ppm each) most effectively decreased the bacterial pathogen to about 97% of the initial concentration.

3.2. Antimycotic and Mycotoxin Neutralizing/Inhibiting Activity

The antimycotic potential of zinc oxide nanoparticles or its composites has been well identified against phytopathogenic fungi belonging to diverse taxonomic groups/classes such as zygomycetous oomycetes genera (*Peronospora tabacina* [141], *Pythium ultimum, Pythium aphanidermatum* [142]), ascomyceteous genera (*Alternaria alternata* [59,62], *Aspergillus flavus/A. fumigatus* [51], *Aspergillus niger* [143], *Botrytis cinerea* [61,62,144,145], *Colletotrichum gloeosporioides* [56,59], *Fusarium graminearum* [146], *Fusarium moniliforme* [40], *Fusarium oxysporum* [66,144,147], *Penicillium expansum* [50,66,144,148]), and basidiomycetous genera (*Erythricium salmonicolor* [68]).

Zinc-derived nanomaterials (nanoparticles/composites) at substantially low working concentrations can kill spores or exhibit inhibition of spore germination (sporostatic/sporicidal activities) besides inhibiting the vegetative mycelial growth of the filamentous fungal plant pathogens, e.g., a significant decrease in fungal growth of *B. cinerea* and *P. expansum* has been observed on ZnO NPs (3 mM L^{-1} concentration) treatment [144]. Likewise, events of spore germination of *Peronospora tabacina* were observed to be completely inhibited on treatment with Zn NPs, ZnO NPs, and $ZnCl_2$ soluble salt at concentrations ranging from 15–20 mg L^{-1} [141].

3.2.1. Mechanism of Antimycotic Activity

Multifarious mechanisms govern the antimycotic activity of the zinc nanomaterials. The primary inhibitory symptoms that appear postincubation of an alive culture of fungi with nanoscale zinc/zinc oxide material include adsorption of nanozinc on the hyphal cell surface, hyphal deformation leading to morphological alterations in the cell wall and cell membrane, formation of sunken or swollen mycelia besides extensive thinning, and branching of the mycelia [144]. The same could be or may not be accompanied by suppression of spore or conidia-forming structures or formation of distorted sporangiophore/conidiophore and absence of formation of perennation structures (spores/conidia) or their number is decreased. Fungal spore nanozinc incubation studies have revealed a delay in spore germination, formation of abnormal stout/short germination tubes, or complete inhibition of spore germination indicating the sporistatic to sporicidal properties of nanoscale Zn material [141].

At the cell ultrastructural level, changes in the cell wall and membrane structure epitomized as enhanced thickening of the cell wall, liquefaction of cell membrane, dissolution or disorganization of the cytoplasmic organelles, hypervacuolization, and detachment of cell wall from cytoplasmic contents indicating incipient plasmolysis like features appear [68].

At the molecular biology scale, the nanoscale Zn materials exhibit interactions with a variety of biomolecules leading to complexation with structural and soluble proteins, inactivation of catalytic proteins, ROS-mediated damage to nucleic acid, particularly the scission of DNA strand, and breakage leading to chromosomal aberrations [25,143,149]. Interaction of zinc nanomaterials with the hyphal cell surfaces also specifically elicit synthesis of nucleic acid and/or production/secretion of the carbohydrates as depicted through increased Raman spectra signal intensities corresponding to these biomolecules [144]. The production of these components may indicate the increased expression of genes involved in subduing the ROS damage induced by the nanozinc material, particularly the osmolytes such as trehalose oligosaccharide. Further, the cell growth cycle also gets altered thereby inhibiting cell division.

3.2.2. Mycotoxin Neutralizing/Inhibiting Activity

The effect of nanoscale zinc materials for mycotoxin production by the filamentous fungal hyphae have also been evaluated [150,151]. Mycotoxins exhibit enormous structural and chemical diversity. Several fungal genera produce different types of mycotoxins primarily including aflatoxins (B1, B2, G1, G2, and M10), ochratoxins, deoxynivalenol, trichothecenes produced by ascomycetous genera *Aspergillus* (sexual stage name: *Eurotium*) [152]. Likewise, various species of another ascomycetous fungus, *Penicillium* (sexual stage name: *Eupenicillium*), synthesizes and secretes a variety of secondary molecules considered as mycotoxins such as penicillic acid, brevianamide A, griseofulvin, patulin, citreoviridin, citrinin, roquefortine, cyclopiazonic acid, PR-toxin, fumitremorgin B, penitrem A, luteoskyrin, ochratoxin A, rugulosin, verrucosidin, verruculogen, viridicarumtoxin, and xanthomegnin [153,154]. Ascomycetous member belonging to order Hypocreales, *Fusarium*, produces trichothecenes (including fumonisins, zearalenone, deoxynivalenol, and diacetoxyscirpenol) besides fusaproliferin, beauvericin, enniatins, and moniliformin [155]. Alkaloids of *Claviceps* sp. are also considered mycotoxins and include clavines, lysergic acids and their amides, and ergopeptides [156–158]. Besides these genera, *Alternaria* sp. produces diverse types of mycotoxins such as altenuene, alternariol, and its methyl ether, altertoxin, and tenuazonic acid [159].

The engineered NPs including ZnO NPs can control mycotoxin production by the mycotoxigenic fungi besides neutralization or adsorption of already formed/secreted mycotoxins [160] (Figure 4). The antimycotic potential of the nano-Zn formulations has already been discussed in Section 3.2.1. The other two mechanisms that are directly responsible for alteration in mycotoxin production by the mycotoxigenic fungi on supplementation of nanozinc formulations in culture/growth media will be dealt with here. Metal oxide nanoparticles exhibit classical size quantization effect such that discrete energy state appears and the number of surface atoms to bulk ratio gets altered besides the changes in the surface topology, which result in enhancing the reactive surface area [161]. Likewise, the thermodynamics of chemical reactivity is varied due to variations in the surface free energy of the NPs. These features adorn NPs the excellent adsorption characteristics. Though classically, carbon nanomaterials, including the amorphous carbon, graphene oxide, carbon nanotubes, and carbon fullerol nanoparticles [150], carbon nanocomposites [162], and inorganic nanocomposites such as MgO-SiO_2 nanocomposite [163] and organo-silicate composites [164], exhibit higher potential for mycotoxin adsorption. However, a recent study on the application of fullerol nanoparticles (FNP) on the aflatoxin biosynthetic pathway in *Aspergillus flavus* has been performed which suggested a concentration-dependent eliciting effect of FNP on aflatoxin synthesis after 120 h of incubation [165]. Therefore, other nanoadsorbent alternatives including the metal and metal oxides particularly the iron, copper, silver, and the zinc NPs [150,166] can be evaluated for mycotoxin adsorption and removal. A research study on flower-shaped zinc nanostructures (Znstr) revealed that supplementation of low concentrations of Znstr (1.25, 2.5, and 5.0 mM) in the liquid growth media led to substantial suppression (97%) of aflatoxin biosynthesis by *Aspergillus flavus* besides reducing the content of aflatoxin (69%) in maize grains [167].

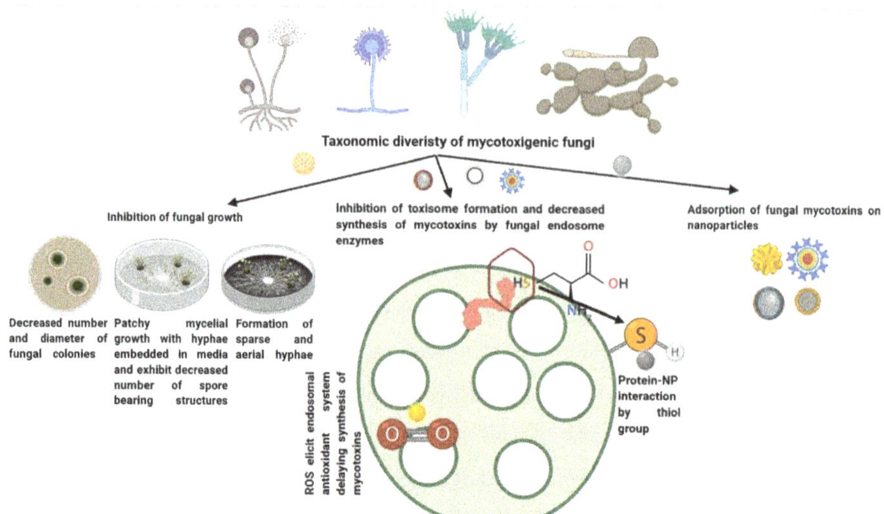

Figure 4. Zinc nanomaterials can exhibit a threefold impact on the production and neutralization of mycotoxins produced by mycotoxigenic fungi.

Apart from the use of nanomaterials for adsorption of mycotoxins, a recent study deciphering the mycotoxin inhibition mechanism of the AgNPs reported a fungus-growth-independent decrease in the aflatoxin B1 production in *Aspergillus parasiticus* [160,168,169]. Unlike the above study, a report documented inhibition of both growth and mycotoxin production potential of *Fusarium graminearum* on the application of biogenic zinc oxide nanoparticles [170]. However, Savi et al. [168] have reported appreciable antifungal and antimycotoxigenic potential of various zinc compounds against *Fusarium graminearum*, *Aspergillus flavus*, and *Penicillium citrinum*. Therefore, zinc nanomaterials have great potential for curbing the growth and mycotoxin contamination of food and feed material [171].

3.2.3. Zinc Nanomaterials for Curbing Plant Viruses/Viroid Diseases

Viruses and viroids cause diverse diseases in crop plants on infection and are responsible for enormous losses posing a great threat to crop productivity and food security. Further, there is a lack of an effective plant viral disease control strategy besides the occurrence of a few commercial antiviral formulations, which enhance the threat for effective control of plant viral diseases. The use of nanomaterials for curbing the spread and disease severity of plant viruses is rather in its incipient stage and research reports on the use of silver [102,103], silver-graphene composite [105], iron oxide [172], and Fe_3O_4 [106] nanomaterials have been published. However, there is one recent report on the application of zinc oxide nanoparticles on the plant foliage to curb Tobacco mosaic virus infection in *Nicotiana benthamiana* [45]. The details regarding the antimicrobial potential of various zinc nanomaterials against plant pathogens have been summarized and presented in Table 1.

Table 1. Antimicrobial potential of zinc nanomaterials on plant pathogenic microbes.

Type of Zn-Nanomaterial Used	Zn-Nanomaterial Characterization	Working Concentration	Study Conditions (Exposure Technique)	Zn-Nanomaterial Application Method	Pathogen Inoculation Technique	Pathogen Studied	Impact	References
				Bacterial pathogens				
Zinkicide SG4, Zinkicide SG6	2-D nanoplate-like structure (dimensions: 0.2–0.5 mm, thickness: ~10.0 nm) nanoparticulate (size: 4–6 nm)	2000 to 1.96 mg/mL	In vitro assay (broth microdilution technique)	Addition in broth at different working concentrations	Broth inoculation	*X. alfalfae* subsp. *citrumelonis*	Two-fold and 7/8-fold lower MIC for Zinkicide SG4 and SG6, respectively	[116]
ZnO NPs	Commercial formulation (size <100 nm)	0.1 mg mL^{-1}	In planta assay	Foliar spray of ZnO NPs suspension (10 mL per lentil plant) under pot culture conditions	Nutrient broth culture (10 mL of 1.2×10^5 CFU mL^{-1}) added around the seedling	*Xanthomonas axonopodis* pv. *phaseoli*	Reduction in disease severity on pathogen challenge	[118]
Zinkicide SG4, Zinkicide SG6	2-D nanoplate-like structure (dimensions: 0.2–0.5 mm, thickness: ~10.0 nm) nanoparticulate (size: 4–6 nm)	Zn (30% *w/v*)	In planta assay	-Foliar spray of Zn formulation (10 mL per grapefruit seedling) using air-brush in greenhouse assay -Foliar spray of Zn formulations (3.0 L per grapefruit tree) with a handgun sprayer	Broth culture (10^4 CFU mL^{-1}) in PBS injection-infiltrated in midrib of leaf 3 each site at both surfaces	*Xanthomonas citri* subsp. *citri*	-Reduction in citrus canker disease -Effective disease control comparable or better than Cu_2O/Cu_2O-ZnO bactericides (no phytotoxicity)	[116]
ZnONPs	TEM: 41–51 nm	4, 8, and 16 μg mL^{-1}	In vitro assay	Variable concentrations of ZnO NPs (10 μL each) dropped on 1-day old bacterial lawn culture	Lawn growth obtained by spread plating of (100 μL, 10^8 cfu mL^{-1}) broth culture followed by incubation for 24 h	*Xanthomonas oryzae* pv. *oryzae* (strain GZ 0003)	Effective antimicrobial agent for bacterial leaf blight of rice	[81]
Cu-Zn hybrid NPs	TEM: 40–100 nm	1000, 500, 200, and 100 μg mL^{-1}	In vitro assay	NP formulations added to broth at different concentrations	Broth culture (20 μL, 10^5 CFU mL^{-1})	*Xanthomonas perforans* (Cu-tolerant GEV485)	Complete inhibition of growth till 24 h of incubation	[117]
Cu-Zn hybrid NPs	TEM: 40–100 nm	500, 200, 100, and 50 μg mL^{-1}	In planta assay	Foliar spray on 4-week old seedlings of tomato variety FL 47 under growth chamber conditions	Pathogen inoculation-foliar spray	*Xanthomonas perforans* (Cu-tolerant GEV485)	Statistically highest decrease in disease symptoms at 500 μg/mL	[117]
				Fungal pathogens				
ZnO NPs	Commercial formulation (< 50 nm particles size)	0, 1, 10, 100, 500, and 1000 μg/mL	In vitro assay (poison food technique)	Supplementation of PDA with different working concentrations	Mycelial plug (5 mm) cut from master culture PDA plate (4-day old growth from edge)	*Alternaria alternata*	-Mean inhibition rate (EC_{50}) range 235 and 848 μg/mL -higher efficacy compared to $ZnSO_4$	[59]

Table 1. *Cont.*

Type of Zn-Nanomaterial Used	Zn-Nanomaterial Characterization	Working Concentration	Study Conditions (Exposure Technique)	Zn-Nanomaterial Application Method	Pathogen Inoculation Technique	Pathogen Studied	Impact	References
ZnO NPs/CS-Zn-CuNPs	DLS: 1.5–20 nm TEM: 6–21 nm	0, 30, 60, and 90 μg mL^{-1}	In vitro assay (poison food technique)	Addition various working concentrations of prepared nanomaterials in PDA media	Mycelial plug (5 mm) cut from edge of 1-week old fungal growth on PDA media	*Alternaria alternata, B. cinerea, R. solani*	-Highest mycelial inhibition by chitosan mixed Zn-Cu nanocomposite	[62]
3D flower-shaped nanostructured ZnO	FE-SEM: 700–800 nm XRD: crystallite size—42.0 ± 0.8 nm	0.3125–5.0 mM	In vitro assay (broth culture experiment)	Supplementation of broth with different concentrations of Zn nanomaterial	Aqueous conidial suspension (125 μL, 4×10^6 spores mL^{-1}) added to Sabouraud dextrose broth (100 mL)	*Aspergillus flavus* Link (UNIGRAS-1231)	-For 1.25–5.0 mM concentrations -78.0% decrease in mycelial growth -99.7% decrease in aflatoxin synthesis	[167]
Metallic (Au/Ag) and ZnO NPs	Commercial formulation DLS: 7 and 477 nm, respectively	50:10 μg/mL	In vitro assay (A. broth microtiter plate test, B. Kirby-Bauer disk diffusion technique)	A. NP suspension (20 μL in 75 μL SDB) B. NP impregnated on sterilized filter paper disks (6 mm diameter)	A. Spore suspension (5 μL, 1×10^5 spores/well) B. Spread plating of spore suspension	*Aspergillus flavus* (NRRL 3518)/*A. fumigatus* (ATCC 1022)	-combination of mix metallic NPs and ZnO-NPs effectively inhibited the fungal growth	[51]
ZNPs	DLS: 30–40 nm TEM: 15–20 nm (average particle size)	50, 100, 250, and 500 ppm	In vitro assay (poison food technique)	Different ZnO NPs concentrations mixed in sterilized PDA media	Fungal spore suspension (3 μL, ~10^4 mL^{-1}) spot plated in center of PDA media plate	*Aspergillus niger*	-dose-dependent decrease in radial growth diameter	[143]
ZnO NPs	Commercial formulation (TEM: 70 ± 15 nm)	0, 3, 6, and 12 mM L^{-1}	In vitro assay (poison food technique)	ZnO NPs mixed in different concentrations in PDA media	Aqueous spore suspension (~10^4 mL^{-1})	*Aspergillus niger* (MTCC-10180), *Fusarium oxysporum* (NCIM-1043, NCIM-1072)	-Significant inhibition in hyphal growth at concentration of 3 mM L^{-1}	[144]
ZnO NPs	Leaf extract of derived NPs	200, 300 and 400 μg mL^{-1}	In vitro assay (poison food technique)	Supplementation of PDA with different working concentrations of NPs	Fungal disc (5 mm diameter) from 5-day old culture growth	*Alternaria alternata, Botrytis cinerea*	-Concentration-dependent decrease in fungal growth	[145]
A. ZnO NPs, B. ZnO:MgO NPs C. ZnO:$Mg(OH)_2$ composite	A. TEM: 22–37 nm B. TEM: 23–30 nm C. TEM: 23–49 nm	Serial dilution ranging from 5 to 0.002 mg mL^{-1}	In vitro assay (broth microdilution and agar-media based poison food technique)	DMSO dissolved NPs were diluted with PDB in a geometric progression	Aqueous spore suspension (1×10^6 conidia mL^{-1}) added in PDB	*Colletotrichum gloeosporioides*	-ZnO NPs alone exhibited highest inhibition of the hyphal growth -Addition of MgO diminished the antifungal potential of ZnO NPs	[56]
ZnO NPs	TEM: 20 nm (spherical), 37 nm (acicular)	3, 6, 9, and 12 mM L^{-1}	In vitro assay (poison food technique)	Supplementation of PDA with different working concentrations of NPs	Mycelial plug (1.5 cm diameter) from 16-day old fungal culture	*Erythricium salmonicolor*	-substantial mycelial growth inhibition at 6 mmol L^{-1}	[68]

Table 1. *Cont.*

Type of Zn-Nanomaterial Used	Zn-Nanomaterial Characterization	Working Concentration	Study Conditions (Exposure Technique)	Zn-Nanomaterial Application Method	Pathogen Inoculation Technique	Pathogen Studied	Impact	References
ZnO NPs	Commercial formulation (size <100 nm)	0, 100, 250, and 500 mg [Zn] L^{-1}	In vitro assay (poison food technique)	Different concentrations of ZnO NPs supplemented in mung bean agar media	Mycelial plugs (~0.5 × 1.0 cm) cut from the margins of the 5-day old fungal growth	*Fusarium graminearum*	-dose-dependent inhibition of fungal growth	[146]
ZnO NPs	TEM: 30–40 nm SEM: triangular- to hexagonal-shaped particles XRD: crystallite size—35.69 nm	25, 50, 75, 100, 125, and 140 µg mL^{-1}	In vitro assay (broth culture experiment)	Different concentrations of ZnO NPs supplemented in Czapek Dox broth	Spore suspension (10 µL, 10^6 spores mL^{-1} in peptone water + 0.01% Tween 80) in Czapek Dox broth (100 mL)	*Fusarium graminearum*	In dose-dependent manner -ROS accumulation in treated mycelial -reduction in deoxynivalenol and zearalenone production	[170]
ZnO NPs	TEM: spherical-shaped 30 nm size NPs XRD: wurtzite crystal nature	10, 25, 50, and 100 mM	In vitro assay (poison food technique)	-Variable concentrations added to PDA -Highest Zn-compounds concentration added to PDA	Mycelial disc (6 mm) obtained from 7-day-old fungal cultures from edge	*Fusarium graminearum, Aspergillus flavus, Penicillium citrinum*	-concentration-dependent decrease in hyphal growth -significant decrease in deoxynivalenol and aflatoxin B1 only by ZnO NPs compared to control	[173]
ZnO NPs	DLS: 111.53 ± 1.3 nm TEM: < 100 nm ζ-potential: −15.89 mV	100–800 ppm	In vitro assay (poison food technique)	-Different concentrations of ZnO NPs added to Czapek Dox agar	Mycelial disc (5 mm diameter) was cut from 5-day old culture	*Fusarium moniliforme*	-Less hyphal growth inhibition due larger sized particles	[40]
ZnO NPs	Commercial formulation (size: 70 ± 15 nm)	0, 2, 4, 6, 8, and 12 mg L^{-1}	In vitro assay (poison food technique)	Different concentrations of ZnO NPs with autoclaved PD agar medium	Fungal mycelia plug (1 cm diameter) taken from the edge of one-week old culture	*Fusarium oxysporum*	-19.3–77.5% hyphal growth inhibition corresponding to for 2–12 mg L^{-1} ZnO NP concentration	[66]
ZnO NPs	Commercial formulation (spherical-shaped 20–30 ± 10 nm NPs)	25, 50, and 100 ppm	In vitro assay (poison food technique)	Working concentrations of ZnO NPs derived from 1000 ppm stock solution added to sterilized PDA medium	Fungal disc (0.5 cm diameter) obtained from 7-old culture	*Fusarium oxysporum f. sp. betae*	-49.3% inhibition of radial hyphal growth at 100 ppm	[147]
ZnO NPs	Commercial formulation (size: <50 nm)	0–15 mM equivalent to 0–1221 ppm	In vitro assay (automated turbidimetric assay)	ZnO NPs suspension-soaked filter papers	Spore suspension (1.73 × 10^3 conidia mL^{-1}) were serially diluted	*Penicillium expansum*	-MIC: 9.8 mM (798 ppm) and NIC: 1.8 mM (147 ppm)	[148]

Table 1. *Cont.*

Type of Zn-Nanomaterial Used	Zn-Nanomaterial Characterization	Working Concentration	Study Conditions (Exposure Technique)	Zn-Nanomaterial Application Method	Pathogen Inoculation Technique	Pathogen Studied	Impact	References
A. Zn NPs B. ZnO NPs	A. TEM: mean diameter 264 nm; hydrodynamic diameter: 615.8 nm; ζ-potential: −1.6 ± 3.7 B. TEM: mean particle diameter 19.3 nm; hydrodynamic diameter: 453.3; ζ-potential: 23.3 ± 5.0	0–65 mg L^{-1}	In vitro spore germination and infectivity tests	Different concentrations of nano-Zn formulations incubated with fungal spore suspension	Spore suspension (10^6 spores mL^{-1}) mixed with DI	*Peronospora tabacina*	-Inhibition of spore germination frequency spore by Zn NPs, ZnO NPs, and $ZnCl_2$ (<10 mg L^{-1}) -Significantly higher inhibition by ZnO NPs compared to bulk ZnO -Reduction in leaf infection in tobacco leaf assay	[141]
ZnO and CuO NPs	Commercial formulation	50, 100, 250, and 500 mg L^{-1}	In vitro assay (poison food technique)	Different concentrations of NPs amended in autoclaved PDA media	Fungal growth plug (0.5 cm^2) placed in center of PDA media	*Pythium ultimum*, *Pythium aphanidermatum*	-Inhibition of growth at low concentrations -morphological changes in the hyphae	[142]
				Viral pathogens				
ZnO NPs	TEM: 18 nm spherical-shaped particles	A. 100 µg mL^{-1} B. 100 µg mL^{-1} (5 mL NP solution foliar spray for 3, 7, and 12 days)	A. In vitro assay B. In planta assay (*Nicotiana benthamiana*)	A. ZnO NP suspension mixed with purified TMV particles B. Foliar spray of NPs suspensions	A. Purified TMV particles mixed with NPs B. Inoculation by rubbing infected leaves onto the oldest leaf	Tobacco mosaic virus	A. aggregation or breakage of tobacco mosaic virus particles B. marked suppression (35.33%) of TMV invasion in the inoculated leaves	[45]

4. Zinc Nanoformulations: In Planta Studies and Crop Plant Responses to Pathogen Attacks

Zinc nanoformulations have been evaluated to curb phytopathogenic infections in various crop plants. The major test crop plants that have been utilized as models to evaluate the antimicrobial potential of the nanozinc products include tomato [67], tobacco [141], pepper [145], rice, and wheat [174]. The antibacterial potential of ZnO NPs against *Pseudomonas syringae* pv. tomato DC3000 that causes bacterial speck disease in tomato [67] has been reported. In planta greenhouse study performed with *Lycopersicon esculentum* cv. Pantelosa transplants involved foliar spray treatment of ZnO NPs (100 $\mu g\ mL^{-1}$) at a five-leaf stage, which significantly reduced the disease severity as compared to untreated control post-1 week of inoculation of the bacterial pathogen. Further, the researchers also indicated elicitation of the plant's innate defense system through physiological and biochemical studies including antioxidant enzyme activities and profound vegetative growth [67]. Another interesting study involving the effect of ZnO NPs on synthesis and secretion of signal compounds (siderophores-pyoverdine) by plant growth-promoting rhizobacteria-*Pseudomonas chlororaphis* O6 improved the lateral root formation in wheat plants besides enhancing the immunity of the treated plants [174]. The use of ZnO quantum dots (QDs) surface-functionalized with kasugamycin antibiotic has been evaluated for on-demand pH-responsive release of the loaded antibiotic in a greenhouse study to effectively control *Acidovorax citrulli* and alleviate the disease severity symptoms of bacterial fruit blotch in watermelon seedlings [175].

The mixed formulation developed as zinc/copper nanocomposites have also been evaluated for their antimicrobial efficacy under field conditions. Suppression of disease symptoms caused by the Citrus canker causative agent, *Xanthomonas citri* subsp. citri were investigated under field conditions on the application of a ZnO-nanoCu-loaded silica gel (ZnO-nCuSiO$_2$ composite) nanocomposite. Young et al. [176] investigated the ZnO-nCuSi for controlling citrus canker disease under field conditions and found that this was effective in suppressing disease at less than half the metallic rate of the commercial cuprous oxide/zinc oxide pesticide, and no phytotoxicity was observed.

Antifungal activities of ZnO NPs biosynthesized from leaf extracts of *Olea europaea* and *Origanum majorana* plants were evaluated. These NPs significantly reduced the appearance of gray and black mold disease symptoms on artificial inoculation with *Botrytis cinerea* and *Alternaria alternata* in test pepper plants compared to chemically synthesized ZnONPs and untreated control plants [145]. Likewise, a comparative in vivo efficacy study for suppression of *Botrytis cinerea* causing gray mold disease on plum fruits (*Prunus domestica*) by treatment with Ag, Cu, and ZnO NPs at two different concentrations (100 and 1000 $\mu g\ mL^{-1}$) was performed [59]. The researchers observed complete inhibition of disease symptoms by AgNPs only while ZnO and CuNPs could help control disease symptoms numerically higher or equivalent to copper hydroxide treatment. A simulation study conducted by Wagner et al. [136] on tobacco leaves revealed the high antifungal potential of Zn nanomaterial against *Peronospora tabacina* primarily through inhibition of the spore germination process. An interactive protective effect of nano-ZnO particle seedling spray/seed soaking followed by seedling spray treatments along with the biocontrol agent, *Trichoderma harzianum*, improved plant's resistance against the causative agent of damping-off disease (*Rhizoctonia solani*) in sunflower seedlings [177].

Zinc nanomaterials also possess elaborate antiviral properties though the reports on in planta studies involving management of the plant viral diseases are recent and incipient. Hence, little literature is available on this aspect. An in vivo experiment on *Nicotiana benthamiana* involved marked inhibition of replication of the Tobacco mosaic virus on foliar spray treatment of ZnO NPs for approximately 2 weeks (12 days). The replication inhibition process may be attributed to improved growth and induction of plant defense responses as indicated by an escalation in accumulation of ROS, and activity of the ROS mitigating enzyme besides upregulation of pathogenesis resistance-related genes [45].

5. Zinc-Derived Nanomaterials for the Development of Tools/Devices for Plant Disease Diagnosis

Pathogenic disorders or diseases in plants can be identified through various imaging, spectroscopy, and conjugate imaging and spectroscopy techniques [178]. Most likely, the role of diagnostic techniques is to achieve quick, early, sensitive, simple, in situ, reliable, and automated high throughput identification and quantification of the causative agent so that the extent of virulence can be obtained before the appearance of the actual visual symptoms of the disease [179]. Nanomaterial-based sensor technologies provide flexible and diverse sensing platforms or methods for elucidation/quantification of the single or multiple analytes [180] and can help ensure early, rapid, and sensitive identification of the plant pathogen [181].

The plant produces a myriad of signal molecules in response to a pathogen attack. Few abundant and signature signal molecules including specific enzymes, gaseous molecules (e.g., nitrous oxide, volatile organic compounds), reactive oxygen species, secretory compounds such as oxylipins and expression of a crucial gene (pathogenesis-related proteins-PRPs, PAMPs) can be aptly utilized as key biomarkers for the development of nanobiosensor platforms [178]. As discussed in Section 3.2.2., several mycotoxigenic fungi produce diffusible exotoxins, which can also be utilized as markers for the identification and confirmation of phytopathogenic fungi. A nano-ZnO film-indium-tin oxide electrochemical impedance sensor was developed by coimmobilization of antibodies and BSA protein to detect ochratoxin-A in produce and other plant-derived products [182]. Likewise, DNA aptamer-functionalized ZnO/ZnS quantum dots can help in easy detection of plant pathogens [181].

Sensors systems based on zinc nanomaterials primarily include the semiconductor quantum dot (core–shell, CdSe/CdTe core ZnS shell QDs, and ZnTe or ZnSe QDs)-enabled optical (fluorescence-based) sensors [183]. High luminescence QDs are fascinating nanomaterials that can be used to develop protein–protein/protein–ligand detection assays including the fluorescence resonance energy transfer technique [184]. In fixed cell systems, the QDs can be extensively used as immunohistochemical labels [185].

The protein—antibody immunofluorescence-based biosensors are finding sensing applications for plant virus pathogens [186]. Medintz et al. [187] have developed a CdTe/ZnS core-shell QD-based sensor by labelling NeutrAvidin on the surface of biotinylated Cowpea mosaic virus (CPMV) and avidin-decorated QDs, which interacted through the biotin-avidin groups. Further, CdTe/ZnSe core–shell QDs can also be utilized for easy detection of DNA sequence change mutation events [188]. The nano-Zn-based QDs exhibit low cytotoxicity and produce high-intensity fluorescence signals, which have resolutions far beyond the diffraction limit of light [183]. Therefore, these can also be utilized for in planta or in vivo assays. Early and sensitive detection (detection limit of 25 $\mu g\ mL^{-1}$) of plant pathogenic *Fusarium oxysporum* has been reported through the use of 3-Mercaptopropionic acid-functionalized CdSe/ZnS QD in a fluorescence-based assay [189].

Other than fluorescence-based sensors, nano-Zn enabled optical biosensors have also been developed. One of the most promising applications of these nano-Zn-enabled optical biosensors is quick and sensitive detection of plant pathogenic viruses. A sensitive immune-optical biosensor was developed, which involved immobilization of antibodies against Grapevine virus A-type (GVA) antigenic proteins on a ZnO thin film prepared by the atomic layer deposition technique [190].

As zinc nanomaterials exhibit electron-hole generation due to their semiconductor behavior; these have also been used to develop another category of a sensor system, the electrochemical sensors. Zinc oxide nanorod cyclic voltammetry-based electrochemical sensor has been developed as a disposable sensor for a rapid, cost-effective, and label-free detection of *E. coli* in food matrices [191]. Tahir et al. [192] have investigated the potential of a ZnO-nanocomposite prepared by decorating zinc nanoparticles (25–500 nm) on the surface of multiwall carbon nanotubes to immobilize probe DNA strands having complementarity to Chili leaf curl virus beta satellite. They have assessed the electrochemical performance of this DNA biosensor through the binding of the DNA by cyclic and differential pulse voltammetry scans. A similar kind of electrochemical DNA biosensor has been reported to be developed involving ZnO nanoparticles-chitosan membrane-doped gold electrode to conveniently identify *Trichoderma harzianum* biocontrol fungus [193].

6. Potential Application of Zn-Based Nanomaterials and Future Use

Zinc nanomaterials have found elaborate applications in diverse fields of agrirelevance such as for fertilizer nutrient delivery [194] through foliar application [195] or sustained release of nutrient from a nanodelivery vehicle [196,197], as novel antimicrobial agent [40,198,199], pesticide [200], and for environmental remediation [201–203]. The role of zinc nanomaterials in nanodiagnostics has been already dealt with in Section 5. Several reports delineating the role of zinc nanomaterials for elicitation of the systemic acquired immunity (SAR) in plants to combat and curb attack by various phytopathogens have been indicating towards the gross positive impact of their use in plant crops [204]. The specific aspects that need to be delved on regarding the voluminous and wide-spread usage of zinc nanomaterials for management of phytopathogens include the development of stable nanozinc formulations and their environmental impacts in the soil food-web on nontarget organisms.

6.1. *Ecosafety Issues of Nanozinc-Derived Products and Devices*

Agriculture is a pivotal global enterprise thrusting the economies of most of nations. Therefore, the products or chemicals utilized for improving the nutritional status (fertilizers) [194,196,197] and for management of the plant pathogenic infections (pesticides) are anticipated to be utilized in quantum amounts. Therefore, a cautious and critical approach is desirable considering the atypical behavior in open, dynamic, and complex multicomponent systems. Further, the ecological nanotoxicity concerns of these materials need to be identified before approving the use of zinc nanomaterial-based agriproducts [205–207].

A pride and prejudice dilemma exists as zinc nanomaterials, particularly the ZnO NPs, are being exponentially synthesized due to amenability for easy and low-cost production processes [208,209]. Further, the functional versatility of nanomaterials renders them affordable for applications or use in diverse fields spanning over electronics, biomedicine [17,18], environment remediation [210], catalysis [211], agriculture [40], and cosmetics industries. However, the release of Zn nanomaterial through municipal wastewater/sewage water, industrial effluent, and surface wash water drifts these nanomaterials to contaminate diverse soil and water ecosystems posing gradual and subtle to drastic effects on soil and aquatic biota thereby exacerbating the health and sanctity of the contaminated eco-niches [212]. The semiconductor (oxidative stress-inducing) properties and heavy metal nature of the zinc nanomaterials (bioaccumulation) further complicate their ecotoxicity concerns [213] besides the fundamental nanoscale aspects (quantum size effects-size, shape, surface charge-dependent properties, and agglomeration/complexation processes), which lead to diverse cyto-/genotoxic and onco-/mutagenic effects [214]. The nano-Zn material and their dissolution product, i.e., Zn^{2+} ions exhibit toxicity to all types of organisms or biotic components of all trophic levels [215]. Further, the occurrence of other pollutants may enhance the pernicious effects of nano-Zn-based products [214]. Therefore, long-term field studies need to be designed besides improvement in the in silico simulation modeling studies to well predict the aftermaths of the rampant use of nano-Zn-based products in agriculture.

6.2. *Improved Nanozinc Formulations: The Scar and Sanctity of Stability and Biosafety*

Zinc nanomaterials can be synthesized using physical and chemical techniques [216]. However, several reports have considered the biologically synthesized nanozinc formulations to be cost-effective, ecosafe, and stable even under ambient storage conditions [217]. Further, higher antimicrobial efficacy and improved photocatalytic activity were reported for the zinc oxide nanoparticles synthesized from the neem leaf extracts [217]. Although the researchers reported a slight difference in the mean size of the ZnO NPs (sol–gel: 33.20 nm and biosynthesized: 25.97 nm), they have argued that the improved efficacy of the neem extract-derived ZnO NPs was due to greater stability of the dispersion owing to surface functionalization by the leaf phenolics or terpenoids.

The stability of nanozinc formulations is governed by size-dependent phenomena. Further, the zeta potential and the surface charge ensure the aggregation, flocculation, or sedimentation of the nanoparticles [218]. Most likely, the zinc nanoformulations are made stable by altering either the charge (charge-stabilized dispersions) or the steric hindrance (sterically stabilized dispersions). The former mechanism slows down the rate of aggregation of the nanoparticles due to electrostatic repulsion forces [219], whereas the latter involves grafting of polymer coating due to the addition of polymers acting as steric stabilizers (e.g., polyvinyl pyrrolidone, polysorbate 80, polyethylene glycol, and many more) on the surface of the dispersed nanoparticles inducing thermodynamic stability [220,221]. However, the surface charge of the ZnO nanomaterial suspensions also decide for the eco- and cytotoxicity of these nanomaterials [222]. The nano-ZnO particle dispersion bearing positive charge at cell physiological pH exhibits an enhanced ability to penetrate the cells than the vice versa [223].

7. Conclusions

The nano-Zn products, particularly the nanoformulations developed for suppression of bacterial, phytoplasma, fungal, or viral diseases in crop plants, can have a gross impact on decreasing the extent of voluminous use of conventional metal(s)-based pesticides. These formulations can be designed for the management of diseases in both open field and closed greenhouse/screen-house conditions and can be applied to crop plants through several application modes. The prior research has shown high effectivity of nano-Zn formulations to curb phytopathogen owing to versatile antimicrobial mechanism of action including photo-oxidation leading to generation of reactive oxygen species, destabilization of the cell membrane, organelles, and other cellular macromolecules, and toxicity due to the release of zinc ions. The zinc nanomaterials have also been utilized for the development of affordable sensor systems for sensitive and early detection of pathogen attack that can be used for predicting the crop losses and for surveillance purposes. Although there are apparent advantages of the use of zinc nanomaterials for diverse benefits, however, their proficient use is limited due to rising concerns about ecohealth deterring aspects of nanomaterials and the bio-/econanotoxicity issues that need to be addressed. The problems such as bioaccumulation across the food chain and food web, complexities of events and components of the plant-soil-atmosphere-pathogen continuum, photo-oxidation properties, and the unprecedented fate of applied nanomaterials in the environment depreciate, comprise, or even negate the advantages of zinc nanomaterials as novel plant disease suppression or eradication agents. Carefully designed protocols and assays dissecting the dimensions and role of nanoscale particles/materials on pathogen and plant can improve our know-how and may direct novel paradigms for adaptation and application of zinc nanomaterials to overt the global food production challenges posed by phytopathogens.

Author Contributions: Conceptualization, A.K.; writing—original draft preparation, A.K.; writing—review and editing, K.A.A.-E. and K.K.; visualization, A.K.; funding acquisition, A.K. and K.K. All authors have read and agreed to the published version of the manuscript.

Funding: This research was funded by Rashtriya Krishi Vikas Yojana (RKVY) scheme, Government of India. It was also supported by UHK (VT2019–2021).

Acknowledgments: A.K. thanks the Head, Department of Soil Science, CoA, for providing the necessary infrastructural facilities and the Director of Research, Punjab Agricultural University, Ludhiana, Punjab, for allocation of funds through RKVY scheme.

Conflicts of Interest: The authors declare no conflict of interest. The funders had no role in the design of the study; in the collection, analyses, or interpretation of data; in the writing of the manuscript, or in the decision to publish the results.

References

1. Savary, S.; Willocquet, L.; Pethybridge, S.J.; Esker, P.; McRoberts, N.; Nelson, A. The global burden of pathogens and pests on major food crops. *Nat. Ecol. Evol.* **2019**, *3*, 430–439. [CrossRef]
2. FAO. The State of Food and Agriculture 2019. In *Moving Forward on Food Loss and Waste Reduction*; FAO: Rome, Italy, 2019.
3. Rajasekaran, P.; Kannan, H.; Das, S.; Young, M.; Santra, S. Comparative analysis of copper and zinc-based agrichemical biocide products: Materials characteristics, phytotoxicity and in vitro antimicrobial efficacy. *AIMS Environ. Sci.* **2016**, *3*, 439–455. [CrossRef]
4. U.S. EPA/OPPTS (Environmental Protection Agency- Office of Prevention, Pesticides and Toxic Substances). Reregistration Eligibility Decision (R.E.D.) Facts-Zinc Salts. EPA 738-R-98-007, 1992, Washington, DC, U.S. Available online: https://archive.epa.gov/pesticides/reregistration/web/pdf/zinc_salt.pdf (accessed on 12 October 2020).
5. Curry, A.S.; Price, D.E.; Tryhorn, F.G. Absorption of zinc phosphide particles. *Nature* **1959**, *184*, 642–643. [CrossRef]
6. Hood, G.A. Zinc Phosphide-A new look at an old rodenticide for field rodents. In Proceedings of the 5th Vertebrate Pest Conference, Fresno, CA, USA, 7–9 March 1972; Volume 5, pp. 85–92.
7. US-EPA. Zinc oxide: Exemption from the requirement of a tolerance. *Fed. Regist.* **2018**, *83*, 42783–42787. Available online: https://www.govinfo.gov/content/pkg/FR-2018-08-24/pdf/2018-18402.pdf (accessed on 12 October 2020).
8. Almoudi, M.M.; Hussein, A.S.; Abu Hassan, M.I.; Mohamad Zain, N. A systematic review on antibacterial activity of zinc against *Streptococcus mutans*. *Saudi Dent. J.* **2018**, *30*, 283–291. [CrossRef]
9. Burgess, J.; Prince, R.H. Zinc: Inorganic & Coordination Chemistry. In *Encyclopedia of Inorganic Chemistry*, 1st ed.; King, R.B., Ed.; John Wiley & Sons, Ltd., Wiley: Hoboken, NJ, USA, 2006; pp. 1–26.
10. Dos Santos, R.A.A.; D'Addazio, V.; Silva, J.V.G.; Falqueto, A.R.; Barreto da Silva, M.; Schmildt, E.R.; Fernandes, A.A. Antifungal Activity of Copper, Zinc and Potassium Compounds on Mycelial Growth and Conidial Germination of *Fusarium solani* f. sp. *piperis*. *Microbiol. Res. J. Int.* **2019**, *29*, 1–11. [CrossRef]
11. Goodwin, F.E. Zinc Compounds. In *Kirk-Othmer Encycl. Chem. Technol.*; Kroschwitz, J., Howe-Grant, M., Eds.; John Wiley & Sons, Inc.: New York, NY, USA, 1998; pp. 840–853.
12. Qureshi, S.A.; Shafeeq, A.; Ijaz, A.; Butt, M.M. Development of algae guard façade paint with statistical modeling under natural phenomena. *Coatings* **2018**, *8*, 440. [CrossRef]
13. Gupta, S.; Sharma, D.; Gupta, M. Climate change impact on plant diseases: Opinion, trends and mitigation strategies. In *Microbes for Climate Resilient Agriculture*; Kashyap, P.L., Srivastava, A.K., Tiwari, S.P., Kumar, S., Eds.; John Wiley & Sons, Inc.: Hoboken, NJ, USA, 2018; pp. 41–56.
14. Luck, J.; Asaduzzaman, M.; Banerjee, S.; Bhattacharya, I.; Coughlan, K.; Debnath, G.; De Boer, D.; Dutta, S.; Forbes, G.; Griffiths, W.; et al. *The Effects of Climate Change on Pests and Diseases of Major Food Crops in the Asia Pacific Region*; Final Report for APN Project (Project Reference: ARCP2010-05CMY-Luck); Asia Pacific Network for Global Change Research: Kobe, Japan, 2010.
15. Lurwanu, Y.; Wang, Y.P.; Abdul, W.; Zhan, J.; Yang, L.N. Temperature-mediated plasticity regulates the adaptation of *Phytophthora infestans* to azoxystrobin fungicide. *Sustainability* **2020**, *12*, 1188. [CrossRef]
16. Ishii, H. Impact of fungicide resistance in plant pathogens on crop disease control and agricultural environment. *Jpn. Agric. Res. Q.* **2006**, *40*, 205–211. [CrossRef]
17. Kalia, A.; Manchanda, P.; Bhardwaj, S.; Singh, G. Biosynthesized silver nanoparticles from aqueous extracts of sweet lime fruit and callus tissues possess variable antioxidant and antimicrobial potentials. *Inorg. Nano-Metal Chem.* **2020**, *50*, 1053–1062. [CrossRef]
18. Kaur, G.; Kalia, A.; Sodhi, H.S. Size controlled, time-efficient biosynthesis of silver nanoparticles from *Pleurotus florida* using ultra-violet, visible range, and microwave radiations. *Inorg. Nano-Metal Chem.* **2020**, *50*, 35–41. [CrossRef]
19. Munir, M.U.; Ihsan, A.; Javed, I.; Ansari, M.T.; Bajwa, S.Z.; Bukhari, S.N.A.; Ahmed, A.; Malik, M.Z.; Khan, W.S. Controllably biodegradable hydroxyapatite nanostructures for cefazolin delivery against antibacterial resistance. *ACS Omega* **2019**, *4*, 7524–7532. [CrossRef]
20. Kalainila, P.; Ravindran, R.S.E.; Rohit, R.; Renganathan, S. Anti-bacterial effect of biosynthesized silver nanoparticles using *Kigelia africana*. *J. Nanosci. Nanoengn.* **2015**, *1*, 225–232.

21. Jangra, S.L.; Stalin, K.; Dilbaghi, N.; Kumar, S.; Tawale, J.; Singh, S.P.; Pasricha, R. Antimicrobial activity of zirconia (ZrO_2) nanoparticles and zirconium complexes. *J. Nanosci. Nanotechnol.* **2012**, *12*, 7105–7112. [CrossRef]
22. Satyavani, K.; Gurudeeban, S.; Ramanathan, T.; Balasubramanian, T. Biomedical potential of silver nanoparticles synthesized from calli cells of *Citrullus colocynthis* (L.) Schrad. *J. Nanobiotechnol.* **2011**, *9*, 43. [CrossRef]
23. Huang, F.; Long, Y.; Liang, Q.; Purushotham, B.; Swamy, M.K.; Duan, Y. Safed Musli (*Chlorophytum borivilianum* L.) Callus-Mediated Biosynthesis of Silver Nanoparticles and Evaluation of their Antimicrobial Activity and Cytotoxicity against Human Colon Cancer Cells. *J. Nanomater.* **2019**, 1–8. [CrossRef]
24. Azizi, S.; Mohamad, R.; Shahri, M.M.; McPhee, D.J. Green microwave-assisted combustion synthesis of zinc oxide nanoparticles with *Citrullus colocynthis* (L.) schrad: Characterization and biomedical applications. *Molecules* **2017**, *22*, 301. [CrossRef]
25. Sánchez-López, E.; Gomes, D.; Esteruelas, G.; Bonilla, L.; Lopez-Machado, A.L.; Galindo, R.; Cano, A.; Espina, M.; Ettcheto, M.; Camins, A.; et al. Metal-based nanoparticles as antimicrobial agents: An overview. *Nanomaterials* **2020**, *10*, 292. [CrossRef]
26. Singh, J.; Vishwakarma, K.; Ramawat, N.; Rai, P.; Singh, V.K.; Mishra, R.K.; Kumar, V.; Tripathi, D.K.; Sharma, S. Nanomaterials and microbes' interactions: A contemporary overview. *3 Biotech* **2019**, *9*, 68. [CrossRef]
27. Díez-Pascual, A.M. Antibacterial activity of nanomaterials. *Nanomaterials* **2018**, *8*, 359. [CrossRef]
28. Mostafa, M.; Almoammar, H.; Abd-Elsalam, K.A. Zinc-based nanostructures in plant protection applications. In *Nanobiotechnology Applications in Plant Protection, Nanotechnology in the Life Sciences*; Abd-Elsalam, K.A., Prasad, R., Eds.; Springer Nature Switzerland AG: Cham, Switzerland, 2019; pp. 49–83. ISBN 9783030132965.
29. Vollath, D. *Nanomaterials: An Introduction to Synthesis, Properties, and Applications*, 2nd ed.; Wiley-VCH, Verlag GmbH & Co. KGaA: Weinheim, Germany, 2013; ISBN 9780470927076.
30. Siegel, R. Nanostructured materials. In *Advanced Topics in Materials Science and Engineering*; Morán-López, J.L., Sanchez, J.M., Eds.; Springer: Boston, MA, USA; New York, NY, USA, 1993; pp. 273–288. ISBN 9788578110796.
31. Murr, L.E. Handbook of materials structures, properties, processing and performance. In *Handbook of Materials Structures, Properties, Processing and Performance*; Murr, L., Ed.; Springer International Publishing: Cham, Switzerland, 2015; pp. 719–746. ISBN 9783319018157.
32. Khan, I.; Saeed, K.; Khan, I. Nanoparticles: Properties, applications and toxicities. *Arab. J. Chem.* **2019**, *12*, 908–931. [CrossRef]
33. Mourdikoudis, S.; Pallares, R.M.; Thanh, N.T.K. Characterization techniques for nanoparticles: Comparison and complementarity upon studying nanoparticle properties. *Nanoscale* **2018**, *10*, 12871–12934. [CrossRef]
34. Roduner, E. Physics and Chemistry of Nanostructures: Why nano is different. *Encycl. Life Support Syst.* **2009**, 1–35.
35. Roduner, E. Size matters: Why nanomaterials are different. *Chem. Soc. Rev.* **2006**, *35*, 583–592. [CrossRef] [PubMed]
36. de Voorde, M.; Tulinski, M.; Jurczyk, M. Engineered nanomaterials: A discussion of the major categories of nanomaterials. In *Metrology and Standardization for Nanotechnology: Protocols and Industrial Innovations*; Mansfield, E., Kaiser, M., Fujita, D., de Voorde, M., Eds.; Wiley-VCH, Verlag GmbH & Co. KgaA: Weinheim, Germany, 2017; pp. 49–73.
37. Christian, P. Nanomaterials: Properties, Preparation and Applications. In *Environmental and Human Health Impacts of Nanotechnology*; Lead, J.R., Smith, E., Eds.; Wiley-Blackwell Publishing Ltd.: Chichester, UK, 2009; pp. 31–77. ISBN 978-1-405-17634-7.
38. Sun, C.Q. Size dependence of nanostructures: Impact of bond order deficiency. *Prog. Solid State Chem.* **2007**, *35*, 1–159. [CrossRef]
39. Andrievskii, R.A. Size-dependent effects in properties of nanostructured materials. *Rev. Adv. Mater. Sci.* **2009**, *21*, 107–133.
40. Kalia, A.; Kaur, J.; Kaur, A.; Singh, N. Antimycotic activity of biogenically synthesised metal and metal oxide nanoparticles against plant pathogenic fungus *Fusarium moniliforme* (*F. fujikuroi*). *Indian J. Exp. Biol.* **2020**, *58*, 263–270.

41. Khan, M.; Shaik, M.R.; Khan, S.T.; Adil, S.F.; Kuniyil, M.; Khan, M.; Al-Warthan, A.A.; Siddiqui, M.R.H.; Nawaz Tahir, M. Enhanced Antimicrobial Activity of Biofunctionalized Zirconia Nanoparticles. *ACS Omega* **2020**, *5*, 1987–1996. [CrossRef]
42. Van Der Wal, A.; Norde, W.; Zehnder, A.J.B.; Lyklema, J. Determination of the total charge in the cell walls of Gram-positive bacteria. *Colloids Surf. B Biointerfaces* **1997**, *9*, 81–100. [CrossRef]
43. Chen, M.; Zeng, G.; Xu, P.; Lai, C.; Tang, L. How Do Enzymes 'Meet' Nanoparticles and Nanomaterials? *Trends Biochem. Sci.* **2017**, *42*, 914–930. [CrossRef]
44. Kaur, M.; Kalia, A. Role of salt precursors for the synthesis of zinc oxide nanoparticles and in imparting variable antimicrobial activity. *J. Appl. Nat. Sci.* **2016**, *8*, 1039–1048. [CrossRef]
45. Cai, L.; Liu, C.; Fan, G.; Liu, C.; Sun, X. Preventing viral disease by ZnONPs through directly deactivating TMV and activating plant immunity in *Nicotiana benthamiana*. *Environ. Sci. Nano* **2019**, *6*, 3653–3669. [CrossRef]
46. Alum, A.; Rashid, A.; Mobasher, B.; Abbaszadegan, M. Cement-based biocide coatings for controlling algal growth in water distribution canals. *Cem. Concr. Compos.* **2008**, *30*, 839–847. [CrossRef]
47. Dizaj, S.M.; Lotfipour, F.; Barzegar-Jalali, M.; Zarrintan, M.H.; Adibkia, K. Antimicrobial activity of the metals and metal oxide nanoparticles. *Mater. Sci. Eng. C* **2014**, *44*, 278–284. [CrossRef]
48. Raghupathi, K.R.; Koodali, R.T.; Manna, A.C. Size-dependent bacterial growth inhibition and mechanism of antibacterial activity of zinc oxide nanoparticles. *Langmuir* **2011**, *27*, 4020–4028. [CrossRef]
49. El-Sayed, A.S.A.; Ali, D.M.I. Biosynthesis and comparative bactericidal activity of silver nanoparticles synthesized by *Aspergillus flavus* and *Penicillium crustosum* against the multidrug-resistant bacteria. *J. Microbiol. Biotechnol.* **2018**. [CrossRef]
50. Jamdagni, P.; Rana, J.S.; Khatri, P.; Nehra, K. Comparative account of antifungal activity of green and chemically synthesized Zinc Oxide nanoparticles in combination with agricultural fungicides. *Int. J. Nano Dimens.* **2018**, *9*, 198–208.
51. Auyeung, A.; Casillas-Santana, M.Á.; Martínez-Castañón, G.A.; Slavin, Y.N.; Zhao, W.; Asnis, J.; Häfeli, U.O.; Bach, H. Effective control of molds using a combination of nanoparticles. *PLoS ONE* **2017**, *12*, 1–13. [CrossRef] [PubMed]
52. Lamsal, K.; Kim, S.W.; Jung, J.H.; Kim, Y.S.; Kim, K.S.; Lee, Y.S. Inhibition effects of silver nanoparticles against powdery mildews on cucumber and pumpkin. *Mycobiology* **2011**, *39*, 26–32. [CrossRef] [PubMed]
53. Park, H.-J.; Kim, S.H.; Kim, H.J.; Choi, S.-H. A new composition of nanosized silica-silver for control of various plant diseases. *Plant Pathol. J.* **2006**, *22*, 295–302. [CrossRef]
54. Kim, H.; Kang, H.; Chu, G.; Byun, H. Antifungal effectiveness of nanosilver colloid against rose powdery mildew in greenhouses. *Solid State Phenom.* **2008**, *135*, 15–18. [CrossRef]
55. Li, J.; Sang, H.; Guo, H.; Popko, J.T.; He, L.; White, J.C.; Parkash Dhankher, O.; Jung, G.; Xing, B. Antifungal mechanisms of ZnO and Ag nanoparticles to *Sclerotinia homoeocarpa*. *Nanotechnology* **2017**, *28*, 155101. [CrossRef] [PubMed]
56. De La Rosa-García, S.C.; Martínez-Torres, P.; Gómez-Cornelio, S.; Corral-Aguado, M.A.; Quintana, P.; Gómez-Ortíz, N.M. Antifungal activity of ZnO and MgO nanomaterials and their mixtures against colletotrichum gloeosporioides strains from tropical fruit. *J. Nanomater.* **2018**, *2018*. [CrossRef]
57. Karimiyan, A.; Najafzadeh, H.; Ghorbanpour, M.; Hekmati-Moghaddam, S.H. Antifungal Effect of Magnesium Oxide, Zinc Oxide, Silicon Oxide and Copper Oxide Nanoparticles Against *Candida albicans*. *Zahedan J. Res. Med. Sci.* **2015**, *17*, 2–4. [CrossRef]
58. Roy, A.; Gauri, S.S.; Bhattacharya, M.; Bhattacharya, J. Antimicrobial activity of CaO nanoparticles. *J. Biomed. Nanotechnol.* **2013**, *9*, 1570–1578. [CrossRef]
59. Malandrakis, A.A.; Kavroulakis, N.; Chrysikopoulos, C.V. Use of copper, silver and zinc nanoparticles against foliar and soil-borne plant pathogens. *Sci. Total Environ.* **2019**, *670*, 292–299. [CrossRef]
60. Choudhary, M.A.; Manan, R.; Aslam Mirza, M.; Rashid Khan, H.; Qayyum, S.; Ahmed, Z. Biogenic Synthesis of Copper oxide and Zinc oxide Nanoparticles and their Application as Antifungal Agents. *Int. J. Mater. Sci. Eng.* **2018**, *4*, 1–6. [CrossRef]
61. Hao, Y.; Cao, X.; Ma, C.; Zhang, Z.; Zhao, N.; Ali, A.; Hou, T.; Xiang, Z.; Zhuang, J.; Wu, S.; et al. Potential Applications and Antifungal Activities of Engineered Nanomaterials against Gray Mold Disease Agent Botrytis cinerea on Rose Petals. *Front. Plant Sci.* **2017**, *8*, 1–9. [CrossRef]

62. Al-Dhabaan, F.A.; Shoala, T.; Ali, A.A.; Alaa, M.; Abd-Elsalam, K.; Abd-Elsalam, K. Chemically-produced copper, zinc nanoparticles and chitosan-bimetallic nanocomposites and their antifungal activity against three phytopathogenic fungi. *Int. J. Agric. Technol.* **2017**, *13*, 753–769.
63. Vahedi, M.; Hosseini-Jazani, N.; Yousefi, S.; Ghahremani, M. Evaluation of anti-bacterial effects of nickel nanoparticles on biofilm production by *Staphylococcus epidermidis*. *Iran. J. Microbiol.* **2017**, *9*, 160–168.
64. Srihasam, S.; Thyagarajan, K.; Korivi, M.; Lebaka, V.R.; Mallem, S.P.R. Phytogenic generation of NiO nanoparticles using stevia leaf extract and evaluation of their in-vitro antioxidant and antimicrobial properties. *Biomolecules* **2020**, *10*. [CrossRef] [PubMed]
65. Bogdan, J.; Zarzyńska, J.; Pławińska-Czarnak, J. Comparison of Infectious Agents Susceptibility to Photocatalytic Effects of Nanosized Titanium and Zinc Oxides: A Practical Approach. *Nanoscale Res. Lett.* **2015**, *10*. [CrossRef]
66. Yehia, R.; Ahmed, O.F. In vitro study of the antifungal efficacy of zinc oxide nanoparticles against *Fusarium oxysporum* and *Penicilium expansum*. *Afr. J. Microbiol. Res.* **2013**, *7*, 1917–1923. [CrossRef]
67. Elsharkawy, M.; Derbalah, A.; Hamza, A.; El-Shaer, A. Zinc oxide nanostructures as a control strategy of bacterial speck of tomato caused by *Pseudomonas syringae* in Egypt. *Environ. Sci. Pollut. Res.* **2018**, *27*, 19049–19057. [CrossRef] [PubMed]
68. Arciniegas-Grijalba, P.A.; Patiño-Portela, M.C.; Mosquera-Sánchez, L.P.; Guerrero-Vargas, J.A.; Rodríguez-Páez, J.E. ZnO nanoparticles (ZnO-NPs) and their antifungal activity against coffee fungus *Erythricium salmonicolor*. *Appl. Nanosci.* **2017**, *7*, 225–241. [CrossRef]
69. Rana, P.; Abdullah, M.; Hameed, H.Q.; Hasan, A.A. The effect of *Olea europea* L. leaves extract and ZrO_2 nanoparticles on *Acinetobacter baumannii*. *J. Pharm. Sci. Res.* **2019**, *11*, 2019.
70. Joshi, S.M.; De Britto, S.; Jogaiah, S.; Ito, S.I. Mycogenic selenium nanoparticles as potential new generation broad spectrum antifungal molecules. *Biomolecules* **2019**, *9*, 419. [CrossRef]
71. Srivastava, N.; Mukhopadhyay, M. Green synthesis and structural characterization of selenium nanoparticles and assessment of their antimicrobial property. *Bioprocess Biosyst. Eng.* **2015**, *38*. [CrossRef]
72. Khiralla, G.M.; El-Deeb, B.A. Antimicrobial and antibiofilm effects of selenium nanoparticles on some foodborne pathogens. *LWT Food Sci. Technol.* **2015**, *63*, 1001–1007. [CrossRef]
73. Shakibaie, M.; Forootanfar, H.; Golkari, Y.; Mohammadi-Khorsand, T.; Shakibaie, M.R. Anti-biofilm activity of biogenic selenium nanoparticles and selenium dioxide against clinical isolates of *Staphylococcus aureus, Pseudomonas aeruginosa*, and *Proteus mirabilis*. *J. Trace Elem. Med. Biol.* **2015**, *29*, 235–241. [CrossRef]
74. Abo Elsoud, M.M.; Al-Hagar, O.E.A.; Abdelkhalek, E.S.; Sidkey, N.M. Synthesis and investigations on tellurium myconanoparticles. *Biotechnol. Rep.* **2018**, *18*, e00247. [CrossRef] [PubMed]
75. Brown, C.D.; Cruz, D.M.; Roy, A.K.; Webster, T.J. Synthesis and characterization of PVP-coated tellurium nanorods and their antibacterial and anticancer properties. *J. Nanopart. Res.* **2018**, *20*, 254. [CrossRef]
76. Siddiqi, K.S.; ur Rahman, A.; Tajuddin; Husen, A. Properties of Zinc Oxide Nanoparticles and Their Activity Against Microbes. *Nanoscale Res. Lett.* **2018**, *13*. [CrossRef] [PubMed]
77. Jaffri, S.B.; Ahmad, K.S. Foliar-mediated Ag:ZnO nanophotocatalysts: Green synthesis, characterization, pollutants degradation, and in vitro biocidal activity. *Green Process. Synth.* **2019**, *8*, 172–182. [CrossRef]
78. Mohamed, M.A.; Abd-Elsalam, K.A. Nanoantimicrobials for Plant Pathogens Control: Potential Applications and Mechanistic Aspects. In *Nanobiotechnology Applications in Plant Protection-Nanotechnology in the Life Sciences*; Abd-Elsalam, K.A., Prasad, R., Eds.; Springer International Publishing AG: Cham, Switzerland, 2018; pp. 111–135. ISBN 978-3-319-91160-1.
79. Agarwal, H.; Menon, S.; Venkat Kumar, S.; Rajeshkumar, S. Mechanistic study on antibacterial action of zinc oxide nanoparticles synthesized using green route. *Chem. Biol. Interact.* **2018**, *286*, 60–70. [CrossRef]
80. Al-Shabib, N.A.; Husain, F.M.; Ahmed, F.; Khan, R.A.; Ahmad, I.; Alsharaeh, E.; Khan, M.S.; Hussain, A.; Rehman, M.T.; Yusuf, M.; et al. Biogenic synthesis of Zinc oxide nanostructures from Nigella sativa seed: Prospective role as food packaging material inhibiting broad-spectrum quorum sensing and biofilm. *Sci. Rep.* **2016**, *6*, 1–16. [CrossRef]
81. Ogunyemi, S.O.; Abdallah, Y.; Zhang, M.; Fouad, H.; Hong, X.; Ibrahim, E.; Masum, M.M.I.; Hossain, A.; Mo, J.; Li, B. Green synthesis of zinc oxide nanoparticles using different plant extracts and their antibacterial activity against *Xanthomonas oryzae* pv. oryzae. *Artif. Cells Nanomed. Biotechnol.* **2019**, *47*, 341–352. [CrossRef]
82. Galié, S.; García-Gutiérrez, C.; Miguélez, E.M.; Villar, C.J.; Lombó, F. Biofilms in the food industry: Health aspects and control methods. *Front. Microbiol.* **2018**, *9*, 1–18. [CrossRef]

83. Kim, S.W.; Kim, K.S.; Lamsal, K.; Kim, Y.J.; Kim, S.B.; Jung, M.; Sim, S.J.; Kim, H.S.; Chang, S.J.; Kim, J.K.; et al. An in vitro study of the antifungal effect of silver nanoparticles on oak wilt pathogen *Raffaelea* sp. *J. Microbiol. Biotechnol.* **2009**, *19*, 760–764. [CrossRef]
84. Lamsal, K.; Kim, S.W.; Jung, J.H.; Kim, Y.S.; Kim, K.S.; Lee, Y.S. Application of silver nanoparticles for the control of *Colletotrichum* species in vitro and pepper anthracnose disease in field. *Mycobiology* **2011**, *39*, 194–199. [CrossRef]
85. Elmer, W.; White, J.C. The Future of Nanotechnology in Plant Pathology. *Annu. Rev. Phytopathol.* **2018**, *56*, 111–133. [CrossRef] [PubMed]
86. Park, S.J.; Park, H.H.; Kim, S.Y.; Kim, S.J.; Woo, K.; Ko, G.P. Antiviral properties of silver nanoparticles on a magnetic hybrid colloid. *Appl. Environ. Microbiol.* **2014**, *80*, 2343–2350. [CrossRef] [PubMed]
87. Galdiero, S.; Falanga, A.; Vitiello, M.; Cantisani, M.; Marra, V.; Galdiero, M. Silver nanoparticles as potential antiviral agents. *Molecules* **2011**, *16*, 8894–8918. [CrossRef]
88. Zeedan, G.S.G.; Abd El-Razik, K.A.; Allam, A.M.; Abdalhamed, A.M.; Abou Zeina, H.A. Evaluations of potential antiviral effects of green zinc oxide and silver nanoparticles against bovine herpesvirus-1. *Adv. Anim. Vet. Sci.* **2020**, *8*, 433–443. [CrossRef]
89. Bekele, A.Z.; Gokulan, K.; Williams, K.M.; Khare, S. Dose and Size-Dependent Antiviral Effects of Silver Nanoparticles on Feline Calicivirus, a Human Norovirus Surrogate. *Foodborne Pathog. Dis.* **2016**, *13*, 239–244. [CrossRef] [PubMed]
90. Shionoiri, N.; Sato, T.; Fujimori, Y.; Nakayama, T.; Nemoto, M.; Matsunaga, T.; Tanaka, T. Investigation of the antiviral properties of copper iodide nanoparticles against feline calicivirus. *J. Biosci. Bioeng.* **2012**, *113*, 580–586. [CrossRef] [PubMed]
91. Kerry, R.G.; Malik, S.; Redda, Y.T.; Sahoo, S.; Patra, J.K.; Majhi, S. Nano-based approach to combat emerging viral (NIPAH virus) infection. *Nanomed. Nanotechnol. Biol. Med.* **2019**, *18*, 196–220. [CrossRef]
92. Ben Salem, A.N.; Zyed, R.; Lassoued, M.A.; Nidhal, S.; Sfar, S.; Mahjoub, A. Plant-derived nanoparticles enhance antiviral activity against coxsakievirus B3 by acting on virus particles and vero cells. *Dig. J. Nanomater. Biostructs* **2012**, *7*, 737–744.
93. Rai, M.; Deshmukh, S.D.; Ingle, A.P.; Gupta, I.R.; Galdiero, M.; Galdiero, S. Metal nanoparticles: The protective nanoshield against virus infection. *Crit. Rev. Microbiol.* **2016**, *42*, 46–56. [CrossRef] [PubMed]
94. Singh, L.; Kruger, H.G.; Maguire, G.E.M.; Govender, T.; Parboosing, R. The role of nanotechnology in the treatment of viral infections. *Ther. Adv. Infect. Dis.* **2017**, *4*, 105–131. [CrossRef] [PubMed]
95. Milovanovic, M.; Arsenijevic, A.; Milovanovic, J.; Kanjevac, T.; Arsenijevic, N. Nanoparticles in Antiviral Therapy. *Antimicrob. Nanoarchit. Synth. Appl.* **2017**, 383–410. [CrossRef]
96. Nikaeen, G.; Abbaszadeh, S.; Yousefinejad, S. Application of nanomaterials in treatment, anti-infection and detection of coronaviruses. *Nanomedicine* **2020**. [CrossRef]
97. Itani, R.; Tobaiqy, M.; Al Faraj, A. Optimizing use of theranostic nanoparticles as a life-saving strategy for treating COVID-19 patients. *Theranostics* **2020**, *10*, 5932–5942. [CrossRef]
98. Haggag, E.G.; Elshamy, A.M.; Rabeh, M.A.; Gabr, N.M.; Salem, M.; Youssif, K.A.; Samir, A.; Bin Muhsinah, A.; Alsayari, A.; Abdelmohsen, U.R. Antiviral potential of green synthesized silver nanoparticles of *Lampranthus coccineus* and *Malephora lutea*. *Int. J. Nanomed.* **2019**, *14*, 6217–6229. [CrossRef] [PubMed]
99. Meléndez-Villanueva, M.A.; Morán-Santibañez, K.; Martínez-Sanmiguel, J.J.; Rangel-López, R.; Garza-Navarro, M.A.; Rodríguez-Padilla, C.; Zarate-Triviño, D.G.; Trejo-Ávila, L.M. Virucidal activity of gold nanoparticles synthesized by green chemistry using garlic extract. *Viruses* **2019**, *11*, 1111. [CrossRef] [PubMed]
100. Kumar, R.; Sahoo, G.; Pandey, K.; Nayak, M.K.; Topno, R.; Rabidas, V.; Das, P. Virostatic potential of zinc oxide (ZnO) nanoparticles on capsid protein of cytoplasmic side of chikungunya virus. *Int. J. Infect. Dis.* **2018**, *73*, 368. [CrossRef]
101. Abdul, W.; Muhammad, A.; Atta Ullah, K.; Asmat, A.; Abdul, B. Role of nanotechnology in diagnosing and treating COVID-19 during the Pandemi. *Int. J. Clin. Virol.* **2020**, *4*, 65–70. [CrossRef]
102. El-Dougdoug, N.K.; Bondok, A.M.; El-Dougdoug, K.A. Evaluation of Silver Nanoparticles as Antiviral Agent Against ToMV and PVY in Tomato Plants. *Middle East J. Appl. Sci.* **2018**, *8*, 100–111.
103. Shafie, R.M.; Salama, A.M.; Farroh, K.Y. Silver nanoparticles activity against Tomato spotted wilt virus. *Middle East J. Appl. Sci.* **2018**, *7*, 1251–1267.

104. Elbeshehy, E.K.F.; Elazzazy, A.M.; Aggelis, G. Silver nanoparticles synthesis mediated by new isolates of *Bacillus* spp., nanoparticle characterization and their activity against Bean Yellow Mosaic Virus and human pathogens. *Front. Microbiol.* **2015**, *6*, 1–13. [CrossRef]
105. Elazzazy, A.M.; Elbeshehy, E.K.F.; Betiha, M.A. In vitro assessment of activity of graphene silver composite sheets against multidrug-resistant bacteria and Tomato Bushy Stunt Virus. *Trop. J. Pharm. Res.* **2017**, *16*, 2705–2711. [CrossRef]
106. Cai, L.; Cai, L.; Jia, H.; Liu, C.; Wang, D.; Sun, X. Foliar exposure of Fe3O4 nanoparticles on *Nicotiana benthamiana*: Evidence for nanoparticles uptake, plant growth promoter and defense response elicitor against plant virus. *J. Hazard. Mater.* **2020**, *393*, 122415. [CrossRef] [PubMed]
107. Hao, Y.; Yuan, W.; Ma, C.; White, J.C.; Zhang, Z.; Adeel, M.; Zhou, T.; Rui, Y.; Xing, B. Engineered nanomaterials suppress Turnip mosaic virus infection in tobacco (*Nicotiana benthamiana*). *Environ. Sci. Nano* **2018**, *5*, 1685–1693. [CrossRef]
108. Hanley, C.; Layne, J.; Punnoose, A.; Reddy, K.M.; Coombs, I.; Coombs, A.; Feris, K.; Wingett, D. Preferential killing of cancer cells and activated human T cells using ZnO nanoparticles. *Nanotechnology* **2008**, *19*, 1–7. [CrossRef] [PubMed]
109. Premanathan, M.; Karthikeyan, K.; Jeyasubramanian, K.; Manivannan, G. Selective toxicity of ZnO nanoparticles toward Gram-positive bacteria and cancer cells by apoptosis through lipid peroxidation. *Nanomed. Nanotechnol. Biol. Med.* **2011**, *7*, 184–192. [CrossRef] [PubMed]
110. Reddy, K.M.; Feris, K.; Bell, J.; Wingett, D.G.; Hanley, C.; Punnoose, A. Selective toxicity of zinc oxide nanoparticles to prokaryotic and eukaryotic systems. *Appl. Phys. Lett.* **2007**, *90*, 1–8. [CrossRef]
111. Sirelkhatim, A.; Mahmud, S.; Seeni, A.; Kaus, N.H.M.; Ann, L.C.; Bakhori, S.K.M.; Hasan, H.; Mohamad, D. Review on zinc oxide nanoparticles: Antibacterial activity and toxicity mechanism. *Nano-Micro Lett.* **2015**, *7*, 219–242. [CrossRef]
112. Tiwari, V.; Mishra, N.; Gadani, K.; Solanki, P.S.; Shah, N.A.; Tiwari, M. Mechanism of anti-bacterial activity of zinc oxide nanoparticle against Carbapenem-Resistant *Acinetobacter baumannii*. *Front. Microbiol.* **2018**, *9*, 1–10. [CrossRef]
113. Sun, T.; Hao, H.; Hao, W.T.; Yi, S.M.; Li, X.P.; Li, J.R. Preparation and antibacterial properties of titanium-doped ZnO from different zinc salts. *Nanoscale Res. Lett.* **2014**, *9*, 1–11. [CrossRef]
114. Jiang, J.; Pi, J.; Cai, J. The Advancing of Zinc Oxide Nanoparticles for Biomedical Applications. *Bioinorg. Chem. Appl.* **2018**, *2018*. [CrossRef]
115. Pinto, R.M.; Lopes-De-Campos, D.; Martins, M.C.L.; Van Dijck, P.; Nunes, C.; Reis, S. Impact of nanosystems in *Staphylococcus aureus* biofilms treatment. *FEMS Microbiol. Rev.* **2019**, *43*, 622–641. [CrossRef] [PubMed]
116. Graham, J.H.; Johnson, E.G.; Myers, M.E.; Young, M.; Rajasekaran, P.; Das, S.; Santra, S. Potential of Nano-Formulated Zinc Oxide for Control of Citrus Canker on Grapefruit Trees. *Plant Dis.* **2016**, *100*, 2442–2447. [CrossRef] [PubMed]
117. Carvalho, R.; Duman, K.; Jones, J.B.; Paret, M.L. Bactericidal Activity of Copper-Zinc Hybrid Nanoparticles on Copper-Tolerant *Xanthomonas perforans*. *Sci. Rep.* **2019**, *9*, 1–9. [CrossRef] [PubMed]
118. Siddiqui, Z.A.; Khan, A.; Khan, M.R.; Abd-Allah, E.F. Effects of zinc oxide nanoparticles (ZnO NPs) and some plant pathogens on the growth and nodulation of lentil (*Lens culinaris* medik.). *Acta Phytopathol. Entomol. Hungarica* **2018**, *53*, 195–212. [CrossRef]
119. Khan, M.; Siddiqui, Z.A. Zinc oxide nanoparticles for the management of *Ralstonia solanacearum*, *Phomopsis vexans* and *Meloidogyne incognita* incited disease complex of eggplant. *Indian Phytopathol.* **2018**, *71*, 355–364. [CrossRef]
120. Alves, M.M.; Bouchami, O.; Tavares, A.; Córdoba, L.; Santos, C.F.; Miragaia, M.; De Fátima Montemor, M. New Insights into Antibiofilm Effect of a Nanosized ZnO Coating against the Pathogenic Methicillin Resistant *Staphylococcus aureus*. *ACS Appl. Mater. Interfaces* **2017**, *9*, 28157–28167. [CrossRef]
121. Fontecha-Umaña, F.; Ríos-Castillo, A.G.; Ripolles-Avila, C.; Rodríguez-Jerez, J.J. Antimicrobial activity and prevention of bacterial biofilm formation of silver and zinc oxide nanoparticle-containing polyester surfaces at various concentrations for use. *Foods* **2020**, *9*, 442. [CrossRef]
122. Jindal, S.; Anand, S.; Huang, K.; Goddard, J.; Metzger, L.; Amamcharla, J. Evaluation of modified stainless steel surfaces targeted to reduce biofilm formation by common milk spore formers. *J. Dairy Sci.* **2016**, *99*, 9502–9513. [CrossRef]

123. Espitia, P.J.P.; Otoni, C.G.; Soares, N.F.F. *Zinc Oxide Nanoparticles for Food Packaging Applications*; Elsevier Inc.: Amsterdam, The Netherlands, 2016; ISBN 9780128007235.
124. Kaur, M.; Kalia, A.; Thakur, A. Effect of biodegradable chitosan–rice-starch nanocomposite films on post-harvest quality of stored peach fruit. *Starch/Staerke* **2017**, *69*, 1–12. [CrossRef]
125. Naskar, A.; Khan, H.; Sarkar, R.; Kumar, S.; Halder, D.; Jana, S. Anti-biofilm activity and food packaging application of room temperature solution process-based polyethylene glycol capped Ag-ZnO-graphene nanocomposite. *Mater. Sci. Eng. C* **2018**, *91*, 743–753. [CrossRef] [PubMed]
126. Gundersen, D.E.; Lee, I.M.; Rehner, S.A.; Davis, R.E.; Kingsbury, D.T. Phylogeny of mycoplasmalike organisms (phytoplasmas): A basis for their classification. *J. Bacteriol.* **1994**, *176*, 5244–5254. [CrossRef] [PubMed]
127. Bové, J.M.; Garnier, M. Walled and wall-less eubacteria from plants: Sieve-tube-restricted plant pathogens. *Plant Cell. Tissue Organ Cult.* **1998**, *52*, 7–16. [CrossRef]
128. Lee, I.; Davis, R.E.; Dawn, E. Phytoplasma: Phytopathogenic Mollicutes. *Annu. Rev. Microbiol.* **2000**, *54*, 221–255. [CrossRef] [PubMed]
129. Jurga, M.; Zwolińska, A. Phytoplasmas in Poaceae species: A threat to the most important cereal crops in Europe. *J. Plant Pathol.* **2020**, *102*, 287–297. [CrossRef]
130. Rao, G.P.; Madhupriya; Thorat, V.; Manimekalai, R.; Tiwari, A.K.; Yadav, A. A century progress of research on phytoplasma diseases in India. *Phytopathog. Mollicutes* **2017**, *7*, 1. [CrossRef]
131. Kumari, S.; Nagendran, K.; Rai, A.B.; Singh, B.; Rao, G.P.; Bertaccini, A. Global status of phytoplasma diseases in vegetable crops. *Front. Microbiol.* **2019**, *10*, 1–15. [CrossRef]
132. Namba, S. Molecular and biological properties of phytoplasmas. *Proc. Jpn. Acad. Ser. B Phys. Biol. Sci.* **2019**, *95*, 401–418. [CrossRef]
133. Fletcher, J.; Wayadande, A.; Melcher, U.; Ye, F. The phytopathogenic mollicute-insect vector interface: A closer look. *Phytopathology* **1998**, *88*, 1351–1358. [CrossRef]
134. Bendix, C.; Lewis, J.D. The enemy within: Phloem-limited pathogens. *Mol. Plant Pathol.* **2018**, *19*, 238–254. [CrossRef]
135. Cagliari, D.; Dias, N.P.; Galdeano, D.M.; dos Santos, E.Á.; Smagghe, G.; Zotti, M.J. Management of pest insects and plant diseases by non-transformative RNAi. *Front. Plant Sci.* **2019**, *10*. [CrossRef]
136. Liu, S.; Jaouannet, M.; Dempsey, D.A.; Imani, J.; Coustau, C.; Kogel, K.H. RNA-based technologies for insect control in plant production. *Biotechnol. Adv.* **2020**, *39*, 107463. [CrossRef] [PubMed]
137. Yang, C.; Powell, C.A.; Duan, Y.; Shatters, R.; Zhang, M. Antimicrobial nanoemulsion formulation with improved penetration of foliar spray through citrus leaf cuticles to control citrus huanglongbing. *PLoS ONE* **2015**, *10*, 1–14. [CrossRef] [PubMed]
138. Yang, C.; Zhong, Y.; Powell, C.A.; Doud, M.S.; Duan, Y.; Huang, Y.; Zhang, M. Antimicrobial Compounds Effective against Candidatus Liberibacter asiaticus Discovered via Graft-based Assay in Citrus. *Sci. Rep.* **2018**, *8*, 1–11. [CrossRef]
139. Gabiel, D.W.; Zhang, S. Use of Aldehydes Formulated with Nanoparticles and/or Nanoemulsions to Enhance Disease Resistance of Plants to Liberibacters. US Patent (US20170006863), 12 January 2017.
140. Ghosh, D.K.; Kokane, S.; Kumar, P.; Ozcan, A.; Warghane, A.; Motghare, M.; Santra, S.; Sharma, A.K. Antimicrobial nano-zinc oxide-2S albumin protein formulation significantly inhibits growth of *"Candidatus Liberibacter asiaticus" in planta*. *PLoS ONE* **2018**, *13*, 1–20. [CrossRef]
141. Wagner, G.; Korenkov, V.; Judy, J.D.; Bertsch, P.M. Nanoparticles composed of Zn and ZnO inhibit *Peronospora tabacina* spore germination in vitro and *P. tabacina* infectivity on tobacco leaves. *Nanomaterials* **2016**, *6*, 50. [CrossRef]
142. Zabrieski, Z.; Morrell, E.; Hortin, J.; Dimkpa, C.; McLean, J.; Britt, D.; Anderson, A. Pesticidal activity of metal oxide nanoparticles on plant pathogenic isolates of *Pythium*. *Ecotoxicology* **2015**, *24*, 1305–1314. [CrossRef] [PubMed]
143. Patra, P.; Mitra, S.; Debnath, N.; Goswami, A. Biochemical-, biophysical-, and microarray-based antifungal evaluation of the buffer-mediated synthesized nano zinc oxide: An in vivo and in vitro toxicity study. *Langmuir* **2012**, *28*, 16966–16978. [CrossRef] [PubMed]
144. He, L.; Liu, Y.; Mustapha, A.; Lin, M. Antifungal activity of zinc oxide nanoparticles against *Botrytis cinerea* and *Penicillium expansum*. *Microbiol. Res.* **2011**, *166*, 207–215. [CrossRef] [PubMed]

145. Hassan, M.; Zayton, M.A.; El-feky, S.A. Role of green synthesized ZnO nanoparticles as antifungal against post-harvest gray and black mold of sweet bell. *J. Biotechnol. Bioeng.* **2019**, *3*, 8–15.
146. Dimkpa, C.O.; McLean, J.E.; Britt, D.W.; Anderson, A.J. Antifungal activity of ZnO nanoparticles and their interactive effect with a biocontrol bacterium on growth antagonism of the plant pathogen *Fusarium graminearum*. *BioMetals* **2013**, *26*, 913–924. [CrossRef]
147. El-argawy, E.; Rahhal, M.M.H.; Elshabrawy, E.M.; Eltahan, R.M. Efficacy of some nanoparticles to control damping-off and root rot of sugar beet in El-Behiera Governorate. *Asian J. Plant Pathol.* **2016**, *11*, 35–47. [CrossRef]
148. Sardella, D.; Gatt, R.; Valdramidis, V.P. Assessing the efficacy of zinc oxide nanoparticles against *Penicillium expansum* by automated turbidimetric analysis. *Mycology* **2018**, *9*, 43–48. [CrossRef] [PubMed]
149. Shoeb, M.; Singh, B.R.; Khan, J.A.; Khan, W.; Singh, B.N.; Singh, H.B.; Naqvi, A.H. ROS-dependent anticandidal activity of zinc oxide nanoparticles synthesized by using egg albumen as a biotemplate. *Adv. Nat. Sci. Nanosci. Nanotechnol.* **2013**, *4*. [CrossRef]
150. Horky, P.; Skalickova, S.; Baholet, D.; Skladanka, J. Nanoparticles as a solution for eliminating the risk of mycotoxins. *Nanomaterials* **2018**, *8*, 727. [CrossRef]
151. Gacem, M.A.; Gacem, H.; Telli, A.; Ould El Hadj Khelil, A. Mycotoxins: Decontamination and nanocontrol methods. In *Nanomycotoxicology: Treating Mycotoxins in the Nano Way*; Rai, M., Abd-Elsalam, K.A., Eds.; Elsevier Inc.: Amsterdam, The Netherlands, 2020; pp. 189–216. ISBN 9780128179987.
152. Tsang, C.C.; Tang, J.Y.M.; Lau, S.K.P.; Woo, P.C.Y. Taxonomy and evolution of Aspergillus, Penicillium and Talaromyces in the omics era—Past, present and future. *Comput. Struct. Biotechnol. J.* **2018**, *16*, 197–210. [CrossRef] [PubMed]
153. El-banna, A.A.; Pitt, J.I.; Leistner, L. Production of mycotoxins by *Penicillium* species. *Syst. Appl. Microbiol.* **1987**, *10*, 42–46. [CrossRef]
154. Frisvad, J.C. A critical review of producers of small lactone mycotoxins: Patulin, penicillic acid and moniliformin. *World Mycotoxin J.* **2018**, *11*, 73–100. [CrossRef]
155. Jimenez-Garcia, S.N.; Garcia-Mier, L.; Garcia-Trejo, J.F.; Ramirez-Gomez, X.S.; Guevara-Gonzalez, R.G.; Feregrino-Perez, A.A. Fusarium mycotoxins and metabolites that modulate their production. In *Fusarium—Plant Diseases, Pathogen Diversity, Genetic Diversity, Resistance and Molecular Markers*; InTech: London, UK, 2018. [CrossRef]
156. Hulvová, H.; Galuszka, P.; Frébortová, J.; Frébort, I. Parasitic fungus *Claviceps* as a source for biotechnological production of ergot alkaloids. *Biotechnol. Adv.* **2013**, *31*, 79–89. [CrossRef]
157. Bennett, J.W.; Klich, M. Mycotoxins. *Clin. Microbiol. Rev.* **2003**, *16*, 497–516. [CrossRef]
158. Schardl, C.L. Introduction to the toxins special issue on ergot alkaloids. *Toxins* **2015**, *7*, 4232–4237. [CrossRef]
159. Ostry, V. *Alternaria* mycotoxins: An overview of chemical characterization, producers, toxicity, analysis and occurrence in foodstuffs. *World Mycotoxin J.* **2008**, *1*, 175–188. [CrossRef]
160. Jesmin, R.; Chanda, A. Restricting mycotoxins without killing the producers: A new paradigm in nano-fungal interactions. *Appl. Microbiol. Biotechnol.* **2020**, *104*, 2803–2813. [CrossRef] [PubMed]
161. Kaur, A.; Saini, S.S. Nanoadsorbents for the preconcentration of some toxic substances: A minireview. *Int. Lett. Chem. Phys. Astron.* **2013**, *21*, 22–35. [CrossRef]
162. Zahoor, M.; Ali Khan, F. Adsorption of aflatoxin B1 on magnetic carbon nanocomposites prepared from bagasse. *Arab. J. Chem.* **2018**, *11*, 729–738. [CrossRef]
163. Moghaddam, S.H.M.; Jebali, A.; Daliri, K. The use of MgO-SiO_2 nanocomposite for adsorption of aflatoxin in wheat flour samples. In Proceedings of the NanoCon 2010, Olomouc, Czech Republic, 12–14 October 2010; pp. 10–15.
164. Daković, A.; Tomašević-Čanović, M.; Dondur, V.; Rottinghaus, G.E.; Medaković, V.; Zarić, S. Adsorption of mycotoxins by organozeolites. *Colloids Surf. B Biointerfaces* **2005**, *46*, 20–25. [CrossRef]
165. Kovač, T.; Borišev, I.; Crevar, B.; Čačić Kenjerić, F.; Kovač, M.; Strelec, I.; Ezekiel, C.N.; Sulyok, M.; Krska, R.; Šarkanj, B. Fullerol C60(OH)24 nanoparticles modulate aflatoxin B1 biosynthesis in *Aspergillus flavus*. *Sci. Rep.* **2018**, *8*, 1–8. [CrossRef]
166. Asghar, M.A.; Zahir, E.; Shahid, S.M.; Khan, M.N.; Asghar, M.A.; Iqbal, J.; Walker, G. Iron, copper and silver nanoparticles: Green synthesis using green and black tea leaves extracts and evaluation of antibacterial, antifungal and aflatoxin B1 adsorption activity. *LWT Food Sci. Technol.* **2018**, *90*, 98–107. [CrossRef]

167. Hernández-Meléndez, D.; Salas-Téllez, E.; Zavala-Franco, A.; Téllez, G.; Méndez-Albores, A.; Vázquez-Durán, A. Inhibitory effect of flower-shaped zinc oxide nanostructures on the growth and aflatoxin production of a highly toxigenic strain of *Aspergillus flavus* Link. *Materials* **2018**, *11*, 1265. [CrossRef]
168. Mitra, C.; Gummadidala, P.M.; Merrifield, R.; Omebeyinje, M.H.; Jesmin, R.; Lead, J.R.; Chanda, A. Size and coating of engineered silver nanoparticles determine their ability to growth-independently inhibit aflatoxin biosynthesis in *Aspergillus parasiticus*. *Appl. Microbiol. Biotechnol.* **2019**, *103*, 4623–4632. [CrossRef]
169. Mitra, C.; Gummadidala, P.M.; Afshinnia, K.; Merrifield, R.C.; Baalousha, M.; Lead, J.R.; Chanda, A. Citrate-Coated Silver Nanoparticles Growth-Independently Inhibit Aflatoxin Synthesis in *Aspergillus parasiticus*. *Environ. Sci. Technol.* **2017**, *51*, 8085–8093. [CrossRef]
170. Lakshmeesha, T.R.; Kalagatur, N.K.; Mudili, V.; Mohan, C.D.; Rangappa, S.; Prasad, B.D.; Ashwini, B.S.; Hashem, A.; Alqarawi, A.A.; Malik, J.A.; et al. Biofabrication of zinc oxide nanoparticles with *Syzygium aromaticum* flower buds extract and finding its novel application in controlling the growth and mycotoxins of *Fusarium graminearum*. *Front. Microbiol.* **2019**, *10*, 1–13. [CrossRef] [PubMed]
171. Mohd Yusof, H.; Mohamad, R.; Zaidan, U.H.; Abdul Rahman, N.A. Microbial synthesis of zinc oxide nanoparticles and their potential application as an antimicrobial agent and a feed supplement in animal industry: A review. *J. Anim. Sci. Biotechnol.* **2019**, *10*, 1–22. [CrossRef] [PubMed]
172. Jabar, A.K.; Aldhahi, H.H.K.; Salim, H.A. Effect of manufactured iron oxides in control of tomato yellow leaf curl virus (TYLCV). *Plant Arch.* **2020**, *20*, 2131–2134.
173. Savi, G.D.; Bortoluzzi, A.J.; Scussel, V.M. Antifungal properties of Zinc-compounds against toxigenic fungi and mycotoxin. *Int. J. Food Sci. Technol.* **2013**, *48*, 1834–1840. [CrossRef]
174. Anderson, A.J.; McLean, J.E.; Jacobson, A.R.; Britt, D.W. CuO and ZnO nanoparticles modify interkingdom cell signaling processes relevant to crop production. *J. Agric. Food Chem.* **2018**, *66*, 6513–6524. [CrossRef]
175. Liang, Y.; Duan, Y.; Fan, C.; Dong, H.; Yang, J.; Tang, J.; Tang, G.; Wang, W.; Jiang, N.; Cao, Y. Preparation of kasugamycin conjugation based on ZnO quantum dots for improving its effective utilization. *Chem. Eng. J.* **2019**, *361*, 671–679. [CrossRef]
176. Young, M.; Ozcan, A.; Myers, M.E.; Johnson, E.G.; Graham, J.H.; Santra, S. Multimodal generally recognized as safe ZnO/Nanocopper composite: A novel antimicrobial material for the management of citrus phytopathogens. *J. Agric. Food Chem.* **2018**, *66*, 6604–6608. [CrossRef]
177. Lahuf, A.A.; Kareem, A.A.; Al-Sweedi, T.M.; Alfarttoosi, H.A. Evaluation the potential of indigenous biocontrol agent *Trichoderma harzianum* and its interactive effect with nanosized ZnO particles against the sunflower damping-off pathogen, *Rhizoctonia solani*. In *IOP Conference Series: Earth and Environmental Science*; IOP Publishing: Bristol, UK, 2019; Volume 365. [CrossRef]
178. Li, Z.; Yu, T.; Paul, R.; Fan, J.; Yang, Y.; Wei, Q. Agricultural nanodiagnostics for plant diseases: Recent advances and challenges. *Nanoscale Adv.* **2020**. [CrossRef]
179. Khiyami, M.A.; Almoammar, H.; Awad, Y.M.; Alghuthaymi, M.A. Plant pathogen nanodiagnostic techniques: Forthcoming changes? *Biotechnol. Biotechnol. Equip.* **2014**, *28*, 775–785. [CrossRef]
180. Giraldo, J.P.; Wu, H.; Newkirk, G.M.; Kruss, S. Nanobiotechnology approaches for engineering smart plant sensors. *Nat. Nanotechnol.* **2019**, *14*, 541–553. [CrossRef]
181. Kumar, V.; Arora, K. Trends in nano-inspired biosensors for plants. *Mater. Sci. Energy Technol.* **2020**, *3*, 255–273. [CrossRef]
182. Ansari, A.A.; Kaushik, A.; Solanki, P.R.; Malhotra, B.D. Nanostructured zinc oxide platform for mycotoxin detection. *Bioelectrochemistry* **2010**, *77*, 75–81. [CrossRef] [PubMed]
183. Martynenko, I.V.; Litvin, A.P.; Purcell-Milton, F.; Baranov, A.V.; Fedorov, A.V.; Gun'Ko, Y.K. Application of semiconductor quantum dots in bioimaging and biosensing. *J. Mater. Chem. B* **2017**, *5*, 6701–6727. [CrossRef]
184. Willard, D.M.; Carillo, L.L.; Jung, J.; van Orden, A. CdSe-ZnS Quantum Dots as Resonance Energy Transfer Donors in a Model Protein-Protein Binding Assay. *Nano Lett.* **2001**, *1*, 469–474. [CrossRef]
185. Chen, F.; Gerion, D. Fluorescent CdSe/ZnS nanocrystal-peptide conjugates for long-term, nontoxic imaging and nuclear targeting in living cells. *Nano Lett.* **2004**, *4*, 1827–1832. [CrossRef]
186. Hong, S.; Lee, C. The current status and future outlook of quantum dot-based biosensors for plant virus detection. *Plant Pathol. J.* **2018**, *34*, 85–92. [CrossRef]
187. Medintz, I.L.; Sapsford, K.E.; Konnert, J.H.; Chatterji, A.; Lin, T.; Johnson, J.E.; Mattoussi, H. Decoration of discretely immobilized cowpea mosaic virus with luminescent quantum dots. *Langmuir* **2005**, *21*, 5501–5510. [CrossRef]

188. Moulick, A.; Milosavljevic, V.; Vlachova, J.; Podgajny, R.; Hynek, D.; Kopel, P.; Adam, V. Using CdTe/ZnSe core/shell quantum dots to detect DNA and damage to DNA. *Int. J. Nanomed.* **2017**, *12*, 1277–1291. [CrossRef]
189. Rispail, N.; De Matteis, L.; Santos, R.; Miguel, A.S.; Custardoy, L.; Testillano, P.S.; Risueño, M.C.; Pérez-De-Luque, A.; Maycock, C.; Fevereiro, P.; et al. Quantum dot and superparamagnetic nanoparticle interaction with pathogenic fungi: Internalization and toxicity profile. *ACS Appl. Mater. Interfaces* **2014**, *6*, 9100–9110. [CrossRef]
190. Tereshchenko, A.; Fedorenko, V.; Smyntyna, V.; Konup, I.; Konup, A.; Eriksson, M.; Yakimova, R.; Ramanavicius, A.; Balme, S.; Bechelany, M. ZnO films formed by atomic layer deposition as an optical biosensor platform for the detection of Grapevine virus A-type proteins. *Biosens. Bioelectron.* **2017**, *92*, 763–769. [CrossRef]
191. Al-Fandi, M.G.; Alshraiedeh, N.H.; Oweis, R.J.; Hayajneh, R.H.; Alhamdan, I.R.; Alabed, R.A.; Al-Rawi, O.F. Direct electrochemical bacterial sensor using ZnO nanorods disposable electrode. *Sens. Rev.* **2018**, *38*, 326–334. [CrossRef]
192. Tahir, M.A.; Hameed, S.; Munawar, A.; Amin, I.; Mansoor, S.; Khan, W.S.; Bajwa, S.Z. Investigating the potential of multiwalled carbon nanotubes based zinc nanocomposite as a recognition interface towards plant pathogen detection. *J. Virol. Methods* **2017**, *249*, 130–136. [CrossRef] [PubMed]
193. Siddiquee, S.; Rovina, K.; Yusof, N.A.; Rodrigues, K.F.; Suryani, S. Nanoparticle-enhanced electrochemical biosensor with DNA immobilization and hybridization of *Trichoderma harzianum* gene. *Sens. Bio-Sens. Res.* **2014**, *2*, 16–22. [CrossRef]
194. Kalia, A.; Kaur, H. *Agri-Applications of Nano-Scale Micronutrients: Prospects for Plant Growth Promotion*; Raliya, R., Ed.; CRC Press: Boca Raton, FL, USA, 2019; ISBN 9781315123950.
195. Bala, R.; Kalia, A.; Dhaliwal, S.S. Evaluation of Efficacy of ZnO Nanoparticles as Remedial Zinc Nanofertilizer for Rice. *J. Soil Sci. Plant Nutr.* **2019**, *19*, 379–389. [CrossRef]
196. Kalia, A.; Sharma, S.P.; Kaur, H.; Kaur, H. Novel nanocomposite-based controlled-release fertilizer and pesticide formulations: Prospects and challenges. In *Multifunctional Hybrid Nanomaterials for Sustainable Agri-food and Ecosystem*; Abd-Elsalam, K.A., Ed.; Elsevier Inc.: Amsterdam, The Netherlands, 2020; pp. 99–134.
197. Kalia, A.; Sharma, S.P.; Kaur, H. Nanoscale Fertilizers: Harnessing Boons for Enhanced Nutrient Use Efficiency and Crop Productivity. In *Nanotechnology in the Life Sciences*; Abd-Elsalam, K.A., Prasad, R., Eds.; Springer International Publishing: Cham, Switzerland, 2019; ISBN 978-3-030-13295-8.
198. Kairyte, K.; Kadys, A.; Luksiene, Z. Antibacterial and antifungal activity of photoactivated ZnO nanoparticles in suspension. *J. Photochem. Photobiol. B Biol.* **2013**, *128*, 78–84. [CrossRef] [PubMed]
199. Sun, Q.; Li, J.; Le, T. Zinc Oxide Nanoparticle as a Novel Class of Antifungal Agents: Current Advances and Future Perspectives. *J. Agric. Food Chem.* **2018**, *66*, 11209–11220. [CrossRef] [PubMed]
200. Jameel, M.; Shoeb, M.; Khan, M.T.; Ullah, R.; Mobin, M.; Farooqi, M.K.; Adnan, S.M. Enhanced Insecticidal Activity of Thiamethoxam by Zinc Oxide Nanoparticles: A Novel Nanotechnology Approach for Pest Control. *ACS Omega* **2019**. [CrossRef]
201. Medina-Pérez, G.; Fernández-Luqueño, F.; Vazquez-Nuñez, E.; López-Valdez, F.; Prieto-Mendez, J.; Madariaga-Navarrete, A.; Miranda-Arámbula, M. Remediating Polluted Soils Using Nanotechnologies: Environmental Benefits and Risks. *Polish J. Environ. Stud.* **2019**, *28*, 1013–1030. [CrossRef]
202. Das, S.; Chakraborty, J.; Chatterjee, S.; Kumar, H. Prospects of biosynthesized nanomaterials for the remediation of organic and inorganic environmental contaminants. *Environ. Sci. Nano* **2018**, *5*, 2784–2808. [CrossRef]
203. Guerra, F.; Attia, M.; Whitehead, D.; Alexis, F. Nanotechnology for Environmental Remediation: Materials and Applications. *Molecules* **2018**, *23*, 1760. [CrossRef]
204. Fu, L.; Wang, Z.; Dhankher, O.P.; Xing, B. Nanotechnology as a new sustainable approach for controlling crop diseases and increasing agricultural production. *J. Exp. Bot.* **2020**, *71*, 507–519. [PubMed]
205. Kookana, R.S.; Boxall, A.B.A.; Reeves, P.T.; Ashauer, R.; Beulke, S.; Chaudhry, Q.; Cornelis, G.; Fernandes, T.F.; Gan, J.; Kah, M.; et al. Nanopesticides: Guiding principles for regulatory evaluation of environmental risks. *J. Agric. Food Chem.* **2014**, *62*, 4227–4240. [CrossRef] [PubMed]
206. Walker, G.W.; Kookana, R.S.; Smith, N.E.; Kah, M.; Doolette, C.L.; Reeves, P.T.; Lovell, W.; Anderson, D.J.; Turney, T.W.; Navarro, D.A. Ecological risk assessment of nano-enabled pesticides: A perspective on problem formulation. *J. Agric. Food Chem.* **2017**. [CrossRef] [PubMed]

207. Kah, M.; Tufenkji, N.; White, J.C. Nano-enabled strategies to enhance crop nutrition and protection. *Nat. Nanotechnol.* **2019**, *14*, 532–540. [CrossRef]
208. Beegam, A.; Prasad, P.; Jose, J.; Oliveira, M.; Costa, F.G.; Soares, A.M.V.M.; Gonçalves, P.P.; Trindade, T.; Kalarikkal, N.; Thomas, S.; et al. Environmental Fate of Zinc Oxide Nanoparticles: Risks and Benefits. In *Toxicology-New Aspects to This Scientific Conundrum*; Soloneski, S., Larramendy, M.L., Eds.; IntechOpen: London, UK, 2016. [CrossRef]
209. Charitidis, C.A.; Georgiou, P.; Koklioti, M.A.; Trompeta, A.F.; Markakis, V. Manufacturing nanomaterials: From research to industry. *Manuf. Rev.* **2014**, *1*, 11. [CrossRef]
210. Kalia, A.; Singh, S. Myco-decontamination of azo dyes: Nano-augmentation technologies. *3 Biotech* **2020**, *10*, 384. [CrossRef]
211. Kaur, P.; Taggar, M.S.; Kalia, A. Characterization of magnetic nanoparticle–immobilized cellulases for enzymatic saccharification of rice straw. *Biomass Convers. Biorefinery* **2020**. [CrossRef]
212. Huang, Y.; Ding, L.; Li, C.; Wu, M.; Wang, M.; Yao, C.; Yin, X.; Zhang, J.; Liu, J.; Zhang, Y.; et al. Safety Issue of Changed Nanotoxicity of Zinc Oxide Nanoparticles in the Multicomponent System. *Part. Part. Syst. Charact.* **2019**, *36*, 1–14. [CrossRef]
213. Nel, A.; Grainger, D.; Alvarez, P.J.; Badesha, S.; Castranova, V.; Ferrari, M.; Godwin, H.; Grodzinski, P.; Morris, J.; Savage, N.; et al. Nanotechnology Environmental, Health, and Safety Issues. In *Nanotechnology Research Directions for Societal Needs in 2020*; Roco, M.C., Hersam, M.C., Mirkin, C.A., Eds.; Springer: Dordrecht, The Netherlands, 2011; pp. 159–220. ISBN 9789400711686.
214. Ali, A.; Phull, A.R.; Zia, M. Elemental zinc to zinc nanoparticles: Is ZnO NPs crucial for life? Synthesis, toxicological, and environmental concerns. *Nanotechnol. Rev.* **2018**, *7*, 413–441. [CrossRef]
215. Paul, S.K.; Dutta, H.; Sarkar, S.; Sethi, L.N.; Ghosh, S.K. Nanosized Zinc Oxide: Super-functionalities, present scenario of application, safety issues, and future prospects in food processing and allied industries. *Food Rev. Int.* **2019**, *35*, 505–535. [CrossRef]
216. Naveed Ul Haq, A.; Nadhman, A.; Ullah, I.; Mustafa, G.; Yasinzai, M.; Khan, I. Synthesis Approaches of Zinc Oxide Nanoparticles: The Dilemma of Ecotoxicity. *J. Nanomater.* **2017**, *2017*, 1–14. [CrossRef]
217. Haque, J.; Bellah, M.; Hassan, R.; Rahman, S. Synthesis of ZnO nanoparticles by two different methods and comparison of their structural, antibacterial, photocatalytic and optical properties. *Nano Express* **2020**, *1*, 010007. [CrossRef]
218. Marsalek, R. Particle size and Zeta Potential of ZnO. *Procedia-Soc. Behav. Sci.* **2014**, *9*, 13–17. [CrossRef]
219. Chai, M.H.H.; Amir, N.; Yahya, N.; Saaid, I.M. Characterization and Colloidal Stability of Surface Modified Zinc Oxide Nanoparticle Characterization and Colloidal Stability of Surface Modified Zinc Oxide Nanoparticle. *IOP Conf. Ser. J. Phys. Conf. Ser.* **2018**, 1123. [CrossRef]
220. Zhulina, E.B.; Borisov, O.V.; Priamitsyn, V.A. Theory of steric stabilization of colloid dispersions by grafted polymers. *J. Colloid Interface Sci.* **1990**, *137*, 495–511. [CrossRef]
221. Fiedot, M.; Rac, O.; Suchorska-Woźniak, P.; Karbownik, I.; Teterycz, H. Polymer-surfactant interactions and their influence on zinc oxide nanoparticles morphology. In *Manufacturing Nanostructures*; Ahmad, W., Ali, N., Eds.; One Central Press: Manchester, UK, 2014; pp. 108–128.
222. Meibner, T.; Oelschlagel, K.; Potthoff, A. Implications of the stability behavior of zinc oxide nanoparticles for toxicological studies. *Int. Nano Lett.* **2014**, *4*, 116. [CrossRef]
223. Hsiao, I.; Huang, Y. Effects of various physicochemical characteristics on the toxicities of ZnO and TiO_2 nanoparticles toward human lung epithelial cells. *Sci. Total Environ.* **2011**, *409*, 1219–1228. [CrossRef]

Journal of
Fungi

Article

Fungus *Aspergillus niger* Processes Exogenous Zinc Nanoparticles into a Biogenic Oxalate Mineral

Martin Šebesta [1,*], Martin Urík [1], Marek Bujdoš [1], Marek Kolenčík [2,3], Ivo Vávra [4], Edmund Dobročka [4], Hyunjung Kim [5] and Peter Matúš [1]

1 Institute of Laboratory Research on Geomaterials, Faculty of Natural Sciences, Comenius University in Bratislava, Mlynská dolina, Ilkovičova 6, 842 15 Bratislava, Slovakia; prif.ulg@uniba.sk (M.U.); marek.bujdos@uniba.sk (M.B.); peter.matus@uniba.sk (P.M.)
2 Department of Soil Science and Geology, Faculty of Agrobiology and Food Resources, Slovak University of Agriculture in Nitra, Trieda A. Hlinku 2, 949 76 Nitra, Slovakia; marek.kolencik@uniag.sk
3 Nanotechnology Centre, VŠB Technical University of Ostrava, 17. listopadu 15/2172, 708 33 Ostrava, Czech Republic
4 Institute of Electrical Engineering, Slovak Academy of Sciences, Dúbravská cesta 9, 841 04 Bratislava, Slovakia; ivo.vavra@savba.sk (I.V.); elekdobr@savba.sk (E.D.)
5 Department of Mineral Resources and Energy Engineering, Jeonbuk National University, 567, Baekje-daero, Deokjin-gu, Jeonju, Jeonbuk 54896, Korea; kshjkim@jbnu.ac.kr
* Correspondence: martin.sebesta@uniba.sk; Tel.: +421-2-602-91-557

Received: 22 September 2020; Accepted: 6 October 2020; Published: 8 October 2020

Abstract: Zinc oxide nanoparticles (ZnO NPs) belong to the most widely used nanoparticles in both commercial products and industrial applications. Hence, they are frequently released into the environment. Soil fungi can affect the mobilization of zinc from ZnO NPs in soils, and thus they can heavily influence the mobility and bioavailability of zinc there. Therefore, ubiquitous soil fungus *Aspergillus niger* was selected as a test organism to evaluate the fungal interaction with ZnO NPs. As anticipated, the *A. niger* strain significantly affected the stability of particulate forms of ZnO due to the acidification of its environment. The influence of ZnO NPs on fungus was compared to the aqueous Zn cations and to bulk ZnO as well. Bulk ZnO had the least effect on fungal growth, while the response of *A. niger* to ZnO NPs was comparable with ionic zinc. Our results have shown that soil fungus can efficiently bioaccumulate Zn that was bioextracted from ZnO. Furthermore, it influences Zn bioavailability to plants by ZnO NPs transformation to stable biogenic minerals. Hence, a newly formed biogenic mineral phase of zinc oxalate was identified after the experiment with *A. niger* strain's extracellular metabolites highlighting the fungal significance in zinc biogeochemistry.

Keywords: biotransformation; biomineralization; metal oxide nanoparticles; nanoparticle dissolution; nanoparticle mobility; fungal leaching

---

1. Introduction

Engineered nanoparticles (ENPs) have been increasingly used in diverse applications concerning renewable energy, electronics, material science, medicine, and agriculture [1]. The increase in ENPs' release to the environment also greatly raises the environmental risks in both frequency and severity [2]. This is especially true for soils and sediments, which represent the major sinks for released ENP. There, the mobility and transformation of ENP are strongly influenced by biogeochemical processes and, thus, they have been increasingly studied in the last two decades [3,4]. This also includes zinc oxide nanoparticles (ZnO NPs) of which, by estimate, up to 8.7 kt end up in soils annually [5].

Zinc oxide, in both bulk and nanosized form, has been extensively used in the industrial and commercial sphere with the estimated global annual production of 1337 and 30 kt in 2014, respectively [6]. Since then, the production of both bulk ZnO and ZnO NPs has shown a rising trend [7,8].

Two main pathways are important to ZnO NP entry to soil environments—(1) unintentional applications with activated sludges that are used as fertilizers, and (2) intentional direct application as a nanofertilizer to supply plants with Zn. Nanotechnology increases the growth and productivity of plants and is used to protect plants from pathogens, and is, thus, increasingly used in agriculture. Therefore, it is expected that the intentional application of ZnO NPs will play a larger role in soil contamination [9–14].

The mobility of ZnO NPs and other Zn forms in the soil is influenced by factors such as pH, the content and quality of clay minerals, oxyhydroxides of Al, Fe, Mn, and organic matter [15–19]. Living organisms also have a direct effect on bioavailability and the transformation of ZnO NPs. ZnO NPs have antimicrobial and antifungal properties and, hence, can affect the composition of microbial communities in soils and affect the growth of soil fungi [17,20–22].

Fungi may dissolve and transform ENPs, including ZnO NPs, and, under the right conditions, recreate them through biomineralization [23]. They have a big influence on the cycling of elements and the transformation of organic matter, with both processes heavily affecting the bioavailability of the elements in soils. Together with other soil organisms, they mechanically and chemically interact with soil particles and enhance weathering. This results in the translocation of trace elements from the solid phase to soil solutions. Thus, these elements become bioavailable to plants and other soil organisms.

Thanks to their ubiquitous activity, fungi have a considerable role in biogeochemical cycles of different elements, including Zn [24]. Fungal-mediated mineralization is ubiquitous in nature and has been used in industrial, biotechnological, and environmental applications [25]. Biomineralization also affects the ZnO NPs released to soil environments and has not been thoroughly studied.

Many fungi, including *Aspergillus niger*, were observed to produce extracellular metabolites capable of complexolysis and/or ligand-promoted dissolution [26,27]. The amino acids, carboxylic acids, phenolic compounds, and siderophores are some known metabolites with the aforementioned properties [28]. Among the strong chelating agents, oxalic acid was shown to be produced in large quantities by *A. niger* and other fungi [29,30]. The extraction efficiency of oxalic acid is pH-dependable and increases with decreasing pH [31]. Thus, it is possible that, in soil microenvironments close to the fungi, oxalic acids and protonation of the environment dissolve Zn, and the Zn is locally redistributed and reprecipitates as Zn oxalate elsewhere [32].

To find out how ZnO NPs interact with filamentous fungus *A. niger* and how they differ from bulk and ionic Zn forms, *A. niger* was grown statically for 7 days in growth media enriched by 6.5 mg $Zn \cdot L^{-1}$ in form of $ZnSO_4$, ZnO NPs or bulk ZnO. Changes in the pH of the growth media and dry biomass were recorded. The influence of extracellular metabolites of *A. niger* was observed in a separate experiment, where the metabolites were applied on ZnO NPs and after 5 days, a transformation into biominerals was recorded.

2. Materials and Methods

2.1. Preparation of ZnO Nanoparticle Suspensions and $ZnSO_4$ Solution

For the experiment with fungal growth, three Zn forms were used, ionic Zn in the form of a solution of $ZnSO_4$, suspension of ZnO NPs, and bulk ZnO. Bulk ZnO was acquired in the form of powder (p.a. quality, Chemapol, Prague, Czech Republic). ZnO NP dispersion used in the experiment was purchased from Sigma Aldrich, St. Louis, MO, USA (<100 nm particle size (TEM), ≤40 nm Avg. part. size (APS), 20 wt. % in H_2O). ZnO NPs used in this work were also used in the article by Kolenčík et al. [16], and additional characterization of the nanoparticles can be found there. Right before the experiment, the suspension of ZnO NPs with a concentration of 65 $mg \cdot L^{-1}$ (1 mmol $Zn \cdot L^{-1}$) was prepared by adding 65 μL of ZnO NP dispersion to a 200 mL volumetric flask that was filled to mark with distilled water. The suspension was then sonicated for 15 min in an ultrasonic bath. An ionic zinc solution of 65 mg·L^{-1} (1 mmol Zn·L^{-1}) $ZnSO_4$ was prepared by dissolving 0.2876 g $ZnSO_4·7H_2O$ (p.a. quality, CentralChem, Bratislava, Slovakia) in 1 L of distilled water.

2.2. Cultivation of *Aspergillus niger*

Microscopic filamentous fungus *Aspergillus niger* (Tiegh.), strain CBS 140837, originally isolated from the mercury-contaminated soil [33], was grown in Sabouraud growth medium (HiMedia, Mumbai, India) in 250 mL Erlenmeyer flasks using a 7-day static cultivation in a growth chamber (dark, 25 °C). Four different types of growth media were created with three forms of Zn at 6.5 mg Zn·L^{-1} (0.1 mmol Zn·L^{-1}), $ZnSO_4$, ZnO NP, and bulk ZnO, and one control without added Zn. The concentration of 6.5 mg Zn·$L^{-}1$ was selected in a preliminary experiment with $ZnSO_4$ (Supplementary Material Table S1, Figure S1), where a concentration of 13 mg Zn·L^{-1} in the form of $ZnSO_4$ prolonged sporulation with negligible effect on dry biomass weight; a concentration of 26 mg Zn·L^{-1} was inhibitory for fungal growth and no compact mycelium was formed after 7 days.

In the case of $ZnSO_{4,}$ and ZnO NPs, 5 mL of either 1 mmol Zn·L^{-1} $ZnSO_4$ solution or ZnO NP suspension were added to 45 mL of Sabouraud growth medium in Erlenmeyer flask. Bulk ZnO in the form of 0.0033 g of ZnO powder was added to 45 mL of growth medium and 5 mL of sterilized distilled water put into Erlenmeyer flask. The control experiment was done in Erlenmeyer flasks filled with 45 mL of the growth medium and 5 mL of sterilized distilled water. Each of the Zn forms and control had six replicates. All the Erlenmeyer flasks with the growth media were then put into the ultrasonic bath for 15 min.

After the aforementioned procedure, each of the growth media in Erlenmeyer flasks was inoculated with 50 μL of *A. niger* spore suspension and grown in the dark in the growth chamber for 7 days. After a 7-day growth period, the weight of dry biomass, the concentration of Zn in dry biomass, pH in the growth media, and the concentration of Zn in growth media in form of ionic Zn and Zn bound in organic or inorganic colloids was measured.

The biomass grown on the top of the growth media was collected and washed several times with distilled water. Afterwards, it was dried out at 60 °C, then weighed, and transferred into PTFE containers and 5 mL of 65% HNO_3 was added to digest the biomass. The PTFE containers were put into high-pressure acid digestion vessels, and the vessels were closed afterwards. Then, the vessels were placed into an oven heated to 150 °C for 4 h to digest the biomass.

To discern between Zn bound to colloidal form and ionic Zn, the removed growth medium was centrifuged at 700 g for 1 min, to remove big clusters of residual biomasses bigger than 1000 nm. Then, part of the supernatant was removed and analyzed for Zn concentration (C_{1000}). The concentration of ionic Zn (< 1 nm, C_1) was acquired after the ultrafiltration of the supernatant, 6 mL of supernatant was transferred to ultrafiltration centrifugation units (Sartorius Vivaspin® 6 mL, 3 kDa, Goettingen, Germany) which were centrifuged at 3500 g for 20 min. 0.5 mL of filtrate was collected, stabilized with HNO_3, and analyzed for Zn concentration. The concentration in filtrates, supernatants, and digested biomass was measured by flame atomic absorption spectrometry (Perkin-Elmer 1100, Perkin-Elmer, Rodgau, Germany). A concentration of Zn bound to colloidal forms (1–1000 nm, $C_{1\text{-}1000}$) was calculated by subtracting the concentration of ionic Zn from the concentration of Zn in colloidal supernatant ($C_{1\text{-}1000} = C_{1000} - C_1$).

The dry weight of mycelia and pH was compared for all applications via a two-tail *t*-test at a significance level $\alpha = 0.05$. Before the *t*-test, data were analyzed for differences in variances by F-test, and then a *t*-test for either equal or unequal variances was used. The statistical evaluation was done with Analysis ToolPak add-in for Microsoft Excel (Redmond, WA, USA).

2.3. Transformation of ZnO Nanoparticles by Extracellular Metabolites of *Aspergillus niger*

In our experiments with *A. niger* mycelia, the used concentration was very low, and most, if not all, the ZnO NPs and bulk ZnO were dissolved. Therefore, the experiment with fungal extracellular metabolites, which were also used in other studies of ZnO NPs synthesis [22], was undertaken to find out if these metabolites are able to dissolve ZnO NPs and transform them into Zn biominerals.

To achieve this, the *A. niger* was statically grown for 7 days on Sabouraud growth media in the growth chamber. It was grown the same way as the control group was grown in the previous experiment.

After the 7-day cultivation period, the mycelium was removed from the Erlenmeyer flask and was put for 3 days into 50 mL of sterilized distilled water. After 3-day cultivation, the mycelium was removed, and the acquired solution was filtered through 0.45 μm membrane filter paper. A small amount of the solution was removed and used for the analysis of oxalic acid produced by the fungal mycelium.

An Erlenmeyer flask was filled with 47.5 mL of a solution of extracellular fungal metabolites, and 500 mg of ZnO NPs was added in the form of 2.5 mL of ZnO NP dispersion. After 5 days of static reaction, the solution was decanted and analyzed for the content of oxalic acid and the white sediment was dried out and sent to X-ray powder diffraction (XRD) analysis and analysis by transmission electron microscopy (TEM). The solutions containing extracellular metabolites were analyzed for the content of oxalic acid by isotachophoresis (ZKI-1, Villa Labeco, Spišská Nová Ves, Slovakia) in itp-itp mode. The acquired isotachopherograms were evaluated by a software suite supplied with the analyzer [34].

2.4. Characterization of ZnO Nanoparticles and Bulk ZnO

Crystalline phases in ZnO NPs, bulk ZnO, and ZnO NPs transformed by the extracellular metabolites of *A. niger* were identified by XRD [35,36].

Surface morphology and the size of ZnO NPs, and bulk ZnO was examined by TEM. TEM images were collected on instrument JEOL-1200 EX (JEOL Ltd., Tokyo, Japan) operating at accelerating voltages of 120 kV. Samples of ZnO NPs and bulk ZnO were diluted in distilled water and ultrasonicated in order to break up large aggregates. A drop of the suspension was placed onto a carbon-coated grid and then air-dried at room temperature overnight. The ZnO NPs and bulk ZnO used here were also used and characterized in our previous study [9].

3. Results

3.1. Interactions of *Aspergillus niger* with ZnO and Aqueous Zn

Fungal mycelia of *Aspergillus niger* were grown statically in Erlenmeyer flasks with Sabouraud growth media spiked with either $ZnSO_4$, ZnO NPs, or bulk ZnO and control without added Zn. The overproduction of protons (H^+) by microscopic filamentous fungus *A. niger* during the 7-day cultivation period generated a substantial decrease of pH from 5.6 to 2.4. It was even more intensified in the presence of Zn where the final pH values decreased to 2.0, 2.1, and 2.2 for $ZnSO_4$, ZnO NPs, and bulk ZnO, respectively (Figure 1).

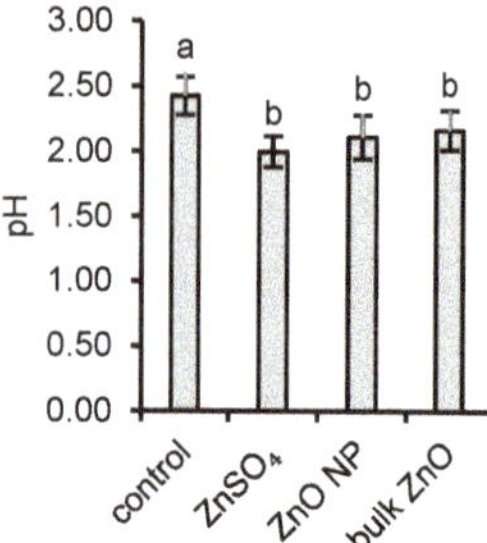

Figure 1. pH of the growth media after the experiment; a and b represent the statistically similar means between different groups at $\alpha = 0.05$ (two-tail *t*-test).

Zn was determined primarily as dissolved in the growth media (Figure 2) regardless of the Zn form applied. Only a negligible fraction was bound to colloidal particles. Thus, the distribution of Zn in colloidal, dissolved and bioaccumulated fractions were similar for $ZnSO_4$, ZnO NPs, and bulk ZnO (Figure 2). The bioaccumulation of Zn by *A. niger* was very effective for all Zn forms (Figure 2) and represented up to 76% of the total Zn added to the cultivation system.

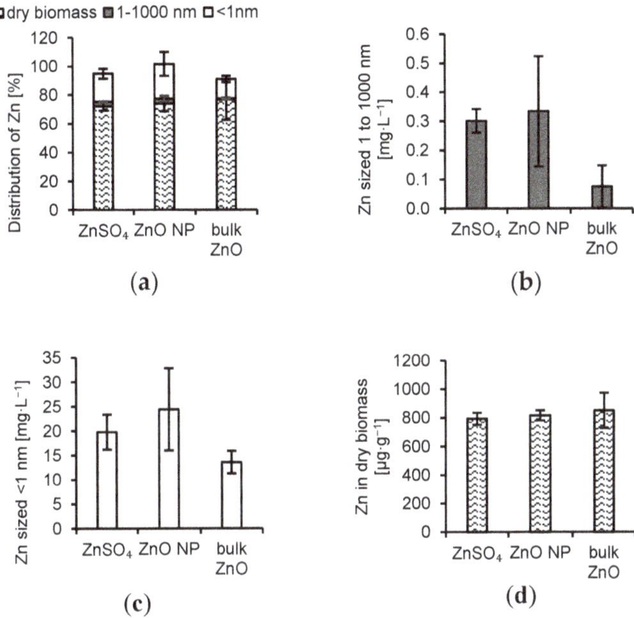

Figure 2. Distribution of Zn in the experimental system (**a**) and concentration of Zn in (**b**) particles sized 1 to 1000 nm, (**c**) dissolved fraction, and (**d**) dry biomass.

The applied concentration of 6.5 mg $Zn{\cdot}L^{-1}$ had no adverse effects on fungal growth and a positive effect of Zn on the growth of the fungal biomass was observed (Figure 3). The application of all Zn form resulted in a higher dry weight of mycelia compared to control. The highest mean dry weight of mycelia was observed for ZnO NP application. The weight was significantly higher than both dry weights in bulk ZnO application and in the control experiment without additional Zn.

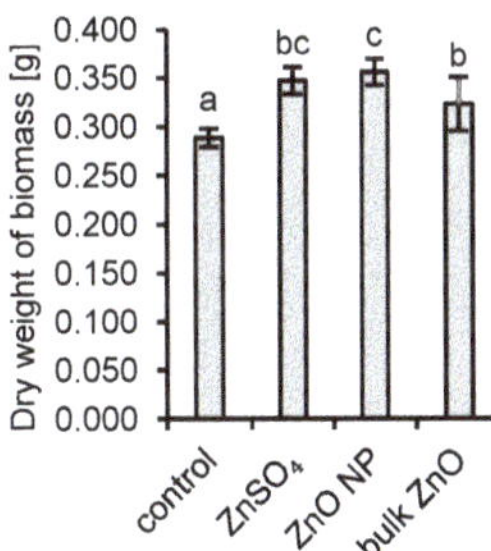

Figure 3. The dry weight of biomass of *Aspergillus niger* after the experiment; a, b and c represent the statistically similar means between different groups at $\alpha = 0.05$ (two-tail t-test).

3.2. Transformation of ZnO Nanoparticles by Fungal Extracellular Metabolites

To experimentally confirm the transformation of ZnO NPs into zinc oxalate, extracellular metabolites of *A. niger* were applied to ZnO NPs. A partial transformation of ZnO NPs was observed, and a new mineral phase—zinc oxalate dihydrate—was identified (Figure 4).

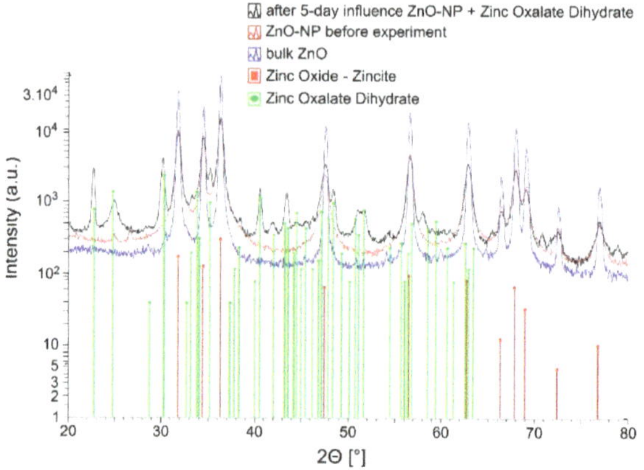

Figure 4. Diffractogram of the partially transformed ZnO NP to Zn oxalate dihydrate.

At the end of the experiment, the oxalic concentration decreased from the initial 11.6 mmol·L^{-1} to 1.2 mmol·L^{-1}. Thus, a 90% decrease in the initial oxalic acid concentration was observed, further confirming the ZnO NPs transformation into the zinc oxalate dihydrate by *A. niger*. Quantitatively, approximately 43 mg or 8.6% of ZnO NPs was transformed into the zinc oxalate dihydrate in the process. Only a partial transformation was observed. The recrystallization did not result in a significant change in the size of the most particles, and it ranged between 40 and 100 nm (Figure 5).

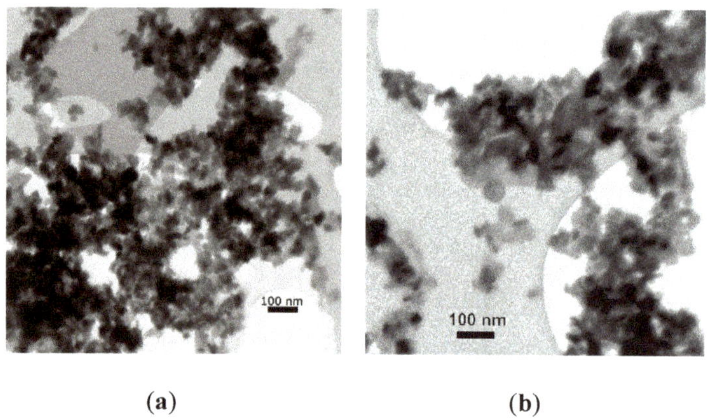

Figure 5. Transmission electron microscopy image of (**a**) Pristine ZnO nanoparticles (**b**) Partially transformed ZnO nanoparticles to Zn oxalate dihydrate.

4. Discussion

4.1. Interactions of Aspergillus niger with ZnO and Aqueous Zn

ZnO NPs are relatively easily dissolvable in soils, especially in more acidic ones [9,37,38]. The dissolution of ZnO NPs is also enhanced by soil organisms, including filamentous fungi, via exudation of various acidic and chelating metabolites. The microbially induced acidification led to the dissolution of ZnO NPs, and bulk ZnO [39] and, therefore, nearly all of the Zn in the solutions of growth media was dissolved regardless of the Zn form.

Since Zn is an essential micronutrient [40,41] and our selected concentration was below the threshold of growth inhibition, all forms of Zn at the applied concentration had a positive effect on the dry weight of biomass after 7 days of cultivation. It was reported that the Zn concentrations as low as 0.065 mg $Zn \cdot L^{-1}$ enhanced the fungal growth [42]. The minimal inhibitory concentration of Zn for microscopic filamentous fungi were reported to be as high as 100 mg $Zn \cdot L^{-1}$ in the literature [43,44].

4.2. Transformation of ZnO Nanoparticles by Fungal Extracellular Metabolites

Filamentous fungi are capable of releasing extracellular metabolites that chemically deteriorate natural mineral Zn phases in soils and sediments via processes of protonation and chelation. However, under specific conditions (e.g., high biogenic chelate concentrations, alkalic pH), the bioextracted Zn can be reprecipitated from the soil solution to form new mineral phases. It has been reported that fungi react to the presence of ZnO NPs by increasing the production of the extracellular metabolites in order to detoxify and immobilize excessive Zn, and thus the transformation of ZnO NPs occurs in the soil environment [39,41]. Immobilization of Zn dissolved by fungi is facilitated by precipitation with oxalates [39].

We used extracellular metabolites of *A. niger* to transform ZnO NPs into a zinc oxalate biomineral. *A. niger* is well known for the high rate of extracellular metabolite production. Oxalic acid is the most produced of organic acids by the fungus when grown on Sabouraud growth media [30].

In our experiment, at the given applied volume and concentration of oxalic acid applied, only a partial transformation of ZnO NPs to zinc oxalates was observed. No significant change in the size of particles was observed. It is most probable that the transformation occurred on the surface of the nanoparticles, creating a surface of Zn oxalate with the core still being ZnO, or the newly formed zinc oxalates did not have time to grow to larger sizes in five days. If a larger volume of the organic acids was applied, both larger amounts of ZnO NPs might have been transformed, and the size of the particles could have seen a bigger change.

A change in Zn minerals into Zn oxalates had been observed before, albeit the minerals were not nanoparticles [32,39,45]. When silicate and sulfide minerals containing Zn were transformed by fungal extracellular metabolites, the resulting crystals of zinc oxalates had various shapes and sizes, with most of the crystals measuring between 50 and 100 μm [39]. When ZnO microparticulate powder was transformed into zinc oxalates by *Aspergillus* species, the biominerals had a different habitus compared to the minerals formed abiotically. The size of these biominerals was above 50 μm.

The mineral size and shape are also dependent on the place of formation. Minerals of zinc oxalate associated with a fungus have a different shape compared to both biomineral and abiotically formed Zn oxalate. However, the biominerals formed farther away from the fungal mycelium had a shape that was more similar to the abiotically formed Zn oxalate [32]. Since our experiment was only done with extracellular metabolites, the smaller change in the size and shape upon partial transformation may also result from the same processes that led to the increase in similarities between the biomineral formed further away from the mycelium and the abiotically formed mineral.

Aspergillus spp. and other fungi can biomineralize inorganic nanoparticles [46] and transform these nanoparticles into more stable oxalate compounds [25,47]. They play a large role in the cycling of elements in the soil environment, and their abilities can also be used for the bioremediation of contaminated areas [48].

5. Conclusions

Fungi, through their ability to acidify their environment, producing strong chelating agents, and the ability to bioaccumulate trace elements in their mycelia, can considerably affect the mobility of elements in soils and locally increase their bioavailability. Our observations affirm that the transformation of

ZnO NPs into biogenic mineral phases occurs under the influence of ubiquitous soil fungus *A. niger* and that these processes may happen, under the right circumstances, in the soil environment. The role of soil fungi is seminal in the distribution of ionic and nanoparticulate forms of Zn. Therefore, fungi should be taken into consideration when models of bioavailability and mobility are constructed for soil environments. Furthermore, these fungal abilities can help us develop new methods of bioremediation in the future.

Supplementary Materials: The following are available online at http://www.mdpi.com/2309-608X/6/4/210/s1, Figure S1: Photographs of Aspergillus niger mycelia after 7-day cultivation in the preliminary test, Table S1: Weight of dry biomass after 7-day static cultivation in the preliminary test.

Author Contributions: Conceptualization, M.Š. and M.U.; Data curation, M.Š.; Formal analysis, M.Š.; Funding acquisition, M.U.and P.M.; Investigation, M.Š.; Methodology, M.Š., M.U. and P.M.; Project administration, P.M.; Resources, M.K., I.V., E.D., H.K. and M.B.; Supervision, M.U., M.K. and P.M.; Validation, M.B.; Visualization, M.Š.; Writing—original draft, M.Š.; Writing—review and editing, M.U. All authors have read and agreed to the published version of the manuscript.

Funding: This research was funded by the Scientific Grant Agency of the Ministry of Education, Science, Research and Sport of the Slovak Republic and the Slovak Academy of Sciences (Vedecká grantová agentúra MŠVVaŠ SR a SAV) under the contracts Nos. VEGA 1/0153/17, VEGA 1/0146/18, and by Cultural and Educational agency of the Ministry of Education, Science, Research and Sport of the Slovak Republic (Kultúrna a edukačná grantová agentúra MŠVVaŠ SR) under the contract No. KEGA 013SPU-4/2019.

Conflicts of Interest: The authors declare no conflict of interest.

References

1. Hochella, M.F.; Mogk, D.W.; Ranville, J.; Allen, I.C.; Luther, G.W.; Marr, L.C.; McGrail, B.P.; Murayama, M.; Qafoku, N.P.; Rosso, K.M.; et al. Natural, incidental, and engineered nanomaterials and their impacts on the Earth system. *Science* **2019**, *363*, eaau8299. [CrossRef] [PubMed]
2. Sun, T.Y.; Mitrano, D.M.; Bornhöft, N.A.; Scheringer, M.; Hungerbühler, K.; Nowack, B. Envisioning Nano Release Dynamics in a Changing World: Using Dynamic Probabilistic Modeling to Assess Future Environmental Emissions of Engineered Nanomaterials. *Environ. Sci. Technol.* **2017**, *51*, 2854–2863. [CrossRef] [PubMed]
3. Peijnenburg, W.; Praetorius, A.; Scott-Fordsmand, J.; Cornelis, G. Fate assessment of engineered nanoparticles in solids dominated media—Current insights and the way forward. *Environ. Pollut.* **2016**, *218*, 1365–1369. [CrossRef] [PubMed]
4. Šebesta, M.; Matúš, P. Separation, determination, and characterization of inorganic engineered nanoparticles in complex environmental samples. *Chem. List.* **2018**, *112*, 583–589.
5. Keller, A.A.; McFerran, S.; Lazareva, A.; Suh, S. Global life cycle releases of engineered nanomaterials. *J. Nanopart. Res.* **2013**, *15*, 1–17. [CrossRef]
6. MarketsandMarkets. *Zinc Oxide Market by Application & by Region—Global Trends and Forecasts to 2020*; MarketsandMarkets™ Research Private Ltd.: Hadapsar, India, 2015.
7. Future Markets Inc. *The Global Market for Zinc Oxide Nanoparticles*; Future Markets Inc.: Edinburgh, UK, 2020.
8. MarketsandMarkets. *Zinc Oxide Market by Process (French Process, Wet Process, American Process), Grade (Standard, Treated, USP, FCC), Application (Rubber, Ceramics, Chemicals, Agriculture, Cosmetics & Personal Care, Pharmaceuticals), Region—Global Forecast to 2024*; MarketsandMarkets™ Research Private Ltd.: Hadapsar, India, 2019.
9. Šebesta, M.; Nemček, L.; Urík, M.; Kolenčík, M.; Bujdoš, M.; Vávra, I.; Dobročka, E.; Matúš, P. Partitioning and stability of ionic, nano- and microsized zinc in natural soil suspensions. *Sci. Total Environ.* **2020**, *700*, 134445. [CrossRef]
10. Raliya, R.; Saharan, V.; Dimkpa, C.; Biswas, P. Nanofertilizer for Precision and Sustainable Agriculture: Current State and Future Perspectives. *J. Agric. Food Chem.* **2018**, *66*, 6487–6503. [CrossRef]
11. Prasad, R.; Bhattacharyya, A.; Nguyen, Q.D. Nanotechnology in Sustainable Agriculture: Recent Developments, Challenges, and Perspectives. *Front. Microbiol.* **2017**, *8*, 1014. [CrossRef]

12. Medina-Velo, I.A.; Barrios, A.C.; Zuverza-Mena, N.; Hernandez-Viezcas, J.A.; Chang, C.H.; Ji, Z.; Zink, J.I.; Peralta-Videa, J.R.; Gardea-Torresdey, J.L. Comparison of the effects of commercial coated and uncoated ZnO nanomaterials and Zn compounds in kidney bean (*Phaseolus vulgaris*) plants. *J. Hazard. Mater.* **2017**, *332*, 214–222. [CrossRef]
13. Raliya, R.; Tarafdar, J.C.; Biswas, P. Enhancing the Mobilization of Native Phosphorus in the Mung Bean Rhizosphere Using ZnO Nanoparticles Synthesized by Soil Fungi. *J. Agric. Food Chem.* **2016**, *64*, 3111–3118. [CrossRef]
14. Liu, R.; Lal, R. Potentials of engineered nanoparticles as fertilizers for increasing agronomic productions. *Sci. Total Environ.* **2015**, *514*, 131–139. [CrossRef] [PubMed]
15. Kabata-Pendias, A.; Szteke, B. *Trace Elements in Abiotic and Biotic Environments*; CRC Press: Boca Raton, FL, USA, 2015; pp. 355–366.
16. Kolenčík, M.; Ernst, D.; Komár, M.; Urík, M.; Šebesta, M.; Dobročka, E.; Černý, I.; Illa, R.; Kanike, R.; Qian, Y. Effect of foliar spray application of zinc oxide nanoparticles on quantitative, nutritional, and physiological parameters of foxtail millet (*Setaria italica* l.) under field conditions. *Nanomaterials* **2019**, *9*, 1559. [CrossRef] [PubMed]
17. García-Gómez, C.; Fernández, M.D.; García, S.; Obrador, A.F.; Letón, M.; Babín, M. Soil pH effects on the toxicity of zinc oxide nanoparticles to soil microbial community. *Environ. Sci. Pollut. Res.* **2018**, *25*, 28140–28152. [CrossRef] [PubMed]
18. Šebesta, M.; Kolenčík, M.; Matúš, P.; Kořenková, L. Transport and distribution of engineered nanoparticles in soils and sediments. *Chem. List.* **2017**, *111*, 322–328.
19. Šebesta, M.; Kolenčík, M.; Urík, M.; Bujdoš, M.; Vávra, I.; Dobročka, E.; Smilek, J.; Kalina, M.; Diviš, P.; Pavúk, M.; et al. Increased Colloidal Stability and Decreased Solubility—Sol-Gel Synthesis of Zinc Oxide Nanoparticles with Humic Acids. *J. Nanosci. Nanotechnol.* **2019**, *19*, 3024–3030. [CrossRef]
20. Erazo, A.; Mosquera, S.A.; Rodríguez-Paéz, J.E. Synthesis of ZnO nanoparticles with different morphology: Study of their antifungal effect on strains of *Aspergillus niger* and *Botrytis cinerea*. *Mater. Chem. Phys.* **2019**, *234*, 172–184. [CrossRef]
21. Rajput, V.D.; Minkina, T.M.; Behal, A.; Sushkova, S.N.; Mandzhieva, S.; Singh, R.; Gorovtsov, A.; Tsitsuashvili, V.S.; Purvis, W.O.; Ghazaryan, K.A.; et al. Effects of zinc-oxide nanoparticles on soil, plants, animals and soil organisms: A review. *Environ. Nanotechnol. Monit. Manag.* **2018**, *9*, 76–84. [CrossRef]
22. Nisar, P.; Ali, N.; Rahman, L.; Ali, M.; Shinwari, Z.K. Antimicrobial activities of biologically synthesized metal nanoparticles: An insight into the mechanism of action. *JBIC J. Biol. Inorg. Chem.* **2019**, *24*, 929–941. [CrossRef]
23. Mohamed, A.A.; Fouda, A.; Abdel-Rahman, M.A.; Hassan, S.E.-D.; El-Gamal, M.S.; Salem, S.S.; Shaheen, T.I. Fungal strain impacts the shape, bioactivity and multifunctional properties of green synthesized zinc oxide nanoparticles. *Biocatal. Agric. Biotechnol.* **2019**, *19*, 101103. [CrossRef]
24. Gadd, G.M. Metals, minerals and microbes: Geomicrobiology and bioremediation. *Microbiology* **2010**, *156*, 609–643. [CrossRef]
25. Qin, W.; Wang, C.; Ma, Y.; Shen, M.; Li, J.; Jiao, K.; Tay, F.R.; Niu, L. Microbe-Mediated Extracellular and Intracellular Mineralization: Environmental, Industrial, and Biotechnological Applications. *Adv. Mater.* **2020**, *32*, 1907833. [CrossRef] [PubMed]
26. Burford, E.P.; Fomina, M.; Gadd, G.M. Fungal involvement in bioweathering and biotransformation of rocks and minerals. *Mineral. Mag.* **2003**, *67*, 1127–1155. [CrossRef]
27. Polák, F.; Urík, M.; Bujdoš, M.; Matúš, P. *Aspergillus niger* enhances oxalate production as a response to phosphate deficiency induced by aluminium(III). *J. Inorg. Biochem.* **2020**, *204*, 110961. [CrossRef] [PubMed]
28. Fomina, M.; Burford, E.P.; Gadd, G.M. Fungal dissolution and transformation of minerals: Significance for nutrient and metal mobility. In *Fungi in Biogeochemical Cycles*; Gadd, G.M., Ed.; Cambridge University Press: Cambridge, UK, 2006; pp. 236–266. ISBN 9780511550522.
29. Polák, F.; Urík, M.; Matúš, P. Low molecular weight organic acids in soil environment. *Chem. List.* **2019**, *113*, 307–314.
30. Boriová, K.; Urík, M.; Bujdoš, M.; Pifková, I.; Matúš, P. Chemical mimicking of bio-assisted aluminium extraction by *Aspergillus niger*'s exometabolites. *Environ. Pollut.* **2016**, *218*, 281–288. [CrossRef]
31. Ström, L.; Owen, A.G.; Godbold, D.L.; Jones, D.L. Organic acid behaviour in a calcareous soil implications for rhizosphere nutrient cycling. *Soil Biol. Biochem.* **2005**, *37*, 2046–2054. [CrossRef]

32. Sutjaritvorakul, T.; Gadd, G.M.; Whalley, A.J.S.; Suntornvongsagul, K.; Sihanonth, P. Zinc Oxalate Crystal Formation by Aspergillus nomius. *Geomicrobiol. J.* **2016**, *33*, 289–293. [CrossRef]
33. Urík, M.; Hlodák, M.; Mikušová, P.; Matúš, P. Potential of Microscopic Fungi Isolated from Mercury Contaminated Soils to Accumulate and Volatilize Mercury(II). *Water Air Soil Pollut.* **2014**, *225*, 2219. [CrossRef]
34. Sádecká, J.; Polonský, J. Determination of organic acids in tobacco by capillary isotachophoresis. *J. Chromatogr. A* **2003**, *988*, 161–165. [CrossRef]
35. Bačík, P.; Ozdín, D.; Miglierini, M.; Kardošová, P.; Pentrák, M.; Haloda, J. Crystallochemical effects of heat treatment on Fe-dominant tourmalines from Dolní Bory (Czech Republic) and Vlachovo (Slovakia). *Phys. Chem. Miner.* **2011**, *38*, 599–611. [CrossRef]
36. Ondruška, J.; Trnovcová, V.; Štubňa, I.; Bačík, P. Depolarization currents in illite. *J. Therm. Anal. Calorim.* **2018**, *131*, 2285–2289. [CrossRef]
37. Du, W.; Sun, Y.; Ji, R.; Zhu, J.; Wu, J.; Guo, H. TiO_2 and ZnO nanoparticles negatively affect wheat growth and soil enzyme activities in agricultural soil. *J. Environ. Monit.* **2011**, *13*, 822–828. [CrossRef] [PubMed]
38. Kool, P.L.; Ortiz, M.D.; van Gestel, C.A.M. Chronic toxicity of ZnO nanoparticles, non-nano ZnO and $ZnCl_2$ to *Folsomia candida* (Collembola) in relation to bioavailability in soil. *Environ. Pollut.* **2011**, *159*, 2713–2719. [CrossRef] [PubMed]
39. Wei, Z.; Liang, X.; Pendlowski, H.; Hillier, S.; Suntornvongsagul, K.; Sihanonth, P.; Gadd, G.M. Fungal biotransformation of zinc silicate and sulfide mineral ores. *Environ. Microbiol.* **2013**, *15*, 2173–2186. [CrossRef] [PubMed]
40. Couri, S.; Pinto, G.A.S.; de Senna, L.F.; Martelli, H.L. Influence of metal ions on pellet morphology and polygalacturonase synthesis by *Aspergillus niger* 3T5B8. *Braz. J. Microbiol.* **2003**, *34*, 16–21. [CrossRef]
41. Tarafdar, J.C.; Agrawal, A.; Raliya, R.; Kumar, P.; Burman, U.; Kaul, R.K. ZnO nanoparticles induced synthesis of polysaccharides and phosphatases by *Aspergillus* fungi. *Adv. Sci. Eng. Med.* **2012**, *4*, 324–328. [CrossRef]
42. Wold, W.S.M.; Suzuki, I. The citric acid fermentation by *Aspergillus niger*: Regulation by zinc of growth and acidogenesis. *Can. J. Microbiol.* **1976**, *22*, 1083–1092. [CrossRef]
43. Sardella, D.; Gatt, R.; Valdramidis, V.P. Assessing the efficacy of zinc oxide nanoparticles against *Penicillium expansum* by automated turbidimetric analysis. *Mycology* **2018**, *9*, 43–48. [CrossRef]
44. Sawai, J.; Yoshikawa, T. Quantitative evaluation of antifungal activity of metallic oxide powders (MgO, CaO and ZnO) by an indirect conductimetric assay. *J. Appl. Microbiol.* **2004**, *96*, 803–809. [CrossRef]
45. Sayer, J.A.; Gadd, G.M. Solubilization and transformation of insoluble inorganic metal compounds to insoluble metal oxalates by *Aspergillus niger*. *Mycol. Res.* **1997**, *101*, 653–661. [CrossRef]
46. Hulkoti, N.I.; Taranath, T.C. Biosynthesis of nanoparticles using microbes—A review. *Colloids Surf. B Biointerfaces* **2014**, *121*, 474–483. [CrossRef] [PubMed]
47. Gadd, G.M. Fungi, Rocks, and Minerals. *Elements* **2017**, *13*, 171–176. [CrossRef]
48. Akhtar, N.; Mannan, M.A. Mycoremediation: Expunging environmental pollutants. *Biotechnol. Rep.* **2020**, *26*, e00452. [CrossRef] [PubMed]

Journal of *Fungi*

Article

Mycosynthesis of ZnO Nanoparticles Using *Trichoderma* spp. Isolated from Rhizosphere Soils and Its Synergistic Antibacterial Effect against *Xanthomonas oryzae* pv. *oryzae*

Balagangadharaswamy Shobha [1], Thimappa Ramachandrappa Lakshmeesha [1,*], Mohammad Azam Ansari [2], Ahmad Almatroudi [3,*], Mohammad A. Alzohairy [3], Sumanth Basavaraju [1], Ramesha Alurappa [1], Siddapura Ramachandrappa Niranjana [4] and Srinivas Chowdappa [1,*]

1 Department of Microbiology and Biotechnology, Bangalore University, Jnana Bharathi Campus, Bengaluru 560056, India; shobhahonnaganga@gmail.com (B.S.); simplesumanth007@gmail.com (S.B.); Ramesha.bio@gmail.com (R.A.)

2 Department of Epidemic Disease Research, Institute for Research and Medical Consultations (IRMC), Imam Abdulrahman Bin Faisal University, Dammam 31441, Saudi Arabia; maansari@iau.edu.sa

3 Department of Medical Laboratories, College of Applied Medical Sciences, Qassim University, Qassim 51431, Saudi Arabia; dr.alzohairy@gmail.com

4 Department of Studies in Biotechnology, University of Mysore, Manasagangotri, Mysore 570006, India; niranjanasr@rediffmail.com

* Correspondence: lakshmeesha@bub.ernet.in (T.R.L.); aamtrody@qu.edu.sa (A.A.); csrinivas@bub.ernet.in (S.C.)

Received: 28 July 2020; Accepted: 17 September 2020; Published: 20 September 2020

Abstract: The Plant Growth Promoting Fungi (PGPF) is used as a source of biofertilizers due to their production of secondary metabolites and beneficial effects on plants. The present work is focused on the co-cultivation of *Trichoderma* spp. (*T. harzianum* (PGT4), *T. reesei* (PGT5) and *T. reesei* (PGT13)) and the production of secondary metabolites from mono and co-culture and mycosynthesis of zinc oxide nanoparticles (ZnO NPs), which were characterized by a UV visible spectrophotometer, Powder X-ray Diffraction (PXRD), Fourier Transform Infrared Spectroscopy (FTIR) and Scanning Electron Microscopy (SEM) with Energy Dispersive Spectroscopy (EDAX) and Transmission Electron Microscope (TEM) and Selected Area (Electron) Diffraction (SAED) patterns. The fungal secondary metabolite crude was extracted from the mono and co-culture of *Trichoderma* spp. And were analyzed by GC-MS, which was further subjected for antibacterial activity against *Xanthomonas oryzae* pv. *Oryzae*, the causative organism for Bacterial Leaf Blight (BLB) in rice. Our results showed that the maximum zone of inhibition was recorded from the co-culture of *Trichoderma* spp. rather than mono cultures, which indicates that co-cultivation of beneficial fungi can stimulate the synthesis of novel secondary metabolites better than in monocultures. ZnO NPs were synthesized from fungal secondary metabolites of mono cultures of Trichoderma harzianum (PGT4), Trichoderma reesei (PGT5), Trichoderma reesei (PGT13) and co-culture (PGT4 + PGT5 + PGT13). These ZnO NPs were checked for antibacterial activity against Xoo, which was found to be of a dose-dependent manner. In summary, the biosynthesized ZnO NPs and secondary metabolites from co-culture of *Trichoderma* spp. are ecofriendly and can be used as an alternative for chemical fertilizers in agriculture.

Keywords: fungal nanotechnology; *Trichoderma* spp.; ZnO nanoparticles; antibacterial activity *Xanthomonas oryzae* pv. *oryzae*; co-cultivation; secondary metabolites

1. Introduction

Rice (*Oryzae sativa* L.) is one of the most vital and essential nourishment sources for half of the world's population, it belongs to the family Poaceae, and it is the most widely cultivated food crop in the world [1]. Across the world, annually, about 40% of rice crops are lost due to biotic stresses such as insects, pathogens, pests and weeds [2]. Some of the most important of these diseases are Bacterial Leaf Blight (BLB) caused by (*Xanthomonas oryzae* pv. *oryzae*), Blast (*Magnaporthe grisea*), Sheath Blight (*Rhizoctonia solani*), Sheath Rot (*Sarocladium oryzae*) and Tungro virus [3].

Rice's BLB is one of the most damaging causes of disease, which is caused by *Xanthomonas oryzae* pv. *Oryzae* (Xoo) [4]. This bacteria restricts annual production of rice in both tropical and temperate regions of the world [5]. In the tropics, the damage is more severe than in the temperate regions [6]. The BLB disease incidence has been recorded in various parts of Asia, USA, Africa, and northern Australia [7]. Various disease management strategies are used to minimize BLB damage, such as chemical control, host–plant resistance, crop system modification, and biological control [8]. In the 1950s, chemical management of BLB in rice fields started with the preventive application of the Bordeaux mixture, other chemicals such as phenazine oxide, tecloftalam and nickel dimethyl dithiocarbamate directly sprayed on plants [9]. Synthetic organic bactericides were also recommended, such as nickel dimethyl dithiocarbamate, phenazine noxide and dithianon phenazine [10]. Overuse of chemical substances often adversely affects the environment, farmers and consumer health [11]. Biological control is an alternative method, which is ecologically sensitive, cost-effective and sustainable in BLB management [12]. An effective means of managing plant diseases can be by using antagonistic microorganisms [13]. Interaction between plant pathogens and biocontrol agents has been extensively studied, and the use of biocontrol agents is promising in protecting some commercially valuable crops [14]. Plant growth-promoting rhizobacteria (PGPR) are the most widely studied group of plant growth-promoting bacteria (PGPB) and plant growth-promoting fungi (PGPF), which colonize root surfaces and closely adhere to the soil interface—the rhizosphere—and can also be used for plants [15,16]. Species of *Trichoderma* were identified as potentially environmentally safe biofertilizer and are non-toxigenic [17]. *Trichoderma* species are effective in agriculture as biological control agents and their frequent addition to soil leads to increased crop yields and control of soil-borne pathogens worldwide [18]. Plant growth-promoting rhizobacteria are root-colonizing, free-living bacteria with beneficial effects on crop plants which work by reducing disease incidence and increasing yields [19]. It contributes to the suppression of disease by various modes of action such as antagonism, space and nutrient rivalry and induction of systemic resistance (ISR) [20]. By eliciting induced systemic resistance, PGPR indirectly mediates biological control in a variety of plant diseases [21].

The co-culture of two or more beneficial fungi can interact, stimulate or enhance the production of secondary metabolites, which are not presented in the mono or single cultures when grown separately in in vitro conditions. The co-culture also triggers certain genes which are not activated in mono or single culture, it can also stimulate various pathways when grown together in in vitro conditions [22].

The introduction of nanotechnology to agricultural science seems to offer promising solutions including the release of modified fertilizers and pesticides [23]. The unique and different properties of nanoparticles such as electrical conductivity, active area, hardness and chemical reactivity can be achieved by reducing the size to nanometers [24].

Biological production of nanoparticles based on natural resources has recently attracted scientific interest. Nanoparticles synthesized using natural resources are called green synthesis or biosynthesis [25]. Nanomaterial biosynthesis has provided a common point between nanotechnology and biotechnology and has led to the development of new materials used in many fields [26]. Fungi have become one of the choices in nanotechnology because of its wide variety of advantages over the bacteria, actinomycetes, plants and other physic-chemical properties. The capability of tolerance and metal bioaccumulation in fungi has made fungi a significant branch in the biosynthesis of nanoparticles [27]. Non-toxic and safe reagents are used in the green synthesis of nanoparticles, which makes them cost effective and environmentally friendly [28].

Surface atomic arrangements influence the antibacterial properties [29]. The specific arrangements of atoms on the surface are selected inorganic oxides and work by a fine-tuning of the morphology. By modifying the conditions and by examining the morphology, synthesis of inorganic oxides can be controlled morphologically [30]. Nano-particles are an alternative method that has gained significant attention in the field of plant defense [31]. In comparison, compared with other metal-NPs, the ZnO NPs have been found to be less harmful to plants and beneficial to soil micro flora [32].

Our study is mainly focused on a biological method for the management of the BLB caused by Xoo, a rice disease using co-culture and monoculture of *Trichoderma* spp., which is eco-friendly. The studies were carried out in vitro using the biosynthesis of Zinc Oxide Nanoparticles synthesized from *Trichoderma* spp., which is a new approach in agriculture for the management of the disease. To our best knowledge, this work is reported for the first time.

2. Materials and Methods

2.1. Collection and Isolation of Bacteria from Infected Rice Leaf

The infected leaf samples were collected from different places of Karnataka and were subjected for the isolation of Xoo. The infected parts of the leaves were cut into 0.5 cm^2, the surface was sterilized with 1% sodium hypochlorite for one minute followed by 3–4 sterile distilled water washes and was then blot dried. The sterilized leaf segments were inoculated onto an agar medium of yeast extract dextrose calcium carbonate (YDC) incubated at 28 ± 2 °C for 72 h. The plates were observed for convex, mucoid and yellow color [33].

2.2. Identification of Isolated Bacteria by Biochemical and Molecular Characterization

The bacterial isolates were identified based on morphological, microscopic, biochemical and molecular characterization. Biochemical tests such as gram staining, catalase, oxidase test, KOH test, gelatin liquefaction, starch hydrolysis, casein hydrolysis and pectin hydrolysis were carried out as described in [34]. Molecular characterization was carried out to identify the organism at a molecular level. Using a Chromous bacterial genomic DNA isolation kit, the genomic DNA of bacterial samples were isolated following the standard protocol. PCR amplification was carried out using universal 16s rRNA primers. The obtained sequences were deposited in an NCBI GenBank and an accession number was obtained [35]. The GenBank accession numbers are as follows: *Xanthomonas oryzae* pv. *oryzae* (MBXoo53) MF787295.1, *Xanthomonas oryzae* pv. *oryzae* (MBXoo70) MF787294.1, *Xanthomonas oryzae* pv. *oryzae* (MBXoo69) MF579736.1.

2.3. Collection and Isolation of Trichoderma spp. from Rhizosphere Soil

Rhizospheric soil (soil around the root zone) samples of different crops were collected from different parts of Karnataka. Five grams of rhizosphere soil was collected by uprooting the plant with the soil attached to the roots. The collected soil samples were preserved in polythene bags and stored at 4 °C until further use [36]. The collected soil samples of the rhizosphere were diluted into various concentration solutions, were well-vortexed and 0.1 mL of the supernatants were poured onto PDA medium (potato dextrose agar, with chloramphenicol antibiotics) plates and incubated at 28 °C for 7 days. Colonies that appeared on the plates were isolated and reinoculated on a new PDA petri plate. After 7 days of subculturing, single-spore colonies were obtained by incubating the plates at 28 °C, and the fungal colonies were further assayed for morphological and physiological characteristics [37].

2.4. Morphological and Molecular Characterization of Trichoderma spp.

The identification of fungi was carried out based on the cultural and microscopic properties using standard manuals [38]. By using a Chromous genomic DNA isolation kit, the genomic DNA of the microbial sample was isolated following the protocol as described by the manufacturer. PCR amplification was carried using universal 18S primers. The obtained sequences were deposited in

NCBI GenBank and an accession number was obtained [39]. The GenBank accession numbers are as follows: *Trichoderma harzianum* (PGT4) MH429899.1, *Trichoderma reesei* (PGT5) MH429901.1, *Trichoderma reesei* (PGT13) MH429900.1.

2.5. In Vitro Screening of Plant Growth-Promoting Fungi (PGPF) Strains of Trichoderma spp. for Its Antibacterial Activity by Agar Plug Method against Xanthomonas oryzae pv. oryzae (Xoo)

The mycelial disc of *Trichoderma* spp. fungi isolated from rhizosphere soil was screened for its antibacterial activity against plant pathogenic bacterial strain *Xanthomonas oryzae* pv. *oryzae* (Xoo) isolated from the infected leaf parts of rice. The different Plant Growth-Promoting Fungi (PGPF) strains of *Trichoderma* spp. were grown on a PDA petri plate for 5–7 days at 24 ± 2 °C with alternate light and dark periods. From seven-day-old, culture the fungal discs were pierced using a sterile cork borer of 5 mm in diameter. The fungal discs were transferred to a Mueller-Hinton agar (MHA) plate which was previously swabbed with the bacteria Xoo. The MHA plates were kept for incubation at 28 °C for 24 h and after incubation, the results were observed [40].

2.6. Trichoderma–Trichoderma Interactions through Co-Culture

In our study, co-culture was carried out to evaluate the ability of selected *Trichoderma* spp. cultures to produce secondary metabolites. This was carried out on solid culture medium potato dextrose agar (PDA) for three selected fungi: *T. harzianum* (PGT4), *T. reesei* (PGT5) and *T. reesei* (PGT13) based on its antibacterial activity against *Xanthomonas oryzae* pv. *oryzae*. These fungi were cultured on PDA and incubated at 25 °C for 7 days. After incubation, 10 mm of fungal discs were taken from actively growing margins of *Trichoderma* spp. on PDA plates. Two different *Trichoderma* spp. cultures of 10 mm were placed on PDA plates at the opposite ends of the petri plates, making a total of 3 dual cultures (pairwise combinations). Triplicates of both dual and monocultures were made and incubated under dark conditions for 9 days at 25 °C [41].

2.7. Production of Secondary Metabolites from Trichoderma spp.

The agar plugs of *Trichoderma* spp., measuring 7 mm in diameter, were taken from actively growing margins of *T. harzianum* (PGT4), *T. reesei* (PGT5) and *T. reesei* (PGT13) cultures grown on PDA media and were inoculated individually: dual cultures (*T. harzianum* (PGT4) and *T. reesei* (PGT5); *T. reesei* (PGT5) and *T. reesei* (PGT13) and *T. reesei* (PGT13) and *T. harzianum* (PGT4)) and co-culture sample PGTA (*T. harzianum*, *T. reesei* and *T. reesei*) were inoculated into 250 mL Erlenmayer flasks containing 100 mL of potato dextrose broth medium (PDB, HIMEDIA) supplemented with chloramphenicol antibiotics, incubated in static condition for 9 days at 25 °C [42]. After the incubation period, to avoid fragmentation of the mycelium, it was removed using a microbial loop and the cultures were filtered under vacuum through filter paper (Whatman No. 4). The final filtrate was called a crude extract of secondary metabolites. A control assay was performed to ensure an optimum filtration (to check for the absence of conidia and mycelia) procedure by spreading 30 µL of the final filtrate on petri plates under sterile conditions containing PDA medium. The plates were examined for fungal growth [41].

2.8. Extraction and Identification of Secondary Metabolites

The secondary metabolites were extracted by a solvent extraction method from filtrates, where ethyl acetate and filtrate in a ratio of 1:1 (*v*/*v*) was used to extract exhaustively. The upper layer of the solvent contains the compounds which were collected separately from the aqueous layer (PDB medium) using a separation funnel. The solvent ethyl acetate was evaporated from the filtrate using a vacuum rotary evaporator at 40 °C, 70 rpm until the extract got reduced to 4 mL; it was maintained at −20 °C in the deep freezer until further use. Ethyl acetate extracts were analyzed by Gas Chromatography Mass Spectroscopy (GC-MS) analysis. The GC-MS analysis was performed using Thermo Scientific, Ceres 800, MS DSQ II (Waltham, MA, USA) and the silica column was packed with Elite-5MS (5% biphenyl 95% dimethylpolysiloxane, 30 m × 0.25 mm ID × 250 µm df). For the separation of components, helium gas was used as a carrier gas, which maintained the constant flow of 1 mL/min. The temperature of the injector was maintained at 260 °C, which was set for the chromatographic run. The sample of 1 μL of extract was injected into the instrument where the temperature of the oven was 60 °C (2 min), followed by 300 °C at the rate of 10 °C·min^{-1} and 300 °C for 6 min. The conditions for the mass detector were a temperature of 230 °C for the transfer line, a temperature of 230 °C for the ion source, an ionization mode electron impact at 70 eV, a scan time of 0.2 s and a scan interval of 0.1 s. The comparison of spectrum of components were carried out with the database of spectrum of known components stored in the GC-MS NIST (2008) library [43].

2.9. In Vitro Screening of Secondary Metabolites Produced from Mono and Co-Culture of Trichoderma spp. for Its Antibacterial Activity by Agar Well Diffusion Method against Xanthomonas oryzae pv. oryzae (Xoo)

The *Xanthomonas oryzae* pv. *oryzae* (Xoo) bacterial cultures were inoculated to 100 mL conical flasks containing nutrient broth and were incubated at 28 °C overnight. The petri plates were swabbed with Xoo cultures whose concentration was adjusted to 10^8 CFU/mL. The agar petri plates were made with the required number of wells using a sterile cork borer, ensuring the proper distribution of wells in the periphery and one in the center. Tetracycline was used as a positive control and distilled water as a negative control. The secondary metabolite crude extracts were loaded in each well. The plates were incubated at 28 °C for 24 h and after incubation, the plates were observed, and the results were recorded [44].

2.10. Mycoynthesis of Zinc Oxide Nanoparticles (ZnO NPs)

The green synthesis method was used to prepare ZnO NPs using the following fungi: *T. harzianum*, *T. reesei*, *T. reesei* and a co-culture of *Trichoderma* spp. The zinc nitrate hexahydrate ($Zn(NO_3)_2{\cdot}6H_2O$) was utilized within the test without any further purification, which was procured from Sigma-Aldrich analytical grade. The $Zn(NO_3)_2{\cdot}6H_2O$ taken in 1 g were dissolved in 10 mL of double-distilled water to the above mixture of 2 mL of fungal extract *T. harzianum*, *T. reesei*, *T. reesei* and co-culture of *Trichoderma* spp. was added and stirred for ~5–10 min using a magnetic stirrer. The mixture obtained was kept in a preheated muffle furnace, which was maintained at 400 ± 10 °C at which temperature the reaction mixture boils, froths, and heat-forming foam dehydrates in less than 3 min. The product obtained was calcinated at 700 °C for 2 h, and the obtained final product was used for further studies [30,45].

2.11. Characterization of Mycosynthesized Zinc Oxide Nanoparticles

The UV-visible spectrophotometer (SL 159 ELICO) was used for recording the UV-visible absorption in the samples. The particles were then characterized by evaluating their chemical composition through FTIR spectroscopy. Powder X-ray diffractometer (Shimadzu) using Cu Kα (1.5418 Å) radiation with a nickel filter was used to examine the phase purity and the crystallinity of the superstructures. The surface morphology of the nanoparticles was examined by SEM (Hitachi Tabletop TM-3000) along with EDAX. Studies were carried out using JEOL JEM 2100 TEM transmission electron microscopy (TEM), high-resolution transmission electron microscopy (HRTEM), and selected area electron diffraction (SAED) [46].

2.12. Antibacterial Activity of Mycosynthesized ZnO NPs against Xanthomonas oryzae

A disc diffusion method was carried out to assess the presence of antibacterial activity of ZnO NPs. A bacterial culture of 0.5 McFarland standard was used to lawn Muller Hinton agar plates evenly using a sterile swab. The discs loaded with ZnO NPs 1 mg/mL (PGT4, PGT5, PGT13 and PGTA) were placed on the Mueller Hinton agar plates. Each test plate had a positive control, a negative control and four treated ZnO NP (PGT4, PGT5, PGT13 and PGTA) discs. The standard tetracycline (0.5 mg/mL) was used as a positive control, and the negative control used was distilled water. The plates were incubated at 28 ± 2 °C for 24 h. After the incubation, the plates were examined and results were recorded in mm [47].

The synthesized ZnO NPs were subjected for antibacterial activity by minimum inhibitory concentration (MIC), which was determined using a broth microdilution method with minor modifications [30]. The desired different concentrations in 100 µg/mL, 50 µg/mL, 25 µg/mL, 12.5 µg/mL and 6.25 µg/mL was obtained by diluting the ZnO NP stock solution (1 mg/mL) with dilution in Mueller Hinton Broth (MHB) medium, which was added to a 96-well sterile microtiter plate. The bacterial suspension measuring 10 µL was added to each well, which were then incubated for 24 h at 28 °C. MHB served as a negative control, the positive control being the tetracycline at the concentration of 100 µg/mL; all these tests were conducted in triplicates. After 24 h of incubation, the addition of 20 µL of iodonitrotetrazolium chloride dye (INT) (0.5 µg/mL) to each well determined the MIC values of the synthesized ZnO NPs. The microtiter plates were incubated for 60 min at 28 °C. MICs determine the lowest concentration of the drug that prevents the change of color from being colorless to red, where colorless tetrazolium salt acts as an electron acceptor and gets reduced to a red-colored formazan product by biologically active organisms [30].

2.13. Statistical Analysis

All data of antibacterial experiments were analyzed statistically, using SPSS software (version 20.0) and Microsoft Excel. The obtained data were further subjected to analysis of variance (ANOVA), and the means were analyzed using Duncan's new multiple range post test at $p \leq 0.05$.

3. Results and Discussion

3.1. Collection and Isolation of Bacteria from Diseased Rice Leaf

The infected leaf samples were collected from different districts of Karnataka like Kolar, Chikkaballapura, Mysuru, Tumkur, Mandya and Bellary; thirty-five bacterial isolates were isolated from samples of 85 different infected samples collected from different rice fields of Karnataka (Table 1). The Xoo was recovered from the samples collected, showing typical Xoo bacterial colony characteristics such as a yellow color and mucoid, convex colonies on plating the samples as explained by Jabeen et al. [33] (Table S1)

Table 1. Antibacterial activity of *Trichoderma* spp. isolated from rhizosphere soil on screening against *Xanthomonas oryzae* pv. *oryzae* (Xoo) (Mean ± standard deviation).

Trichoderma spp. Culture	Xoo Isolates (Zone of Inhibition)		
	MBXoo69 (mm in Diameter)	MBXoo70 (mm in Diameter)	MBXoo53 (mm in Diameter)
PGT1	20.33 ^{mno} ± 0.577	22.33 ^{qrs} ± 0.577	19.33 ^{lmn} ± 1.155
PGT2	19.67 ^{lmn} ± 0.577	21.00 ^{mno} ± 1.732	19.33 ^{mno} ± 1.155
PGT3	20.00 ^{mnop} ± 0.000	20.33 ^{lmn} ± 1.528	19.00 ^{lmn} ± 1.000
PGT4	18.33 ^{jk} ± 0.577	22.67 ^{pqr} ± 1.155	24.67 ^{v} ± 0.577
PGT5	19.67 ^{lmn} ± 0.577	22.00 ^{pq} ± 1.000	19.67 ^{lmn} ± 0.577
PGT6	19.33 ^{lmn} ± 0.577	21.33 ^{pq} ± 0.577	21.33 ^{pqr} ± 0.577
PGT7	19.67 ^{lmn} ± 0.577	19. 67 ^{lmn} ± 0.577	20.33 ^{mno} ± 0.577
PGT8	19.67 ^{lmn} ± 0.577	19.67 ^{lmn} ± 0.577	19.67 ^{lmn} ± 0.577
PGT9	19.67 ^{lmn} ± 0.577	19.67 ^{lmn} ± 0.577	20.00 ^{mno} ± 0.000
PGT10	19.67 ^{lmn} ± 0.577	17.67 ^{ijk} ± 0.577	19.67 ^{lmn} ± 0.577
PGT11	19.67 ^{lmn} ± 0.577	21.33 ^{pq} ± 0.577	19.67 ^{lmn} ± 0.577
PGT12	19.67 ^{lmn} ± 0.577	16.67 ^{gh} ± 0.577	20.00 ^{mno} ± 0.000

Table 1. *Cont.*

Trichoderma spp. Culture	Xoo Isolates (Zone of Inhibition)		
	MBXoo69 (mm in Diameter)	MBXoo70 (mm in Diameter)	MBXoo53 (mm in Diameter)
PGT13	20.00 ^{mno} ± 0.000	21.67 ^{pqr} ± 0.577	19.67 ^{lmn} ± 0.577
PGT14	19.67 ^{lmn} ± 0.577	21.67 ^{pqr} ± 0.577	19.67 ^{lmn} ± 0.577
PGT15	16.00 ^{fg} ± 0.000	18.33 ^{jk} ± 0.577	20.00 ^{mno} ± 0.000
PGT16	17.67 ^{ijk} ± 0.577	18.67 ^{kl} ± 0.577	19.67 ^{lmn} ± 0.577
PGT17	19.67 ^{lmn} ± 0.577	17.00 ^{hi} ± 0.000	19.33 ^{lm} ± 0.577
PGT18	17.67 ^{ijk} ± 0.577	14.67 ^{cd} ± 0.577	18.33 ^{jk} ± 0.577
PGT19	19.67 ^{lmn} ± 0.577	18.33 ^{jk} ± 0.577	18.00 ^{jk} ± 0.000
PGT20	19.67 ^{lmn} ± o.577	18.33 ^{jk} ± 0.577	19.67 ^{lmn} ± 0.577
PGT21	17.33 ^{hij} ± 0.577	18.33 ^{jk} ± 0.577	18.33 ^{jk} ± 0.577
PGT22	19.67 ^{lmno} ± 0.577	21.67 ^{pqr} ± 0.577	18.33 ^{jk} ± 0.577
PGT23	18.33 ^{jk} ± 0.577	19.67 ^{lmno} ± 0.577	19.67 ^{lmno} ± 0.577
PGT24	20.67 ^{nop} ± 0.577	19.67 ^{lmn} ± 0.577	20.00 ^{mno} ± 0.000
PGT25	14.00 ^{c} ± 0.000	21.67 ^{pqr} ± 0.577	14.67 ^{cd} ± 0.577
PGT26	12.33 ^{b} ± 0.577	18.33 ^{jk} ± 0.577	15.67 ^{ef} ± 0.577
PGT27	11.00 ^{a} ± 0.000	18.00 ^{jk} ± 0.000	15.00 ^{de} ± 0.000
PGT28	11.00 ^{a} ± 0.000	18.33 ^{jk} ± 0.577	15.00 ^{de} ± 0.000
PGT29	12.33 ^{b} ± 0.577	17.67 ^{ijk} ± 0.577	19.67 ^{lmn} ± 0.577
PGT30	24.67 ^{v} ± 0.577	24.67 ^{v} ± 0.577	20.67 ^{nop} ± 0.577
PGT31	22.67 st ± 0.577	23.67 ^{u} ± 0.577	21.00 ^{opq} ± 0.000
PGT32	24.67 ^{v} ± 0.577	25.00 ^{v} ± 0.000	21.67 ^{pqr} ± 0.577
PGT33	22.67 st ± 0.577	25.33 ^{v} ± 0.577	23.00 ^{uv} ± 0.000
PGT34	21.00 ^{opq} ± 0.000	21.67 ^{pqr} ± 0.577	19.67 ^{lmno} ± 0.577
PGT35	22.67 st ± 0.577	20.00 ^{mn} ± 0.000	20.67 ^{nop} ± 0.000
Positive	25.00 ^{v} ± 0.577	26.00 ^{v} ± 0.000	25.00 ^{rs} ± 0.000
Negative	0 ^{a}	0 ^{a}	0 ^{a}

The mean values are recorded and the values are not significantly different according to Duncan's Multiple Range Test indicated by different alphabets in the superscripts at $p < 0.05$.

3.2. Identification of Isolated Bacteria by Biochemical and Molecular Characterization

Morphologically identified samples were pure cultured and further used for biochemical and physiological identification methods according to [33] for the tests such as Gram's reaction, oxidase test, catalase test, KOH test, starch hydrolysis test, casein hydrolysis test, gelatin liquefaction test and pectin hydrolysis test. The isolated bacteria were Gram-negative, short rods producing yellow-colored pigment. The bacterial isolates tested were positive for catalase, oxidase, 3% KOH, gelatin liquefaction, starch hydrolysis and pectin hydrolysis. Our results correlate with the results of Arshad et al. [48] The isolates were positive for all the tests except for the Gram's reaction, which showed it to be negative, indicating that it is a gram-negative organism. The identification of the isolated bacteria was confirmed to be *Xanthomonas oryzae pv. oryzae* (Xoo) by molecular analysis. The isolates were sub cultured and maintained for further antibacterial studies.

3.3. Collection of Rhizospheric Soil Samples

A total of 180 rhizospheric soil samples were collected from different crop-growing regions of Karnataka districts like Chikkaballapura, Kolar, Ramanagara and Tumkur and are shown in Table S2.

3.4. Isolation and Identification of Selected Trichoderma spp. by Molecular Characterization

The colonies of *Trichoderma* spp. were grown on PDA plates by incubating for 6–7 days at 28 °C. A total of 35 isolates of *Trichoderma* spp. were isolated from 197 different rhizosphere soils collected from different parts of Karnataka. Standard techniques were used for DNA isolation, agarose gel electrophoresis and PCR amplification. Three fungal cultures of *Trichoderma* spp. were identified by molecular identification and were found to be *Trichoderma harzianum* (PGT4) isolated from *Cucumis sativus* (cucumber) rhizosphere soil, *Trichoderma reesei* (PGT5) isolated from *Solanum melongena* (brinjal) rhizosphere soil and one more strain of *Trichoderma reesei* (PGT13) isolated from *Coriandrum sativum* (coriander) rhizosphere soil sample collected from the Chikkaballapura district. These isolates were further confirmed by database searches which were carried out with the BLAST programs available at the National Centre for Biotechnology Information (Bethesda, MD, USA), which showed the identification matching to 99%, 100% and 100%, respectively. Our findings co-related with that of Ru and Di [37], where the isolates were isolated from rhizosphere soil of potato (Table S3).

3.5. In Vitro Screening of PGP Strains of Trichoderma spp. for Its Antibacterial Activity by Agar Plug Method against Xanthomonas oryzae pv. oryzae (Xoo)

The control of a broad range of plant pathogens, including fungal, bacterial and viral diseases, through elicitation of Induced Systemic Resistance (ISR) by *Trichoderma* spp. or localized resistance has been reported. Some *Trichoderma* spp. rhizosphere-competent strains have been shown to have direct effects on plants, increasing their growth potential and nutrient uptake, fertilizer use efficiency, percentage and rate of seed germination and stimulation of plant defenses against biotic and abiotic damage by Hermosa et al. [49]. A total of 35 *Trichoderma* spp. were subjected for preliminary antibacterial activity screening on solid media. The plates were pre swabbed with Xoo and the agar plugs of 7 mm of fungi taken from an actively growing region of 7-day-old culture were placed on the Mueller Hinton agar (MHA). The plates were kept for incubation in the incubator for 24 h at 28 °C. After the incubation period, the plates were observed for zone of inhibition and in each plate, a standard tetracycline was used as a positive control and distilled water was taken as a negative control. All the isolates showed activity against the tested three plant pathogenic bacteria Xoo. The majority of these isolates showed a wide range of inhibition of the Xoo cultures. The zone of inhibition ranged from 11 mm to 25 mm. There are reports of *Trichoderma harzianum* isolate which showed strong antagonism against fungal species by Leelavathi et al. [50] (Figure S1) (Table 1). Based on the zone of inhibition, three *Trichoderma* spp. were selected for further analysis.

3.6. Trichoderma–Trichoderma Interactions through Co-Culture

The co-culture of selected *Trichoderma* spp. isolates showed the production of secondary metabolites on incubation, showing growth compatibility when grown on solid media. However, no dual cultures were found to be incompatible. Our findings co-related with the results of that of the development of secondary metabolites on a laboratory scale; fungal isolates containing the largest areas of secondary metabolite accumulation on mono and dual cultures on semi-solid media were used by Ortuno et al. [41] (Figure 1).

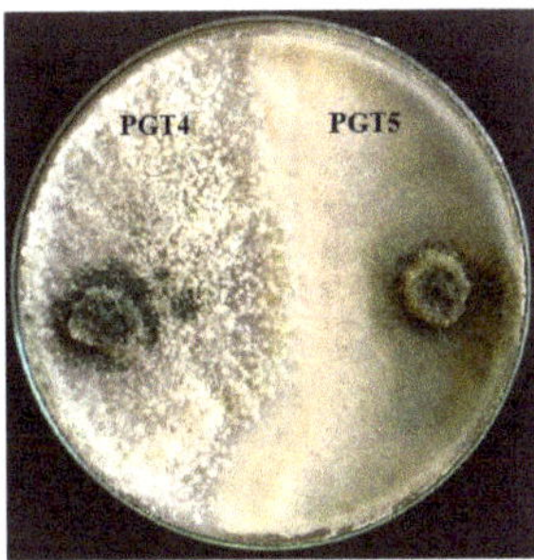

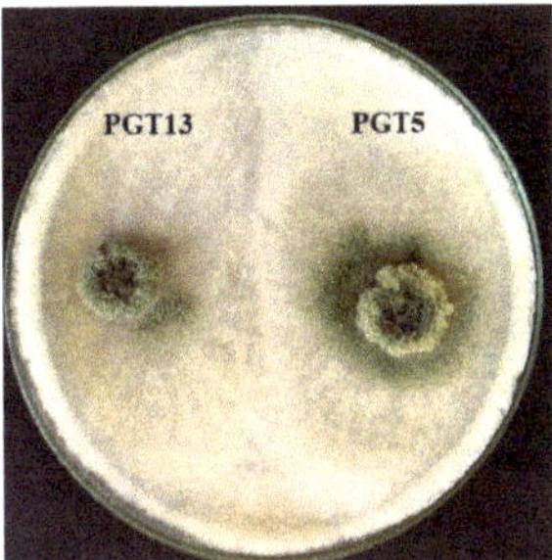

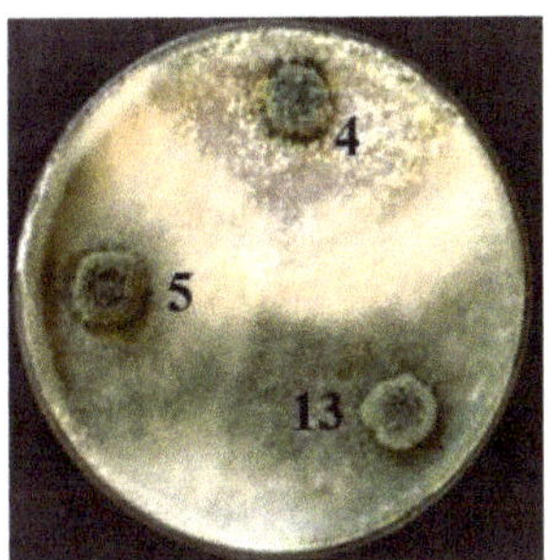

Figure 1. Co-culture of *Trichoderma* spp. plates showing the compatibility: 4—*Trichoderma harzianum*, 5—*Trichoderma reesei* and 13—*Trichoderma reesei*.

3.7. Identification of Secondary Metabolite Compounds by GC-MS Analysis

The ethyl acetate extract of the monocultures of *T. harzianum* (PGT4), *T. reesei* (PGT5) and *T. reesei* (PGT13) and co-cultures of Sample A (PGT4, PGT5, PGT13) were analyzed by GC-MS and the analysis has led to the identification of different compounds present in mono and co-culture of *Trichoderma* spp. Co-cultivation of beneficial fungi can stimulate the synthesis of novel secondary metabolites rather than in monocultures. The significant compounds that were found in the ethyl acetate solvent extract for PGT4 (Table 2), PGT5 (Table 3) and PGT13 (Table 4) and Sample A (Table 5). Gas chromatographic (GC) methods are usually done for the determination of volatile fungal metabolites for different fungi such as *Aspergillus, Fusarium, Mucor, Penicillium* and *Trichoderma* (Siddiquee et al. [51] (Figure 2)).

Table 2. List of compounds of *Trichoderma harzianum* (PGT4) detected by Gas Chromatography Mass Spectroscopy (GC-MS). RT—Retention Time.

SL. No.	RT (mins)	Name of the Compound *Trichoderma harzianum* (PGT4)	Molecular Formula	Molecular Weight
1	10.70	4-Propylbenzaldehyde	$C_{10}H_{12}O$	148
2	13.37	6-Pentyl-2H-pyran-2-one	$C_{10}H_{14}O_2$	166
3	14.02	2,4-Di-tert-butylphenol	$C_{14}H_{22}O$	206
4	15.76	4,6-*O*-Furylidene-d-glucopyranose	$C_{11}H_{14}O_7$	258
5	16.55	Trimethyl-3,4-undecadiene-2,10-dione	$C_{14}H_{22}O_2$	222
6	17.37	1,5-Diphenyl-3-(3-cyclopentylpropyl)pentane	$C_{25}H_{34}$	334
7	18.10	1-[2-Methyl-2-(-4-methyl-3-pentenyl)cyclopropyl]ethanol	$C_{12}H_{22}O$	182
8	21.44	Phthalic acid, diisobutyl ester	$C_{16}H_{22}O_4$	278
9	22.36	Dibutyl phthalate	$C_{16}H_{22}O_4$	278
10	22.77	Phthalic acid, butyl 2-pentyl ester	$C_{17}H_{24}O_4$	292
11	23.38	Phthalic acid, 6-ethyl-3-octyl butyl ester	$C_{22}H_{34}O_4$	362
12	24.17	Dibutyl phthalate	$C_{16}H_{22}O_4$	278
13	31.79	3-Ethyl-3-hydroxyandrostan-17-one	$C_{21}H_{34}O_2$	318
14	32.44	Mono(2-ethylhexyl) phthalate	$C_{16}H_{22}O_4$	278
15	35.87	Digitoxin	$C_{41}H_{64}O_{13}$	764

Table 3. List of compounds of *Trichoderma reesei* (PGT5) detected by GC-MS. RT—Retention Time.

SL. No.	RT (mins)	Name of the Compound *Trichoderma reesei* (PGT5)	Molecular Formula	Molecular Weight
1	8.02	n-Nonaldehyde	$C_9H_{18}O$	142
2	10.82	p-propylbenzaldehyde	$C_{10}H_{12}O$	148
3	12.95	4-(2-Hydroxyethyl)phenol	$C_8H_{10}O_2$	138
4	14.13	Phenol, 2,4-di-tert-butyl	$C_{14}H_{22}O$	206
5	19.14	(3E)-3-Octadecene	$C_{18}H_{36}$	252
6	20.65	Phthalic acid, butyl isobutyl ester	$C_{16}H_{22}O_4$	278
7	22.52	Dibutyl phthalate	$C_{16}H_{22}O_4$	278
8	22.93	Phthalic acid, butyl 2-pentyl ester	$C_{17}H_{24}O_4$	292
9	23.54	1,2-Benzenecarboxylic acid, bis(2-methylpropyl) ester	$C_{16}H_{22}O_4$	278
10	24.34	Dibutyl phthalate	$C_{16}H_{22}O_4$	278
11	27.11	1-Hydroxy-4-methylanthra-9,10-quinone	$C_{15}H_{10}O_3$	238
12	28.82	3-Chloro-6-(phenylsulfsnyl)bicycle(3.1.1)hept-2-ene	$C_{13}H_{13}ClS$	236
13	32.57	Mono(2-ethylhexyl) phthalate	$C_{16}H_{22}O_4$	278
14	33.27	(10E)-10-Henicosene	$C_{21}H_{42}$	294
15	21.59	Phthalic acid, diisobutyl ester	$C_{16}H_{22}O_4$	278

Table 4. List of compounds of *Trichoderma reesei* (PGT13) detected by GC-MS. RT—Retention Time.

SL. No.	RT (mins)	Name of the Compound *Trichoderma reesei* (PGT13)	Molecular Formula	Molecular Weight
1	7.93	5,6-Dimethylundecane	$C_{13}H_{28}$	184
2	9.36	1-Methylene-1H-indene	$C_{10}H_8$	128
3	10.72	Benzaldehyde, 4-propyl	$C_{10}H_{12}O$	148
4	14.03	Phenol, 2,4-di-tert-butyl	$C_{14}H_{22}O$	206
5	14.35	Maleic acid, dibutyl ester	$C_{12}H_{20}O_4$	228
6	15.31	Phthalic acid, ethyl 2-methylbutyl ester	$C_{15}H_{20}O_4$	264
7	19.65	Tetradeconic acid, 1-methylethyl ester	$C_{17}H_{32}O_2$	270
8	21.44	Phthalic acid, butyl isobutyl ester	$C_{16}H_{22}O_4$	278
9	22.34	Dibutyl phthalate	$C_{16}H_{22}O_4$	278
10	22.77	Phthalic acid, butyl 2-pentyl ester	$C_{17}H_{24}O_4$	292
11	24.15	Phthalic acid, butyl hexyl ester	$C_{18}H_6O_4$	306
12	26.92	Anthraquinone, 1-hydroxy-2-methyl	$C_{15}H_{10}O_3$	238
13	30.73	Cis-4,7,10,13,16,19-Docosahexanoic acid, tert-butyldimethyldilyl ester	$C_{28}H_{46}O_2Si$	442
14	32.46	9-t-Butyltricyclo[4.2.1.1(2,5)]decane-9-10-diol	$C_{14}H_{24}O_2$	224
15	36.62	(5E,7E)-25-[(Trimethylsilyl)oxy]-9, 10-secocholesta-5,7,10-triene-1,3-diol	$C_{30}H_{52}O_3Si$	488
16	38.21	2,6-Ditert-butyl-4-methylphenyl 2-methylcyclopropanecarboxylate	$C_{20}H_{30}O_2$	302

Table 5. List of compounds of *Trichoderma* spp. co-culture (PGTA) detected by GC-MS. RT—Retention Time.

SL. No.	RT (mins)	Name of the Compound Sample A (4,5,13)	Molecular Formula	Molecular Weight
1	10.71	Benzaldehyde, 4-propyl	$C_{10}H_{12}O$	148
2	18.11	1-[2-Methyl-2-(-4-methyl-3-pentenyl)cyclopropyl]ethanol	$C_{12}H_{22}O$	182
3	5.79	4-Hydroxybenzenephophoric acid	$C_6H_7O_4P$	174
4	7.94	Nonanal	$C_9H_{18}O$	142
5	14.03	Phenol, 2,4,-di-tert-butyl	$C_{14}H_{22}O$	206
6	21.55	1-Oxa-spiro[4,5]deca-6,9-diene-2,8-dione, 7,9-di-tert-butyl	$C_{17}H_{24}O_3$	276
7	22.28	Hexadecanoic acid	$C_{16}H_{32}O_2$	256
8	23.91	Chrysophanic acid anthranol	$C_{15}H_{12}O_3$	240
9	25.58	Trans-13-Octadecenoic acid	$C_{18}H_{34}O_2$	282
10	26.95	1-Hydroxy-4-methylanthra-9, 10-quinone	$C_{15}H_{10}O_3$	238
11	29.29	10,12-Pentacosadiynoic acid	$C_{25}H_{42}O_2$	374
12	30.35	4-(2-Oxiranyl)-9H-fluoren-9-ol	$C_{15}H_{12}O_2$	224
13	32.47	9-t-Butyltricyclo[4.2.1.1(2,5)]decane-9,10-diol	$C_{14}H_{24}O_2$	224
14	33.14	6,9-Octadecadiynoic acid, methyl ester	$C_{19}H_{30}O_2$	290
15	36.61	Ledene oxide-(II)	$C_{15}H_{24}O$	220

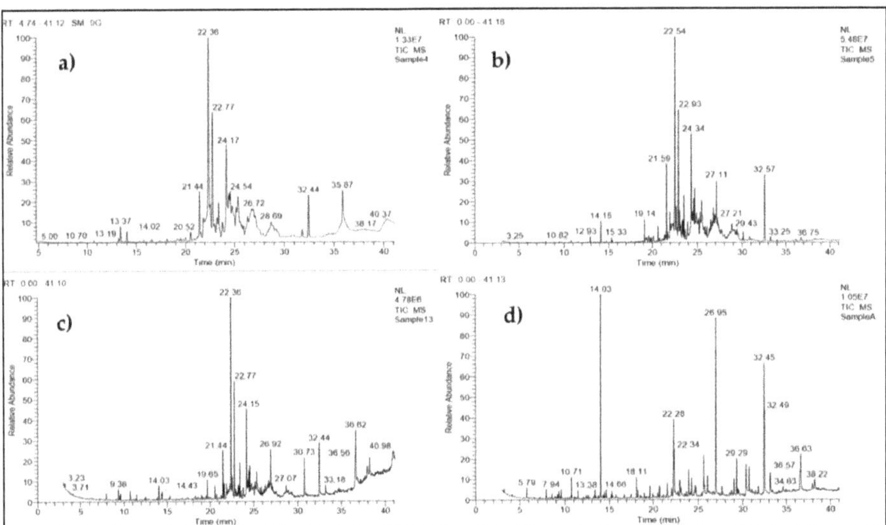

Figure 2. (**a**–**d**) Gas Chromatography Mass Spectroscopy (GC-MS) of compounds identified from secondary metabolite crude extracts of (**a**) *T. harzianum* (PGT4), (**b**) *T. reesei* (PGT5), (**c**) *T. reesei* (PGT13) and (**d**) co-culture of *Trichoderma* spp. (PGTA).

Interpretation of Mass Spectrum Gas Chromatography Mass Spectroscopy (GC-MS) was carried out using the database of the National Institute of Standard and Technology (NIST) with more than 62,000 patterns. The unknown spectrum of components was compared with that of the known components contained in the NIST library. The name, molecular weight and structure of the components of the test material were ascertained.

3.8. *In Vitro Screening of PGPR Trichoderma spp. Co-Culture Secondary Metabolites for Its Antibacterial Activity by Agar Well Diffusion Method against Xanthomonas oryzae* pv. *oryzae* (Xoo)

In our study, we analyzed the effects of growing the fungal cultures in single cultures or in combination cultures of *T. harzianum* (PGT4), *T. reesei* (PGT5) and one more culture of *T. reesei* (PGT13) for the production of fungal secondary metabolites in liquid media.

The zone of inhibition was found to range from 26 mm to 29 mm in diameter for the co-culture of (PGTA) followed by *T. harzianum* (PGT4) and *T. reesei* (PGT13) ranging from 23 mm to 26 mm in diameter, followed by a mono culture of *T. reesei* (PGT13). The zone of inhibition for *T. reesei* (PGT5) and *T. reesei* (PGT13) ranged from 20 mm to 26 mm. The zone of inhibition was found to be highest in the co-culture rather than in the mono cultures. Our findings co-related with the results of *T. harzianum* M10 and *T. pinophilus* F36CF on the production of fungal secondary metabolites in the liquid culture both in single and combined treatment by Vinale et al. [22] (Figure 3) (Table 6).

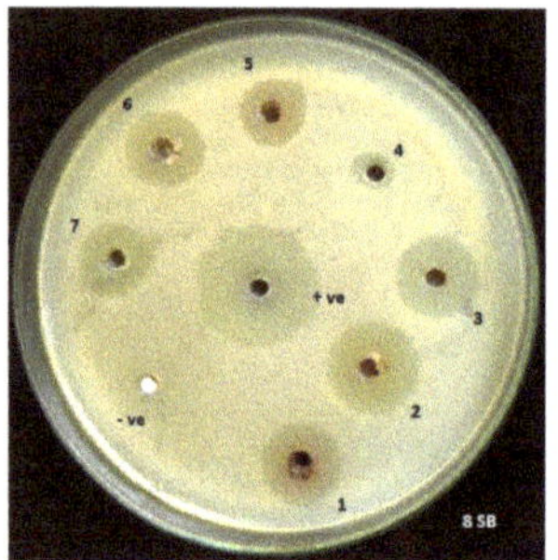

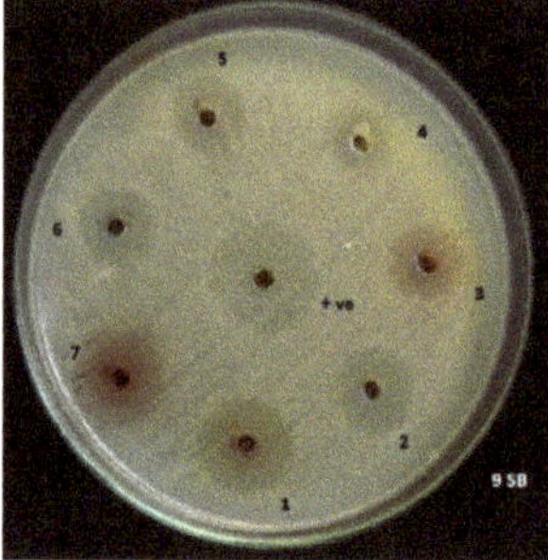

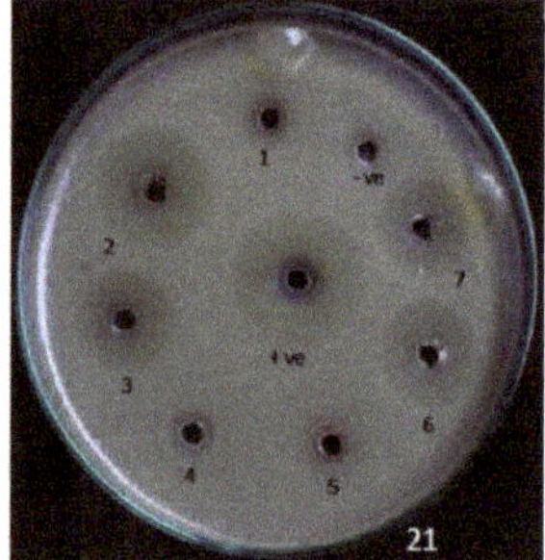

Figure 3. Screening of *Trichoderma* spp. co-culture secondary metabolites for its antibacterial activity against plant pathogen *Xanthomonas oryzae* pv. *oryzae* (Xoo); 8SB, 9SB and 21 represent strains MBXoo69, MBXoo70 and MBXoo53, respectively.

Table 6. Antibacterial activity of *Trichoderma* spp. co-culture secondary metabolite against plant pathogen Xoo (Mean ± standard deviation).

SL. No.	*Trichoderma* spp. Coculture Secondary Metabolite	Xoo Isolates (Zone of Inhibition)		
		MBXoo69 (8SB) (mm in Diameter)	MBXoo70 (9SB) (mm in Diameter)	MBXoo53 (21) (mm in Diameter)
1	PGT13	$24^{defgh} \pm 1.00$	$25.33^{gh} \pm 1.15$	$24.67^{fgh} \pm 0.577$
2	PGT4,5,13	$22.33^{cde} \pm 1.528$	$28^{i} \pm 1.00$	$21.67^{cd} \pm 1.528$
3	PGT4,13	$22^{cd} \pm 2.000$	$24.67^{gh} \pm 0.577$	$22.00^{cd} \pm 1.000$
4	PGT4	$15.67^{b} \pm 1.155$	$15.67^{b} \pm 1.155$	$13.33^{a} \pm 1.528$
5	PGT5	$20.33^{c} \pm 0.577$	$23.33^{def} \pm 1.155$	$24.33^{efg} \pm 0.577$
6	PGT5,13	$20.67^{c} \pm 0577$	$23.33^{def} \pm 1.528$	$21.67^{cd} \pm 1.528$
7	PGT4,5	$21.67^{cd} \pm 0.577$	$21.67^{cd} \pm 0.577$	$20.67^{c} \pm 0.577$
Positive	Positive	$26.00^{gh} \pm 1.732$	$29.33^{hi} \pm 1.155$	$27.67^{i} \pm 0.577$
Negative	Negative	0^{a}	0^{a}	0^{a}

The mean values are recorded and the values are not significantly different according to Duncan's Multiple Range Test indicated by different alphabets in the superscripts at $p < 0.05$.

3.9. *Characterization of Synthesized Nanoparticles*

The UV-visible absorption spectra was observed within the range of 372–374 nm. The obtained results matched with that of the UV–vis spectra of ZnO NPs for olive leaves (*Olea europaea*), chamomile flower (*Matricaria chamomilla* L.) and red tomato fruit (*Lycopersicon esculentum* M.) and showed strong absorption bands at 384, 380 and 386 nm respectively according to Ogunyemi et al. [52] (Figure 4a).

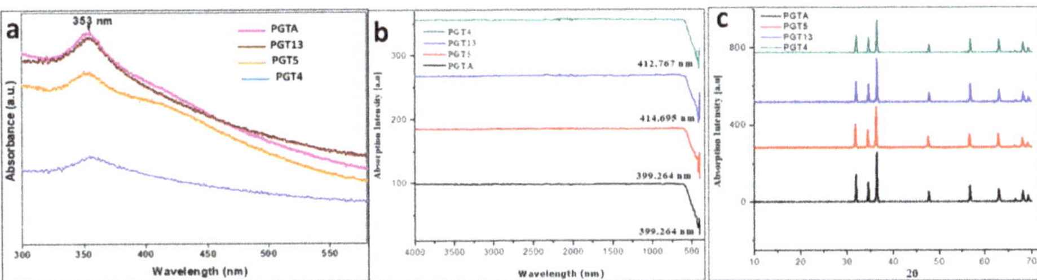

Figure 4. (**a**). UV–vis spectra of co-culture of *Trichoderma* sp. (PGTA), *Trichoderma reesei* (PGT5), *Trichoderma reesei* (PGT13) and *Trichoderma harzianum* (PGT4); (**b**) FT-IR Spectrogram of synthesized ZnO NPs *Trichoderma harzianum* (PGT4), *Trichoderma reesei* (PGT13), *Trichoderma reesei* (PGT5) and co-culture of *Trichoderma* sp. (PGTA); (**c**) PXRD patterns of ZnO NPs of *Trichoderma* spp.; co-culture of *Trichoderma* spp. (PGTA), *Trichoderma reesei* (PGT5), *Trichoderma reesei* (PGT13) and *Trichoderma harzianum* (PGT4).

The FTIR spectrum showed the absorption at 400 cm $^{-1}$ to 600 cm $^{-1}$ which further confirms the presence and formation of ZnO nanoparticles by using *Trichoderma* spp. Similar results were observed from nanoparticles synthesized by the biological method using plant extracts Ogunyemi et al. [52] At 700 °C, the obtained fungal secondary metabolites were converted to their respective oxides, leading to the formation of ZnO NPs. This implies that most of the compounds present in the sample do not have a high thermal stability. Hence there were no other vibration modes detected in the FT-IR spectra as shown in (Figure 4b) other than ZnO NPs.

The biosynthesized ZnO NPs PXRD patterns showed noticeable peaks and it was well-matched to JCPDS No. 75-576. Similarly, nanoparticles from *Trichoderma* spp. are synthesized with ZnO. The biosynthesized ZnO-NPs of the crystalline structure was confirmed by stiff and narrow diffraction peaks with no significant variance in the diffraction peaks, suggesting that the crystalline product was free of impurities. Similarly, Lakshmeesha et al. [30] reported the green synthesis of *Nerium oleander* ZnO-NPs with no impurities in the obtained crystalline product. The size of the present study's crystalline particles of green synthesized ZnO-NPs was calculated using Scherrer's formula, which was within a range of 12–35 nm. Accordingly, Dobrucka and Dlugaszewska [53] reported the biosynthesis of ZnO nanoparticles using *Trifolium pratense*, with a hexagonal wurtzite shape and the sharp peaks calculated using Scherrer's formula were 60–70 nm according to Murali et al. [54] (Figure 4c).

The formed shapes of ZnO NPs were displayed in SEM images with different surface morphology. The elements involved in the formation of nanoparticles were subjected for the EDAX analysis to know the qualitative difference as well as the quantitative difference. The analysis revealed the highest proportion of zinc (50.36%) in nanoparticles and oxygen (49.64%) in all the synthesized nanoparticles according to Prasad et al. [55] (Figures 5–8).

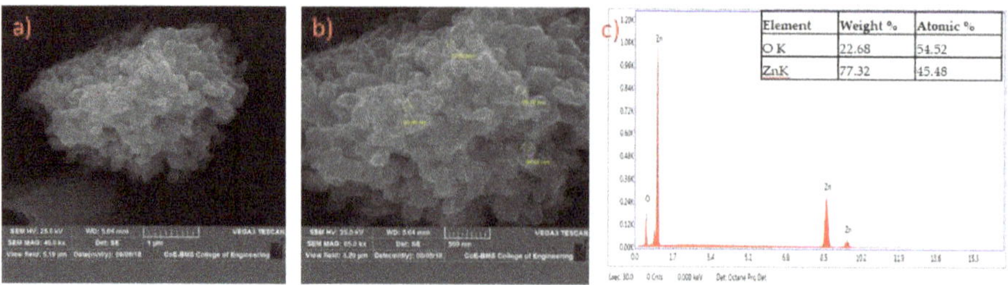

Figure 5. SEM image of zinc oxide nanoparticles synthesized from *Trichoderma harzianum* (PGT4) at lower (**a**) and higher magnification (**b**); (**c**) represents energy-dispersive X-ray spectroscopy (EDAX) analysis.

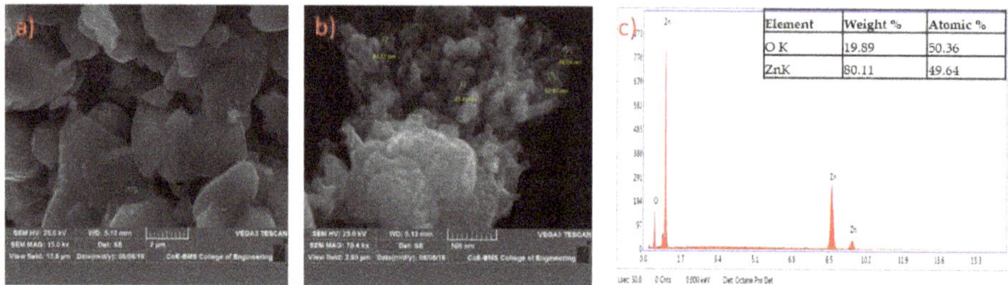

Figure 6. SEM image of zinc oxide nanoparticles synthesized from *Trichoderma reesei* (PGT5) at lower (**a**) and higher (**b**) magnification; (**c**) represents energy-dispersive X-ray spectroscopy (EDAX) analysis.

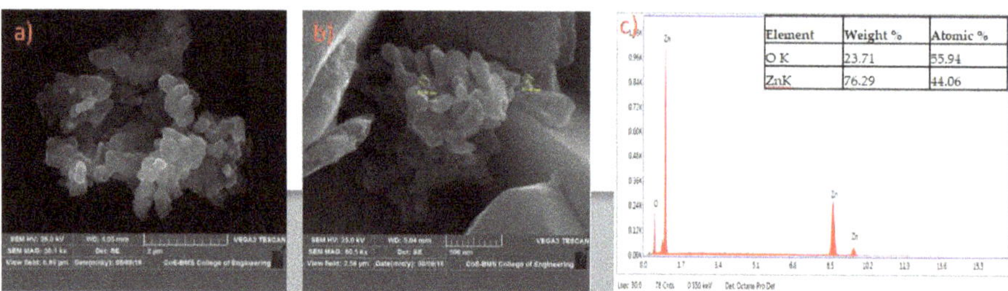

Figure 7. SEM image of zinc oxide nanoparticles synthesized from *Trichoderma reesei* (PGT13) at lower (**a**) and higher (**b**) magnification; (**c**) represents energy-dispersive X-ray spectroscopy (EDAX) analysis.

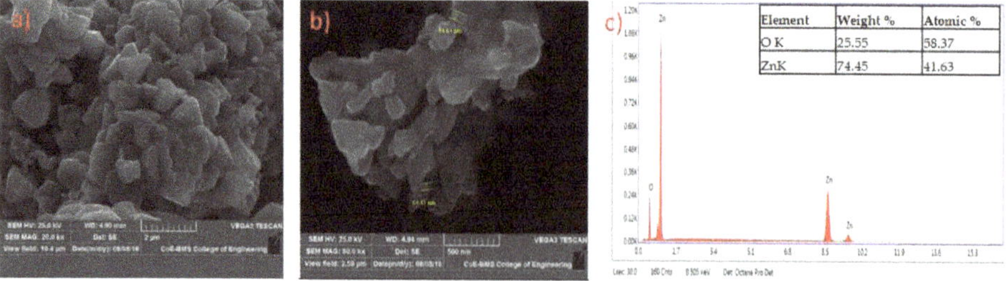

Figure 8. SEM image of zinc oxide nanoparticles synthesized from *Trichoderma* spp. co-culture of (PGTA) at lower (**a**) and higher (**b**) magnification; (**c**) represents energy-dispersive X-ray spectroscopy (EDAX) analysis.

TEM and SAED patterns correspond to ZnO compounds were obtained. The TEM image shows the agglomerated small particles of ZnO. The high resolution TEM image shows the well-defined crystal planes. The SAED patterns are well matched with the (hkl) values corresponding to the prominent peaks of the PXRD profiles (Figure 9a–d).

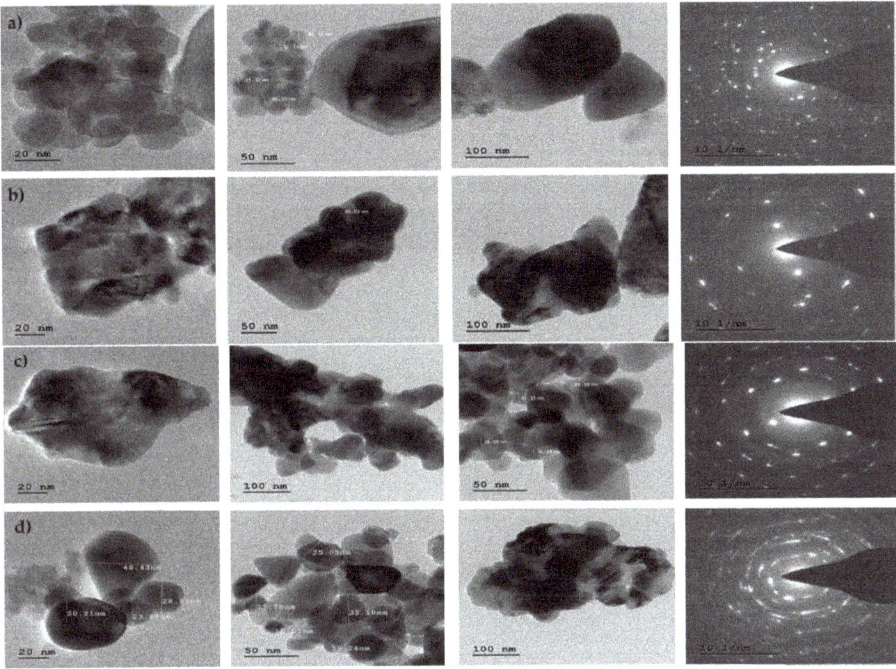

Figure 9. TEM micrographs (at different magnifications) and SAED patterns of Zno NPs synthesized from (**a**) *Trichoderma harzianum* (PGT4); (**b**) *Trichoderma reesei* (PGT5); (**c**) *Trichoderma reesei* (PGT13); and (**d**) *Trichoderma* spp. co-culture (PGTA).

3.10. Antibacterial Activity

The antibacterial activity of ZnO NPs has been evaluated by measuring the zone of inhibition around the disc. The antibacterial activity of biosynthesized ZnO NPs was tested by an agar disc diffusion method, which was placed on the pre-swabbed Mueller-Hinton agar plate. The zone of inhibition is represented in Figure S2 and tabulated in Table 7. Further MIC values were determined for the biosynthesized ZnO NPs by the 96 well plate method, which is tabulated in Table 7 and is represented in Figure S3. The pronounced antibacterial activity of ZnO NPs can be due to its relatively small size and high surface-to-volume ratio. The present study clearly signifies the potentiality of ZnO NPs as antibacterial agents against Xoo (Figure S3) (Table 7). Our results correlated with the results of Ogunyemi et al. [52], where the zone of inhibition was recorded and the antibacterial activity of ZnO NPs was checked against Xoo when used in different concentrations.

Table 7. Evaluation of the bactericidal activity of biosynthesized zinc oxide nanoparticles from different species of *Trichoderma* against different strains of *Xanthomonas oryzae* pv. *oryzae* (Xoo) (Concentration expressed in μg/mL).

ZnO NPs	Disc Diffusion Values (in mm)	MIC Values (μg/mL)
PGT 4	00	50
PGT5	14.33 ± 0.33	25
PGT13	00	50
PGTA	15.67 ± 0.33	25
Positive	13.67 ± 0.33	25
Negative	00	00

4. Conclusions

Biological control provides an alternative for chemical fertilizers and reduces costs as well as environmental pollution. *Trichoderma* spp. are a good source of secondary metabolites. Co-cultivation of these beneficial fungi can stimulate the synthesis of novel secondary metabolites better than in monocultures. These *Trichoderma* cultures in combination can be used in field trials as they are able to inhibit the growth of the *Xanthomonas oryzae* pv. *oryzae* in in vitro conditions. The maximum zone of inhibition was recorded from the co-cultures rather than the monocultures. Biosynthesized ZnO NPs were also able to inhibit the growth of the *Xanthomonas oryzae* pv. *oryzae* in in vitro conditions, which was found to be dose-dependent.

Supplementary Materials: The following are available online at http://www.mdpi.com/2309-608X/6/3/181/s1, Figure S1. Screening of fungal discs (*Trichoderma* spp.) for antibacterial activity against *Xanthomonas oryzae* pv. *oryzae* A- MBXoo69, B- MBXoo53 (+- tetracycline - distilled water). Figure S2: Antibacterial activity of zinc oxide nanoparticles (ZnO NPs) against *Xanthomonas oryzae* pv. *oryzae*(Xoo) by disc diffusion method (+ ve = tetracycline; – ve = distilled water). Figure S3: Bacterial sensitivity test of biosynthesized zinc oxide nanoparticles from *Trichoderma harzianum* (PGT4) (MH429899.1), *Trichoderma reesei* (PGT5) (MH429901.1), *Trichoderma reesei* (PGT13) (MH429900.1) and co-culture of *Trichoderma* sp. (PGTA) against different strains of plant pathogen *Xanthomonas oryzae* pv. *oryzae* (MF579736.1) (concentration in μg/ml) (a- Trail 1, b- Trail 2, and c- Trail 3). Table S1: Place of collection of infected samples from different crop-growing regions of Karnataka. Table S2: Places of Rhizospheric soil sample collection. Table S3: Morphological and physiology characters of Rhizosphere Trichoderma spp. fungi.

Author Contributions: Conceptualization, T.R.L. and S.C.; Data curation, M.A.A. (Mohammad Azam Ansari), A.A., M.A.A. (Mohammad A. Alzohairy), S.B., R.A. and S.R.N.; Formal analysis, T.R.L., A.A., M.A.A. (Mohammad A. Alzohairy) and S.R.N.; Investigation, B.S. and S.B.; Methodology, B.S.; Resources, T.R.L., M.A.A. (Mohammad Azam Ansari), A.A., M.A.A. (Mohammad A. Alzohairy), S.B. and R.A.; Software, M.A.A. (Mohammad Azam Ansari) and M.A.A. (Mohammad A. Alzohairy); Supervision, S.C.; Visualization, S.B.; Writing—original draft, B.S. and T.R.L.; Writing—review and editing, M.A.A. (Mohammad Azam Ansari), R.A., S.R.N. and S.C. All authors have read and agreed to the published version of the manuscript.

Funding: This research was funded by UGC SAP-DRS II program and University Grants Commission (UGC)-New Delhi, India for providing financial support under UGC Rajiv Gandhi National Fellowship for SC (No. 201516-RGNF-2015-17-SC-KAR-19382).

Conflicts of Interest: The authors declare no conflict of interest. The funders had no role in the design of the study; in the collection, analyses, or interpretation of data; in the writing of the manuscript, or in the decision to publish the results.

References

1. Szareski, V.J.; Carvalho, I.R.; da Rosa, T.C.; Dellagostin, S.M.; de Pelegrin, A.J.; Barbosa, M.H.; Aumonde, T.Z. Oryza Wild Species: An alternative for rice breeding under abiotic stress conditions. *Am. J. Plant Sci.* **2018**, *9*, 1093. [CrossRef]
2. Seck, P.A.; Diagne, A.; Mohanty, S.; & Wopereis, M.C. Crops that feed the world 7: Rice. *Food Sec.* **2012**, *4*, 7–24. [CrossRef]

3. Latif, M.A.; Badsha, M.A.; Tajul, M.I.; Kabir, M.S.; Rafii, M.Y.; Mia, M.A.T. Identification of genotypes resistant to blast, bacterial leaf blight, sheath blight and tungro and efficacy of seed treating fungicides against blast disease of rice. *Sci. Res. Essays* **2011**, *6*, 2804–2811.
4. Banerjee, A.; Roy, S.; Bag, M.K.; Bhagat, S.; Kar, M.K.; Mandal, N.P.; Maiti, D. A survey of bacterial blight (Xanthomonas oryzae pv. oryzae) resistance in rice germplasm from eastern and northeastern India using molecular markers. *J. Crop Prot.* **2018**, *112*, 168–176. [CrossRef]
5. Jonit, N.Q.; Low, Y.C.; Tan, G.H. Xanthomonas oryzae pv. oryzae, Biochemical Tests, Rice (Oryza sativa), Bacterial Leaf Blight (BLB) Disease, Sekinchan. *Appl. Environ. Microbiol.* **2016**, *4*, 63–69.
6. Hoque, M.E.; Mansfield, J.W.; Bennett, M.H. Agrobacterium-mediated transformation of Indica rice genotypes: An assessment of factors affecting the transformation efficiency. *Plant Cell Tiss. Org.* **2005**, *82*, 45–55. [CrossRef]
7. Chukwu, S.C.; Rafii, M.Y.; Ramlee, S.I.; Ismail, S.I.; Hasan, M.M.; Oladosu, Y.A.; Olalekan, K.K. Bacterial leaf blight resistance in rice: A review of conventional breeding to molecular approach. *Mol. Biol. Rep.* **2019**, *46*, 1519–1532. [CrossRef]
8. Naqvi, S.A.H.; Umar, U.D.; Hasnain, A.; Rehman, A.; Perveen, R. Effect of botanical extracts: A potential biocontrol agent for Xanthomonas oryzae pv. oryzae, causing bacterial leaf blight disease of rice. *Pak. J. Agric. Sci.* **2018**, *32*, 59–72. [CrossRef]
9. Nasir, M.; Iqbal, B.; Hussain, M.; Mustafa, A.; Ayub, M. Chemical Management Of Bacterial Leaf Blight Disease In Rice. *J. Agric. Res.* **2019**, *57*, 99–103.
10. Gnanamanickam, S.S.; Priyadarisini, V.B.; Narayanan, N.N.; Vasudevan, P.; Kavitha, S. An overview of bacterial blight disease of rice and strategies for its management. *Curr. Sci.* **1999**, 1435–1444.
11. Sattari, S.Z.; Van Ittersum, M.K.; Bouwman, A.F.; Smit, A.L.; Janssen, B.H. Crop yield response to soil fertility and N, P, K inputs in different environments: Testing and improving the QUEFTS model. *Field Crops Res.* **2014**, *157*, 35–46. [CrossRef]
12. Naqvi, S.A.H. Bacterial leaf blight of rice: An overview of epidemiology and management with special reference to Indian sub-continent. *Pak. J. Agric. Sci.* **2019**, *32*, 359. [CrossRef]
13. Kohl, J.; Kolnaar, R.; Ravensberg, W.J. Mode of action of microbial biological control agents against plant diseases: Relevance beyond efficacy. *Front. Plant Sci.* **2019**, *10*, 845. [CrossRef] [PubMed]
14. Shyamala, L.; Sivakumaar, P.K. Integrated control of blast disease of rice using the antagonistic Rhizobacteria *Pseudomonas fluorescens* and the resistance inducing chemical salicylic acid. *Int. J. Res. Pure Appl. Microbiol.* **2012**, *2*, 59–63.
15. Zhang, S.; Gan, Y.; Xu, B. Application of plant-growth-promoting fungi Trichoderma longibrachiatum T6 enhances tolerance of wheat to salt stress through improvement of antioxidative defense system and gene expression. *Front. Plant Sci.* **2016**, *7*, 1405. [CrossRef]
16. Tariq, M.; Noman, M.; Ahmed, T.; Hameed, A.; Manzoor, N.; Zafar, M. Antagonistic features displayed by plant growth promoting rhizobacteria (PGPR): A review. *J. Plant Sci. Phytopathol.* **2017**, *1*, 38–43.
17. Nurfalah, A.; Ayuningrum, N.; Affandi, M.R.; Hastuti, L.D.S. Promoting Growth of Tomato (*Solanum lycopersicum* L.) by using Trichoderma—Compost—Rice Bran based Biofertilizer. In *Proceedings of the IOP Conference Series: Earth and Environmental Science, Medan, Indonesia, 8–9 December 2018*; 2019; IOP Publishing: Bristol, UK; p. 012074.
18. Lee, S.; Yap, M.; Behringer, G.; Hung, R.; Bennett, J.W. Volatile organic compounds emitted by Trichoderma species mediate plant growth. *Fungal Biol. Biotechnol.* **2016**, *3*, 7. [CrossRef]
19. Prasad, R.; Kumar, M.; Varma, A. Role of PGPR in soil fertility and plant health. In *Plant-Growth-Promoting Rhizobacteria (PGPR) and Medicinal Plants*; Springer: Cham, Germany, 2015; pp. 247–260.
20. Rahman, A.; Korejo, F.; Sultana, V.; Ara, J.; Ehteshamulhaque-Haque, S. Induction of systemic resistance in cotton by the plant growth promoting rhizobacterium and seaweed against charcoal rot disease. *Pak. J. Bot.* **2017**, *49*, 347–353.
21. Gouda, S.; Kerry, R.G.; Das, G.; Paramithiotis, S.; Shin, H.S.; Patra, J.K. Revitalization of plant growth promoting rhizobacteria for sustainable development in agriculture. *Microbiol. Res.* **2018**, *206*, 131–140. [CrossRef]
22. Vinale, F.; Nicolett, R.; Borrelli, F.; Mangoni, A.; Parisi, O.A.; Marra, R.; Lorito, M. Co-Culture of Plant Beneficial Microbes as Source of Bioactive Metabolites. *Sci. Rep.* **2017**, *7*, 14330. [CrossRef]

23. Guilger, M.; Pasquoto-Stigliani, T.; Bilesky-Jose, N.; Grillo, R.; Abhilash, P.C.; Fraceto, L.F.; De Lima, R. Biogenic silver nanoparticles based on Trichoderma harzianum: Synthesis, characterization, toxicity evaluation and biological activity. *Sci. Rep.* **2017**, *7*, 44421. [CrossRef] [PubMed]
24. Sun, K.G.; Hur, S.H. Highly sensitive non-enzymatic glucose sensor based on Pt nanoparticle decorated graphene oxide hydrogel. *Sens. Actuators B Chem.* **2015**, *210*, 618–623.
25. Mazhar, T.; Shrivastava, V.; Tomar, R.S. Green synthesis of bimetallic nanoparticles and its applications: A review. *J. Pharm. Sci. Res.* **2017**, *9*, 102.
26. Zare, E.; Pourseyedi, S.; Khatami, M.; Darezereshki, E. Simple biosynthesis of zinc oxide nanoparticles using nature's source, and it's in vitro bio-activity. *J. Mol. Struct.* **2017**, *1146*, 96–103. [CrossRef]
27. El Sayed, M.T.; El-Sayed, A.S. Tolerance and mycoremediation of silver ions by Fusarium solani. *Heliyon* **2020**, *6*, e03866. [CrossRef]
28. Iravani, S.; Varma, R.S. Greener synthesis of lignin nanoparticles and their applications. *Green Chem.* **2020**, *22*, 612–636. [CrossRef]
29. Zheng, K.; Setyawati, M.I.; Leong, D.T.; Xie, J. Antimicrobial silver nanomaterials. *Coord. Chem. Rev.* **2018**, *357*, 1–17. [CrossRef]
30. Lakshmeesha, T.R.; Sateesh, M.K.; Prasad, B.D.; Sharma, S.C.; Kavyashree, D.; Chandrasekhar, M.; Nagabhushana, H. Reactivity of crystalline ZnO superstructures against fungi and bacterial pathogens: Synthesized using Nerium oleander leaf extract. *Cryst. Growth Des.* **2014**, *14*, 4068–4079. [CrossRef]
31. Lahuf, A.A.; Kareem, A.A.; Al-Sweedi, T.M.; Alfarttoosi, H.A. Evaluation the potential of indigenous biocontrol agent Trichoderma harzianum and its interactive effect with nanosized ZnO particles against the sunflower damping-off pathogen, Rhizoctonia solani. In *Proceedings of the IOP Conference Series: Earth and Environmental Science, Banda Aceh, Indonesia, 21–21 August 2019*; 2019; Volume 365, IOP Publishing: Bristol, UK; p. 012033.
32. Chavali, M.S.; Nikolova, M.P. Metal oxide nanoparticles and their applications in nanotechnology. *SN App. Sci.* **2019**, *1*, 607. [CrossRef]
33. Jabeen, R.; Iftikhar, T.; Batool, H. Isolation, characterization, preservation and pathogenicity test of *Xanthomonas oryzae* pv. *oryzae* causing BLB disease in rice. *Pak. J. Bot.* **2012**, *44*, 261–265.
34. Garrity, G.M. Bergey's manual of Systematic bacteriology. The Proteobacteria, 2nd ed.; In *Bergey's Manual Trust*; Department of Microbiology and Molecular Genetics: East Lansing, MI, USA, 2005.
35. Sandhu, A.F.; Khan, J.A.; Ali, S.; Arshad, H.I.; Saleem, K. Molecular Characterization of Xanthomonas oryzae pv. oryzae Isolates and its Resistance Sources in Rice Germplasm. *PSM Microbiol.* **2018**, *3*, 55–61.
36. Goswami, D.; Dhandhukia, P.; Patel, P.; Thakker, J.N. Screening of PGPR from saline desert of Kutch: Growth promotion in Arachis hypogea by Bacillus licheniformis A2. *Microbiol. Res.* **2014**, *169*, 66–75. [CrossRef] [PubMed]
37. Ru, Z.; Di, W. Trichoderma spp. from rhizosphere soil and their antagonism against Fusarium sambucinum. *Afr. J. Biotechnol.* **2012**, *11*, 4180–4186.
38. Barnett, H.L.; Hunter, B.B. *Illustrated Genera of Imperfect Fungi*; APS Press: St. Paul, MN, USA, 1998; p. 218.
39. Shahid, M.; Singh, A.; Srivastava, M.; Rastogi, S.; Pathak, N. Sequencing of 28SrRNA gene for identification of Trichoderma longibrachiatum 28CP/7444 species in soil sample. *Int. J. Biotechnol. Wellness Ind.* **2013**, *2*, 84–90. [CrossRef]
40. Santos, I.P.D.; Silva, L.C.N.D.; Silva, M.V.D.; Araújo, J.M.D.; Cavalcanti, M.D.S.; Lima, V.L.D.M. Antibacterial activity of endophytic fungi from leaves of Indigofera suffruticosa Miller (Fabaceae). *Front. Microbiol.* **2015**, *6*, 350. [CrossRef]
41. Ortuno, N.; Castillo, J.A.; Miranda, C.; Claros, M.; Soto, X. The use of secondary metabolites extracted from Trichoderma for plant growth promotion in the Andean highlands. *Renew. Agr. Food Syst.* **2017**, *32*, 366–375. [CrossRef]
42. Vinale, F.; Nicoletti, R.; Lacatena, F.; Marra, R.; Sacco, A.; Lombardi, N.; Woo, S.L. Secondary metabolites from the endophytic fungus Talaromyces pinophilus. *Nat. Prod. Res.* **2017**, *31*, 1778–1785. [CrossRef]
43. Dubey, S.C.; Tripathi, A.; Dureja, P.; Grover, A. Characterization of secondary metabolites and enzymes produced by Trichoderma species and their efficacy against plant pathogenic fungi. *Indian J. Agric. Sci.* **2011**, *81*, 455–461.
44. Irshad, S.; Mahmood, M.; Perveen, F. In vitro antibacterial activities of three medicinal plants using agar well diffusion method. *Res. J. Biol. Sci.* **2012**, *2*, 1–8.

45. Shelar, G.B.; Chavan, A.M. Myco-synthesis of silver nanoparticles from Trichoderma harzianum and its impact on germination status of oil seed. *Biolife* **2015**, *3*, 109–113.
46. Nayak, S.; Bhat, M.P.; Udayashankar, A.C.; Lakshmeesha, T.R.; Geetha, N.; Jogaiah, S. Biosynthesis and characterization of Dillenia indica-mediated silver nanoparticles and their biological activity. *Appl. Organomet. Chem.* **2020**, *34*, e5567. [CrossRef]
47. Mohammed, H.A.; Al Fadhil, A.O. Antibacterial activity of Azadirachta indica (Neem) leaf extract against bacterial pathogens in Sudan. *Afr. J. Med.Sci.* **2017**, *3*, 246–2512.
48. Arshad, H.M.I.; Naureen, S.; Saleem, K.; Ali, S.; Jabeen, T.; Babar, M.M. Morphological and biochemical characterization of Xanthomonas oryzae pv. oryzae isolates collected from Punjab during 2013. *Adv. Life Sci.* **2015**, *2*, 125–130.
49. Hermosa, R.; Viterbo, A.; Chet, I.; Monte, E. Plant-beneficial effects of Trichoderma and of its genes. *Microbiology* **2012**, *158*, 17–25. [CrossRef] [PubMed]
50. Leelavathi, M.S.; Vani, L.; Reena, P. Antimicrobial activity of Trichoderma harzianum against bacteria and fungi. *Int. J. Curr. Microbiol. Appl. Sci.* **2014**, *3*, 96–103.
51. Siddiquee, S.; Cheong, B.E.; Taslima, K.; Kausar, H.; Hasan, M.M. Separation and identification of volatile compounds from liquid cultures of Trichoderma harzianum by GC-MS using three different capillary columns. *J. Chromatogr. Sci.* **2012**, *50*, 358–367. [CrossRef] [PubMed]
52. Ogunyemi, S.O.; Abdallah, Y.; Zhang, M.; Fouad, H.; Hong, X.; Ibrahim, E.; Li, B. Green synthesis of zinc oxide nanoparticles using different plant extracts and their antibacterial activity against Xanthomonas oryzae pv. oryzae. *Artif. Cells Nanomed. Biotechnol.* **2019**, *47*, 341–352. [CrossRef]
53. Dobrucka, R.; Długaszewska, J. Biosynthesis and antibacterial activity of ZnO nanoparticles using Trifolium pratense flower extract. *Saudi J. Biol. Sci.* **2016**, *23*, 517–523. [CrossRef]
54. Murali, M.; Mahendra, C.; Rajashekar, N.; Sudarshana, M.S.; Raveesha, K.A.; Amruthesh, K.N. Antibacterial and antioxidant properties of biosynthesized zinc oxide nanoparticles from Ceropegia candelabrum L.—an endemic species. *Spectrochim. Acta A Mol. Biomol. Spectrosc.* **2017**, *179*, 104–109. [CrossRef]
55. Prasad, K.S.; Pathak, D.; Patel, A.; Dalwadi, P.; Prasad, R.; Patel, P.; Selvaraj, K. Biogenic synthesis of silver nanoparticles using Nicotiana tobaccum leaf extract and study of their antibacterial effect. *Afr. J. Biotechnol.* **2011**, *10*, 8122–8130.

Journal of *Fungi*

Article

Copper-Chitosan Nanocomposite Hydrogels Against Aflatoxigenic *Aspergillus flavus* from Dairy Cattle Feed

Kamel A. Abd-Elsalam [1,*], Mousa A. Alghuthaymi [2], Ashwag Shami [3,*], Margarita S. Rubina [4], Sergey S. Abramchuk [4], Eleonora V. Shtykova [5] and Alexander Yu. Vasil'kov [4]

1 Plant Pathology Research Institute, Agricultural Research Center (ARC), 9-Gamaa St., Giza 12619, Egypt
2 Biology Department, Science and Humanities College, Shaqra University, Alquwayiyah 19245, Saudi Arabia; malghuthaymi@su.edu.sa
3 Biology Department, College of Sciences, Princess Nourah bint Abdulrahman University, Riyadh 11543, Saudi Arabia
4 A.N. Nesmeyanov Institute of Organoelement compounds (INEOS) of Russian Academy of 13 Sciences, 119454 Moscow, Russia; margorubina@yandex.ru (M.S.R.); abr@polly.phys.msu.ru (S.S.A.); alexandervasilkov@yandex.ru (A.Y.V.)
5 V. Shubnikov Institute of Crystallography of Federal Scientific Research Centre "Crystallography and Photonics" of Russian Academy of Sciences, 119333 Moscow, Russia; shtykova@ns.crys.ras.ru
* Correspondence: kamelabdelsalam@gmail.com (K.A.A.-E.); AYShami@pnu.edu.sa (A.S.); Tel.: +20-10-910-49161 (K.A.A.-E.); +966-11-823-3175 (A.S.)

Received: 29 June 2020; Accepted: 16 July 2020; Published: 21 July 2020

Abstract: The integration of copper nanoparticles as antifungal agents in polymeric matrices to produce copper polymer nanocomposites has shown excellent results in preventing the growth of a wide variety of toxigenic fungi. Copper-chitosan nanocomposite-based chitosan hydrogels (Cu-Chit/NCs hydrogel) were prepared using a metal vapor synthesis (MVS) and the resulting samples were described by transmission electron microscopy (TEM), X-ray fluorescence analysis (XRF), and small-angle X-ray scattering (SAXS). Aflatoxin-producing medium and VICAM aflatoxins tests were applied to evaluate their ability to produce aflatoxins through various strains of *Aspergillus flavus* associated with peanut meal and cotton seeds. Aflatoxin production capacity in four fungal media outlets revealed that 13 tested isolates were capable of producing both aflatoxin B1 and B2. Only 2 *A. flavus* isolates (Af11 and Af 20) fluoresced under UV light in the *A. flavus* and *parasiticus* Agar (AFPA) medium. PCR was completed using two specific primers targeting aflP and *aflA* genes involved in the synthetic track of aflatoxin. Nevertheless, the existence of *aflP* and *aflA* genes indicated some correlation with the development of aflatoxin. A unique DNA fragment of the expected 236 bp and 412 bp bands for *aflP* and *aflA* genes in *A. flavus* isolates, although non-PCR fragments have been observed in many other Aspergillus species. This study shows the antifungal activity of Cu-Chit/NCs hydrogels against aflatoxigenic strains of *A. flavus*. Our results reveal that the antifungal activity of nanocomposites in vitro can be effective depending on the type of fungal strain and nanocomposite concentration. SDS-PAGE and native proteins explain the apparent response of cellular proteins in the presence of Cu-Chit/NCs hydrogels. *A. flavus* treated with a high concentration of Cu-Chit/NCs hydrogels that can decrease or produce certain types of proteins. Cu-Chit/NCs hydrogel decreases the effect of G6DP isozyme while not affecting the activity of peroxidase isozymes in tested isolates. Additionally, microscopic measurements of scanning electron microscopy (SEM) showed damage to the fungal cell membranes. Cu-Chit/NC_S hydrogel is an innovative nano-biopesticide produced by MVS is employed in food and feed to induce plant defense against toxigenic fungi.

Keywords: aflatoxins; *Aspergillus* section *Flavi*; chitosan; feeds; nanocomposites

1. Introduction

The contamination of agricultural and dairy products with aflatoxins is a major problem for economic and public health. Aflatoxins (AFs) are fungal subsidiary products mainly developed by *Aspergillus flavus* and *Aspergillus parasiticus* strains on cereals, nuts, dried fruits, dairy, and animal feed under warm and humid conditions [1,2]. The significant source of AFs spoilage is *A. flavus*, especially aflatoxin B1, which has received a lot of attention in the food and feed industry [3]. High concentrations of aflatoxin could even prompt the disease of aflatoxicosis, an infection that affects serious disease and can lead to cancer in severe cases [1,4,5]. Additionally, chronic absorption of aflatoxins causes various adverse effects, such as increased susceptibility to various pathogens, loss of production, and a decrease in milk production yield and quality in dairy cattle [6]. The nanotechnology approach seems to be an encouraging, effective, and affordable way to reduce the health problems of mycotoxins in humans and animals. There are three different approaches to reduce mycotoxin risks: effects on mold and flour retention, mycotoxin, and minimization of toxic effects by various nanomaterials [7–9]. Chitosan and self-assembled benzoic acid polymers were synthesized, and it was found that the encapsulation of CS-BA nanogels significantly enhanced the half-life and antifungal activity properties of thyme oil against *A. flavus* strains [10]. The antifungal efficacy of mycogenic silver nanoparticles hybridizing with simvastatin against three species of the *Aspergillus Flavi* group was measured. Some nano-formulations regulated the development of the toxigenic *Aspergillus* species [11]. Plant-mediated CuO NPs were synthesized from *Cissus quadrangularis* and applied as antifungal agents against *A. niger* and *A. flavus*. The produced nano-copper showed a better performance than the carbendazim fungicide [12]. In addition, hybrid nanocomposites based on organic polymeric and inorganic matrices as effective anti-aflatoxigenic strains were explored [3,4,13,14]. Chitosan-based nanocomposite film vapor assays were applied to hybrids between thyme-organo, thyme-tea tree, and thyme-peppermint EO mixtures and demonstrated strong antifungal action against some toxic fungi, including *A. flavus*, *A. parasiticus*, and *P. chrysogenum*, limiting their production ranged from 51 to 77% [15]. There is a direct association between the concentration of aflatoxin M1 (AFM1) in milk and aflatoxin B1 (AFB1) in dairy cattle feed which results in AFM1 being found in the milk of animals on contaminated feeds with AFB1 [16]. To our understanding, antifungal action of copper-chitosan nanocomposite-based chitosan hydrogels (Cu-Chit/NC_S hydrogels) against *A. flavus* strains from animal feed samples is not previously studied. Present study aimed to: (1) recognize *alfP* and *aflA* as two essential genes that lead to development of aflatoxin in animal feed via *Aspergillus* genus. (2) determination of AFB1 and AFB2 frequency and distribution of *A. flavus* strains in relation to feed delivered to dairy cows in small farms. (3) copper-chitosan nanocomposite was produced utilizing metal vapor synthesis (MVS), the physicochemical characteristics of the nanocomposites formed were described by electron microscopy (TEM) transmission, X-ray fluorescence analysis (XRF), and X-ray scattering (SAXS) small angle. (4) The fungicidal effect of the hydrogel Cu-Chit/NCs were screened against three *A. flavus* strains. (5) Protein, isozymes, and DNA fragmentations were investigated using two electrophoresis techniques, finally, scanning electron microscope was used to assess morphological changes in NCs-treated fungi.

2. Materials and Methods

2.1. Chemicals and Reagents

High-quality Acetone with a special purity 99.5% was used as a solvent for the production of metal nanoparticles via metal vapor synthesis (MVS) technique. Prior to the synthesis solvent, it was dried under molecular sieves (4 Å) and degassed in a vacuum pump under 10^{-1} Pa by freezing and thawing cycles. The metal source was Cu foil (99.99 percent) with a surface pre-treated with concentrated HNO_3 and diluted H_2O to remove oxide film.

In this work, two types of chitosan were used. Chitosan with a high molecular weight (ChitHMW) was purchased from ACROS Organics (Wheaton, IL, USA). Chitosan with a low molecular weight (ChitLMW) was bought from Wirud (Hamburg, Germany). Until impregnation, the chitosan powder was degassed at 40 °C for 12 h under a vacuum of 10^{-1} Pa. Oxalic acid dihydrate was of analytical consistency.

2.2. Preparation of Chitosan Powder Modified with Cu NPs

Chitosan powder decorated with Cu NPs was prepared according to the procedure described in the previously published works [17–20]. For the preparation of organosol, 0.56 g of Cu foil (about 200 μm of thickness) was resistively dispersed from the tantalum boat (90 mm × 5 mm) and co-condensed with 160 mL of acetone on the liquid nitrogen-cooled walls of a quartz 5 L vessel. This procedure was carried out at a residual pressure of 10^{-5} within 1 h. Then the cooling was removed and the reactor was filled with pure argon. Under these conditions, the cryomatrice warmed to room temperature naturally within 15 min. As a result, the Cu-acetone organosol was obtained. The calculated solvent-to-metal molar ratio in the synthesis was 250:1 and the concentration of the copper organosol was 5 10^{-2} M. Cu-acetone organosol was then infiltrated with chitosan powder (HMW or LMW) in an evacuated Schlenk vessel. During the deposition procedure, the flask was stirred manually to obtain homogeneous material. Thereafter, the solvent was removed and the chitosan powder containing Cu NPs was dried in a vacuum of 10^{-1} Pa at 40 °C for 6 h. As a result, two types of powdered Cu-carrying composites based on ChitHMW and ChitLMW were prepared.

2.3. Preparation of Chitosan Gels Modified with Cu NPs

The following technique was used to prepare chitosan gels filled with Cu NPs. Cu-carrying chitosan powder (1.32 g, 5 percent *w/w*) was dissolved in an oxalic acid solution (25 mL, 1 M) with vigorous stirring at 80 °C for 30 min. The mixture was poured into the cylindrical molds (20 mm × 10 mm) and put in the water bath below 22 °C to maintain a steady gel formation temperature. Each mold had 4 g of chitosan solution. After 12 h, the prepared gels were soaked in a beaker filled with distilled water and cleaned thoroughly from excess acid until neutral pH was reached. With the methods described above, two types of gels were prepared:

Cu@ChitHMW—chitosan hydrogel doped with Cu NPs based on ChitHMW

Cu@ChitLMW—chitosan hydrogel doped with Cu NPs based on ChitLMW

For best solidification, chitosan gels were stored in a water/isopropanol (6:1, v/v) bath at room temperature (RT).

2.4. Characterization Techniques

2.4.1. Transmission Electron Microscope (TEM)

TEM images were performed with a transmission electron microscope LEO 912AB OMEGA, Zeiss (Oberkochen, Germany) at an acceleration voltage of 100 kV. Cu-carrying chitosan composites for measurements were previously suspended in deionized water (resisting 18 MΩ) and sonicated in an ultrasonic bath for 15 min at RT. Then, a small drop of the suspension was dripped onto a copper grid (200 mesh) previously coated with formvar film. Then, the samples were dried at RT for 15 min and placed under the microscope.

2.4.2. X-Ray Fluorescence (XRF) Analysis

To determine metal concentration (% *w/w*) in the composites, a VRA 30 X-Ray fluorescent analyzer (Leipzig, Germany) was used. To excite XF, an X-Ray tube with a Mo anode was used at the acceleration of 50 kV and current of 20 μA. For analysis, Cu-chitosan powders or gels in an amount of 10–12 mg was thoroughly ground and pressed into pills. Then the XRF spectra of the composites and reference samples were recorded. The standard buffer solution for spectrometer calibration was composed of a mixture of polysterene/metal salt. Quantity analysis was conducted through comparison of the peak intensity of Cu Kα line in XRF spectrum of the composite with the values of the calibration curve obtained previously.

2.4.3. Conventional Small-Angle X-Ray Scattering (SAXS) Analysis

SAXS measurements were done on laboratory diffractometer "AMUR-K" (developed in A. V. Shubnikov Institute of Crystallography, Moscow [21]). Wavelength of X-rays $\lambda = 0.1542$ nm was used, applying Kratky type geometry covered the range of scattering vector modulus $0.12 < s < 6.0$ nm^{-1} ($s = 4\pi sin\theta/\lambda$; 2θ is the scattering angle). Experimental data were normalized to the intensity of the incident beam, and then a correction on collimation error was made according to standard procedure [22]. Further data processing and interpretation was done using the program suit ATSAS [23].

Volume size distribution functions $D_V(R)$ of heterogeneities in the specimens and distance distribution functions $p(r)$ were computed by means of the regularization technique realized in program GNOM [24]. The low-resolution shapes of the Cu nanoparticles in the Cu-carrying chitosan were reconstructed ab initio using distance distribution function $p(r)$ and program DAMMIN [25]. The program utilizes a simulated annealing algorithm to build models fitting the experimental data $I_{exp}(s)$ that minimizes the discrepancy:

$$\chi^2 = \frac{1}{N-1}\sum_j \left[\frac{I_{\exp}(s_j) - cI_{calc}(s_j)}{\sigma(s_j)}\right]^2$$

where N is the number of experimental points, c is a scaling factor and $I_{calc}(s_j)$ and $\sigma(s_j)$ are the calculated intensity from the model and the experimental error on the intensities at the momentum transfer s_j, respectively. The program DAMMIN was run of about a dozen separate calculations to identify the most typical models.

2.5. *Aflatoxin-Producing Ability Medium*

Selected isolates from *A. flavus* (21 isolates) associated with dairy cattle feed samples collected from different governorates in Egypt were screened for aflatoxins production by four solid media. Four culture media recipes including *A. flavus* and *parasiticus* Agar (AFPA) (20 g/L yeast extract; 10 g/L bacteriological peptone; 0.5 g/L ferric ammonium citrate; and 15 g/L agar) [26], coconut agar (CA), PDA+ 20% NaCl, and PDA were employed to confirm aflatoxins production. All tested media were equipped as described earlier by some mycologists [1,27]. Aflatoxins production was detected via UV fluorescent light, light brown circle closed to fungal colonies was appeared after one-week incubation at 28 °C.

2.6. *VICAM Aflatoxins Assay*

Culture broth filtrates of 21 aflatoxigenic *Aspergilii* collected from peanut cake and cotton seeds were analyzed for aflatoxins production (B1, B2, G1, and G2) (Figure 1) using AflaTest Immunoaffinity (VICAM) Chromatography assay that quantifies total aflatoxin concentrations according to the manufacturer's instructions. Briefly, 5 mL of diluted sample extract (2:3, extract:water) was filtered through the column with immunological properties with a drop per second. The column was cleaned with 10 mL of water by running two drops per second through the column. Aflatoxin was eluted by transferring 1 mL of methanol at one drop per second from the column. A volume of 1 mL of bromine developer was added to the methanol elute and the total aflatoxin concentration was read in a pre-calibrated VICAM Series-4 Fluorometer set at 360 nm absorption and 450 nm emissions with a detection limit of 2 ppb (2µg/kg). Results for each sample were averaged and reported in ppb [28].

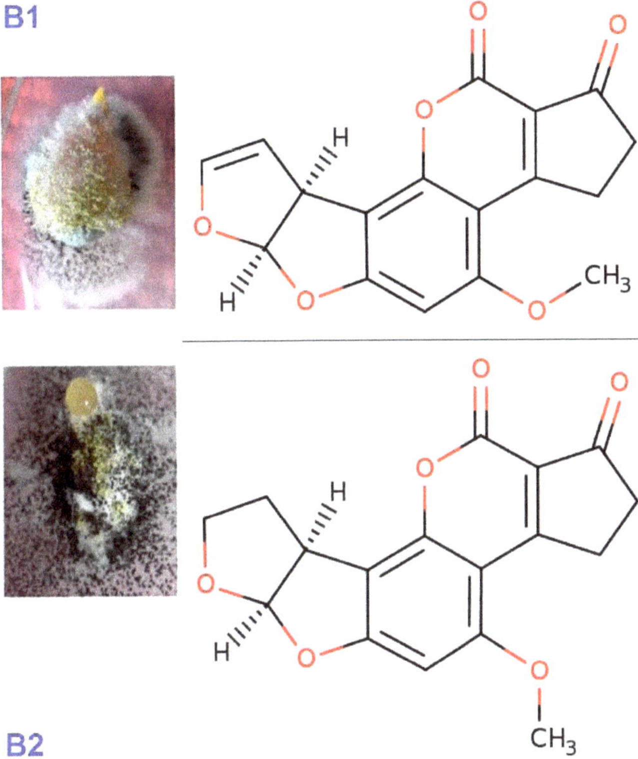

Figure 1. Structural formula of two type of aflatoxins produced by *A. flavus* isolated from peanut meal and cotton seed. Aflatoxin (B1) $C_{17}H_{12}O_6$ (6aR,9aS)-4-Methoxy-2,3,6a,9a-tetrahydrocyclopenta[c] furo[3′,2′:4,5]furo[2,3-h]chromen-1,11-dion. Aflatoxin (B2), $C_{17}H_{14}O_6$ (6aS,9aR)-4-Methoxy-2,3,6a,8,9,9a hexahydrocyclopenta[c]furo[3′,2′:4,5]furo[2,3-h]chromen-1,11-dion. Formula available online from: http://www.chemspider.com.

2.7. DNA Extraction

For the PCR amplification and DNA degradation assays, *Aspergillus* mycelium was grown in 20 mL of potato dextrose broth liquid medium (24 g/L of potato dextrose broth (Difco Laboratories, Detroit, ML, USA)). Fungal mats were gathered by separation through mesh sieves (40 mm), finally washed using sterile deionized, and dropped inside a Whatman filter paper to eliminate extra water. For homogenization, fungal mycelium was milled to acceptable powder in a mortar employing liquid nitrogen. The DNA protocol modified by Bahkali et al. [29] was used to obtain a highly purified DNA amplicon.

2.8. PCR Assay for A. flavus Detection

For specific detection of *A. flavus*, 64 PCR reaction contained 2 µL of the extracted DNA and 23 µm PCR mix containing 11.5 µL Taq DNA polymerase (Jena, Germany), 5 mM of two specific primers for *A. flavus aflP* (F-5′-CATGCTCCATCATGGTGACT-3′), (R-5′ CCGCCGCTTTGATCTAGG-3′) [30], *aflA* (F-5′-GGTGGT GAAGAAGTCTATCTAAGG-3′), and (R-5′AAGGCATAAAGGGTGTGGAG-3′) [31]. PCR thermal cycler program was adjusted as follows: 7 min at 94 °C tracked by 40 amplification cycles at 94 °C for 30 s, annealing temperature 62 for 30 s, 72 °C for 30 s, and then 72 °C for 3 min for the final extension. The amplified DNA was separated via 2% agarose gel electrophoresis containing ethidium bromide at 90 V for 30 min. Agarose gel was detected in UV transilluminator light via Gel Documentation System (Uvitec, Cambridge, UK).

2.9. Antifungal Assay

To determine the inhibition of mycelial growth of *A. flavus*, four Cu-Chit/NCs gel concentrations (60, 120, 180, and 240 ppm) were prepared as hydrogel discs for every disc containing 60 ppm. The anti-fungal efficacy of Ag-Chit-NCs was assessed by determining the reduction in fungal growth of *A. flavus* using agar-well diffusion assessment [32]. Each concentration of Cu-Chit/NCs gel has been applied to PDA plate, and petri dishes were inoculated with *A. flavus* fungal disks. Flavus isolates were incubated at 28 °C for 10 days. The inhibition factor of growth was estimated and evaluated by the equation below. [33]. Growth Inhibition (percent) = (R1−R2)/R1 × 100 where R1 was the control's radial growth and R2 for each therapy was the radial growth. After one week, the photographic record and the development of the radial colony was calculated. The experiments were conducted in three folds.

2.10. Protein Profile Degradation Assay

To investigate the Cu-Chit/NCs gel mediated protein expression in *A. flavus*, SDS-PAGE analysis was performed by Laemmli method [34]. The extracted protein from the fungal mycelium was treated with 180 ppm of Cu-Chit/NCs gel concentrations and incubated for 8 h. SDS-PAGE was performed using a 5–10% gradient of polyacrylamide gels containing 0.1% SDS. Proteins were investigated in 1.5 mm and 15 cm gels that work in dual vertical electrophoresis glass plates (Hoefer Scientific Instruments, San Francisco, CA, USA). Twenty microliters of the extracted protein were inoculated into polyacrylamide gels. SDS-PAGE samples were differentiated by separating the acrylamide gel at a stable electrical current of 30 mA, and by using a separate gel at room temperature at 90 mA, the gel was stained with silver staining. [34]. The typical molecular weight used for gel analysis was the Sigma protein marker, which is between 66,000, 45,000, and 22,000 kDa.

2.11. Native PAGE Isozyme Assay

Fungal isozymes were purified by grinding 100 mg of fungal mats in 1.0 mL extraction buffer (0.1 M Tris-HCl + 2 mM EDTA, pH 7.8). The extracted enzyme from fungal mats was treated with 180 ppm of Cu-Chit/NCs gel concentrations and incubated for 8 h. Native-PAGE was used to separate two enzymatic activities under native conditions [35]. Electrophoretic technique were conducted with the electrode buffer Tris/Glycine (pH 8.3) using 5 percent of the stacking gel and 6 percent of the separating gels. The 5 μL enzyme samples were placed over each well of the stacking gel and the gel was initially run at 60 V replaced by 100 V later. After running native acrylamide gel, and for staining glucose 6-phosphate dehydrogenase (G6PD) (EC.1.1.1.49), native protein gel was incubated in a staining solution (0.1 mM tris-HCL buffer, pH8.8, 7.5 glucose 6-phosphate (di-sodium salt), 20 mg NADP, 10 mg MTT, 10 mg PMS, 0.2 M $MgCl_2$) in the dark at 37 °C until dark blue band appear. To stop the reactions, the isozyme gel was washed and fixed in 50% ethanol [36]. Peroxidase isozyme (EC 1.11. 1.7) was stained with incubated gels in a staining solution (50 mM phosphate buffer, pH5.0, 50 mg benzidine dihydrochloride, and 3% hydrogen peroxide), and the gel was washed in water and fixed in 50% glycerol [37].

2.12. Binding/Degradation of Genomic Fungal DNA

To check DNA quality, 10 microliters of *A. flavus* DNA were treated with Cu-Chit/NCs hydrogel (180 ppm) for 2 h at 37 °C. The DNA amplicon treated NCs gel was separated on 1.5% (*w*/*v*) agarose gels prepared in 1× Tris-acetate-EDTA (TAE) and stained with ethidium bromide (EtBr, 10 mg/mL). Five microliters of extracted and treated DNA from every pattern, along with 1 μL DNA loading dye, become loaded into the wells. Agarose gel was run for 30 min at 90 V and visualized to check for DNA degradation inside the GelDoc (Uvitec, Cambridge, UK).

2.13. Scanning Electron Microscopy (SEM)

Agar disks of *A. flavus* had been inserted into sterile cellophane films and transferred to PDA mixed with 180 ppm of Cu-Chit/NCs hydrogel in Petri dishes, with one plugin keeping in the middle of the plate, and 3 replicates were used for every strain. The Petri dishes were incubated at 28 °C for 7 days. Cellophane coated with *A. flavus* was transferred and stuck in 2% (*w*/*v*) glutaraldehyde at 4 °C for 12 h in 0.1 M phosphate buffer (PB; pH 7.0). Aspergillus specimens will be put in the desiccator before further use. Upon drying, the samples prepared are assembled into the Poutron SEM coating system using standard double-sided adhesives with $\frac{1}{2}$ inch SEM nozzles and with a gold-palladium gold coating (60 s, 1.8 mA, 2.4 kV). All samples were tested with JEOL JXA-480 SEM (JEOL, Tokyo, Japan) at the National Research Center in Giza, Egypt.

3. Results

3.1. Preparation and Characterization of Cu-Carrying Chitosan Powders

Chitosan modified with metal NPs was prepared in two steps. Firstly, Cu NPs were prepared by interaction of metal vapors with acetone vapors according to the MVS protocol (see experimental part). During the second step, freshly prepared Cu-acetone organosol was deposited in situ onto chitosan powders. During the deposition procedure, the flask was stirred manually to obtain homogeneous material. Discoloration of organosol indicated the completeness of the nanoparticle's deposition on the biopolymer support and the color of powders changed from beige to dark-green. XRF analysis shows that the metal proportion of the Cu-carrying powders based on ChitLMW and ChitHMW is 0.5 and 0.83%, respectively. Previously, the same method for preparing Cu-carrying chitosan powders with a high copper concentration of 3–5% *w*/*w* was used [20].

The TEM images in the bright field and the selected area diffraction pattern (SAED) for the newly prepared Cu-acetone organosol have been seen in Figure 2A. As can be shown, SAED has diffuse reflexes suggesting the production of a significant number of very small particles. NPs have a mainly spherical shape and blurry boundaries. The NPs' sizes estimated from Figure 2B are in the range of $1 < d < 4$ nm. The good solvating properties of acetone for preparing Cu NPs were shown in previously published works [38,39].

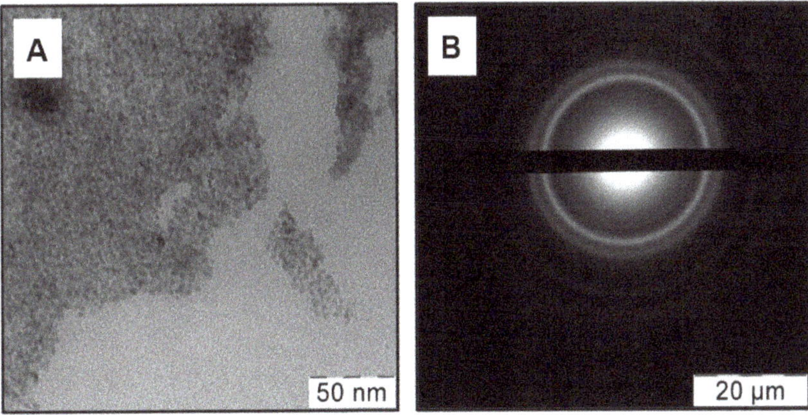

Figure 2. TEM image in bright field (**A**) and selected area diffraction pattern (SAED) (**B**) from a region for Cu-acetone organosol.

Five rings correspond to the lattice planes of Cu and Cu_2O. As a result of the proximity of some interplanar distances of Cu and Cu_2O and relatively broad rings, their superposition was observed. The formation of a core-shell structure of copper NPs with metallic copper as a core and copper oxide

(I) as a shell can be assumed. A similar structure of Cu NPs in organosols prepared with different solvents via MVS was detected [40].

In Figure 3 TEM images in bright/dark field and SAED of powdered chitosan doped with Cu NPs using the impregnation step are shown. It was detected that Cu-carrying chitosan composite contains Cu and Cu_2O phases as well as Cu-acetone organosol (Figure 3B). The crystallite sizes estimated from dark field image are in the range of 2–4 nm (Figure 3D).

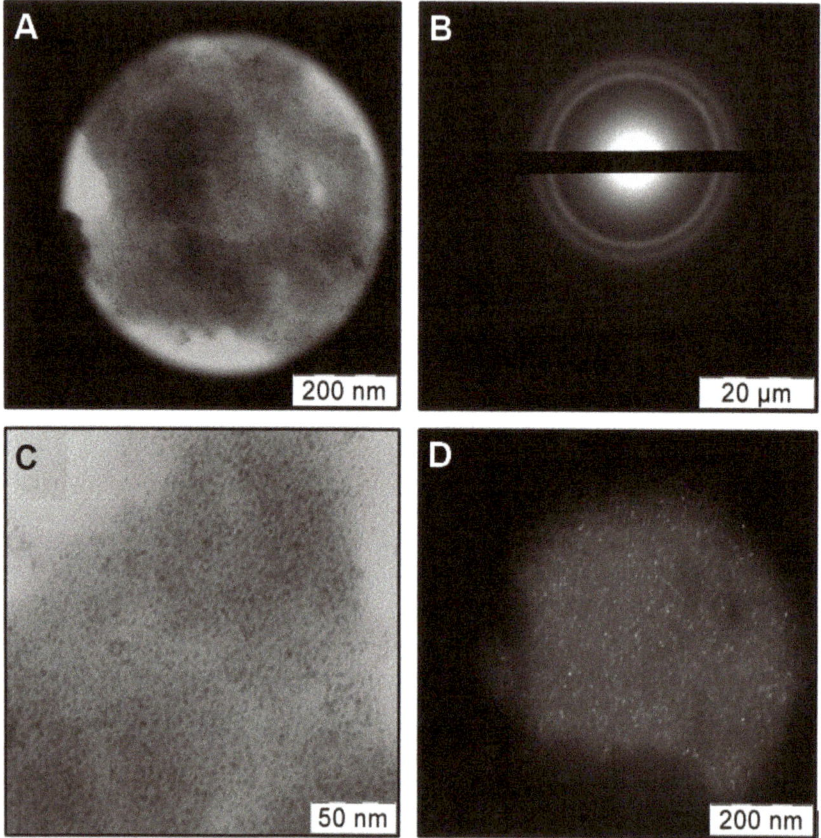

Figure 3. TEM images in bright (**A**,**C**) and dark fields (**D**) of chitosan with a high molecular weight (ChitHMW) doped with copper nanoparticles (Cu NPs) as well as SAED (**B**) of highlighted field.

It was previously demonstrated that the surface of the composite prepared with Cu-acetone organosol contains two oxidized copper states Cu^{2+} and Cu^{+}, with concentrations (at. %) of 10.7 and 3.6%, respectively [20]. Experimental SAXS curves of Cu-carrying chitosan and pristine non-modified chitosan (ChitLMW) are described in Figure 4A.

Volume size distribution functions $D_V(R)$ of heterogeneities presented in pristine non-modified chitosan and that of Cu nanoparticles embedded in the chitosan are shown in Figure 4B. To obtain $D_V(R)$ only for the Cu nanoparticles, a difference SAXS curve was calculated by subtraction of the scattering of the pristine non-modified chitosan from the scattering of the Cu-carrying chitosan (Figure 4A insert). As we can see from Figure 4B, Cu nanoparticles in this system are practically monodisperse and more compact (average size is about 1.5–2 nm) to compare with the sizes of the pores of the pristine non-modified chitosan (about 3 nm), where the Cu nanoparticles are located. A small detectable number of larger particles (possible aggregates or clusters of Cu nanoparticles) is also present in the sample. Due to the practically monodisperse character of the $D_V(R)$ function for Cu nanoparticles, one can reconstruct an average shape of the Cu nanoparticles [21]. For the ab initio restoration by the program DAMMIN [24], distance distribution function p(r) was calculated. A shortened curve with no initial part at the range of momentum transfer < 0.7 nm^{-1} was used for the calculation to minimize the influence of scattering from large aggregates on the result of the shape restoration. The distance distribution function $p(r)$ is shown in Figure 5 (insert on the top right) along with a model scattering curve from a restored shape and with a smoothed curve after the of collimation corrections.

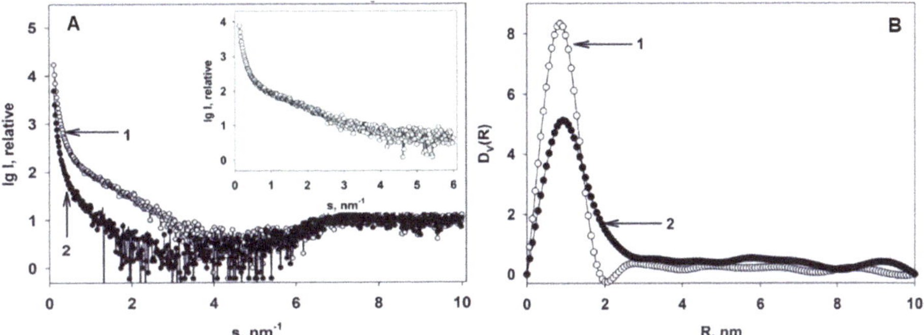

Figure 4. (**A**) Experimental (SAXS) curves: 1—Cu-carrying chitosan; 2—pristine non-modified chitosan (chitosan low molecular weight (ChitLMW)). Insert—difference SAXS curve for the embedded Cu nanoparticles. (**B**) Volume size distribution functions $D_V(R)$: 1—Cu nanoparticles; 2—pristine non-modified chitosan (ChitLMW).

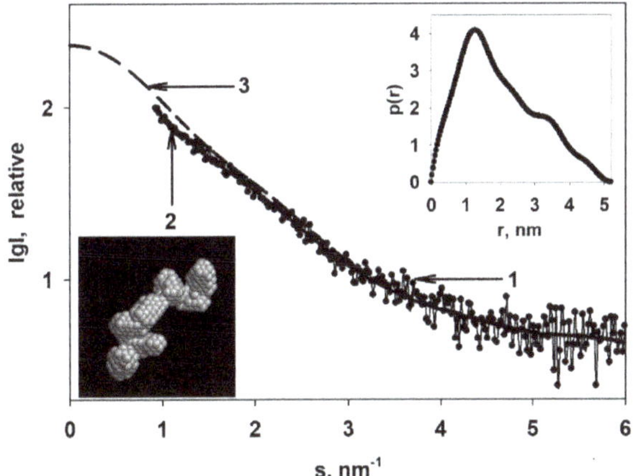

Figure 5. Reconstruction of the shape of the Cu nanoparticles in the Cu-carrying chitosan: 1—difference SAXS curve; 2—a model scattering curve calculated from the restored shape of the Cu nanoparticles; 3—extrapolated to zero angles smoothed scattering curve after the introduction of collimation corrections. Inserts: top right—distance distribution function $p(r)$; bottom left—restored shape of the Cu nanoparticles.

The restored shape of the Cu nanoparticles is a cluster consisting of 5–6 individual Cu nanoparticles with the average sizes of about 1.5–2.0 nm and with the length of the cluster of about 5 nm. Due to the presence of some amount of large aggregates it is impossible to restore the shape of the individual

Cu nanoparticles. However, the shape of the cluster is restored with very good accuracy: $\chi^2 = 0.92$, and the separate nanoparticles in the cluster are clearly visible.

Chitosan gels have been produced by ionic physical gelation of the oxalic acid-biopolymer [41,42]. Two types of Cu-carrying chitosan gels were obtained with the consecutive procedures of dissolution in oxalic acid at high temperature, gelation and thorough washing procedures (Figure 6).

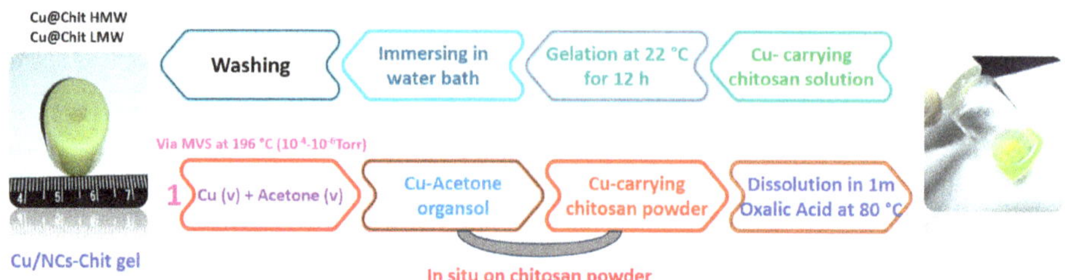

Figure 6. Scheme of preparation of chitosan hydrogels from Cu-carrying chitosan powders.

3.2. *Aflatoxins Production Ability*

The production ability of toxigenic isolates was screened on four solid media. Beige rings seen without light are observed in aflatoxigenic fungal cultures. It is also possible to visualize the blue fluorescence ring that surrounds the aflatoxigenic colony under ultraviolet light. The aflatoxigenic *A. flavus* °C. The detected beige ring diameter and the strength of its fluorescence emission were improved under UV over time with the maximum observation by the end of the week (Table 1). It is possible to use AFPA media that is suitable for fast screen aflatoxigenic fungi associated with feeds. Four forms of aflatoxin activity have been quantified and measured (B1, B2, G1, and G2). Production patterns of AFs by aflatoxigenic *A. flavus* isolates are presented in Table 1. Thirteen of these were producers of AFB_1 and AFB_2 aflatoxins and 8 were nonproducers of aflatoxins. Thirteen of the isolates produced aflatoxin B1 ranging from 4.50 to 19.44 ppb, while B2 was produced in the same isolates with a 0.02–5.29 ppb. *A. flavus* (Af1) produced the highest AFB1 concentration (19.44 ppb) while *A. flavus* (Af13) produced the intermediate quantity of AFB1 (9.10 ppb) and Af13 produced the lowest quantity of AFB1 (4.50 ppb). None of the tested isolates produced aflatoxins G1 and G2.

Table 1. Fast screen aflatoxins by various cultural media and concentration of aflatoxins assayed by VICAM test in 21 *Aspergillus flavus* isolates collected from peanut meal and cottonseeds.

Isolate Code	Feed	Fluorescence Detection under UV Light (365 nm)				The Concentration of Aflatoxins (ppb)		
		AFPA	CA	PDA + Na Cl	PDA	AFB1	AFB2	Total Aflatoxins
Af1	Peanut meal	+	−	+	−	19.44	0.03	19.47
Af2	Peanut meal	+	+	+	−	13.54	0.02	13.56
Af3	Peanut meal	+	+	−	−	10.13	0.05	10.18
Af4	Peanut meal	+	+	+	−	14.22	5.29	19.51
Af5	Peanut meal	+	−	−	−	10.14	1.30	11.44
Af6	Peanut meal	+	+	−	−	ND	ND	ND
Af7	Peanut meal	−	+	+	−	12.10	3.20	15.30
Af8*	Peanut meal	+	+	+	−	12.10	3.56	15.66
Af9	Peanut meal	+	+	−	−	ND	ND	ND
Af10	Peanut meal	+	+	+	−	10.13	0.05	10.18

Table 1. *Cont.*

Isolate Code	Feed	Fluorescence Detection under UV Light (365 nm)				The Concentration of Aflatoxins (ppb)		
		AFPA	CA	PDA + Na Cl	PDA	AFB1	AFB2	Total Aflatoxins
Af11	Cotton seeds	+	−	−	−	ND	ND	ND
Af12	Cotton seeds	+	+	+	−	7.20	2.13	9.33
Af13	Cotton seeds	+	+	+	−	9.10	2.50	11.60
AF14	Cotton seeds	+	+	+	−	8.14	1.34	9.48
AF15	Cotton seeds	+	−	+	−	ND	ND	ND
AF16	Cotton seeds	+	+	+	−	ND	ND	ND
AF17	Cotton seeds	+	+	−	−	10.64	1.36	12.00
AF18	Cotton seeds	−	+	+	−	ND	ND	ND
AF19	Cotton seeds	+	+	+	−	4.50	1.30	5.80
AF20	Cotton seeds	+	−	−	−	ND	ND	ND
AF21	Cotton seeds	−	−	−	−	ND	ND	ND

AF: Aflatoxin, +: Positive fluorescence, −: Negative fluorescence, ND: Not detected, *Aspergillus flavus* and *parasiticus* Agar (AFPA), Coconut agar (CA), potato dextrose agar (PDA), *Af4, highly producer isolate. Af13, Intermediate producer isolate. Af19, Low producer isolate.

3.3. *Aspergillus flavus PCR Detection*

A total of 21 *Aspergillus flavus*, and *Aspergillus* related isolates such as two isolates for both *Aspergillus clavatus*, *Aspergillus ochraceous*, *Aspergillus niveus*, *Aspergillus terreus*, *Aspergillus fumigatus*, *Aspergillus versicolor*, *Penicillium paneum*, *Penicillium expansum*, *Penicillium citrinum*, *Penicillium verrucosum* in addition to one isolate from *Alternaria alternata* were used for testing specificity of primers. Figure 7 indicates the findings of the PCR product inspection of the agarose gel assay; the presence of a band at 236 bp for primers *aflP* and 412 bp *aflA* primers, respectively and suggested the same predicted PCR product. The samples were prepared independently from individual strains, with an identical amount of DNA (60 ng/μL). Figure 7 shows the results for screening the specificity of the tested primers. Thus, two sets of primer pairs (*aflP* and *aflA*) are specific for the detection of *A. flavus* (Figure 7A,C). The results show that only *A. flavus* DNA can be amplified (lines 1 to 21), no PCR products from other *Aspergillus* species and Penicillium were obtained from lane 1 to 21 (Figure 7B). For rapid and accurate detection of *A. flavus* isolates tested in the current report, two PCR primer sets (*aflP* and *aflA*) were employed. The applicable primers amplified a PCR fragment sited near the 18SrRNA region with 95.3% efficiency and 100% specificity. Ultimately, existing examinations suggest the powerful specificity of the *aflA* PCR primer over different typically available diagnostic primers for correct, speedy, and large-scale identification of *A. flavus* isolated from feeds. Most *A. flavus* strains were detected with the aid of specific-PCR by using the examined primer sets.

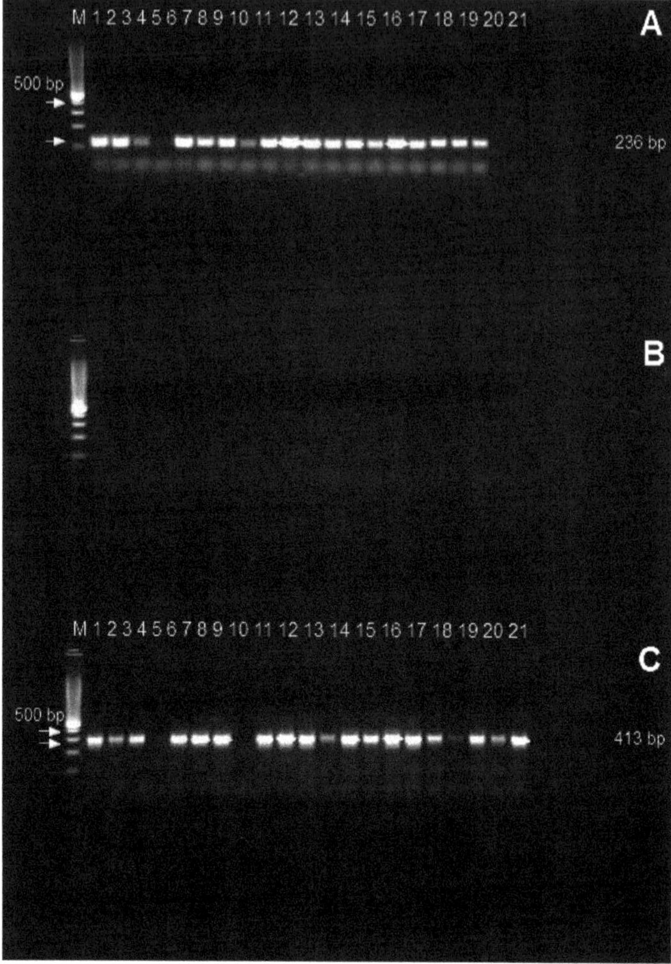

Figure 7. PCR amplicons obtained using primer pairs developed for the *aflP* (*omtA*) and *aflA* genes in 21 *Aspergillus flavus* (Lane1–21) tested with *aflP* primers (**A**), and an *Aspergillus* related isolate such as two isolates for both of *Aspergillus clavatus, Aspergillus ochraceous, Aspergillus niveus, Aspergillus terreus, Aspergillus fumigatus, Aspergillus versicolor, Penicillium paneum, Penicillium expansum, Penicillium citrinum, Penicillium verrucosum* and one *Alternaria alternate* isolate (Lane 1–21) tested with *aflP* primers (**B**). A total of 21 *Aspergillus flavus* (Lane1–21) tested with *aflA* primers (**C**).

3.4. Antifungal Activity of NCs

Various concentrations of Cu-Chit/NCs hydrogels were used to study the inactivation of *A. flavus* mycelia growth. The antifungal activity of the synthesized Cu/NCs chit gel was assessed by measuring the radial growth of mycelium for all treatments (Figure 8). The highest mycelial growth inhibition was found at a concentration of 240 ppm followed by 180, 120, and 60 ppm concentrations of Cu-Chit/NCs hydrogel. The antifungal activity of Cu-Chit/NCs hydrogel did not increase with increasing concentrations ranging between 60 and 120 ppm, while the inactivation rate constant increases with the concentration of 240 ppm of the nanocomposite. The mycelial growth inhibition varies from 100% to 5.14% in different concentrations of prepared nanocomposites (Table 2).

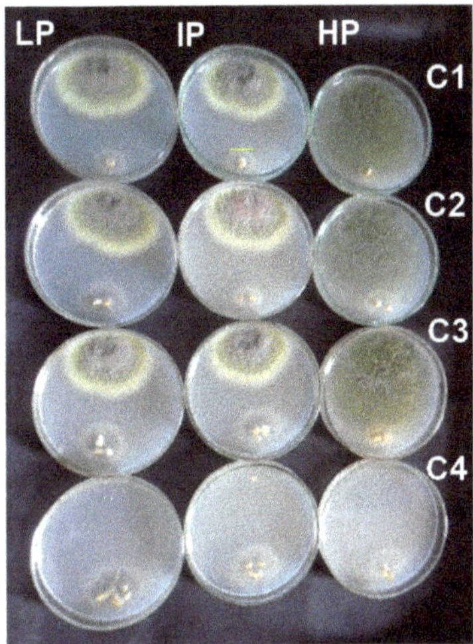

Figure 8. Antifungal activity for different concentration of Cu-Chit/NCs gel (C1 = 60, C2 = 120, C3 = 180, and C4 = 240 ppm) against *A. flavus* isolated from feeds by the plate assay. All Petri dish treatments were incubated at 28 °C for one week.

Table 2. Antifungal activity of Cu-Chit/NCs gel against three *Aspergillus flavus* strains.

Serial Number	NCs Gel Concentrations (ppm)	Strain Code	% Inhibition
1	C1 = 60	LP	27.89
2		IP	28.66
3		HP	5.14
4	C2 = 120	LP	25.67
5		IP	28.43
9		HP	6.48
7	C3 = 180	LP	29.75
8		IP	29.20
9		HP	6.57
10	C4 = 240	LP	100
11		IP	100
12		HP	100

3.5. Protein and Isozymes Profile Degradation

SDS-PAGE analysis was carried out to evaluate the change in gene expression of *A. flavus* handled with a hundred and eighty ppm of Cu-Chit/NCs gel. Some of the protein bands in the *A. flavus* isolates were not seen in the Cu-Chit/NCs gel treatment. In the control group, the protein pattern gave enhanced protein bands. In particular, there were 3 principal bands within the protein maker, inclusive of 66, 45, and 22 kDa, respectively. In high producer (HP) isolate from *A. flavus* treated with 180 ppm of Cu-Chit/NCs gel, five bands completely disappeared with molecular weights of 12, 17, 26, 38, and 55 kDa, respectively. While in intermediate producer (IP) and low producer (LP), isolates were treated with the same concentration of nanocomposites, this resulted in the induction of three newly expressed proteins with approximate molecular weights of 10 kDa, 32 kDa, and 40 kDa (Figure 9A). Native-PAGE results observed that the activity of tested isozymes changed in the treated mycelium. For G6PD activity, six different banding patterns of enzymes appeared in untreated *A. flavus* isolates while enzyme activity in treated isolates with nanocomposites decreased to four isozymes (Figure 9B). Analysis of peroxidase isozymes revealed that from 4–6 peroxidase isozyme loci in both treated and untreated isolates. As a result, Cu-Chit/NCs gels have no effect on peroxidase isozymes activity in fungal mats (Figure 9C). In the present report, we divide *A. flavus* isolates based on their aflatoxins ability into three types: high producer (HP), intermediate producer (IP), and low producer (LP). The same isolates were treated with Cu-Chit/NCs gel as a fellow T1, T2, and T3, respectively.

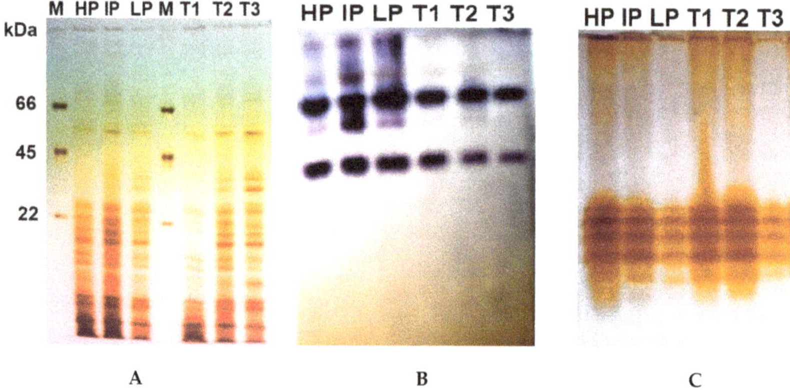

Figure 9. (**A**) Protein expression profile of SDS-PAGE extracted from *A. flavus* mycelium treated with a high Cu-Chit/NCs gel concentration. Lane M shows a standard protein molecular weight marker. Protein marker including three molecular bands ranging from 66, 45, and 22 kDa was used. Isoenzymes electrophoresis of G6PD (**B**) and peroxidase (**C**) isozymes extracted from *A. flavus* mycelium treated with a high Cu-Chit/NCs gel concentration.

3.6. DNA Binding and Degradation

Separation of genomic DNA of fungi treated with Cu-Chit/NC gel by agarose electrophoresis is broken and DNA band are faint for selected samples, while no serious harm has occurred for untreated DNA. On the other side, the fungi treated with 180 ppm of nanocomposites a slightly less intense band can be observed compared to the untreated sample (Figure 10). The genotoxic effects for fungal mycelium were investigated after treatment with Cu-Chit/NCs gel DNA degradations was separated especially at high concentrations of Cu-Chit /NCs gel.

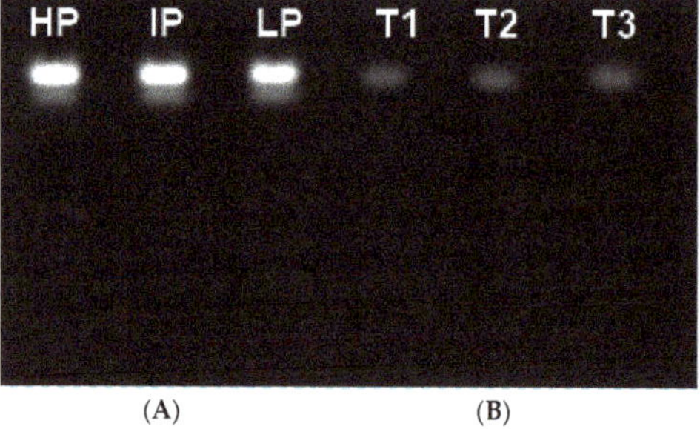

Figure 10. Agarose gel electrophoretic pattern of the fungal genomic DNA treated with 180 ppm of Cu-Chit/NCs gel, (**A**) Lanes 1–3: DNA for untreated *A. flavus* isolates, Lane 1: *A. flavus* (high producer (HP) isolate), Lane 2: *A. flavus* (intermediate producer (IP) isolate), Lane 3: *A. flavus* (low producer (LP) isolate). (**B**) Lanes T1, T2, and T3, three *A. flavus* isolates DNA treated with Cu-Chit/NCs gel, showed total damage to fragmented DNA bands.

3.7. Fungal SEM

Concerning the morphological structure of *A. flavus* investigated by SEM, adjustments in conidiophore attributes were watched. *A. flavus* was refined on PDA corrected with 180 ppm of Cu-Chit/NCs hydrogel caused slight changes in mycelial structure, highlighted by hyphal twisting and by decreasing of regenerative structures, for example, conidia and conidiophores.

SEM investigation of conidiophore changes showed that fungal spores turned into glaringly extraordinary, in which mycelia and conidiophores were contracted in comparison to untreated controls (Figure 11). Alterations of hyphae structure have been determined as proven in Figure 11A–C along with lower of cytoplasmic content material and adjustments of membrane integrity. whilst, in untreated control, development of mycelium and conidiophore was regular with considerable conidia (Figure 11D–F).

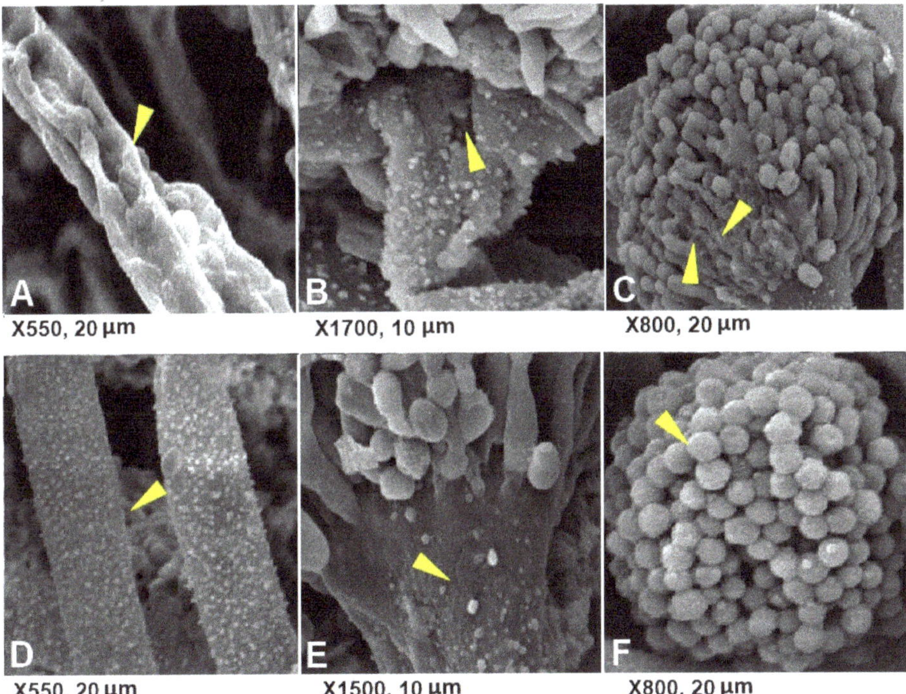

Figure 11. Scanning electron micrographs showing shriveled conidiophores (**A–C**) and the healthy mycelium or conidiophores (**D–F**) of *A. flavus* on PDA treated with Cu-Chit/NCs hydrogel. The yellow arrows mean hyphae or conidiophore.

4. Discussion

There is an enormous interest to expand effective and eco-friendly fungicides anti-toxigenic fungi with low level or zero mycotoxin residues without affecting the plant growth and crop productivity of the essential agriculture ingredients [9]. The primary target of the current examination is to evaluate the antifungal impacts of bio-polymers like chitosan cross breed with copper metals against *A. flavus* aflatoxin-creating strains. Measures of aflatoxin formed in cultural media show some alarms on the toxigenic capability of various fungal isolates to deliver a high quantity of aflatoxin in agricultural supplies [43]. In the present work, analysis of aflatoxin-producing ability by fluorescence in CA an AFPA medium showed a good correlation with the biochemical examination of aflatoxins, a cutting-edge finding in a harmony with Monda et al. [44]. However, for most purposes, we found the CA media screening technique to be simpler, faster, and much cheaper than any of the different techniques examined [45]. Although the data indicate that the aflatoxins distinguished producer media such as AFPA is not completely persistent in differentiating between aflatoxin-producing and nontoxigenic strains of *A. flavus*, it is important that the fungal medium did not yield false-positives [46].

Our results show that 62% percent of *A. flavus* is aflatoxin-producing isolates. Fifty percent of the screened isolates of *A. flavus* collected from discolored rice grains in India can produce aflatoxin B1 [47]. These results are in agreement with Abbas et al. [48] who investigated more noteworthy producer *A. flavus* strains. Lai et al. [49] indicated that more than 35% of *A. flavus* strains secreted various quantities of aflatoxins in the rice grain. The diverse aflatoxin production capacities of the *A. flavus* isolates would be affected by the various resources of the strains and also ecological problems. The aflatoxin production pathway includes roughly 30 genes, some having unsure functions in aflatoxin biosynthesis [50].

In the present work, *A. flavus* isolates were tested for the presence of gene *alfA*, which is code for fatty acid synthases, while structural gene *aflP* is one of the main genes responsible for transforming ST into O-methylsterigmatocystin [51]. Our findings show that the two primers sets are specific for fast detection of *A. flavus*. Based on specific target genes such as *aflA* and *aflP*, the present findings confirmed the applicability of PCR assays for the detection of *A. flavus* isolated from the feeds. Researchers reported strong antifungal activity of Cu NPs and chitosan nanocomposites against *A. flavus*, for example, benzoic acid nanogel (CS-BA) [10], CuO NPs [12], nanocomposites anti-aflatoxigenic [3,5,13,14]. In the current work, Cu-Chit/NC_S hydrogel showed complete inhibition of growth against *A. flavus* strains at the highest concentration (240 ppm). Furthermore, we reported that the antifungal efficacy is influenced not only by nanocomposites concentration but also by type of tested strain. The antifungal efficacy of copper oxide nanoflowers as an antifungal agent against some phytopathogenic fungi like, *A. niger, A. flavus, Penicillium notatum*, and *A. alternata* were reported [52]. In addition, CS-Cu and CSZn NCPs show strong in vitro antifungal activity against *A. alternata, Rhizoctonia solani*, and *A. flavus* and are introduced as potential materials for innovative antimicrobials in cosmetics, foodstuffs, and textiles [53]. In in vitro assays, Cu-chitosan NPs were found to be effective in inhibiting fungal growth of some plant pathogens such as *Alternaria solani* and *Fusarium oxysporum* [54]. The antifungal activity of CS NPs against two aflatoxin producers such as *A. flavus* and *A. parasiticus* was demonstrated [55] and CS NPs succeeded in reducing total aflatoxin production and inhibiting the extent of fungal growth. The main protein composition in the absence or presence of Cu-Chit/NCs gel was analyzed by SDS-PAGE in comparison with protein markers and depending on the amino acid composition. Many protein bands have not been seen in nanocomposite treatment. Subsequently, the treatment of chitosan nanocomposites generates some biological reactions, such as oxidative stress-induced metabolic changes, which in turn affect the protein synthesis rate [56]. The toxicity of nanocomposites in fungal cells is due to severe metabolic changes, in particular protein synthesis, which resulted in a maximum protein reduction, as verified by the absence of the most important protein synthesis [57].

G6PD is a housekeeping enzyme that primarily regenerates adenine dinucleotide phosphate (NADPH) nicotinamide to sustain cellular redox homeostasis. Since NADPH is necessary for NADPH oxidase (NOX), synthase of nitrogen oxides to generate reactive oxygen species, and for signaling nitrogen, several new cellular functions have been established for G6PD [58]. Lack of glucose-6-phosphate dehydrogenase isozyme can make nanoparticles more susceptible to oxidative stress [59]. Several experiments can be checked that they explain the impact of metals on G6PDH activity, also Cd^{++} greatly influences on G6PDH activity in bacteria, fungi, and vertebrates; Ni^{++} inhibits the enzyme 's kinetic properties in mammals; Zn^{++} has extreme effects from a variety of influences on G6PDH; Cu^{++} has severe effects from bacteria and animals on G6PDH [60].

G6PD stimulates xenobiotic metabolism via the Nof2 signaling pathway and impacts the xenobiotic-metabolizing expression of the enzyme [61]. The full sense is that the inhibition of *A. aculeatus* G6PD activity by zinc and many other metal nanoparticles may be reinforced by potential production or otherwise formulation of polyketide mycotoxins in toxigenic fungi, including Aspergillus [62]. Additional attempts and modes of action study are needed to examine the molecular mechanisms on which G6PD interacts with the Nrf2 pathway. This is the first report showing G6PD isozymes activity in *A. flavus* strains treated with prepared nanocomposites to understand the antifungal mechanisms. SEM images of the treated pathogen above show that the hyphae also had a swollen appearance, damaging the plasma membrane of both fungal spores and mycelium. Similar outcomes were investigated by Rubina et al. [20], who discovered that Cu-chitosan nanocomposites deteriorate fungal mycelia of *R. solani* from cotton and also *S. rolfsii* pathogenic to onion. Gold nanoparticles may alter and disturb the fungal cell membranes of *A. flavus*, *F. verticillioides*, and *P. citrinumdue* [63]. Weak sporulation with shrinking spores and defects was found in all *A. versicolor* strains treated with the modified nanocomposites [64]. More omics tools such as functional genomics, transcriptomics, proteomics, and metabolomics are required for the identification of different antifungal mechanism pathways for various nanomaterials that can be used against aflatoxigenic strains of Aspergillus and also suppress their aflatoxins production.

5. Conclusions

Dairy cattle feed is prone to fungal infections and major fungus infecting peanut meals, and cotton seeds are *A. flavus* with aflatoxins producing nature. Therefore, an urgent need to produce novel and safer nano-biocides to prevent fungal contamination of food and feed. Current research shows that these two media can only distinguish between aflatoxigenic and non-aflatoxigenic isolates from *A. flavus*. Most *A. flavus* strains react positively with *aflP* primers and *aflA* covering regions 236 bp and 412 bp, respectively. Current results suggest that the prepared nanocomposites hydrogel could be used not only as an effective fungicide against plant pathogens but also can be effectively used for the management of toxigenic fungi. In addition, omics technologies can be extended to improve our knowledge of toxigenic fungi, classify fungal species, predict fungal contamination, and may also facilitate the progress of plant breeding by gene insertion technologies to improve host plant tolerance, deter or minimize contamination of mycotoxins in feed. In addition, omics technologies can be extended to improve our knowledge of toxigenic fungi, classify fungal species, predict fungal contamination, and may also facilitate the progress of plant breeding by nanobiotechnologies to improve host plant tolerance, deter, or minimize contamination of mycotoxins in feed.

Author Contributions: K.A.A.-E., project administration, resources, conceptualization; M.A.A., biology methodology; A.S., methodology, data analysis, resources; M.S.R., nanomaterials synthesis; S.S.A. and E.V.S., material characterizations; A.Y.V., methodology, validation; K.A.A.-E., A.S., M.S.R., S.S.A., E.V.S., A.Y.V., writing—review and editing, supervision. All authors have read and agreed to the published version of the manuscript.

Funding: Current research was supported by the Science and Technology Development Fund (STDF), Joint Egypt (STDF)-South Africa (NRF) Scientific Cooperation, Grant ID. 27837 to Kamel Abd-Elsalam.

Acknowledgments: The authors would like to thank Ministry of Science and Higher Education of the Russian Federation for Physicochemical characterizations of copper-chitosan nanocomposite.

Conflicts of Interest: The authors declare no conflict of interest. The funders had a main role in the design of the study; in the collection, analyses, or interpretation of data; in the writing of the manuscript, or in the decision to publish the results.

References

1. Almoammar, H.; Bahkali, A.H.; Abd-Elsalam, K.A. A Polyphasic method for the identification of aflatoxigenic'aspergillus' species isolated from camel feeds. *Aust. J. Crop. Sci.* **2013**, *7*, 1707.
2. Omeiza, G.K.; Kabir, J.; Kwaga, J.K.P.; Kwanashie, C.N.; Mwanza, M.; Ngoma, L. A Risk assessment study of the occurrence and distribution of aflatoxigenic *Aspergillus flavus* and Aflatoxin B1 in dairy cattle feeds in a Central Northern State, Nigeria. *Toxicol. Rep.* **2018**, *5*, 846–856. [CrossRef]
3. Li, Z.; Lin, S.; An, S.; Liu, L.; Hu, Y.; Wan, L. Preparation, characterization and Anti-Aflatoxigenic activity of chitosan packaging films incorporated with turmeric essential oil. *Int. J. Biol. Macromol.* **2019**, *131*, 420–434. [CrossRef] [PubMed]

4. Mateo, E.M.; Gómez, J.V.; Domínguez, I.; Gimeno-Adelantado, J.V.; Mateo-Castro, R.; Gavara, R.; Jiménez, M. Impact of bioactive packaging systems based on EVOH films and essential oils in the control of aflatoxigenic fungi and aflatoxin production In Maize. *Int. J. Food Microbiol.* **2017**, *254*, 36–46. [CrossRef]
5. Kumar, A.; Singh, P.P.; Prakash, B. Unravelling the antifungal and Anti-Aflatoxin B1 Mechanism of Chitosan Nanocomposite Incorporated with *Foeniculum vulgare* essential oil. *Carbohydr. Polym.* **2020**, *236*, 116050. [CrossRef] [PubMed]
6. Loeffler, S.H.; de Vries, M.J.; Schukken, Y.H. The effects of time of disease occurrence, milk yield, and body condition on fertility of dairy cows. *J. Dairy Sci.* **1999**, *82*, 2589–2604. [CrossRef]
7. Abd-Elsalam, K.A.; Hashim, A.; Alghuthaymi, M.A.; Bahkali, A.H. Nanobiotechnological strategies for molud and mycotoxin control. In *Nanotechnology in Food Industry, Volume VI: Food Preservation*; Grumezescu, A.M., Ed.; ELSEVIER: Amsterdam, The Netherlands, 2016; pp. 337–364.
8. Horky, P.; Skalickova, S.; Baholet, D.; Skladanka, J. Nanoparticles as a solution for eliminating the risk of mycotoxins. *Nanomaterials* **2018**, *8*, 727. [CrossRef] [PubMed]
9. Rai, M.; Abd-Elsalam, K.A. *Nanomycotoxicology Treating Mycotoxins in Nano Way*, 1st ed.; Academic Press: Amsterdam, The Netherlands, 2019; p. 450. Available online: https://www.elsevier.com/books/nanomycotoxicology/rai/978-0-12-817998-7 (accessed on 19 April 2020).
10. Khalili, S.T.; Mohsenifar, A.; Beyki, M.; Zhaveh, S.; Rahmani-Cherati, T.; Abdollahi, A.; Bayat, M.; Tabatabaei, M. Encapsulation of Thyme Essential Oils In Chitosan-Benzoic Acid Nanogel with Enhanced Antimicrobial Activity Against *Aspergillus flavus*. *LWT-Food Sci. Technol.* **2015**, *60*, 502–508. [CrossRef]
11. Bocate, K.P.; Reis, G.F.; de Souza, P.C.; Junior, A.G.O.; Durán, N.; Nakazato, G.; Furlaneto, M.C.; de Almeida, R.S.; Panagio, L.A. Antifungal Activity of silver nanoparticles and simvastatin against toxigenic species of *Aspergillus*. *Int. J. Food Microbiol.* **2019**, *291*, 79–86. [CrossRef]
12. Devipriya, D.; Roopan, S.M. *Cissus quadrangularis* Mediated Ecofriendly Synthesis of Copper Oxide Nanoparticles and its Antifungal Studies Against *Aspergillus niger, Aspergillus flavus*. *Mater. Sci. Eng. C* **2017**, *80*, 38–44. [CrossRef]
13. Jampílek, J.; Kráľová, K. Nanocomposites: Synergistic nanotools for management of mycotoxigenic Fungi. In *Nanomycotoxicology*; Academic Press: Amsterdam, The Netherlands, 2020; pp. 349–383.
14. Tamayo, L.; Azócar, M.; Kogan, M.; Riveros, A.; Páez, M. Copper-Polymer nanocomposites: An excellent and cost-effective biocide for use on antibacterial surfaces. *Mater. Sci. Eng. C* **2016**, *69*, 1391–1409. [CrossRef]
15. Hossain, F.; Follett, P.; Salmieri, S.; Vu, K.D.; Fraschini, C.; Lacroix, M. Antifungal activities of combined treatments of irradiation and Essential Oils (Eos) Encapsulated chitosan nanocomposite films In Vitro and In Situ conditions. *Int. J. Food Microbiol.* **2019**, *295*, 33–40. [CrossRef] [PubMed]
16. Xiong, J.L.; Wang, Y.M.; Nennich, T.D.; Li, Y.; Liu, J.X. Transfer of dietary Aflatoxin B1 to Milk aflatoxin M1 and effect of inclusion of adsorbent in the diet of dairy cows. *J. Dairy Sci.* **2015**, *98*, 2545–2554. [CrossRef] [PubMed]
17. Vasil'kov, A.Y.; Rubina, M.S.; Naumkin, A.V.; Zubavichus, Y.V.; Belyakova, O.A.; Maksimov, Y.V.; Imshennik, V.K. Metal-Containing systems based on chitosan and a Collagen-Chitosan composite. *Russ. Chem. Bull.* **2015**, *64*, 1663–1670. [CrossRef]
18. Vasil'kov, A.Y.; Rubina, M.S.; Gallyamova, A.A.; Naumkin, A.V.; Buzin, M.I.; Murav'eva, G.P. Mesoporic material from microcrystalline cellulose with gold Nanoparticles: A new approach to metal-carrying polysaccharides. *Mendeleev Commun.* **2015**, *25*, 358–360. [CrossRef]
19. Rubina, M.S.; Kamitov, E.E.; Zubavichus, Y.V.; Peters, G.S.; Naumkin, A.V.; Suzer, S.; Vasil'kov, A.Y. Collagen-Chitosan scaffold modified with Au and Ag nanoparticles: Synthesis and structure. *Appl. Surf. Sci.* **2016**, *366*, 365–371. [CrossRef]
20. Rubina, M.S.; Vasil'kov, A.Y.; Naumkin, A.V.; Shtykova, E.V.; Abramchuk, S.S.; Alghuthaymi, M.A.; Abd-Elsalam, K.A. Synthesis and characterization of chitosan-copper nanocomposites and their fungicidal activity against two sclerotia-forming plant pathogenic Fungi. *J. Nanostructures Chem.* **2017**, *7*, 249–258. [CrossRef]
21. Mogilevskiy, L.Y.; Dembo, A.T.; Svergun, D.I.; Feygin, L.A. Small-angle X-ray diffractometer with single coordinate detector. *Crystallography* **1984**, *29*, 587–591.
22. Feigin, L.A.; Svergun, D.I. *Structure Analysis by Small-Angle X-ray and Neutron Scattering*; Plenum Press: New York, NY, USA, 1987.

23. Franke, D.; Petoukhov, M.V.; Konarev, P.V.; Panjkovich, A.; Tuukkanen, A.; Mertens, H.D.T.; Kikhney, A.G.; Hajizadeh, N.R.; Franklin, J.M.; Jeffries, C.M.; et al. ATSAS 2.8: A Comprehensive Data Analysis Suite for small-angle scattering from macromolecular solutions. *J. Appl. Crystallogr.* **2017**, *50*, 1212–1225. [CrossRef]
24. Svergun, D.I. Determination of the regularization parameter in indirect-transform methods using perceptual criteria. *J. Appl. Crystallogr.* **1992**, *25*, 495–503. [CrossRef]
25. Svergun, D.I. Restoring low resolution structure of biological macromolecules from solution scattering using simulated annealing. *Biophys. J.* **1999**, *76*, 2879–2886. [CrossRef]
26. Pitt, J.I.; Hocking, A.D.; Glenn, D.R. An improved medium for the detection of *Aspergillus flavus* and *A. parasiticus*. *J. Appl. Bacteriol.* **1983**, *54*, 109–114. [CrossRef] [PubMed]
27. Samson, R.A.; Visagie, C.M.; Houbraken, J.; Hong, S.B.; Hubka, V.; Klaassen, C.H.; Perrone, G.; Seifert, K.A.; Susca, A.; Tanney, J.B.; et al. Phylogeny, identification and nomenclature of the genus aspergillus. *Stud. Mycol.* **2014**, *78*, 141–173. [CrossRef] [PubMed]
28. Kaaya, A.N.; Eboku, D. Mould and aflatoxin contamination of dried cassava chips in eastern uganda: Association with traditional processing and storage practices. *J. Biol. Sci.* **2010**, *10*, 718–729. [CrossRef]
29. Bahkali, A.H.; Abd-Elsalam, K.A.; Guo, J.R.; Khiyami, M.A.; Verreet, J.A. Characterization of Novel Di-, Tri-, and tetranucleotide microsatellite primers suitable for genotyping various plant pathogenic Fungi with special emphasis on *Fusaria* and *Mycospherella graminicola*. *Int. J. Mol. Sci.* **2012**, *13*, 2951–2964. [CrossRef]
30. Al-Shuhaib, M.B.S.; Albakri, A.H.; Alwan, S.H.; Almandil, N.B.; AbdulAzeez, S.; Borgio, J.F. Optimal Pcr primers for rapid and accurate detection of *Aspergillus flavus* Isolates. *Microb. Pathog.* **2018**, *116*, 351–355. [CrossRef]
31. Trinh, H.L.; Anh, D.T.; Thong, P.M.; Hue, N.T. A Simple PCR for Detection of *Aspergillus flavus* in Infected Food. *Qual. Assur. Saf. Crop. Foods* **2014**, *7*, 375–383. [CrossRef]
32. Perez, C.; Paul, M.; Bazerque, P. Antibiotic assay by agar-well diffusion method. *Acta Biol. Med. Exp.* **1990**, *15*, 113–115.
33. Pariona, N.; Mtz-Enriquez, A.I.; Sánchez-Rangel, D.; Carrión, G.; Paraguay-Delgado, F.; Rosas-Saito, G. Green-Synthesized copper nanoparticles as a potential antifungal against plant pathogens. *RSC Adv.* **2019**, *9*, 18835–18843. [CrossRef]
34. Laemmli, U.K. Cleavage of structural proteins during the assembly of the head of bacteriophage T4. *Nature* **1970**, *227*, 680–685. [CrossRef]
35. Rabilloud, T.; Vuillard, L.; Gilly, C.; Lawrence, J.J. Silver-staining of proteins in polyacrylamide Gels: A General overview. *arXiv* **2009**, arXiv:0911.4458.
36. Agarwal, S.; Nath, A.K.; Sharma, D.R. Characterisation of Peach (*Prunus Persica* L.) cultivars using isozymes as molecular markers. *Sci. Hortic.* **2001**, *90*, 227–242. [CrossRef]
37. Vallejos, C.E. Enzyme activity staining. In *Tanksley SD, Isozymes in Plant Genetics and Breeding*; Orton, T.S., Ed.; Elsevier: Amsterdam, The Netherland, 1983; p. 469.
38. Evangelisti, C.; Vitulli, G.; Schiavi, E.; Vitulli, M.; Bertozzi, S.; Salvadori, P.; Bertinetti, L.; Martra, G. Nanoscale Cu Supported catalysts in the partial oxidation of cyclohexane with molecular oxygen. *Catal. Lett.* **2007**, *116*, 57–62. [CrossRef]
39. Vitulli, G.; Bernini, M.; Bertozzi, S.; Pitzalis, E.; Salvadori, P.; Coluccia, S.; Martra, G. Nanoscale Copper particles derived from solvated Cu atoms in the activation of molecular oxygen. *Chem. Mater.* **2002**, *14*, 1183–1186. [CrossRef]
40. Ponce, A.A.; Klabunde, K.J. Chemical and catalytic activity of copper nanoparticles prepared via metal vapor synthesis. *J. Mol. Catal. A Chem.* **2005**, *225*, 1–6. [CrossRef]
41. Yamaguchi, R.; Hirano, S.; Arai, Y.; Ito, T. Chitosan salt gels thermally reversible gelation of chitosan. *Agric. Biol. Chem.* **1978**, *42*, 1981–1982.
42. Rubina, M.S.; Elmanovich, I.V.; Shulenina, A.V.; Peters, G.S.; Svetogorov, R.D.; Egorov, A.A.; Naumkin, A.V.; Vasil'kov, A.Y. Chitosan aerogel containing silver nanoparticles: From Metal-Chitosan powder to porous material. *Polym. Test.* **2020**, *86*, 106481. [CrossRef]
43. Fakruddin, M.; Chowdhury, A.; Hossain, M.N.; Ahmed, M.M. Characterization of aflatoxin producing *Aspergillus flavus* from food and feed samples. *SpringerPlus* **2015**, *4*, 159. [CrossRef]
44. Monda, E.; Masanga, J.; Alakonya, A. Variation in occurrence and aflatoxigenicity of *Aspergillus flavus* from two climatically varied regions in Kenya. *Toxins* **2020**, *12*, 34. [CrossRef]

45. Davis, N.D.; Iyer, S.K.; Diener, U.L. Improved mmethod of screening for aflatoxin with a coconut agar medium. *Appl. Environ. Microbiol.* **1987**, *53*, 1593–1595. [CrossRef]
46. Wicklow, D.T.; Shotwell, O.L.; Adams, G.L. Use of Aflatoxin-Producing ability medium to distinguish aflatoxin-producing strains of *Aspergillus flavus*. *Appl. Environ. Microbiol.* **1981**, *41*, 697–699. [CrossRef]
47. Reddy, K.R.N.; Surendhar Reddy, C.; Nataraj Kumar, P.; Reddy, C.S.; Muralidharan, K. Genetic Variability of Aflatoxin B1 producing *Aspergillus flavus* strains isolated from discolored rice grains. *World J. Microbiol. Biotechnol.* **2008**, *25*, 33–39. [CrossRef]
48. Abbas, H.K.; Weaver, M.A.; Zablotowicz, R.M.; Horn, B.W.; Shier, W.T. Relationships between Aflatoxin production and sclerotia formation among isolates of aspergillus section flavi from the mississippi delta. *Eur. J. Plant Pathol.* **2005**, *112*, 283–287. [CrossRef]
49. Lai, X.; Zhang, H.; Liu, R.; Liu, C. Potential for Aflatoxin B1 and B2 Production by *Aspergillus flavus* strains isolated from rice samples. *Saudi J. Biol. Sci.* **2015**, *22*, 176–180. [CrossRef] [PubMed]
50. Yu, J.; Ehrlich, K. *Aflatoxins-Biochemistry and Molecular Biology*; USDA/ARS, Southern Regional Research Center: New Orlenas, LA, USA, 2011.
51. Caceres, I.; Khoury, A.A.; Khoury, R.E.; Lorber, S.; Oswald, I.P.; Khoury, A.E.; Atoui, A.; Puel, O.; Bailly, J.-D. Aflatoxin Biosynthesis and genetic regulation: A review. *Toxins* **2020**, *12*, 150. [CrossRef]
52. Mageshwari, K.; Sathyamoorthy, R. Flower-Shaped CuO nanostructures: Synthesis, characterization and antimicrobial activity. *J. Mater. Sci. Technol.* **2013**, *29*, 909–914. [CrossRef]
53. Kaur, P.; Thakur, R.; Barnela, M.; Chopra, M.; Manuja, A.; Chaudhury, A. Synthesis, Characterization andin vitroevaluation of cytotoxicity and antimicrobial activity of chitosan-metal nanocomposites. *J. Chem. Technol. Biotechnol.* **2014**, *90*, 867–873. [CrossRef]
54. Saharan, V.; Sharma, G.; Yadav, M.; Choudhary, M.K.; Sharma, S.S.; Pal, A.; Raliya, R.; Biswas, P. Synthesis and In Vitro antifungal efficacy of Cu–Chitosan nanoparticles against Pathogenic Fungi of tomato. *Int. J. Biol. Macromol.* **2015**, *75*, 346–353. [CrossRef]
55. Mekawey, A.I.I. Effects Of Chitosan Nanoparticles as antimicrobial activity And on mycotoxin production. *Acad. J. Agric. Res.* **2018**, *6*, 101–106.
56. Kulatunga, D.; Dananjaya, S.; Godahewa, G.; Lee, J.; De Zoysa, M. Chitosan silver nanocomposite (CAgNC) as an antifungal agent against *Candida Albicans*. *Med. Mycol.* **2016**, *55*, 213–222. [CrossRef]
57. Alghuthaymi, M.A.; Abd-Elsalam, K.A.; Shami, A.; Said-Galive, E.; Shtykova, E.V.; Naumkin, A.V. Silver/Chitosan nanocomposites: Preparation and characterization and their fungicidal activity against dairy cattle toxicosis *Penicillium expansum*. *J. Fungi* **2020**, *6*, 51. [CrossRef] [PubMed]
58. Wu, Y.H.; Lee, Y.H.; Shih, H.Y.; Chen, S.H.; Cheng, Y.C.; Chiu, D.T.Y. Glucose-6-Phosphate dehydrogenase is indispensable in embryonic development by modulation of Epithelial-Mesenchymal transition via the NOX/Smad3/miR-200b Axis. *Cell Death Dis.* **2018**, *9*, 1–14. [CrossRef] [PubMed]
59. Ho, H.; Cheng, M.; Chiu, D.T. Glucose-6-Phosphate Dehydrogenase- from oxidative stress to cellular functions and degenerative diseases. *Redox Rep.* **2007**, *12*, 109–118. [CrossRef] [PubMed]
60. De Lillo, A.; Cardi, M.; Landi, S.; Esposito, S. Mechanism(s) of Action of heavy metals to investigate the regulation of plastidic Glucose-6-Phosphate dehydrogenase. *Sci. Rep.* **2018**, 13481. [CrossRef]
61. Lin, H.-R.; Wu, C.-C.; Wu, Y.-H.; Hsu, C.-W.; Cheng, M.-L.; Chiu, D.T.-Y. Proteome-Wide dysregulation by Glucose-6-Phosphate Dehydrogenase (G6PD) reveals a novel protective role for G6PD in Aflatoxin B1-Mediated cytotoxicity. *J. Proteome Res.* **2013**, *12*, 3434–3448. [CrossRef]
62. Ibraheem, O.; Adewale, I.O.; Afolayan, A. Purification and properties of glucose 6-Phosphate dehydrogenase from aspergillus aculeatus. *BMB Rep.* **2005**, *38*, 584–590. [CrossRef]
63. Savi, G.D.; Scussel, V.M. Inorganic compounds at regular And nanoparticle size and their Anti-Toxigenic Fungi activity. *J. Nanotechnol. Res.* **2014**, *97*, 589–598.
64. Ammar, H.A.; Rabie, G.H.; Mohamed, E. Novel fabrication of gelatin-encapsulated copper nanoparticles using aspergillus versicolor and their application in controlling of rotting plant pathogens. *Bioprocess Biosyst. Eng.* **2019**, *42*, 1947–1961. [CrossRef]

MDPI
St. Alban-Anlage 66
4052 Basel
Switzerland
Tel. +41 61 683 77 34
Fax +41 61 302 89 18
www.mdpi.com

Journal of Fungi Editorial Office
E-mail: jof@mdpi.com
www.mdpi.com/journal/jof